Block by Block

Block by Block

The Historical and Theoretical Foundations of Thermodynamics

Robert T. Hanlon

Illustrations by Robert Hanlon and Carly Sanker

OXFORD
UNIVERSITY PRESS

OXFORD
UNIVERSITY PRESS

Great Clarendon Street, Oxford, OX2 6DP,
United Kingdom

Oxford University Press is a department of the University of Oxford.
It furthers the University's objective of excellence in research, scholarship,
and education by publishing worldwide. Oxford is a registered trade mark of
Oxford University Press in the UK and in certain other countries

© Robert T. Hanlon 2020

The moral rights of the author have been asserted

Published in the United States of America by Oxford University Press
198 Madison Avenue, New York, NY 10016, United States of America

British Library Cataloguing in Publication Data

Data available

Library of Congress Control Number: 2019956700

ISBN 978–0–19–885154–7 (hbk.)
ISBN 978–0–19–885155–4 (pbk.)

Printed and bound by
CPI Group (UK) Ltd, Croydon, CR0 4YY

For mom, dad, and Kevin, my constant inspirations

Acknowledgements

This eighteen-year (and counting) journey afforded me the opportunity to meet, either in person or via email, a wide range of people much more knowledgeable than I about the specific topics covered in my book. I relied on this group very much to check my work and validate the direction I had set for myself. This journey also afforded me the opportunity to benefit from the support of those not familiar with the technical content but very familiar with who I am and my reasons for taking on this project. To this entire generous group, I say, "Thank you."

For Part I (Big Bang) I relied on the very strong technical support of Katherine Holcomb in helping me understand certain cosmological concepts, and then for Part II (Atom), I relied on Milton Cole, John Durso, and Mark Schure. I thank all four for their kindness and shared expertise.

Regarding Parts III (Energy) and IV (Entropy), I split my acknowledgements into two groups, the historians and the scientists. Starting with the former, as I sought to learn more about the history that led to both energy and entropy, I discovered an entire community of historians that I hadn't really known about before, first through the books and articles they wrote and then through my attendance at a History of Science Society meeting in Philadelphia. Out of all of this, I began communicating with some members of this group, which in turn led me to others. I especially thank Stephen Brush and Robert Fox. I had the good fortune to meet both in person and thoroughly enjoyed our conversations and follow-up email exchanges. Both Stephen and Robert read sections of my manuscript and provided very helpful comments. Others who provided helpful feedback on selections of my manuscript were: Helge Kragh, Kenneth Caneva, David Harriman, and Phil Hosken. I additionally thank Robert Deltete, Francis Everitt, Ken Jolls, Thomas Settle, and Crosbie Smith.

As for the technical support I received for Parts III and IV, I especially thank Arieh Ben-Naim, Ken Dill, Randy Field, Sudha Jayaraman, Ken Jolls, Themis Matsoukas, John O'Connell, Roger Riehl, Kim Sharp, Jefferson Tester, Darrell Velegol, and Hans von Baeyer. I additionally thank Tom Degnan, Steve Fields, Derek Mess, Raymond Smith, and Harry Watson.

As for transforming my ideas and concepts into illustrations, I thank Carly Sanker, whose creativity and intellectual curiosity helped make this part of book writing an enjoyable and energizing experience.

All support received from the experts listed above helped ensure historical and technical accuracy. Any mistakes in the resulting product are all mine.

On a more personal note, many who were otherwise unfamiliar with the details of what I was doing but very familiar with my larger intent provided full, ongoing moral support, starting with a note written to me on a scrap sheet of paper years ago with the simple directive, "Write that book!" I've kept that scrap in my wallet ever since. Deep thanks to my family, my relatives, and my close friends from high school, university, graduate school, and my professional life. Special additional thanks to Koko for being there at the beginning and to Thomas Bartos for being the first to critically read the entire manuscript from beginning to end.

Finally, it is a rare and fortuitous occasion when an author can marry his best friend who also happens to be a professional writer. The emotional and technical support my wife, Dr. Colleen Terrell, has provided me over the past many years of my writing journey has been invaluable toward creating a loving, home-based writing environment that enabled me to maintain my focus, my sanity, and, most importantly, my sense of humor. For all of this and much more, I say, "Thank you, dear."

Contents

List of Figures

Afterword

Introduction

For those of us who didn't get it the first time around.

Prologue

Why?

I found myself thinking this question one day at work while calculating the temperature rise of a gas undergoing compression. It's not that I was questioning how to do the calculation, for I had done it numerous times before, but for some reason this time I suddenly realized that I didn't *really* understand why the temperature should rise in the first place. Here I was, a successful engineer with a graduate degree, doing a straightforward calculation we were taught over and over again in school, and, to my embarrassment, I never bothered to learn on a molecular level why the temperature increases.

Frustrated, I went to my chemical engineering textbooks and learned that…well…learned that they didn't have an answer for me. They provided the equations, showed how they were derived from fundamental classical thermodynamic principles, but they didn't explain at a physical level what happens to the gas to cause its temperature to rise.

Frustrated even more, I then started working this through in my own mind.

The gas molecules must somehow be picking up speed, and hence temperature, during the compression. The only thing that's moving is the piston. Hmmm. When molecules strike a stationary wall, they rebound with the same speed. However, when molecules strike the moving piston, they rebound at a higher speed. This must be it.

Right around this time I started flipping through the pages of Richard Feynman's *Lectures on Physics*. Lo and behold, I finally found someone who explained adiabatic compression on the fundamental level I wanted. And, to my joy, I found that I was correct in my thinking.

Feynman wrote,

> *Suppose a piston moves inward, so that the atoms are slowly compressed into a smaller space. What happens when an atom hits the moving piston? Evidently it picks up speed from the collision. You can try it by bouncing a ping-pong ball from a forward-moving paddle, for example, and you will find that it comes off with more speed than that with which it struck. (Special example: if an atom happens to be standing still and the piston hits it, it will certainly move.) So the atoms are "hotter" when they come away from the piston than they were before they struck it. Therefore all the atoms which are in the vessel will have picked up speed. This means that when we compress a gas slowly, the temperature of the gas increases. So, under slow compression, a gas will increase in temperature, and under slow expansion it will decrease in temperature.*[1]

---

[1] (Feynman et al., 1989a) Vol. I, p. 1-4.

Block by Block: The Historical and Theoretical Foundations of Thermodynamics,
Robert T. Hanlon, Oxford University Press (2020). © Robert T. Hanlon.
DOI: 10.1093/oso/9780198851547.001.0001

At this point, I started to reflect on all of the science theories, concepts and equations I had learned in school, challenging myself with the question, *do I truly understand this*? I was reasonably good at using the equations; I thought I understood them. But now I wasn't so sure. I realized that I had learned what the equations were and how to use them, but hadn't learned the physical and historical underpinnings behind them that would explain why they worked or why they existed as they did in the first place. I had learned the *what* but not the *why*.

My Motivation – Too Many Unanswered Questions

> *As problems get more and more difficult, and as you try to understand nature in more and more complicated situations, the more you can guess at, feel, and understand without actually calculating, the much better off you are!…Once you get the idea that the problems are not mathematical problems but physical problems, it helps a lot.*
>
> – Richard Feynman[2]

At the heart of many fields—physics, chemistry, engineering—lays thermodynamics. As a chemical engineering student, I was continually frustrated in my attempts to deeply understand this subject. Such concepts as flowing heat, free energy, maximum work and multiple versions of the 2nd Law didn't make sense to me. Attempts by the teacher to present a cause–effect chain of logic rarely, if ever, got to the foundational bedrock. We were taught the concepts and associated equations we needed to know, but rarely the physical meaning and historical origins behind them. Admittedly, it was difficult for the teachers to drill down so deeply in the classroom, where the amount of material to teach overwhelmed the time available. But even after I graduated and sought to educate myself, I faced significant challenges in finding the relevant material. It was at this point that I became motivated to find the answers, put them in a book, and so share them with others.

I succeeded in my professional career—two Fortune 500 companies and MIT's Department of Chemical Engineering—because I knew how to solve problems using the skills I acquired during my formal education, including those associated with thermodynamics. But I could have achieved even more had I had a greater intuitive understanding (à la Feynman) of the physical nature of the problems involved, for such an understanding would have provided me with more confidence, more strength and more creativity in using these skills to solve them.

My technical background combined with my researching and writing this book has placed me at the interface between three worlds: practitioner, academician and historian. I offer an original perspective on thermodynamics from this interface. I have spent my career working with other practitioners and students and am aware that many of them, including me at the

---

[2] (Feynman et al., 2013) p. 72–73. Along these same lines, on August 19, 2014, Dr. Samuel Fleming recounted a story to me about a lecture he attended while a student in MIT's Department of Chemical Engineering in which "T. K. Sherwood himself, no slouch when it came to mathematics and models in chemical engineering, observed at a doctoral seminar, grinning broadly, 'Well, the trouble with these mathematical models is that if they begin to work, pretty soon you start to believe them.' And then after the laughter died down, he turned deadly serious and asked the student presenter, 'What are the molecules doing?'" Sam told me that he was sitting right behind TKS and never forgot the moment or the "Sherwood Axiom."

time, had challenges in deeply understanding the concepts central to thermodynamics. In fact, I vividly recall one conversation in which a fellow PhD student told me, "I tried reading Gibbs and couldn't get past the third page." To conquer the challenge of Gibbs and many other challenges—what *is* entropy?—I searched through much relevant literature to answer my repeated *why* questions, stopping short of going too deep into material irrelevant to my objectives, such as quantum physics. I then extracted the answers I sought and combined them with my own practitioner insights and perspectives into a coherent, seamless whole.

This book targets the "student" who is asking the same questions I asked, whether the student is in the classroom, the laboratory, the office or the home reclining chair, and whether the student is a practitioner, an academician or a historian. In addition to offering a different way of approaching the subject, one that enables greater confidence and creativity in approaching and ultimately solving the thermodynamic problems, this book also offers the student inspiration by historical example. When students read about Galileo rolling up his sleeves to finally conduct an experiment to quantify free fall, they may be inspired to roll up their own sleeves and go into the laboratory or the chemical plant to conduct their own experiments, gather their own data and so challenge their own hypotheses.

General Approach

The book is structured in four parts. The first describes the physics behind the origins of the elements starting with the Big Bang; the second, the atom and its features; the third, energy and the conservation laws; and lastly, the fourth, entropy and the laws of thermodynamics. I included the first two parts since I felt it relevant to include the science on the origin and behaviors of atoms in a book on thermodynamics. I structure each part to lay the foundational building blocks of both the *what* and the historical and theoretical *why* of thermodynamics. The first chapter of each of the four parts explains the physical science involved, while the subsequent chapters explain the history that led to this understanding. Regarding the science, I strove to include the physical mechanisms involved at their most basic levels and with the most basic language. Regarding the histories, I selected both major and minor contributors to provide a logical thread, recognizing that such a thread is only evident in hindsight. Regarding the content, I strove to effectively engage the reader by including relevant stories and original illustrations.

To achieve my aims, I have synthesized and gathered into one accessible volume a wide range of specialized topics built using my own practitioner's perspective on what students— classroom or otherwise—will find most valuable. Specifically regarding the history, I share how the concepts and terminologies we use today rose from their historical origins in the belief that such historical understanding, when combined with the scientific understanding, can enhance one's mastery of thermodynamics. Additionally, regarding the history, I share the struggles many encountered when attempting to understand nature and why some ideas succeeded while others did not, believing that the failures are just as educational as the successes. Learning what something is and is not; learning why some scientists succeed while others do not; learning why some ideas succeed when they shouldn't—these lessons enhance the depth of education. They also show that while the technologies of the past are much different than those of the present, the people aren't. The same highs and the same lows occur for the same reasons today as they did yesterday. All impact the ways by which science is conducted. The full history adds richness and dimensionality and enables one to consider subjects from different perspectives and so better ingrain the science in one's mind.

The Four Parts

Part I – The Big Bang and the Synthesis of the Elements in the Stars

The historical timeline of the universe in this book starts when hydrogen and helium initially formed and the four fundamental interactions—strong, electromagnetic, weak, gravitation—came into being. I share the current thinking about what happened during the Big Bang, how hydrogen and helium formed and spread throughout the universe, and then how they came back together again in the stars and combined to create the periodic table of elements.

I also share the history of how a very small number of very smart individuals discovered the Big Bang (as it became known) and its aftermath. The fact that the stars are all moving away from us in a structured velocity-versus-distance relationship as first discovered by Edwin Hubble is simply fascinating, almost as fascinating as learning that just prior to this, most everyone—including Einstein himself—assumed that the stars were static, non-moving objects in the sky, just sitting there.

The macroscopic world of thermodynamics is governed by the microscopic world of atoms. A natural question to ask is, where did the atomic world even come from? In Part I, I answer this question. In Part II, I describe how atoms behave.

Part II – The Atom: Hard Spheres that Attract and Repel

From afar an atom looks and behaves like a billiard ball; up close it looks and behaves much differently. Constructed from the building blocks of protons, neutrons, and electrons, and from the four fundamental interactions acting within and between them, the atom is the starting point for understanding the array of chemistry surrounding us and for understanding thermodynamics. As it was for the Big Bang, the history behind the atom's discovery was one of finding many pieces to a jigsaw puzzle of an unknown final design lying on a table and slowly over time connecting them together to solve the puzzle.

Part III – Energy and the Conservation Laws

Energy has a visceral feel to it. We conceptually understand that energy is conserved. We know that engines can't run without fuel. We know that we'll never witness rocks rolling uphill. We know that events just don't happen, that there must be some cause behind the effect, and that there must be something lost for something gained. We see energy around us in the basketball's trajectory as it heads toward the basket, in the collisions between cars shown on the nightly news, and in the lifting of an air conditioning unit by a crane to the top of a building.

We grasp most of this and yet, do we *really* understand how a simple lever works? Do we really understand that the Moon is continuously falling toward the Earth with the same acceleration as does the dropped apple once corrected for the distance of each to the center of the Earth? And do we understand why water heats up when a spinning paddle agitates it? We may *think* we know, but do we really? For a long time, many didn't even believe that agitation could heat water; it didn't make sense. And then along came James Joule, and he ever-so-methodically quantified this, showing how the work associated with a falling weight is equivalent to the heat generated by the paddle it spun. It was Joule's demonstration of work–heat equivalence that helped break open the dam and so enable Rudolf Clausius to both correct the single flaw in Sadi Carnot's theoretical analysis of the heat engine and to then formally introduce the concept of energy. In Part III are the details behind this breakthrough and how it led to the 1st Law of Thermodynamics.

But these final steps can only be understood in the context of the fact that the individuals involved, as Newton recognized, stood on "the shoulders of others." Many steps preceded these final steps: Archimedes' lever, Galileo's free fall, Newton's laws of motion and universal gravitation, the arrival of calculus to bring structured quantifiable equations to bear on the analysis. And then the realization of what heat was and wasn't, a journey that required both strong experimental evidence provided by, for example, the thermometer, Joseph Black's heat capacity, and the extremely accurate measurements of gas properties by such individuals as Henri Victor Regnault in his Paris laboratory, and a group of scientists willing to challenge the prevailing caloric theory that had "consistency, elegance, and other genuine merits"[3] but was otherwise incorrect and therefore stifled progress. All of these different pieces funneled together in the mid-1800s to result in the discovery of energy and its conservation and also in the realization that the long sought-after perpetual motion machine is indeed impossible.

Part IV – Entropy and the Laws of Thermodynamics

As one of the most challenging concepts to comprehend in thermodynamics, entropy started in an unusual location, the steam engine, or more accurately, the piston. The evolution of the piston-in-cylinder device and its use in the early "fire engines" and the later "steam engines" takes us to Britain where these machines were used to pump water out of mine shafts, thus enabling deeper access to the underground coal. The accomplishments of the British led to the subsequent theoretical analysis by a Frenchman, the aforementioned Sadi Carnot, who sat in his small Parisian apartment and wondered, *is the motive power of heat unbounded*? His resulting essay found its way (only barely) into the hands of theoreticians in both Britain and Germany, and their analyses led to the founding of classical thermodynamics and the discovery of entropy. This fascinating journey brings forth the contributions of many individuals and cultures. You'll meet the engine engineers of Britain, Sadi Carnot of course, the all-star Scottish team of William Thomson (before he became Lord Kelvin), William Rankine and James Clerk Maxwell, along with the adopted Joule (England) and Hermann von Helmholtz (Germany), Rudolf Clausius in Germany, and finally and ultimately J. Willard Gibbs at Yale University. The subsequent founding of statistical thermodynamics that connected the micro-world of the

[3] (Fox, 1971) p. 5.

atom with the macro-world of classical thermodynamics began with the development of the kinetic theory of gases by Clausius and Maxwell followed by the development of statistical mechanics by Maxwell, Ludwig Boltzmann and Gibbs yet again.

Included in Part IV is a broader discussion of the scientific method, what good science looks like and how it should be conducted. Those who made significant contributions to thermodynamics generally adhered to a similar process: gather data, induce a set of hypotheses, deduce the consequences, test against reality and identify the hypothesis that (best) explains reality. This methodology is naturally not limited to thermodynamics.

History

> While hindsight can detect the great pattern of concept and theory that was being woven at that time, it would be wrong to assume that it was equally apparent to contemporary observers. A closer study of the period conveys a bewildering impression of scientific conjectures and refutations. Hypotheses that were later shown to be wrong were acclaimed while much sounder ones were ignored. More confusing still, the right ideas were sometimes put forward for the wrong reasons. Confusions of this sort are no doubt inseparable from the advance of science in any age, and the historian must duly report them, without emphasizing them.
>
> The historian has to describe certain quite unrelated activities which were going on at the same time because in later years these activities merged in one great synthesis of knowledge. But he must be careful to avoid giving the impression that those who engaged in these unrelated researches envisaged the grand synthesis which he, the historian, knows took place much later.
>
> – D. S. L. Cardwell[4]

I draw attention to this quote by D. S. L. Cardwell to address the challenge of writing history. I fully realize that I selected specific people and events to create timelines for each of the four parts in this book to achieve my objectives. And I fully recognize that such timelines may suffer from being so-called "false narratives." As excellently discussed by Nassim Taleb in *The Black Swan*,[5] such narratives occur when historians gather facts about past events and then create cause–effect connections that lead to a conclusion. This is only natural, since no one would really want to buy a history book filled only with disconnected facts. But such narratives can at times be more fictional than real. To counter this tendency, Taleb argues that the historian should seek the diaries, letters and primary sources of the people involved in the events, as such materials reflect the thinking of someone who doesn't know what tomorrow will look like. I attempted this where possible and indeed found that history is messier than what some history books suggest.

---

[4] (Cardwell, 1971) p. 121.

[5] The primary theme of Taleb's *The Black Swan* is how many of the significant events in our history were caused by events that were not predicted, events that caught us totally by surprise, just like the sight of a black swan would. Taleb says that we tend to wrongly believe we can actually predict the occurrence of "black swans" in the future based on the fact that we believe we can explain their occurrence in the past. Taleb's conversation inexorably leads into the study of history itself, which is why I'm bringing it up here as this book—the one in your hand—is heavy on the history.

The Silent Evidence

People say, Think if we hadn't discovered Emily Dickinson. I say, Think of all the Emily Dickinsons we've never discovered
    – Catherine Thomas Hale, character in Mark Helprin's *In Sunlight and in Shadow*[6]

Before ending this Introduction, I want to point out another reality discussed by Taleb. Silent evidence. Consider that the history we read addresses only those artifacts or "fossils" that we have discovered, or only those stories of the survivors or the winners. In other words, consider that we only become aware of a small subset of the large sample pool of individuals gathered around a common concept. For example, when you read about professional gamblers, you'll likely learn that each had good luck early on that enabled them to eventually succeed. But you're not reading about all those who started out to be professional and lost everything. They don't end up in the history books. No one takes into account the "silent evidence" of those who lost, those who died, the books that were never published, the books that were published but disappeared, the stone tablets that erosion erased—and the scientific findings, ideas and dead ends[7] that never saw the light of day.[8] In fact, as we'll see in this book, some of the breakthrough ideas that we do learn about and that changed the course of science only barely survived. Sadi Carnot wrote his famed book on the steam engine and hardly anyone read it. But for the stroke of good fortune his book might never have been discovered by Thomson and Clausius. Where would we have been then? Who else was prepared to discover the 1st and 2nd Laws of Thermodynamics? Would someone have discovered an entirely different set of laws that achieved the same solution, based on other properties of matter that we're unaware of?

This Book is for the Curious Mind

I believe that anyone who has taken high school science, and especially anyone who may have further taken an introductory college course in thermodynamics, will enjoy at least some if not all of this book, especially since the science is accompanied by an intriguing history. The lever—who would have thought that such a simple machine would prove so hard to explain? The steam engine—who would have thought that the discovery of entropy would lay inside such a rough and rugged machine? Starlight—who would have thought that we could learn so much about the universe from the tiny specks of light reaching us from faraway stars? Aristotle, Clausius, Archimedes, Carnot, Thomson, Hubble, Lavoisier, Boltzmann, Gibbs. So many scientists, so many characters, so much progress. I have tried to capture both the science and the history in a way to interest and engage any who are curious about the foundations of thermodynamics.

---

[6] (Helprin, 2012) p. 121–122.
[7] Regarding an extensive piece of experimental work by James Joule that went nowhere, D.S.L. Cardwell wrote, "It was to prove one of the dead ends that litter the record of science, that receive little or no mention in the history books and that are completely unknown to the practicing scientist." (Cardwell, 1989) p. 73.
[8] As Abraham Lincoln's character said in the 2012 movie *Abraham Lincoln: Vampire Hunter*, "History only remembers a fraction of the truth."

This book is not a textbook. I intentionally chose not to delve into the use of equations to solve problems; this is covered very well and very extensively in many high-quality textbooks. However, while I do not discuss how to solve problems, I do provide the foundations on which to build solutions. Also covered very well in textbooks is how to calculate the physical properties of matter. Thermodynamics is based on the interrelationships between these properties. While I do not indicate how to calculate these properties, I do show how and why they're connected.

In the end, I wrote this book for the curious mind, regardless of whether that mind is inside or outside the classroom.

So with this Introduction now complete, I share with you my findings.

Robert T. Hanlon
Philadelphia
January 2019

PART I

The Big Bang

THE BIG BANG DISCOVERY MAP

THE DISCOVERY OF THE BIG BANG
TOOK A VARIETY OF EXPERIMENTAL + THEORETICAL
APPROACHES, FAILURES + SMALLER DISCOVERIES

PTOLEMY (150 AD)
- Earth-centered model of the universe

COPERNICUS (1543)
- Sun-centered model of the universe

KEPLER (1609)
- discovered elliptical planet orbits

GALILEO (1610)
- improved the telescope 20x
- helped launch Copernican Revolution
- endorsed Sun-centered model
- observed imperfections in the sky

THE COSMOLOGISTS

observed expanding universe

assumed single point of origin

EINSTEIN (1915)
- General Theory of Relativity (GTOR)
- assumed static universe; inserted cosmological constant to counteract gravitational pull

NEWTON (1687)
- law of universal gravitation
- laws of motion

LEMAÎTRE (1927)
- unknowingly repeated Friedmann's work: proposed that universe is expanding
- first to propose origin of universe as compact region: the primeval atom

FRIEDMANN (1922)
- proposed GTOR valid without cosmological constant if universe is expanding
- early death / low profile slowed his theory's spread to public

THEORY ⟶ **DATA**

HUBBLE (1929)
- used Doppler Shift plus Cepheid variables to determine that stars are moving away from Earth at speeds that increase with distance

LEAVITT (1900)
- discovered how to use Cepheid variable stars to calculate distance to stars from Earth

SLIPHER (1917)
- first to measure line shifts and so quantify direction and speed of stars relative to Earth

THE ASTRONOMERS

HUGGINS (1868)
- husband and wife team
- turned the spectroscope toward the stars
- observed Doppler Effect

BUNSEN + KIRCHOFF (1860)
- developed spectroscope to provide more accurate line positions

FRAUENHOFER (early 1800s)
- using Prism, detected lines in spectrum of starlight

THE BIG BANG THEORY

PARTICLE PHYSICISTS ENTER THE WORLD OF COSMOLOGY

GAMOW, ALPHER, HERMAN (1940s + 1950s)

- first group to quantify conditions of the early universe
- predicted existence of cosmic background radiation (CMB)
- low profile slowed theory's spread into public

HOYLE (mid 1950s)

- solved 8-particle gap in nucleosynthesis reaction network
- along with others, modeled evolution of stars, including synthesis of elements

PEEBLES + DICKE

- unknowingly repeated work of Gamow, Alpher, Herman
- predicted existence of CMB originating from Big Bang

PENZIAS + WILSON (1964)

- using radio telescope, detected noise EVERYWHERE THEY LOOKED!
- sought meaning from Peebles + Dicke

CMB THEORY VALIDATED

SMOOT (mid 1970s)

- predicted presence of trace imperfections in early universe which would later cause clumping and formation of stars
- proposed Satellite to test hypothesis

COSMIC BACKGROUND EXPLORER (1989)

- (COBE) Smoot's non-uniformity detected— 1 part in 100,000
- final validation of the Big Bang Theory

NON-UNIFORMITY IN EARLY UNIVERSE VALIDATED

1 The Big Bang: science

The Inflation Theory of How the Universe Began and the Atoms Arrived

> *Ten or twenty billion years ago, something happened – the Big Bang, the event that began our universe. Why it happened is the greatest mystery we know. That it happened is reasonably clear.*
>
> – Carl Sagan[1]

> *So after a few of the most productive hours I had ever spent at my desk, I had learned something remarkable...[and began] to call it inflation*
>
> – Alan Guth[2]

About 13.75 billion years ago our entire universe occurred as an infinitesimally small region, more than a billion times smaller than a proton, with a near-infinite temperature and density of 10^{29} °C and 10^{80} gm/cc, respectively.[3] Staggering, unfathomable numbers.

Stop here and think about this. This region, very much smaller than the period at the end of this sentence, was all that there was. There was *nothing* else, at least nothing that influenced our universe.

Within the first second of what became the start of time arose a tremendous repulsive force, the "bang" in the Big Bang, that drove space itself, per Alan Guth's inflation theory,[4] to expand at tremendous rates. The cause of the repulsion? Nobody knows, but presumably some form of "dark energy" repulsion having a different underlying physical process than the first form. Both forms of repulsion contribute to the repulsive force contained within Einstein's cosmological constant (Chapter 2) and provide a counter-force to gravity in Einstein's Theory of General Relativity.

This was not an explosion as we commonly know it. There was no center to this event, no edge, no boundary. It occurred simultaneously everywhere. Everything rushed apart from everything else. Space itself rapidly swelled.

During this brief flicker, which ended within the first 10^{-30} seconds of time, the "dark energy" repulsion caused the universe to expand exponentially, doubling in size once

---

[1] (Sagan, 1980) p. 246.
[2] (Guth, 1997) p. 176.
[3] (Guth, 1997) pp. 86, 109, 170, 185.
[4] (Guth, 1997) p. 186. If Guth's theory is indeed correct, then the observed universe is only a "minute speck" in a universe that is many orders of magnitude (10^{23} times) larger.

Block by Block: The Historical and Theoretical Foundations of Thermodynamics,
Robert T. Hanlon, Oxford University Press (2020). © Robert T. Hanlon.
DOI: 10.1093/oso/9780198851547.001.0001

every 10^{-37} seconds, more than 100 such doublings in all, a huge factor, before extinguishing. The rapid decay in the repulsive force slowed the expansion and transferred energy into the rapid formation of particles. Photons, protons, electrons, neutrons, and all sorts of anti-matter. Their creation occurred everywhere, almost perfectly uniformly filling the expanding universe, which at the end of inflation was only about one meter radius. The entire mass of the universe existed inside this small, inflated sphere.

Particles collided with each other so quickly that uniformity prevailed throughout the expansion. The collisions sometimes resulted in energy transfer, other times in the annihilation of the colliding particles and the creation of others. Photons collided with protons, neutrons with electrons, matter with anti-matter.

As the universe expanded, space itself stretched and, in turn, stretched the wavelength of the photons, causing their energy to decrease and the universe to cool. The cooling of the universe is sometimes confusingly attributed to the universe's expansion from the standpoint that as a gas expands, it cools. But this cause–effect logic doesn't work. While adiabatic expansion in a piston does lead to cooling, this is caused by the fact that the piston is doing work, thereby meaning that the gas molecules rebound from the retracting piston at a slower rate than when they struck the piston. But there is no such piston at work in the universe. There is no boundary that the universe is expanding against. So why then did the universe cool? The increasing wavelength of photons with increasing expansion of the universe follows from application of Einstein's General Theory of Relativity. And this naturally begs another question: how does the resulting energy balance work? Where does the energy lost by the photons end up going? It appears that Einstein's equations are not consistent with the conservation of energy. The actual mechanism by which the stretching of space causes the stretching of photon wavelength is unknown and not currently consistent with the conservation of energy.[5]

The stretching of space caused photon wavelength to increase and energy to decrease. Prior to recombination (when electrons combined with the nuclei), photons and matter existed in thermal equilibrium on account of the high collision frequency between all particles. Matter cooled as photons cooled. However, after recombination, radiation decoupled from matter and the former continued to cool with the expanding universe while the latter continued to cool by other pathways. After 1/10th of a second, the universe stood at 300 billion °C. After a second, 10 billion °C. After three minutes, one billion °C.

After the First Three Minutes – One Billion °C and the Formation of Heavier Nuclei

The decreasing temperatures shifted the relative reaction rates between the particles such that at the end of three minutes, mostly photons remained, 1 billion per each nuclear particle. The

---

[5] For a good discussion of this topic see (Hawley and Holcomb, 2005) pp. 414–415. Where does the energy lost by the photons go? It seems to just disappear, but that is incompatible with energy conservation. However, we do not currently understand whether the law of conservation of energy applies to the universe as a whole or what a consistent definition of the total energy of the universe might be. Perhaps further research into the nature of space and time will explain this mystery.

cooling universe also opened the door for the formation of heavier nuclear particles, starting with the reaction of protons (protium) and neutrons to form deuterium nuclei; the electrons remained unattached in these high temperature conditions (Figure 1.1). The delay of this reaction, which became known as the "deuterium bottleneck," was caused by the fact that deuterium is not stable at high temperature. After about three minutes, the cooled universe broke open the bottleneck, leading to the production of deuterium and then the subsequent rapid reactions involving deuterium to form the even heavier nuclei of tritium, helium-3, and ordinary helium, helium-4 (He-4). It was these primordial reactions that led to the large abundance of helium in the galaxy.[6]

Hydrogen Isotope Nuclei			
Protium	1 proton	0 neutron	dominant isotope
Deuterium	1 proton	1 neutron	
Tritium	1 proton	2 neutrons	
Helium Isotope Nuclei			
Helium-2 (He-2)	2 protons	0 neutron	unstable[7]
Helium-3 (He-3)	2 protons	1 neutron	
Helium-4 (He-4)	2 protons	2 neutrons	dominant isotope

He-4 was where Big Bang nucleosynthesis hit a roadblock. Everything piled up on a dead-end street with nowhere else to go. Further addition of single protons to He-4 and the reaction of He-4 with itself were both blocked since lithium-5 and beryllium-8 are both unstable. These bottlenecks became known as the "5-particle gap" and the "8-particle gap," respectively.

So after the first three minutes, the nuclear mass in the universe consisted of about 75% unattached protons (protium), 25% He-4, and a vast sea of photons. Free neutrons no longer existed. With a half-life of only 10 minutes, they could only exist in the early minutes of the universe. Their capture by protons effectively saved these Big Bang neutrons from extinction; additional neutrons would form later from proton–proton reactions inside the stars. Unattached electrons also existed, one for each proton; the universe was still too hot for them to establish orbits around the nuclei.

For the next 300 000 years, the universe continued to expand and cool, little happening in the way of further nuclear reactions as the density of matter was too low. Photons continued to collide with the electrically charged electrons, preventing the formation of atoms (electrons in orbit around nucleus) while also creating an opaque universe due to their resulting scattered lines of travel.

---

[6] (Weinberg, 1993) p. 113. "There is, of course, extremely little helium on Earth, but that is just because helium atoms are so light and so chemically inert that most of them escaped the Earth ages ago."

[7] Proton–proton nuclei don't form because they are not stable; the electrostatic repulsion between them needs to be lowered by the presence of a neutron in order for the strong nuclear interaction (force) to dominate.

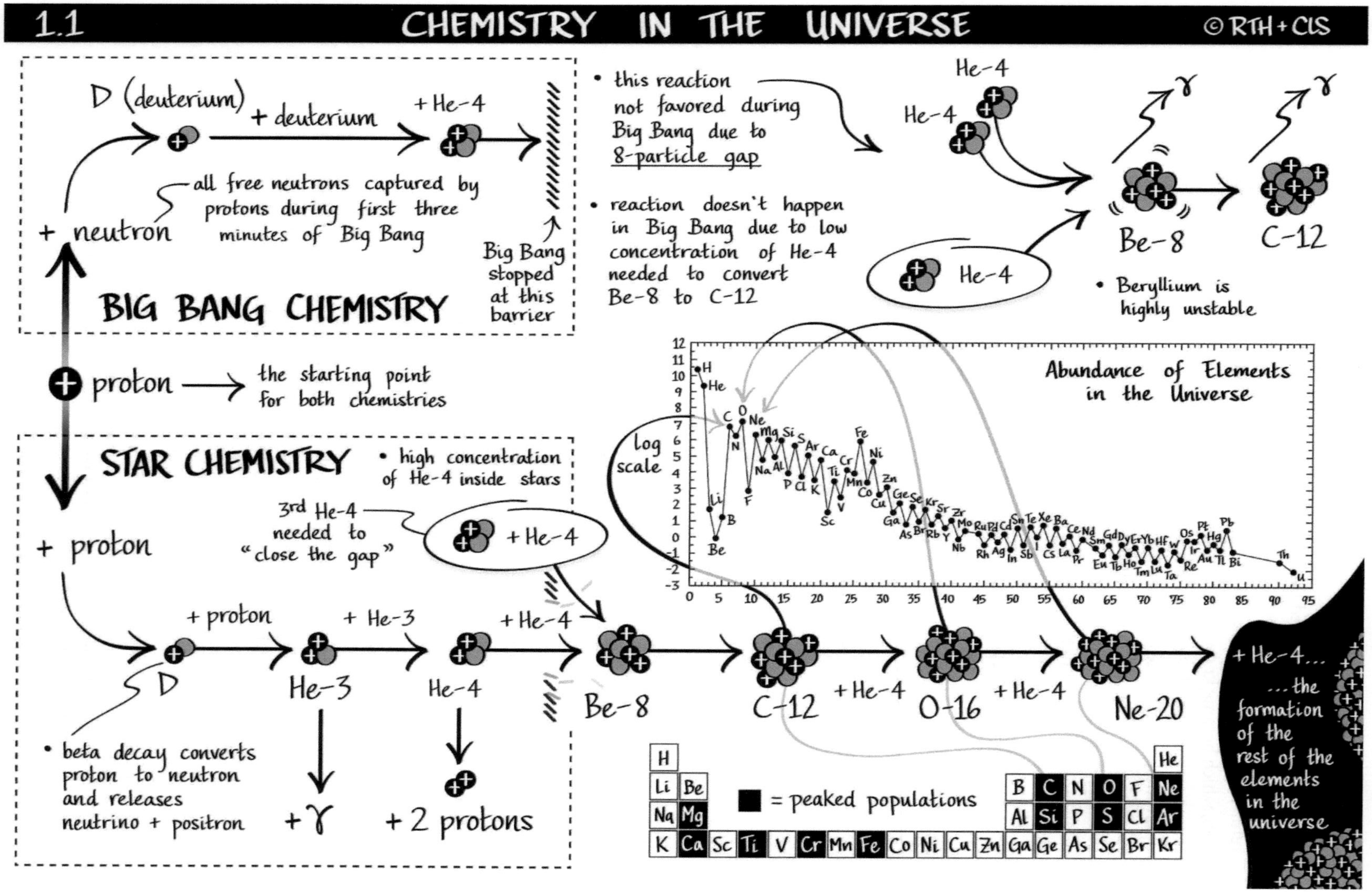

Figure 1.1 Chemistry in the universe

After the First 300 000 Years – 3000 °C and Recombination

At 300 000 years a transition occurred. By then the universe had finally cooled enough (3000 °C) for electrons to stabilize in orbits around the nuclei and thus form hydrogen (protium) and helium (He-4) atoms. Since bound electrons do not scatter light like free electrons do, recombination[8] also made the universe transparent to light. No longer influenced by the presence of the electrically charged electrons, the photons commenced an unimpeded journey through the universe.

At this point in time, the universe was uniform, with photons and the newly formed atoms spread evenly throughout. Well, almost evenly. Small variations in density, on the order of 1 part in 100 000, created the smallest of imperfections, the slightest of wrinkles. And from such small imbalances, atoms started to be pulled more in one direction than the other by the force of gravity.

Slowly, during the ensuing millions of years as the universe continued to expand, atoms found other atoms and clouds formed. Gravity increasingly pulled more and more atoms into the clouds and increasingly pulled the clouds in on themselves, so causing the transformation of potential energy into kinetic energy and the corresponding increase in temperature and pressure. Electrons were stripped off atoms, leaving bare nuclei (packed at much higher density) to start fusing together inside the early stars. Free protons fused together to form deuterium, one of the reactant protons having converted to a neutron in the process, and the resulting deuterium nuclei fused together to form He-4 (Figure 1.1). This "hydrogen burning" process generated tremendous radiant energy and served to push matter outwards, against the inward pull of gravity.[9] A kind of push–pull, contract–expand balance existed for many more years, with the pull of gravity slowly overtaking the decreasing push of radiation as the hydrogen fuel was consumed, leading to smaller, denser, and even hotter stars.

The Formation of the Elements

The stars are the seat of origin of the elements
– Burbidge et al.[10]

The increasing temperature and density inside these early stars (10^8 degrees and 10^5 gm/cc) eventually enabled a leap over the 8-particle barrier mentioned earlier. This barrier couldn't be breached in the early universe since by the time He-4 was created, the density was too low, around that of water, to promote this breakthrough. While He-4 can react to form beryllium-8, consisting of 4 protons and 4 neutrons, this product is <u>very</u> unstable, having a half-life of less

---

[8] While this word is generally used to describe the historical event when electrons entered into orbits about nuclei to establish atoms, it's a bit of a misnomer as the electrons were not previously in such orbits.

[9] Even though photons are massless, they still carry momentum in their electromagnetic field and can "collide" or otherwise interact with electrons with sufficient force to cause the electrons to recoil, thus resulting in pressure.

[10] (Burbidge et al., 1957) p. 550.

than a millionth of a billionth of a second. In the early universe, the fleeting formation of such nuclei led to nothing. However, in the aged universe when dense new stars came into existence, the high-density center ensured that He-4 nuclei were physically close enough to beryllium-8 nuclei to react with them before they had a chance to break apart. The result? Carbon-12. While slow, such "helium burning"[11] when viewed over billions of years was significant.

Once carbon-12 started to form inside the early stars, the floodgates opened and synthesis of the elements began (Figure 1.1). Successive addition of He-4 led to peaked production of oxygen-16, neon-20, magnesium-24, silicon-28, and so on, while successive addition of protons filled in the nuclei between these numbers. The relative rates and concentrations involved in these reactions shifted over time as the composition of the stars shifted over time. The first stars were naturally composed solely of hydrogen and He-4. As the hydrogen exhausted itself by burning to additional He-4, the outward radiation push dropped off, causing the stars to collapse further due to gravity; internal temperatures then rose, triggering the next reaction in line, He-4 to carbon-12. As He-4 then exhausted itself, the stars collapsed even further, reaction conditions became more severe and the next reactions in line were triggered, this time the He-4 addition reactions. The progression of these contractions and reactions were not always as smooth as written here. At times, the speed at which they occurred was such that sudden releases of energy resulted, leading to explosions and sometimes super explosions, especially during the rapid collapse of very massive stars. During such violent events, the contents of the stars spewed back out into the universe, only to be pulled back into other clouds and other stars, starting the whole cycle over again, with one critical exception. Each new generation of stars began with heavier elements contaminating the hydrogen, leading to a different set of element-building reactions. In this way, the temperature and density inside the stars grew over time as progressively heavier nuclei participated in the reactions, all the way up to iron, the most stable of nuclei.

Alongside these reactions was a separate set of reactions. Production of lithium, beryllium, and boron was very low as these rare light elements were not primary products of either hydrogen or helium burning processes. Some lithium production occurred during primordial nucleosynthesis but was very limited due to the aforementioned 5-particle barrier. Additional lithium production together with beryllium and boron production resulted from "spallation" reactions in which cosmic rays (mostly protons) struck heavier nuclei such as carbon and oxygen nuclei with so much energy that the nuclei broke apart, generating the light elements. The rarity of these reactions accounts for the relative scarcity of these elements in the universe today. Nuclei heavier than iron (atomic number 26) originated when exploding stars shot

---

[11] (Burbidge et al., 1957) p. 550–551. "Except at catastrophic phases a star possesses a self-governing mechanism in which the temperature is adjusted so that the outflow of energy through the star is balanced by nuclear energy generation…[when] hydrogen becomes exhausted as stellar evolution proceeds, the temperature rises until helium becomes effective as a fuel. When helium becomes exhausted the temperature rises still further until the next nuclear fuel comes into operation, and so on. The automatic temperature rise is brought about in each case by the conversion of gravitational energy into thermal energy."

neutrons into existing star nuclei, resulting in the formation of elements all the way up to uranium (atomic number 92), the heaviest naturally occurring element.[12]

Observed elemental population distributions of the universe are consistent with these reactions.[13]

---

[12] Uranium atom = U-238 = 92 protons + 146 neutrons + 92 electrons. This is the most common isotope of uranium. Heavier elements do exist, up to atomic number of 118 (oganesson) as of 2019, but do not occur naturally. They are produced in man-made accelerators and are not stable.

[13] For a wonderfully detailed discussion of all the reactions that led to the formation and population of the periodic table, along with the supporting data, read (Burbidge et al., 1957).

2 The Big Bang: discovery

We take it for granted that the Earth spins on its axis and revolves around the Sun, that we are a part of a solar system, that our solar system is but one of billions in the Milky Way galaxy, and that the Milky Way is but one of billions of galaxies in the universe. We take it for granted that everything is in motion relative to everything else and that the net direction of motion is 'away,' with everything rushing away from everything else in our still expanding universe. For many, we take it for granted that this all started about 14 billion years ago with the Big Bang.

However, picture yourself *not* knowing this. Picture yourself standing on the Earth hundreds of years ago, seeing what you see with your own eyes, feeling what you feel with your own body. What do you see? What do you feel? Think about it. You see the Sun rise and the Sun set and the Moon rise and the Moon set. You see a very large number of stars moving as one entity across the night sky, each standing absolutely still relative to the others, motionless. Among them you see several wanderers and eventually call them planets. And you feel...nothing. No sense of movement at all.

What do you conclude from this? You could easily conclude that you are located at the center of everything and that the Sun, Moon, and stars revolve around you. Not knowing anything else, not having any information to contradict this, you'd be content to sit inside this paradigm.

But then data would start to arrive, data that challenged your beliefs. And the tide would slowly turn. The curious amongst us would look at the data and wonder why they don't fit with their beliefs. And then the brave would propose new beliefs, new ways of seeing things, new theories, new paradigms and set about testing them by obtaining more data. Does the new theory agree with the data? No? Then try another theory. Yes? Maybe we're on to something.

And this is exactly what happened with the development of the Big Bang theory, one of the most wondrous of human achievements. Ever. A fascinating story of discovery, drama, and the human spirit, as scientists handed off the baton of discovery, one to the other, across hundreds years.

Copernicus – the Return to a Sun-centered Universe and the Onset of the Scientific Revolution

> *For a long time I was in great difficulty as to whether I should bring to light my commentaries written to demonstrate the Earth's movement*
>
> – Nicholas Copernicus (1543)[1]

[1] (Copernicus, 1995) p. 4.

Block by Block: The Historical and Theoretical Foundations of Thermodynamics, Robert T. Hanlon, Oxford University Press (2020). © Robert T. Hanlon. DOI: 10.1093/oso/9780198851547.001.0001

Enter Nicholas Copernicus (1473–1543), one of the first to carry the baton in this story, a great astronomer. Born in Poland, Copernicus wrote but one and a half publications that hardly anyone read during his lifetime. In particular, his main publication, *De revolutionibus* (1543 – published just after he died), proposed that the Earth revolves around the Sun and not the Sun around the Earth. His internal debate on whether or not to publish was understandable given that his theory ran counter to the deeply held beliefs taught by the church and to the well-entrenched beliefs proposed by the ancient Greeks.

Copernicus was the first to suggest a Sun-centered model into Western culture. While the ancient Greeks also proposed this, they later rejected it in favor of the Earth-centered model proposed by Ptolemy in 150 AD since this model was more consistent with three observations: people do not feel movement, objects fall toward Earth, and stars appear as a stationary backdrop moving across the sky. The belief was that each of these observations would not occur if the Earth were revolving around the Sun. And so the Ptolemy model became the paradigm for the next fifteen hundred years. The Catholic Church likewise supported the Earth-centered model, citing Biblical references regarding the motionless Earth and moving Sun.

Copernicus believed otherwise. Based on his own observations of the Sun, Moon, stars, and planets, Copernicus proposed a paradigm shift, a Sun-centered universe, in which the Earth spins on its axis once per day and, together with the planets, orbits the Sun in perfect circles at constant speeds. The Copernican model better fit the data and arguably triggered the onset of the scientific revolution.

Brahe and Kepler – the Elliptical Orbit

It was as if I had awakened from sleep
— Johannes Kepler (1596) on discovering Mars' elliptical orbit[2]

Some 50 years later in the late 1500s, with surprisingly no recorded action in between, Copernicus' baton was finally picked up by Tycho Brahe (1546–1601) and Johannes Kepler (1571–1630). Born in Germany and Denmark, respectively, the two proved to be a fortuitous pairing of complementary skills, Brahe the data collector and Kepler the interpreter. Based on many hours of tedious devotion to highly accurate readings, Brahe prepared the best collection of astronomical readings at that time. Kepler then spent eight years (!) analyzing these data until his own paradigm shifted. Initially, he assumed that the Copernican model was accurate. However, Brahe's data didn't fit this model, especially the data for the path of Mars, which showed the greatest error. Kepler's leap occurred when he abandoned the assumptions handed to him and made new ones, the main one being that the planets move in ellipses as opposed to perfect circles. This seemingly simple and yet profound change in assumptions led to a much better fit of the data to the model.

Why were Kepler's findings so momentous? First, they established a much higher degree of agreement between the model predictions and data. Second, they shifted the orbits from the

[2] Cited in (Gingerich, 2005) p. 47.

perfection of circles to the seeming imperfection of ellipses, a shift that ran counter to the dogma of the day.

The main problem limiting further progress beyond Copernicus and Kepler was, simply put, lack of more substantive evidence. Enter Galileo.

Galileo – One Data Point is Worth One Thousand Opinions

Galileo proposed to strike out on a different course—to drop all Aristotelian talk of why things moved, and focus instead on the how, through painstaking observations and measurements

– Dava Sobel[3]

Born in Italy, Galileo Galilei (1564–1642) was the first to see the richness of the heavens by developing an improved means of collecting data—the telescope—and pointing it skyward. While Galileo did not actually invent the telescope, his improvements, especially 20× magnification,[4] allowed him to see the universe in more detail than anyone else had seen before.

He collected data and evidence, both in his efforts here and in other endeavors, and through his embrace of the scientific method pushed the scientific revolution forward even further. Observe, hypothesize, test, observe, hypothesize, test. He pointed his telescope to the night sky and saw…imperfections! Craters on the Moon and dark spots on the Sun, moons around Jupiter and phases of Venus. Galileo provided the much needed evidence to show that the objects in the universe are not perfectly circular and spherical and smooth. And he showed that not everything orbits the Earth. He used these findings to strongly endorse both the Sun-centered model and what became known as the Copernican Revolution.

To convince the obstinate who care only for the vain applause of the stupid herd?…The testimony of the stars would not suffice, even if the stars came to Earth and spoke for themselves.

– Galileo[5]

In their battle to establish the heliocentric theory, the new scientists provided a philosophic lesson that changed the course of history. They broke the stranglehold of religious dogma, at least for a while, and demonstrated that man can know the world—if he uses the method of observation, measurement and logic.

– Leonard Peikoff[6]

Unfortunately for Galileo, the book he wrote and finally printed in 1632 to capture his thinking was not well received by the Roman Catholic Church. In 1616, the church banned the Sun-centered model and Copernicus' book with it, citing Copernican astronomy as being "false and contrary to Holy Scripture."[7] The intensity of the church's view only increased in 1618 when the Thirty Years' War began between the Protestant and Roman Catholic states in Germany, eventually spreading to many other countries. The arrival of Galileo's book, *Dialogue Concerning*

---

[3] (Sobel, 2000) p. 301.
[4] (Harriman, 2000) p. 8.
[5] Cited in (Harriman, 2000) p. 10.
[6] Cited in (Harriman, 2010) Preface, p. 1.
[7] (Sobel, 2000), p. 79.

the Two Chief World Systems,[8] in 1632 amidst the war was too much. It forced the Church to act. Pope Urban VIII, already stretched thin by the war with worry and growing financial debt, could not allow publication of such a contrary book to go unpunished. So in 1633, at sixty-nine years of age, Galileo was tried. Facing severe punishment, he succumbed and testified against his own work, against his own theories, against the Copernican model.[9] In the end, Galileo was sentenced to what amounted to house arrest,[10] which lasted until the end of his life in 1642— the same year that Sir Isaac Newton (1642–1727) was born. In addition, *Dialogue* was banned, eventually being placed on the published Index of Prohibited Books in 1664, where it remained for the next 200 years.[11]

> *Already the minds of men are assailing the heavens, and gain strength with every acquisition. You have led in scaling the walls, and have brought back the awarded crown. Now others follow your lead with greater courage.*
>
> – Mark Welser in letter to Galileo[12]

But Galileo was not finished. While under house arrest he continued to write, creating what is thought to be his greatest original contribution to the field of science, summing up his full life's experience as an observer, thinker, mathematician, and experimenter. He wrote about motion, which we'll visit in Chapter 7. How things move, their speed, their acceleration, their direction, the resistance they experience. He quantified time, distance, and acceleration. He became the father of modern physics. Published in 1638, *Discourses and Mathematical Demonstrations Relating to Two New Sciences*[13] was his final work. And in these efforts Galileo explained something that troubled the ancient Greeks. Their final belief in the Earth-centered universe rested in part on the fact that humans cannot feel any sense of motion, which meant that Earth must be fixed in space. Galileo showed that we can most certainly travel through space at high constant speeds and not feel a thing, so long as everything around us is moving exactly as we are. He noted that when an object is dropped from a mast on a moving ship, the people on the ship observe it moving straight down while those on the shore observe more of a parabolic motion. Our sense of motion depends on an external reference that's not moving as we are. Without such a reference, we have no idea whether we are moving or not. Motion is relative, a theme that would be more fully developed by Albert Einstein.

---

[8] Dialogue Concerning the Two Chief World Systems (1632)—despite taking care to adhere to the Inquisition's 1616 instructions, the claims in the book favoring the Copernican Sun-centered model of the solar system led to Galileo being tried and banned on publication.

[9] (Harriman, 2000) p. 23. Galileo: "After having been instructed by the Holy Office to abandon completely the false opinion that the sun is in the center of the world…I abjure, curse, and detest [my] errors and heresies."

[10] (Sobel, 2000) p. 277. After his sentencing, rumor has it that Galileo rose from his knees while muttering under his breath, "Eppur si muove" (But still it moves). Unfortunately, no records exist to suggest that Galileo made such a statement on that day, at that time. Perhaps later, with friends, but not then. Whether rumor or not, the image does make for a great and powerfully defiant moment in time.

[11] (Harriman, 2000) p. 24. "Italy had long held a role of intellectual leadership, but that changed when the center of the Renaissance put reason on trial and found it guilty…Progress would continue in England and France, but astronomy was dead in Italy."

[12] Cited in (Harriman, 2000) p. 26.

[13] Discourses and Mathematical Demonstrations Relating to Two New Sciences (1638)—summarized work he had done some forty years earlier on the two sciences now called kinematics and strength of materials.

Newton and Gravity

What Galileo started, Newton completed. Newton created the laws of motion that we learn today in the classroom (Chapter 8). He brought structure and order to the phenomena around us. He created ideas far too voluminous to pay justice to here. He defined and standardized. Force. Mass. Momentum. Action. Reaction. Calculus. Optics. And he discovered the force of gravity. All was eventually published in 1687 as *The Principia*, one of the most influential books in history.

In an amazing leap of intuition, Newton realized that the same force that pulls the apple toward the center of the Earth also pulls the moons and planets toward each other. He quantified this force by proposing that it arises from mass and is simply proportional to the product of the masses of two bodies and inversely proportional to the square of the distance between them. Copernicus, Kepler and Galileo contributed to a quantified understanding of the Sun-centered model and Newton pulled it all together by showing where the numbers and relationships come from. As the sole force at work over the large distances in the cosmos, gravity explained the reason behind the motions of the "Copernican model" and became a critical component of the Big Bang model that was yet to come.

The Twentieth Century

We skip from the 1687 publication of *The Principia* to the early twentieth century, a time owned by Albert Einstein (1879–1955). The Special Theory of Relativity, the General Theory of Relativity, Brownian Motion, the Photoelectric Effect, Quantum Physics. Einstein pushed back the frontier of physics to such an extent that we are still amazed and paying homage some 100 years later, with *Time* magazine honoring Einstein as "Person of the Century." As with Newton, Einstein's work and, more importantly, the analysis and influence of his work fill volumes of written material that simply can't be addressed here. So instead, we focus on the one element of his work that is critical to this story, his General Theory of Relativity.

Einstein and the Static Universe

In this theory, Einstein proposed a new model of how gravity works. Newton's model rests on the assumption that space and time are fixed and well-defined entities. Einstein's model assumes otherwise, that space and time are flexible and linked in a term called spacetime, and that it is spacetime that is the underlying cause of gravity. While both Newton's and Einstein's models agree with each other for conditions found on Earth, they depart from each other under certain conditions in outer space, such as extreme gravitational force. An experiment was needed to decide between the two, preferably an experiment that forced each model to predict something that had not yet been observed.

> The most exciting event I can recall in my own connection with Astronomy is the verification of Einstein's prediction of the deflection of light at the eclipse of 1919... Three days after the

eclipse, as the last lines of the calculation were reached, I knew that Einstein's theory had stood the test and that the new outlook of scientific thought must prevail.

– Arthur Eddington[14]

Einstein outlined a defining experiment. His model predicted that light emitted by stars sitting behind the Sun should bend twice as much around the Sun as predicted by Newton's model. Since the light from such stars could only be observed when the Sun's own light was blocked out, the measurements had to be taken during a full solar eclipse. And so it was that Arthur Eddington, member of the Royal Astronomical Society and an early supporter in Britain of Einstein's relativity theories, organized and led a data-collecting expedition to watch the solar eclipse of 1919. He traveled with his team to the island of Príncipe off the west coast of Africa while a second team traveled to Brazil as a contingency plan. The resulting photographic plates taken at the two locations showed Einstein's theory to be correct.[15] By effectively overthrowing Newton, Einstein became the world's first science superstar, a celebrity resulting in part by the need for the world to celebrate the genius of man after the darkness of The Great War.

It was around this time that things started to get interesting for our story. As he was wont to do, Einstein decided to test his model on a grand scale by applying it to the entire universe and what he found disturbed him, for his equation suggested that the universe is unstable and should collapse.

This was an amazing moment in the history of the Big Bang theory and demonstrated how a paradigm can stifle even the most powerful of minds. Up until the early 1900s, no one, including Newton and Einstein, seemed to have conceived that the universe could be anything other than static, likely influenced by the general consensus among astronomers that this was indeed the case. When humans looked to the skies, all they saw was a stationary ceiling of stars rotating en masse across the night sky. Why assume otherwise? And so when Einstein considered the implications of his model on the universe, he had to conclude that unless the universe were perfectly uniform and homogeneous, then gravity would eventually pull the universe together and cause its eventual collapse, a conclusion that Einstein couldn't accept. So instead of considering that the universe was *not* static, Einstein added a new term to his equation, a term that predicted a repulsive force in the universe to exactly balance gravity's attractive force, thereby resulting in a static universe.

Einstein later admitted that this was the "biggest blunder"[16] he had ever made in his life. He relinquished the beauty of his General Theory of Relativity equation by adding an unneeded term, a cosmological constant, a fudge factor, to balance gravity and so stabilize the universe. This modified equation was short lived, as it was only a matter of time until other scientists weighed in on the matter, some with theory and some, more importantly, with data.

[14] Cited in (Chandrasekhar, 1998) p. 10.

[15] The story behind these measurements is quite a tale, and the conclusions drawn were not as straightforward as history suggests. For a more detailed account of this defining moment in history, see (Waller, 2002) Chapter 3. For additional interesting discussion, see (Chandrasekhar, 1998) pp. 110–141, which includes a good history of Einstein's cosmological constant.

[16] As cited in (Isaacson, 2007) p. 355–6, George Gamow recalled the following: "When I was discussing the cosmological problems with Einstein, he remarked that the introduction of the cosmological term was the biggest blunder he ever made in his life." For a contrary opinion, see (Livio, 2013) Chapter 10 in which Mario Livio concludes that while Einstein indeed had regrets about using the cosmological constant, his use of the word "blunder" was almost certainly a "hyperbole" created by Gamow.

Friedmann, Lemaître, and the Expanding Universe

Enter Alexander Friedmann (1888–1925). Born in Russia, educated in mathematics and interested in cosmology, Friedmann proposed that Einstein's equation is valid without the cosmological constant if one were to assume a dynamic universe, one that is expanding against the counteracting pull of gravity. Unfortunately, Friedmann's early death in 1925 and low profile relegated his profound theory to the backwoods of history.

Next up, Georges Lemaître (1894–1966). Born in Belgium, Lemaître studied to become both an ordained priest and, to pursue his strong interest in cosmology, a physicist. He began developing his own models of the universe and unbeknownst to him reinvented Friedmann's model of the expanding universe a decade after Friedmann. This was an unfortunate event in history, an event often repeated in science, where one person's discovery is overlooked when recognizing another person's identical discovery later in time.

Lemaître took his model to a higher level. While Friedmann was a mathematician, Lemaître was a cosmologist and so contemplated the real-world significance of his model. In another leap of intuition, and there were many such leaps in the history of the Big Bang, Lemaître realized that if the universe were indeed expanding, then yesterday must have been smaller than today and the day before yesterday smaller still. Taking this thought process to its logical conclusion, Lemaître ran the clock of time backwards and proposed that at some point in our far past, all the stars of the universe were condensed into a very compact region, which he called the *primeval atom*. The moment of creation of the universe then occurred when this small region suddenly destabilized, ejected matter outwards, and created the expanding universe.

Lemaître was the first person to accurately propose the origin of our universe.

So why did other scientists resist Lemaître's theory? Well, they took their lead from Einstein himself. Einstein was aware of Friedmann's earlier theory when he became aware of Lemaître's similar theory and rejected both, not being able to accept the concept of an expanding universe when he *knew* the universe to be static. Since Einstein was preeminent at the time that Lemaître presented his theory to the public in 1927, other scientists deferred to him and similarly rejected Lemaître, a devastating blow to Lemaître, made all the more so by Einstein's rather infamous statement: "Your calculations are correct, but your physical insight is atrocious."[17]

Deciphering Starlight

Enter Edwin Powell Hubble (1889–1953). Yes, the same Hubble for which that magnificent telescope in outer space was named. In a body of work that led to his becoming one of the greatest astronomers in the early 1900s, Hubble was the one who finally pointed his telescope toward the sky and obtained the data showing that the universe is indeed expanding, although it wasn't quite as simple as that. So let's take a short detour in our story.

The only means we have of determining what's happening in the universe is light. Light is the messenger. Without the light, we simply wouldn't have a clue. At first glance, one wonders

---

[17] (Deprit, 1984) p. 370.

how light can tell us anything. Light is light, right? Well, no. Light contains much information. And the use of light to break the code of the heavens had a path of development all its own.

In the early 1800s, a German optician, Joseph von Frauenhofer observed that a prism could spread out the light from the Sun into a range of colors. Now this was not new as many others, including Newton and most children, had similarly observed such a phenomenon. As light passes through a prism or any transparent object that doesn't have parallel entry and exit surfaces, like water droplets (think rainbow), a spread in colors results. This is because light consists of a multitude of photons, each with its own characteristic wavelength. As light travels through the transparent object, the photons of different wavelengths travel at different speeds and upon exit become separated according to wavelength. The red colors (long wavelength) go here. The blue colors (short wavelength) go there.

But what Frauenhofer noticed in 1814–1815 that was different was that lines of *darkness* occurred at certain wavelengths in the resulting spectrum. He and others soon concluded that atoms in the Sun's atmosphere must be absorbing specific wavelengths of light. Robert Bunsen and Gustav Kirchhoff took this one step further around 1860 by developing a more accurate means of collecting the data: a specially designed instrument they invented called a spectroscope. They turned this instrument toward the Sun and compared the observed pattern of dark lines to the known patterns of elements obtained in laboratory experiments. In this way, they discovered a range of elements in the Sun's atmosphere, a huge finding in and of its own right, but not the reason to bring it up here.

In 1868, Sir William Huggins and his wife Margaret turned Bunsen and Kirchhoff's spectroscope toward the other stars and noticed that their absorption lines did not exactly line up with those from the Sun. Instead, they were slightly shifted, leading them to conclude that the stars must be in motion. They knew of the Doppler effect, discovered by Johann Christian Doppler in 1842, in which the wavelengths of sound and light are shifted to either shorter or longer wavelengths as the object generating the waves is moving toward or away from the observer, respectively, the amount of the shift being proportional to the speed of motion. So they applied this theory to their observations and concluded that lines shifted toward the red end of the spectrum (where longer wavelengths reside) must be caused by stars moving away from Earth while lines having a blue shift must be caused by stars moving toward Earth.

Astronomers subsequently realized the opportunity offered by this new spectroscope. In 1912 Vesto Slipher (1875–1969), an American astronomer at the Lowell Observatory in Flagstaff, Arizona, became the first to start measuring the line shifts in the starlight from far galaxies.[18] Slowly, painstakingly slowly, measurements were made, one galaxy at a time. Between 1912 and 1917 he obtained 25 measurements, of which 21 were red shifted, a rather shocking result, certainly not a random mix. Unfortunately, Slipher received little recognition for this work, a situation likely caused in part by his lack of desire for such recognition. He became one of the "unsung heroes of astronomy"[19] for it was his groundbreaking work that later inspired Hubble. At that point in time, the Milky Way was thought to have been the entire universe, but based on the very high speeds of retreat he was observing, Slipher suggested an

[18] (Brooks, 2008) p. 11–12.
[19] (Brooks, 2008) p. 11.

even more stunning theory: what we thought were Milky Way's stars were instead very far away galaxies. Unfortunately, he didn't have the means to test this hypothesis.

While such findings drove a stake through the heart of the stationary universe, they didn't make their way to Einstein quickly enough to forestall him from adding his cosmological constant in 1917, a situation indicative of the communication gaps that can occur between different bodies of scientists, in this case the astronomers and the physicists, a theme that would play itself out again.

The great precision of the spectroscope to analyze light from the stars together with the increasing power of the telescopes to collect the light opened the door for the astronomers to map the speeds of many more stars in the universe around them. But there was another variable that was needed to resolve this mystery, a variable that proved more difficult to measure. Distance. While there is nothing inherent about light that lends itself to quantifying the distance of stars, there is a feature of stars that does. Their brightness.

Henrietta Leavitt's Cepheid Variables

In the late 1800s the Harvard College Observatory developed a reputation for photographing stars and leaving piles of images for others to analyze. Sensing opportunity, Edward Pickering, director of the observatory starting in 1877, recruited a team to mine these data. One on this team was Henrietta Leavitt (1868–1921). Operating around the turn of the century, Leavitt compared two photographic plates of the same stars over a period of time and observed a cyclical variation in the stars' brightness, swinging from high to low and back again, over and over, with the peaks separated by a range of days or months. These stars became known as Cepheid variables.

In a fascinating piece of focused scientific work, Leavitt further determined that the inherent brightness of the Cepheid is related to the speed of this variation in brightness. All stars have an inherent, intrinsic brightness. However, what we observe on Earth is *apparent* brightness. The further away a star is, the lower its apparent brightness, even though its inherent brightness remains the same. This is similar to when someone with a flashlight walks away from us. We see the brightness of the flashlight fall off even though it's the same flashlight with the same inherent brightness.

Leavitt was able to determine that the inherent brightness of the Cepheids is related to their speed of variation; thus, by measuring this speed, one can calculate inherent brightness. By further measuring apparent brightness, one can then calculate distance, since she observed a direct correlation between the fall off in brightness with distance. In this way Leavitt created a means of measuring distance to the stars.

Hubble Provides Evidence of the Expansion

Returning now to the mystery of the red shift, Hubble had at his disposal two very powerful tools and demonstrated his genius by bringing both together, a spectroscope to measure the speed of stars relative to Earth, following in the footsteps of Slipher, and Leavitt's technique for

The Doppler Shift

FRAUENHOFER
BUNSEN + KIRCHOFF
HUGGINS

- spectroscopy of starlight
- calculate star's velocity based on shift

SLIPHER (1917)
- did not have the means to calculate distance

HUBBLE (1929)

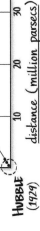

star's velocity (km/sec)

500   1000

star's distance (parsecs)

10^6 2×10^6

1 parsec = 3.26 light years

HUBBLE (1931)

velocity (km/sec)

5000   10,000   15,000   20,000

distance (million parsecs)

10   20   30

HUBBLE (1929)

LEAVITT

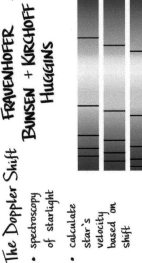

maximum brightness of stars

minimum brightness of stars

period (days)

12 13 14 15 16

0 20 40 60 80 100 120

inherent brightness

- Cepheid variable star displays cyclic variation in brightness that is correlated with inherent brightness
- measure period to determine inherent brightness
- calculate distance to the star based on the difference between the measured brightness and the inherent brightness

Figure 2.1 Hubble provides evidence that universe is expanding

quantifying the distance to those same stars (Figure 2.1). He also had one of the most powerful telescopes on Earth at the Mt. Wilson observatory in California. With these tools, he set down to work, which consisted of the tedious, arduous, physically demanding work of nighttime observation in the cold nights on top of the mountain. His efforts paid off. The data arrived and spoke to him. Forty-six data points in all. When he plotted them in 1929, the world was never the same, for this graph became the first true observational evidence for the Big Bang theory that was yet to come. A critical turning point in history was defined, for a correlation was there. How many of us have eagerly plotted data looking for a correlation to help explain some phenomenon? Remember the joy at seeing that a correlation existed? Think about how Hubble must have felt, or Leavitt for that matter. The sheer joy of discovery, of pulling back the cloak just a little bit further on the hidden world of nature around us.

What many could see in Hubble's famous plot was this. As the galaxies became more distant from the Earth, their receding velocities became faster. And the conclusion was clear. The universe is expanding.[20] The further away the galaxies, the faster they are moving, just what one would expect from an explosion that started from a center. The material that is blown the furthest is traveling at a higher speed.[21] Hubble's graph showing galaxies racing away at speeds proportional to their distance naturally led to the conclusion that all the galaxies must have started from the same common point some time far in the past.[22]

Einstein credits Lemaître

In a wonderful coincidence of timing, Lemaître's theory from 1927, and Friedmann's before him, perfectly matched Hubble's observations from 1929. Finding Hubble's data very convincing, Einstein publicly renounced his own belief in the static universe and fully endorsed the model of the expanding universe, crediting Lemaître in the process for the "most beautiful and satisfactory explanation of creation" that he had ever heard, thus marking the start of Lemaître's life as a celebrity.[23] In so doing, Einstein reconsidered his equation for general relativity and removed the ugly cosmological constant. With an expanding universe, the need for such a constant that was originally included to explain a static universe no longer existed.[24]

You would think that this would be the end of the story, but it wasn't. Many questions remained to be resolved. And this required the meeting of two separate groups of people, the cosmologists/astronomers and the particle physicists.

[20] (Singh, 2004) p. 248. Hubble did not necessarily leap to this conclusion as fast as others. His primary purpose was embodied in his mantra: "Not until the empirical results are exhausted need we pass on to the dreamy realms of speculation." Although he had collected the data, he did not personally get involved in the cosmological debate around the Big Bang.

[21] Once again, the Big Bang was not an explosion like we think of explosions today. The galaxies aren't rushing away from us, per se. They are moving with the expanding space.

[22] In conducting his research, Hubble also validated Slipher's hypothesis by determining that the famed Andromeda Nebula was not a cluster of stars and gas within the Milky Way but instead an entirely separate galaxy laying far beyond. See (Singh, 2004) Chapter 3.

[23] (Singh, 2004) p. 276.

[24] The cosmological constant has since re-appeared in Einstein's equation to account for the more recently discovered presence of dark energy in the universe.

The Meeting of Two Worlds: Cosmology and Particle Physics

The issue that was still to be resolved concerned the population of the elements in the universe. The arrival of the spectroscope allowed astronomers to quantify relative abundances of the elements outside Earth and they found that hydrogen was by far the most abundant element, followed by helium. The two together account for approximately 99.9% of all observable atoms in the universe. Whatever theory was proposed to account for the origin of the universe had to account for this as well.

And this is where particle physicists and cosmologists started to come together, a fascinating overlap since cosmological distances are on the order of light-years while particle physics distances are on the order of angstroms and less. That the two would have reason for talking with one another was never anticipated. However, in hindsight, it was inevitable. Running time backwards shrinks the universe from extremely large to extremely small distances, bringing the cosmologists toward the particle physicists as time approaches zero.

Gamow, Alpher, and Herman Calculate Conditions of the Early Universe

Enter George Gamow (1904–1968). Born in the Ukraine, Gamow was one of the first to combine nuclear physics and cosmology. Intensely unhappy under the Soviet regime and the associated Communist ideology, Gamow fled the Ukraine in 1933 and eventually arrived in America to pursue his interest in the Big Bang and the original formation of the atomic nuclei, known as nucleosynthesis. He started his research in the early 1940s at George Washington University and since most other American scientists were out of sight in Los Alamos working behind the secrecy of the Manhattan Project, he worked alone. His previous life in the Soviet Union, even though as a dissenter, precluded him from joining the ranks of his fellow physicists at Los Alamos.

Gamow chose to start his calculations by starting with the universe as it is today and then running the clock backwards. Using simple mathematics, he quantified the rise in both temperature and density as the universe shrunk and potential energy was transformed into kinetic energy. The hotter and more compact universe eventually stripped the electrons from the atoms and blasted apart the resulting nuclei into protons and neutrons. The Big Bang in reverse. As Gamow approached the starting point of the universe (t = 0), temperature and density rose very strongly as did his need to draw on his background in particle physics. Thus did the largeness of cosmology and the smallness of nuclear physics first overlap.

At this point, Gamow, together with his students Ralph Alpher (1921–2007) and later Robert Herman (1914–1997), began piecing together what must have happened in the early moments of the universe. Using known reaction data obtained from particle physicists and some basic assumptions about temperature, density and uniformity in the early universe, and spending a lot of time doing calculations—three years' worth, back in the days before computers—Gamow, Alpher, and Herman showed that only hydrogen and helium nuclei were created in the early universe, in a ratio of about 10 to 1. Consisting of only one proton, the hydrogen nuclei were stable and could exist alone. However, being unstable with a half-life of only about 10 minutes, free neutrons were predicted to not exist as they would decay to protons.[25] Instead, neutrons were predicted to exist only in a combined form, alongside protons inside helium-4 (He-4) nuclei.

These never-before-done calculations were remarkable. That three scientists could sit down and make estimates about what went on during the first 300 seconds of the universe and arrive at answers consistent with how we observe the universe today, especially regarding the currently observed hydrogen to helium ratio, was simply amazing.

The 5- and 8-Particle Gaps

Alas, Gamow, Alpher, and Herman eventually hit a significant roadblock. They couldn't show a way for elements heavier than He-4 to form during the Big Bang. They could show how protons and neutrons combined in the early moments of the universe to form the 2-, 3-, and 4-particles nuclei of deuterium, tritium, and He-4. But realizing that 5- and 8-particle nuclei are not stable, they could not find a way to grow He-4 nuclei via the addition of either one proton or another He-4 nucleus. Neither route worked. And the likelihood of He-4 reacting with two protons simultaneously to form the 6-particle lithium nucleus was extremely low, especially in the low-density conditions rapidly established after the Big Bang.

While the 5- and 8-particle gaps prevented further work on the formation of heavier elements during the Big Bang, there was another aspect the trio pursued that was equally important. Photons. Once again, using their knowledge of particle physics, these scientists realized that photons scatter strongly in the presence of charged particles, especially free electrons, but don't scatter in a neutral environment such as exists in gases consisting of fully developed atoms in which the charged electrons orbit the charged nucleus.[26] So during the early universe, when both photons and free electrons were present, they proposed that the universe was opaque since the photons couldn't travel far without being scattered by the plasma. However, once the electrons combined with the protons to form atoms, the photons could travel in uninterrupted straight lines, making the universe as clear as it is today. This transition should have occurred when the universe temperature cooled to 3000 °C, a temperature low enough for electrons to form stable orbits in the atom. This temperature corresponded to a 300 000 year-old universe. Their concluding calculation in 1948 suggested that the continued expansion of space and associated stretching of the photon wavelengths from this point on should have further cooled these photons down from the 3000 °C starting point to about 5 K today—a remarkably accurate estimate considering that recent measurements put this number at 2.7 K—and further suggested that these photons should be observable today as the fossil record of the Big Bang.

Predicting something that had not yet been observed is such a rare and very exciting occasion, especially such a prediction as this, since it offered a means to definitively and finally validate the Big Bang model during a time in the late '40s and early '50s when the model had not yet been fully accepted by the scientific community. This was a great moment in the history

---

[25] Neutrons played a significant role in the initial three minutes of the universe as explained in (Weinberg, 1993) Figure 9, p. 111.

[26] The reasons for this are discussed in Chapter 1.

of cosmology that ultimately led…nowhere. Gamow and his team were nuclear physicists, not cosmologists. What they had proposed required validation by the latter, but the wide gap between the two groups significantly hindered communications and collaborations. Very few such people with such a broad interest in both worlds existed at that time. This fact combined with the statement mentioned above, that not all astronomers were convinced, even in the '40s and '50s, that the Big Bang theory was valid, unfortunately led Gamow, Alpher, and Herman to bring their research program to a close in 1953 so that they could move on with their lives. And yet progress on a different front continued.

Hoyle Shows the Bridge across the 8-Particle Gap

Enter Fred Hoyle (1915–2001). Born in England, Hoyle was a mathematician who evolved into a cosmologist. His curiosity drove him to theorize about the stars and set him to work modeling their evolution through the various stages of life, describing the slow contraction and resultant rise in temperature and pressure, leading to the fusion of protons and the eventual formation of heavier elements. Building on the work of Fritz Houtermans and Hans Bethe, two exiled German physicists who in the late '20s and early '30s conducted the first theoretical calculations detailing the mechanism of stellar fusion, thereby explaining the seemingly endless source of energy emanating from the Sun,[27] Hoyle eventually pieced together a map of how, starting from carbon-12, all the heavier elements could form in the stars[28] and thus showed these reactions as being distinct and different from those that solely formed hydrogen and helium in the Big Bang. The formation of the heavier elements required the higher pressures offered by the collapsing stars. He explained the relative abundances of almost all the elements in the periodic table and showed that successive addition of He-4—analogous to a He-4 polymerization reaction—accounted for the large amounts of oxygen and iron and the small amounts of gold and platinum. His study of star evolution enabled him to explain the variable Cepheids. He explained much. And yet, as with the Gamow team, he too was stymied by the 5- and 8-particle gaps. He could not explain how He-4 could react to form either lithium-5 or beryllium-8, and thus could not explain how carbon-12 came into being.

In another leap of genius that was to characterize the major steps forward in the story of the Big Bang, Hoyle used the *anthropic principle* to state that, since we exist, then the universe must allow for human existence, and so there must have been a route to carbon-12. Using this thought as motivation, Hoyle relentlessly pursued discovery of this route and eventually achieved a breakthrough. His theory? Two He-4 nuclei could react to form beryllium-8 and further react with another He-4 to carbon-12, but only if the beryllium-8 could be struck quickly enough by He-4 to interrupt its rapid decay (millionth of a billionth of a second). The only way for this to happen was in an extremely dense and high-temperature (200 million °C)

---

[27] (Singh, 2004) Chapter 4. Fritz Houtermans along with Robert d'Escourt Atkinson in 1929 were the first to detail the fusion of two protons, while Bethe, enabled by Chadwick's discovery of the neutron in 1932, completed this work in 1939 by detailing the further fusion reactions that form He-4.

[28] The entire mapping of heavy element formation in the stars and all of the associated nuances therein was thankfully pulled together into one single famous paper (Burbidge et al., 1957).

environment in which He-4 is immediately available to intercept the beryllium-8 and convert it to carbon-12. While such conditions were not present during the Big Bang—the density was not high enough as it fell too fast—they were present in the stars.[29]

Hoyle's calculations showed the conversion of He-4 to carbon-12 to be a very slow process. But multiplying this slow process by two very large numbers, the age of the universe and the amount of matter in the universe, resulted in a large amount of carbon, enough to form the heavier elements we see today, both here on Earth and in the universe around us. While amounts are small compared with hydrogen, such heavier elements are indeed present and measurable. With these calculations, Hoyle showed that the path to carbon was across the 8-particle gap as opposed to the 5-particle gap.

Hoyle also showed that the large abundance of helium in the universe could only have originated in the early moments of the Big Bang and not in the stars where the milder conditions resulted in slower fusion of protons to helium.

Hoyle's and others' work describing the formation of the heavier elements in the stars culminated in the mid '50s, right around the same time as Gamow, Alpher, and Herman were closing shop on their own work. Interestingly enough, Hoyle did not believe in the Big Bang theory,[30] preferring instead the alternative theory of the time, the Steady State theory, in which the universe always was and always will be.[31] The eternal universe. So even though his work helped put another piece into the Big Bang puzzle, the one showing that the heavier elements did not need to originate at the beginning but could have instead evolved later in the stars, he still didn't believe in the Big Bang.

At this time in the '50s, cosmology was in the middle of a paradigm shift with two very different theories competing against each other—a single Big Bang expansion event vs. a steady-state expansion with continuing formation of mass—neither clearly the leader, both eagerly awaiting definitive data that would tip the scale toward one over the other. Opinions and egos were all at play[32] as the different theories competed for the top prize. A defining set of data was needed to resolve the competition. And the data could again come from the only messenger available to Earth-bound scientists: light from the universe.

Up Until Now, Only Theory. Then Serendipity Arrived.

> *Penzias said that the measurements were going fine, but that there was something about the results he didn't understand. Burke suggested to Penzias that the physicists at Princeton might have some interesting ideas on what it was that his antenna was receiving.*
>
> – Steven Weinberg[33]

---

[29] Many scientists contributed to Hoyle's success and really to the entire discovery of the Big Bang. Read the key references of (Weinberg, 1993), (Guth, 1997), (Kragh, 1996), and (Singh, 2004).

[30] (Singh, 2004) pp. 352–3. It was rather ironic that it was Hoyle who, during a 1950 radio program, coined the term "Big Bang" in light of the fact that he was one of the greatest critics of the theory.

[31] The two competing theories of the universe, Big Bang (an evolutionary universe with a beginning in time) and Steady State (a stationary universe of infinite age), are well discussed in (Kragh, 1996).

[32] (Singh, 2004) Chapter 5.

[33] (Weinberg, 1993) p. 49.

Light, both visible and invisible, is the only avenue for us to study the universe. All sorts of data are available in such light: the distribution of wavelengths, the spectral lines, the period of change in variation. Without any other means to understand the universe, we study light. It tells us what's happening, or, better put, what happened. By the time it reaches us, whatever it was that generated the light happened long ago. And so it was that the study of one aspect of light proved to be the deciding factor here. Enter Arno Penzias (1933–) and Robert Wilson (1936–).

As radio astronomers conducting research at Bell Laboratories in New Jersey in 1964, Penzias and Wilson conducted experiments with a large telescope that measured light in the wavelength of radio waves as opposed to visible light. They sought to measure radio noise coming from the Milky Way galaxy in order to expand the range of data made available by accessing this wavelength of light.

The critical step for them was to first 'zero' their equipment. Because the radio signals from distant galaxies are weak to begin with, they had to make sure that their equipment was sensitive enough to detect them. This meant that they had to ensure that when no radio waves were present, there was no signal detected by their telescope. Many scientists and engineers have been fooled by equipment and instruments that didn't read 'zero' when they should have.

So in 1964 Penzias and Wilson pointed their telescope at a location in the emptiest part of the sky where no signal should have existed…and found a signal. Thinking this to be a mistake, they tried and tried again to eliminate all possible extraneous sources for such a signal, such as bad wiring or nesting pigeons. They broke down and cleaned out their equipment again and again. And no matter what part of the empty sky they pointed their telescope at, the signal remained. And so they did the only thing they could do, measure the signal, specifically two critical features of the signal: wavelength and intensity. From these points, Penzias and Wilson calculated that this light had an equivalent temperature of about 3.5 K.[34]

At this point, Penzias and Wilson did not understand their finding and its significance… but their neighbors did. In a magnificent demonstration of the power of networking, Penzias and Wilson and their unrecognized discovery were eventually put in contact with theoretical astronomy professors at nearby Princeton by the names of P.J.E. Peebles (1935–) and Robert H. Dicke (1916–1997). Unknowingly repeating Gamow, Alpher, and Herman's earlier work, Peebles and Dicke in the mid-60s predicted that there ought to be a background of radio noise coming from the sky, a relic of the Big Bang, signifying the time at around 300 000 years of age in the universe when electrons combined with nuclei to form electrically neutral elements and thereby leaving the photons to travel unscattered in straight lines ever since. The temperature at which the photons started their straight-line journey was around 3000 °C. As time marched on and the universe continued to expand, the photons' wavelengths got stretched down to lower temperatures, which Peebles and Dicke calculated to be around 10 K (versus the appr. 3.5 K measured by Penzias and Wilson, since revised to 2.7 K). Once Penzias and Wilson learned of Peebles and Dicke's theory, their discovery became clear.

Penzias and Wilson's discovery of this Cosmic Microwave Background (CMB) radiation effectively brought to an end the development and validation of the Big Bang theory. No other

---

[34] The conversion of a light's wavelength and intensity to an equivalent temperature relies on the blackbody radiation theory.

theory could account for such background radiation. This radiation was always present, ready to be observed by those on Earth. It just took a long time before advanced equipment was available to find it.

There was an unfortunate side of this major final piece to the Big Bang puzzle. The scientists who originally proposed the existence of the CMB radiation, Gamow, Alpher, and Herman, were effectively left out of the celebrations, receiving virtually no recognition for their first-to-publish efforts. This was not deliberate but instead reflected how different camps of scientists, such as cosmologists and particle physicists, were simply unaware of each other. This was neither the first time nor the last. Knowledge can die or at least go into hibernation.

The discovery of the CMB radiation validated the Big Bang theory, and in so doing, validated that the universe had a beginning, a moment of creation, a finite age.

Epilogue: The Validating Imperfection

Of course, there is an epilogue to this story, a final piece that fully completed the puzzle. This piece had to do with the reason how our Earth or any other clumped body of matter exists in the universe. Back at the beginning of time, the universe was perfectly uniform, well almost so. For if it were perfectly uniform, then mass never would have clustered, being pulled equally in all directions by gravity, with no inclination toward one direction over another. But how could such imperfection be detected?

Enter George Smoot (1945–). A professor at the University of California at Berkeley in the mid '70s, Smoot started his long and passionate quest to find these imperfections by taking more measurements of the CMB radiation. With the Big Bang theory now firmly established, scientists at the time realized that the CMB radiation represented the earliest available view of the universe. Whatever happened prior to 300 000 years is lost forever in the opaqueness that existed then, at least from the standpoint of our seeing it in the light that hits Earth. We can naturally *infer* and predict what happened prior to 300 000 years as discussed earlier through theories and calculations and data of the observable universe. But we can't observe it.

So detecting variation in the early universe translated into detecting variation in the CMB radiation, since this was the earliest data point available. If variation existed, it had to exist in this radiation.

Unfortunately, detecting such variation from Earth was next to impossible as the expected variations were too small to be detected with the associated interferences from the atmosphere. So a satellite was needed.

After years of effort, Smoot finally convinced NASA to launch a satellite to collect the data. In 1989 this satellite, called the Cosmic Background Explorer (COBE) satellite, was launched. At first, nothing. The CMB radiation was perfectly consistent, no variation from one point to another. An amazing confirmation of the Big Bang theory in a way, since it was impossible to otherwise explain how the CMB radiation from points in the sky separated by billions of miles could possibly be the same. But this perfection had to have some small flaw, some small wrinkle, or how else could we be here?

The COBE team at NASA continued to take data and ever so slowly, very small variations began to emerge: only 1 part in 100 000, but that was all that was needed to start the formation of the stars and the galaxies…and of us. The resulting CMB map of the sky that the team created from one of the most sensitive measurements in the history of science awed the audience when it was presented in 1992.

The COBE map effectively ended the story of the Big Bang. My writing here hardly does justice to the scores of people involved at all stages in the building of this theory, one of the most magnificent constructions in the history of science.

On to the Atom

I find the history fascinating and the theory awing, for lack of a better word. The numbers involved are incredibly large and small, far outside the comfort zone of human thinking. As humans, we tend to seek comprehension by comparing new information with information we've seen or experienced and stored in our brain. We understand to a large extent by analogy. We live in a world centered around the number 1. We have finite fingers to count with, finite people to talk with. Every day things occur to us as whole numbers. I'm 46 years old. You're 6 feet tall. We created measurements and definitions consistent with this.

So when numbers are presented to us far outside of the world of our experience, we simply can't fathom them. What does 10^{80} gm/cc even mean? I don't know. I can't relate to this number. But it doesn't mean that this wasn't the initial density of the universe. Just because I can't comprehend something, doesn't mean it wasn't so.

Why did I include the story of the Big Bang in this book? First, because it explains the origins and populations of the elements in the periodic table. Second, because the world of astronomy rarely overlaps with the world of physics, chemistry, and engineering, and perhaps something, some opportunity to gain insight, is lost because of this. Our division of science and engineering into different camps serves a purpose—we can't study everything—but perhaps this does us an injustice as it narrows our view, and hence our understanding, and hence our ability to explore and create. Look at Gamow's story. He discovered something new based on combining particle physics with cosmology; it's just unfortunate that his groundbreaking work didn't receive proper recognition simply because he wasn't part of the cosmology community. Life at the interface between two sciences can be fertile ground and also rather exciting. We'll see this theme again in some of the stories to follow, such as those involving Galileo.

So what's next? Well, in Part II we go from the large to the small, from the universe to the atom, on which many of our thermodynamic theories are ultimately based.

PART II

The Atom

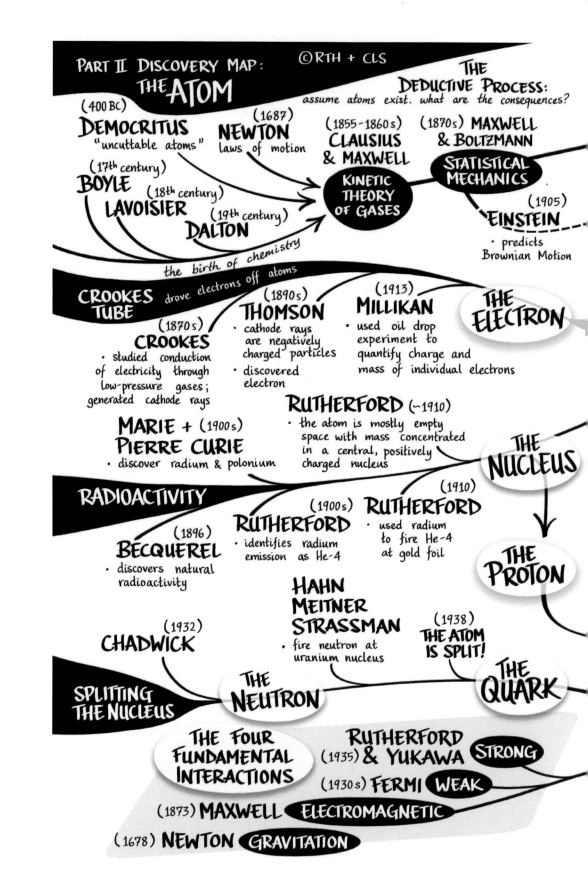

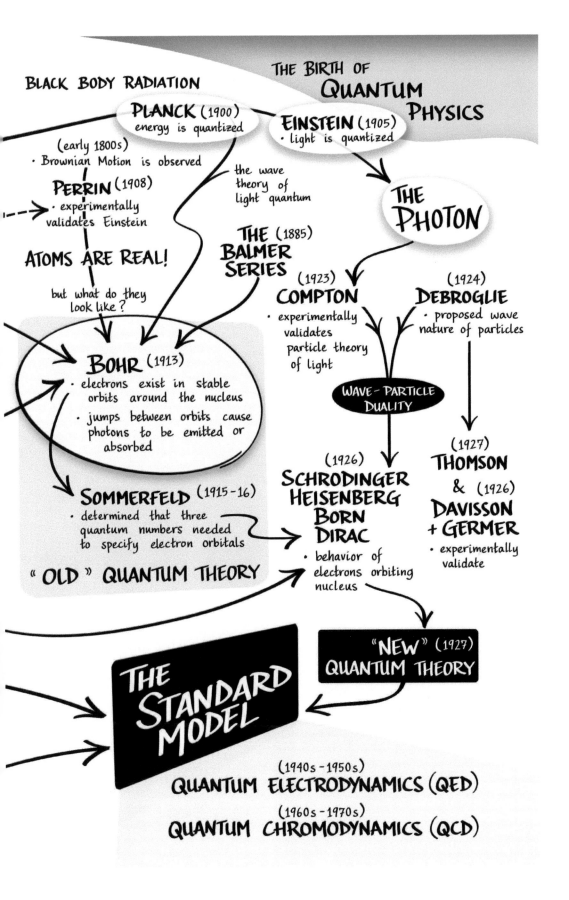

BLACK BODY RADIATION

THE BIRTH OF
QUANTUM
PHYSICS

PLANCK (1900)
energy is quantized

EINSTEIN (1905)
· light is quantized

(early 1800s)
· Brownian Motion is observed

PERRIN (1908)
· experimentally
validates Einstein

the wave
theory of
light quantum

THE
PHOTON

ATOMS ARE REAL!

THE (1885)
BALMER
SERIES

but what do they
look like?

(1923)
COMPTON
· experimentally
validates
particle theory
of light

(1924)
DEBROGLIE
· proposed wave
nature of particles

BOHR (1913)
· electrons exist in stable
orbits around the nucleus
· jumps between orbits cause
photons to be emitted or
absorbed

WAVE-PARTICLE
DUALITY

SOMMERFELD (1915-16)
· determined that three
quantum numbers needed
to specify electron orbitals

(1926)
SCHRODINGER
HEISENBERG
BORN
DIRAC
· behavior of
electrons orbiting
nucleus

(1927)
THOMSON
& (1926)
DAVISSON
+ GERMER
· experimentally
validate

"OLD" QUANTUM THEORY

"NEW" (1927)
QUANTUM THEORY

THE
STANDARD
MODEL

(1940s-1950s)
QUANTUM ELECTRODYNAMICS (QED)
(1960s-1970s)
QUANTUM CHROMODYNAMICS (QCD)

3 The Atom: science

If, in some cataclysm, all of scientific knowledge were to be destroyed, and only one sentence passed on to the next generations of creatures, what statement would contain the most information in the fewest words?...all things are made of atoms – little particles that move around in perpetual motion, attracting each other when they are a little distance apart, but repelling upon being squeezed into one another. In that one sentence, you will see, there is an enormous amount of information about the world, if just a little imagination and thinking are applied.

– Richard Feynman[1]

Forming the Elements – Review

Atoms weren't there at the beginning. The ingredients were but the early universe was too hot for them to combine. It took a little over three minutes for the universe to cool enough (about 1 billion °C) for protons and neutrons to combine (1:1) into deuterium nuclei. Once this bottleneck was broken, the production of an array of two-, three-, and four-particle nuclei soon followed. These reactions bound almost all free neutrons into helium nuclei, thus preventing their disappearance from the universe on account of their short (~10 minute) half-life. It would take another 300 000 years for the universe to cool further enough (3000 °C) for the electrons to fall into stable orbits around these nuclei.

The mixture resulting from the Big Bang, largely atomic hydrogen (75 wt%) and helium-4 (25 wt%), became the raw materials for a second stage of reactions that took place inside the early stars due to the severe conditions present: high temperature, high pressure, and tremendous particle-spewing explosions. The repeating star cycle of formation–collapse–explosion with increasingly heavier elemental starting materials led to the creation of the periodic table of elements at the universe-wide populations we observe today. After each cycle was completed and after matter cooled, the electrons settled back into their orbits, one electron for each proton, thus preserving electrical balance.

Some Staggering Numbers

So what is an atom and what are its properties? A seemingly simple question that becomes increasingly more complicated as one delves further. Let's start with the simple first. An atom consists

---

[1] (Feynman et al., 1989a) Volume I, p. 1–2.

Block by Block: The Historical and Theoretical Foundations of Thermodynamics, Robert T. Hanlon, Oxford University Press (2020). © Robert T. Hanlon.
DOI: 10.1093/oso/9780198851547.001.0001

of protons, neutrons (except for the simplest atom, atomic hydrogen), and electrons.[2] That's it. Just three fundamental building blocks behind the chemistry of the periodic table. The protons and neutrons, collectively called nucleons, comprise the nucleus, giving the atom nearly all its mass, while the electrons move in orbit around the nucleus, giving the atom nearly all its volume.

As with the Big Bang, when discussing the atom, the numbers used are enormously far removed from 1, the number we're most used to. For example, the atom is $1–2 \times 10^{-8}$ centimeters in radius, a number so small that it's difficult for us to grasp. In a cubic centimeter of a solid, about the size of the end of your thumb joint, there are approximately 10^{22} atoms, a number so large that it's also difficult to grasp. How big is this number? Well, if you spread this many peas across the continental United States, you would have a pile about 10 miles deep.

The sizes naturally get even smaller *inside* the atom. It turns out that the nucleus has a radius of only 10^{-13} centimeters, a hundred thousand times smaller than the atom itself. In fact, much of an atom contains no mass. Sticking with the peas, imagine placing a pea on the pitcher's mound inside a baseball stadium. The pea to the stadium is similar in size ratio as the nucleus to the orbiting electrons.[3] And since the mass of the protons and neutrons far exceeds the mass of the electrons, nearly all of the weight of the atom is concentrated in a central region less than 0.0001% of the atom's volume.

Because the atom's mass is concentrated in the infinitesimally small region of the nucleus, the density of the nucleus is huge. About 3×10^{14} gm/cc. One teaspoon of nuclear matter weighs as much as 50 billion tons. The particles of nuclear matter, namely the proton, the neutron, and thus the nucleus, have this density.[4]

Strange Behavior at Small Scales

> *The behavior of matter on a small scale… is different from anything that you are used to and is very strange indeed*
>
> – Richard Feynman[5]

While the atom may seem like a simple, quiet entity, to borrow a cliché, still waters run deep. The strong proton–proton repulsive forces at play inside the nucleus lead to the high-speed

---

[2] The atomic number defines the element and equals the number of protons, while the atomic weight is equal to the combined weight of protons and neutrons. The presence of isotopes reflects the fact that varying neutrons—and thus varying atomic weights—can exist for a given element. A number following the element name denotes an isotope. For example, carbon consists of 6 protons and so carbon-12 consists of 6 protons plus 6 neutrons.

[3] Other effective analogies include: If the atom were the size of a soccer ball, its nucleus would be less than the size of the period at the end of this sentence. Or, if the nucleus were the size of a golf ball, its electron would reside one mile away.

[4] While the atomic densities of the elements range from a minimum of 0.09 to a maximum of 22.5 gm/cc for hydrogen and osmium, respectively, the densities of the nuclei are constant at about 3×10^{14} gm/cc. The range in atomic densities reflects the fact that density equals mass divided by volume. As each nucleon is added to the nucleus, tight packing is maintained such that the increase in volume is directly proportional to the increase in mass, thus leaving the nucleus density the same. For the atom, however, as each nucleon is added to the nucleus, mass increases while the volume of the atom doesn't really change that much. For example, the radius of hydrogen is about 0.8×10^{-8} cm while that of osmium is 1.4×10^{-8}. So the much higher mass of osmium (76 protons + 116 neutrons) relative to hydrogen (1 proton) is squeezed into a similar volume (only about 8 times larger), resulting in a much higher atomic density as shown above. See (Emsley, 1990).

[5] (Feynman et al., 1989b) Volume II, p. 35–1.

motions of protons and neutrons relative to each other, and the electrons move around the nucleus in excess of about 1300 miles/sec, achieving quadrillions of rotations every second in a plane that itself is in motion, performing billions of full rotations per second.[6] During each billionth of a second, the electron sweeps across the entire outer surface area of the atom.

Such terms as orbit, particle, and velocity are convenient terms for us to use. But they're not accurate, for at the small scales involved the strange world of quantum physics takes over and matter behaves unlike anything we're used to. The electron, for example, collides with other mass, just like a particle, and yet it has a wavelength and frequency, just like a wave. It's these wave-like characteristics that explain why the electron's motion about the nucleus is like an orbit but really isn't and also why the electron is pulled very strongly toward the proton but doesn't crash into it as one would expect. When working to understand or interpret such strange concepts, we often reach for words we're familiar with to set up x-is-like-y analogies. As Feynman wrote, "Because of the weakness of the human brain, we can't think of something really new; so we argue by analogy with what we know."[7] But at the small dimensions of the atomic world, such approaches break down and at times create confusion. As we'll see later, this topic about language comes up time and time again in science. Nature does what she does, regardless of our way of thinking, and we struggle to describe her. Unfortunately we can't force her to conform to our words. Now, having said all this, I have to stay with such inexact words like orbit and particle because they do fill a role and because I don't have any other words to use.

The Atomic Building Blocks and the Interactions Between Them

In the following I address the science of the atom (Figure 3.1) by introducing the four relevant particles—proton, neutron, electron, photon—along with the interactions[8] between them, but I don't do this one-by-one. While each individual particle and interaction is interesting, it's really their participation in two larger structures that's relevant to this book, the first such structure being the nucleus where the protons and neutrons reside, held together by the strong nuclear interaction and stabilized by proton:neutron ratio adjustments enabled by the weak interaction. While the properties and behaviors of the nucleus aren't critical to understanding the basic thermodynamic chemistry covered in this book, they did play a significant role in the historical steps leading to the discovery of the atom as discussed in Chapter 4.

The second larger structure is the atom itself, comprised of electrons held in orbit around a positively charged nucleus by the electromagnetic interaction. Free electrons together with other fundamental particles naturally played a significant role in the early moments of the Big Bang, but it's the bound electrons that give us chemistry and thus most interests us now. I'll explain how the orbiting electrons create a seemingly hard (but comparatively massless) sphere around the nucleus, a sphere that can't be penetrated by the orbiting electrons of neighboring

---

[6] (Simhony, 1994) Chapter 5.

[7] (Feynman et al., 1989b) Volume II, p. 28–12.

[8] The four fundamental interactions—strong, electromagnetic, weak, gravitation—are sometimes referred to as the four fundamental forces. In general, I use "force" when referring to the attraction or repulsion that occurs as a result of the interaction.

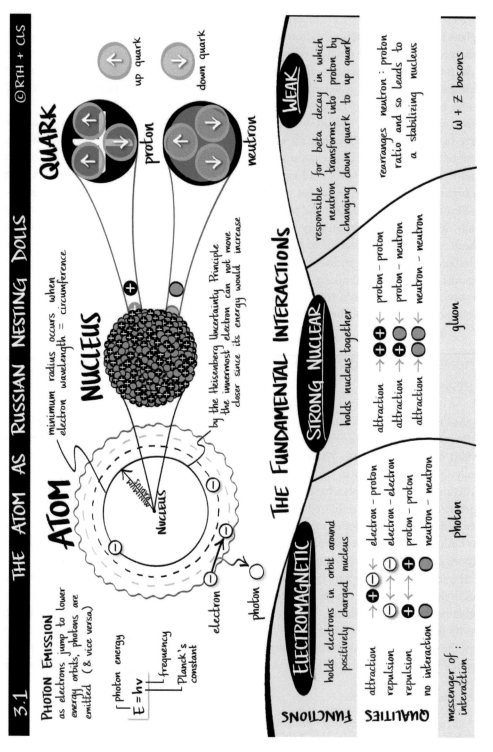

Figure 3.1 The atom as Russian nesting dolls

(non-reacting) atoms, and how these properties create atomic volume. I'll then explain what it is about the orbiting electrons that leads to intermolecular forces.

I'll conclude this chapter by introducing a final fundamental particle, the photon, which serves as the messenger of energy between atoms through its interaction with the orbiting electrons.

While admittedly brief, I will periodically return to the following science background throughout the remainder of this book as a means of connecting the microscopic world with our macroscopic theories and equations.

> The Nucleus
> Protons, neutrons, and the strong interaction
> holding them together

The atom's nucleus is comprised of both protons and neutrons. Since the neutrons are neutral and the protons are positively charged, thus strongly repelling each other via the electromagnetic interaction, the question must be asked, what is holding the nucleus together? How can it even exist?

Gravity and electromagnetism—like charges repel, unlike charges attract—represent two of the four fundamental interactions in nature. Operating inside the nucleus is a third interaction acting between the protons and neutrons. It's this interaction that holds the nucleus together, an interaction that acts at very small distances and is much larger than the electromagnetic interaction. It's called the strong nuclear interaction.

The strong interaction is attractive, pulling all of the nucleons toward each other—protons to protons, protons to neutrons, neutrons to neutrons. It acts equally strongly between these particles and over very small distances, about the same size as the diameter of the particles themselves, and serves to counter the repulsive electromagnetic interaction between the protons. It's the presence of neutrons and the contributing attractive strong interactions they bring that makes possible atoms heavier than hydrogen.

The strong interaction falls off extremely rapidly with distance, much more so than the electromagnetic interaction, halving for each 10^{-13} cm increment, which is about the diameter of a nucleon. Because of this, the strong interaction is confined solely to the nucleus and plays no role in the interaction between the nucleus and the orbital electrons or neighboring atoms.

Consider a proton at the surface of a nucleus. The strong interaction holding it to the surface is greater than the combined electromagnetic repulsive interaction of the other protons in the nucleus. If you tried to pull this single proton away from the surface, you'd need about the same force as that required to lift a 50-lb suitcase.[9] However, as you continued to pull and eventually move the proton just a small distance away from the surface, on the order of the radius of the proton itself, the *net* force dramatically switches from attractive to repulsive, and once this switch occurs, the repulsive force is so large that the proton explodes away from the nucleus at extremely high velocities. This phenomenon—positively charged nucleus fragments separating from the remaining positively charged nucleus—is the source of the tremendous energy released in atomic fission.

[9] (Frisch, 1973) p. 135.

For the elements with low atomic number, the nucleus contains about an equal number of protons and neutrons. For example, carbon-12 has six neutrons and six protons. With increasing atomic number however, a higher fraction of neutrons is required to hold the nucleus together. Iron-56 has 26 protons and 30 neutrons, gold-197 has 79 protons and 118 neutrons, and uranium-238 has 92 protons and 146 neutrons. The reason for this shift is that each nucleon, whether proton or neutron, feels the strong attractive interaction only from its immediate neighbors, while each proton feels the repulsive interaction of all the other protons inside the nucleus. Two protons on opposite sides of the nucleus aren't attracting each other at all; the sole interaction acting between them is repulsive. The increasing neutron:proton ratio with increasing atomic weight is needed to increase the attractive interactions inside the nucleus and so maintain stability.

This balancing act between protons and neutrons hits a breaking point at uranium-238, the heaviest naturally occurring element.[10] Barely stable, this element is primed to explode given the least disturbance. Its nucleus packs so much repulsive force from the 92 positively charged protons that even with so many additional neutrons the strong interaction is only just strong enough to hold the nucleus together. If the nucleus is disturbed ever so slightly, it breaks apart—well, really it explodes apart—into two pieces that are each positively charged. The strong electromagnetic repulsion between them causes them to fly apart at tremendous velocities.

Quarks – Source of the Strong Interaction

The source of the strong interaction lies inside the protons and neutrons in the form of quarks. The atom is somewhat like a Russian nesting doll. Inside the atom is the nucleus. Inside the nucleus are nucleons. Inside the nucleons are quarks. Some theorize that inside the quark are "strings." We won't discuss this further other than to say that six types of quarks exist, of which two, one labeled "up" and the other labeled "down," are relevant to this discussion. Each nucleon has three of these quarks, two up and one down for the proton, and two down and one up for the neutron. Being the actual source of the attractive strong interaction, these quarks only occur in tightly bound groups. An interesting aspect of the quarks is that the attractive interaction between them counterintuitively *increases* with increasing distance, meaning that they can't be pulled away from each other, further meaning that free quarks don't exist. Evidence of their existence is necessarily indirect.

Nuclear Decay – Alpha, Beta and Gamma

Nuclear decay, also known as radioactive decay since the emission of particles from a decaying source 'radiates' outward in all directions, reveals itself through detection of the decayed products, namely alpha particles, beta rays, and gamma rays (Figure 3.2).

---

[10] Heavier elements do exist, but they are man-made using special equipment and their nuclei are so unstable that they live a very short life.

3.2 NUCLEAR DECAY MECHANISMS ©RTH + CLS

ALPHA DECAY

since the nucleus is a very dynamic environment, there are moments when an alpha particle (He-4) will vibrate just far enough away from the main nucleus that the attractive strong interaction is overwhelmed by the repulsive electromagnetic interaction

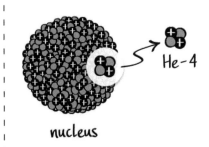

nucleus

He-4

BETA DECAY

during beta decay, which is caused by the weak interaction, a neutron transforms to a proton and emits an electron (together with an anti-neutrino)

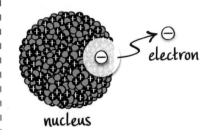

nucleus

electron

GAMMA DECAY

when a nucleus moves from an excited state to a stable state, it emits a gamma ray which is a high-energy photon

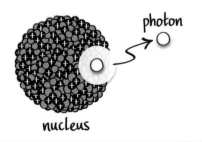

photon

nucleus

NUCLEAR FISSION

when certain "unstable" nuclei vibrate, one part may slightly separate from the other such that the attractive force holding them together is overwhelmed by the repulsive electromagnetic interaction.

"Fission" occurs and the two parts fly away from each other at extremely high speeds

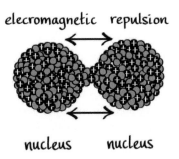

elecromagnetic repulsion

nucleus   nucleus

Figure 3.2 Nuclear decay mechanisms

Alpha decay. An "alpha" particle consisting of two protons and two neutrons (He-4 nuclei) experiences a moment where its electromagnetic repulsion of the main nucleus overcomes its attractive strong interaction toward its nearest neighbors in the same main nucleus and thus explodes away at about 5% of the speed of light.

Beta decay. A beta ray (high-energy electron) and a neutrino are emitted from an atomic nucleus when a neutron transforms into a proton to stabilize the nucleus.

Gamma decay. A gamma ray (high-energy photon) is emitted when an excited nucleus decays to a lower energy state.

I introduce these three decay processes for two reasons, the first because each played a significant role in the discovery of the atom, as I'll recount in Chapter 4, and the second to introduce the weak interaction. During beta decay, the weak interaction occurs within the quarks and thus operates at much shorter distances than even the strong interaction. It allows quarks to exchange fundamental particles such that a down quark within a neutron changes to an up quark, thereby converting the neutron to a proton plus the emission of a beta ray (electron) plus a neutrino.[11]

$$\text{Beta decay}: \text{Neutron} \rightarrow \text{Proton} + \text{electron} + \text{anti} - \text{neutrino}^{12}$$

> The Atom
> Electrons, protons, and the electromagnetic
> interaction holding them together

Of the four fundamental interactions, I'm primarily interested in the electromagnetic interaction, because in the end it's this interaction that governs the relationship between the electrons and protons and thus governs chemistry. Here's why. First consider the four fundamental interactions along with their dramatically different (relative) strengths.

Strong interaction = 1	acts inside nucleus; not relevant to this book's thermodynamics
Electromagnetic interaction = 10^{-2}	the root cause of chemistry
Weak interaction = 10^{-13}	acts inside nucleus; not relevant to this book's thermodynamics
Gravitation interaction = 10^{-38}	conservation of mechanical energy – the lever & free fall

The strong and weak interactions act across very short distances, much less than the diameter of the very small nucleus where they operate. While these interactions are critical to our world, serving to keep the nucleus intact, we don't sense them in action. This is left to the particle physicists who intentionally probe their inner workings.

Of the other two interactions, we are more familiar with gravitation. We see direct evidence of gravity every second of every day. It's the dominant force shaping our universe, embedded in space, reaching beyond the atom all the way to the most distant stars. It is the cause of motion in the lever and during free fall, the realization of which led early scientists to an early version of the conservation of mechanical energy.

---

[11] The reverse process is called electron capture and can occur inside atoms when the nucleus absorbs an inner bound electron.

[12] This particle was initially termed "neutrino" and was later changed to "anti-neutrino."

While gravity is part of our everyday language, the electromagnetic interaction is more obscure. We don't think of it. People on the street don't talk about it. It's a more difficult concept to understand. But, when all is said and done, it's the electromagnetic interaction (and resulting forces) that makes our life so interesting. Our body's senses evolved to be triggered by this interaction. It governs, either directly or indirectly, how we see, feel, hear, smell, and touch. It's the electromagnetic relationship between electrons and protons and the associated separation of charges that governs the chemistry of the periodic table. It's a seemingly simple interaction that provides an abundance of wonderful and colorful complexity and beauty.

The Electromagnetic Interaction

The electromagnetic interaction originates from charges—opposite charges attract, like charges repel, acceleration generates photons—and overwhelms gravity as shown by their relative strengths above. Put two grains of sand thirty meters apart from each other. They can barely feel the attractive interaction of gravitation between them. But if all of the mass in one of the grains were associated with a positive charge and all of the mass in the other grain were associated with a negative charge, the resultant attractive force between them would be three million tons.[13] We don't talk about the electromagnetic interaction because we live in a paired world of electrons and protons—it's tough to keep them apart. But this seemingly neutral world is an illusion. It's only neutral from a distance. Like an Impressionist painting, an atom is smooth and continuous only from a distance. Move up close and things start to get fuzzy as the neutrality disappears and the granularity shows itself. It's up close where the electromagnetic interaction does what it does, governing the motions of the granular parts and creating the atom's volume along with the tendencies of atoms to attract and react.

Why the Atom has Volume – the Quantized Orbit

What happens when a body is far removed from Earth? Gravitation interaction pulls the body toward Earth, causing it to accelerate. The body's total energy remains constant, as its potential energy is transformed 1:1 into kinetic energy. Depending on the situation, it can fly past Earth (with a change in direction), it can go into orbit, or it can crash.

What happens when an electron is far removed from a proton? Something similar but only up to a point. The electromagnetic interaction pulls the negative electron toward the positive proton. But this is where similarity ends and Feynman's world of "strange behavior" begins. Yes, if the electron is initially moving fast enough, as it was prior to the first 300 000 years of the universe, it will fly past with a change in direction. So this isn't that strange. And if it's moving slowly enough, it can go into orbit. So this isn't strange either. But it's what happens next that's strange.

---

[13] (Feynman et al., 1989b) Volume I, p. 2–4.

A body orbiting Earth will remain at whatever radius it established based on its initial conditions (momentum plus location) and will then stay in this orbit (almost) forever. Think of the planets orbiting the Sun. The total energy of the orbiting body stays the same; the swing back and forth from potential to kinetic and back to potential occurs on average in a 1:1 relationship. But an orbiting electron experiences something different because it can lose energy (via photon ejection) while the orbiting body can't. Once the electron establishes an orbit at a certain maximum or outermost radius, it starts moving in discrete jumps closer and closer toward the proton, shedding energy as it does so by ejecting photons. The electron moves inward from one orbital to the next, as if it were descending a staircase, sometimes one step at a time and sometimes skipping steps altogether, releasing one photon per each step equal in energy to the energy difference between the orbitals, until it can't go any further. Thus, as opposed to a free-falling body experiencing gravitational transformation of potential to kinetic energy, a "falling" electron experiences electromagnetic transformation of potential energy to photon emission.

For each discrete "quantized" step of the electron toward the proton, a photon is ejected. This process can be reversed if a photon of the same energy returns to the system to be absorbed by the electron and so energize it back up the stairs. Thus the electron moves down and down until it reaches the lowest possible radius. This movement is also sometimes referred to as an "energy minimization" process since the inward movement and corresponding photon ejection ultimately minimizes the total energy of the electron.[14]

Why a Minimum Radius?

The above still leaves us with the question, what stops the electron's descent? Why doesn't it keep on going and crash into the proton? Classical physics says it should. As we'll see later, this historic question was asked by Neils Bohr and the simple answer goes back to the electron's wave characteristics. In short, even though the electron is a particle, it has a wavelength. The early quantum mechanical rulebook, prior to the more complicated rulebook of Schrödinger and Heisenberg and others which isn't necessary to get into here, states that the electron can only establish stable orbits around the nucleus when the circumference (as determined by the radius) is equal to an integer multiple of the electron's wavelength. Thus, as it descends from orbital to orbital, the electron moves toward the nucleus in very precise integer steps, or multiple-integer steps if it makes larger jumps, from radius to radius, with each accessible radius corresponding to an orbital circumference equal to an integer multiple of the electron's wavelength, ejecting one photon for each step taken, until it gets to a final orbital whose circumference equals the length of a single wavelength of the electron. This "ground state" thus occurs at the minimum orbital circumference and thus the minimum orbital radius. For hydrogen, this radius is called the Bohr radius.

---

[14] The total energy of the electron, comprised of both kinetic and potential (electromagnetic) terms, is quantified in a mathematical function known as the Hamiltonian. The mathematics here are complicated especially when more than a single electron is involved. The Hamiltonian played a critical role in the development of quantum theory.

Heisenberg Uncertainty

But the question still remains, why does such a minimum exist? We return to Feynman's concept of "strange behavior." While the electromagnetic interaction pulls the electron toward the proton, the electron's wavelike character keeps it from crashing fully inward. Very advanced mathematical equations show that as the electron starts moving closer to the nucleus than its minimum radius, its wave is compressed into a smaller region, causing its total energy to increase, which makes this step highly improbable. Thus, the electron remains at the minimum radius where its total energy is minimized.

At the heart of these advanced mathematics is the famed Heisenberg uncertainty principle. What this principle roughly says is that as a particle's location becomes more defined, its momentum becomes less defined. This is one of those strange parts of nature for which there's no easy explanation. It's just so. Thus, as the electron moves closer to the nucleus than the minimum radius, it moves within a tighter space, making its location more certain but its momentum less certain, and this leads to its energy increase. The uncertainty principle explains why electrons don't crash into nuclei and thus why atoms can't be compressed. It was quantum mechanics that explained this.

*   *   *

Because the bound electron's potential energy decreases as it moves toward the proton (and ejects photons along the way), it has a lower potential and a lower total energy than it did at infinite distance. The energy of an orbital electron is typically quantified by the energy required to pull it away from the atom to an infinite distance. This energy is thus called the electron's "binding energy." Since the energy at infinite distance is arbitrarily set to zero, the energy of orbital electrons is necessarily negative. The ground state electron requires the highest removal energy; its "binding energy" is thus the most negative.

Pauli Exclusion

When we're dealing with atoms having more than one electron and one proton, an additional principle steps in to govern things. Modifying our previous examples, imagine releasing multiple electrons near a nucleus containing an equal number of protons (and associated neutrons). Once released, the electrons cascade down the orbital energy levels, ejecting photons while doing so, until they eventually get to the point where they can't go any further. Now one might think that all the electrons would end up in the same bottom-most orbital of minimum radius. But this doesn't happen. In yet another point of "strangeness," it turns out that electrons have a property called spin,[15] which has something to do with angular momentum, and this property has one of two possible values, +1/2 and −1/2. According to something called the Pauli exclusion principle,

---

[15] The term "spin" was coined during early research on electrons when they demonstrated properties associated with angular momentum, which implied that they must be spinning. However, in order for something to spin, it must have an internal structure. To the best of physicists' understanding, electrons do not have an internal structure, which means they can't actually spin. They must truly be fundamental, irreducible particles without structure but with an inherent angular momentum. Even if they were standing still, orbiting nothing, they would have an angular momentum, a property as fundamental as charge and mass.

two electrons of the same spin can't co-exist in the same orbital, thus meaning that the maximum number of electrons in a given orbital is two, and they must have opposite spin. So once the bottom-most orbital fills with two electrons, each of opposite spin, the next orbital up (larger radius) gets filled, again with two opposite-spin electrons, and this is repeated, again and again, until all the electrons have placed themselves in a bottom-up filling of the orbitals.

The Behavior of Orbiting Electrons Enables Modeling the Atoms as Hard Spheres

While the Pauli exclusion principle explains the electron housing rules, it also helps explain something else and brings us back to the start of this section. It contributes to our understanding of how the behavior of orbiting electrons mathematically turns atoms into billiard balls. Why do atoms "collide" with each other? Why don't their electron clouds just pass through each other? The answer? Pauli exclusion, which roughly says that no more than two electrons (of opposite spin) can be in the same place. The clouds aren't allowed to overlap. As Feynman said, "almost all the peculiarities of the material world hinge on this wonderful fact."[16] But note that while Pauli exclusion prevents overlap, it doesn't create the rebound effects of the collision. Pauli exclusion is not a force itself; it's a principle. The forces at work behind the rebound are the electromagnetic repulsion forces between electrons and between protons.

Now it is possible for electron clouds from two atoms to overlap in a special way. Consider two hydrogen atoms. As they move toward each other, if their respective electrons have opposite spin, the two electron clouds can indeed overlap. In fact, such overlap opens the door to chemical (covalent) bonding as a new orbital can form, a bonding orbital, in which the two electrons can reside while orbiting both protons. The two protons are pulled toward each other by the presence of electrons between them.

That the atom conceptually exists as a stable hard sphere is rather amazing since the electromagnetic interaction seeks both to explode the nucleus into fragments (proton–proton repulsion) and to collapse the electrons into the nucleus (electron–proton attraction). Fortunately for us, the presence of the countering strong interaction prevents the nucleus explosion and the quirkiness of the quantum world prevents both collapse (Heisenberg uncertainty principle) and atom overlap (Pauli exclusion principle). Acting together these principles explain why matter, despite being over 99.9999% empty of mass, has such strength and thus why atoms can be treated as hard, incompressible spheres, an extremely simplifying feature that opens the door to very powerful mathematical modeling. The additional fact that mass is concentrated in the center and not spread evenly throughout the volume provides a further simplifying assumption: the thermal energy associated with rotation for a monatomic atom is effectively zero. This fact impacts the heat capacity of monatomic elements as we'll see in later chapters.

```
Atoms
Why they attract
```

---

[16] (Feynman et al., 1989c) Volume III, p. 4–13.

Picture two of the simplest of atoms, atomic hydrogen, each containing one electron and one proton, sitting far enough apart from each other such that the only force between them is the attractive interaction of gravity (Figure 3.3). While they each contain both positive and negative charges, they appear neutral to each other since the distance between the internal charges is small relative to the distance between the two atoms. The attraction and repulsion forces between the two charges of one and the two charges of the other cancel each other out. Gravity's all that's left. And even though gravity is such a small force, when it's the only game in town, it dominates.

However, as the hydrogen atoms approach each other and the distance between them shrinks toward the scale of the atoms themselves, the electromagnetic forces between the two no longer cancel out. The charges from one start to "see" the charges from the other and so begin influencing each other's motions.

When the electron (E1) of the first hydrogen atom moves between the two protons (P1 and P2), it repels the electron (E2) of the second, forcing it to move away and behind P2. E1 is attracted to P2 resulting in net attraction between the two atoms. Since E1 is closer to P2 than P1 is, the attractive force between E1 and P2 is greater than the repulsive force between P1 and P2. One could argue though, that E1 will eventually circle around behind P1, thus reversing this situation. But one can't ignore E2 in the process. When E1 departs, E2 comes back out front again and creates a new net attractive force with P1. In this way, the two electrons develop a synchronized motion, resulting in a net–net attractive force between the atoms that switches back-and-forth and back-and-forth at very high frequency. This attractive force, called the van der Waals force or London dispersion, is a natural response between any two atoms. The strength of attraction varies, but it's always there.

In essence, each atomic hydrogen atom is a microscopic magnet containing a positive (proton) and a negative (electron) pole. The opposite poles from the two magnets attract each other. In the real atoms, even though the poles are spinning at very high speeds, the atoms sense their presence and so respond to each other, aligning their motions such that the opposite poles, although moving, can still line up in a synchronized way and create a net attractive force.

Between all atoms, not just atomic hydrogen, the attractive van der Waals attraction is always present. If atoms approach each other slowly enough, these attractive forces will lead to weak van der Waals bonding (not covalent bonding) and can result in such physical phenomena as condensation. When additional attractive forces are involved, such as exist in polar molecules where certain nuclei attract electrons more than others, an extreme case being polar water molecules in which the oxygen nuclei have much stronger attraction of the electrons than the hydrogen nuclei, the weak bonding tendency (again, not covalent bonding) increases. In these situations, however, if the approach speed is too high, the atoms and molecules collide and bounce off each other like billiard balls. With increasing speeds, the bounce is stronger and prevents weak-bonding attachment from occurring, much like two "sticky" objects will bounce off of each other if you throw one at the other with high enough speed. Replace the word "speed" here with "temperature" and this explains why changes in temperature cause transitions from solid to liquid to vapor.

Atoms
Why they react

3.3   THE ATOM AS HARD SPHERES ©RTH + CLS

ATTRACT FROM A DISTANCE | **AND REPEL UP CLOSE**

ATTRACTION | **REPULSION**

ATTRACTION | **ATTRACTION**

E1 P1 → ← E2 P2   P1 E1 → ← P2 E2

- E1 and E2 synchronize motions to always result in a net attractive force between atoms

- this force is often referred to as London dispersion force or Van der Waals force

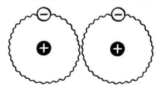

- when atoms get close enough to each other, their electron clouds repel each other, resulting in a "collision" — hence the modeling assumption of "hard spheres."

- by the Pauli Exclusion Principle, the clouds cannot overlap because no two electrons can have the same quantum number.

- however if an orbital offers a lower energy state to an electron from each atom, then a reaction can occur

Figure 3.3   The atom as hard spheres

Returning to atomic hydrogen for now, as the two atomic hydrogen atoms attract each other and move closer to each other, they react since each electron has a new lower-energy orbital accessible to entry. This new orbit is called a bonding orbit. The sharing of the electron pair between the two atoms results in a very strong bond, called a covalent bond, and leads to the formation of a hydrogen molecule, denoted as H_2. In a covalent bond, two electrons, one from each atom, occupy the same orbital, called a molecular orbital. The orbital is shaped like an hourglass, with each proton located in its own bulb. When the electrons' orbit takes them to the center of the hourglass, they serve to pull the two nuclei toward each other. The length of the resulting bond is determined by an equilibrium balance between the attraction of the two nuclei toward the centrally located electrons and the repulsion of the two protons for each other.

The covalent bond is the strongest chemical bond. There is none stronger because the Pauli exclusion principle limits the maximum number of electrons in the space between the nuclei to two—and they must be of opposite spin. All other types of bonding, such as van der Waals bonding, hydrogen bonding, ionic bonding, and so on, involve the occurrence of less than two electrons between the nuclei.

The Incomplete Outer Shell

To generalize for more complicated atoms, the reason why a bond could form between the two hydrogen atoms in the last example is that each had an unfilled outer orbital and the Pauli exclusion principle allows both electrons to form a new bonding orbital (lower energy than the two separate orbitals) so long as they are of opposite spin. This generalization also applies to atoms comprised of multiple protons and thus multiple electrons but things get more complicated since increasing the number of electrons opens the door to increasing orbital shapes. Orbitals are typically categorized by energy level, which is sometimes referred to as a shell. The innermost energy shell, which is characterized by a principal quantum number and which I don't need to get more into here, is the ground state and is comprised of a single orbital having a spherical shape. The second energy shell is comprised of four orbitals—one sphere, three hourglasses—while the third shell is comprised of nine orbitals—one sphere, three hourglasses, five cloverleaves. (Orbitals having the same energy level but different characteristic shapes are called degenerate.) So it's really more accurate to say that when the outermost energy shell isn't full, meaning that one or more of the degenerate orbitals is not full, then the atom can react. But if the outermost energy shell is full, then the element is inert. There are chemistry subtleties here that I don't want to get into as it would necessarily lead to a lengthy discussion of the rather complicated physics of electron orbital structures and chemical reactions. But again, in general, because the housing rules are based on bottom-up filling, whether or not an element is reactive depends on whether or not the outermost energy shell is full.

These issues explain the structure behind the periodic table. The chemical properties of the elements are governed largely by the extent of saturation (or valence) of the outermost energy shell. The inner orbitals aren't relevant because they're full. For example, elements of similar valence behave similarly in chemistry and are thus positioned in vertical columns in the periodic table. Elements having a valence of zero have full outer shell(s) and are thus chemically inert. These elements are the inert elements (noble gases) that populate their own column.

Given this, one can see why the discovery of the atomic number played such a crucial role in the development of the periodic table. The chemical properties of an atom are largely dictated by the number of electrons it has, which is equivalent to its atomic number as defined by the number of protons it has.

Alas, the wonderful description of the elements as one marches through the periodic table is beyond the scope of this book but does fill many books and makes for fun reading. The interested reader is encouraged to explore the works of, for example, John Emsley or Eric R. Scerri.

*   *   *

The totality of the above phenomena lies at the core of chemistry. It's the fact that negative electrons and positive protons are separated in space within a bound atom that gives rise to attractive forces between atoms and molecules. It's the fact that negative electrons continually seek to minimize total energy by moving as closely as possible to the positive protons that gives rise to chemical reactions. It's the fact that a negative electron can only get so close to the positive proton combined with the fact that orbiting electrons of colliding atoms can't overlap that gives the atom volume. And it's the combination of all of these facts and others that give us chemistry.

The Photon

Our eyes detect photons as colored light. Our skin feels photons as heat from a red hot metal bar across the room. Doctors fire photons called x-rays through our body to detect if a bone is broken. We heat dinner using microwaves. In each case, the photons, whether in the form of light, thermal radiation, x-rays, or microwaves, transfer energy between matter, typically through their interaction with orbital electrons.

Photons are discrete particles of electromagnetic energy that don't sit still. Ever. They always move at the speed of light (186 000 miles/sec in a vacuum) and have zero mass.

As an elementary particle, a photon is generated or destroyed as a single particle and once generated, remains this way, travelling vast distances in a straight line, not spreading out over distance and time as a wave would.

A sea of photons exists all around us. On a sunny day, about a thousand billion (10^{12}) photons of sunlight fall on a pinhead *each second*. We don't sense these discrete particles of energy; their overwhelming presence makes them appear to us as a continuum. We can't see the gaps.

Yet, while such language suggests that photons are particles, don't be fooled. Like the electron, or even the proton and neutron for that matter, each photon has both particle and wave properties, a wave–particle duality that we haven't yet captured in a word. Like a particle, each photon carries energy, linear momentum, and angular momentum. Like a wave, each photon displays both wavelength and frequency, the two linked together by the constant wave speed equal to the speed of light. Many discrete photons added together form an electromagnetic wave. Wave or particle? It all depends on one's perspective, neither perspective being more truthful than the other. Once again, our language fails us in this small and strange world.

Why should we care about photons? Because when photons are absorbed (or emitted), they increase (or decrease) the energy of matter. Just because they have negligible mass doesn't mean they have negligible influence. Each photon is an electromagnetic wave that interacts with charged particles and causes them to move; thus, each photon has momentum; they cause motion; they affect the thermal energy of matter; they play an important role in any energy balance such as those based on radiative heat transfer. When bound electrons move from orbit to orbit or when energy transitions occur inside the nucleus, photon absorption or emission is involved. At the most fundamental level, photons are messengers of electromagnetic interactions, playing a key role in the events that occur between charged particles. They really aren't any different from the mass-carrying constituents of the atom. Each shows wave–particle duality, some more so than others. At these small scales, there is no distinction between wave and particle.

When photons are emitted or absorbed by matter, the rule of interaction is very specific. The photon's energy, as defined by its frequency, is exactly equal to the difference in energy between the two states involved. There is no partial emission or partial absorption. The science of spectroscopy is based on such phenomena as these.

*   *   *

The process that led to the discovery of the atom was rather fascinating and drew on the talents of a wide array of scientists from a wide array of fields. It started with the assumption that invisible structures must be causing certain observed chemical and gas property behavior phenomena and then evolved as both theory and experiment combined to make the invisible visible. In the following chapter we'll cover the highlights and the stories behind them.

4 The Atom: discovery

The curious mind wants to know
    – Michael Munowitz[1]

The evolution of physics does not always follow logical paths
    – Abraham Pais[2]

Now you know what the final puzzle looks like. It's all put together. You can study it in the high school classroom and think to yourself, *this is relatively easy*. But stop and really think about this. Many things look easy once they're there. But at the beginning, there weren't even any pieces available to this puzzle. Just pure thought and logic. No path. No equipment.

Yes, over time the pieces started to appear. But they didn't come with assembly instructions. There was no idea how to fit the pieces together. No easy way forward. No edge pieces. No corner pieces. Not even all the pieces.

The discovery of the atom and its large infinitesimal wonder was a journey like that of the discovery of the Big Bang. No one individual made the discovery working alone. Rather, it was a collective journey of thinking and experiments and more thinking and more experiments, a false turn here and there, but eventually leading in the direction of success, one step at a time, each success building on the previous success.

We were destined to discover the atom by the simple fact that the questions *Why does that happen?* and *How does that work?* are an integral part of what makes us human beings. Our never-ending quest to understand is what led us to the conclusion that atoms must exist, even in the absence of high technology proof. Pure logic told us that the process of going smaller and smaller and smaller must lead to a final indivisible end-point where no additional "smaller" is possible. This thought process started over 2000 years ago. It just took us until the twentieth century to develop the tools to prove it.

First with the ancient Greeks (Leucippus and Democritus) around 400 BC and then with the ancient Romans (Lucretius) around 50 BC rose the proposal that atoms, of Greek origin meaning "uncuttable," must exist and that matter must have some inner structure. But in later centuries this idea was met with resistance since another integral aspect of the human being is, unfortunately, that new ideas run right into reflexive resistance.[3] In this case, the proposal that atoms exist and

---

[1] (Munowitz, 2005) p. 285.

[2] (Pais, 1988) p. 11.

[3] (Johnson, 2009) p. 44. In reference to the feud that developed between Sir Isaac Newton and Robert Hooke, as will be discussed in Chapter 8, "The merciless dissection of new ideas would become a normal part of science."

Block by Block: The Historical and Theoretical Foundations of Thermodynamics,
Robert T. Hanlon, Oxford University Press (2020). © Robert T. Hanlon.
DOI: 10.1093/oso/9780198851547.001.0001

travel in straight lines (gas phase) until they collide with other atoms implied determinism, which in turn implied, with a small leap in thought but a large leap in meaning, atheism. Embracing determinism in which atoms in motion continue in motion with certain speeds and directions, their futures fixed and entirely predictable, suggested the absence of an intervening God.

This notion, with its implied atheism, didn't sit well with those who believed that there is a higher being, deity, or mythical god imposing purpose on the world. This conflict between the scientists on the one hand and the philosophers and religious adherents on the other split these groups a long time ago and resulted in an ongoing conflict throughout history and even up to the present as manifested in the debates between the Creationists and the Evolutionists. This split was so strong that it, along with the stifling hold of the ancient (not-always-correct) Greek philosophers on science, impeded development of the atomic theory for hundreds of years.

The Rise of Chemistry from Alchemy

In their search for gold, the alchemists discovered other things—gunpowder, china, medicines, the laws of nature.

– Arthur Schopenhauer[4]

When the atomic theory started to rise again, it did so from an unexpected and unplanned source: man's search for a way to turn metals into gold. Magic and sorcery prevailed for the early chemists, named Alchemists. For many centuries "frenzied alchemy held the world in its grip."[5] The possibility to transmute—turn one metal into another, preferably gold—held center stage for many centuries, up into the 1700s. Even the best and the brightest, such as, for example, Roger Bacon (1214–1292) and Sir Isaac Newton, believed in the possibility of transmutation. While naturally unsuccessful, this quest did yield one extremely fortunate and beneficial byproduct: the birth of chemistry.

Slowly over time, the mystical pursuit of trans*mutation* gave way to the more purposeful pursuit of trans*formation*—the conversion of one chemical into another. Thus did alchemy give way to chemistry. The chemists built a body of knowledge about chemical reactions and developed rules based on their experiences. Learning became accepted and ingrained from generation to generation. Certain metals such as iron, copper, and gold, were deemed elemental, not able to be broken down into other metals. Other elements were discovered. Reactions were found to occur in strict proportions. Exactly two parts of hydrogen and one part of oxygen comprise water. Exactly three parts of hydrogen and one part of nitrogen comprise ammonia.

Scientists such as Robert Boyle (1627–1691) and Antoine Lavoisier (1743–1794) contributed greatly, Boyle by raising the level of accuracy in measurements and Lavoisier by introducing the seemingly simple but extremely powerful concept of the mass balance to chemistry. Lavoisier then coined the term "element" to denote a substance which could not be decomposed into two or more different substances, another subtle yet powerful introduction that enabled chemists to start re-organizing their thinking, studies, and knowledge. Overall,

[4] (Schopenhauer, 2008) p. 13.
[5] (Jaffe, 1934) p. 9.

Lavoisier brought organization and structure into the language of chemistry, which had previously been rather awkward, and so made possible faster, more efficient, and better-understood communications amongst chemists.

The first person to most clearly put these pieces together into the modern version of the atomic theory was John Dalton (1766–1844). In the early 1800s Dalton interpreted these earlier results by proposing that each element consists of atoms, which are all alike, immutable, and different from those of other elements and that elements combine in definite proportions and have masses generally close to multiples of the mass of the hydrogen atom.

The False Clarity of History

> While hindsight can detect the great pattern of concept and theory that was being woven [in science in the early 1800s], it would be wrong to assume that it was equally apparent to contemporary observers. A closer study of the period conveys a bewildering impression of scientific conjectures and refutations.
>
> – D. S. L. Cardwell[6]

> Once we know something, we find it hard to imagine what it was like not to know it
>
> – Chip Heath and Dan Heath[7]

This history seems obvious to us in hindsight, a very logical conclusion given the data and our current understanding of the atomic theory. But putting together all the pieces of the puzzle back then was challenging, especially since communication was slow between different countries and disciplines. And remember that what we read today about this history makes everything appear nice and neat, suggesting a logical and smooth progression of science. But this wasn't the case at all. History has been largely filtered for our ease of learning. What we don't see are all the articles and communications following false leads, erroneous theories, bad experiments, and dead ends. In general, we don't see the mess that was; instead, we only see the good stuff and a clear line of progress.[8]

For example, we generally aren't exposed to the early attempts to explain the fundamental theories of combustion and heat transfer, concepts that were not understood at all before the nineteenth century. Fire itself was thought to be a definite chemical entity ("phlogiston"), while heat transfer was thought to be the movement of a material substance ("caloric") between bodies. The resulting paradigms that resulted from these false leads created much confusion and stifled much progress. We know now, based on lessons taught in elementary school, that burning involves the reaction between oxygen and combustible species like paper or gasoline, which releases heat, and we also now know, based on lessons in later years, that heat transfer is caused by the motions and collisions of the atoms comprising matter. Well, scientists didn't know all this back then. They couldn't, even though they spent years studying these phenomena. It wasn't until the likes of Boyle, Lavoisier (who eventually clarified combustion in 1774),

---

[6] (Cardwell, 1971) p. 121.
[7] (Heath and Heath, 2008) p. 20.
[8] For a good discussion of this topic I refer you to (Taleb, 2010).

and Dalton that science started to break through the paradigms then in existence. We generally don't learn much about such history today, leaving it hard for us to grasp how it was that people *didn't* know such things yesterday. Perhaps the absence of this kind of history from our education leaves us lacking appreciation for the journey as a result.

The Rise of Modern Physics

Alongside these advances in chemistry stood mathematical advances in natural philosophy, which today we call physics.

In the 1600s, Sir Isaac Newton formulated the laws of motion in his masterpiece, the *Principia*, which will be discussed in more detail in Chapter 8. These laws provided a quantitative description of what happens when particles move and collide. A body in motion stays in motion; a change in motion is caused by an outside force; an outside force causes a body to accelerate; acceleration is proportional to the force divided by the mass. And so it was only a matter of time until the thinkers started to consider, *Let's assume that there are atoms. What happens when we apply Newton's laws to their motions? Can we predict what's happening around us?*

The arrival of the *Principia* opened the door to the use of theoretical physics for exploring the atomic theory. In a great demonstration of deductive logic—make assumption, see if consequences align with nature, if so maybe assumption is valid, if not it isn't—physicists created mechanical worlds of invisible colliding spheres, like billiard balls on a three-dimensional pool table, applied Newton's laws to the collisions and made estimates of select resulting bulk properties. Such models stood on their own as mathematically sound. It was then that they compared their predictions against the ideal gas laws that were rising in parallel by the experimentalists. Agreement meant that the starting assumptions that went into the model might reflect some truth about nature, such as the existence of atoms.

Ideal gas pressure and temperature were two of the main bulk property targets for these predictive equations. Most everyone was familiar with these two quantities. *We feel pressure. We feel temperature. There must be something there that's hitting our skin. How small are these objects? How fast are they moving?* As I'll discuss later, pressure, being more mechanical in nature, was the easier of the two. Temperature was different, in part because it was tied to the nebulous concepts of hot and cold and fire and heat and the aforementioned caloric. People didn't really know what heat was and so attempting to quantify an ill-defined concept of heat in terms of invisible tiny hard balls required too great a leap of imagination. But not for everyone. There are always thinkers at the upper end of the bell curve who can put together a puzzle before others, even though they're not always listened to. This select group started using Newton's laws of motion to make predictions about what happens when small objects move and bounce off each other, in particular gaseous objects since much attraction was put on the expansion and contraction of gases, especially with the onset of the piston-driven steam engine in the late 1700s and early 1800s.

By the mid-1800s, this collective effort evolved into the kinetic theory of gases and incorporated the significant assumption that heat is nothing but the motion of particles through space. To some, this assumption or hypothesis seemed right and made sense. But, being so outside contemporary thinking, the theory attracted little attention. Actually, it was worse than this. John Waterston's (1811–1883) 1845 paper on the topic was rejected by the Royal Society of London. The problem, once again, was that because people couldn't see atoms and lacked the leap-of-faith required to believe in their existence, and because they had other paradigms to hold onto, they resisted.

Not to over generalize, but over time, truth usually wins, especially since over time, the nay-sayers, the holders of legacy, eventually die and are thus removed from the debate. The younger generation, neither bound to any legacy beliefs nor vested in past theories, sees and absorbs the truth rather quickly. As Max Planck (1858–1947) wrote, "A new scientific truth does not triumph by convincing its opponents and making them see the light, but rather because its opponents eventually die, and a new generation grows up that is familiar with it."[9]

Things finally took a turn when a physicist of influence, Rudolf Clausius (1822–1888), focused his mind on this subject. In his historic work *The Nature of the Motion which we call Heat* (1857),[10] Clausius proposed that a volume of gas is comprised of tiny atoms and that its temperature and pressure are both related to the motion of these atoms. He then went ahead and quantified these relationships and so provided some of the early estimates of the speed of gaseous atoms. While the concept was not entirely novel, Clausius' strong reputation as an established professor of physics in Berlin and later Zurich together with his grounded mathematical approach led to his being recognized as the first to truly bring the atomic theory of heat into existence. While I address this in greater detail in Chapter 40, the reason for bringing it up now is that Clausius' work furthered the notion that atoms exist.

Subsequent scientists developed Clausius' kinetic theory of gases, most notably James Clerk Maxwell (1831–1879) and Ludwig Boltzmann (1844–1906). In the 1860s, Maxwell brought statistics into physics to quantify not only the speed but also the distribution of speeds of gas phase atoms. In the 1870s, Boltzmann advanced Maxwell's statistical mechanics to an even higher level. The collective ability of these new techniques to accurately predict real-world phenomena such as gas viscosity, gas heat capacity, and even the impact of rising elevation on the composition of the atmosphere, all supported the starting assumption that gas is comprised of constantly moving and colliding particles.

If only life were this easy. As Boltzmann continued to drive progress on the kinetic theory of gas, culminating in his famed *Lectures on Gas Theory* (1896–8),[11] the disbelievers fought back, which greatly distressed the sensitive Boltzmann and contributed to his eventual suicide in 1906.

For some, it was obvious. The existence of atoms made complete sense. Predictions could be made of real, measurable properties based on the assumption that atoms exist and follow Newton's laws. And these predictions were shown to be accurate. *Just because we can't see them, doesn't mean they aren't there. What more is needed? Let's move on!*

But for others, and especially the "old school" physicists as exemplified by, for example, Ernst Mach (1838–1916) and Wilhelm Ostwald (1853–1932), things weren't so clear. "I don't believe that atoms exist!"[12] Mach stated as late as 1897 at a scientific conference. He, and others like him, more so in Germany than in England, partly because of England's heritage of Newton

---

[9] (Kuhn, 1996) p. 151. Another version of Planck's quote is: "Science advances one funeral at a time."
[10] (Clausius, 2003a).
[11] (Boltzmann, 1995).
[12] (Lindley, 2001) p. VII.

and Dalton, understood the math and the predictions and yet refused acceptance since, at that time, it was impossible to definitively prove the existence of atoms. This group's philosophical approach to science was inherently skeptical, challenging any concept or theory that was based on anything other than what can be directly experienced or measured. The fact that a theory led to accurate predictions was not enough for them to accept the assumption on which the theory was based.[13] The history here as captured well in David Lindley's *Boltzmann's Atom*[14] makes for a fascinating read and shows how two highly intelligent schools of thought can differ so strongly in their thinking on a given subject matter. Today the atomic theory is taught as if it were obvious, but back then, even for the leading scientists, such was clearly not the case.

And so it was left to the giant, Albert Einstein (1879–1955), to bring an end to the discussion by employing Boltzmann's theories to quantify Brownian motion, thereby providing the final proof, short of detection, of the existence of atoms.

Brownian Motion – Manifestation of Atoms in Motion

> *My major aim in this [analysis of Brownian motion] was to find facts which would guarantee as much as possible the existence of atoms of definite finite size. In the midst of this I discovered that, according to atomistic theory, there would have to be a movement of suspended microscopic particles open to observation, without knowing that observations concerning the Brownian motion were already long familiar.*
>
> – Albert Einstein[15]

In the early 1800s, Robert Brown (1773–1858), a curious botanist, pointed his microscope at a collection of small pollen grains in water. Surprisingly, he couldn't get perfect, stable images. The grains kept moving around, jiggling and drifting, this way and that.

Following Brown's observations, scientists attempted explanations of this phenomenon, which was eventually called Brownian motion. Many correctly perceived that the underlying cause was the constant collision of tiny, invisible particles of water with the larger grain particles. Bam, bam, bam! Small objects hammering away at the larger object, with minute variations on one side or the other leading to the jostling motion. So the *qualitative* part made sense and scientists bought into it. It was the *quantitative* part that was the challenge.

And here's where Einstein's genius kicked in. Einstein fully bought into Boltzmann's statistical approach to physics and imagined that the statistical distribution of velocities for the invisible water molecules could lead to momentary imbalances of local pressure and thus cause motion in larger particles like pollen grains suspended in the water. He further speculated that if he could bring quantitative tools to the problem, then his results would provide a direct validation of the atomic hypothesis.

---

[13] (Darrigol, 2018) p. 364. Even Planck, early in his career, was hostile toward the atomic theory of matter. Based on his belief that the molecular hypothesis contradicted the 2nd Law of Thermodynamics, he wrote "…in spite of the great successes of the atomic theory in the past, we will finally have to give it up and to decide in favor of the assumption of continuous matter."

[14] (Lindley, 2001).

[15] Cited in (Lewis, 2007) p. 144.

Brownian motion was not a trivial problem to solve. Einstein started by first assuming that atoms exist and then set out to obtain facts to prove it. But instead of quantifying the jiggling, he made the critical decision to quantify the particle's net movement, the "random walk" as it was called due to its zigzag motion, a seemingly simple change in approach and yet one that no one else had considered. He made these quantitative predictions in 1905, amazingly enough before even knowing anything about Brownian motion. So, in a way, all of the data obtained in earlier Brownian motion studies later became the "proof" of Einstein's theory. Subsequent experimental work by Jean Perrin (1870–1942) in 1908 provided the final validation.

Einstein was, at heart, a determinist. His inclusion of statistics to analyze large groups of atoms was entirely consistent with this since at the most fundamental level of this analysis, the atoms obeyed Newton's laws of motion and determinism prevailed. I point this out since, as we'll see, this use of probability and statistics was not in any way the same as their later use to define the unpredictable nature of quantum physics.

Starting in the 1860s with Clausius and Maxwell, and continuing with Boltzmann and then with Einstein, theoretical physics came into its own. Previously, theory and experiment were linked. Propose a theory, validate with experiment. But with the atomic theory, this was no longer possible since the existence of atoms couldn't be directly validated with experiment. The tools just weren't there. So this rising breed shifted validation from the theory to the *if–then* deductive consequences of the theory. While this new approach could no longer provide definitive proof, it did provide substantive support. Its success regarding Brownian motion provided a strong win for theoretical physics and encouraged an even stronger presence of mathematics in this field.

To conclude, in the early 1900s Einstein's work on Brownian motion together with the collective work of many chemists "proved" the existence of atoms, even though the atoms couldn't be seen. From 1908 onwards, atoms were considered real.[16] But while this provided a significant piece to the atomic puzzle, a corner piece if you will, much remained to be done in finding and organizing all the other pieces. *What was it that actually differentiated the elements?* The answers to such questions and the resulting puzzle pieces they created started to arrive very quickly.

Discovering What's Inside the Atom

There must be something simpler than ninety-two separate and distinct atoms of matter
– Joseph John (J.J.) Thomson (circa 1897)[17]

It's one thing to know that atoms and elements exist. It's an entirely other thing to know what they're made of and how they work. The main entities involved are few in number: proton, neutron, and electron, with the photon to send messages between them. Who could have imagined that from such simplicity would come so much complexity?

---

[16] (Lewis, 2007) p. 145. Quoting Stephen Brush, "In fact the willingness of scientists to believe in the "reality" of atoms after 1908, in contrast to previous insistence on their "hypothetical" character, is quite amazing. The evidence provided by the Brownian-movement experiments of Perrin and others seems rather flimsy, compared to what was already available from other sources."

[17] Cited in (Jaffe, 1934) p. 271.

It was the discovery of these different entities and the description of their behavior that eventually led to the completion of the atomic puzzle. One by one, each particle was discovered, following no set path other than pure discovery. When the time was finally right, in the early 1900s, all the pieces of the puzzle came together in a rush, one discovery after the other, completing the story.

Crookes → Thomson → Röntgen → Becquerel → Curie → Rutherford → Planck → Einstein → Bohr → Schrödinger → Heisenberg → Born

Since the onset of the modern atomic theory in the early 1800s, Dalton's atom was thought to be indivisible. A century later, English chemists proved otherwise.

> *Looking back on that period, one is amazed at the number of times the correct conclusion was missed…On reading their papers now [electric sparks in containers], one marvels at the wealth of ingenious experiments, devised to answer the wrong questions, and the number of wrong ideas seemingly supported by the results.*
>
> – Otto Frisch[18]

Starting in the mid to late 1800s, scientists, notably including Sir William Crookes (1832–1919) and then Joseph John (J.J.) Thomson (1856–1940), commenced pulling the atom apart, although they didn't realize it at the time. In his London home, Crookes, looking to study the conduction of electricity through low-pressure gases, connected the positive (anode) and negative (cathode) leads from a high voltage power supply to the ends of a glass tube (Figure 4.1). The generator was basically an electron pump, pulling electrons from the anode and forcing them to the cathode where the repulsion between them resulted in a voltage high enough to drive the electrons off the cathode and across the gap in the glass tube toward the electron-starved anode. But Crookes did not know about any of this then.

Crookes initially couldn't detect a flow of electricity because of the presence of residual gaseous atoms in the glass tube. As the electrons jumped off the cathode, they didn't get very far as they quickly struck and energized these gaseous atoms, exciting their electrons to higher energy levels. The orbital electrons eventually returned to their initial energy levels, re-emitting the energy in the form of photons. Crookes observed these photons as light and this process of converting a directional electrical current into a non-directional spread of light eventually became the process used today in neon signs and fluorescent lights.

Interestingly enough, the glow in the tube became secondary to the real breakthrough that occurred when Crookes started using more advanced air pumps, such as those developed by Johann Heinrich Geissler (1815–1879) and later by his assistant Charles A. Gimingham (1853–1890), to lower the pressure in the tube much further, so creating a stronger vacuum, void of essentially all remaining gaseous atoms such that all that was left were the cathode, anode, and the strong voltage across the empty space between them. When he did this, the

---

[18] (Frisch, 1973) pp. 54–55.

WILLIAM CROOKES (mid to late 1800s)

- used advanced air pumps to create vacuum in tubes to enable unhindered flow of electrons from cathode to anode

- the high speed electrons shot past the anode and hit the glass wall, exciting the atoms which gave off yellow-green light (fluoresce)

J.J THOMSON (1847-1899)

- studied impact of magnetic and electric fields on cathode rays

- cathode rays are made of negatively charged particles

- determined charge-to-mass ratio of particles

- separately worked with Wilson and his cloud chamber to quantify charge of particle and thus its mass

- thus did Thomson discover the electron

ROENTGEN (1895)

- discovered x-rays (which were caused by electrons jumping orbitals)

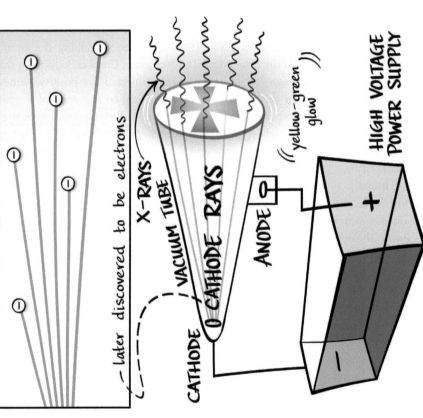

- later discovered to be electrons

X-RAYS

VACUUM TUBE

CATHODE

CATHODE RAYS

(yellow-green glow)

ANODE

HIGH VOLTAGE POWER SUPPLY

+

−

Figure 4.1 Crookes' tube reveals electrons and photons

light disappeared and the tubes went dark, all except for a greenish light near the anode. What he didn't know was that this effect was caused by the fact that removal of air provided unhindered flow of the electrons from cathode to anode. The high-speed electrons were traveling so fast that many missed the anode and hit the glass wall, where they excited the glass atoms and made them give off a yellow-green light (fluoresce). Crookes' use of high vacuum in his eponymous tube provided the means by which a pure stream of electrons could be detected and studied. He naturally didn't realize what he had done. He simply noticed the green light and reported the result.

It was J.J. Thomson who finally figured out what was happening. Others came close. But Thomson spoke out clearly and convincingly, quantifying the magnitudes of the electron's mass and charge, and so became the discoverer of the electron and the winner of the Nobel Prize.[19]

Crookes → Thomson → Röntgen → Becquerel → Curie → Rutherford → Planck → Einstein → Bohr → Schrödinger → Heisenberg → Born

As head of the famed Cavendish Laboratory in London, Thomson picked up Crookes' line of research and for the next 20 years studied and characterized what became known as "cathode rays" generated inside the "Crookes tube." He noted, for example, that both magnetic and electric fields could be used to bend the rays, thus initiating the eventual development of the television tube, and so concluded in an 1897 announcement to the Royal Society that because deflection occurred and because light (photons), being without mass or charge, aren't affected by such fields, then the beam was not light, as the Germans believed,[20] but was instead negatively charged matter. Based on his measurements of the deflection of the electron beam, he calculated the charge-to-mass ratio (e/m) of these particles, the two values inextricably linked since the amount of deflection of a moving charged mass depends on the magnitude of *both* the charge and the mass, higher charge leading to more deflection and higher mass leading to less.

Had Thomson stopped here he may not have won the Nobel Prize for the discovery of the electron. But he didn't stop, for in his Cavendish lab, one of his students, C. T. R. Wilson (1868–1959), a Scotsman who grew up around fogs and clouds, had developed an experimental method called the cloud chamber that could be used to quantify the size of charges. The science behind the cloud chamber was based on the fact that supersaturated water vapor

[19] While J.J. Thomson won the Nobel Prize in Physics in 1906 for proving the electron is a particle, his son, George Paget Thomson (1892–1975), won the Nobel Prize in Physics in 1937 for showing that the electron is also a wave.

[20] At this time there was great rivalry between German and British researchers. Specifically regarding the cathode ray, the Germans were quick to denounce Crookes' work due to his misinterpretation of some of the experiments while the British accepted Crookes' main findings while simply acknowledging and then discarding the errors. Ultimately, the Germans tended to believe that Crookes' cathode rays were a wave (like light), whereas the British tended to believe that the rays were particles. As events unfolded over the next few decades and theory of the wave-particle duality of nature arose, both would be proven correct.

condenses around charged particles. A moving charged particle thus leaves a trail through a cloud, which can be photographed in the chamber with a bright flash of light.

In 1899 Thomson shared results based on this use of the cloud chamber to measure the charge of the electron and was thus able to also calculate the mass since he separately knew the charge-to-mass ratio. And it was for this total discovery of the electron, the initial determination of the charge-to-mass ratio combined with the separate determination of the charge, that Thomson won the Nobel Prize.[21] In parallel, the arrival of the cloud chamber out of this effort did much to convince everyone of the reality of subatomic particles.[22] The existence of invisible particles became evident by the trail they left behind.

While not highly accurate, Thomson's numbers were the right order-of-magnitude. It was left to Robert Andrews Millikan (1868–1953) at the University of Chicago to more accurately quantify the two numbers in his famed oil drop experiments (published 1913), similar in concept to the cloud chamber experiments, but this time with oil as opposed to water. In these experiments, Millikan determined the value of the charge and thus the mass of the *individual* electrons, showing as a result that *all* electrons have the same charge, a finding not to be overlooked.[23] Simple to grasp today, this concept was amazing back then, for who really knew what the electrons were? When a Millikan droplet was seen to change its charge, it always gained or lost the charge of an entire electron. So the idea of the electron as a smeared-out cloud of charge didn't work.

Returning to Thomson, based on his initial findings, he took a rather large leap and proposed that these negatively charged particles, which he referred to as "corpuscles," were building blocks of the atom. This proposal went directly against the wisdom at that time that the Daltonian atom was indivisible. This was really the first suggestion that the atom itself is comprised of building blocks. The atom had been split! Well, sort of…

This hypothesis naturally opened the door to many other questions, primary among them: *if the atom is neutral and if a part of the atom is negative, what's the part that's positive?*

*   *   *

There was no structured path of subsequent discovery here. Results started coming from every direction. So it is difficult now to write the history to make it look like a fixed path. How could someone keep everything straight? Who to believe? What to believe? So very fast. To the person who put all of these facts together into a coherent picture, the prize. One thing led to another. No direction. Only discovery. Hand off. Ideas to ideas. The turn of the century in the late 1800s and early 1900s. What an exciting time to have lived and worked as a physicist and a chemist, a striking journey of hypothesis and discovery.

*   *   *

---

[21] (Pais, 1988) p. 86.

[22] (Weinberg, 2003) p. 86.

[23] (Frisch, 1973) p. 58. We take it for granted that the building blocks of atoms (protons, neutrons, electrons) and thus the atoms themselves (of same element, same isotope) are identical. But it's only because we know such things that we take them for granted. If we didn't know such things, then we'd live in a different paradigm where such things weren't even an idea in someone's mind. That all of these particles are identical is truly remarkable.

The invention of the cathode-ray tube enabled the discovery of the electron and soon became a valuable tool in studying particle physics, eventually and indirectly enabling the discovery of the atomic nucleus, among other discoveries.

Crookes → Thomson → Röntgen → Becquerel → Curie → Rutherford → Planck → Einstein → Bohr → Schrödinger → Heisenberg → Born

In 1895, Wilhelm Röntgen (1845–1923), a German physicist, found that when cathode rays hit the glass wall of his Crookes tube, highly penetrating radiation was emitted in the form of a faint green light against the far wall. Röntgen learned that these x-rays, a name he used since he didn't know what the rays actually were, could travel through anything put in front of them. As he continued to explore this, putting various objects in front of the generator to see what happened, he noticed the outline of the bones from his hand displayed on the wall.[24]

What Röntgen couldn't have known at the time is this. When high-energy, cathode-ray electrons strike atoms, such as those comprising the glass wall in the Crookes tube, they penetrate deeply into their electron shells and knock-out electrons from the inner-most shells. When the resulting vacancies are re-populated by electrons from the outer shells, the large gap traversed by the electrons causes the emission of high-energy photons in the form of x-rays.

Crookes → Thomson → Röntgen → Becquerel → Curie → Rutherford → Planck → Einstein → Bohr → Schrödinger → Heisenberg → Born

In the following year of 1896, Professor Antoine Henri Becquerel (1852–1908) of the Ecole Polytechnique in Paris, France, attempted to determine whether phosphorescent substances would emit Röntgen's x-rays after exposure to sunlight. In the course of his experiments, he inadvertently put one of these substances, an otherwise unremarkable uranium ore, on a photographic plate lying on a table in a dark room. Days later, when he went back into the room, he noticed that the area of the plate underneath the ore was exposed, even though visible light was not present. So what could be causing exposure? After many repeated experiments, Becquerel proposed that some substance contained within the uranium ore, but not the uranium itself, was causing the exposure. Becquerel had unknowingly discovered radioactivity.

One of the great assignments in history then took place when Becquerel turned to one of his students and recommended that the identification of this substance become her doctoral thesis.

---

[24] Röntgen first began figuring out the medical use for these x-rays when he saw a picture of the bones of his wife's hand on a photographic plate that formed due to the rays. This x-ray image of the human body was the first ever and served to open the door to a whole new medical science.

Crookes → Thomson → Röntgen → Becquerel → ⬚Curie⬚ → Rutherford → Planck → Einstein → Bohr → Schrödinger → Heisenberg → Born

For hours at a time she stood beside the boiling pots stirring the thick liquids with a great iron rod almost as large as herself

– Jaffe[25]

In 1891, out of fear of the Russian police, Marie Skłodowska (1867–1934) fled her native Poland, then under Russian rule, to Paris. Raised by her father, a professor of mathematics and physics in Warsaw, she sought to continue her academic life by studying and teaching. In 1894 she met and later married Pierre Curie (1859–1906), a French physicist, and eventually found her way into the laboratories of the Sorbonne where her experimental skills caught the attention of Becquerel. Becquerel presented her with the problem he faced, *What was the powerful unknown element in uranium ore?*, and asked her if she would undertake this as a research project. After conferring with Pierre, both became excited by the opportunity to discover a new element and decided to drop all of their other work and focus on answering this question.

Starting with a ton of pitchblende, the uranium-containing ore obtained from the Austrian government that demonstrated the most power for exposure, Marie and Pierre worked under rather challenging conditions—abandoned shed, leaky roof, boiling cauldrons, stifling fumes, bitter winter cold—to isolate the active species, slowly concentrating it by physical and chemical means, along the way discovering a new element they called polonium for Marie's native country. In the challenge, they found joy. Recalled Marie years later, "We lived in a preoccupation as complete as that of a dream."[26]

After two years (!) of repeated crystallizations, all the while using an electrometer to track which way the radioactivity had gone, ensuring they didn't lose the product down the drain, the Curies isolated a very tiny amount of extremely active salt crystals, small enough to fit into laboratory glassware. Finally, all of their hard physical work paid off. They identified in this salt the new element radium (1899). Of all the discoveries of elements, this discovery of radium became one of the most impactful as it became the starting point for a whole new area of research that ultimately led to an entirely new way of thinking about the atom.

All of this work that Curie completed, including the discovery of polonium, the discovery of radium, the quantification of properties of each, including the astounding fact that radium gave off 250 000 times the amount of heat as the burning of an equivalent weight of coal, all of this information, for which she, Pierre, and Becquerel were eventually awarded the Nobel Prize (1903),[27] was compiled into a single document, her doctoral thesis. Unbelievable. The professors at her thesis defense didn't even know what questions to ask in the face of such a mass of

---

[25] (Jaffe, 1934) p. 251.

[26] (Jaffe, 1934) p. 251.

[27] Marie Curie was the first woman to be awarded the Nobel Prize. She would later go on to win yet another Nobel Prize in 1911 for isolating pure radium from the radium salt. She did this on her own after Pierre was killed after being hit by a horse-drawn vehicle on the streets of Paris.

original work. Some considered her thesis "the greatest single contribution of any doctor's thesis in the history of science."[28]

But why is this story important to share in the context of this discussion? Well, this new element became the source of bullets that could be fired into the depths of atoms and so became a tremendously powerful tool for advancing the science of atomic physics. And the person who recognized this was Ernest Rutherford (1871–1937), a New Zealander who became a student of J.J. Thomson's at the Cavendish lab in 1895. It was Rutherford's genius that identified this opportunity as he proceeded to bring his strong, enthusiastic, persistent, hands-on, intuitive curiosity to bear on the study of the atom, becoming in the process one of the greatest experimental physicists in the world.

The shared learning between chemists like Curie and physicists like Rutherford began blurring the walls separating the two disciplines and so strengthened the hunt for the atomic theory.

* * *

Recall that the atomic nucleus is not static. Protons and neutrons in the nucleus are in constant motion governed by two fundamental and opposing forces: electrostatic repulsion between the protons and strong nuclear attraction between the nucleons, proton and neutron alike. As the nucleus moves and slightly deforms, the balance between these forces shifts ever so slightly. While both forces decrease with distance, the strong interaction (force) holding the nucleus together falls off at a much faster rate. So when deformation reaches a certain point, the dominant force switches from attractive to repulsive, splitting the nucleus apart and releasing a range of very high-energy products, including what became known as the aforementioned alpha, beta, and gamma rays, which eventually were determined to be He-4 nuclei, electrons, and photons, respectively. This "fission" process occurs much faster for certain elements, such as uranium and radium due to their large and unstable nuclei, than others. And it was the study of these radiation phenomena and their associated high-energy emissions that ultimately enabled discovery of the final missing pieces of the atomic theory puzzle.

Crookes → Thomson → Röntgen → Becquerel → Curie → Rutherford → Planck → Einstein → Bohr → Schrödinger → Heisenberg → Born

Then I remember two or three days later Geiger coming to me in great excitement and saying, "We have been able to get some of the alpha particles coming backwards…" It was quite the most incredible event that has ever happened to me in my life. It was almost as incredible as if you fired a 15-inch shell at a piece of tissue paper and it came back and hit you.

– Ernest Rutherford[29]

Rutherford's remarkable career spanned two continents, starting at the Cavendish Laboratory under Thomson, moving to McGill University in Montreal in 1898, back to England in 1906 as chair of physics at Manchester University, then back to the Cavendish Laboratory in 1919

[28] (Jaffe, 1934) p. 254.
[29] Cited in (Weinberg, 2003) p. 114.

where he succeeded Thomson as the Cavendish Professor of Experimental Physics at Cambridge, residing there until his death in 1937. At each location, Rutherford made significant strides in pushing back the frontiers of atomic physics, guided to a large degree by a question that kept nagging him. *If the atom contains Thomson's negatively charged electrons, then what positively charged constituents does it also contain to render the complete atom neutral?*

Rutherford's accomplishments can be grouped into two broad categories: 1) clarifying the nature of "natural" radioactivity, and then 2) using radioactivity as a tool in the discovery of the structure of the atom. Regarding (1), in his early years at the Cavendish Laboratory and then McGill, where he worked with Frederick Soddy (1877–1956), his research partner, Rutherford started untangling the different emissions of radioactive substances, which he coined alpha rays, beta rays, and gamma rays. The range of these different emissions was causing great complications at the time.

Research by Bequerel, among others, showing that the beta rays could be deflected by electric and magnetic fields much in the same way as Thomson's electron corpuscles could be, correctly concluded that the beta rays were indeed high-energy electrons, traveling at much higher speeds than Thomson generated in his Crookes tube.

Applying similar techniques, Rutherford showed that the alpha rays could be similarly deflected, but that the deflection was much more difficult on account of the alpha rays being more massive than the beta rays. After much work to refine these measurements, Rutherford, working with students, colleagues, and additional tools, eventually isolated the alpha rays, measured their spectral lines, and concluded in 1908 that the alpha rays were the nucleus of helium-4, which we now know contains two protons and two neutrons, a fact Rutherford didn't, couldn't, know at the time.

In 1903 Rutherford closed the book on the three types of radioactive emission by naming the third as gamma rays. First observed in France by Paul Villard (1860–1934) in 1900, Rutherford speculated that these gamma rays were similar to x-rays, i.e., high-energy photons of short wavelength. But it wasn't until 1914 that he was able to prove them as such.

* * *

It was one thing to identify and classify the various radioactive emissions, but an entirely different thing to speculate on their origins. Rutherford's identification of the nature of the emissions was a tremendous accomplishment in and of its own right. But it's what he did next that shook the world of physics to its core.

Based on his studies into the nature of radioactivity, Rutherford and Soddy speculated in a 1903 paper that the process of radioactivity causes the change of one element into another. This was a significant proposal at the time as atoms were then considered "immutable." But Rutherford measured the radioactive emissions and could not conclude otherwise. The Rutherford–Soddy "disintegration theory" was so dramatic a shift in thinking that it was not wholly welcomed.[30] Rutherford noticed that the radioactive decay of atoms was a one-atom process, unlike a normal chemical reaction in which collisions of atoms and molecules are required and in which temperature, which affects such collisions, can impact the reaction rate. He found no such evidence of this in radioactive decay. The process was inherent to the atom, nothing else, and

---

[30] One has to wonder how many dramatic shifts in thinking have ever been welcomed. How many scientists truly welcome change? Must they constantly be on guard against false change?

led to the concept of half-life decay laws. And then he started quantifying half-lives: 1700 years for radium, 6 billion years for uranium. His rather remarkable conclusion: atoms are not stable.

The next critical step for Rutherford, continuing to reflect his genius, was to shift his view of the alpha particles as something to study to something to use, a tool. While the emitted alpha particles were small, they were also heavy (8000 times the mass of the electron) and very fast (12 000 miles/second when emitted from radium). The electron shells defending the nucleus didn't stand a chance. Since large particle accelerators had not yet been constructed at the time, such an energetic particle emitted by naturally radioactive substances became the only available probe and an excellent one for Rutherford to use in his search for the positive particles in the atom that countered the presence of the negatively charged electron discovered by Thomson.

Working in his Manchester laboratory with student Ernest Marsden (1889–1970) and post-doctoral student Hans Wilhelm Geiger (1882–1945), and with a piece of radium provided by the Curies and small enough to fit on the head of a pin, Rutherford built a device to shoot the alpha rays emitted from the radium at a piece of gold foil, thin enough to permit complete penetration. By detecting deflections in the alpha rays exiting the foil, Rutherford reasoned that he would gain an understanding of the nature of the charge and mass distribution of the atoms in the foil.

Plum Pudding

Before going further, let's first recognize the prevailing atomic paradigm at the time, as put forth by Thomson himself. Based on his own studies and thinking, Thomson proposed the "plum pudding" model of the atom in which a sphere of smoothly spread positive charge (pudding) contained embedded negatively charged corpuscles (plums), which were assumed to be the electrons. And since the mass of the electron corpuscles was so small, the mass of the atoms must therefore lie in the pudding and be spread *uniformly* across the entire atom.

With this "plum pudding" paradigm strongly present in his mind, Rutherford didn't anticipate anything other than small deflections of the alpha rays as they passed through the atom field in the gold foil. In 1909, chasing some fleeting thought that he didn't fully comprehend himself, Rutherford asked Geiger and Marsden to conduct a "damn fool experiment"[31] and set up their alpha ray detector to measure large deflections. The rest is history. Geiger's report back that a low but measurable fraction of alpha rays did indeed bounce backwards shook Rutherford deeply with excitement.[32] There was no possible way for such massive and high energy alpha particle to bounce straight back, unless…

And here's where Rutherford's theory of the atomic structure came into being. Based on his continued measurements, calculations, and thinking, Rutherford proposed that the only way that such scattering could occur is if the atom consisted of mostly empty space, with all of its mass concentrated in a central, positively charged nucleus and with the negatively charged corpuscles at a comparatively great distance away. After deliberating over all of these findings and the significance of what his team had discovered, Rutherford published his results and

---

[31] (Lightman, 2005) p. 89.

[32] It's difficult to truly appreciate the significance and the shock and the amazement of Rutherford's experiment. Reading it now, I can't feel this because I wasn't present to the mindset that existed back then when no one had a clue. All the theories pointed in a very different direction. And then, bam! The alpha particle bounced off the gold foil and came backwards.

theories in 1911. As sometimes happens, the scientific community mostly ignored this theory, that is until a post-doctoral student of Rutherford's, a Danish theoretical physicist by the name of Niels Bohr (1885–1962), was able to explain it.

Blackbody Radiation

But before continuing this story, another piece to the puzzle must be revealed first, because Bohr relied on both Rutherford's findings and this other piece to propose his new theory of the quantum atom.

Recall the photon. By the early 1900s, while most scientists believed light to be a continuous wave of energy as accurately described by the famous Maxwell equations, data started arriving that could not be explained by these equations, the most disrupting data coming from studies involving blackbody radiation.[33]

It was well known in the early 1900s that hot objects glowed in colors that shifted from one to another with increasing temperature. Think of the red-hot coils on your stove and then the much hotter white bars of iron being worked in a steel mill. To better study such phenomena, scientists prepared specially enclosed hollow furnaces containing a very small sampling hole in the side. As the furnace was heated, the thermal blackbody radiation emitted from the hole was collected and analyzed, using various devices such as prisms to dissect the radiation into its component parts, making graphs of the resulting data, plotting the energy density against frequency. The resulting distribution graphs showed a smooth curve that started low, increased to a maximum, and then decreased back down. One curve existed for each temperature, the curve remaining the same for a given temperature regardless of what the furnace was constructed of, just so long as it behaved as an ideal blackbody.

Scientists were bothered by these graphs for quite a while as they could not explain their shapes. They could model one end or the other but not the entire graph, mainly because the underlying physics was not known. Something clearly was not understood with such blackbody radiation graphs. And when something is not understood, it becomes a magnet for the curious.

Of course, what scientists didn't know at the time is that as heat is added to a solid, atoms and the charges they contain start moving faster and faster, vibrating back and forth, not in any specific direction, just moving back and forth, faster and faster with increasing temperature. Such motion of charges generates photons, but not any kind of photons, only very specific photons. Just as bound electrons only exist in certain energy "shells" surrounding the atom, bound nuclei in solid matter only exist in certain vibrational energy states relative to neighboring nuclei, meaning that the vibrations between nuclei are not arbitrary. As more energy in the form of heat is pumped into a solid and motion of the charged nuclei increases, the vibrating nuclei, together with the electrons surrounding them, jump up into their well-defined higher energy states. Once there, they can then decay back down to lower energy states, emitting a single photon for each "jump," equal in energy to the energy difference between the two states. The gaps between these energy states are equal to a whole number multiple of a constant called Planck's constant.

---

[33] For an excellent overview of the historical role of blackbody theory on the rise of quantum discontinuity, see (Kuhn, 1987).

The heat added to the solid by thermal motion of the atoms (from the furnace flame) is thus radiated away in the form of photons, which is called thermal radiation. The distribution profile of the radiated photons thus takes on the shape of the distribution profile of the vibrating nuclei. At each temperature of the furnace, the vibrating nuclei take on a range of energy levels, some high, some low, most in the middle, a typical shape of a distribution curve. As temperature increases, the nuclei continue to take on a range of energies, with the average moving upwards. Since the photons are generated as a result of the motion, the number and energy of the photons mirrors the number and the energy of the vibrating nuclei. More nuclei vibrating at a given energy level translates into more photons emitted at that energy level. The distribution shapes trend each other.

Note that the description here is for thermal motion leading to thermal radiation. Well, the opposite happens as well. Thermal radiation leads to thermal motion, which is why you feel the heat from sunlight. When in equilibrium, thermal motion causes thermal radiation just as much as thermal radiation causes thermal motion. Each is a cause of the other.

Crookes → Thomson → Röntgen → Becquerel → Curie → Rutherford → Planck → Einstein → Bohr → Schrödinger → Heisenberg → Born

> The questions viewed from this standpoint led me to consider the relation between energy and entropy, by taking up again Boltzmann's viewpoint. After a few weeks, which were certainly occupied by the most dogged work of my life, a flash of lightning lit up the dark in which I was struggling.
>
> – Max Planck[34]

> The beginning of the end of classical mechanics
>
> – Richard Feynman[35]

The scientist who really cracked this case open was Max Planck (1858–1947), a theoretical physicist teaching at Berlin University. In 1900, on the very day he learned of some new results in this area called blackbody radiation, Planck created and soon published an empirical mathematical formula that fit the new data and has stood ever since as Planck's radiation law.[36] Having wrestled with radiation since 1894, he was aware of the various attempts to model the phenomenon and so was primed to realize upon seeing the new data the mathematical correction needed. But he arrived at the correct answer without knowing why. Hence, use of the word "empirical."

Bothered by the absence of an underlying theory, bothered by the need to know *why*, because surely such exactness of fit doesn't just happen, Planck spent the next two months deriving this formula based on theory, using Boltzmann's work in statistical mechanics as a guide. The

---

[34] Cited in (Cercignani, 2006) p. 219.
[35] (Feynman et al., 1989a) Volume I, p. 41–6.
[36] (Pais, 2005) p. 368. Also see (Kuhn, 1987) Chapter IV, (Lindley, 2001) pp. 174–7, and (Cercignani, 2006) Section 12.3. Also see Chapter 42 in this book.

theory he arrived at was very unconventional. Planck found that he could only explain his formula by proposing that the total energy of the "vibrating resonators" comprising the solid walls in the furnace[37] resulted from the addition of a well-defined number of equal parts, i.e., the whole is the sum of the parts, and that there is a smallest part, later called a quantum, below which energy can no longer be divided. Wow. Such a seemingly simple theory. And yet so very profound since Planck was, in effect, proposing that energy is not continuous but is instead the sum of discrete parts, a theory that clearly ran counter to the prevailing paradigm that energy is continuous and can be infinitely divided into smaller and smaller pieces.

On December 14, 1900 Planck delivered his findings to the German Physical Society in Berlin. I specifically note this date as many now consider it to be the birth date of quantum physics. But at that time, no one really thought of it as such; history has only written it as such, for while Planck's excellent mathematical findings were quickly accepted, his underlying theoretical proposal was "scarcely noticed."[38]

Planck was the first to propose the quantized form of energy, for which he won the Nobel Prize in 1918. As Einstein later said of Planck, "he showed convincingly that in addition to the atomistic structure of matter there is a kind of atomistic structure of energy."[39] And so began a revolution, one that no one seemed to notice, including Planck himself, who didn't fully appreciate what he himself had done. And he became a "reluctant revolutionary"[40] with what happened next.

Crookes → Thomson → Röntgen → Becquerel → Curie → Rutherford → Planck → Einstein → Bohr → Schrödinger → Heisenberg → Born

Five years after the publication of Planck's findings, Einstein stepped in (where *wasn't* Einstein?) and, also influenced by Boltzmann's work, concluded that not only did the resonators exist in certain energy states but so did the radiation emitted and absorbed *by* the resonators. Einstein was the first to truly recognize the essence and implications of Planck's findings. Inspired by the idea that blackbody radiation data must provide some glimpse into the fundamental nature of matter and energy—*it must tell us something!*—Einstein took Planck's proposal about the quantized nature of energy and applied it to the radiation itself.

In his own publication on this matter, during his famous year of 1905,[41] Einstein suggested that certain phenomena such as blackbody radiation were better explained by assuming a *discontinuous* nature of light, a very radical idea since the continuous nature of light was very well established and proven at that time, as best manifested in the continuity equations of James Clerk Maxwell, which accurately described all aspects of electromagnetic radiation. Not one to

---

[37] At this point, Planck doubted the existence of atoms. See (Kragh, 2000).
[38] (Kragh, 2000) p. 33.
[39] (Lightman, 2005) p. 14
[40] (Kragh, 2000).
[41] Like Newton, Einstein had his own *Annus mirabilis* in 1905 when he published four articles that contributed substantially to the foundation of modern physics. In these papers he explained the photoelectric effect, Brownian motion, special theory of relativity and mass–energy equivalence and thus changed our views on space, time, mass, and energy.

be intimidated by prevailing thinking, Einstein began his paper with the assumption that blackbody light behaves like a system of gas particles. And from here, from this entirely new direction, he re-derived Planck's formula, and in so doing determined that each light particle, which he called "energy quanta," has an energy (E) related to its frequency (ν) and that the relationship constant (h), which we now naturally call Planck's constant, a fundamental constant of nature, is exactly the same constant as Planck determined in his analysis.

$$E = h\nu$$

Planck never proposed the existence of photons. In fact, his theory didn't even apply to any fundamental nature of light itself. Instead he proposed that matter emitted energy in discrete units, and that the feature of discreteness was simply due to how radiation was emitted and absorbed by a surface rather than a feature of the actual light wave as it propagates through space.[42] Planck's view on light was fully aligned with the continuous wave approach so well expressed by Maxwell. This was a critical distinction as it revealed the depth of this paradigm and thus accentuated the height and impressiveness of Einstein's leap in thinking. It was Einstein who basically said, *hold on, perhaps our senses are misleading us, perhaps this radiation energy only and always comes in discrete packages or quanta, an inherent characteristic of photons and not simply a result of the way matter emitted light, and it's only the fact that so very many of these packages are present that our senses mislead us into thinking of them as a continuous flow.*

The true father of quantum theory, Einstein was the first to open Pandora's box regarding the wave–particle duality of light by being the first to propose the particle nature of light. He was the first to propose that light inherently occurs in discrete packets of energy, which eventually acquired the name of photons, which can only be produced and absorbed as complete units. And unlike a wave that spreads out with distance, he theorized that the photon remains intact as an indivisible unit for however far it travels.

In this same paper, Einstein took his findings one step further by using them to explain another phenomenon called the photoelectric effect. In this phenomenon, a beam of light strikes a metal surface, delivering sufficient minimum energy, one photon particle at a time, to eject the metal's orbital electrons and so cause an electric current to flow. The point at which the current starts flowing depends not on the number of photons striking it but on the energy of the photon as measured by its frequency. This becomes the threshold frequency. The wave theory of light could not explain this. The particle theory of light could. Einstein clarified this. With Robert Andrews Millikan's experimental verification of the photoelectric effect at the University of Chicago during 1914–1916, both physicists won Nobel Prizes, Einstein in 1921[43] and Millikan in 1923.

The particle nature of light was not an easy sell in the early 1900s. Planck couldn't "see" the discrete photons and thus never fully accepted their existence, and in fact resisted their existence,

[42] (Isaacson, 2007) p. 156.

[43] Einstein received his Nobel Prize in 1922, the delay resulting from disagreement within the Swedish Academy on whether Einstein deserved an award for a highly speculative theory such as general relativity. Alfred Nobel dictated in his will that the prize should go to "the most important discovery or invention." The Academy felt that relativity was neither. And so advocates of Einstein, committed to awarding Einstein the Nobel Prize, which he could have won several times over for his massive impact on physics, chose a different path, pushing to give the prize to Einstein for "the discovery of the law of the photoelectric effect," which was not technically the same as the discovery of light quanta, the dominant feature of his 1905 paper. Bohr won the 1922 Nobel Prize for his model of the atom, which was built on the laws describing the photoelectric effect laws as clarified by Einstein. See (Isaacson, 2007) pp. 309–315.

only admitting toward the end of his life the futility of his attempt to explain the quantization of energy by classical means.[44] Ironically, Bohr too couldn't believe that light quanta were truly fundamental particles and actually fought fiercely against this theory, being one of the last to accept the photon as a real particle,[45] even though he was to use Einstein's $E = hv$ equation to explain how atoms emit and absorb packets of light at specific frequencies. Even Millikan, and Einstein himself, had a difficult time accepting that light quanta were real and fundamental. While he was the first to propose light quanta,[46] Einstein too believed in Maxwell's wave nature of light, and the idea of discontinuous light would basically invalidate Maxwell's hugely successful theories. Furthermore, as we shall see, the theory of light quanta eventually brought the concepts of spontaneity and probability into physics, which bothered Einstein for the rest of his life, as reflected by his writing, "All these fifty years of pondering have not brought me any closer to answering the question, 'What are light quanta?'"[47]

This entire story once again shows that what we take for granted today was not at all taken for granted yesterday, even by the major figures in physics.

It would take until May, 1923 for this issue to be resolved when Arthur Compton (1892–1962), at Washington University in St. Louis, published results that solidly and decisively demonstrated the particle-like nature of radiation when he discovered that light in the form of x-rays bounces off electrons exactly as one would expect if the x-rays exist as photons.[48] The Compton Effect or Compton Scattering "stands in the history books as the crucial evidence that light quanta had to be taken seriously."[49]

Just When it Seemed that Physics was Ending

While it is never safe to affirm that the future of Physical Science has no marvels in store… it
seems probable that most of the grand underlying principles have been firmly established
– Albert A. Michelson (1894)[50]

The discovery that light behaves as if it were a stream of particles caused enormous confusion at the time. As often happens, just when the book on physics was closing, just when physicists thought they had things all figured out, that their understanding of the fundamental laws were essentially complete, a new revolution came into being, spurred by all sorts of new experimental results and thinking.

---

[44] (Isaacson, 2007) p. 100.

[45] (Pais, 1988) p. 212.

[46] Newton actually proposed the existence of light in the form of corpuscles and so this is a matter of dispute. Perhaps Einstein was the first modern physicist to propose the particle nature of light.

[47] (Isaacson, 2007) p. 101.

[48] The effect was observed in the increase in wavelength (lower energy) of the photon.

[49] (Lindley, 2001) p. 96.

[50] Full quote: "While it is never safe to affirm that the future of Physical Science has no marvels in store even more astonishing than those of the past, it seems probable that most of the grand underlying principles have been firmly established and that further advances are to be sought chiefly in the rigorous application of these principles to all the phenomena which come under our notice. It is here that the science of measurement shows its importance—where quantitative work is more to be desired than qualitative work. An eminent physicist remarked that the future truths of physical science are to be looked for in the sixth place of decimals." (Michelson, 1896) p. 159.

Indeed, around 1890 many physicists believed their task was largely done.[51] Every phenomenon had been identified and explained, except for one or two small issues that were essentially wrapped up. Or were they?

Crookes → Thomson → Röntgen → Becquerel → Curie → Rutherford → Planck → Einstein → Bohr → Schrödinger → Heisenberg → Born

> *As soon as I saw Balmer's formula, the whole thing was immediately clear to me.*
> – Niels Bohr[52]

We now return to Niels Bohr. Bohr had two pieces of information in front of him. First was Rutherford's discovery that essentially all the mass of the atom resides in a very small positively charged central core or nucleus, surrounded by negatively charged electrons. Second was Einstein's discovery that radiation is emitted from matter in the form of quanta and that its energy is directly proportional to its frequency ($E = h\nu$).

What Bohr didn't have was an understanding of how the electrons could possibly exist in orbit about the nucleus. From a classical physics perspective, the electrons should simply spiral right down into the nucleus, radiating photons the whole way down. Furthermore, he also didn't understand why the atom had the specific radius of 10^{-8} cm and why the elements had a periodically repeating set of chemical properties as reflected in Mendeleyev's periodic table. None of these could be explained. He needed yet one more piece of information, something to crystallize his own discovery of how the atom worked. And this piece arrived in the form of the Balmer series.

The Balmer Series

In 1885, Johann Jacob Balmer (1825–1898), a Swiss physicist and mathematician, created a mathematical formula to describe the patterns in spectral lines of atomic hydrogen. When the complete range of sunlight passes through atomic hydrogen, photons of very specific frequencies are absorbed by the electrons, causing them to jump to higher energy orbitals. The energy of the absorbed photon, as quantified by Einstein's equation relating a photon's energy to its frequency ($E = h\nu$), corresponds to the difference in energy levels between the two energy states involved in the electron's jump. Similarly, when the energized electron eventually drops back down to its ground state, either jumping straight down or stopping along the way at various energy levels, photons are emitted, one photon for each jump with a frequency equal to, once again, the energy gap covered by the jump.

Well, Balmer didn't know this. He simply saw a pattern in the frequencies where others did not. And he wrote a very simple but powerfully exact formula to describe these frequencies.

---

[51] (Frisch, 1973) p. 196.
[52] Cited in (Lightman, 2005) p. 155.

His formula was well recognized as a creation of beauty, even though no one knew why it worked. The discovery of spectral lines and Balmer's subsequent development of a simple yet powerful equation to describe them provided early evidence that nature is not continuous and that atoms exist in certain very specific energy states.[53]

Bohr had not heard of Balmer's formula. And so when he saw it, and believing that beautiful patterns do not happen by chance, all became immediately clear to him. His ideas crystallized and within hours he delivered atomic physics to the world.

* * *

Bohr created his model based two main assumptions.[54] This first sounds simple, but in actuality represented "one of the most audacious hypotheses ever introduced in physics."[55] Bohr postulated that an electron has a stationary and stable state of lowest energy, later called the ground state, which does not radiate. He thus proposed that a charged electron can orbit about a charged nucleus without emitting radiation, a proposal that ran directly counter to the electromagnetic theories of that day, specifically those of Maxwell and Hendrik Lorentz (1853–1928), who extended Maxwell's theories to charged particles. The classical theories apparently failed at the small dimensions of the atom.

In the second assumption, Bohr further proposed that the electron orbits the nucleus with an angular momentum equal to a whole number multiple of h/2π (later termed ħ), which represented yet another huge leap in thinking, truly reflecting Bohr's genius intuition. Bohr recognized that the units or dimensions of angular momentum (mass × radius × velocity) were the same as those of Planck's constant. And so he theorized that the angular momentum of the electrons must be some multiple of Planck's constant, with the lowest energy state, the ground state, having an angular momentum of h/2π. Playing with this line of thinking and using the physical data available at the time, Bohr discovered something rather striking. He showed this assumed "multiple" to occur as whole numbers and not fractions, a finding consistent with the concept of quantum behavior.[56]

Working with Newton's equations to describe the stable orbit of a body, i.e., the electron, around another body, i.e., the proton, but with Coulomb's law of electromagnetic attraction substituted for Newton's law of gravitational attraction, Bohr then developed a mathematical model of the atom describing a series of stable orbits where the electron can exist. The innermost orbit (the ground state) would have the lowest energy. With increasing radius, additional stable orbits of increasing energy levels would exist. At infinite radius, the electron would be free and have an energy level of zero, a reference point chosen for convenience. Because of this, the energy levels of electrons bound in atoms are all negative.

Bohr then proposed that an electron jump from an upper to a lower orbit would cause the radiation of a discrete amount of energy equal to the energy difference between the two orbits. And since the radiation energy is defined by its frequency as described by the classical wave

---

[53] The existence of such spectral lines suggested an internal complexity inconsistent with the simple assumptions being used by Maxwell and Boltzmann in the theoretical calculation of specific heat capacities, an issue that both were aware of and concerned with.
[54] (Frisch, 1973) pp. 84–85.
[55] (Pais, 1988) p. 199.
[56] (Frisch, 1973) p. 84.

theory, then only specific frequencies of emissions are involved, as shown by the Balmer series. Voila! (Important to note here is that Bohr rejected the photon concept until 1925.[57])

Bohr ran the calculations, showed how the electrons exist in stable orbits, determined that the smallest orbit for atomic hydrogen was in rough agreement with the known size of atoms at that time and in 1913 finally achieved a significant "breakthrough"[58] by using his model to derive Balmer's formula from first principles. Professor Arnold Sommerfeld (1868–1951) of Munich, master of applied mathematics, then furthered Bohr's work by constructing the more complicated and sophisticated math behind the orbits, showing in the process that three "quantum" numbers are required to specify the size, ellipticity, and orientation of these orbits.

Thus arrived the Bohr–Sommerfeld atom. The atom consists of a central positively charged nucleus containing essentially all of the mass, surrounded by electrons in specific, well-defined orbits and corresponding energy levels, differing by multiples of some basic quantum. A tremendous step forward from Thomson's Plum Pudding model, the Bohr–Sommerfeld atom explained a great many things such as the size of the atom and all sorts of spectroscopic data and provided the structure upon which to understand the repeating patterns in the periodic table.[59]

One problem remained, however. Just as Balmer's equation beautifully described the spectral lines of the hydrogen atom but without any underlying theory, the Bohr–Sommerfeld atom also beautifully explained the atomic model behind Balmer's equation but without any underlying theory. Their theory was eventually considered to represent the "old" quantum theory in that they proposed the electron's motion to follow the rules of motion set out by Newton, with some adjustments made for Einstein's relativistic effects and some quirky "quantum" rules defining the range of permitted shapes and sizes and alignments for the electron orbitals. But the underlying physics behind the quantum rules were still not known. Their model could not explain such seemingly simple questions as, *why doesn't the orbiting electron radiate energy (charges moving in relation to each other) and simply spiral down into the nucleus?*, and *how do electrons "jump" from one orbital to another?*

Crookes → Thomson → Röntgen → Becquerel → Curie → Rutherford → Planck → Einstein → Bohr → Schrödinger → Heisenberg → Born

After a time gap in scientific progress during the Great War (WWI) of 1914–1918, which redirected physicists' efforts toward government service, the 1920s brought in the next critical advances in atomic physics, especially concerning the behavior of electrons in atoms. By mid-1924, there was still no solid theory accounting for their behavior as reflected by the gaps between the Bohr–Sommerfeld model and experimental data. Among other discrepancies, while the model was based on planetary-motion physics and integer-related momenta, the

---

[57] (Stachel, 2009)
[58] (Frisch, 1973) p. 85.
[59] (Frisch, 1973) p. 86–88. (Lindley, 2007) pp. 58–61.

electrons' behaviors did not align with either, especially for atoms having two or more elec-trons.[60] Attempts to resolve this situation led to the formation of two camps with competing theories, one representing perhaps more of an extension of the old quantum theory and the other representing an entirely new quantum theory. Let's start with the former.

Particles Behave Like Waves

After long reflection in solitude and meditation, I suddenly had the idea, during the year 1923, that the discovery made by Einstein in 1905 should be generalized by extending it to all material particles and notably to electrons

– Louis de Broglie[61]

In November 1924, Louis de Broglie (1892–1987), a doctoral student at the University of Paris, proposed a very simple yet profound idea. If waves show properties associated with particles, then perhaps particles show properties associated with waves. Starting with this assumption, de Broglie calculated a wavelength for the electron particle, showing it to be inversely proportional to the electron's speed. He then looked at the implications of this calculation and found the rather amazing result that the wavelength of an electron existing in the innermost orbit of an atom exactly matches the circumference of the orbit. And for an electron in the next higher orbit, a distance of two wavelengths exactly matches the longer circumference for the larger radius, and so on, in simple progression, with the allowed radii being defined by a whole number increase in wavelength. The conclusion to be made was that only electrons having waves that linked up by coming back to the same starting point could exist. De Broglie's theory that electrons behave as standing waves about the nucleus had a certain appeal since standing waves, such as those cre-ated when violin strings are plucked, were well understood at the time. Einstein welcomed the theory—"I believe it is a first feeble ray of light on this worst of our physics enigmas."[62]—and in 1927 Clinton J Davisson and Lester H. Germer at Bell Labs, and separately Sir George Thomson, son of J.J., demonstrated the diffraction of electron beams, just as waves are diffracted, thus earn-ing De Broglie (1929), Thomson (1937), and Davisson (1937) Nobel Prizes.

Upon learning of de Broglie's work, Erwin Schrödinger (1887–1961), a physicist working out of Zurich, became enamored of the concept that electron orbits are really standing waves. Challenged by a colleague who asked, *if electrons are waves, then where is the wave equation?*, Schrödinger developed just such a wave equation that captured de Broglie's theory in a more formal manner. His published findings in 1926 were gratefully received by the physics com-munity, and especially by Einstein since, being based largely on classical physics and its inher-ent cause–effect determinism, the findings were much more understandable than a competing theory that developed during the same time period.

Schrödinger's wave mechanics provided the means to a much better understanding of how light is emitted during electron transitions, and in so doing cleared up Bohr's unacceptable sudden jump concept. His wave-function model proposed electrons to behave as diffuse, pul-sating, standing-wave clouds of a certain energy level as characterized by their orbits. During

---

[60] (Pimentel and Spratley, 1969) pp. 16–17.
[61] Cited in (Pais, 1988) p. 252.
[62] (Pais, 1988) p. 252.

a transition, one wave function overlaps with the other, a behavior allowed in wave mechanics, the first gradually weakening with the second gradually strengthening, until the first gives way completely to the second. Schrödinger calculated the time of transition to be finite, on the order of 10^{-9} seconds, not zero as in Bohr's model.[63]

Crookes → Thomson → Röntgen → Becquerel → Curie → Rutherford → Planck → Einstein → Bohr → Schrödinger → Heisenberg → Born

Now everything is in Heisenberg's hands... to find a way out of the difficulties.
– Niels Bohr[64]

I remember discussions with Bohr which went through many hours till very late at night and ended almost in despair; and when at the end of the discussion I went alone for a walk in the neighboring park I repeated to myself again and again the question: Can nature possibly be so absurd as it seemed to us in these atomic experiments?
– Werner Heisenberg[65]

When one wishes to calculate 'the future' from 'the present' one can only get statistical results since one can never discover every detail of the present
– Werner Heisenberg[66]

The development of the competing electron theory started around 1920 when Sommerfeld accepted two new students, Wolfgang Pauli (1900–1958) and Werner Heisenberg (1901–1976), who approached physics with fresh and highly intelligent eyes. Like Einstein before, neither showed a vested or biased interest in preserving the theories of the past, each willing to challenge any and all old rules with new theories, not being strong adherents of or well-enough versed in the old theories to see why their new theories wouldn't work. The shifting of a paradigm requires this, requires new thinking that starts with a clean sheet of paper, building up new theories from a whole new empty space.[67] And it certainly helped that Sommerfeld gave Pauli and Heisenberg the room, space, and freedom to explore and propose controversial ideas into the public arena. The ideas from this group grew significantly once Heisenberg travelled to Copenhagen in 1924 to begin collaborating with Bohr.

The critical issue faced by this group, and especially by Heisenberg and Bohr, concerned phenomena that were abrupt, spontaneous, and discontinuous. Electrons "jumped" from one atomic orbit to another with no apparent cause, no apparent reason, no apparent smoothness of change. At that time, the mathematics, based largely on differential calculus that describes incremental change, and the philosophies, especially concerning determinism, simply didn't

[63] (Frisch, 1973) pp. 94–95.
[64] Cited in (Lindley, 2007) p. 108.
[65] (Heisenberg, 2007) p. 16.
[66] Cited in (Isaacson, 2007) p. 333.
[67] Thomas Kuhn discusses all aspects of this issue in his *The Structure of Scientific Revolutions*. See especially the 1970 edition as it includes a valuable postscript section.

work to explain such behavior. An atom was in one energy state, and then in another. How did this happen? Why did this happen?

In an act of profound struggle, obsession, and creation, Heisenberg finally arrived at a solution, effectively launching the "new" quantum mechanics. In June 1925, attempting to heal himself of severe hay fever, Heisenberg isolated himself on an island off the north coast of Germany, and there had his own eureka moment—as he later recalled, "Well, something has happened"[68]— and developed an extremely complex mathematical description of the quantized energy of a mechanical system like the atom, an approach that would be hard to sufficiently describe in this book here. The math behind the approach was later discovered to be matrix algebra, which had already been invented years previously and was thus effectively re-invented by Heisenberg.

*   *   *

So two mathematical models existed in 1926 to describe the motion of an electron around a nucleus, Schrödinger's based on wave mechanics and Heisenberg's based on matrix mechanics. While both were complicated, the first was more easily interpreted, usable, and thus welcomed by the physics community. In fact, of the two, wave mechanics is still more widely, but not exclusively, taught and applied today.

Interestingly enough, once the two versions of the new quantum mechanics were laid before the public, Schrödinger and later Paul Dirac demonstrated their mathematical equivalence, an exciting finding were it not for the fact that this just furthered the confusion. How could two very different approaches end up with equivalent answers?

Crookes → Thomson → Röntgen → Becquerel → Curie → Rutherford → Planck → Einstein → Bohr → Schrödinger → Heisenberg → Born

Max Born (1882–1970), a theoretical physics professor at Göttingen, solved this dilemma by bringing the concept of probability into quantum mechanics. While physicists were enamored of Schrödinger's math, they, and notably Einstein, also recognized its limitations. The model predicted that over time, the wave could mathematically spread over a large area, which couldn't really be what was happening.[69] Born brought clarity to this problem by pointing out that Schrödinger's wave equations didn't describe actual particles and classical waves, but something entirely new based on probabilities. In the case of the electron orbiting a nucleus, for example, Born said that the wave equation doesn't represent a physically spread-out mass or charge but instead represents the *probability* of finding it in one location or another. And this description worked well with Heisenberg's matrix mechanics in a deeply philosophical and mathematical way since Heisenberg's approach was based on what an electron *might* be doing as opposed to a specific indication of what it *was* doing. While both the science and the math here are difficult to follow, the key is the advent of true, real probability in physics. This was a revolutionary theory that, bitterly to Born, received little immediate recognition, as reflected by the long delay in his being awarded the Nobel Prize in 1954.

---

[68] (Lindley, 2007) p. 114.
[69] (Isaacson, 2007) p. 330.

The Historic Solvay Conference of 1927

During this time, that culminated in the historic 5th Solvay Conference in Brussels in 1927, where the world's top physicists gathered around the theme of "Electrons and Photons," physics took a sharp turn from classical to quantum philosophies. In classical physics, probability exists as shown, for example, by James Clerk Maxwell's and Ludwig Boltzmann's use of probability distributions in the late 1800s to describe the behaviors of a very large numbers of particles, such as gaseous atoms, since trying to calculate such behaviors one atom at a time would be near impossible. But at the core of these cases existed a deterministic belief in the cause–effect behavior of nature as characterized and quantified by Newton's rule book. Indeed, the principle of causality was at the heart of classical physics. The only probability that then arose concerned the accuracy of the measuring devices used. But with increasing accuracy of measuring device came increasing power to predict the future effect of a present cause.

In quantum physics, the concept of probability is very different in nature. It's inherent to the system, an inescapable condition that can't be resolved with expensive analytical devices. No matter how hard one tries, one can only specify the probabilities of outcomes. Effect doesn't exactly follow cause. Things just happen one way or the other with no underlying reason why. The quantum mechanical particle follows no set path. While the classical world is crystalline and certain, the quantum world is fuzzy and indeterminate.

Things were to get even more complicated when Heisenberg furthered his theory. Starting with Born's logic that a collision between particles doesn't result in a definite outcome but instead in a range of probable outcomes, Heisenberg showed that it was impossible to exactly and simultaneously determine both the position and momentum of a particle. For example, in the classical world, deliberately colliding a photon into an electron, as Compton showed was possible, would allow exact determination of the location and momentum of the electron. You would exactly know the change in direction and energy of the photon caused by the collision and thus be able to exactly calculate where the electron must have been, how fast it was moving, and in what direction it was moving. The use of light in this way is a common approach in determining both position and speed of a material object. And it works quite well in the classical world.

But if Born were indeed correct, then a range of collisions could produce a given outcome, which in turn would mean that a given outcome could have been caused by a range of collisions. This then makes it impossible to determine the exact collision, as defined by the electron's position and momentum, at cause in the matter. You can focus on either position or momentum but not both simultaneously. The more you accurately determine one, the less accurately you determine the other. And it's not as simple as this statement implies. It's not that it's either-or. It's an inherent fundamental limitation in exactness that's always present. While Heisenberg called this inherent limitation "inexactness," Bohr later applied the term "uncertainty" and hence was born the Heisenberg uncertainty principle.

In short, quantum mechanics says that identical states measured in identical ways can yield different results. One can't predict the future. While most practicing scientists accepted this Copenhagen Interpretation, as it would soon be called since it was based largely on the influence of Bohr and Heisenberg, few were happy with it. Einstein, for one, the true father of quantum physics, never accepted it, arguing that quantum mechanics is an incomplete theory, masking an underlying reality with hidden variables still to be exposed.

Quantum mechanics' removal of determinism from physics greatly disturbed Einstein, as reflected in his famous statement, "I for one am convinced that He does not throw dice."[70] He continued fighting against the concept of probability in physics but could find no way to eliminate it. No thought experiment or real experiment could find a flaw in this new theory and no theorist could come up with something better, even to this day. Einstein felt that the source of the "uncertainty" was man's lack of understanding of the world as opposed to some fundamental aspect of nature itself. His arguments with Bohr against the Copenhagen Interpretation were legendary. The fact that he couldn't knock it down helped lead to its ultimate validation. It was a theory that kept standing in the face of attack from a giant.[71]

The debate continues to this day, as some number of physicists remain unconvinced that quantum theory depends on pure chance. They continue to work toward developing quantum models that are deterministic underneath the seeming uncertainty.

Pauli Exclusion, the Zeeman Effect, and Electron Spin

Further modeling of the behavior of the electron about the nucleus would provide yet one more major breakthrough in our understanding of the atom. By the early 1920s, three quantum numbers had been accepted as the requirements to describe an electron's orbit around a nucleus, the first for radius (energy), the second for shape (sphere, dumbbell, cloverleaf), and the third for orientation and direction of the shape.[72] In 1924, Pauli proposed a fourth number to account for his belief in the existence of two "types" of electrons. Furthermore, he said that in an atom there can never exist two electrons having the same four quantum numbers. This rule eventually became known as the Pauli exclusion principle since it effectively meant that no more than two electrons are permitted in any given orbit. But Pauli knew that the story was not yet complete, writing "We cannot give a more precise reason for this rule."[73] This was left to others, namely two young Dutch physicists, Samuel Goudsmit (1902–1978) and George Uhlenbeck (1900–1988), who had the original theoretical breakthrough in 1925, and then Paul Dirac (1902–1984), who wrapped it all up into his own major mathematical breakthrough, which he published in 1928.

Spectroscopy

But before sharing this story, a short detour is needed to revisit the science of spectroscopy. When light passes through a gas, certain photons are absorbed by the electrons, which excites them to

---

[70] (Lindley, 2007) p. 137.

[71] (Isaacson, 2007) pp. 324–6.

[72] The first quantum number corresponding to total energy is called the principle quantum number. In quantum mechanics, an energy level is called degenerate if it corresponds to two or more different measurable states of a quantum system. In other words, an energy level is degenerate if there exist different combinations of variables (2nd, 3rd, 4th quantum numbers) that result in the same energy (1st quantum number). This concept becomes important in statistical mechanics, for example, in light of the fact that different combinations of x,y,z-velocity components can result in the same speed (u) and thus the same kinetic energy. $u = \sqrt{(u_x^2 + u_y^2 + u_z^2)}$.

[73] (Pais, 1988) p. 274.

higher energy levels. After a very brief moment, the atoms return to their ground state, perhaps stopping momentarily at other energy levels along the way. For each gap traversed by the electron between orbits, either on the way up or on the way down, the photon that is absorbed or emitted, respectively, has a frequency (v) matching the energy difference ($E = hv$) between the gaps. The resulting spectra, for either photon absorption (dark lines on light background) or emission (bright lines on dark background), are very sharp and specific, unique for each element.

Analysis of these spectral lines—beautiful patterns of very specific spectral lines—led to the conclusion by many, long before the conclusion could be finalized, that there must be some very detailed structure contained within the atom itself, inherent to the atom, seemingly unaffected by the surrounding environment. The world of spectroscopy soon opened up, helping us to both identify atoms by their spectral fingerprints and to study their workings. As I wrote in Chapter 2, such a tool helped in the study of the light coming to Earth from the universe, thus enabling our determination of the composition of the universe and our unlocking of the secrets of the Big Bang.

Scientists, and especially the Dutch physicist Pieter Zeeman (1865–1943), soon brought new techniques to the world of spectroscopy. In 1896, building on Michael Faraday's (1791–1867) intuition, Zeeman demonstrated that a magnetic field causes spectral lines to broaden and eventually split. Zeeman's countryman, Hendrik Lorentz, applying Maxwell's electromagnetic equations to charges and currents carried by fundamental particles in ways too complicated to address here, was able to explain *most*, but not all, of these effects, which was rather amazing in light of the fact that Bohr's model of the atom had not yet been created. Zeeman and Lorentz shared the Nobel Prize in 1902 for their combined work on the influence of magnetism upon such spectral lines.

Two types of "Zeeman effects" were observed. The first, or "normal" effect, was characterized by the magnetic-induced splitting of a single spectral line into three spectral lines, which came to be known as the Lorentz triplet. The second, or "anomalous" effect, was discovered by Alfred Marie Cornu (1841–1902) in 1898 and manifested itself by any pattern of splitting other than the Lorentz triplet. While Lorentz's math worked wonders in describing how the force of a magnetic field on the moving negative charges could result in the normal effect, it couldn't fully resolve the underlying reason for the anomalous effect.

Well versed in Pauli's exclusion principle, which had added a fourth quantum number to describe each electron, the team of Goudsmit and Uhlenbeck were primed to solve this problem. As Goudsmit recounted, "Our luck was that the idea [of spin] arose just at the moment when we were saturated with a thorough knowledge of the structure of atomic spectra, had grasped the meaning of relativistic doubling, and just after we had arrived at the correct interpretation of the hydrogen atom."[74] And as Uhlenbeck recounted, "It was then that it occurred to me that since (as I had learned) each quantum number corresponds to a degree of freedom of the electron, the fourth quantum number must mean that the electron had an additional degree of freedom – in other words the electron must be rotating!"[75]

In 1925 Uhlenbeck and Goudsmit published their discovery of "spin," which accounted for the two values—spin in either one direction or the other—for the fourth quantum number. It was

---

[74] (Pais, 1988) p. 277.
[75] (Pais, 1988) p. 277.

clear that the spirit behind their discovery was a combination of classical and old quantum theories. They and others literally believed the electron to be a rigid body, spinning about its axis. The term "spin" was coined because the electrons demonstrated properties associated with angular momentum, which implied that they must spin. Physicists have since learned that things aren't as simple as this. In order for something to spin, it must have an internal structure. And to the best of physicists' understanding, electrons do not have an internal structure, which means they can't actually spin. Instead, they must truly be fundamental, structureless, irreducible particles with an inherent angular momentum. This means that even if the electrons were standing still, orbiting nothing, they would still have an angular momentum, a property as fundamental as charge and mass.

Paul Dirac

You may wonder: Why is nature constructed along these lines? One can only answer that our present knowledge seems to show that nature is so constructed. We simply have to accept it. One could perhaps describe the situation by saying that God is a mathematician of a very high order, and He used very advanced mathematics in constructing the universe.

– Paul Dirac[76]

The discovery of spin resolved much if not all of the spectral lines observed, including those resulting from splitting by magnetic fields. But left unresolved was the reason behind the spin. This was left to Paul Dirac, a mathematician.

In 1928, starting with the intent to simply improve Schrödinger's model by including Einstein's relativistic effects to account for the high speed of the electron's movement, Dirac and his powerful mathematical approach uncovered the fact that some energy levels of the electrons were not single levels but double levels very close together. This was a fascinating development, a huge moment in the history of physics, as Dirac had not started out by assuming spin. Spin arose directly from his math. Dirac used pure mathematical reasoning to fit Schrödinger's model into Einstein's relativity theory, and so discovered the last major clue in understanding atomic structure prior to the physical discovery via experimentation.

The combined work of Pauli, Uhlenbeck and Goudsmit, and Dirac served to complete the description of an electron's behavior inside an atom. Since then, the theory has been extended and improved, but not basically changed.

The Neutron

"The complacency of 1930—the widespread belief that physics was nearly complete—was shattered by the discovery of subatomic particles."

– Otto Frisch[77]

---

[76] (Dirac, 1963) p. 53.
[77] (Frisch, 1973) p. 199.

With the arrival of quantum mechanics in 1928, it would seem that this chapter has reached an end, with the atom defined and the electron behavior described. Yes, further developments in quantum mechanics did continue, but they are far outside the scope of this book.

But upon reflection, you'd realize that we couldn't have yet reached the end, for as you may have already realized since I didn't bring it up, the neutron had not yet been discovered by 1928. And so all the significant depths of mathematics applied to quantum theory, mostly regarding the behavior of electrons about the nucleus, were completed without there being an accurate understanding of the composition and inner workings of the atomic nucleus.

* * *

In the twenty years following Rutherford's discovery, physicists thought that the nucleus contained only protons (hydrogen nucleus)[78] and electrons. So, for example, the nucleus of helium-4 was believed to consist of four protons and two electrons, thus matching the atomic weight of 4 (electrons have negligibly low mass relative to protons) and the net atomic charge of +2 needed to neutralize the −2 charge of the orbital electrons. Since most experiments involving the disintegration of the nucleus resulted in the emission of either alpha particles (He-4 nucleus) or beta particles (electrons), there was no reason to think that any particles other than protons and electrons comprised the atom.

And yet...

In 1932, James Chadwick (1891–1974) at Cavendish laboratory, a student and the colleague of Rutherford's at Manchester and the Cavendish laboratory, respectively, noticed that, upon bombardment with very fast alpha particles, beryllium and other light elements emitted highly penetrating radiation, much more penetrating than the protons emitted from other nuclear disintegrations. Chadwick eventually determined that these rays were actually particles of weight nearly identical to the weight of the proton. He also deduced that, because these particles showed high penetrating power, they must be electrically neutral, thus enabling their ability to pass through the electrically charged environment within atoms.

While Chadwick believed this "neutron" particle to be comprised of a proton and an electron, other physicists soon realized that the neutron itself was an elementary particle, although the exact point at which this shift occurred is difficult to identify. The realization of the neutron as the final significant elementary particle of the atom, until the discovery of the quark in the 1970s, quickly solidified the terminology of atoms and helped clarify the differences between atomic number, atomic weight, and isotopes. Much confusion in the early stages of nuclear physics was caused by the unrecognized presence of isotopes in which the same element could be comprised of a different number of neutrons. Chlorine, having an atomic weight of 35.45, was a classic example. Its weight occurred not as an integer but as a fraction, which wasn't consistent at all with atomic theory. We know today that chlorine has an atomic number of 17 and consists of a blend of two isotopes, 76% Cl-35 (18 neutrons) and 24% Cl-37 (20 neutrons), thus resulting in a weight of 35.45.

From here the final major pieces of the atomic puzzle, including discovery of the weak interaction and the strong interaction, were rapidly put together, opening the door to the modern era of nuclear physics. The starting point for this final push? Enrico Fermi (1901–1954), Italy, 1933.

[78] (Romer, 1997) In 1920 Rutherford, suspecting that the hydrogen nucleus contained one particle, named this particle the *proton*.

The Weak Interaction

Fermi contributed two important ideas. The first concerned beta radioactivity, a very elusive phenomenon since it seemed to defy the law of conservation of energy, something we'll address in more detail in Chapter 23. Suffice to say here that Fermi, building on ideas set forth by Pauli, brought great clarity to this process by explaining that the origin of the beta rays (electrons) from the nucleus did not mean that the electrons were in the nucleus all along, as many had thought. Just as the emitted photon is not actually present in the atom until an electron jumps orbits, the emitted electron is not present until a neutron disintegrates via beta decay into a proton, an electron, and the yet-to-be-discovered neutrino. This groundbreaking work showed that neutrons could be converted into protons, and vice versa, meaning that neither could be elementary particles themselves, and thus brought much clarification to physics.

Fermi's work here in this regard revealed the action of an entirely new interaction, called the weak interaction, now considered as one of the four fundamental interactions of nature. Whereas the electromagnetic interaction causes the emission of photons during electron jumps, the weak interaction causes the emission of electrons during beta decay. The weak interaction lays behind the adjustment that occurs between neutrons and protons depending on the surroundings they're in. For example, while free protons are stable, free neutrons are not, having a half-life of about 10 minutes. Yet in some nuclei, while neutrons are stable, protons are not. The weak interaction and resultant beta decay are the source of this behavior.

For his second contribution, Fermi proposed that the newly discovered neutrons would make even better atomic projectiles than alpha particles, especially when trying to penetrate the nucleus. The problem with the alpha particle is its positive charge, which makes it easily repelled by the positively charged nucleus. Only high-energy alpha particles stand a chance of penetrating the nucleus. The neutron, on the other hand, has no charge and can easily pass through the nucleus' defenses without the need for high speed. And so a new branch of physics commenced based on the use of neutron sources, which were easily created by physically mixing radium (alpha emitters) and light elements such as beryllium.

The Splitting of the Atom

Soon physicists, notably Otto Hahn (1879–1968) and Lise Meitner (1878–1968), latched on to Fermi's idea and started firing neutrons at different elements to study the radioactive disintegration process, uranium being a favorite target since it contained the nucleus with the largest naturally occurring atomic number. Hahn and Meitner, colleagues in Berlin since 1907, teamed with Fritz Strassmann (1902–1980) in 1935 for their historic experiments that they initially anticipated would last but "a few weeks"[79] to check on some of Fermi's results. Hahn's strength in analytical chemistry contributed much to their efforts since such experimental skills were essential in identifying the many fragments of isotopes and elements resulting from radioactive disintegration. Over the course of their program, they surprisingly found that the reactions

---

[79] (Frisch, 1978) p. 428.

seemed more probable the slower (less energetic) the neutron projectiles, exactly opposite what they expected. At this time, most physicists felt that the disintegration resulting from such bombardments was caused by the physical smashing of one particle into another, much like a high speed bullet would shatter a rock. But their results suggested otherwise. It was the slow speed bullet that was the most effective.

In a repeat of trends that occurred during WWI, scientific progress slowed in 1933 when Adolf Hitler rose to power in Germany and occupied Austria in March 1938. Collaborative efforts between the close-knit British and European community of physicists began to unravel. Meitner, a Viennese of Jewish ancestry, fled Germany and eventually resided in Sweden. Hahn together with Strassmann continued their experiments in Berlin, sharing results with Meitner as they dared.

During one such dare in November 1938, Hahn secretly met Meitner in Copenhagen to discuss his most recent experimental results showing that the firing of slow neutrons at uranium produced fragments which he analyzed to be radium…or so he thought. The proposed reaction pathway and energy balance didn't make sense to Meitner and she requested further study and analysis to check the initial findings. So Hahn did this, repeated the experiments and came up with a different answer. Staggeringly different.

If you look at the periodic table, radium sits immediately below barium, thus meaning that they share the same electron structure in the outer shells and hence share similar chemical properties. It was easy in Hahn's case to assume that a single small neutron moving at slow speed didn't have the energy to chip off anything more than tiny pieces of the huge uranium nucleus, thus leaving as the main product a nucleus close in size to uranium. His chemical analysis pointed to radium (atomic number = 88), which was quite close to uranium (atomic number = 92). What he didn't realize until Meitner's strongly urged re-analysis was that his analytical tests weren't picking up radium at all, but instead were picking up barium (atomic number = 56), an element half the size of uranium! How could this possibly be? "It was as if a stone from a slingshot had cracked open a mountain."[80] The thought that he had actually split the atom went against all of Hahn's experience, so much so that he couldn't bring himself to fully embrace this in his publication of these data. It was Meitner who did. The world had shifted.

Meitner's realization that uranium had indeed been split occurred when she was discussing Hahn's results with her nephew, Otto Frisch (1904–1979), yet another physicist of Jewish descent who fled Germany. The discussion occurred during their historic walk in the snow in Sweden during Christmas in 1938. They concluded that uranium could be split, not by the physical cracking or chipping caused by the neutron's impact, but by the slight deformation caused by the presence of an additional neutron inside the nucleus. This theory would account for the observed time delay in the fission process since the former chipping process would be near instantaneous while the latter deformation process would require a very short but finite time for the unstable nucleus to break apart.

Meitner and Frisch realized that if they envisioned the uranium nucleus to be a sphere, or a drop of liquid water as imagined by Bohr, and since a sphere represents the tightest packing of matter, then any deformation, however slight, such as that caused by the arrival of an extra neutron in the nucleus, would lead to less tight packing. So if the uranium nucleus were sitting

---

[80] (Lightman, 2005) p. 305.

right on the edge of a stability cliff,[81] then a single neutron might be enough to cause it to tip. They figured that there must be some force holding the nucleus together, a force we now know to be due to the strong interaction that overwhelmed the very powerful repulsive force caused by the presence of all those protons inside the nucleus. How else could the nucleus be stable? However, once the uranium nucleus started jiggling and deforming, it was only a matter of time until the net force inside the nucleus would shift from attractive to repulsive, causing the nucleus to split, just like a water droplet does once it starts to deform. The two smaller nuclei would then fly apart at tremendously high speeds. Using straightforward pre-quantum physics to calculate the repulsive force caused by the protons, Meitner determined that 200 million electron volts of energy would be released during a single fission! Fissioning one gram of uranium would produce about three million times as much energy as burning a gram of coal.

Ultimately, Meitner and Frisch realized what had happened. The atom had been split! The news from these experimental findings and accompanying theories were released in January 1939 and spread like wildfire. Frisch told Bohr and Bohr pretty much told the world since within a month he travelled across the Atlantic to the United States and shared these results with many. Within a year, more than one hundred papers on fission had been published and the nuclear age was fully underway.

Hahn alone won the Nobel Prize in Chemistry for this finding, a severe discredit to the contribution of Meitner. Several forces likely influenced this decision, including the fact that the Nobel Prize is traditionally awarded based on experimental discoveries rather than theoretical interpretations. Hahn did the experiments and Meitner interpreted the results. It wasn't the first time, nor the last, that the Nobel Prize was awarded incompletely.

The Strong Interaction

Remaining to be answered was the question, what held the nucleus together? How was this coiled force contained? The beginning of the answer was addressed by Rutherford. Based on his work with radioactivity, Rutherford speculated that there must exist some attractive force keeping the nucleus together, some force to counteract and indeed overwhelm the repulsive electromagnetic force caused by the close packing of the protons. As mentioned earlier, throughout the early 1900s, most physicists believed that electrons existed inside the nucleus and so served as the cohesive force. But it wasn't clear why some electrons should be bound in the nucleus while others orbited the nucleus. Additionally, "no one had any idea anyway of what sort of force might be operating at the extremely short distances separating particles within a nucleus."[82]

---

[81] In a sequence similar to the cocking of a mousetrap, uranium atoms were formed billions of years ago when the tremendous packing force of gravity inside stars was sufficient to overcome the tremendous repulsive force of the protons, packing them tighter and tighter together until the strong nuclear attractive force was sufficient to take over, thus setting the trap. During star explosions, more nuclear matter in the form of neutrons were added to this foundation, building up elements heavier than iron, unstable elements that behave like a cocked mousetrap in that the slightest disturbance causes each to explode. In an interesting back-and-forth flow of energy, the Big Bang forced matter apart, gravity brought matter back together, and nuclear fission forced matter back apart again.

[82] (Weinberg, 2003) p. 133.

So after the discovery of the neutron and the demise of the electron-in-nucleus theory, speculation increased rapidly that a strong interaction must exist between the protons and the neutrons. Clarity arrived in 1935 when Japanese theorist Hidekei Yukawa (1907–1981) proposed an answer.[83] While the interaction couldn't be electromagnetic in nature, as the neutron had no charge, Yukawa theorized another interaction at work, a new kind of interaction similar in nature to the electromagnetic interaction, but involving a new particle. While the electromagnetic interaction involves the exchange of photons between charged particles in ways too difficult to go into here, this new interaction was proposed to involve the exchange of entirely new and not-yet-detected particles he labeled mesons[84] between the protons and neutrons. The subsequent discovery of this particle in 1937 solidified Yukawa's theory involving the mesonic force, i.e., the strong interaction.

*   *   *

While WWII stifled shared scientific discovery between 1936 and 1945, secret discovery escalated on the use of the newly discovered fission for the atomic bomb, especially in the United States. Most all of the physicists available, including those who fled Europe such as Fermi, were pulled together to rapidly scale-up Meitner and Hahn's findings and create the atomic bomb, which helped to bring an end to the Pacific War with Japan.[85]

Soon thereafter, as the doors to science opened once again, more sub-atomic particles were discovered and more theories developed, too many and too complicated to go into here, with one exception.

The Quark

In 1968, physicists at the Stanford Linear Accelerator Center (SLAC) took a page out of Rutherford's notebook and fired high-energy electrons at protons. While most passed right through the protons with some deflection, a small fraction suffered...large scattering angles! Déjà vu! Just as Geiger and Marsden's alpha particle scattering results with gold foil suggested the presence of a nucleus within the atom, the SLAC results suggested the presence of some type of unknown particle *inside* the proton. The opening of the Russian nesting doll continued.

Originally thought to be fundamental particles, the proton and neutron were now themselves observed to consist of building blocks, which became known as "quarks."[86] While scientists couldn't isolate quarks, due to the counter-intuitive fact that the force binding the quarks together *increases* with distance, they could infer their presence based on the SLAC experiments and soon developed theories to explain their nature. It became apparent that just as electrons fall into two categories, $+ \frac{1}{2}$ and $- \frac{1}{2}$ spin, these quarks also fall into two categories, up and down. The proton and neutron each contain three of these quarks, two up and one

---

[83] (Pais, 1988) p. 430.

[84] Mesons were later to be more clearly identified as pi-mesons or pions.

[85] (Frisch, 1978) p. 428. Recollection of Meitner's nephew, Otto Frisch: "She had worried about the development of nuclear weapons and refused to take part in it when invited to do so. Their success distressed her greatly, but she hoped for the peaceful use of fission."

[86] (Munowitz, 2005) p. 357. "The term quark is a whimsical usage inspired by Murray Gell-Mann's reading of a line from James Joyce's Finnegan's Wake: 'Three quarks for Muster Mark!'"

down for the former, and one up and two down for the latter. Furthermore, scientists proposed that the quarks are held together by exchange of elementary particles they named "gluons."

The Standard Model

Today, the full mathematics and science of these sub-atomic particles and the interactions between them are contained in what is called the Standard Model. Several tasks remain to complete the model, including further validation of the 2012 Higgs boson discovery and inclusion of the graviton, which is the hypothesized messenger particle of gravity.

The Standard Model is synthesized from two sub-models involving three of the four fundamental interactions. One of these sub-models, called Quantum Chromodynamics (QCD), provides the full theory of the strong interaction involving, for example, quarks and gluons. The other sub-model, called Quantum Electrodynamics (QED), provides the full theory of the electromagnetic and the weak interactions involving, for example, electrons and protons, or more accurately, the electrically charged quarks within the protons.

More relevant to our interests in this book, QED is the theory of all chemistry. Considered one of the greatest success stories in physics, QED unified the electromagnetic interaction and the weak interaction into the *electroweak* interaction based on the learning that the two interactions share a common origin, specifically that moment shortly after the Big Bang when the universe was hot enough for the two interactions to exist together as a combined interaction. The development of this fundamental theory of the interaction between light and matter, or between electric field and charge, contains the basic rules for all the known electrical, mechanical and chemical laws.

Of the four fundamental interactions of nature, only gravity remains unaccounted for in the Standard Model. Pursuit of a unified theory to account for everything by integrating gravity into the Standard Model is considered the Holy Grail, a journey famously and futilely undertaken by Einstein, among others.[87] The difficulty to overcome is the fact that gravity is so very weak compared to the other three interactions, making it difficult to incorporate into a unified model. The other interactions simply overwhelm it. So the only opportunity to study and develop a truly unified model would be to run an experiment in which gravity plays an equal role with the other interactions. Well, such an experiment has already been done, billions of years ago. The Big Bang. This was the one event in our history where all four interactions played significant roles and is now the focal point of current study regarding a unified theory of everything. (Alas, validation experiments for this scenario will naturally be difficult.)

---

[87] (Krauss and McCarthy, 2012) pp. 311–2. Richard Feynman: "People say to me, 'Are you looking for the ultimate laws of physics?' No, I'm not. I'm just looking to find out more about the world. If it turns out there is a simple, ultimate law which explains everything, so be it; that would be very nice to discover. If it turns out it's like an onion, with millions of layers, and we're sick and tired of looking at the layers, then that's the way it is. But whatever way it comes out, it's nature, and she's going to come out the way she is."

Conclusion

At the most fundamental level, the computer is based on very simple logic. Either the switch is off or it is on. Either it's 0 or 1. Binary logic. That's it. The complexity only arises as layers upon layers of these zeros and ones are built up upon each other, making it very hard for the human mind to follow.

So too with physics and chemistry. At the most fundamental level, the atom behaves rather simply, following some rather straightforward laws. Each individual interaction between particles, atoms, molecules, and photons is determinable and quantifiable, ignoring for now the complexities that arise in the depths of quantum mechanics. It's when the full variety of atoms in the periodic table are considered and when billions and billions of this range of atoms are present that things become complicated, leading to the complexity we see around us today. But just because it's difficult for us to comprehend doesn't mean that it's not simple. At its core, it is simple. Really.

Using the core understanding provided in this chapter as our starting point, our bedrock, we can start moving forward in the direction of billions and billions. But before moving too far, we need to learn about a concept called energy. We spoke of single atoms and their characteristics and how two atoms interact with each other in various ways, such as attraction, repulsion, collision, and reaction. And we also discussed the role of the photon as a means of transferring energy. We now need to discuss the concept of energy itself, which we use to keep track of all these interactions. Energy provides an accounting system for us to keep track of credits and debits, in the end knowing that the two must balance because there's a law that says they must balance. This law lies at the core of the 1st Law of Thermodynamics and is called the conservation of energy.

PART III

Energy and the Conservation Laws

ARISTOTLE
(incorrect theories
for both)

THE LEVER
- weight × velocity
  is important

HERO (50 AD)
- weight × displacement
- 5 simple machines

JORDANUS (1200s)
- weight × vertical displacement

$$w\Delta h$$

further study
of machines

perpetual motion
is impossible

MOTION
- free fall influenced by weight
- all that moves contains
  something else doing
  the moving

GALILEO (1638)
- broke from Aristotle's paradigm
- law of fall: $h \propto t^2$

DECARTES (mid 1600s)
- focus on weight × speed
- first attempt at
  conservation law for motion

HUYGENS (1847)
- delivered a new variable
  to the world: wv^2

But since heat
can generate work,
how to account for heat
in a conservation law?

early version of
CONSERVATION OF MECHANICAL ENERGY

$$\tfrac{1}{2}mv^2 + mgh = \text{constant}$$

LEIBNIZ (1686 - 1695)
- showed logical connection
  between $w\Delta h$ and wv^2

NEWTON
- laws of motion

(1732 - 36)
BERNOULLI / EULER
- unite **NEWTON / LEIBNIZ**

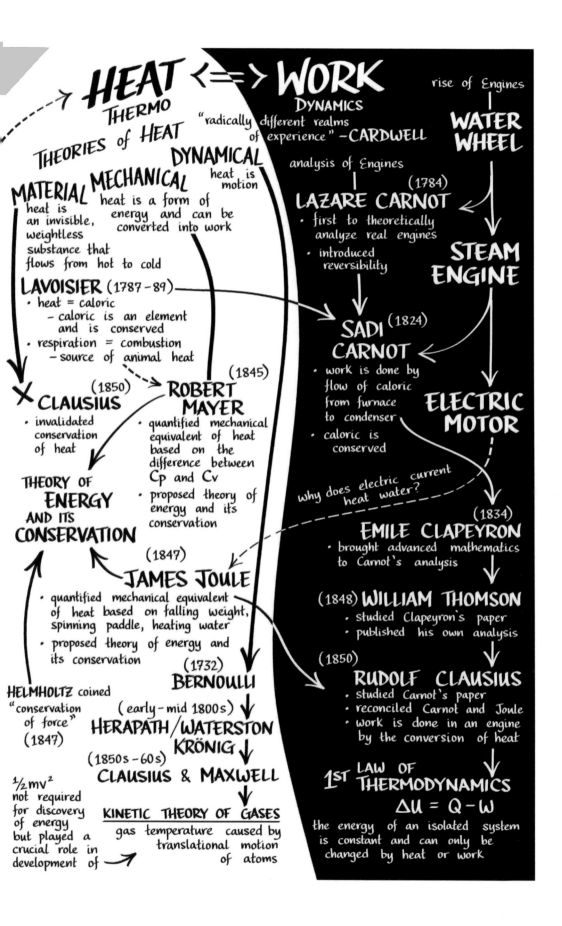

HEAT <==> WORK

THERMO **DYNAMICS**

THEORIES of HEAT

"radically different realms of experience" –CARDWELL

DYNAMICAL heat is motion

MATERIAL **MECHANICAL**

heat is an invisible, weightless substance that flows from hot to cold

heat is a form of energy and can be converted into work

LAVOISIER (1787-89)
- heat = caloric
  - caloric is an element and is conserved
- respiration = combustion
  - source of animal heat

✗ **CLAUSIUS** (1850)
- invalidated conservation of heat

ROBERT MAYER (1845)
- quantified mechanical equivalent of heat based on the difference between Cp and Cv
- proposed theory of energy and its conservation

THEORY OF ENERGY AND ITS CONSERVATION

JAMES JOULE (1847)
- quantified mechanical equivalent of heat based on falling weight, spinning paddle, heating water
- proposed theory of energy and its conservation

HELMHOLTZ coined "conservation of force" (1847)

BERNOULLI (1732)

HERAPATH/WATERSTON (early-mid 1800s)
KRÖNIG

CLAUSIUS & MAXWELL (1850s-60s)

½mv² not required for discovery of energy but played a crucial role in development of ➔

KINETIC THEORY OF GASES
gas temperature caused by translational motion of atoms

rise of Engines

WATER WHEEL

analysis of Engines

LAZARE CARNOT (1784)
- first to theoretically analyze real engines
- introduced reversibility

STEAM ENGINE

SADI CARNOT (1824)
- work is done by flow of caloric from furnace to condenser
- caloric is conserved

ELECTRIC MOTOR

why does electric current heat water?

EMILE CLAPEYRON (1834)
- brought advanced mathematics to Carnot's analysis

(1848) **WILLIAM THOMSON**
- studied Clapeyron's paper
- published his own analysis

(1850)

RUDOLF CLAUSIUS
- studied Carnot's paper
- reconciled Carnot and Joule
- work is done in an engine by the conversion of heat

1ST LAW OF THERMODYNAMICS

$$\Delta U = Q - W$$

the energy of an isolated system is constant and can only be changed by heat or work

5 Energy: science (and some history)

There is a fact, or if you wish, a law, governing all natural phenomena that are known to date. There is no known exception to this law – it is exact so far as we know. The law is called the conservation of energy. It states that there is a certain quantity, which we call energy, that does not change in the manifold changes which nature undergoes.

– Richard Feynman[1]

Energy Invented to Quantify Change

Left alone, atoms would experience no change. If still, they would remain still. If moving, they would remain moving. But in our universe, being left alone is impossible. The four fundamental interactions (or forces) ensure this. They were there at the beginning in the extreme smallness immediately after the "bang" in the Big Bang. While we don't know exactly what form of dark energy repulsion caused the "bang", we do know pretty well what happened afterwards. As space inflated and continued to expand over the next 13.75 billion years of the universe's existence, the four fundamental interactions were in action. The strong interaction pulled neutrons and protons together, the electromagnetic interaction pulled electrons to the resulting nuclei, and then gravity pulled the resulting atoms together, packing them tightly, fusing them, igniting reactions, releasing heat, and causing explosions, starting an in-and-out cycle of star formation and destruction, like a swinging pendulum, over and over again. Throughout all of this, the weak interaction caused neutrons to convert to protons, thus enabling a path to stabilize the growing nuclei. Not to be ignored, the photons. Left alone, photons would travel in a straight line at the speed of light, forever. However, in our universe, photons interact with each other and with matter—disappearing, reappearing, changing direction.

The four fundamental interactions pervade the universe and are at the root of change. It's the change in the potential energy originating from these forces over distance that causes a change in the kinetic energy of particles over time. We invented the concept of energy as a means to account for the change in both.

Energy is a very abstract man-made concept. Although we invented it, we're not really sure what it or its underlying mechanisms are. It's a quantifiable number that we calculate by adding up all the different energies manifested in matter and radiation, based on all of the forces and motions

---

[1] (Feynman et al., 1989a) Volume I, p. 4–1.

Block by Block: The Historical and Theoretical Foundations of Thermodynamics,
Robert T. Hanlon, Oxford University Press (2020). © Robert T. Hanlon.
DOI: 10.1093/oso/9780198851547.001.0001

involved. While the calculation of this number may not be terribly exciting to consider, there's an aspect of it that is. It never changes. This number always remains the same. Energy is always conserved.[2]

Events Happen but Total Energy Remains the Same

Matter and photons have been interacting with each other since the beginning of time. During each interaction, such as when an atom collides with another atom, or an electron collides with a photon, or a photon is emitted or absorbed, the conservation of energy applies. If you sum the total energy of all the participating particles before the event, then this number remains constant throughout the entire event and then after the event is over. Not almost constant. Exactly constant. Because this has never been known to fail, we call it a law. We can't prove a law. But when something never *has* failed, we take it for granted that it never *will* fail and so call it a law of nature. And we use it as a basis for much of our science.

Because the total energy remains constant at each and every instant of time during an event, we can state that the energy *before* the event is equal to the energy *after* the event and actually skip over what happens *during* the event. This makes our lives easier since our ability to measure energy before and after is much easier than during, when the changes can be very fast. It's easier to calculate the energy of the before and after collision of two billiard balls than the energy during the collision when the forces between many atoms are at play.

It's the Change in Energy that Matters

Physics was founded on the study of motion of both the seen (moving bodies) and the unseen (heat). As we sought to control motion and so progress civilization, we invented the concept of energy as a means to help guide our efforts and understand our limitations. *How much coal do we need to burn to generate steam to drive the pump to move the water up into a storage tank?* Such a line of questioning that follows the cause–effect conversion of these different forms of energy and associated motions from one to another cuts to the heart of the matter, which is our core belief that you can't get something from nothing, that water doesn't simply flow uphill of its own accord, and that we need energy to cause the change we want. Our core belief is that total energy doesn't change. What this total was or is, we don't care. What matters is that it doesn't change. And so regardless of what event happens, when we add up the total energy before and after, we believe that the two numbers must be the same. And if they're not, such as occurred during the discovery of the neutrino in the early-to-mid 1900s, which we'll visit in Chapter 28, then we believe that we've calculated wrong. We trust the law—hence our calling it a law—more than we trust ourselves.

Fortunately, we're only interested in energy as a means to quantify the *difference* in energy between two systems, for it's this difference that quantifies the energy required to cause the change. We're not interested in the absolute energy of either, for each individual number would

---

[2] I'm choosing to ignore here the complications of cosmological energy conservation for the universe.

tell us nothing, and besides, there is no such thing as absolute energy. So when we discuss energy, we discuss a quantity that represents a difference. Even when energy is presented to us as an absolute, it's typically done so with the understanding that it's referenced against some set of standard temperature and pressure conditions; thus, even in this situation, the energy quantifies a difference.

The fact that we're only interested in energy differences is fortuitous since this allows us to ignore all the different manifestations of energy that don't change. These remain largely constant in most of the events we're interested in, the same before as after, and so when calculating changes in energy, subtracting the before from the after, they cancel out. As an object falls toward Earth and potential energy transforms to kinetic energy, we don't need to know anything about the strong nuclear or weak interaction internal forces since they're not changing. We only need to know about gravitational potential energy and kinetic energy. When considering energy, only the differences matter.

Force, Energy, Terminology, History, and Theory – All Intertwined

The Big Bang was an event, the atom an object. Both lent themselves to discovery. Energy was different. We *invented* energy to help us understand the changes around us.[3] We observed these changes, quantified these changes, and then discovered that when we added up all of the quantified terms, the sum itself didn't change. The parts did but not the sum. Because of this, it's difficult to write here what energy is without also including brief mention of the relevant history behind the theory and without also including some of the terminology we use to discuss energy, terminology that can be confusing since it was created prior to energy's arrival. For example, for a long time we readily used such concepts as heat and work in our everyday lives. During the long discovery of energy, these colloquialisms crept into the technical language without a vetting process. Much as the delivery of strict definitions by Lavoisier helped forward the world of chemistry, the delivery of strict definitions by others of such terms helped forward the world of energy. But these definitions and their early histories remain a challenge to us. It is also necessary to introduce the concept of force in this chapter since force and energy are intertwined in history, which can lead to further confusion, especially since at one point force was used to describe energy. In this Part III, I address all of these concepts and their respective histories in detail and hopefully bring clarity to you in the process. In this chapter, I show you the larger picture of how they're all connected, believing that understanding this will help you absorb the later chapters.

The Deep Dive into the Four Fundamental Interactions

To address the science of energy, we first start with a deep dive into the most fundamental world where quantum mechanics governs the behavior and motion of the elementary particles. From

---

[3] As you'll see throughout this book, it is sometimes difficult to define the boundary between discovery and creation.

this micro-world, we then move to the macro-world in which the motions and related energies of bodies of matter are governed largely by Newton's laws of motion combined with simplified force equations. While the development of classical thermodynamics didn't require an understanding of the connection between the two worlds, the subsequent development of more advanced thermodynamics did.

At the quantum-mechanical scale of the elementary particles exist both motion and the potential for motion. The two are intertwined, inseparable and ever present. Each contributes to total energy, the former as kinetic energy, the latter as potential energy. Each causes change in the other. The cause–effect dance is exact. Any loss in one form of energy is exactly balanced by an identical gain in another.

Force Results from the Change in Potential Energy with Distance

Particles interact with each other through one or more of the following four fundamental classifications—strong, electromagnetic, weak, gravitation (Figure 5.1). These interactions are currently thought to be fundamental in the sense that they can't be reduced to more basic interactions.

While the four fundamental interactions are not completely understood, we do know that, as described in the Standard Model, they arise through the emission and absorption of elementary particles, called exchange or messenger particles, between other elementary particles. For example, the emission of a photon by one electron and its subsequent absorption by another electron causes repulsion between the two electrons. How this all works is described in the very deep layers of mathematics comprising the QED and QCD theories of the Standard Model, which will hopefully someday include the yet-to-be-unified gravity theory.[4] These four forces are mathematically described as fields; a field is a region in which each point is affected by an interaction (or force). A classic field is considered for gravitation interaction, quantum fields for the others. Fortunately for you the reader and me the writer, all of this is beyond the scope of this book.

But given this complexity, we do know certain critical aspects of the exchange particles associated with these four fundamental interactions. We know that the exchange particles populate space and that their density determines potential energy. We also know that their density falls off with distance from the emitter, meaning that the magnitude of the potential energy is dependent on location. And finally, we know that it's the *change* in this magnitude with location that results in attraction or repulsion for any particle placed at that location. It's the change in potential energy with distance that results in force. The particle feels the attraction or repulsion *force* and accelerates in response to it, either by a change in speed or direction or both. And this is where the mathematics becomes complex.

---

[4] (Munowitz, 2005) pp. 320–332. (Feynman et al., 1989a) Chapter 2: Feynman wrote this prior to the completion of the Standard Model. Gravity cannot yet be explained at the quantum level by the exchange of force-carrying particles (gravitons), although note that as of 2015 gravity waves have been observed. Currently, gravity is characterized by Einstein's Theory of General Relativity.

As two elementary particles move relative to each other, the attractive or repulsive forces they experience change as the distance between them changes, which in turn changes the movement of the two particles relative to each other. Cause leads to effect, with each effect becoming the next cause. Modeling such integrated motion is complicated but can be done through the use of calculus, which breaks down single events into infinitesimal changes, thus allowing one to accurately track and predict the cause–effect progression over the course of the event. In fact, Newton and Leibniz invented calculus for this very reason. By combining classical mechanics with calculus, scientists conquered the world of dynamics by enabling the prediction of a body's motion under the influence of an external force, such as gravity, during which both force and motion change with time. NASA uses such an approach today, for example, in calculating the launching and positioning of satellites in space.

While the above cause–effect nature of the fundamental interactions and motions is indeed complex, further complexity awaits. For one thing, the interactions themselves are not instantaneous since the speed of the exchange particles is limited to the speed of light,[5] and for another, since the particles are moving so fast, any equations used to describe their motions must be corrected for relativistic effects.

To help handle such complexities involving motions and potentials, the concept of "field" was created. Fields and the associated quantum electrodynamics (QED) and quantum chromodynamics (QCD) field equations, introduced in Chapter 4, describe the presence of the exchange particles and so can have, to some degree, a life of their own, absent the emitter and/or absorber. Particles produce fields and lead to equations of potential energy; fields act upon particles and lead to equations of motion.

It's the Change in Potential Energy with Distance that Causes Acceleration

Each of the four fundamental interactions contributes to the total potential energy existing at each location in space. Since potential energies are additive, the total is proportional to the sum of all of the exchange particles.[6] It's the change in the total potential energy from one point to another in space that ultimately results in attraction or repulsion and thus results in accelerated motion.

Simply put,

$$- \, d \, (P.E.) / d\mathbf{s} = m\mathbf{a} \qquad\qquad [5.1]$$

As potential energy decreases from one point to another in space, where the spatial coordinates are defined by "s," a particle of mass "m" will accelerate "$\mathbf{a}$" in that direction. Here, the

---

[5] If the Sun were to suddenly disappear right now, we here on Earth would not feel the effect until just over eight minutes later.

[6] A specific force arises directly from each of the four fundamental interactions. For example, a proton adjacent to another proton inside the nucleus of an atom experiences the electromagnetic force of repulsion, the strong nuclear force of attraction and the gravitational force of attraction. Exchange particles associated with these interactions are all involved. It just so happens that in this particular case, the strong nuclear force dominates and the protons remain adjacent.

ENERGY FUNDAMENTALS AND FOUR INTERACTIONS

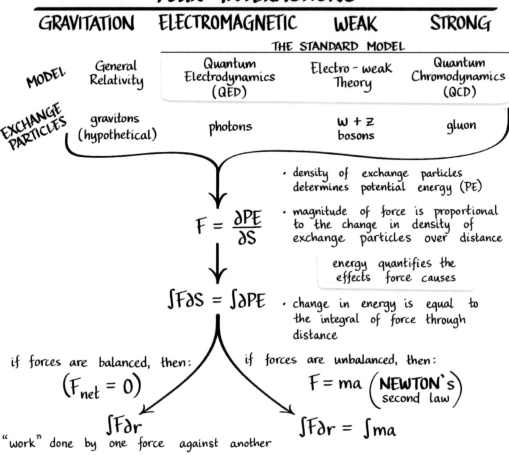

	GRAVITATION	ELECTROMAGNETIC	WEAK	STRONG
		THE STANDARD MODEL		
MODEL	General Relativity	Quantum Electrodynamics (QED)	Electro-weak Theory	Quantum Chromodynamics (QCD)
EXCHANGE PARTICLES	gravitons (hypothetical)	photons	W + Z bosons	gluon

- density of exchange particles determines potential energy (PE)

$$F = \frac{\partial PE}{\partial S}$$

- magnitude of force is proportional to the change in density of exchange particles over distance

energy quantifies the effects force causes

$$\int F \partial S = \int \partial PE$$

- change in energy is equal to the integral of force through distance

if forces are balanced, then:

$$(F_{net} = 0)$$

$$\int F \partial r$$

"work" done by one force against another

if forces are unbalanced, then:

$$F = ma \left(\begin{array}{c} \textbf{NEWTON's} \\ \text{second law} \end{array}\right)$$

$$\int F \partial r = \int ma$$

THE LEVER

examples of gravitational force

FREE FALL

$$m_1 g \Delta h_1 = m_2 g \Delta h_2$$

$$mg\Delta h = \frac{1}{2} mv^2$$

potential energy transforms to potential energy

potential energy transforms to kinetic energy

it was the understanding that these two terms quantified the same kind of energy change (the energy required to lift is of the same nature as the energy lost when falling) that led to the early version of the conservation of mechanical energy (1750)

$$\frac{1}{2} mv^2 + mgh = constant$$

Figure 5.1 Energy fundamentals and thermodynamic models

FUNDAMENTAL MODELS

- works well in particle physics but does not easily apply to Classical Thermodynamics

CLASSICAL THERMODYNAMICS

- makes no assumptions about nature
- science based on macroscopic properties

EMPIRICAL MODELS

KINETIC ENERGY $+$ POTENTIAL ENERGY $=$ CONSTANT

temperature results from kinetic energy of matter

<u>matter</u>
$\frac{1}{2} mv^2$

<u>photons</u>   Planck's constant
$h\nu$ — frequency

<u>thermal</u>
CT — heat capacity

gravitational
electromagnetic

MOLECULAR ENERGIES

translational
rotational
vibrational

vibrational
intermolecular

← ORBITAL ELECTRONS →

exhibit both kinetic and potential energies

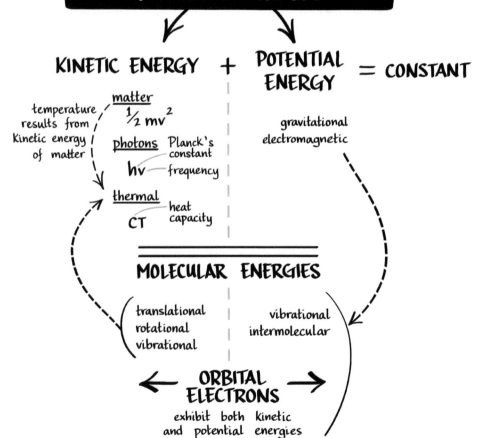

bolded terms "**s**" and "**a**" are vectors to account for the need to specify direction. Note also the need for the negative sign to signify that a particle experiences positive acceleration in the direction of decreasing potential energy. Why particles move in such a direction is not known. They just do. At a certain point, we just can't go any deeper in our understanding.

"But wait!" you say, "I thought that mass times acceleration equals force, not some complicated differential involving energy!?" And you'd be right. This is what we're taught in school. Newton's famous 2nd Law of Motion. $F = ma$. So what's going on here? This is where we need to return to our discussion of force.

What is Force?

F = ma by itself does not provide an algorithm for constructing the mechanics of the world
– Frank Wilczek[7]

The concept of force according to $F = ma$ was actually discovered by Newton before the arrival of potential energy. In his 2nd Law of Motion, Newton stated that change in momentum is proportional to force, which evolved into force being equal to mass times acceleration. If a body is accelerating, then an external force must be present and acting on it. While revolutionary, this definition in and of itself did not provide an independent meaning of what force actually is. It defined force by the effect it caused. As Feynman said, it's basically a definition going around in a circle.[8]

Newton's equation provided a definition, a way to link a particle's mass and acceleration to external interactions, and so became a very useful tool for analyzing nature. But it didn't define those external interactions. It provided no independent meaning for force, leaving the 2nd Law of Motion incomplete while also leaving many scientists incomplete in their understanding of what force truly is.

Complete force laws only started to form after separate laws were developed to describe the source of the "F" term on the left side of Newton's equation, starting with Newton's discovery of the law of gravitational force between two bodies and followed by Coulomb's discovery of the law of electrostatic force between two charged particles.

Our modern understanding of force has since evolved. We now understand force to be an independent entity of material origin; there must be some *thing* in the neighborhood that is a source of the force. Having both magnitude and direction, force arises from the presence of exchange particles in space, and specifically the change in their density with distance. Place a particle at a given point in space and it will accelerate in response to the change in density. The product m**a** quantifies the magnitude and direction of the force, but not the force itself, which instead is described at the fundamental level by the term $d(P.E.)/ds$ in Equation 5.1, thus necessitating knowledge of the potential energy sources.

---

[7] (Wilczek, 2004a)
[8] (Feynman et al., 1989a) Volume I, p. 12-1.

Technically, the term force quantifies the change in potential energy with location in space; it's the space derivative, otherwise known as the gradient,[9] of potential energy.[10] The greater the change with location, the stronger the force. This concept is similar to that displayed in topographical maps in which lines of constant elevation are drawn. The closer the iso-elevation or contour lines are to each other on the map, the steeper the mountain. The closer the iso-potential lines are to each other, the stronger the force. At the quantum-mechanical scale, we define and quantify force in this way.

Empirically Derived Force Equations as a Reasonable Assumption

In theory, the natural phenomena we study can be derived from Equation 5.1. The behavior of elements, atoms, molecules, gases, liquids, and solids, the behaviors that become viscosity, conductivity, vapor pressure, solubility, reactivity—all of these phenomena are, in theory, entirely and completely predictable from the above relationship combined with Pauli's exclusion principle.

Ah, if life were only this simple. The above goal would be doable except for two HUGE barriers. First, we don't even fully understand all the fundamental potential energy functions at the elementary particle level. We're close, having created the Standard Model, but we're not quite there, as reflected, for example, by the fact that gravity is not yet unified into this model. And second, even if we did know all the fundamental functions, we'd be stopped dead in our tracks since any phenomenon that we study typically involves many millions, billions, and trillions of particles, all in accelerated motion relative to each other, attracting, repulsing, colliding, reacting, appearing, disappearing, and so on. Even if we knew all the fundamental equations involved, the math would overwhelm our capacity to compute.

So we are left to make simplifying approximations based on our knowledge of the fundamentals of the interactions, developing Newtonian force models based on Equation 5.2 to shield ourselves from the "vastly complex"[11] details inherent to the potential energy, details that we really don't need to know.

$$F = ma \qquad\qquad [5.2]$$

Early physicists used empirical methods to derive force models devoid of potential energy functions. They didn't need them because the models worked quite well and also because the concept of potential energy fields didn't appear until the 1900s. Examples of such models include Newton's Theory of Universal Gravitation and Coulomb's Theory of Electrostatic Force. Another excellent example is the more complicated Lennard–Jones (L-J) potential

---

[9] In physics, a gradient is an increase or decrease in the magnitude of a property (e.g., temperature, pressure, or concentration) observed in passing from one point or moment to another.

[10] As we'll discuss later, force is also the time derivative of momentum, which is the product of a particle's mass and velocity: $d\,(P.E.) / d\mathbf{s} = F = \mathbf{ma} = d\,(mv) / dt$

[11] (Feynman et al., 1989a) Volume I, p. 12-6. Referring to the forces of attraction between two atoms, Feynman wrote, "they are due to the vastly complex interactions of all the electrons and nuclei in one molecule with all the electrons and nuclei in another."

written below in terms of potential energy V(r) between two atoms or molecules which can be readily converted to force by differentiation with respect to distance (r).[12]

$$V(r) = 4\varepsilon\,[(\sigma/r)^{12} - (\sigma/r)^{6}] \tag{5.3}$$

The use of potential energy as a variable in the L-J model (1924) was not based on development of QED quantum mechanics (circa 1950) but instead on the combination of physical insight and empirical data.

Equation 5.3 quantifies the interactions between neutral atoms and molecules by capturing the long-range attractive van der Waals force, also known as the London dispersion force, in the $1/r^6$ term and the short-range electromagnetic repulsion between orbiting electrons in the $1/r^{12}$ term. The fundamental physics at play here are quite complex, dealing with balance between the induced polarity of atoms and the Pauli exclusion principle that results in only electromagnetic repulsion between electron clouds and not a competing low-energy opportunity for cloud overlap. Attempts to model such energy interactions and resultant forces at the most fundamental QED level are extremely complicated. The approximation developed by Lennard-Jones provided an extremely simplified but highly useful alternative. The parameters eters can be fit using experimental data for a wide range of atoms and molecules. Similar approaches have been used to develop other force models.

When developing simplified force models, we also make other approximations as physicists, chemists, and engineers by choosing to ignore the weak interactions, the strong nuclear interactions and relativistic effects, thus leaving gravity and electromagnetic interactions. For us, it really all comes down to only these two. I earlier highlighted the role of the electromagnetic force in chemistry. Gravity becomes important to us on the macroscopic scale where we consider separation processes based on density and energy calculations based on performing work.

Force and Conservation of Energy

> The total energy of the world, kinetic plus potential, is a constant when we look closely enough... [If we see energy not conserved, then] this is due to a lack of appreciation of what it is that we see.
>
> – Richard Feynman[13]

At the level of the fundamental particles as embodied by QED and QCD mathematics, the change in potential energy over distance causes force, and force causes the change in kinetic energy[14] over time. Because energy is conserved, the transformation of potential energy into kinetic energy (or vice versa) is exact. Even with the classically derived gravitational force mathematics, the response of particles to forces always occurs such that total energy is conserved. Hence, these four fundamental interactions are called conservative, meaning that no

[12] (Brush, 2003c) pp. 468–476. Brush's commentary of the origin of this law, which is written here as a potential but which Brush describes based on the corresponding force.
[13] (Feynman et al., 1989a) Volume I, p. 14-6.
[14] Technically, force causes the change in momentum over time, by definition. I'm using the energy equivalent in this discussion to align with the overall energy theme.

approximations are found in their definitions. All is accounted for, nothing left out, no other interactions at work, no further reductions possible. So long as all the particles, including the photons, are accounted for in whatever event occurs due to the presence of conservative forces, energy is implicitly conserved and no special attention to the energy conservation law is required.

A full understanding of this concept sheds light on the reason why Newton never had to address the conservation of energy—not because he didn't want to, but because the subject of his study didn't require it. Like the other fundamental interactions, the gravitational interaction is conservative at the fundamental level. And fortunately for Newton, this interaction is also conservative at the higher level of our larger scale world. So when Newton linked his Universal Law of Gravitation (cause) to his Laws of Motion (effect), and because no other forces like friction were significant, then energy was inherently and implicitly conserved and he was able to successfully explain both celestial and terrestrial motions. As the apple falls to the earth, its potential energy is converted exactly into its kinetic energy. It's just that he never used nor had to use such terms.

As we move from the quantum world of the fundamental interactions and conservative forces to the macro-world of, for lack of a better phrase, the pseudo-forces, we shift from strict definitions to simplified approximations. In the macro-world of trillions of atoms, calculating force as the gradient of potential energy just isn't feasible. We don't know enough to do this, and even if we did, we wouldn't have the computing power to handle it. So instead we develop models, as Lennard-Jones did, based on combining the essence of physical insight with empirical data. In so doing, sometimes something is missed and energy conservation appears to be false. But, as we've learned, energy conservation is a law, and so when it appears that it doesn't work, then something else must be happening, something that we're not seeing, something that we're neglecting, something that we're not accounting for. We must be studying only a part of the system and not the whole system. For when we look closely enough, when we zoom from the macro-world down into the micro-world on which it is ultimately based, the fundamental interactions govern and so total energy as comprised of both kinetic and potential energies is constant.

In light of this, when we create empirical force laws based on macroscopic phenomena as opposed to potential energy gradients, then we classify such a force as non-conservative. A classic example of this is friction. Friction hinders motion and so causes deceleration. Thus, according to Newton's 2nd Law, friction occurs as a negative force. You feel this force when you push a table across the floor and see this force when you let go of the table and it comes to a halt. At the most fundamental level, the friction force is quite complicated, originating as it does from the fundamental electromagnetic interactions of the many atoms involved at the contact points between the two moving surfaces. As the table moves, the atoms at the surfaces are displaced by these interactions until they eventually snap back toward their anchor and start vibrating around this point, which results in the heating of the two surfaces. During each and every one of these interaction events between atoms and molecules, energy is conserved. But since we can't capture all of these interactions in a model, we take a short-cut by inventing a concept called friction along with a simple mathematical formula to describe it. Even though it has no relation to potential energy gradients, the formula works quite well. Scientists and engineers can develop a formula for a friction event based on correlating the force needed to overcome friction, which is treated as an opposing force and which can be quantified from the weight of the object. The resulting correlation coefficient will vary with the type of elements involved and the physical characteristics of the surface. One can use such force models to predict, for example, how quickly a moving car will slow down once its brakes are locked.

There's just one small problem with the formula for friction force. It doesn't account for the energy lost to heating the interface surfaces. So when viewed from a far distance, it appears that energy is not conserved during a friction event, which is why this force is non-conservative. However, this apparent failure of the conservative of energy occurs only because we're so far away, for when we look closely enough, energy is always conserved and forces are always conservative. There are really no non-conservative forces.[15] There are simply approximated pseudo-forces that appear non-conservative only until someone looks closely enough at the fundamental underlying mechanisms at work and then such forces show themselves to be conservative.

Energy Lies at the Core of Thermodynamics

Now that we've learned the core science of energy, we need to shift gears. The concepts we just learned play a critical role in particle physics, as they are extremely powerful to use in studying the motion of individual particles. However, such concepts don't play such a critical role in thermodynamics. Yes, they still govern the behavior of individual particles, but in the world of trillions, we simply can't follow all of these particles. Fortunately, when thermodynamics was created in the mid-1800s, this wasn't an issue since no one knew about individual particles then anyway as the atomic theory had not yet been born. In spite of this, the early physicists were still able to create thermodynamics. How? By discovering energy and its conservation in the world of trillions. They didn't need to model the individual particles involved. With energy, they could instead rely on macroscopic properties.

The science of energy as applied to thermodynamics reflects the history by which it was created. Since the history and the theory are intertwined, I now begin to cover both together, completing this chapter with a high-level overview and then using the subsequent chapters of Part III to go into the details.

Energy of Motion of a Single Body

The concept of conservation demands quantification. We start with energy associated with motion. We're most familiar with kinetic energy, if not by name then by experience. We are always moving relative to something else, and something else is always moving relative to us. The moving car, the falling rock, the running athlete, the flying cannon ball. Each has kinetic energy that can be quantified as,

$$\text{Kinetic Energy} \left(\text{K.E.} \right) = \tfrac{1}{2} \, m \, v^2 \qquad [5.4]$$

---

[15] (Feynman et al., 1989a) Volume I, p. 14-7. "…it has been discovered that all the deep forces, the forces between the particles at the most fundamental level, are conservative."

where m and v represent the mass and the speed of the body relative to a stationary frame of reference. Kinetic energy is a scalar number, meaning that there is no direction of motion attached to it, hence the use of speed and not velocity in the equation since velocity is technically a vector. [Note: having said this, for simplicity's sake, unless otherwise stated, I use v to quantify scalar speed.]

Each moving particle or body has the capacity to affect an object it strikes. Kinetic energy quantifies this capacity in a way that makes sense to us since our everyday experience tells us that this capacity increases with both the body's mass and its speed. We know to be more afraid of the moving rock than the moving sponge, even though they're each moving at the same speed. We know to be more afraid of the stone dropped from the top of a tall building than a slower moving stone tossed to us by a friend, even though they're each of the same mass. The quantification of kinetic energy captures the influence of both components. You may be wondering why v^2 and not some other function of v like just plain v, and you wouldn't be the first. Such wonderment goes all the way back to Galileo when he first attempted to define motion and was to become the subsequent source of many heated debates between the v-followers of Descartes and the v^2-followers of Huygens and Leibniz, as we shall see.

We can calculate this simple quantity for individual objects we can see, like a bullet or an asteroid, and those we can't, like an atom or a molecule. But what happens when we're dealing with trillions? How do we quantify the energy of motion contained within such a system? This is where we open the door to the realm of thermodynamics, starting with the concept of temperature.

Temperature Quantifies the Average Kinetic Energy of Many Bodies

When we place a thermometer into a glass of water, the water molecules collide with the thermometer, transferring energy with each collision, first to the glass atoms and then to the mercury atoms, causing the bulk mercury to either expand (hot) or contract (cold) to some level which we read as temperature. While simple to measure, temperature comprises a complicated concept and is the foundation on which thermodynamics is built. I'll discuss this in more detail in later chapters but need to address it now in the context of energy.

Temperature quantifies the average kinetic energy of an ensemble of atoms and molecules that are too many to count and too small to see. It reduces all of the complexities involved with motion to a single value. We convert this single value to units of energy by multiplying it by Boltzmann's constant as follows:

Thermal Energy (per particle or molecule) = ½ kT (for each degree of freedom)   [5.5]

Thermal energy is related to the motion energy of a particle or molecule in a system. The ( ) on the right-hand side of the equation says that thermal energy applies to each independent direction of translational motion available. For example, if we were dealing solely with translational movement of a gaseous atom or molecule from one location to the other, three directions are possible as represented by the x,y, and z coordinates. Since in the absence of external forces none of these is favored over the other, they are independent. Thus, translational motion has three independent directions of motion, or, as it's more frequently stated, it has three "degrees of freedom" (DOF). There's something called the equipartition theorem that says that the quantity of energy applies *equally* to each degree of freedom; if there's movement in any degree of freedom, there's the same movement in all degrees of freedom. So, assuming we have an ideal gas comprised of monatomic atoms, and since each of these atoms thus has three degrees of freedom, then the average kinetic energy of the atoms is 3/2 kT and the total energy is simply this number times the total number of atoms. From this number, one can calculate the average speed of an atom by equating 3/2 kT to ½ mv² where m is the atom's mass.

If the ideal gas is comprised of molecules containing two atoms, like molecular oxygen, additional degrees of freedom need to be accounted for. The higher the degrees of freedom, the higher the energy for a given temperature. Think of someone throwing an object at you comprised of two balls connected by a strong spring. First think of the person just simply throwing it at you. This is translational motion. Then think about someone first imparting a high-speed spin to the object before throwing it at you. Then think about someone first causing the balls to vibrate back-and-forth very fast and then imparting a high-speed spin before throwing it at you. For each scenario, the speed of the throw is the same. But in the sequence just listed, your level of fear increases significantly. There's much more energy in a translating, rotating, vibrating molecule than in a molecule only translating and this is accounted for through the fact that these additional motions contribute to the total degrees of freedom value and thus to total energy.

Now it becomes a little more complicated than this as reflected in the above caveats. The molecule may be comprised of many atoms, each with its own motions including vibration, which contributes both kinetic and potential energies to the total. The molecules themselves may be attracted to other molecules and so contribute another quantity of potential energy to the overall total. The molecules could be in a gas phase, a liquid phase, or a solid phase. Then there may be quantum mechanical phenomena such as the fact that the number of degrees of freedom may be temperature dependent. For example, vibration motion in certain molecules isn't accessible until sufficient energy is available. At some point things become too complicated to use Equation 5.5.

We'll discuss this more in later chapters, but the reason to bring it up now is to say that we *do* have a theoretical means to quantify the total energy of a system based on temperature plus knowledge about the degrees of freedom in that system. Ah. But there's the rub. We don't always know in advance what the degrees of freedom are and we certainly didn't have a handle on this in the early days of thermodynamics. But the beauty of those early days is that we didn't *need* to know because we had an alternative means to quantify thermal energy, one based on the world of trillions, involving the direct measurement of a macroscopic property called "specific heat capacity" (designated C) of the material that captures all forms of changing energy, both kinetic and potential (except for isothermal phase change which is treated separately), contained in the system.

$$\text{Thermal Energy} \left(\text{per mass}\right) = CT \tag{5.6}$$

Using a calorimeter, we can quantify the specific heat capacity C of a substance referenced against, say, water, which has been given an arbitrary heat capacity of 1.0 Btu per pound per degree Fahrenheit (or 1.0 calories per gram per degree Celsius). We can do this by, for example, starting each off at different temperatures and then allowing them to contact each other. Since energy is conserved, the *change* in total thermal energy of the substance as quantified by mCΔT must equal the negative of the change in total thermal energy of the water, also quantified by mCΔT. Assuming no losses (good insulation), the conservation of energy demands this equality. We finally calculate the relative specific heat capacity of the substance based on the ratios of the masses and the temperature changes of the two systems (Equation 5.7).

$$\text{Specific Heat Capacity} \left(C \right) \text{of Substance} \left(\text{per mass} \right) = \left(mC\Delta T \right)_{water} / \left(m\Delta T \right)_{substance} \qquad [5.7]$$

I cover much more on this subject in Chapter 14 since Joseph Black's use of the calorimeter to create the concept of heat capacity as separate from temperature became an important milestone in the history of energy. In addition to quantifying heat capacity, the calorimeter would also be used to quantify the heat generated by chemical reaction. In Chapter 39 I discuss this expanded role to highlight how the resulting data aided in the analysis of reaction spontaneity and the concept of "free energy."

Radiant Energy of Photons

So far our discussion has centered on quantifying energy associated with the motion of particles containing mass. But what about moving particles that don't have mass, specifically the photon? As discussed in Chapter 2, we determine the photon's energy by multiplying its frequency (ν) by Planck's constant (h).

$$\text{Photon Energy} = h\nu \qquad\qquad\qquad [5.8]$$

Free Fall Theory – the Fundamental Connection between Δh and v^2

The discovery of potential energy started with the quantity Δh, the change in elevation or height of a body in a gravitational field. As man contemplated the lever, this term (combined with weight) kept appearing in the analysis. But the lever involves only the transformation of potential energy into potential energy; kinetic energy is not involved. It wasn't until physicists separately discovered the relation between Δh and v^2—and made the logical tie between the Δh associated with raising a weight and the Δh associated with a falling body—that the early concept of energy conservation started to appear. The fact that these two terms are connected is best seen in the science of free fall. Indeed, it was Galileo's Law of Fall that first started shedding light on this subject and eventually led to an early form of the conservation of mechanical energy law in 1750, an early form since energy wouldn't be discovered (or created) until around 1850. This pre-law served to embolden later physicists in their belief that a more encompassing conservation law must exist, one that included heat. Given this, let's review the science of free fall.

The potential energy associated with gravitational force between two bodies is written as

$$\text{Gravitational Potential Energy} = -G\, m_1 m_2\, /r \qquad [5.9]$$

where G equals the gravitational constant, m equals mass and r equals the distance between the two bodies 1 and 2. While not based on a similar quantum-level analysis as the other three interactions, especially since a quantum-mechanical model of gravity has not yet been discovered, Equation 5.9 is highly accurate and considered conservative. What does it tell us? As the distance between two bodies goes to infinity, gravitational potential energy goes to zero, which is the arbitrarily defined energy reference point for gravity, having no fundamental significance behind its use. Since potential energy decreases (becomes a larger negative number) with decreasing "r," then per Equation 5.1, the two bodies accelerate toward each other. Bodies want to move in a direction of decreasing potential energy. Because the gravitational potential energy causes matter to move toward matter, we say that it is an attractive force.

Let's now quantify this attractive force by calculating the derivative of the potential function. (Note: the negative sign in Equation 5.1 is removed to reflect fact that the gravitational force acts toward the emitter. Also, both variables s and r quantify the same distance in this example and can be substituted for each other.)

$$\begin{aligned}
\text{Gravitational Force} &= d\,(\text{P.E.})\,/\,ds \\
&= d\,(-G\, m_1 m_2\, /r)\,/\,dr \\
&= -G\, m_1 m_2\,(-1/r^2) \\
&= G\, m_1 m_2\, /r^2 \qquad [5.10]
\end{aligned}$$

This is Newton's famous equation for gravitational force and shows how his inverse-square law develops from the potential energy function, or, since he discovered his law prior to any such concept as energy, vice versa.[16] Gravitational force is indeed positive and thus attractive. Also, it's never zero; it reaches across the entire universe.

Now you can put this force equation into Newton's 2nd Law of Motion to define how fast a body, m_1, accelerates in a gravitational potential energy field toward another body, m_2.

$$F = d\,(\text{P.E.})\,/\,ds = G\, m_1 m_2\, /r^2 = m_1 \mathbf{a}$$

What's interesting here is that m_1 appears on both sides of the equality sign. If we eliminate it we get

$$\mathbf{a} = G\, m_2\, /r^2$$

and see that the acceleration of a body toward another is independent of the body's own mass, a phenomenon demonstrated, as legend has it (discussed more in Chapter 7), around 1590 when Galileo dropped two balls of different masses off the Leaning Tower of Pisa and saw them both

---

[16] When the distance between two bodies is large, potential energy is high and the attractive force low. As the bodies accelerate toward each other, potential energy decreases as the attractive force (slope of the P.E. versus distance curve) increases.

hit the ground at the same time. A more updated version of this beautiful visual experiment occurred in 1971 when Commander David Scott of Apollo 15 dropped a hammer and a feather while standing on the Moon. They each hit the ground at the same time.[17]

Let's keep going. What is acceleration? It's the change in velocity (vector) with time, dv / dt. Let's make this substitution and then multiply everything by differential distance, ds, which we'll convert to dr since attraction acts along the radius (spherical coordinates).

$$F \, dr = d \, (P.E.) = (G \, m_1 m_2 \, / \, r^2)dr = (m_1 \mathbf{a})dr = (m_1 (dv / dt))dr$$

Since dr / dt is velocity

$$F \, dr = d(P.E.) = (G \, m_1 m_2 \, / \, r^2)dr = (m_1 v)dv$$

Let's integrate this to see what happens when a body is dropped from a height above the Earth's surface and allowed to free fall without resistance or constraint. Setting r and m_2 to the radius and mass of the Earth, respectively, and then setting v to zero at the starting height above the Earth, and finally assuming that h is very small relative to r, the equations for free fall of a body under the force of gravity are

$$\int F \, dr = - \, \Delta P.E. = g m_1 \Delta h = \tfrac{1}{2} \, m_1 v^2 = \Delta K.E. \qquad [5.11]$$

where

Δ h is the distance fallen (written as a positive number)

$$g = G m_2 \, / \, r^2$$

In this example, g is called the local acceleration due to gravity at the Earth's surface. When you plug in the value for G together with the values for Earth's properties, g equals approximately 32.2 ft/sec^2. When you drop something, regardless of how big it is, after one second it's falling at 32.2 ft/sec, after two seconds it's falling at 64.4 ft/sec, and so on.[18]

We call the "mg" term weight, w ($w = mg$). It's the downward force of Earth's gravity acting on the mass of a body. When we step on a scale, we're measuring our weight and not our mass. As we move further above the Earth's surface, r increases and thus g decreases and so our weight decreases but our mass remains constant. Mass is a conserved quantity, not weight. We convert weight to mass by dividing by g (32.2 ft/sec^2). But this leaves us with a weird set of units, lb_f-sec^2/ft, where lb_f denotes pounds associated with force or weight. To convert this to units of mass that we're more familiar with, we multiply by a term called the gravitational conversion factor, g_c, which is equal to 32.2 lb_m-ft/lb_f-sec^2. This makes the conversion of weight to mass quite easy. At the Earth's surface, one pound force (lb_f) conveniently equals one pound mass (lb_m). If only life were always so simple!

The fundamental development behind Equation 5.11 provided, long after the fact, the theoretical justification for the conservation of mechanical energy, an early version of which was established around 1750. As a body free falls, its potential energy decreases by mgΔh while its

---

[17] If you consider that each infinitesimal part of m_1 is attracted toward m_2 and accelerates accordingly, then you realize that it makes no difference whether the m_1 parts are attached to each other or not. Nothing magic happens when they are attached.

[18] The other equation used in this free fall calculation is $\Delta h = \tfrac{1}{2} \, gt^2$. Substitution into Eqn. 5.11 yields $v = gt$.

kinetic energy increases by $\frac{1}{2}m_1v^2$. This trade-off is exact; it's an equality; energy is conserved. Such equations and their derivatives showed that many types of celestial and terrestrial phenomena could be predicted and explained, and thus they strongly supported the belief that we live in a deterministic world governed by Newtonian machinery.

Newton developed the essence of this relationship in one of the examples he used in the *Principia* without the need to rely on the concept of energy since gravitational interaction is conservative. His path to this equation was relatively simple once he discovered his laws and in fact was so simple that he didn't really dwell on its significance. He wasn't thinking "energy" even though it was he who provided the theoretical basis for the above relationship.

Because of the complexity involved in linking Newton's force with energy-related analyses, others didn't latch on to Newton as the path to energy. Sure, isolated incidences of genius, such as Daniel Bernoulli and Leonard Euler, saw the connection, but such occasions were rare. The *Principia* was hard enough to understand as it was, let alone build upon. So other scientists, who pursued energy without really knowing what they were pursuing, arrived at the basic relationship between h and v^2 shown in Equation 5.11 along a totally different path.

The Lever – It's All About w∆h

Looking first at the "mg∆h" term (or w∆h) in Equation 5.11, the discovery of this term, which quantifies the change in gravitational potential energy, rose from analysis of the lever. For a lever that is perfectly balanced and thus in a state of equilibrium, the change in potential energy of a mass at one end, as quantified by this product, exactly equals the negative change in this same term for a mass at the other end such that the net change in potential energy at the two ends equals zero, leaving total energy conserved. The path to this momentous result started with the ancient Greek philosophers, Aristotle and Archimedes. This concept later found its way into the analysis of engines, especially those involving the fall of water to do work, which is simply the same process as what a lever does except using water instead of solid bodies. And it was in this context that the concept of potential energy and really the concept of energy conservation first started. The ability of a water wheel to do work corresponded to the "potential" offered by the change in height of the falling water.

Free Fall History – Galileo Discovers the Relationship between h and v^2

We showed previously the theoretical path to quantifying kinetic energy as $\frac{1}{2}mv^2$. The historical path was different and equally fascinating. Here we start with Galileo (Chapter 7) who made the tremendous break-through of discovering the proportional trade-off during free fall between vertical height and the square of velocity.

$$h \propto v^2$$

Much has been written about this famous discovery. Through a combination of genius, logic, and experimentation, Galileo derived his famous Law of Fall stating that the distance travelled by a free-falling body is proportional to the square of the time, $h \propto t^2$. By further asserting that velocity increases linearly with time, $v \propto t$, he then arrived at $h \propto v^2$.

While Galileo connected h to v^2 experimentally, Leibniz, building on the work of Huygens, followed by connecting them theoretically, ingeniously employing another discovery, the impossibility of perpetual motion that also rose from the above efforts. He incorporated weight into the relationship, something that Galileo didn't need to address since free fall is not influenced by weight, and in so doing suggested the fundamental importance of both $mg\Delta h$ and mv^2. The origins of potential and kinetic energy theories rested on these two terms.

*　*　*

The mathematical derivation leading to Equation 5.11 quantifies how the potential energy of one body is transformed to the potential energy of another in the operation of a lever, and how potential energy is converted into kinetic energy for a body's free fall in a gravitational field. Similar examples can be derived for other particles in other potential energy fields. In each example, the same rules apply. The change in the motion-energy of the particle—as quantified by change in either speed or direction or both—is exactly balanced by the change in the potential-motion-energy of the particle—as quantified by its location in the potential energy field. Such equations and studies concern particle physicists. In these situations, the physicists say that the field does "work" on the particle and the integral of force through distance—$\int F dr$—quantifies the work.

The general concepts used to derive Equation 5.11 also apply to thermodynamics since this subject was founded on the conversion of kinetic energy in the form of heat ("thermo") to potential energy in the form of work ("dynamics"). But because thermodynamics deals with trillions of particles, some of the concepts and definitions are different than for particle physics since the behaviors of individual particles can't be followed. For fluids, the term $\int F dr$ is used but in a different form, $\int P dV$, which still quantifies work but not the work done on a single particle. Instead it quantifies the work done on one system by another due to a moving boundary between them. We'll get into this later, but it's important to understand such differences.

Also important to realize is that while the potential energy field and resulting force lead to acceleration at the molecular scale, the phenomenon of acceleration is absent in the mathematics of thermodynamics. In thermodynamics, processes are "reversible" in that a near-exact balance of opposing forces is maintained throughout. Such balance results in zero net force and thus zero acceleration—at the macro-scale. For example, the quintessential heat-to-work process in thermodynamics concerns the piston-in-cylinder. Heat added to the cylinder causes an increase in internal pressure, which pushes the piston out and so lifts a weight like a bucket of coal. In this process, the internal force acting on the piston (pressure times area) almost exactly equals the external counter-force applied by the raised bucket. The resulting near-perfect balance means that the reversible process can go in either direction. Only infinitesimal changes are allowed, meaning that the piston moves infinitesimally slowly, either in or out, and thus experiences no acceleration itself, for if it did experience acceleration, it would smash into another piece of process equipment and generate unwanted heat and unwanted energy loss from the system. Such acceleration of the macro-equipment leads to an "irreversible" process since you couldn't move the process in the opposite direction without a finite (not infinitesimal) addition of external energy to compensate for the energy lost.

The Mechanical Theory of Heat – Weight Falls, Paddle Spins, Water Heats

By 1750, an early version of the conservation of mechanical energy based on Δh and v^2 terms was established. But with the rise of the steam engine, it became apparent that heat generated by burning coal could also generate work in the form of $mg\Delta h$. But what was the energy equivalent that was lost in the process? Engineers realized that work didn't simply appear and that the burning of coal was needed. But when the theorists started pondering the larger concept of conservation, one that would account for the fact that heat is needed to generate work, they were stymied for they didn't have a term to account for this, mainly because they lived inside a caloric paradigm that said heat is conserved and thus couldn't be converted to work. The conceptual leap required to move from the 1750 to the 1850 concept of energy was immense.

Prior to 1850, scientists' incorrect belief in the conservation of heat (not energy) prevented them from comprehending the connection between heat and work. It would take years of experiments and overwhelming data to demonstrate that heat is consumed while work is done (and vice versa) and so turn the tide. Success finally arrived when one of the key figures in this story, James Joule, demonstrated that by slowly dropping a weight tied to a string in order to spin a paddle in a bath of water, he could make the temperature of the water rise. In other words, Joule demonstrated that mechanical work (weight times drop in elevation) can be converted to heat (via friction of the moving paddle against the water) at an exact exchange rate, what we today call the mechanical equivalent of heat (MEH) (Figure 5.2). Such a simple experiment; such a profound meaning. Joule's conclusions were paradigm shifting. Heat is not conserved. Instead, heat can be generated by work and vice versa. The concept of heat–work equivalence (or work–heat equivalence; the two are the same and I use them interchangeably) became the core component of the mechanical theory of heat[19] in which heat and work are two sides of the same coin, which was later to be called energy.

The fascinating aspect of this major discovery was that it didn't depend on any fundamental knowledge of what heat actually was. People long thought that heat must be caused by motion, and indeed, Joule's finding was all about motion. As the spinning paddle continually struck the water molecules, the molecules moved faster and temperature rose. The viscous resistance "felt" by the paddle was quantified by the falling-weight work needed to push the paddle through that resistance. The work done by the weight was exactly equal to the temperature gained by the water as quantified by the MEH. But Joule's mechanical theory of heat was not dependent on this understanding. This is why the $\frac{1}{2} mv^2$ term sat on the sidelines during this mid-1800s discovery period. It was a critical and necessary component behind the discovery

---

[19] Also known as the dynamical theory of heat.

5.2 THE MECHANICAL EQUIVALENT OF HEAT (MEH)

WORK <=> HEAT

each can be transformed to
the other by a fixed ratio

$$MEH = \frac{778 \text{ lbs} \cdot \text{ft}}{1 \text{ lb water} - 1°F}$$

— the work expended by dropping
778 lbs of weight through 1ft generates
a quantity of heat sufficient to
raise 1lb water 1 degree Fahrenheit

→ energy is conserved during each atomic collision between the wall and the fluid

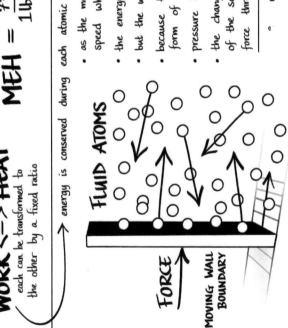

FLUID ATOMS

FORCE

MOVING WALL BOUNDARY

- as the moving wall strikes the atoms, the atoms recoil at a higher
  speed which causes the fluid temperature to increase

- the energy gained by the fluid must be lost by the wall

- but the wall's energy loss is _not_ in the kinetic energy of its own atoms

- because the wall is solid, it collects the change of energy in the
  form of "force," which is pressure times area, acting through distance

- pressure is the macroscopic quantification of the microscopic collisions

- the change in energy of the fluid atoms is caused by the motion
  of the solid wall against them and is equal to the integral of the
  force through distance

$$\text{work done} \atop \text{by wall} = \int F \, dr = {\text{change in energy of fluid as} \atop \text{quantified by temperature change}}$$

- the source of this force can vary

	MOVING WALL BOUNDARY	SOURCE OF FORCE CAUSING WALL TO MOVE
MAYER	flexible bladder used to maintain constant pressure during heat capacity measurements (c_p)	system of gas atoms being heated, which causes expansion at constant pressure
JOULE	spinning paddles in water tank	falling weights strategically selected to enable motion (and spinning paddles) while preventing acceleration

Figure 5.2 The mechanical equivalent of heat (MEH)

of the 1750 early version of the conservation of mechanical energy, but once this was done, it was no longer needed because temperature was available to replace it as revealed by Joseph Black in his work involving the calorimeter.

The Kinetic Theory of Gases

But the ½ mv² term would rise again. As the atomic theory of John Dalton became more accepted, people started applying Newton's mechanics to the unseen world, thereby explaining the linkage of, for example, gas temperature and pressure to the kinetic energies of its constituent atoms and molecules. Such an approach became the developmental basis for the kinetic theory of gases, statistical mechanics, and statistical thermodynamics. We delve into these advanced concepts in Part IV.

*   *   *

As physicists, chemists, and engineers, we work with a range of substances, a range of phases (gases, liquids, solids), a range of conditions (pressure, temperature) and a range of processes (reaction, separation, heating, pumping), and throughout, we embrace conservation of energy. We use it to determine process requirements. We use it to help trouble-shoot problems. *How much energy will I need to operate this process? Why doesn't this energy balance close? Where is the missing energy?* We similarly embrace conservation of mass. Indeed, the mass and energy balance, an approach arguably launched by Lavoisier, is a trademark of our craft.

But given all the above, we still need a way to do the accounting. How do we keep track of kinetic energy and potential energy? How to we monitor the changes between the motions and interactions of all the atoms and molecules? If you were starting from a clean sheet of paper, the task in front of you would be daunting. You could start with temperature, perhaps. But temperature is only part of the story. You also have to consider heat capacity and its dependence on the many intermolecular interactions that lead to attraction, repulsion, phase change, and reaction. You also have to consider how to handle the complexity of a trillion interacting molecules and the potential energy stored in their respective bonds. Naturally you could simplify things by ignoring the strong and weak fundamental interactions along with gravity. But you'd still have the electromagnetic forces together with all the spinning, vibrating, and translational motions to handle. Daunting is an understatement. You simply couldn't take account of all of this detail.

Thank goodness for such geniuses as Rudolf Clausius. It was he who in 1850 set us on the right path by cutting the Gordian knot and telling us not to worry about any of this. He didn't concern himself with what was actually going on inside a given isolated system. Instead, with the newly invented concept of energy and its conservation in hand, he proposed a simple solution by stating that any given system has an energy associated with it and it doesn't matter what this energy is because it's only the change in energy that matters. Further, he said that this energy, whatever it is, remains constant unless the system interacts with the environment, and proposed two options for such interactions: heat flow across the boundary (heat) or movement of the boundary itself (work). These interactions reflected the two means by which energy could change in Sadi Carnot's piston-in-cylinder assembly. As we know now, but Clausius didn't know then, both modes of energy transfer source from the same phenomenon, mechanical collisions at the atomic level. If the atoms in the boundary are moving slightly faster on average than the atoms inside the isolated system, whether due to a higher temperature or an inward moving wall, then the system gains energy.

Note my use of the term "energy transfer." While this is a commonly used expression in thermodynamics, it's important to understand that there is actually no *thing* that is transferred. The term "transfer" along with others such as "heat flow" originated under a different paradigm of heat and so cause confusion today when attempting to understand them based on their common usage. I'll point out such occurrences along the way to help bring clarity.

The 1st Law of Thermodynamics

In 1850, Clausius, in one amazing fell swoop of simplification, resolved all of the complexity involved with energy by creating what would become known as the 1st Law of Thermodynamics. He put the concepts together into the following equation (modified to the structure we're most familiar with):

$$\Delta U = Q - A \times W \tag{5.12}$$

This equation looks so simple and yet says so much.

The term U refers to internal energy and serves to quantify all the different forms—both kinetic and potential—of energy present within a given material. Clausius realized that it was the *change* in energy as opposed to absolute energy that was the issue, hence the 1st Law being written as such. If we know heat (Q) and we know work (W), which can be converted to the same units as U by multiplying by A—Clausius used A which is the reciprocal of the mechanical equivalent of heat—then we know ΔU even though we don't know any of the detailed phenomena happening inside the system. We also know that if both Q and W are zero, then ΔU must also be zero. [Note: In subsequent uses of Equation 5.12, I exclude the 'A' term with the understanding that 'W' is converted into energy units.]

In Equation 5.12, Q quantifies the heat gained by the system (positive value) while W quantifies the work done by the system (positive value). A system gains energy for positive Q and loses energy for positive W (by doing work).

Note that both Q and W quantify *change*. When the environment heats a system, it loses energy while the system gains energy in the same amount; Q quantifies the change in energy. Regarding work, when the environment does work *on* a system, it loses energy while the system gains energy in the same amount; W quantifies the change. Neither heat nor work represents an absolute quantity; each represents the *difference* between two quantities of energy.

Note also that the equation is based on the creation of two separate systems: the system of interest and the environment. Equation 5.12 applies to both. The energy lost by either is the energy gained by the other. Furthermore, for reversible processes the values of Q and W for either are exactly equal to the negative of these values for the other. The sign conventions can become confusing because of this. Again, in Equation 5.12 as written for the system of interest, the thermal energy received by the system and work done by the system are both considered positive; the former increases the system's energy while the latter decreases it. Hence the form of the equation.

We've relied on Equation 5.12 and its role as the 1st Law of Thermodynamics ever since Clausius wrote it down. In essence the equation represents the first energy balance conducted around a body for which the following applies: energy accumulation = energy in – energy out. When Clausius derived his equation, the primary means for energy change in the system of interest, which in this case was the closed piston-in-cylinder assembly inherent to the steam engine, were heat and work. Equation 5.12 was later expanded to include other means such as the flow of mass in and out.

Clausius and later Gibbs used Equation 5.12 to establish the first fundamental equation (based on energy) of thermodynamics based purely on the properties of (fluid) matter and written in differential form (Equation 5.13). You'll note the inclusion of entropy (S) in this equation. This property and its discovery by Clausius in 1865 are discussed in Part IV.

$$dU = TdS - PdV \qquad\qquad [5.13]$$

Years after 1850 and especially once the atomic theory became accepted, scientists naturally started to focus on the physical phenomena inside the isolated system. While in classical thermodynamics based on macroscopic properties this wasn't necessary, in the advanced kinetic theory of gases and statistical thermodynamics it was. The linking of the microscopic behavior of atoms and molecules to the macroscopic properties and the exact relationships between them became and continues to be the subject of much theoretical and experimental research, all focused on the different types of kinetic and potential energies that exist within a system of atoms and molecules.

Both Equations 5.12 and 5.13 are the defining equations around which classical thermo-dynamics was built, the former representing the 1st Law of Thermodynamics and the latter representing both. Neither relies on any understanding of the four fundamental interactions of nature. Both rely on the concept of energy and its conservation. From these equations, and especially from 5.13 as shown by Gibbs, we can use the power of differential calculus to develop a full mathematical description of the variety of ways that a system's energy can change. We'll cover this in Part IV. In the meantime, let's visit the history that led to the discovery of energy and its conservation. Sit back. It's quite a ride.

6 Motion prior to Galileo

History is Only Clear in Hindsight

Our emotional apparatus is designed for linear causality...[but] Linear progression...is not the norm.

– Nassim Taleb[1]

Discovery is like putting a puzzle together. The pieces are laid out on the table, but not all are there. Some are missing. And some are pieces of some other puzzle, serving to confuse and distract. But eventually, as we are "explanation seeking animals,"[2] the human mind is drawn to creating structure, to fitting the pieces together, to discovering the missing pieces, and to slowly watching the puzzle evolve toward completion. The completion can happen in one single "ah-ha!", that breath-taking moment of sudden insight into how nature works, or it can come more slowly over time.

While several huge ah-ha moments occurred during completion of the Big Bang (Part I) and the atom (Part II) puzzles, the discovery of energy and its conservation was not as clean-cut. The Big Bang was an event, the atom an object, and each was a product of nature. Each leant itself to true discovery, however circuitous the route. But energy was different. It was a man-made concept based on many observations relating to motion and change. It wasn't sitting there waiting to be discovered. It was an abstract concept waiting to be created. As each observation was made, it wasn't always evident in that moment that it represented a true piece to the puzzle, largely because no one knew what the final puzzle looked like. In general, as very well discussed by Nassim Taleb in *The Black Swan*, real history as reflected in the personal letters and diaries of those involved is not nearly as clear as revised history, when the re-write suggests a clear, well laid-out path. There was no such path for the conservation of energy; the path was filled with serendipity. The fact that we ended up with such a powerful result is rather amazing.

So given all this, let's look at how the different pieces to this specific puzzle evolved, starting with the first recorded ideas about motion by the ancient Greeks and ending with the independent and simultaneous completion of the final puzzle by different scientists in Western Europe in the mid-1800s. You can form your own opinion on the history of this puzzle. Many others have. Regardless, the importance of learning this history is this. Without understanding the history, one can't appreciate the achievement. Galileo can't be understood without Aristotle,

---

[1] (Taleb, 2010) p. 88–89.
[2] (Taleb, 2010) p. 119.

Block by Block: The Historical and Theoretical Foundations of Thermodynamics,
Robert T. Hanlon, Oxford University Press (2020). © Robert T. Hanlon.
DOI: 10.1093/oso/9780198851547.001.0001

Newton can't be understood without Galileo, and Einstein can't be understood without Newton. Furthermore, understanding history enables one to learn how a given problem was solved, which thus better ingrains the solution into one's own thinking. True understanding of a scientific concept can be significantly enhanced by knowledge of its historical development.

How to Reconcile the Varieties of Motion

While the different pieces of this particular puzzle evolved around the basic concept of motion, it wasn't entirely clear that the pieces actually belonged together in the same puzzle. There was the motion of those things that we could see with our own eyes such as falling objects, colliding balls, soaring projectiles, and the mysterious movements of the Sun, Moon, and wandering planets. And then there was the motion of those things that we couldn't see, like heat and radiation. There was the motion of objects that required human or animal exertion and sweat, like lifting heavy blocks of stone. And then there was the highly abstract motion of things that hadn't yet moved, such as that point in time immediately after a body is dropped, when motion is about to be but has not yet been. And there were other mysterious energy-related phenomena involving concepts of, for example, friction, phase change, and inelastic collisions, that couldn't be fully understood. How could all these seemingly different and abstract concepts be reconciled without creating a ghastly complex array of one *ad hoc* explanation after another for each and every phenomenon?

So let's make this even more daunting, as if this weren't daunting enough. Reconciliation required quantification. But how and of what? What equipment, procedures, and methods would be used to do the quantification to the level of precision and accuracy required to create a law? And exactly *what* would be quantified?

The variables seemed simple, but they weren't. Just re-read long-ago writings and see how confused you become with both the attempts to quantify and the attempts to describe the quantification. There were *seemingly* simple variables like distance and weight. And then there were more complex variables like time and speed. But, you say, time is simple. And I would reply, really? How would you even reliably measure time way-back-when, especially in the range of seconds or less required for quantifying instantaneous speed of falling bodies or shooting projectiles when the necessary timepiece hadn't even been invented yet? And how would you even define time? Many thought of time as being granular, occurring in discrete segments or instants. Even weight itself was complicated as eventually demonstrated when Newton separated weight and mass into two distinct concepts. And then there were even more complex variables like velocity, acceleration, mass, and force. And don't forget all those variables that were tried and didn't work. And don't forget all those different words that were tried and didn't stick. Words in English, French, German, Italian, Latin, Greek, and Arabic. Not to state the obvious, but in the world of science, exact meaning is critical. It's so very easy to become tangled and lost in words. It was truly a daunting task to discover a law out of such a messy and confusing landscape.

So what drove scientists to spend so much time, energy, and frustration pursuing reconciliation of the above? Clearly, as we will discuss, there was the desire to understand and to then control the performance of machines that could help humans protect and progress life. But while the solution of real technological problems was a driver, it wasn't the key driver. Instead, it was man's innate curiosity and desire to problem solve that drove him to complete the puzzle. It was also man's very deep belief that a simple and beautiful structure *must* exist, that there must be an orderly pattern, a logical reason, a cause and effect, and simple equations to tie the two together. It was man's belief that, in the end, that there must ultimately be some *thing* that's conserved. Magic doesn't happen. Rocks don't roll up hill. And this deep belief that there's something there to be discovered became an extremely strong attractive magnet for the curious human mind. That this entire and complex journey did indeed lead to a very simple set of equations involving rather profound concepts is a remarkable testament to both nature's simplicity and man's beliefs, along with man's insatiable curiosity and ego-driven desire to be the first.

Classical Mechanics – the Lever and Free Fall

Before moving further into this chapter, I need to bring you back to our university physics course to remind you of classical mechanics. Today we define mechanics as being concerned with the behavior of physical bodies when subjected to external forces. At the heart of this science is the concept of the *inertial state*, which is the tendency of a body to remain at rest or in uniform motion in a straight line—Newton's 1st Law of Motion—unless that state is changed by an external force. When the forces acting on a body are in balanced equilibrium, the body remains in its inertial state. It experiences no change in motion. When the forces are not balanced, acceleration results—Newton's 2nd Law of Motion.

Two classic "case studies" associated with classical mechanics and involved in this section are the lever and free fall. As you'll see, much of the difficulty encountered when deriving the concept of energy occurred as a result of philosophers trying to find a simple connection between these two phenomena when in fact the connection was very abstract. Today, the two are appropriately situated in two very different disciplines, the lever in statics and free fall in dynamics. Historically, study of the former brought forth the concepts of work, potential energy, and the impossibility of perpetual motion, while study of the latter brought forth the concepts of force and kinetic energy. The concept of momentum, central to Newton's Laws of Motion, rose from an overlap between the two disciplines. These two very different circles of knowledge wouldn't truly come together until the connection between them was recognized, the connection naturally being the rudimentary concept of mechanical energy and its conservation, which preceded the concept of energy that included temperature and heat.

Aristotle Turned Man's Mind toward the Natural World

For lack of more ancient records, history of mechanics starts with Aristotle

– René Dugas[3]

Aristotle may have been the last person to know everything there was to be known in his own lifetime

– Daniel Tunkelang[4]

---

[3] (Dugas, 2012) p. 19.
[4] (Tunkelang, 2009) p. 3.

He brought inquiry to almost every subject possible. He made significant contributions to each. His views continue to influence Western Civilization to this day. A student of Plato, Aristotle (384–322 BC), among other Greek philosophers, indeed played the dominant role in leading the evolution of thought from the internal mind to the natural world, addressing in the process man's need to understand causality. He strove to find connections between phenomena, to see cause–effect relations, to observe and study closely, and to seek organization and classification of the results. And so, of course, we must start with him.

While a recognized genius, Aristotle was not perfect. Many of his ideas were right, but some significantly weren't. Since he was such a powerful and persuasive individual, his ideas were built into the foundation of Western thought and so initially accelerated but eventually inhibited scientific progress.

Specifically regarding the concept of motion, as opposed to the myriad of other topics that he studied if not mastered, Aristotle considered two separate forms: the motion of objects such as projectiles and falling bodies and the motion associated with the use of simple machines such as the lever. The developments of each took different evolutionary paths. We start this section by first reviewing the first path, seeing how the thinking on the motion of projectiles, falling bodies, and moving planets evolved from Aristotle to Galileo to Newton, when this issue was finally resolved through Newton's Laws of Motion and his discovery of one of the four fundamental interactions or forces, Universal Gravitation.

Next we'll review the second path by learning how Aristotle's ideas on the motions of simple machines led to the early concepts of work and the impossibility of perpetual motion, the forerunner to energy and its conservation.

Within these stories we'll further learn how Newton's momentum and Leibniz's kinetic energy evolved as separate and distinct quantifications of motion. This history is fascinating in and of itself in light of the long controversy during the seventeenth century about which was more relevant, velocity or the square of velocity? As we'll learn, both were.

We'll then see how the two paths met again by 1750 when an early form of conservation of mechanical energy based on both kinetic and potential energies was realized. Finally, we'll follow the journey over the next 100 years that would link this law to the concept of heat as a form of energy and so finalize the concept of energy and the law of its conservation.

*　*　*

As summarized by Dijksterhuis,[5] Aristotle's beliefs[6] on motion were:

1. A body removed from all external influences will be at rest.

2. The speed of a falling body increases with weight.

3. All that moves contains something else doing the moving.

Regarding projectile motion, the written record shows that Aristotle really had only vague references and obscure remarks. It is difficult to truly understand his position on acceleration.

---

[5] (Dijksterhuis, 1969) pp. 163–184.

[6] The views of Aristotle originated from both Aristotle and his successors, known as the Peripatetics, from the Latin phrase for walking up and down, which was Aristotle's practice while teaching. In the text, I do not differentiate and simply assign all thoughts to Aristotle.

These beliefs represent the challenge that Aristotle presented to those who followed, the challenge of separating wheat from chaff. While (1) successfully led to the concept of the inertial state once the equivalence between rest and uniform straight-line motion was realized, (2) and (3) were just plain wrong. In fact, regarding (3), Aristotle missed the mark even further by not realizing that something must be conserved. For example, he said that a thrower transfers something, an *impetus*,[7] to the layer of the object that remains with the object during the motion. This something, this external cause residing in or on the object, provides continual action to the body. The motion then simply dissipates when the object lands. Aristotle made no attempt to quantify where the lost motion had gone.

Why do I raise these three notions? Because they evolved into an extremely deep paradigm about motion that subsequent generations first embraced and then rejected. For while (1) set man on the right path, (2) and (3) each contributed to the misleading and stifling paradigm that surrounded scientists for the subsequent hundreds of years. To truly appreciate the genius of Galileo and Newton and their respective breakthroughs in the subject of motion, one must first appreciate the world in which they lived, and this world was significantly influenced by Aristotle.

---

[7] (Dijksterhuis, 1969) p. 173. The concept of *impetus* rose from Jean Buridan in the fourteenth century based on Aristotle's assumption that a projectile's motion had an external cause that resided in the medium.

7   Galileo and the Law of Fall

Galileo, perhaps more than any other single person, was responsible for the birth of modern science

– Steven Hawking[1]

[Galileo] stood with one foot in the neatly ordered cosmos of medieval philosophy and with the other greeted the dawn of the mechanical universe

– Richard Westfall[2]

As we learned by his contributions to the Big Bang's discovery in Chapter 2, Galileo Galilei (1564–1642) was fascinated by motion. He studied motion most of his life, wrote about it in *De motu* (On Motion; 1589–92; unpublished until the nineteenth century), started to re-orient his thinking in 1592, eventually rejected the mechanics he wrote about, and finally, toward the end of his life, reformulated his thinking into his masterpiece *Discourses and Demonstrations Concerning Two New Sciences* (1638), a publication that opened the door to the modern era in physics.

To fully comprehend Galileo's achievements and especially the evolution of his thinking in going from *De motu* to *Discourses*, one must first recognize his early immersion in the existing interpretations of nature at that time, based largely on Aristotelian physics. As a student in Pisa and teacher in Padua, Galileo spent years absorbing the teachings of the Ancients, including Aristotle, Euclid, and Archimedes, and fully embraced Aristotle's and his successors' ideas about motion. Such teachings helped set the stage for the ensuing conflict. Having not been similarly immersed in Aristotelian physics, we can't truly appreciate the internal struggle represented in the difference between the Aristotle-influenced content of *De motu* and the self-created content of *Discourses*. The path blazed by Aristotle frustrated many, including Galileo. But while many were thus stymied, Galileo wasn't. He simply created his own new path.

Galileo's Break from Aristotle

Galileo most likely began his break from Aristotle in 1592,[3] his first year of teaching at Padua University, when he was exposed to, ironically enough, non-academic science. On his way to

---

[1] (Hawking, 1988) p. 179.
[2] (Westfall, 1971) p. 42.
[3] (Renn et al., 2000).

Block by Block: The Historical and Theoretical Foundations of Thermodynamics,
Robert T. Hanlon, Oxford University Press (2020). © Robert T. Hanlon.
DOI: 10.1093/oso/9780198851547.001.0001

Padua from Pisa, Galileo is suggested[4] to have visited his benefactor Guidobaldo del Monte, a man of learning and wealth, at his home. The two had started to exchange letters in 1588, and Guildobaldo, suitably impressed, endorsed Galileo's teaching position at Padua.

During this likely visit, historians further speculate that Galileo and Guidobaldo collaborated in an experiment concerning projectile trajectory. The details here are unclear regarding who owned the key idea, but most believe it was Galileo. Regardless, the outcome was clear. As summarized later by Galileo in *Discourses*, when one rolls a small metal ball, made warm and damp from contact with a person's hand, or alternatively covered in ink, across an inclined metal mirror at an upward angle, beautifully symmetrical parabolas are traced.

Discovery of the Parabolic Shape of a Projectile's Trajectory

This discovery of the parabolic shape of a projectile's trajectory in 1592 was monumental. Cameras obviously did not exist back then. So Galileo's concept of capturing such a dynamic event on such a creative "film" was truly remarkable. It afforded him the opportunity to actually *look* at the shape, frozen in time, and realize its symmetry. Many had wondered and speculated. Galileo took the "photograph."

But what was so monumental about this symmetry? Well, in the absence of fact, opinion dominates. And the prevailing opinion at that time, based on Aristotle's philosophy, was that a projectile's trajectory could *not* be symmetrical since it's determined at the beginning and at the end of its path by different causes. At the beginning, it is dominated by the "violent" impetus impressed into it, and at the end by its "natural" motion toward the center of the Earth.[5] Per this opinion, a trajectory based on two very different causes could not be symmetrical.

Galileo's discovery directly contradicted this. Newsworthy, right? Galileo didn't think so. He didn't make mention of it until he published *Discourses*.[6] Why not? The answer seems clear. Galileo must not have initially understood the full implication of his finding. Two reasons are suspected. First, at this point in his career, Galileo began to shift from pure science to applied technology and wasn't yet prepared or ready to grapple with the theoretical aspects of projectile motion. His focus was on the practical, and indeed, he used these findings to support his teachings on artillery science. Second, Galileo may have explained the symmetry by simply saying that the decrease in violent motion exactly balanced the increase in natural motion. This new insight would have left him comfortably remaining in Aristotle's paradigm. So this discovery may not have been a significant discovery to Galileo, at least not until he reinterpreted it based on his later findings relating to free fall and inertia.

---

[4] (Renn et al., 2000) p. 299. "In spite of being the subject of more than a century of historical research, the question of when and how Galileo made his major discoveries is…still only insufficiently answered." Alas, history can only be based on the fossils we find.

[5] Violent motion was thought to be any forced motion that does "violence" to the natural order. For example, an upward motion would be violent since it violated the "natural" downward motion toward the center of the Earth. Natural motion is motion freed from all constraints.

[6] As a general comment, Galileo was at times frustrating to study since he didn't always record what he did or the thought process behind his conclusions.

Galileo's "Radical Re-orientation" from Pure to Applied Science

While still a significant achievement, this discovery was not the dominant reason behind the start of Galileo's break from Aristotle. Instead, when viewed from a larger perspective, the dominant reason that the year 1592 marked a "radical re-orientation"[7] in Galileo's life was his direct contact with Guidobaldo and the resulting shift in his approach from pure to applied science. Guidobaldo was considered to be an "engineer scientist,"[8] a name given to the group who rose over the previous centuries when the combination of new construction practices and increasingly limited availability of labor stimulated new thinking and innovation. This group of intellectuals gained many real-world experiences, experiences not taught or available in academia, and in so doing they became well-versed in practical mathematics and engineering sciences. Galileo's contact with their experimental techniques and research interests, with their environment filled with technological and industrial problems waiting to be solved, and with their thinking that nature was not only to be contemplated but controlled, his contact with all of their activities which came as a result of his relationship with one of their leading figures, Guidobaldo, is thought to have encouraged him to adopt a new approach to science in which new knowledge is acquired through practical experience. Such a practical turn was, perhaps, reinforced by his childhood background during which, encouraged by his musician father, he experimented with musical instruments and the application of mathematics to their understanding. His later combination of this applied science approach with the pure science approach embodied in the exciting exchanges of ideas in and around the university atmosphere, and especially since at that time many such exchanges were converging on the problem of motion, set Galileo down a very powerful path of discovery. As you'll see several times in this book, discovery often occurs at such interfaces between disciplines.[9]

The Pull toward Experimentation

From 1592 to 1610, Galileo's mind continually pulled him toward experiments and measurements and practical applications, all centered on the motion of falling bodies. He used pendulums, cannons, and balls on inclined planes. He attempted to explain his findings using Aristotle's theories but increasingly arrived at findings in conflict with those theories. The resulting frustration drove him to develop new ideas and methods.

In 1593, he wrote on the fact that, ignoring air resistance, the speed of a free falling body is independent of its weight. Whether or not he actually conducted the supporting experiments off the Tower of Pisa, a subject of continued speculation and controversy,[10] Galileo's

[7] (Renn et al., 2000) p. 349.
[8] (Renn et al., 2000) p. 336.
[9] In all, Galileo operated at four different interfaces: two philosophical ones, Aristotle's physics and modern physics, and two professional ones, academia and craftsmanship.
[10] The story that Galileo discovered this fact by dropping lead balls of different weight from the top of the Tower of Pisa is much disputed. Some historians believe that, more likely, Galileo ran a thought experiment and simply concluded this known fact. The thought could have been something like, two balls tied together by a string don't fall faster

stated belief showed an increasing willingness to challenge Aristotle's own beliefs as outlined in Chapter 6.

The Law of Fall

Then in 1604, Galileo arguably achieved his most significant discovery, known as "The Law of Fall," which states that the vertical distance travelled by a free falling body is proportional to the square of the time.[11] Later in 1608, using a similar apparatus in addition to "thought experiments," he stated that a body moving horizontally would continue uniformly in the absence of external resistance. These findings led him to fully explain the parabolic trajectory of projectiles as he summarized much later in *Discourses*:[12]

- Free projectiles move horizontally with constant speed and fall vertically with constant acceleration. The two motions do not interact but instead combine to produce a perfectly symmetrical parabolic trajectory.
- The vertical fall is independent of weight and covers distance as the square of the time.

Using "h" to represent the height of vertical fall, or the distance travelled from rest, the Law of Fall thus says that,

$$h \propto t^2 \qquad \text{Galileo's Law of Fall}$$

Galileo went on to assert that given the Law of Fall and given that height fallen equals average speed multiplied by time, then speed itself must be proportional to time.

$$v \propto t$$

Combining the two relationships, Galileo arrived at one of the most significant discoveries in history.[13]

$$h \propto v^2$$

Galileo also asserted that "a heavy body falling from a height will, on reaching the ground, have acquired just as much *impeto* as was necessary to carry it to that height," and that, based on his

---

than two balls not tied together; matter is simply comprised of smaller pieces of matter "tied" together. However, an alternative view shared with the author in a 2017 email exchange with science author David Harriman is that Galileo indeed did conduct the experiment based on Viviani's recollection of a conversation in which Galileo spoke of dropping objects from towers early in his career. Viviani did not specifically mention the Leaning Tower, but it is a good guess since Galileo was in Pisa at the time. Dutch physicist Simon Stevin (1548–1620) conducted the first validation experiment of this concept in 1586. See also (Settle, 1983).

[11] (Drake, 1973) p. 84. Some historians feel that Nicole Oresme (1323–1382) discovered this law first through his analysis of accelerated motion using new mathematical concepts on rate-of-change. But he never applied his results to the changing speed of a freely falling object. He did not connect uniform acceleration with free fall.

[12] The delay in publishing his findings on motion was caused in large part by Galileo's shift in focus to telescope astronomy in 1609.

[13] For a good discussion of Galileo's logical development of free fall mathematics, see (Harriman, 2010) Chapter 2 – Experimental Method, pp. 36–80.

observation of pile drivers, that the impact or intensity of blow, which we today quantify as kinetic energy, is proportional to height.

Simply put, these findings were momentous, being the first to capture the mathematics central to the conservation of mechanical energy. You'll learn more on this later, when I discuss their deeper meanings as uncovered by Galileo's successors.

The Law of Fall Determined by an Ingenious Experiment

Galileo arrived at the Law of Fall through a rather ingenious experiment (Figure 7.1), an approach similar to that we repeat in high school physics. Ingenuity was needed, because how could one really quantify free fall? There are three variables involved: time, distance, speed. (Recall that mass plays no role in free fall.) Which ones to use? How to relate them? What to search for? Not as easy as it looks, or else it would have been done much earlier.

During free fall, speed changes with time, and there was simply no way back then to meas-ure this, to capture instantaneous speed. So Galileo focused on what he could directly measure: time and distance. To study the relation between the two, he used the approach he trusted most throughout his life, the proportionality theory of Euclid, based on the earlier work of Eudoxus: x_1 is to x_2 as y_1 is to y_2. With this approach, one only needs precision and not the accuracy offered by calibrated instrument since the use of ratios eliminates the need for absolute units, i.e., the units cancel out. But first he needed good data. And when objects fall, they fall fast. How could one even obtain highly accurate time/distance data for a free falling object? Understanding this challenge, Galileo decided to slow everything down by using a grooved inclined plane as with his earlier studies. He systematically started letting balls roll down the plane and, because the incline reduced the speed of the ball, he was able to mark with high accuracy the distance from rest at fixed time increments. And from these data, Galileo dis-covered the Law of Fall: vertical distance from rest during free fall increases as the square of time.

There are so many themes, meanings, and implications emanating from this experiment, this single experiment that arguably started the modern era of physics.

How Did Galileo Measure a Time-varying Speed?

First let's look at the experiment itself and especially Galileo's measurement of time.[14] Was it real or fiction? Some historians had criticized Galileo for presumably not really having even conducted these experiments to begin with. In *Discourses*, Galileo claimed to have used a "water clock," dripping water out a nozzle and then weighing the result, to quantify time incre-ments. Historians suggested that this provided a too-good-to-be-true accuracy. As late as 1953, Alexandre Koyrè, the highly regarded analyst of the scientific revolution, suggested that Galileo made no use of experiment but instead was led to his conclusion by pure mathematical reason-ing. The issue was finally resolved in a way that Galileo would have approved of. In 1960,

---

[14] (Johnson, 2009) Chapter 1. (Drake, 1975).

7.1   GALILEO: THE LAW OF FALL   ©RTH + CLS

FREE FALL
Galileo used an inclined plane to slow down free fall

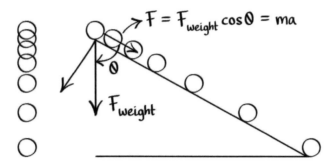

$$F = F_{weight} \cos \theta = ma$$

F_{weight}

GALILEO'S DATA AS DISCOVERED BY STILLMAN DRAKE (1975)

TIME*	DISTANCE
.55	32.9
1.10	131.4
1.65	295.7
2.20	424.7
2.75	821.5
3.30	1182.4
3.85	1609.8
4.40	2103.1

distance (graph: distance vs time², values 500–2500 on y-axis, 0–25 on x-axis)

LAW OF FALL: $h \propto t^2$

GALILEO'S LOGIC: $v \propto t$ given the Law of Fall and given that height fallen equals average speed multiplied by time, then speed itself must be proportional to time

the important conclusion relevant to energy: $h \propto v^2$

* Drake speculated that the fixed time increments reflect Galileo's use of cadence while singing a song to mark time

Figure 7.1 Galileo: The Law of Fall

Thomas B. Settle, historian of science, simply repeated Galileo's experiment with a "water clock" and confirmed its stated accuracy. But the true breakthrough occurred in 1973 when Galileo's original data were discovered, thus providing the proof that he actually did obtain his law from experiments. One would have thought all of Galileo's papers had been thoroughly analyzed by 1973, but Stillman Drake, a leading Galileo expert, discovered otherwise.[15] He was going through Galileo's own notebooks and surprisingly unearthed unpublished experimental data—how could such data remain undetected for so long?!—supporting the Law of Fall. For whatever reason, Galileo published the law and a description of the experiment but never the data. And this is what understandably caused the confusion. Drake acknowledged that, despite Settle's findings, a water clock most likely was *not* initially used due to its inherent inaccuracies but suggested instead a viable and rather unexpected alternative. Galileo was raised in a musical world and so was very likely to have a deep respect for the strong internal rhythm innate to human beings. Drake proposed that Galileo made use of this by singing a song or reciting a poem and using the cadence to mark time with rubber "frets" along the incline during the experiments to create audible bumps when the ball passed by. By adjusting or tuning these frets, Galileo was able to accurately synch the bump sounds with his internal cadence, thus providing a means to achieve equal divisions of small time increments. This proposed approach is strongly supported by the fact that Galileo's data show fixed time increments, as opposed to sets of times and distances like a water clock would provide. To Drake, the only method that would result in fixed time increments would be a fixed cadence. "But wait!," you say. "How could this possibly provide the necessary accuracy?" Well, just observe yourself listening to live music when the drummer is but a fraction of a second off-beat. You cringe, right? This is because your innate rhythm is so strong. Drake proposed that while highly accurate, such an unorthodox approach would hardly be one that Galileo would want to share with the scientific world. And so he most likely used this approach to first discover the Law of Fall and then used the water clock afterwards to demonstrate it in a more respectable and scientific manner.

The Dawn of a New Variable to Science – Time

Now let's look closer at the concept of time itself. Galileo was one of the first to use the concept of time as a dimension in a mathematical relationship. As noted by science historian Charles Gillispie, "Time eluded science until Galileo."[16] Linking time with another dimension, distance, opened the door to developing more complex relationships involving speed and acceleration. Galileo's time flowed uniformly, also a new concept since many viewed time as occurring in increments.

Now let's look at the actual result. The distance measured from rest increases with the square of the time. So during free fall, speed increases uniformly. The continuity implied by these results, reflective of a *constant* acceleration, may seem rather straightforward to us since we readily grasp continuity. But in Galileo's time, as influenced by Aristotle, most felt that

---

[15] (Drake, 1975).
[16] (Gillispie, 1990) p. 42.

*dis*continuity was involved in free fall. Most thought that acceleration in free fall occurred as a series of small and rapid spurts of uniform speed, caused by discrete impressed increments of *impetus* from nature. This concept may have sounded reasonable and Galileo may have started with this philosophy since he was raised in it, but it rapidly crumbled in the face of Euclid–Eudoxus mathematics, leaving many to ask with even further frustration, *so what* is *the true cause of free fall?*, a question that wouldn't be answered until Newton's arrival. All of this marked a significant transition away from Aristotle, with Galileo sitting at the transition point.

What is Speed?

Now let's look at the concept of speed. While certain variables like distance or weight are directly measureable, others are more complicated. Speed, for example, is the ratio of distance to elapsed time. During acceleration, such as with the free fall of a body, speed continuously changes. But how could one possibly measure instantaneous speed in the early 1600s? Such measurements would require something akin to stopwatches, but the arrival of such fine time pieces wouldn't occur until many years later. Men such as Galileo and, later, Christiaan Huygens made significant advances in time keeping through use of the pendulum. But the finesse offered with such devices clearly wasn't compatible with the objective of measuring instantaneous speed during the rapid free fall of objects.

So given all these reasons why Galileo shouldn't have been able to do what he did, how did he end up doing it? He surely wasn't one to let such seeming impossibilities get in the way. He wasn't daunted in the face of this challenge. While he knew that he couldn't measure speed, he didn't ignore its importance.

Galileo mentioned speed or "degrees of speed" in some of his writings but unfortunately generated confusion as a result, mostly because he never really defined what he meant by it. He was stepping into new undefined territory here. When a body moves, there must be some quantity to describe its motion. Today we have fully defined terms, such as speed, velocity, acceleration, momentum, kinetic energy, etc. But such was not the case in Galileo's time. No definition of "speed" during acceleration, a revolutionary concept in and of itself in Galileo's day, had ever been clearly made in terms of proportional ratios of distances and times.[17] Archimedes had only done it for uniform motion, which is much easier since there's no need for "instantaneous." Clearly no experimental device was available in Galileo's time to make such a measurement during acceleration itself. So he was left to define speed, or *velocità* as he called it, in the way that he felt best described nature. At one point, as shared earlier, while observing a pile driver hammering away near his home, Galileo observed that the driver strikes twice as hard when the weight falls twice as far ($\Delta v^2 \propto \Delta h$). Seeking to establish a standard for measuring the force of percussion, a problem that bothered Galileo and many others until Huygens solved it, he thus reasoned that *velocità* represents whatever it is that changes the striking power of a body falling from different heights and so it must be proportional to distance from rest (twice as hard for twice as far). Based on his law of fall ($h \propto t^2$), this meant that *velocità* must then also be proportional to the square of time. As we know today, the effect of a

---

[17] (Drake, 1973).

pile driver is governed not by speed but by its kinetic energy. It's v^2 that's proportional to the square of time and thus creates the percussion effect. So in trying to define speed, Galileo mixed together two different concepts, v and v^2. He eventually resolved this situation by stating that *velocità* is that which is proportional to time and not its square (v ∝ t). The heated debate between different definitions of quantities relating to motion, and *especially* between v and v^2 as influenced by both the law of the lever and free fall, would continue throughout the seventeenth century and into the eighteenth century when it was finally demonstrated that both definitions are valid, a fact that contributed greatly to the confusion.

Galileo's Use of Mathematics to Quantify the Physical World

Now let's look at the use of math. Historically, physicists and mathematicians didn't interact. Physicists resisted the use of math since professors in this area were invariably philosophers and not mathematicians. Galileo, operating at yet another interface between disciplines, joined the two fields together by using a mathematical approach to describe and quantify the physical world and so test the hypotheses he formed. This represented another break from the past. Aristotle did not encourage the use of mathematics in describing natural phenomena since he was primarily interested in the causes, and mathematics did not deal with physical causes. Galileo, on the other hand, made use of a full set of new mathematical tools, such as the Euclid–Eudoxus law of proportions and especially the approach originated by Oresme and more fully developed by René Descartes in which data are graphed on an x–y axis, which today we refer to as the Cartesian coordinate system, thus enabling the decomposition of motion into horizontal and vertical components. Galileo had such tools available based on his education and simply used them in new ways of discovery.

Now let's look at another aspect of math, much more profound in nature. Galileo's philosophy developed to believe that, "[The universe] is written in the language of mathematics."[18] He believed that mathematics could be used to describe all natural phenomena; therefore, he believed that all natural phenomena must follow mathematical behavior. In his search for the Law of Fall, for example, he believed that a simple equation existed and then found the equation. Such a very profound philosophy. As I type this, I wonder whether there are natural phenomena that don't follow mathematics. I haven't studied this but also believe the answer to be "no." And this seems quite profound to me. Galileo and others helped create this philosophy for us.

Galileo Chose to Ignore Cause

Now let's look at what Galileo *didn't* do. Galileo didn't develop an underlying theory for motion, a point critically highlighted by Descartes.[19] He discovered the parabolic trajectory and

[18] (Popkin, 1966) p. 65.
[19] (Galilei, 1974) Stillman Drake's Preface, p. xxi. The eventual grand program which we know as "the mechanical philosophy" was not due to Galileo but instead due to Descartes who flatly rejected Galileo's physics as having been

quantified free fall and chose not to seek the cause of either. Perhaps, though, this reflected Galileo's frustration with Aristotle's approach that favored thinking over action. While Aristotle was driven by his interest to classify, which could not provide anything useful and couldn't tell us any more than what it summarized and so stopped short of quantification, Galileo, not being one to sit idly by, didn't. Galileo chose to ignore "cause," and in fact outright rejected the quest for "cause," and turned to a purely kinematic approach, "to investigate and demonstrate some of the properties of accelerated motion, whatever the cause of this acceleration may be."[20] The lack of supporting theoretical foundation does not detract from Galileo's discoveries. Put into proper context, these discoveries and the experimental–mathematical philosophy behind them are truly remarkable and became the bedrock on which his successors built. But it does highlight the fact that Galileo's contributions required further work by those who followed, such as Huygens, Leibniz, and Newton, before classical mechanics became fully developed. In some sense, one could say that the later creation of classical mechanics provided meaning to Galileo's discoveries. Even Galileo did not immediately understand the implications of his work.

As an aside, it's interesting that Newton attributed his famous 2nd Law of Motion, $F = ma$, to Galileo. Westfall[21] stated that Newton gave Galileo more than his due, saying that the 2nd Law of Motion truly belonged to Newton even though he had to build on the work of others to create it. Looking at Galileo's writing's, one can indeed see glimpses of the 2nd Law there. Newton saw it. But Galileo never did. Newton was primed to see it, but Galileo was not even close to doing so. Because while Galileo broke from Aristotle, the break wasn't complete. Some key beliefs of his were still very strongly linked to Aristotle. For example, Galileo continued to believe in the concepts of "violent" and "natural" motions to the end of his life. With such thinking came an inability to conceive of some external force as being at the root of free fall. Galileo simply wasn't ready to grasp action-at-a-distance, as reflected by his statement that the notion that the Sun and the Moon cause the tides "is completely repugnant to my mind."[22] He never even suggested that free fall was produced by an external force acting on a body. Galileo truly stood at the transition point itself, one foot in Aristotle's world (*De motu*) and the other in the modern era of physics (*Discourses*). He quantified free fall, ascribed it to a "natural" tendency, and then stopped. To him, force, as Newton later defined it, would have been absurd. And none of this discussion detracts from Galileo's accomplishments. One could even argue that his accomplishments were even more remarkable and revolutionary given the depth of the paradigm from which he had to at least partly emerge.

The Scientific Method

Finally, let's look at Galileo's approach. Thinking about this now, all I can say is "Wow!" As scientists, we're taught the scientific method[23] and embrace it, either knowingly or not, in our

---

built without foundation. So while Galileo contributed to the "birth" of classical mechanics, it was left to others to formulate the final structure eventually given by Newton.

[20] Galileo quote cited in (Crombie, 1969) p. 86.

[21] (Westfall, 1971) p. 2.

[22] Galileo quote cited in (Westfall, 1971) p. 46.

[23] (Brush and Segal, 2015) This book provides a good historical overview of the scientific method.

work. But what we don't always recognize is Galileo's role in making it *the* method in science. We work today in a world largely created by Galileo. We make observations of some process or phenomenon and then make a hypothesis, e.g., a mathematical model, to explain it. And then we design experiments to generate data to test the hypothesis. Is it right or wrong? The true power of this approach is that we validate our theory based on its being able to predict that which hasn't yet happened,[24] such as when Einstein used his theory of General Relativity to boldly predict the amount of bending of light around the Sun and then had it later confirmed by Eddington's measurement. Certainty arises neither from mathematics alone nor from experience alone, but only from their agreement. As physicists, chemists, and engineers, we constantly obtain experience, i.e., data, and employ mathematics to develop and validate theories based on these data. We employ these tools in everything we do, from strict science problems such as determining reaction kinetics to all of our quality improvement programs, e.g., Statistical Quality Control, Total Quality Management, 6 Sigma, etc. All are built on Galileo's scientific method that favors data over preconceived ideas.[25]

Galileo and the Launch of the Scientific Revolution

Thus only has the study of nature entered on the secure methods of a science, after having for many centuries done nothing but grope in the dark.
          – Kant in reference to Galileo and others using experiments to understand nature[26]

Galileo was a bridge from the Renaissance to the Age of Reason
                                                              – David Harriman[27]

A brief mention is warranted here on the scientific revolution,[28] within which the scientific method played a critical role. Largely a sixteenth and seventeenth century effort that led science out of superstition, alchemy, and magic and into rationality, experimentation, inductive reasoning, and mathematics,[29] this revolution arguably started, if one had to pick a year, in 1453 with the publication of *De revolutionibus orbium coelestium* (On the revolutions of Heavenly Spheres) in which Copernicus put the Sun back at the center of the solar system

---

[24] The general scientific method follows the path of: employ inductive logic to create a hypothesis based on observations or data, deduce a consequence based on the hypothesis about something not already known, test the prediction, reject the theory if the test fails. The method is not based on confirming what is already known but instead is more powerfully based on being able to predict that which has not yet been observed. For an excellent overview of the use of logic in science, see (Harriman, 2010). This reference is particularly enlightening as it draws on the respective achievements of, among others, Galileo, Newton, and those involved with establishing the atomic theory. See also (Platt, 1964) for a deeper discussion of the scientific method based on "strong inference."

[25] (Taleb, 2010) p. 318. One of Francis Bacon's criticisms of Aristotle is that he made conclusions about natural phenomena prior to conducting experiments.

[26] (Kant, 1896) p. 692.

[27] (Harriman, 2000) p. 1.

[28] I wrote this section based largely on the wonderful set of papers and commentaries delivered at the Institute of the History of Science event held at the University of Wisconsin in 1957 and published in (Clagett, 1969).

[29] (Fromkin, 2000) pp. 132–133.

where it belonged.[30] Galileo's support of Copernicus (Chapter 2) and his overthrow of Aristotle's ideas on motion continued the revolt.

The ancient Greeks clearly progressed Western Civilization by creating science as a means to describe and understand nature. Inherent in their philosophy was the assumption of the existence of an abstract order behind the regular changes observed in the world. Many of their ideas positively influenced succeeding generations, but unfortunately not all were correct as some originated more from unfounded logic than from experiment. So at the same time that they accelerated the positive influence of science on our lives, they also created obstacles that slowed progress.

For a time, though, none of the above really even mattered since the ancient Greek writings were lost to the Western World, due in part to the decline of the Roman Empire in the fifth century. It took until the early Renaissance around the twelfth century for the writings to re-emerge, thanks largely to some dedicated and motivated individuals who took it upon themselves to translate the writings into Latin or into Arabic and then into Latin. Furthering the re-emergence of the Ancients was the Ottoman Empire's capture of Constantinople in 1453, which resulted in the migration of bundles of Ancient manuscripts to Italy in the hands of scholars, and also Gutenberg's invention of the printing press in 1439, which helped spur the wide dissemination of knowledge to an appreciative public. The "reawakening of the human mind"[31] followed.

Along with this re-emergence of the Ancients came the seeds of their overthrow. Society began to change in response to the accelerated application of science to aid civilization. This rise in science led to man's own quest for understanding, instead of having others, such as those in authority, e.g., the Catholic church, tell them what the understanding is in spite of conflicting evidence from nature herself.[32] Society became less accepting and more questioning, more inspired by scientific inquiry to achieve quantifiable proof. This was not an easy endeavor as it ran into the resistance of those who didn't want such questioning, as best illustrated by the resistance Galileo encountered from the Catholic Church (Chapter 2).

These changes occurred slowly. During the favorable Medieval Warm Period of the eleventh and twelfth centuries and after barbarian invaders established settlements in what we now call Western Europe, population increased rapidly, logically accompanied by the need for progress in building, construction, food, water, and power. In response, societies began turning their interest to technology and its use to obtain control over nature. The mid-thirteenth century saw the rise in the specialist and also the enlargement of science in the universities, although schools and universities often failed to adapt to the pace of progress of the applied scientists and early engineers. People started to learn through practical experience, leading to the rise in empiricism based on gaining knowledge through experimentation and observation. The desire to understand cause–effect naturally followed. "*What caused this to happen? How can I prove it?*"

---

[30] Johannes Gutenberg's development of the printing press around this time period also contributed to the scientific revolution.

[31] (Harriman, 2000) p. 1.

[32] A relevant Stillman Drake quote cited in (Harriman, 2000) p. 11: "In an age when authority was everywhere taken for granted, Galileo's watchword was the rejection of authority of any kind. His entire attitude has been well summed up by saying that he was never willing to accept any intermediary between himself and nature."

Further societal improvements relative to what had existed in ancient Greek society rose to spur scientific advancement. The new societies in early Western Europe encouraged greater participation of the people. Man's need for structure led to improved organization of science. Communications improved, thanks in large part to the arrival of the aforementioned printing press, which made literature, publishing, and self-education available to all. Science and technology became valued in and of themselves as a contribution to a welcoming society that sought progress.

In parallel to the above advancements came advances in mathematics as well. The thirteenth and fourteenth centuries saw the rise in new mathematical concepts, such as rate-of-change, functional dependence, and the use of graphing on Cartesian x-y coordinates, as initiated by Oresme and developed by Descartes, to analyze variable quantities, something the Greeks never exploited.

But throughout, society pretty much followed the world according to Aristotle, really only revising aspects that they largely inherited from the Greeks and Arabs. Greek science was absorbed, accepted, and even embraced and built upon. But the science did not always match reality and so it was only a matter of time until the revolt happened.

The stage was set. The motions in the skies were placed at center stage, surrounded by a storm of ideas. Others certainly helped prepare this perfect storm. But a genius was required to unleash it and genius is not additive. More people don't necessarily make an insight more likely.[33] The insight, the paradigm shift, occurs in a single brain when neurons alter their activity all at once, an effect that you've most likely felt at least once yourself. The sum of one thousand people thinking falls well short of one single Einstein thinking. A single genius was needed, someone conversant in the different disciplines of mathematics, contemporary motion theory, applied science, constructing tools for experimentation, experimentation itself, and yes, even music. Someone with the genius and especially the courage to make a break from the past, to create something new. Someone who recognized the key missing ingredient: measurement of motion as actually found in nature.

Galileo was in the right place at the right time to help set science in this new direction. While Copernicus may have started the scientific revolution, Galileo clearly fanned its flames.[34] His conviction that practical and controlled experimental observation, enhanced by mathematical theory and quantification, was the key to a more satisfying understanding of nature became an inspiration to those who followed.

One Data Point is Worth One Thousand Opinions

Between Galileo and Aristotle there were just a lot of guys with theories they never bothered to test
– Helen Monaco's character in Philip Kerr's *Prayer: A Novel*[35]

---

[33] A relevant Galileo quote cited in (Harriman, 2000) p. 11: "I should agree that several reasoners would be worth more than one, just as several horses can haul more sacks of grain than one can. But reasoning is like racing and not like hauling, and a single Arabian steed can outrun a hundred plow horses."

[34] Galileo was naturally not alone in this effort. For example, "Francis Bacon (1561–1626)…was widely regarded in the later seventeenth century as the founding father and patron saint of modern English science because of his vigorous advocacy of an experimental approach to the study of nature." (Lewis, 2007) p. 1.

[35] (Kerr, 2015) p. 73.

It had long since come to my attention that people of accomplishment rarely sat back and let things happen to them. They went out and happened to things

– Leonardo da Vinci[36]

It's interesting to read that some historians suggest that what Galileo did wasn't monumentally new. After all, in the centuries before Galileo, many had moved such similar concepts forward. And, living as we do now in the paradigm that Galileo helped to create, it's difficult for us to look back upon this and truly understand what the fuss was all about. We don't really know deep down the emotions and conflicts of those involved in progressing his science. We repeat Galileo's experiments in high school and don't recognize their importance.

So what did Galileo *truly* do that made the difference?

There are two answers to the question. The first—data! It's fascinating to know that while everyone from Aristotle to those immediately preceding Galileo *thought* about all sorts of things, many of the same things that Galileo was to think about, none of them took any measurements. Galileo measured while others thought. We see this around us today. Much thinking, proposing, and speculating. But without measurements, it really doesn't mean anything. As a former boss of mine once wisely said, "One data point is worth one thousand opinions." Rarely has this been better put.

If there is a persistent incongruity between a man's actual performance and his account of what he does, it is not unreasonable to ask whether the individual does have a clear grasp of what it is he is saying, or alternatively, whether the man's words mean what we suppose them to mean.

– Ernest Nagel[37]

Regarding the second answer, while some claim that thirteenth and fourteenth century scientists already developed what Galileo got credit for, it's critical to recognize, as the above quote indicates, that actions are very telling. If these earlier scientists actually achieved a new understanding of motion before Galileo did, then their achievements would testify to their grasp of such ideas. But the contrary appears to be the case. They may have thought about it, but it doesn't appear by their actions that they truly achieved this understanding. Many may think and speculate and hypothesize. But the critical issue is whether they truly understand. If there is a persistent disconnect between a man's actions and his words, it is reasonable to ask whether the individual truly understands his words. It can seem like such a small thing, the difference between understanding and not understanding. But it's actually quite large. Vast, in fact. Galileo understood. His predecessors didn't.

[36] A popular quote attributed to Leonardo Da Vinci.
[37] (Nagel, 1969) p. 154. In (Taleb, 2010), Taleb also points out (p. 322) that after a successful discovery is made, others rapidly come forward to claim that they had gotten there first and so deserve recognition. But this typically reflects what Taleb refers to as "retrospective distortion" involving the belief that someone knew something in the past but really didn't.

8 Newton and the Laws of Motion

The landscape has been so totally changed, the ways of thinking have been so deeply affected, that it is very hard to get hold of what it was like before. It is very hard to realize how total a change in outlook [Newton] produced

– Hermann Bondi[1]

The quest for "cause" is probably as old as mankind itself. As analytical thinkers, we are obsessed with cause. It's a philosophy that lies deep within us. We were born with this obsession. When we see an effect, we are compelled by our genetic make-up to discover the cause. Such a philosophy strongly manifested itself during the study of motion in the seventeenth century.

In 1609 Johannes Kepler (1571–1630) published the discovery of a monumental "effect" when he quantified the elliptical motion of planets. In 1638 Galileo did likewise when he described the complete parabolic trajectory of projectiles. That both kinematic effects[2] were consequences of the same single set of dynamic principles was not known. They stood apart as descriptive quantifications with no underlying cause–effect equations. In fact, Galileo intentionally stayed away from cause. What was left was for someone to identify the common cause. As we've seen before, such an effort could not be done by committee. A single mind was needed, someone to integrate all of this, someone to unite everything.

It would take another 50 years for this person to step forward and lay claim to solving this great puzzle. This person was indeed a genius, a single mind, who in his formative years worked alone, isolated from the rest of the scientific community. Sometimes a genius simply needs to be left alone to follow the thread of discovery in his or her own mind.

By now you know that this person was Sir Isaac Newton (1642–1727). Okay, he didn't do it entirely alone. Others were involved. Newton was clearly exposed to the ideas of such predecessors as Descartes and Galileo and such contemporaries as Gottfried Wilhelm Leibniz (1646–1716), Christiaan Huygens (1629–1695), Robert Hooke (1635–1703), and Edmund Halley (1656–1742). That this collective influenced Newton is reflected in his own writing, "If I have seen further it is by standing on the shoulders of giants."[3] But in the end, it was Newton who pulled it all together.

* * *

---

[1] (Bondi, 1988) p. 241.

[2] Kinematics is the branch of mechanics concerned with the motion of objects without reference to the forces causing the motion. Dynamics is another branch concerned with the motion of objects but under the action of forces.

[3] (Gleick, 2003) p. 98.

Block by Block: The Historical and Theoretical Foundations of Thermodynamics,
Robert T. Hanlon, Oxford University Press (2020). © Robert T. Hanlon.
DOI: 10.1093/oso/9780198851547.001.0001

Much as the transformation of mechanics from Aristotle to Galileo was not straightforward, neither was its continued transformation from Galileo to Newton.

Started unintentionally by Copernicus in the late sixteenth century, furthered by Galileo in the early seventeenth century, and following no path other than that created in the moment by man's perpetual need to question, the scientific revolution eventually rose to overthrow the Ancient philosophies, especially Ptolemy's Earth-centered model of the universe and the Aristotelian views of motion and, more generally, scientific methodology itself. While Galileo's achievements did much to further the revolution, educational institutions were slow to learn of and embrace his lead. The curriculum remained stagnant. Aristotle remained the traditional singular authority, heavy on qualitative observations, light on quantitative measurements and validating mathematics. But the real world was moving ahead, beyond Aristotle, even if academia couldn't keep up. And that world with its advancing technologies slowly showed Aristotle's mechanics to be "quaint and impotent."[4] Another step forward was needed. But in what direction?

Sir Isaac Newton

Amicus Plato amicus Aristoteles magis amica veritas
Plato is my friend; Aristotle is my friend; but my best friend is truth
Sir Isaac Newton, 1664, from his notebook[5]

Born in England the year Galileo died, Newton grew up an introvert, most likely due to a fatherless and motherless childhood. His father died shortly before his birth and his mother remarried and left him to be raised by her parents. He never married, had few close friends and, being suspicious of others, developed a "paralyzing fear of exposing his thoughts,"[6] a trait that would unfortunately lie at the root of his angry intellectual quarrels with the likes of the same aforementioned contemporaries, especially Hooke and, above all, Leibniz.

In 1661, Newton's developing intellect led him to enroll in Cambridge University where he was initially immersed in Aristotle, just as Galileo had been. But the rise of published scientific communications in the seventeenth century, especially those by Galileo, René Descartes, Pierre Gassendi, and Robert Boyle, presented Newton with alternative and contrary viewpoints. Newton's discovery of these writings in 1664 combined with his own obsessive quest to understand the nature of the world in which we live—and to bring simplicity, order, and quantified proof to such an understanding—led him to realize the failings of the Aristotelians. You can almost hear Newton thinking, *Say what?*, when reading the ancient Greek philosophies. Newton's conversion away from Aristotle and toward the rising mechanical philosophy as advanced by Descartes, in which nature consists solely of matter in motion, was almost immediate.

---

[4] (Gleick, 2003) p. 24.
[5] Newton quote cited in (Cohen and Westfall, 1995) p. 3.
[6] (Cohen and Westfall, 1995) p. 314. Referenced to John Maynard Keynes, "Newton, the Man," in The Royal Society Newton Tercentenary Celebrations (1947). "[Newton's] deepest instincts were occult, esoteric, semantic—with profound shrinking from the world, a paralyzing fear of exposing his thoughts, his beliefs, his discoveries in all nakedness to the inspection and criticism of the world."

Annus Mirabilis – Newton's Miraculous Year

In 1664, the Plague arrived in London and fatefully interrupted Newton's studies, forcing him to escape to his family home in the countryside. He took with him all of his books, all of his learnings, and all of his thoughts, and in this one year of isolation, Newton "voyag[ed] through strange seas of thought alone."[7] Einstein had his patent office, Heisenberg had his beach, and Newton had his isolated home in the countryside. In each instance, solitude provided the necessary environment for true deep insight.[8]

Isolated, his mind focused, Newton started thinking. And then, single-handedly, he moved the entire world of science forward. Like the Ancients, Newton was a very good observer. But unlike the Ancients, thankfully so, Newton was also an excellent experimentalist, capable of creating experiments to intentionally probe his observations. Adding to this mix were his skills in mathematics that provided him the courage to wade into complex calculations, seeking exactness, refusing to turn away from what he could not explain. During his isolation (1665–1666) and with these skills and his bold imagination, Newton began uncovering the logic and laws of many different aspects of nature, including planetary motion. His many resulting accomplishments gained the year 1666 great fame, so great that the year itself became known as *Annus Mirabilis*, Miraculous Year. In a rather humbling aside, Newton was all of 24 years old at the time. What did each of us accomplish by 24?

> Annus Mirabilis
> Highlights

Below are select highlights and other relevant discussion points stemming from Newton's *Annus Mirabilis*. The insights and discoveries he generated during that year would later provide him with the necessary tools to fully develop the Laws of Motion and Universal Gravitation, something he was not in a position to do in 1666. It would take until 1687 for his thinking, expertise, knowledge, and experience to mature before he could create these laws and so create the *Principia*.

Gravity and Action at a Distance

While Descartes' work strongly influenced Newton, it wasn't a blinding influence. The same intelligence that led Newton to see the faults in Aristotle started to lead him to see the faults in Descartes. Newton started willingly rejecting aspects of Descartes' mechanical philosophy

---

[7] (Wordsworth, 1850) Lines 58–63. "And from my pillow, looking forth by light/Of moon or favouring stars, I could behold/The antechapel where the statue stood/Of Newton with his prism and silent face/The marble index of a mind for ever/Voyaging through strange seas of Thought, alone."

[8] My opinion: The tendency in education to involve students in team learning seems at odds with the requirements of the human mind to achieve insight. Team learning, either in the classroom or on social media, prevents sustained focus, without which the mind cannot follow threads of ideas to logical conclusions and insights. When others are present, distraction results and the thread is lost. Isolation provides focus. Also, one can easily hide in a group.

that didn't make sense to him. Especially relevant here was Descartes' strict adherence to matter-in-motion as the cause of *all* phenomena. In Descartes' world, action-at-a-distance was impossible. To Newton, this led to "absurd consequences"[9] involving the invention of imagined mechanisms to explain, for example, the attraction between matter that Newton directly observed in his studies involving the cohesion of bodies, capillary action, and surface tension.

But Newton didn't totally reject the mechanical model. Indeed, when all was said and done, he remained a mechanical philosopher, but just not the one defined by Descartes.[10] His full acceptance of Gassendi's concept of atomism[11] led him to continue to see the world as matter-in-motion. But to Newton, matter-in-motion was only part of the story. The other part had to account for the invisible mechanisms that caused attraction between pieces of matter. During his year of isolation, Newton slowly developed a theory of force involving action-at-a-distance to account for these mechanisms and so shifted from a mechanical philosophy to a dynamic mechanical philosophy in which an external force, a new and abstract concept that he created, became the cause of changes in motion. But, as we'll discuss shortly, the final step to this revolutionary thinking would not truly crystallize until 1679 as a result of his communications with Robert Hooke.

Newton's development of action-at-a-distance was enhanced by another powerful skill, his ability to see links between different phenomena, similar in a way to Galileo's ability to see opportunities at the interface between different disciplines. The apple falls to the Earth. The Moon orbits the Earth. Perhaps the same force is at work behind both? As Newton later recalled, "I began to think of gravity extending to the orb of the Moon."[12] Alas, whether a falling apple truly inspired this tremendous insight is unknown, but it continues to make for fun speculation.[13]

The Rise of Calculus

In addition to achieving this first insight into gravity, Newton also created a whole new approach to mathematics, his first intellectual passion. During this year, Newton taught himself most all of the prior work in math, and then, although he didn't know it at the time, pushed past the frontier, surpassed the Greeks, and became one of the world's leading mathematicians. Out of this effort, calculus was born, thus solving a need to quantify time-varying parameters, such as accelerating velocity and instantaneous rates-of-change, and also thus contributing to what some consider to be the most creative age in mathematics.

[9] (Westfall, 1971) p. 337.
[10] (Westfall, 1971) Chapter 7, "Newton and the Concept of Force."
[11] (Westfall, 1971) Chapter 3, "Mechanics in the Mid-Century." A contemporary of Descartes', Pierre Gassendi (1592–1655) was a French philosopher who revived the atomistic theory of the Ancients—especially Epicurus and Lucretius—involving nature as particles of matter in motion.
[12] (Gleick, 2003) p. 55.
[13] (Stukeley, 1936) p. 20. William Stukeley recorded that after dinner on 15 April 1726, Newton wondered aloud, "Why should that apple always descend perpendicularly to the ground?... Why should it not go sideways or upwards, but constantly to the Earth's centre?"

The Privacy of His Thoughts

A man must either resolve to point out nothing new or to become a slave to defend it
– Sir Isaac Newton[14]

Amazingly and unfortunately, even though Newton captured many of his thoughts and ideas from this year in his own private notebooks, he didn't publish any of it. Seeing no reason to tell anyone what he had discovered, he wrote mostly for himself, leaving others to discover his discoveries on their own. His reluctance to subject himself to criticism even led him to intentionally write the *Principia* in a tightly worded, hard-to-read "abstruse" style, catering only to the most capable mathematicians, and so "avoid being baited by little smatterers in mathematics."[15]

The Dispute with Leibniz

Newton's unpublished discoveries from this period would later lead to the aforementioned conflicts as others began making the same discoveries. Such a situation manifested itself most prominently in Newton's very public disputes with Leibniz regarding the discovery of calculus. Newton invented calculus during 1665–66 but didn't publish it. Leibniz invented his own version of calculus around 1675 and *did* publish it (1684–86). So Newton had priority of discovery while Leibniz was first to publish. You can see the origin of the battle over ownership that subsequently ensued. I won't go further into this here other than to say that while Newton may have won the public battle of opinion, at least in England, Leibniz won the war as his approach to calculus proved much easier to use and was ultimately embraced by scientists. Today we study calculus using Leibniz's symbols and terminology.

The Path to the Principia

After the *Annus Mirabilis*, Newton returned to Cambridge, became Professor of Mathematics in 1669, and delved heavily into a broad range of pursuits, including theology, alchemy, optics, and light. All the while, in the background, a clock was ticking. Unfinished business lay on the table. The genius of his *Annus Mirabilis* work sat idle, a tool waiting to be used.

Robert Hooke and the Challenge of Circular Motion

The most difficult subjects can be explained to the most slow-witted man if he has not formed any idea of them already; but the simplest thing cannot be made clear to the most intelligent

---

[14] Newton quote cited in (Chandrasekhar, 1998) p. 42.
[15] ('Espinasse, 1956) p. 33.

man if he is firmly persuaded that he knows already, without a shadow of doubt, what is laid before him.

– Leo Tolstoy[16]

There exists an infinite and omnipresent spirit in which matter is moved according to mathematical laws

– Sir Isaac Newton[17]

Motion and especially *change* in motion, thanks to Galileo's work, remained the central focus in science during the seventeenth century, and the need to resolve these concepts, especially as they pertained to planetary motion, energized many coffeehouse discussions. Such discussions and communications later confused historians' attempts to assign original ownership to the laws of motion and gravity. Newton's own reluctance to publish naturally didn't help the situation. While he achieved certain breakthrough insights behind these laws as far back as his *Annus Mirabilis*, he didn't finalize them and thus couldn't publish them, as he still had learning to do.

Some claimed that Newton's Laws of Motion and Universal Gravitation should not be solely attributed to him. Clearly, Newton was involved in communications and idea sharing with the critical players. So one could indeed dispute original ownership of the laws.

In this regard, Robert Hooke played a prominent role, for it was he who caused Newton's most significant advancement in thinking. Prior to 1679, Newton was clearly torn between two worlds. He believed in the mechanical philosophy but found it significantly lacking when it came to reconciling certain phenomena he saw in nature. He was ready for a new paradigm. It was Hooke who delivered it.

The issue at hand concerned circular motion, action-at-a-distance, and the relation between the two. Circular motion, which appeared so perfectly natural in Aristotle's world, became a significant puzzle in the mechanical world. As inspired by Descartes' pioneering work, many, including Newton, incorrectly viewed circular motion as an equilibrium, which suggests *non-acceleration*, between an internal "*centrifugal*" force that a circling body exerts *away* from the center and a gravitational force directed toward the center and caused by multiple impacts of invisible bodies. This deep belief in a purely mechanical philosophy became an obstacle for progress.

In his famous exchange of letters with Newton in 1679, Hooke, refusing to be limited by the mechanical philosophy's absolute rejection of attraction, corrected this mistake and properly framed the issue. He proposed that planetary orbits follow motions caused by a central attraction continually diverting a straight-line inertial motion into a closed orbit. To Hooke, a single unbalanced force was at work in circular motion, specifically an *inverse-square attraction* of the Sun for the planets. Frustrated with mechanical philosophy's inability to provide solutions to such problems, Newton immediately latched onto this concept as the critical missing piece to his mechanical philosophy and adopted the concept of attractive and repulsive forces as being at cause for *any* change in a body's straight-line inertial motion, especially change that results

---

[16] (Tolstoy, 1894).
[17] Newton quote cited in (Westfall, 1996) p. 509.

in circular motion. With this Newton completed his shift from Descartes' mechanical philosophy to his own dynamic mechanical philosophy.

Without the insight provided by Hooke, Newton's *Principia* would have been impossible.[18] The crux of the insight was not, as many may think, the famous inverse-square relationship—which Hooke later claimed that Newton had plagiarized, but which, unbeknownst to Hooke, Newton had actually derived (but not published) earlier—but the more subtle clarification of circular motion.

1679 – Newton Conquers Circular Motion

It was Hooke who came along and sliced away the confusion by "exposing the basic *dynamic* factors [of circular motion] with striking clarity."[19] It was Hooke who corrected the misconceptions about circular motion. It was Hooke who properly framed the problem. It was Hooke's conceptual insight and mentoring that "freed Newton from the inhibiting concept of an equilibrium in circular motion."[20] With his newfound clarity, Newton let go of the concept of the fictitious centrifugal force, embraced the concept of his newly created *centripetal* gravitational force pulling *toward* the center, and so changed the world. The year 1679 was a crucial turning point in Newton's intellectual life. And Hooke was the cause.

Why not Hooke?

Given all this, why Newton and not Hooke? Why didn't Hooke author the *Principia*? Why all the acclaim to Newton? The short answer, according to Westfall,[21] is that the bright idea is overrated when compared to the demonstrated theory. While Hooke did indeed provide a critical insight to Newton, the overall problem of motion, including a fundamental understanding of Universal Gravitation, remained unsolved, and the reason was that no one, including Hooke, could work out the math. Without the accompanying mathematics and quantified validations, all such ideas represented mere conjecture. The fact that Halley recognized Newton's genius and so travelled to him in 1684 to call him forward to solve this problem clearly indicates that the puzzle was far from complete.

Thank goodness for Halley! He is one of the heroes in this story. What would have happened had he not been present? He was responsible for lighting the fire within Newton toward solving this problem.[22] And he did it with skill, approaching Newton with ego-stroking flattery as reflected by his direct request in 1687: "You will do your self the honour of perfecting scientifically what all past ages have but blindly groped after."[23]

---

[18] (Westfall, 1971) p. 433.
[19] (Westfall, 1971) p. 426.
[20] (Westfall, 1971) p. 433.
[21] (Westfall, 1971) p. 428.
[22] The acrimony that rose between Robert Hooke and Newton over priority as regards gravity's inverse-square law also inadvertently helped fan the flames of Newton's desire to set the record straight through the *Principia*.
[23] (Gleick, 2003) p. 129.

And so the furnace was lit. Through Halley's gentle but firm push, Newton shifted his intense focus away from his other pursuits and toward the cosmos. Newton drew upon his volume of unpublished *Annus Mirabilis* work and his later Hooke-inspired insights and pursued the answer. Slowly. Methodically. "A fever possessed [Newton], like none since the plague years."[24] And when it was all done, the *Principia* was born.

Clearly, no Hooke, no Halley, no *Principia*. But even more clearly, no Newton, no *Principia*.

The *Principia* (1687) – The Rise of Force

As he wrote, computed, and wrote more, he saw the pins of a cosmic lock tumbling into place, one by one.

– James Gleick[25]

[Newton wrote the Principia] with a speed and a coherence unparalleled in the intellectual history of man

– S. Chandrasekhar[26]

The Principia is written in a style of glacial remoteness which makes no concessions to his readers

– S. Chandrasekhar[27]

It's interesting to follow the logical path Newton took in his famous *Philosophiae Naturalis Principia Mathematica*, since shortened to the *Principia*. Published in 1687 by the 45-yr-old and eventually hailed by the scientific community as a masterpiece, the *Principia* did indeed follow a path, showing that Newton's breakthrough work was not based on a flash of insight but was instead based on a painstakingly methodical, intentional, and mathematically intense process, one that, in keeping with the spirit of this book, is valuable history to understand. Insight doesn't just happen; it comes when hard work combines with a prepared mind. With Newton as the Master Builder, each step in the *Principia* became part of the foundation of a new physics of motion that we now call classical mechanics.

The heart of Newton's *Principia*, and, in fact, the heart of Newton's contribution to science, was his newly created concept of force and its central role in a quantitative dynamics.[28] Prior to Newton, force had been used in a variety of descriptions, including but certainly not limited to Galileo's use in the discussion of the simple machines concerning the "forza" applied to one end of a lever, and Descartes' use to describe the "force" of a body's motion, which he quantified as mass times velocity. From this Cartesian perspective, which Newton initially embraced, force referred to an inherent property of the body.

But these differing definitions eventually frustrated Newton, leading him to take a seemingly simple but actually very revolutionary step. Starting from his study of impact between

---

[24] (Gleick, 2003) p. 124.
[25] (Gleick, 2003) p. 123.
[26] (Chandrasekhar, 1998) p. ix.
[27] (Chandrasekhar, 1998) p. 45.
[28] (Westfall, 1971) p. 466.

two bodies, a common enough phenomenon to study in the seventeenth century, especially for the mechanical philosophers, Newton realized that the force of a body's motion could be seen from another perspective. The force of one body's motion functions as the "external cause" that can alter another body's inertial state of rest or straight-line uniform motion. He realized that each body involved in a collision could be viewed as the external cause of the other's change in inertial motion. In this way, he thus shifted the perspective of force from an inherent property to an external cause. If Body 1 strikes Body 2, the force of Body 1 causes a change in Body 2's motion. He then assigned this change in Body 2's motion to a force external to Body 2. In this way, Newton created a simple, clear, and yet rather abstract and wholly new definition of force. Force is whatever it is that causes a change in motion as quantified by that change in motion it causes. "[Newton] was the first man fully to comprehend the implication of inertia for dynamics."[29] If an inertial state changes, a force must be present.

While Newton initially used force to quantify the change in motion resulting from impact, he later used it, or I should say, he later profoundly used it, to quantify the *rate* of change in motion resulting from the continuously acting gravity, the complexity of which ultimately drove his invention of calculus. The units between these two definitions are naturally different, the former being "mv" and the latter "d (mv) / dt" or simply "ma." But Newton did not dwell on such distinctions. He used both definitions correctly and interchangeably throughout the *Principia*, especially in the problems he solved. I won't delve into this any further, and as Westfall stated,[30] "it is fruitless to torture individual passages for the exact implications of their words," but if you are interested in learning more about such fascinating distinctions, I strongly recommend Westfall's *Force in Newton's Physics*.[31]

Newton melded this new and precise definition of force, which allowed for the possibility of action-at-a-distance since force is *whatever* causes a change in motion, together with his powerful mathematical tools to create an entirely new dynamic mechanics. All sorts of very difficult and different problems regarding motion, and especially those involving attraction, could now be solved. But where to start? The solution of what problem would best embody or manifest the power of Newton's newly created tools and best commence discussion in the *Principia* of the Laws of Motion and Universal Gravitation? Here Newton returned to circular motion.

As already discussed, circular motion was one of the major stumbling blocks toward achieving a sound dynamic mechanics. The mechanical philosophers just couldn't grasp the concepts involved. Hooke grasped the concepts but couldn't do the math. He then effectively pointed Newton down the right path. As the pieces quickly fell together in Newton's mind, he picked up his new tools and went to work.

Circular motion was a perfect starting point for Newton in the *Principia*, beautifully revealing one key aspect of Newton's philosophical approach to science: simplicity. The problem was simple enough to solve but complex enough to demonstrate Newton's new dynamics. Newton fundamentally believed in God and fundamentally believed simplicity to be at the heart of God's creations, however complex they might appear. He felt that all of the different ideas and

---

[29] (Westfall, 1971) p. 344.
[30] (Westfall, 1971) p. 299.
[31] (Westfall, 1971)

theories about motion being offered at that time were overly complex, especially when he saw a new and different cause arising *ad hoc* for each new and different effect. He sought clarity and insight into the complex reality by starting with simplifying assumptions, an approach that we're taught today, believing such a starting point to be ultimately aligned with the simplicity of the yet-to-be-determined end point. His philosophy mirrored Galileo's own belief that all natural phenomena must follow mathematical behavior. That Newton ended up simplifying the many bewildering complexities of terrestrial and celestial motion into a single small set of governing equations is an amazing testament to this philosophy.

In keeping with the simplicity theme, he did not start his analysis of circular motion by look-ing at the planets. Instead he turned to his own observations of two simple systems: a weight attached to the end of a rope and swung in a circle, and a ball rolling around in a circle inside a bowl. In each system, continuing along the path blazed by Hooke, Newton identified the cause of the circular motion. For the rope, it was the man pulling inward. For the ball, it was the bowl pushing inward. In both cases, the uniform circular motion resulted from a constant force directed toward the center of the circle. Without the force, the object would fly off in a straight line. And so evolved the first law.

Newton's 1st Law of Motion

Every body perseveres in its state of being at rest or of moving uniformly straight forward, except insofar as it is compelled to change its state by forces impressed.

While the general concept behind this law had been in print for years before *Principia*, stretch-ing back to Galileo's law of inertia and even earlier, three distinctions are important to high-light for our understanding of the paradigms that existed back then. First, a body at rest is treated as being similar to a body in uniform motion, the former is really but one possibility within the range of the latter, a paradigm-shifting concept that would eventually lead the way to Einstein's relativity theories. Aristotle had earlier proposed that a body in motion contains something keeping it in motion, a thought process that continued for many hundreds of years. Even Kepler never grasped that motion was possible in the absence of force. This law put such incorrect thinking to rest.

Second, Galileo assumed in his own version of this law that uniform horizontal motion implied uniform circular motion around the planet, since he recognized that any horizontal motion on Earth would have to follow the Earth's circular path. But Newton and others saw this differently, clearly identifying straight-line motion as the defining inherent feature of motion. In this way he thus identified circular motion as containing a hidden, previously unrecognized, accelerating component. While a body could travel at the same speed for both straight-line and circular motions, Newton clearly showed that the two are not the same. Some force must be present to cause motion to deviate from the former to the latter. Release this force and the latter returns to the former.

While the first two distinctions were each in print prior to Newton, the third distinction belongs to Newton. The real critical and differentiating concept that Newton introduced into his 1st Law concerned "impressed forces." In earlier formulations of this law by others, deviation from straight-line motion implied, according to the mechanical philosophers, the existence of a physical impediment. The more abstract approach used by Newton based on "impressed forces" opened the door for other types of forces, such as gravity and its implied action-at-a-distance as inspired by Hooke.

Newton next began developing the concept of this "impressed force" by seeking to quantify exactly how it changes motion. Here, Newton took an approach that mirrored Galileo's. Galileo explained parabolic trajectory by decomposing it into horizontal and vertical components, the former being constant speed and the latter being constant acceleration. Newton similarly explained uniform circular motion. By shrinking distances and using the calculus of limits and differentiation, which effectively enabled him to replace complicated curvature with simpler linearity, Newton was able to show that the instantaneous motion of a body moving in a circle is also comprised of both horizontal (tangential) and vertical components. He then isolated the vertical component and so was able to calculate the rate of vertical fall required to convert motion from a straight-line to a circle.

The Inadequacy of Language

Half the problem of dynamics was the ambiguity of words
– Richard Westfall[32]

At this point, the inadequacy of language hampered progress. Building on his study of impact, in which he recognized that for certain impacts involving a change in direction, quantity of motion is not conserved, Newton recognized that for a given body moving at a given speed, the critical difference between straight-line and circular motions is direction. For the former, direction is constant. For the latter, direction constantly and uniformly changes. But at that time in history, both speed and acceleration were scalar quantities and didn't account for changing direction. To Descartes, a change in direction didn't really require any action, a surprising assertion given that Descartes also recognized the importance of the concept of an inertial state. So Newton newly defined the term velocity as a vector containing both magnitude (speed) and direction. He also expanded the concept of acceleration. Originally a scalar, acceleration became a vector equal to the rate of change in velocity. Thus, a change in *either* speed *or* direction would be quantified by the concept of acceleration. While such definitions seem simple in hindsight, they represented a revolutionary step at the time, one that made possible all that followed in the science of dynamics.

Newton next shifted his focus back to the original problem, planetary motion. If the planets are not moving in a straight line but are instead moving in circular motion, then they are accelerating. If they are accelerating, then a force must be present. He didn't speculate as to the nature of the force. He only said that a force must be present that is causing the planets to deviate from a straight line. And then, since the effective center of the planets' circular motions is the Sun, he concluded that it must be the Sun that's exerting the (attractive) force on the planets.

---

[32] (Westfall, 1971) p. 179.

Newton's Early Insights into Universal Gravitation

Building on his early insight into gravity during his *Annus Mirabilis* and the added insight provided by Hooke in 1679, Newton then proceeded using the following logic: "To the same natural effects we must, as far as possible, assign the same causes."[33] Sound or not, such logic led Newton once again to the concept of universal gravity. The planets circle the Sun; the Sun exerts an attractive force on the planets. The Moon circles the Earth; the Earth exerts an attractive force on the Moon. And finally, a free-falling apple (yes, Newton's famous apple!) accelerates toward the Earth; the Earth exerts an attractive force on the apple. In a stroke of genius, seeking links where others didn't, Newton asked, as he originally did back in the *Annus Mirabilis, are all of these attractive forces one and the same?! Are they universal?* It may seem obvious now, but it wasn't at all obvious back then that such gravitation could depend *solely* on mass and not, for example, on the varying "attractions" observed in Newton's chemical studies involving the cohesion of bodies, capillary action, and surface tension.

The only way to answer this question was to perform the calculations, Newton's trademark approach. And so that's what he eventually did. But before he could do this to the accuracy he desired, Newton first had to attain a better understanding of the relationship between force and acceleration by completing his laws of motion. Otherwise, he had no equations to work with.

Newton discovered in his 1st Law that an abstract push or pull force is needed to change a body's velocity, i.e., to accelerate the body. But what was the relationship between the two? To help clarify this line of questioning, Newton decided first to quantify the relation between force and acceleration. Where to even start? Back to Galileo!

In his inclined plane studies, Galileo varied the slope of the incline to slow down the speed of the ball. As we know today, what Galileo was truly doing and not really realizing was that he was changing the force on the ball. The force on the ball in the direction of the roll is simply the component of its weight in the direction of the incline. Newton analyzed Galileo's finding and concluded that the time of the ball's descent varied in direct proportion to force in the direction of motion. Newton used this finding together with other supporting studies involving pendulums as his starting point: acceleration is directly proportional to force.

In this next step, Newton struggled. The above starting point was good for a given body. But what happens when different bodies are used? There must be some intrinsic aspect of the body that creates the force and so causes its acceleration. Something like weight, but not weight, since Newton recognized that weight itself is not an intrinsic property of a body. His earlier work showed that, for example, a body's weight decreases as it moves further away from Earth. So the body must interact with the Earth to result in weight. But what property of the body causes this interaction? This line of questioning opened up a completely new territory in physics.

Mass – a New Concept

The concept of mass (as distinct from weight) had not yet been defined by anyone. The Greeks dabbled in this area, suggesting that all matter consists of an intrinsic property of either

[33] (Newton, 2017) p. 160.

"heaviness" or "lightness." But it took until 1643 for progress to jump forward when Evangelista Torricelli explained why a pump could not lift up water more than 34 feet: a pump acts by creating a vacuum, and it's the weight of the air at sea level that pushes the water up into the pipe on the suction-side of the pump. The conclusion that air has weight was a significant breakthrough since it disproved the Greeks' belief in the existence of any such property as absolute "lightness." All matter is heavy, some is just heavier than others. Another victory for the scientific revolution.

But this led to the next question: What does "heaviness" mean? Typically, heaviness was measured as weight, i.e., the magnitude of the downward pull on a body. But, as discussed above, Newton showed weight to be variable. And why would weight as a downward pull be an important variable in impact studies involving bodies moving horizontally? So in yet another stroke of genius, Newton created a new term, an intrinsic property of matter that he called "quantity of matter" or "mass." He then defined mass as the product of volume and density for an incompressible body. Since density measurements at that time were based on specific gravity, i.e., a ratio of absolute densities, they represented an intrinsic property not influenced by gravitational force like weight is. In Newton's mind, for a given amount of matter, mass remains constant while weight and volume can change. Raise a body relative to Earth, and its weight decreases but its mass remains constant. Increase the density of a body, such as by compacting snow, and its volume decreases but its mass remains constant. In all cases, the body's mass remains constant.

By creating this concept, even though Ernst Mach would later justifiably criticize the above logic as being circular,[34] Newton effectively separated mass from weight and made mass a truly intrinsic property. And more importantly, although Newton didn't explicitly spell this out, mass also became a conserved quantity. Newton inherently assumed total mass to equal the sum of the masses of its building blocks, and further assumed that such building blocks are neither created nor destroyed. Lavoisier later put these assumptions into law. Some consider the "conservation of mass" to be Newton's zeroth law of motion.[35]

Newton next addressed how mass impacts the relation between force and acceleration. Recall that Galileo either demonstrated at the Tower of Pisa or concluded based on others' work and his own thinking that the rate of fall is independent of the weight of the body. If the rates of fall are equivalent, i.e., if the accelerations are equivalent, then what does this say? If you increase the size of a ball, then you increase both its weight (force) and its mass by the identical ratio. If acceleration doesn't change as you increase the size of the ball, and if acceleration is directly proportional to force, then acceleration must also be inversely proportional to mass. In other words, a = F / m, which brought Newton to his 2nd Law of Motion.

Newton's 2nd Law of Motion – "Soul of Classical Mechanics"

A change in motion is proportional to the motive force impressed and takes place along the straight line in which that force is impressed.

[34] (Westfall, 1971) p. 448.
[35] (Wilczek, 2004a) p. 11.

Right now you're probably not understanding the above formal statement, and you'd be correct in thinking so. I too didn't follow this logic because the 2nd Law as Newton wrote it is not exactly equivalent to the famous F = ma.

Newton spent time in the *Principia* creating the concept of "quantity of motion," which today we call momentum, as the product of mass and velocity (vector). His inspiration here was likely Descartes who had defined "quantity of motion" as weight times velocity (scalar). As will be discussed in Chapter 10, Descartes was, in turn, likely inspired by Aristotle's earlier analyses of the lever in which the quantity of weight times "virtual velocity" was isolated as the measure of "force" at each end of the lever, thus signifying the continuing and misleading role of the (static) lever in the study of a dynamic phenomenon.

Although Newton, while not stating so, used his newly defined "quantity of motion" in each of his three laws, he still wrote "motion" and not "quantity of motion" in his 2nd Law. Additionally confusing this situation, he used two different definitions of motion: the change in motion (based on impact) and the rate of change of motion (based on gravity). The important point to recognize here is this. It is in his examples that Newton shows what force truly is. It is in his examples, especially with the continuously acting gravity, that we find the relationship known today as the 2nd Law: F = ma or F = d (mv) / dt. Newton's dynamics are best understood by the problems he solved. It was Euler who would later modernize Newton's 2nd Law using Leibniz's calculus symbols to the form we work with today.

The 2nd Law is considered the "soul of classical mechanics."[36] It represented a tremendous breakthrough that simplified all the confusion over many terms and concepts into one single simple, powerful equation. It brought new meaning to velocity and acceleration by redefining each as vectors dependent on direction. It identified a new concept called mass that is separate from weight. And it (finally!) defined force. With this equation, one could finally quantify the force causing a known motion or the motion resulting from a known force. This proved to be an extremely powerful advance in mechanics.

Newton Turns to Experimentation

Newton was still not finished. One significant question remained. His earlier thoughts on gravity were based on a smaller body being attracted to a larger body, such as the apple to the Earth, or the Earth to the Sun. But what does this mean about the larger body? Is it attracted to the smaller one? Is the Earth attracted to the apple?

Newton resolved this the best way he knew how: experimentation. To do so, he employed a rather ingenious approach involving swinging pendulum bobs to study the forces that occur between two colliding bodies. He varied the mass of the bobs, their composition, e.g., steel, cork, and even tightly wound wool, and their initial amplitudes, and then measured their final amplitudes after the collision. Based on Galileo's earlier work with swinging bobs, Newton knew how to correlate amplitude to speed. The initial amplitude translates into the speed at the bottom of the swing just prior to impact while the final amplitude translates into the speed just after impact.

[36] (Wilczek, 2004a) p. 11.

What Newton found was this: the mass of the first bob multiplied by its change in velocity was equal to the mass of the second bob multiplied by its change in velocity. In other words, the change in momentum of the first bob equaled the change in momentum of the second. Since, per the 2nd Law, the force exerted on each bob at the moment of impact is equal to the change in its momentum, Newton had proven that the bobs exert forces on each other that are equal in magnitude and opposite in direction.

He then extended this thinking to non-contact forces. Place a magnet and a block of iron on a boat. The magnet pulls the iron and yet the boat doesn't move. Why? The iron must be equally pulling the magnet. Newton showed that this worked for a full range of forces: gravitational, magnetic, elastic and inelastic collisions. He found no exceptions and so called this his 3rd Law.

Newton's 3rd Law of Motion

> To any action there is always an opposite and equal reaction; in other words, the actions of two bodies upon each other are always equal and always opposite each other.

In addition to modernizing Cartesian collision theory by bringing much needed quantifiable mathematics to its study, Newton's work demonstrated that conservation of momentum applies to both elastic and inelastic collisions, a finding that can also be derived from his 2nd Law, showing just how interconnected his laws are: if there is no external force on a system of two or more colliding particles, then the change in the total momentum of the particles is zero and so the total momentum of the particles remains constant, even if the collisions are inelastic. The application to a system of more than two colliding particles is made possible since Newton showed that forces are additive.

Force – a New Concept

The development of the concept of force by Newton in his laws of motion was a significant step forward in the history of science. When Newton first began writing, neither he nor anyone else really knew exactly what force was. Some had used the word, or its German counterpart *Kraft*, to define a concept that would later evolve into *energy*. Newton cleared away the confusion. By defining what force is, he thereby defined what force isn't. He separated force as a concept distinct from the yet-to-be-defined energy.

Prior to Newton, the concept of force had grown naturally in colloquial language in response to our need to describe an everyday fact of life: we must exert effort by pushing or pulling or lifting or holding up objects. We feel this exertion. While the definition was loose, the concept stuck as it captured our physical sensory experience. This rock is heavy; we'll need a lot of force to move it. The definition was extended to quantify some property of a body's motion, ranging from Descartes' mass times velocity to Leibniz's mass times velocity squared.

In the 2nd Law, Newton finally provided an equation that both defined and quantified force, and in so doing raised the concept of force to a much higher and more abstract level. Force is whatever it is that causes a change in momentum, and it is equal to the rate of this change. This shifted force from the historical "static" concept associated with the lever to a new "dynamic" concept, a term coined by Leibniz in the 1690s. If a body is in its inertial state, meaning that it is either not moving or moving in a uniform straight line, then the forces acting on that body must be exactly balanced. However, if the body changes its inertial state and thus accelerates, then a force (or unbalanced forces) must be present. The force could be discrete, resulting from direct contact between two bodies, or it could be continuous, resulting from action-at-a-distance, e.g., gravity. Newton didn't specify, instead choosing to rather ingeniously defuse the situation by separating force away from the mechanism by which it acts. And then he provided the equations to quantify force. His equations could work for any geometry and any system, including one that he foresaw: the electromagnetic force. In the *Principia*, Newton speculated about such a force, not by name of course, as being that which coheres matter together.

The concise simplicity of the three Laws of Motion is quite deceptive. Newton incorporated new and very advanced supporting concepts, such as mass, momentum, and a vector-based approach to velocity and acceleration, into these laws—concepts that we continue to use to this day. They represent the consolidation of a vast array of both terrestrial and celestial data. Together with calculus they provide a means to solve most all non-quantum problems associated with motion in nature. I recall my university physics class in which the professor led us through the solution of each problem by first starting with one of Newton's Laws of Motion and deriving the solution from this. He would light his cigarette, take a long drag, write F = ma on the board, and then start deriving the answer from there. This is some 300 years after Newton discovered the laws. Talk about withstanding the tests of time.

But, for Newton, these laws were simply a starting point, a means to an end. These laws enabled him to analyze motion and thus quantify the forces causing the motion. From this analysis, Newton could finalize his theory of Universal Gravitation.

Newton's Law of Universal Gravitation

For all the difficulty of philosophy seems to consist in this—from the phenomena of motions to investigate the forces of nature, and then from these forces to demonstrate the other phenomena
– Sir Isaac Newton[37]

Both Joseph-Louis Lagrange and Pierre-Simon, Marquis de Laplace regretted that there was only one fundamental law of the universe, the law of universal gravitation, and that Newton had lived before them, foreclosing them from the glory of its discovery
– I. Bernard Cohen and Richard S. Westfall[38]

With his new tools in hand, Newton returned to his calculations relating the Earth's attraction for the Moon to the Earth's attraction for the legendary apple. Building on his *Annus Mirabilis* insights, he first sought to explain how gravitational force changes as a function of distance. His initial thoughts suggested that a rapid decrease in gravitational attraction with distance

---

[37] (Newton, 2006) p. x.
[38] (Cohen and Westfall, 1995) p. xiv–xv.

must occur in order to explain the differences between the "fall" of the Moon and the fall of the apple. But what was the exact relationship and how could he prove it?

The first step was to acquire data. He knew the distance of the Earth to the apple since he knew the radius of the Earth. And he knew the distance from the Earth's surface to the Moon based on the work of the ancient Greeks with their mathematical tools. Based on his newfound insight into circular motion, he calculated the acceleration of the Moon toward the Earth. Finally, based on his analysis of Kepler's laws, he proposed that gravitational attraction must be proportional to the inverse-square of distance, a proposal consistent with his simplicity philosophy and also consistent perhaps with his earlier thinking that gravitational intensity might fall off with distance as does the brightness of light; it decreases with the increasing area of the sphere through which it passes. Since the area increases with r^2, then the intensity must decrease with $1/r^2$.

When Newton finished this calculation for the Moon's acceleration and then multiplied it by $(60)^2$, 60:1 being the ratio of the distances of the Moon and the apple to the Earth's center, he arrived at an estimate of the acceleration of the apple at the Earth's surface of 32.2 ft/sec², a value in excellent agreement with Huygens' independent measurement (32.2 ft/sec²) of acceleration at the Earth's surface using pendulums.

The implication was stunning. The Moon orbits for the same reason the apple falls. By identifying that the same cause attracts both Moon and apple, Newton united the Heavens and the Earth. He did not speculate. He did the math. He proposed that Earth exerts a gravitational attraction that decreases with $1/r^2$. His quantification was his proof. The numbers fell into place. Like Lagrange and Laplace, I would have loved to have been the one to do this calculation and see it work.

But the law still wasn't complete.

"A prudent question is one-half of wisdom" – Francis Bacon

Newton then asked the insightfully penetrating question, a question you may be asking yourself: Why is it possible to use the Earth's radius as the distance from the Earth to the apple in the above exercise? This implies that the spherical Earth attracts as if all of its mass is at one single point, the center. In an exercise that many of us repeated in university physics, when you work through the math as Newton did, using the inverse-square relationship and assuming that each infinitesimal part of the Earth gravitationally attracts a body laying anywhere beyond the surface and that the total attraction force is simply the sum of these parts, which implies that total gravitational force is proportional to mass, you find that this is indeed the case.[39] You find that you can use the center of mass in such gravity calculations, effectively shifting (and tremendously simplifying) future calculations from a large body to a single point. The fact that this calculation explains the data strongly supports the starting assumption: the universal gravitational force is proportional to mass. Another pin in the lock tumbled into place.

---

[39] You must also assume that the Earth is a sphere of uniform density.

Taking this logic one step further, if the attractive force of the Earth is proportional to its mass, then the attractive force of any body, terrestrial or celestial—it was becoming apparent that there is no difference, meaning that mass is mass regardless of type, thus further supporting the use of the word "universal"—is also proportional to mass since Newton proved that gravity on Earth is the same as gravity in the skies. Furthermore, based on his 3rd Law, all bodies, large (Earth) and small (apple), must possess this same attractive force since the attractive force of the Earth to the apple must equal the attractive force of the apple to the Earth. Everything attracts everything else. This ultimately meant that the mass of each body must be included in the final equation. Newton thus arrived at Equation 5.10, the complete law of Universal Gravitation:

$$F = G \, m_1 m_2 \, / \, r^2 \qquad\qquad\qquad [5.10]$$

where G became known as the gravitational constant or Newton's constant. Newton didn't actually write this equation as such, instead preferring to write in terms of proportions, similar to Galileo's use of the Euclid–Eudoxus law of proportions. The addition of the constant, G, was all that was required to form the equation. It would take until 1798 for G to be indirectly measured in torsion experiments by Henry Cavendish and then again until around 1873 for G to be more directly calculated from Cavendish's data. Cavendish's original goal was to measure the density of the Earth with his equipment; he never actually calculated G. His data, though, provided the eventual answer, which is an equally interesting history of events by itself.

"The Principia was a Tour-de-Force Demonstration of the Intelligibility of the Universe"

Newton then began systematically building the case for his theory of Universal Gravitation by using his laws and calculus to analyze most all known motions in the sky. With data he obtained from astronomers, he quantified attractive forces, explained motions, and even explained unusual and unexpected behaviors. He explained nearly everything, including the orbits of the moons of Jupiter, Saturn, and especially Earth. He unified Galileo and Kepler by showing that each of their laws could be explained from the same single set of his own laws.

And in a remarkable breakthrough, he explained the tides. Newton came to understand that just as the Earth attracts the Moon, the Moon attracts the Earth, every single part of the Earth, including the oceans. Because of the decrease in force with distance, the Moon attracts the water on the closest side of Earth with more force than it attracts the solid Earth itself, and similarly attracts the solid Earth with more force than the water on the farthest side of the Earth. Thus, the Moon pulls the nearside water away from the Earth, resulting in a high tide, and also pulls the Earth away from the far-side water, resulting in another high tide. The Moon causes two simultaneous high tides. He explained that the tides remain fixed in place respective to the Moon, but move relative to us since the Earth itself is spinning, generating two tides per day for any seaside location.

Newton also explained why the same side of the Moon always faces the Earth. I didn't know this. Did you? There's a small bulge on the side of the Moon facing Earth. This effectively locks the Moon in place and keeps it from spinning.

He explained that the Earth itself must not be perfectly spherical since the dynamics involved with a spinning globe, for which the surface at the equator moves much faster than the surface at the poles, causes the Earth to bulge at the former and flatten at the latter. Newton used his math to quantify this.

For his grand finale, Newton explained the motion of comets, whose utter unpredictability tagged them as being signs of some deeper mysterious God-based forces. Newton's clear, quantified explanation that comets move in accordance with the laws of motion and gravity, orbiting the Sun just as the planets do, removed the myth. The fact that both comet and planet orbits are determined by the same attraction to the Sun was a major step toward the proof of <u>univer-sal</u> gravitation. Once again, this was not at all clear to Newton prior to his work in the *Principia*. It wasn't clear that matter of the same mass but different composition would exert identical gravitational attraction. It took until 1685 to realize this.

While one can't prove laws, one can provide sufficient evidence to support the declaration of laws. The breadth and depth of the math and measurement in Newton's *Principia* more than provided the necessary evidence, solving a host of problems that had seemed impossible to solve at the beginning of the seventeenth century. The *Principia* itself was the proof. It doesn't get much better than when one can explain both the heavens and the Earth with the same simple set of laws (hence the comment by (Harriman 2010)[40] used as the title to this section).

*　*　*

How difficult it must have been for Newton to extract himself from his own Aristotelian-influenced paradigms in creating the *Principia*. We'll see a similar situation with William Thomson when he extracted himself from the caloric paradigm. It makes one wonder what paradigms we're living in today that we simply don't recognize.

*　*　*

The Principia Broke Physics Away from Philosophy

> *The scientific revolution of the seventeenth century heralded man's coming of age as a rational being.*
>
> – David Harriman[41]

> *For it is to [Newton] who masters our minds by the force of truth, and not to those who enslave them by violence, that we owe our reverence.*
>
> – Voltaire[42]

With the *Principia*, Newton effectively brought the scientific revolution to a close and established the scientific method as *the* new paradigm of science. The Aristotelian approach to motion, the cosmos, and to science in general, the approach that blocked progress for two thousand years, was over. The likes of Copernicus, Kepler, Galileo, Huygens, Leibniz, Hooke, and Newton overthrew the irrational, swept out the mysticism, and left us with reason.

---

[40] (Harriman, 2010) p. 140.
[41] (Harriman, 2000) p. 1.
[42] Voltaire quote cited in (Durant and Little, 2002) p. 24.

The Age of Enlightenment began once Voltaire introduced Newton's mechanics (and Locke's philosophy) to France. As others embraced Newton, amazing progress resulted, like a fire sweeping through a field, burning the old, enabling the new.

Newton's exact mathematical approach enabled a much higher level of discrimination between competing scientific theories. Every systematic discrepancy between observation and theory, no matter how small, could be taken as telling us something important about the world. So powerful was this change, this bringing of reality and exactness and absolutes and determinism to science, that physics broke from philosophy, creating two new disciplines where once there was one. From this time onwards, physics was set on solid ground, explaining the *how* and leaving the *why* for philosophy and religion.

As a society, although we're not always so good about how we go about doing science and making policy decisions, we do embrace Newton as the standard by which work *should* be conducted. We are Newtonians! Let's test our assumptions! Let's test our conclusions! Show me the data!

Epilogue – Completion of the Scientific Revolution

After completing the *Principia*, Newton moved on with his life, preparing second and third editions of the *Principia*, furthering his studies in theology and alchemy, becoming involved with government as Master of the Royal Mint, and continuing to lead science as President of the Royal Society. While he didn't continue to pursue his studies of motion with such passion, others fortunately did. The leading physicists and mathematicians of Europe, including Jean d'Alembert, Charles Coulomb, the Bernoulli family and Leonhard Euler, Lagrange and Laplace, who all read and were deeply inspired if not awed by the *Principia*, eagerly took up, corrected where needed, and extended Newton's work to all aspects of the natural world such that by the end of the eighteenth century, their combined work "left no astronomical problem unsolved."[43] Such efforts further validated Newton's laws of motion and gravity and completed the foundation of classical mechanics.

> *Newton's opponents could not grasp that knowledge is gained by starting with observations and proceeding step by step to the discovery of causes, and eventually to the discovery of fundamental causes. They wished to start with the first causes and deduce the entire science of physics from them. Newton knew that this rationalist method led to the indulgence of fantasy, not to scientific knowledge.*
>
> David Harriman[44]

But as with most any other monumental work, it wasn't as easy as this. Acceptance of Newton's *Principia* was not instantaneous. Opposition ran strong and deep against his work's most critical implication: action-at-a-distance. The glaring absence of a mechanism for gravitational force was deeply disturbing to some, especially the mechanical philosophers such as Huygens and Leibniz. While Newton discovered and quantified Universal Gravitation, he didn't, and

---

[43] (Cohen and Westfall, 1995) p. 67.
[44] (Harriman, 2010) p. 142.

really couldn't, go to the next level down and explain the cause of gravity itself. We're still wrestling with this issue today as physicists attempt to unify Einstein's General Theory of Relativity with the Standard Model.

The mechanical philosophy of matter and motion that dominated the seventeenth century was not of one mind. Of the many possible interpretations, especially regarding the elusive understanding of planetary motion, two dominated. The Continental-based Cartesians, such as Huygens and Leibniz, believed that all motion results from matter in *direct* contact with other matter. Thus, they proposed that the planets orbit the Sun due to an invisible but material "plenum" that spins around the Sun in a vortex, carrying the planets with it. In their mind, since space cannot be empty; rather, *infinitely divisible matter* must extend to fill the entire universe. The ancient atomic approach, on the other hand, as embraced by Gassendi, said that matter consists of *indivisible particles* that move through a void at immense velocity. In this philosophy, space is comprised mostly of a continuous void and only a small amount of matter. Furthermore, as discussed earlier, those atomists in England and especially Robert Hooke, proposed that a central attractive force acts across the void to bind the planets to the Sun and the moons to the planets. The Cartesians strongly rejected the atomists' concept of a void and the implied action-at-a-distance, while the atomists directly rejected the Cartesian's plenum.

Newton created his own philosophy from a mixture of the above. He believed in the Cartesian philosophy of colliding matter and believed this matter to be comprised of atoms even though it was difficult for him to reconcile the discontinuous world of small particles with the continuous world of smooth changes. He also embraced action-at-a-distance, although he did not say so publicly since his scientific approach was to stay away from that which he could not explain. To him, gravity was a mathematical force, not a physical force. He preferred to give no explanation, believing in the existence of gravity but stopping short of speculating to the underlying cause, at least in his published writing.

But while Newton claimed no cause for gravity, he didn't let this stop him, like it did for Huygens and Leibniz, who simply could not accept action-at-a-distance. Newton ingeniously separated the problem into the mathematical and the physical aspects and then worked out the former without having to address the need for a causal mechanism for the latter. He showed that the math worked and so validated the void-based approach of the atomists. The vortex theory, on the other hand, resisted mathematization and simply faded away once the *Principia* was published. That's the nature of paradigms. When others are born and raised in a new paradigm that rings of truth, such as that created by the *Principia*, acceptance is often quite easy. Others raised in the old competing paradigms simply can't let go. Over time, resistance to action-at-a-distance slowly crumbled away as the new scientists entered and the old scientists exited.

Newton's masterpiece of mathematical physics established the foundations of dynamics and the theory of gravity. And yet its reception was mixed. The British generally loved it, but the Continental mathematicians, as you'll see in Chapter 11 on the Bernoulli family, distrusted its physics, initially but unsuccessfully holding out for a revival of the vortex theory of planetary motion, which Newton had refuted. Euler's role would be critical in forwarding the *Principia*'s acceptance. As you'll see, he re-wrote Newton's ideas into the language of the calculus, after which it was much easier for mathematicians to appreciate and extend what Newton had done. While Newton invented calculus, he left it out of the *Principia*, favoring the use of geometric analysis. The challenge involved in understanding this approach, and in fact, the challenge of understanding the *Principia* at all, contributed significantly to its slow acceptance. Euler helped bring much needed clarity to Newton's ideas.

Newton's Laws of Motion and Universal Gravitation would remain the cornerstone of physics for the next two hundred years, until one of their flaws was exposed. In Newton's world, action-at-a-distance occurred instantaneously. It would take until the early 1900s for Einstein to set all of this straight in his theories of relativity that changed our concepts of gravity, space, and time. But this too is for another book. For those of us working largely in the non-quantum, non-relativistic world, Newton's laws remain standing.

Newton's Relevance to Energy

The history of Newton's derivation of the Laws of Motion and Universal Gravitation is both fascinating and worthwhile knowing, especially as an education on the development and progress of our scientific methodology and also as an education on the first discovery of one of the four fundamental interactions in nature. But as you may be asking yourself by now, what does this have to do with energy?

While Newton never really discussed or considered energy and its conservation, such concepts appeared in the *Principia* as a natural consequence of his solution to certain problems. For example, Newton stated the mathematical equivalent of the work–energy equation by showing that the integration of force over distance, which today we define as "work," is equal to the change in the square of velocity, i.e., the kinetic energy, for a body. But unfortunately, as discussed by Westfall,[45] Newton's "lack of interest in any equation except that relating force to change of velocity led him to ignore the significance of the equations he implicitly wrote." The quantity mv^2 simply held no special meaning for him. As we shall see, it was Leibniz who identified this meaning.

But the critical point is that Newton's laws lead directly to the concepts of energy. And this is why Newton is so important to our understanding of energy. He demonstrated how force, distance, mass, and velocity are all related and unknowingly left it to others to use such a structure to ultimately discover and define energy and the conservation laws.[46] Energy came to be viewed through the classical mechanics paradigm created by Newton in the *Principia*. To fully understand and appreciate this, we now return to Aristotle and the simple machines.

---

[45] (Westfall, 1971) p. 490.
[46] (Cardwell, 1971) p. 17. Newton's importance to the science of heat was indirect. He made no authoritative pronouncements about heat.

9 The lever

Give me a place to stand on, and I will move the Earth
– Archimedes, on the lever

There's a set of survival genes in each of us that evolved to put us in motion: to catch food, to build structures, to defend communities. As human beings, these genes compelled us to physically move ourselves and the objects around us from one location to another. Without this tendency to motion, we would have died off as a species. Early in our history, these genes drove us to pick up sticks and stones and use them to build small huts for living and defend ourselves during war. We then started wanting to move larger things, like massive boulders and cut rocks. As we attempted this, we reached the physical limit of what we could do alone or even with a group. And so, spurred by another set of genes, we started inventing tools to help us: strong logs to be used as levers, circular stones to be used as wheels, long roads with gradual inclines to be used to help raise heavy blocks of stone, water wheels to grind grain, and windmills to pump water. Each invention enabled us to do more than we could otherwise do.

Even though the link to this discussion is not obvious, we can't forget fire, that natural phenomenon we harnessed to keep warm, prepare food, process metals, and generate power. There's motion here as well to consider, unseen to our naked eyes.

While our unconscious brain learned through experience how to improve our ability to do work and control the world around us, our conscious brain sought answers. We were able to do things, but didn't know how or why and this bothered us deeply. And so another set of genes representing our innate curiosity led us to think and wonder about all of this. *How is it possible for these tools to work so? How can their performances be quantified? Are the performances linked together in some way?* And finally, *why can't these machines run forever?*

*   *   *

Two paths may be available for a journey. Do I go left or right? Sometimes you don't even know that there's a choice. But in the end, you choose and so set your future. Looking at the history of science, you can see the natural philosophers at the crossroads repeatedly. There were always choices to make, regardless of whether or not they were seen and evident, and each choice led down a separate path. If one particular path was initially successful, or if one scientist was influential enough, then this path influenced succeeding generations, regardless of whether the path was the correct one or not. Sometimes the path was correct, but sometimes it wasn't, and this left us with concepts and words that caused confusion and hindered further progress. Such was the case with Aristotle's incorrect philosophies about motion. It required

Block by Block: The Historical and Theoretical Foundations of Thermodynamics,
Robert T. Hanlon, Oxford University Press (2020). © Robert T. Hanlon.
DOI: 10.1093/oso/9780198851547.001.0001

the significant efforts of such individuals as Galileo and Newton to correct them. Much time was required to change and correct words, meanings, concepts, and paradigms.

Analysis of the Lever led to the Creation of Potential Energy

It's fascinating in this context to read the history of the creation of potential energy. This tremendous breakthrough started with a study between input and output of the simple lever. Who would have thought that study of such a seemingly boring concept would yield such an impact? This analysis was done without the need to discuss motion itself, which explains why this thread is separate from the Galileo–Newton thread. It all had to do with the transformation of one form of potential energy to another and the eventual realization that one can never get more out than what one puts in, or, in other words, that perpetual motion is impossible, which was a critical step toward that eventual realization that energy is conserved.

* * *

The lever. So much discussion throughout history by so many about a single plank resting on a simple wedge. It doesn't get much simpler than this. Place a fulcrum underneath a strong bar. Put a man at one end and a heavy object at the other. So long as the distance from the man to the fulcrum is larger than that from the object to the fulcrum, the man can raise objects heavier than himself. Workers have used levers and the mechanical advantage they provide since forever to help move heavy objects.

But if the lever was such an easy concept to implement, why was it so difficult to comprehend? Well, its simplicity was deceiving. The concepts involved relating to energy were really quite complex and not readily apparent. And to be honest, who really had the time or interest or motivation to care? Progress ruled the day and didn't wait for the theorist.

But the theorist had to come along, right? Who could stare at the lever and not ask, how? And so the theorizing began, naturally with the ancient Greeks, and evolved, with no real path or eureka moment, into the realization that perpetual motion is impossible and that energy is conserved. But I'm getting ahead of myself.

Aristotle and Archimedes each recognized the importance of understanding the lever. But each chose a different path of analysis. This most likely was not a conscious decision but simply a reflection of the way in which the lever occurred to each. Yes, there is more than one way to look at a lever. Aristotle chose one way, Archimedes chose another. And this set the course of history.

Aristotle and Archimedes–Two Different Views of the Lever

Aristotle (384–322 BC)[1] and Archimedes (287–212 BC) assumed an identical starting point: the lever is exactly balanced and in equilibrium. But then they went in profoundly different directions. Aristotle pictured in his mind small up-and-down movements of the balanced

---

[1] The ancient Greek ideas on the lever were captured in the *Mechanica*. While many feel that Aristotle didn't actually write this historic text but instead feel that it was written by his contemporaries or Peripatetic followers, they

9.1   THE LEVER   ©RTH + CLS

HOW DOES IT WORK?

ARISTOTLE: equate the effects at opposite ends

BUT WHAT EFFECTS?

ARISTOTLE: weight x velocity
HERO : weight x displacement
JORDANUS : weight x verticle displacement Δh — inclined plane

THE 5 SIMPLE MACHINES

wheel + axle
lever
pulley
wedge
screw

wΔh
is the correct "effect"
to use when analyzing
the work done by machines

study of perpetual motion by many
led to conclusion that

PERPETUAL MOTION
IS IMPOSSIBLE

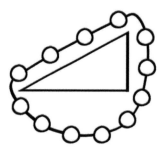

STEVIN (1586)

- Stevin demonstrated the absurdity
of perpetual motion with
his famed illustration

- the upper half suggests that
motion should occur while
the lower half doesn't

in all such machines,
because they are balanced with no net forces present,
Kinetic energy is absent and Input = Output
as quantified by wΔh (potential energy)

Figure 9.1 The lever

lever, neither direction being favored since equilibrium existed, like a teeter-totter balanced by two children at the playground (Figure 9.1). By imagining such "virtual" movements, i.e., movements that don't need to happen for one to understand and quantify the implications, Aristotle concluded, based on simple geometry, that because the heavier weight was closer to the support, its "virtual velocity" was slower by an amount inversely proportional to its weight.

Archimedes, on the other hand, saw something very different when he looked at the lever. To him, the lever was non-moving. Given this, he proceeded to use a rather ingenious approach to analyze the static situation and quantitatively prove his famous inverse-lever-arm law: the ratio of weights is inversely proportional to the ratio of lever arms. In this proof, Archimedes crucially relied upon his self-evident but non-supported concept of center of gravity. A given body resting on a lever could be viewed as having an identically heavy second body resting exactly on top of the fulcrum where it doesn't impact the balance. Since the center of gravity of these two bodies lies exactly at their mid-point, one could combine the two together and place the resulting heavier weight at this mid-point and maintain equilibrium balance. With this conceptual approach, one could move and alter weights along the lever so as to maintain balance and thus arrive at the law showing how the weight must change with distance along the lever to maintain equilibrium. You can move an object half-way toward the fulcrum by doubling its weight, or you can move the object twice as far away from the fulcrum by halving its weight. In either direction, you use the concept of moving counter weights away from or onto the fulcrum. And in this way you arrive at Archimedes' proof of his Law of the Lever. An ingenious approach indeed.

In the analysis of the balanced lever, Aristotle embraced motion but not center of gravity, while Archimedes embraced center of gravity but not motion. In a quirk of history, Aristotle's ideas moved forward through history more continuously than Archimedes'. The reason? Translation. Aristotle's work received timely translations while Archimedes' work lay in wait until around 1300 when it was finally translated from Greek into Latin, by which point the philosophers had already aligned with and built upon Aristotle's dynamic approach to mechanics. And it was this dynamic approach that ultimately led to the foundation of classical mechanics, while Archimedes' approach led to the separate science of Weights and Measures. As Ernest A. Moody remarked: "Ignorance of [Archimedes' work] proved, in the long run, a boon to the progress of mechanics."[2] Isn't history fascinating? What would have happened had Archimedes' approach dominated?

Regardless of whose ideas were correct and whose ideas were most favored by subsequent history, both Aristotle and Archimedes created an inspiring line of thought for civilization by recognizing and acting on the need to bring rigorous and profound inquiry to the seemingly simple lever.

believe that it was written at the time of Aristotle and that it contained his ideas and terminology, and so ascribe it to Aristotle.

[2] (Hiebert, 1981) p. 31. Moody remarks: "In the absence of Archimedes' work, the [thirteenth century authors] undertook to derive the lever principle from Aristotle's general "law of motion" thereby committing themselves to a far more ambitious undertaking than that which Archimedes had attempted, and involving themselves in problems whose solution required nothing less than the discovery and elaboration of what we know as modern classical mechanics. Ignorance of Archimedes [work] proved, in the long run, a boon to the progress of mechanics."

Aristotle—Equate the Effects at the Opposite Ends of the Lever

Let's return now to Aristotle's work. Aristotle's crucial insight was the need to equate the effects of the opposite ends of the lever. If the lever is in equilibrium, then little up-and-down motions favor neither one side nor the other. If this is true, then some quantifiable aspect of each infinitesimal motion must equal the other. What one gains, the other must lose to maintain equilibrium. But what was this aspect?

Clearly, no single measurable parameter, such as weight or geometry, could be the answer. So it must be some combination of such parameters, some combination that showed that a small weight could lift a large weight only if some other parameter decreased by the same proportion, for you can't just simply lift a heavy weight with a light weight without losing something in return, right? You can now see how this line of thinking led to the conservation of energy. Aristotle started down a path and others eagerly followed, because deep down, many felt that there must be *some* relationship between a machine's input and output. If something goes up, then something else must go down. Cause leads to effect. Things don't just happen.

As we now know, all of these inquiries were really inquiries around the concept of conservation of potential energy. The change in potential energy at one end of the lever had to equal the change in potential energy at the other end such that total energy was conserved. The concept of kinetic energy was never a part of this specific story line—free fall of an unbalanced lever was not considered—nor did it need to be since it was never really involved in the study of machines and so developed separately.

Unfortunately, the lever's design doesn't allow for the systematic study of the operating variables involved as its geometry provides but a single degree of freedom. Once you choose one parameter, such as the length of the lever-arm, all the other geometric parameters, such as the length of the arc, become fixed. But it turns out that while it then seems that any variable would be valid, only one is correct as the true fundamental energy-related variable. Alas, Aristotle chose the wrong one. He chose velocity. Since the time of motion was the same for both ends of the lever during any wiggle, and since the heavier weight moved a shorter distance, then the heavier weight must move slower. And so he concluded that the heavier objects move slower than the lighter objects in exact inverse proportion to their weights. It would take hundreds of years to correct this mistake.

Aristotle, along with Archimedes, brought necessary attention to the lever as an important problem to solve. But his solution ended up troubling mechanics all the way into the seventeenth century. For in his solution, he indirectly and incorrectly created a new but unnamed term to quantify motion. In his analysis, Weight 1 is to Weight 2 as Virtual Velocity 2 is to Virtual Velocity 1, which was but one algebraic step away from saying that equality exists between the product of weight and virtual velocity at each end of the lever. While this solution was incorrect and held no relevance in either statics or dynamics, it did eventually lead the way toward creating the correct concepts of both momentum and kinetic energy within the discipline of dynamics, once Newton added direction to convert velocity into a vector for the former and once Leibniz realized the significance of v^2 as opposed to v when considering aspects of the latter. Unfortunately, Aristotle's influence was so strong that many philosophers continually and inappropriately tried to apply his weight times velocity term to dynamic analysis, thus leading to much confusion.

Before moving on, two comments are warranted. First, Aristotle's "virtual" approach to an otherwise static system offered insight without the need to experiment and so could possibly be considered the predecessor of the "thought experiment" that Einstein, among others, made so famous later on.[3] Second, the "virtual" concept became a path toward crystallizing thinking around the nebulous theory of potential energy since both terms capture the concept of what *could* happen but hasn't yet actually happened. When "virtual" was finally and explicitly used in scientific writings, it was taken from other writings common during Aristotle's time as a way to accurately capture the need to distinguish between potential and actual. Others would eventually use "virtual" to describe several different types of infinitesimal movements, such as displacement, speed, and work.

Subsequent philosophers built upon the early Greek efforts, seeking to understand the capacity of a machine to perform work in terms of the effort expended. Unfortunately, few written records for such works survived, and so they too fall into the category of Taleb's "silent evidence." While we get glimpses in the references used in the later writings of others, we don't have the originals in our hands. But as discussed before, just because we don't have this history doesn't mean that it didn't happen. Unrecognized others may have had key insights, just as Aristotle and Archimedes did. So anytime the word "first" is used to describe someone's achievement, we must take care to more accurately use the words "first recorded" or perhaps more accurately, "first recorded, survived, and found" as, right or wrong, it is this person who receives recognition.

Hero and his Five Simple Machines

Of the succession of others who followed in this history, and especially those whose writings survived, two stood out, starting first with Hero of Alexandria (10–70 AD). Despite the fact that much of Hero's work was lost,[4] we do have available some of his critical works thanks to the efforts of Qusta ibn Luqa of Baghdad who translated Hero's *Mechanica* in 862–866. Thank goodness for those translators, another group among the unsung heroes of history.

As with Archimedes before him and Galileo after, Hero applied excellent skills in mechanics and mathematics to produce great practical accomplishments that were later highly valued by the Romans. It wasn't all theory to Hero. Building on the work of his predecessors, he wielded Occam's razor—before its invention—to tremendously simplify the analysis of machines by stating that all machines are comprised of but five "simple" machines: wheel & axle, lever, pulley, wedge, screw. He then additionally recognized that the appropriate way to quantify a machine's mechanical output was through the product of weight times displacement as

---

[3] "Virtual" motion and the associated responses like virtual velocity, distance, and work, are a part of classical mechanics and specifically the sub-division called statics. In this field, infinitesimal displacements are imagined, compatible with the constraints of a given rigid structure, and the resulting impacts analyzed.

[4] One of the most famous quotes from Hero's work concerns a very tantalizing concept. Pappas, a Greek around the end of the third century AD, wrote of Hero's work on finding the cause of the efficiency of the simple machines and alludes to a major breakthrough when he wrote: "Hero and Philo also have shown us that these machines are based upon a single natural principle, although they (the machines) seem to differ considerably from one another." (quoted in (Hiebert, 1981) p. 16). Nothing further was available! Ugghhh! So tantalizingly close!

opposed to weight times Aristotle's speed. This was the first known use of what would be later called "virtual work." In the eighteenth century it led to the definition of mechanical work, defined as the product of an acting force and the distance through which the force acts in the direction of this force, and ultimately became *the* standard measure of energy.

Hero got it almost right. His quantification was a huge step forward. The only problem was his focus on the lever. When the lever lifts a weight, it does so in an arc, and so Hero's displacement was the length of this arc. Which brings us to Aristotle's second notable successor.

Jordanus and Vertical Displacement

Jordanus de Nemore, a thirteenth century European mathematician of whom little is actually known, made corrections to Hero's work.[5] While Hero took a great step forward, he never made clear the relevance of the *direction* of the weight's displacement. He never explicitly referred to vertical displacement. Jordanus recognized the need to be more specific in quantifying the output of machines. In a true ground-breaking insight, Jordanus recognized the need to resolve force into two different directions, one along the line pointing toward the center of the Earth and the other perpendicular to this line. And it was displacement along the first that was needed to correctly quantify output. Jordanus brought the concepts of vertical displacement and "positional gravity" to the discussion, two monumental steps forward toward defining potential energy.

But how could he achieve such insight from looking at the lever, which had but a single degree of freedom? He didn't. He shifted the conversation from the lever to the inclined plane. Was this such a big deal? Absolutely.

The Inclined Plane

In yet another fascinating piece of historical trivia, neither Aristotle nor Hero nor Archimedes had succeeded in giving a correct analysis of the inclined plane, because they all failed to see the necessity of resolving the forces of weight into horizontal and vertical components. Just as Galileo would later gain significant insight when he resolved parabolic motion into horizontal and vertical directions, so too did Jordanus gain such insight through resolving force into these same directions. By studying the inclined plane, he realized that it is the vertical distance that is critical to the analysis of a machine and not any of the other terms.

The theorems of Jordanus influenced science for the next several centuries. Mathematicians and physicists of the sixteenth century, especially those in Italy, combined Aristotle's dynamic tradition with Jordanus' static theories to solve new problems related to dynamics, basing their arguments upon virtual work. But in the midst of such progress, all was not peaceful.

[5] Once again, I've had to preface most historical statements with variations of "Building on the work of others..." since it is very difficult to identify from the historical records who owned the critical insights during this time period.

You Can't Prove a Law

At this point in time, the general approach to analyzing machines was to equate the "virtual work" (weight times vertical distance) of a machine's input and output and then to draw conclusions accordingly. Some sought to prove why such an approach was valid, since the concept of proof as initiated by the ancient Greeks was a very powerful driving force. But as the philosophers sought such a proof, they began agonizing, as only philosophers could do. Because how can one prove something when the initial assumption in the proof was that which they wished to prove? They had to assume that the machine was in equilibrium before proving that it was in equilibrium. The fact that this so deeply bothered man, this issue of not being able to explain something, became a very good thing to mankind as it served as an excellent motivator to discovery and invention.

What these early philosophers didn't realize was that this was the only thing they *could* do. They were bumping up against the law of conservation of energy, specifically the equality between the input and output of simple machines based on our modern day concept of "work" (weight times change in vertical distance). While this concept opened up great opportunities for the development of ever more powerful machines, proof of the concept remained elusively impossible. And the reason is that you can't prove a law. It's impossible.

But out of all these entangled ideas, one thing rose that was clear, the eventual understanding that something else must also be impossible. Perpetual motion.

* * *

As an aside, if you're not able to sense a logical progression in this story, don't worry, because the path itself was not clear. Recall Taleb: "linear progression is not the norm."

Perpetual Motion and the March of Analysis

Long sought, never found, perpetual motion became to mechanists what alchemy became to chemists, a very positive driving force for great advances in science based on a false premise. Just as many wanted to convert lead into gold, many also wanted to build machines that could run forever. And every attempt to build such a perpetual motion machine, along with the associated investment strategies, failed. To this day, students are asked in graduate thermodynamics to identify the key flaw in a given proposed perpetual motion machine. It's a great learning exercise because it teaches us that the flaw is always there.[6]

* * *

As philosophers after Jordanus continued to study machines and continued to validate, time after time after time, the starting assumption that "virtual work" in and out equal each other, they began to realize that perpetual motion is impossible. Output can't exceed input. The continual demonstration of this impossibility led to its becoming a law in and of itself. The process was gradual with no eureka moment involved. People simply became very frustrated at ending

---

[6] (Klein, 1967) p. 516. "The problem of finding exactly how and where some particular proposal for a perpetual motion machine violates the laws of thermodynamics can often call for considerable insight and ingenuity, even though one is certain that the flaw in the argument is there to be found."

up with the same answer: perpetual motion is impossible. But the frustration slowly and favorably evolved into realization. Perpetual motion is impossible! And this subsequently became a critical foundation stone in the development of the conservation of energy. Simon Stevin (1548–1620) from Bruges perhaps best captured this frustration and subsequent realization by demonstrating the absurdity of perpetual motion with his classic and famous drawing of a system of interconnected weights around a triangle that suggested motion when viewed on the upper half of the drawing but showed the absurdity of motion when viewed on the lower half (Figure 9.1). Stevin's practical nature as a military engineer most likely contributed to his frustration with the impractical and impossible concept of perpetual motion. So powerful was his argument that the drawing ended up on his tombstone.

Others continued to apply this basic precept along with associated variations, such as John Wallis' (1616–1703) conclusion that "movement results whenever the common center of gravity of many joined weights descends"[7] and, as highlighted by Jordanus and then Galileo, "a body falling from a given height acquires sufficient velocity so that it can raise its own weight or another equal weight to the original height from which it fell [but no further],"[8] to further the study of mechanics as applied to machines, pendulums, and even fluids. Galileo started applying Aristotle's dynamic approach to analyzing fluid flow, and this work was furthered by the likes of Blaise Pascal (1623–1662), who developed theories of hydraulics, and Daniel Bernoulli (1700–1782), who developed his famous equation for fluid flow. Others continued to apply these approaches to more complicated mechanical analysis, such as Evangelista Torricelli (1608–1647) and Jean le Rond d'Alembert (1717–1783), the founder of analytic mechanics, and Joseph-Louis Lagrange (1736–1813), who applied infinitesimal calculus to the evaluation of the mechanical performance of machines and mechanical work in general.

While the above work contributed greatly to the development of Mechanics, when all was said and done, the above work really boiled down to the same concept: assumed exclusion of perpetual motion. So what was wrong with this? Well, it led to continued frustration because the philosophers correctly sensed that something was still missing. The above provided one critical clue, the equal exchange or transformation of "virtual work," which we know today as the change in potential energy, between the input and output of a machine. But people didn't know where to go with this. As Louis Poinsot (1777–1858) said, referring to Lagrange's work but easily extendable to this collective work, it all provided nothing clearer than "the march of analysis."[9]

So what *was* missing? By now you'll have noticed that the above discussions of levers and machines and "virtual work" lacked any mention of the concept of real motion of the kind that we associate with unbalanced forces, kinetic energy, and heat. It was simply not required as a prerequisite to the discovery of the impossibility of perpetual motion. But it *would* be required to arrive at a more comprehensive energy conservation law.

---

[7] (Hiebert, 1981) p. 52.
[8] (Hiebert, 1981) p. 76.
[9] (Hiebert, 1981) p. 56.

10 The rise of mv²

This chapter considers the role of three noted physicists—René Descartes, Christian Huygens, Gottfried Wilhelm Leibniz—and three physical phenomena—free fall, two-body collisions, effects in simple machines—in the rise of mv² as an important quantity in physics during the time period from 1638 (Galileo) to 1686 (Leibniz). In his free-fall studies Galileo determined the famed "Law of Fall" relating distance to the square of time; weight was not involved as it plays no role in free fall. Weight did play a role, though, in the studies of the simple machines such as the lever; weight times vertical distance quantified what would become known as "work." Two-body collisions intuitively involved both weight and speed but the functional form of this involvement was not clear, especially since two forms were involved: conservation of momentum and conservation of kinetic energy (for elastic collisions). The achievement of this group was to find the correct path to connect free fall with the lever and in so doing to connect speed with weight and so arrive at the correct form of energy based on motion (mv²).

As we know today, the critical physics concepts involved in the above concepts are the properties mass (and weight), speed, and velocity, in addition to the overriding concepts of momentum, energy, and the conservation of each. I bring this up here since these properties and concepts weren't known during 1638–1686. Mass and velocity (vector) wouldn't be defined until Newton's *Principia* arrived in 1687 and even then wouldn't be fully digested until years later, and the conservation laws wouldn't start taking shape until around 1750 for mechanical energy and then around 1850 for the totality of energy itself. So while the fundamental theories involve mass, velocity, and the conservation laws, the original theories were written in terms of weight, speed, and rough outlines of conservation ideas. Instead of having you the reader trying to follow such "chaos in terms,"[1] as Westfall put it, I've written the following, unless otherwise stated, using the correct terms and conservation laws we use today.

Important to realize is that the most important findings of the three physicists are revealed in the logic they employed and not the terms they used.

Galileo Connected h with v²

And so now we return to the concept of motion, which means we must return to Galileo, for he was the one who really started the study of motion when he shifted the conversation from

---

[1] (Westfall, 1971) p. 61.

Block by Block: The Historical and Theoretical Foundations of Thermodynamics,
Robert T. Hanlon, Oxford University Press (2020). © Robert T. Hanlon.
DOI: 10.1093/oso/9780198851547.001.0001

qualification to *quantification*. This clearly was not an easy shift. He had to define new terms and then invent his own tools to study and measure those terms. Certainly the terminology challenge in and of itself slowed Galileo's progress and, in fact, slowed the progress of all who attempted to build a quantitative dynamics[2] in the seventeenth century. But the design and execution of the experiments was no less demanding.

Perhaps the most critical term confronting Galileo was velocity. Neither he nor anyone else *really* knew what this term meant. Recall that at one point, he looked at the performance of a pile driver and concluded that its "intensity of blow when stopped," which he surmised to be the appropriate way to define velocity but which we now know to be related to kinetic energy, is proportional to height. *Twice as hard for twice as far*—twice $\Delta\frac{1}{2}mv^2$ for twice Δmgh. But then his thinking evolved and he switched to defining velocity as distance divided by time, thus unknowingly leaving to others the challenge of quantifying the more complicated terms of velocity (including direction), momentum, and the kinetic energy involved with such phenomena as collisions and pile-driving percussion. But this decision presented him with a new problem. How could one even measure velocity back then, especially in an accelerating environment with no stopwatch? He understood that this quantity couldn't be directly measured during free fall, but, as we learned earlier, he pushed through this obstacle and by sheer genius devised experiments, both real and in thought, to discover two seemingly simple but tremendously monumental relationships ("h" represents the change in vertical height):

$h \propto t^2$	The Law of Fall	Galileo's experimental result
$v \propto t$		Galileo's logic-based result
$h \propto v^2$		The important conclusion relevant to energy

It's rather fascinating that when you look at the final relationship, with some small modifications of the plus and minus signs to account for direction, you see the conservation of mechanical energy for a body staring you right in the face.

$$mgh + \frac{1}{2}\, mv^2 = \text{constant}$$

Of course, Galileo worked in the world of ratios and not equations. But still, from our vantage point today, when we look at his findings we realize that he seemed to come so very close to linking kinetic and potential energies. But this is naturally an illusion as he was a huge paradigm shift away from this linkage.

---

[2] As broached in Chapter 8, p. 1, footnote 2, dynamics is the branch of mechanics concerned with the motion of bodies under the action of forces, while statics is the branch of mechanics concerned with bodies at rest and forces in equilibrium. Thus, Newton's force-based Laws of Motion are located in dynamics while Archimedes' analysis of the lever is located in statics. Since Kinematics is concerned with the motion of objects without reference to the forces causing the motion, it encompasses Galileo's Law of Fall.

Descartes – the First to Attempt a Conservation Law based on Motion and...

Descartes deserves the honor of having first formulated a theory for the conservation of motion.

– Erwin Hiebert[3]

I do not accept any difference between violent and natural motions.
– René Descartes[4]

What Galileo started, René Descartes (1596–1650) continued. As an extremely influential French thinker and writer—"*I think, therefore I am*"—Descartes crystallized focus on a broad range of themes in such fields as philosophy, mathematics, and physics. Regarding this specific discussion on energy and motion, his contributions and those of his "Cartesian" followers were more philosophical than experimental, which does not mean to detract from the contributions but more to highlight their critical importance at this stage in history. While Aristotle's influence on the science of motion remained strong during this time, unfortunately so, Descartes found that "the Aristotelian definition of motion [was] so obscure as to be meaningless."[5] Thus motivated, he set out to bring a new conception of motion to science.

Descartes believed deeply in causality, writing that "an effect is always in proportion to the action which is necessary to produce it."[6] Such a strong belief originated, at least in part, from his detailed study of earlier works concerning the fundamental laws of the simple machines, such as the quantification of performance as weight times vertical distance, the equality of this quantity between input and output, and ultimately the impossibility of perpetual motion.

With such a belief as his guide, Descartes became the chief enthusiast of the mechanical philosophy that rose in the early seventeenth century based on the proposition that *all* physical phenomena can be explained solely in terms of collisions between bodies. This philosophy represented a significant break from Aristotle's in that it removed the "occult" from cause–effect discussions—no more "violent" or "natural" motions—and replaced it with a cold philosophy of matter-in-motion that had zero tolerance for magic. This mechanical approach, in turn, stimulated the early thinking about conservation. If cause equals effect, then there can be no effect without cause. Events just don't happen. A change here must be compensated by an opposite change there. What is lost must be gained. Descartes then raised this discussion about conservation to an even higher level when he proposed the existence of a certain permanency of motion based on God's action in the world.[7]

---

[3] (Hiebert, 1981) p. 67.
[4] Descartes quote cited in (Westfall, 1971) p. 56.
[5] (Westfall, 1971) p. 57.
[6] (Hiebert, 1981) p. 49.
[7] (Hiebert, 1981) p. 67.

Descartes –... the First to Propose a Mathematical Characterization of Said Motion

Descartes realized that mathematical quantification would be needed to support conclusions based on conservation and so became one of the first to propose a mathematical characterization of a body's motion. As suggested earlier, this was all part of a larger need that rose in the sixteenth and seventeenth centuries for a more rigorous and precise mathematical framework to formalize and quantify concepts, including a corresponding need for rigorous words and definitions that cut across language and national boundaries. Science begged for such rigor, especially when the concept of conservation was involved and the need to compare input and output, or before and after, on a quantifiable basis was critical. It's one thing to talk and pontificate about such ideas as conservation, it's an entirely other thing to actually formalize and quantify this thinking, especially when very disparate concepts are involved. *How does motion mathematically combine with "virtual work"? And more importantly, why should the two even be related?*

Galileo's quantification of speed was helpful but not sufficient to address these issues. When two balls collide, speed is important but so is mass. Surely the quantity of motion, whatever it means, must be related to both, and Galileo's above proportionality relationships didn't address this—and really didn't need to since free fall is independent of mass. So Descartes and his successors, likely influenced by Aristotle's weight times "virtual velocity" term (incorrectly) derived from his study of the lever, made the profound initial proposal that the appropriate definition of "quantity of motion" is weight times speed (scalar) and that the totality of this quantity remains the same before and after a collision of two bodies. Descartes went so far as to declare that the total quantity of motion in the universe is as constant as the amount of matter that it contains.[8] These ideas represented the first attempt at a conservation law based on motion. While they were not correct, they became reasonable starting points for discussion. Indeed, they attracted the attention of a large number of natural philosophers who followed in Descartes' foot-steps. Sometimes someone simply has to throw something against the wall to see if it will stick, to toss out an idea to get things started, to bring focus, to crystallize thinking. Descartes was the someone who brought focus to the right question for the development of dynamics: *how is motion quantified?*

Descartes' choice of weight times speed to quantify motion was incorrect from two different perspectives: it was not a correct quantification of either momentum or energy. The resulting journey to correct this mistake thus took two different paths. We already followed the first path in which his "quantity of motion" based on speed (scalar) evolved into Newton's *momentum* based on velocity (vector) by incorporating direction. Now let's follow the second path in which his same "quantity of motion" based on speed evolved into *kinetic energy* based on the square of speed.

The Evolution of Kinetic Energy

> *The creation of modern dynamics in the seventeenth century appears as one of the supreme conquests of the human spirit, a triumph wholly worthy of the 'century of genius'*
> – Richard Westfall[9]

---

[8] (Hiebert, 1981) p. 66.
[9] (Westfall, 1971) p. x.

The evolution of kinetic energy as a concept in the seventeenth and eighteenth centuries occurred as an outcome of the convergence of three separate branches of mechanics, namely simple machines, free fall, and collisions. How these three were connected wasn't obvious, nor was the reason why they should even be connected in the first place. It would take the brain-power of a series of geniuses of the seventeenth century to unveil the fundamental link.

Jordanus had thoroughly summarized the performance of simple machines and Stevin the impossibility of perpetual motion that was derived from analysis of the simple machines. Galileo pretty much captured the critical kinematic aspects of free fall except for its link to gravity, which would have to wait. Collision theory was still a work in progress. Scientists of the later Middle Ages believed that something must be conserved in the mechanical world when bodies fell or collided; the conservation of energy was "implicit"[10] in their writings although not explicitly stated. Based on his cause-and-effect philosophy and an incorrect premise derived from Aristotle, Descartes made the first stab at an answer: weight times speed (scalar). All of this work opened the door for the more in-depth analysis that followed, one that brought together these branches of mechanics. Which brings us to Christiaan Huygens.

Huygens – Galileo's Successor

[It was] Huygens, upon whose shoulders the mantle of Galileo fell…
— Ernst Mach[11]

Between Galileo and Descartes at the beginning of the century and Leibniz and Newton at its end, Huygens stood in the middle as the indispensable link whereby the endeavors of the age acquired continuity and retained coherence.
— Richard Westfall[12]

A brilliant Dutch philosopher and mathematician who was exposed to the achievements of Descartes and Galileo while he was still a boy,[13] Christiaan Huygens (1629–1695) pursued a wide range of study in Paris that included such revolutionary concepts as his development of Galilean relativity and the wave theory of light. While he personally knew Descartes and was immersed in Cartesian thinking, as especially reflected by his absolute refusal to accept action-at-a-distance, a refusal that possibly prevented him from beating Newton to the Laws of Motion,[14] Huygens' true inspiration was Galileo and many consider him as Galileo's successor. In the end, it was his understanding of the interplay between Descartes and Galileo that enabled his achievements.

Regarding this specific discussion, one could argue that it all started with a contest. Who would have thought that the collisions between objects would prove so important to the progress of energy? But it did. It had to. It was at the core of the mechanical philosophy. Bodies

---

[10] (Theobald, 1966) p. 28.
[11] (Mach, 1911) p. 28.
[12] (Westfall, 1971) p. 146.
[13] (Westfall, 1971) p. 146.
[14] (Westfall, 1971) p. 188.

colliding with bodies. People couldn't figure out the rules—or rather, their minds couldn't rest without figuring out the rules—governing the impact between bodies, ones that would enable the prediction of direction and velocity of bodies following two-body collisions. Descartes first raised the issue that something must be conserved. But what? As we know today, *both* momentum and kinetic energy are the conserved quantities for elastic collisions (only momentum is conserved for inelastic collisions[15]). The fact that there were actually two correct solutions, which sowed seeds of confusion into the situation, shows just how difficult the problem was to solve.

Collision Theory – the First Recorded Usage of mv²

In response to a contest created by the Royal Society of London in 1668 seeking papers on collision theory—which really served to highlight how important this issue was at that time— three notable physicists, Huygens, John Wallis, and Christopher Wren (1632–1723), each prepared a response. While I won't go into each historic paper here, especially since collision theory in and of itself did not involve potential energy and as such was not critical to the development of the conservation of energy, I do draw attention to Huygens' paper since this was the first recorded usage of the quantity mv^2 as a measure of motion.

Based on his use of Galileo's findings and his own independent thinking and research on pendulums and other mechanical devices, Huygens stated in this paper, but without proof, that the quantity mv^2 is conserved during elastic collisions. In this paper and in other subsequent writings, especially his *Horologium Oscillatorium* (1673), Huygens clearly built on Galileo's work involving free fall. Time and time again, the seventeenth-century philosophers relied on this work of Galileo's to help uncover the nature of motion. Huygens was the first to truly latch onto the significance of this relationship when he had the insight to turn it around by saying that if the relationship applies to a free-falling body, then it must also apply to its subsequent rising. What Huygens suggested here is that one means of quantifying the horizontal motion of a body is to change its direction from horizontal to vertical and then measure how high it rises, realizing that the measured height (h) is proportional to the square of the horizontal velocity (v^2). This shift in logic sounds simple but was actually rather profound, reflecting Huygens' insight that free fall and *not* the lever was the right path to relate motion and ascent. He properly viewed the simple machines in terms of equilibrium; there is no path to relate motion and ascent with the lever. (Note: I intentionally do not use the word *work* here as it had not yet been defined; its definition would come later.)

Building on this insight, Huygens next turned to Torricelli, who had proposed the new and very important principle that two or more bodies joined together can be treated as a single body concentrated at their center of gravity, and that the conclusions established for single bodies, including those regarding the inertial state of the bodies, can be applied as well to the motion of this center of gravity. Taking this one step further, Huygens suggested that the

---

[15] Using Newtonian mechanics, the net force (F_{net}) acting on a system of bodies equals the rate of change in the net momentum of that system (d $\sum (mv)_i$ / dt for i = 1 to n, the total number of bodies. If F_{net} = 0, then $\sum (mv)_i$ must be a constant.

same logic must also apply to the motion of bodies *not* joined together, asserting that two bodies in impact can be considered as one concentrated at their center of gravity and that "the center of gravity of bodies taken together continues always with a uniform motion in the same direction and is not disturbed by any impact of the bodies."[16] Thus, elastic impact could be viewed purely as a kinematic event (no net force acting on the system) and not a dynamic event since no change in the inertial state of the center of gravity occurs. The insight became immensely important to the future science of mechanics and today's students continue to learn about collisions from this same center-of-gravity perspective.

Center of Gravity Calculations Provided a Means to Connect Weight (Mass) with Speed

Huygens then extended this thinking even further to the analysis of the fall and subsequent rise of any number of weights. While I won't go into the details of his argument here, as shared previously, he considered the relationship between a body's speed and its ability to achieve a vertical rise in elevation ($h \propto v^2$). By considering multiple bodies in a given system, say a collection of swinging pendulums for example, and quantifying the movement of the center of gravity of these bodies, specifically proposing that this center of gravity cannot rise *on its own* as a result of anything that goes on inside the system, he established the connection between the weight of each body—the individual weights are needed to calculate the center of gravity—to the terms in the $h \propto v^2$ relationship, and in so doing connected weight with speed and thus delivered a new variable to the world of physics, mv^2.[17] From this analysis, returning to the study of collisions in general, Huygens stated that during "elastic" collision of two bodies, i.e., ones in which the bodies don't change shape, the total mv^2 of the two before the collision equals the total after. In other words, total mv^2 is conserved and not mv as proclaimed by Descartes.

While Huygens' focus on mv^2 was a significant step forward in the history of mechanics as it connected weight with motion in a way that Galileo's free fall couldn't, to him it was a stand-alone, untethered insight. He did not show a direct interest in pursuing this line of thought any further and perhaps did not possess a clear enough understanding of how his thinking might combine with the concept of mechanical work, i.e., weight times vertical height. Which brings us to Gottfried Wilhelm Leibniz.

Leibniz and the Birth of Modern Physics

In regard to the quantity mv^2, what was a mere number to Huygens was invested by Leibniz with cosmic significance

– Richard Westfall[18]

---

[16] (Westfall, 1971) p. 153.
[17] For more details on this, see (Theobald, 1966) p. 28–31 discussing Huygens' treatment of the pendulum in his *Horologium Oscillatorium* in 1673.
[18] (Westfall, 1971) p. 284.

A German mathematician and philosopher, Gottfried Wilhelm Leibniz (1646–1716) was well acquainted with the intellectual community of Europe. While mechanics was not one of his primary interests, he together with Newton effectively became the two co-creators of modern dynamics, a word coined by Leibniz. Someone had to make sense of all the learning that had been gathered and so bring a systematic analysis to motion. Both Newton and Leibniz succeeded in doing this. The fact that they came up with two very different but equally valid answers, which were eventually reconciled, reflected the difficulty involved.

Leibniz's role in this story was not the addition of any new quantitative relationships but instead the clarification of the relationships already developed, especially those by Galileo, Descartes, and Huygens, who was a contemporary collaborator of Leibniz despite being 17 years older. Motivated as Huygens had been to bring attention to Descartes' error in defining "quantity of motion," Leibniz conducted a rather ingenious inquiry into the impact of bodies free falling from different heights. He sought the correct measure of whatever it is in the motion of a body that leads to the magnitude of its impact and used as his assumption that two bodies would strike with equal effect only if equal effort went into elevating them off the ground prior to dropping them. Recall that Galileo wrote briefly about this—"twice as hard for twice as far"—regarding the percussive performance of pile-drivers but didn't progress his thinking beyond this. Leibniz believed that within this examination lay a hidden truth about a cause–effect equality in nature. The word he used to encompass both cause and effect was *vis*, which later became known as energy.[19]

Leibniz's published results (1686–1695) from his inquiry started with the following assumption. A body falling from a certain height acquires the same *vis* as that required to return it back to its original height. This assumption really stemmed from Galileo's proposal in his 1638 *Discorsi* that a body can rise to the height from which it fell but no further and was embraced by most seventeenth century philosophers, although Leibniz's use of *vis* in this context reflected a higher interpreted understanding, one broaching the concept of conservation, something that surely attracted him.

At face value, the two concepts of free fall and raising a weight differed from each other greatly as they originated from totally different historical contexts, the former from Galileo's study and the latter from the collective study of simple machines. It was the magnitude of this gap between the two that made Leibniz's use of *vis* to link them through his above assumption such a monumental breakthrough. His deep inner belief, building on the thinking of Descartes and others, in the impossibility of perpetual motion, in the equality between cause and effect, and in the conservation of some universal quantity led him to his assumption, which can be re-stated in today's words: the gain in kinetic energy of a falling body—$\frac{1}{2}mv^2$—must equal the mechanical work—$mg\Delta h$—required to raise it back to its starting point.

---

[19] Leibniz's use of the word *vis*, which we now call energy, created later confusion in the literature as this Latin word translates to force. The word force in this context is clearly not the same as the word "force" used by Newton. To Leibniz, *vis* was a property of a body, something that a body had and because of that, it provided the body with the means to act or have an effect. He used *vis* to quantify the effect that a body can produce, whereas Newton used force to quantify an external cause of a change in a body's momentum.

Leibniz's Famed Thought Experiment Provided a Different Means to Connect Weight with Speed

Leibniz's challenge was to find the right path to link, in essence, free fall with the lever. To accomplish this, he proposed a thought experiment, which I paraphrase here, involving two bodies, A and B (Figure 10.1). He assumed that the *vis* required to raise body A (1 pound) to the height of 4 units is the same as that required to raise body B (4 pounds) to the height of 1 unit, reflecting the importance of what others had already recognized, that weight times vertical distance, which we now call work, is the relevant *vis*-related quantity to consider with ascent. He then assumed that the *vis* acquired by A's 4-unit free fall must equal the *vis* acquired by B's 1-unit free fall, again believing that if the two ascents require equal *vis*, then their respective free falls acquire this *vis*. His logic worked as follows:

> Because *vis* of ascent is equal for these two scenarios:
> Body A: *vis* required to lift 1 pound to height of 4 units = $1 \times 4 = 4$
> Body B: *vis* required to lift 4 pounds to height of 1 unit = $4 \times 1 = 4$
>
> Then *vis* of free fall must also be equal for these two scenarios:
> Body A: *vis* acquired by 1 pound after falling 4 units = $1 \times$ f(speed after falling 4 units)
> Body B: *vis* acquired by 4 pounds after falling 1 unit = $4 \times$ f(speed after falling 1 unit)
>
> Solving,
> Body A *vis* from free fall = Body B *vis* from free fall
> $1 \times$ f(speed after falling 4 units) = $4 \times$ f(speed after falling 1 unit)
> f(speed) must be linear with height (h)

Leibniz's above reasoning led him to search for an unknown function of speed acquired during free fall that varies linearly with height. He discovered his answer in the work of Huygens, which was based on the work of Galileo: the square of velocity varies linearly with height of fall ($v^2 \propto h$). In this way Leibniz saw that it must be the square of velocity that quantifies the motive *vis* and not velocity alone as Descartes suggested for his "quantity of motion" that was incorrectly derived by Aristotle for the static (equilibrium) lever for which motion plays no role. Building a clearer explanation than Huygens did but otherwise aligning with Huygens, Leibniz stated the importance of v^2 and the error of Descartes' v. It took a lot to prove a legend wrong, but Leibniz did just that and made sure others knew about it. The title of his publication started with the following, "A brief demonstration of a famous error of Descartes…"

To fully quantify his motive *vis*, Leibniz multiplied v^2 by weight to arrive at mv^2 based on his further insight that two bodies will strike the ground with an equal *vis* only if equal amounts of effort had been expended in raising them to their respective drop heights. Since the raising effort is proportional to weight, then the striking *vis* must also be proportional to weight. Double the weight, double the striking *vis*.

Leibniz raised the significance of motive *vis*, mv^2, which he named *vis viva*, when he proposed, based on argument alone, that it become a measure of all activity, whatever the source— free fall, collision, even the release of a compressed spring. He proposed, as did Huygens, that the total sum of this number remains constant during elastic collision, but unlike Huygens, he even proposed such conservation for inelastic collisions, having the intuition to suggest that

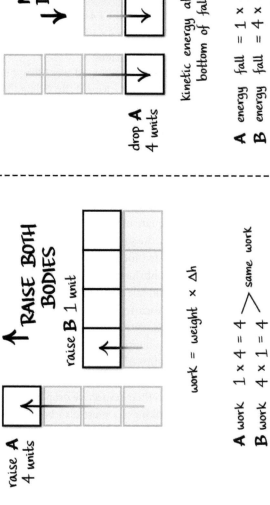

10.1 LEIBNIZ'S THOUGHT EXPERIMENT © RIH + CLS 2019

Leibniz used the term " vis "
which we now call " energy ".

Body **A** [ ] Body **B** [= 4x Body A]

RAISE BOTH BODIES

raise **A** 4 units

raise **B** 1 unit

work = weight × Δh

A work 1 × 4 = 4
B work 4 × 1 = 4 } same work

LEIBNIZ

· the energy required to raise a weight
  must be equal to the energy of the
  weight after it is dropped.
· both **A** and **B** lifted with same energy.
  thus same energy must result when
  A and **B** are dropped
· Leibniz called this energy "vis"

NOW DROP BODIES

drop **A** 4 units

drop **B** 1 unit

kinetic energy at
bottom of fall = weight × f(speed)

A energy fall = 1 × f(V_A) when dropped 4 units
B energy fall = 4 × f(V_B) when dropped 1 unit

$$1 × f(V_A) = 4 × f(V_B)$$

f(v) must be linear with height

since Galileo showed h ∝ v²

then v² must quantify energy (and not v)

Figure 10.1 Leibniz's thought experiment

any loss of *vis viva* from two colliding large bodies is simply gained by the "parts [that] receive it."[20] He did not make the connection between this statement and heat—"he too was unable to leap a century beyond his time"[21]—but this was clearly a preview of what would eventually become in others' hands the full conservation of energy.

1686 – the Birth of dynamics

One could argue that the science of dynamics as distinct from statics was born in 1686 with Leibniz's publication identifying mv² instead of mv as the correct dynamic measure of motion. Like Huygens before, Leibniz arrived at the correct quantity v² based on Galileo's free fall, whereas Descartes arrived at the incorrect quantity v based on Aristotle's lever. Even though he arrived at the same complete answer of mv² as Huygens did, Leibniz's approach was different, being clearer and perhaps on a more solid theoretical foundation since he had the insight to explicitly relate two otherwise unrelated concepts: free fall and simple machines.

Continuing his work in the direction of conservation, Leibniz eventually realized that he needed another term to balance *vis viva* and so created *vis mortua* to account for the *vis* of a body at rest such as a raised weight. Later, he created the more general term *potentia* to encompass any condition of a body that can bring about action or change, such as motion, elevation, and even the stretched condition of an elastic body. These various terms embodied the "germ"[22] of our potential energy concept and reflected Leibniz's efforts to create a cause–effect balance for his *vis viva*.

What Leibniz did here was to take the first step, far beyond the thinking that existed at that time, toward defining the conservation of mechanical energy involving both kinetic and potential quantities. But this was pretty much the extent of it as he did not trouble himself to extensively investigate the full implications of *vis mortua* as a concept within dynamics. For example, he showed no inkling of the continual interplay between *vis mortua* and *vis viva* whereby an increase in one is accompanied by a decrease in the other. For another example, while he understood the recognized importance and special significance of the product of weight times vertical height, he didn't understand this enough to generalize it to a concept of work or a change in energy.

Newton's force and Leibniz's *vis* (energy) added new dimensions to the study of motion. That the concepts complemented each other was not readily apparent and so they sat waiting for someone to see the connection. Since Newton's *Principia* was more immediately relevant to the solution of real problems being faced by science, it was his work that received the more immediate attention. But soon the connection between his force and Leibniz's *vis* would be revealed. And this now brings us to the Bernoulli family and Leonard Euler.

---

[20] (Hiebert, 1981) p. 89.
[21] (Westfall, 1971) p. 295.
[22] (Hiebert, 1981) p. 81.

11 Bernoulli and Euler unite Newton and Leibniz

Entering the eighteenth century we had Newton's Laws of Motion based on his concept of external force, Leibniz's *vis viva* (mv²) and its proposed conservation, and two different versions of calculus. Hidden in the background lay the deeply held axiom: perpetual motion is impossible. The next generation of scientists and mathematicians were left to interpret, judge, and unify these concepts. The difficulty involved was significant since, from a larger perspective, the two approaches represented two very different philosophies of motion: Newton's based on momentum and Leibniz's based on energy. The two wouldn't definitively be united until Einstein showed in his Theory of General Relativity that the conservation of each originates from the same conservation law in which energy is a tensor that incorporates momentum. But I'm really getting ahead of myself. The point is that there was a significant technical challenge to bringing Newton and Leibniz together.

The above unification would have been difficult in the best of conditions. But the conceptual difficulty was confounded by the emotional one caused by the vicious animosity that existed between Newton and Leibniz, largely because each felt himself to be the inventor of calculus. The emotions that flared between them soon brought their respective homelands into intellectual conflict as well, each seeking to defend their own. Furthering this emotional dispute was the Continent's rejection of Newton's theory of universal gravitation. Like Huygens, Leibniz deeply rejected the idea of action-at-a-distance and many others who embraced the mechanical philosophy on the Continent followed suit. Because of this, it was hard for many to remain open-minded enough about Newton's work to see its connection to Leibniz's.

The unification of Newton and Leibniz thus required a genius in his own right, one who was both open-minded enough to see the genius in each body of great work and motivated to bring them together. Of the few who sought to do this, I've selected two of the key persons to discuss here, namely Daniel Bernoulli and Leonard Euler. We'll start with Bernoulli but we really can't start with him without first starting with his father, Johann.

The Bernoulli Family and Johann the Father

> *The primary [principle] is the conservation of living forces, or, as I say, the equality between actual descent and potential ascent.*
>
> – Daniel Bernoulli in *Hydrodynamica*[1]

---

[1] Bernoulli quote cited in (Coopersmith, 2010) p. 115.

Block by Block: The Historical and Theoretical Foundations of Thermodynamics,
Robert T. Hanlon, Oxford University Press (2020). © Robert T. Hanlon.
DOI: 10.1093/oso/9780198851547.001.0001

The Bernoulli family tree was quite remarkable, a true dynasty, one filled with both achievement and drama.[2] The short version of their family lineage is this. Jakob the Elder begat Nikolaus (1623–1708), Nikolaus begat Jakob (1654–1705) and Johann (1667–1748), and Johann begat Nikolaus II (1695–1726) and Daniel (1700–1782).

This story begins with the sixteenth century Reformation during which a group from within broke from the Catholic Church in protest of its rigid doctrines. This group logically became known as the Protestants. Unfortunately, this break was not without consequences and led to Catholic persecution against the Protestants, especially the French Huguenots and the German Calvinists, many of whom fled to safer and more tolerant environments, such as Basel, Switzerland. And so it was that in 1622, Jakob the Elder, as a Huguenot suffering from such persecution in Belgium, fled to Basel and there established the beginnings of his family's 50-year-dominance in mathematics. Here again we see the influence of religion on the progress of science. How many countries lost and how many gained throughout history due to the migration of scientists from intolerance to tolerance? Look at Italy. After Galileo endured what he did, then what? Who wanted to pursue astronomy there? As Harriman noted, "Italy had long held a role of intellectual leadership, but that changed when the center of the Renaissance put reason on trial and found it guilty... Progress would continue in England and France, but astronomy was dead in Italy."[3] Once Galileo died, the science of astronomy indeed shifted away from Italy and toward other regions, such as Germany, France, and England, where the Catholic Church had less influence.

Now Jakob the Elder didn't intentionally establish a dynasty of mathematicians. In fact, his son, Nikolaus, sought practical jobs for his own sons, Jakob and Johann. But how could he win against the influence of a rock star, namely Leibniz? The arrival of Leibniz's publication on calculus in 1684 set the Bernoulli brothers on fire. This was not in dad's plans. While few in the world could comprehend this new concept, the brothers remained undaunted and impassioned. They struggled. They didn't stop. And finally, in their own eureka moment, they saw it, first Jakob and then Johann. They finally understood how the machinery worked. Together they moved forward. Together they learned about the huge power of the "infinitesimal," a huge stumbling block for many. Together they used this phenomenal machine to solve many complicated problems in many fields of study that had been sitting and waiting. The fame and accolades flowed. Alas, if only such flows were equal for all. Unfortunately in this situation (in all situations?) they flowed more in one direction than in the other, an imbalance that only the most disciplined of egos could withstand. You see, Johann's path took him to France to become a tutor of French nobility, and then into direct communication with the rock star himself. Through written correspondence, he developed a close relationship with Leibniz and became a very strong and active public supporter of him in his calculus war against Newton. His association with Leibniz in such a publicly popular and emotional battle led to fame for Johann across the Continent. Unfortunately, Jakob's path didn't lead to the same fame, even though his attainment of the seat of Professor of Mathematics in the University of Basel was quite an achievement in its own right. As an unfortunate result, jealousy set in. When Johann sought to return to Basel to also become a professor, Jacob worked behind the scenes to derail the

[2] (Guillen, 1996) p. 65–117.
[3] (Harriman, 2000) p. 24.

process. Johann learned of the betrayal and the relationship rapidly deteriorated from there, the feud amplifying itself through public taunting in professional journals. So why go into this family drama? Because it was to play itself out again, but this time not as brother-vs-brother but as father-vs-son.

Johann eventually developed his own strong academic career in mathematics. He became chair of the department at the University of Groningen in the Netherlands, at the invitation of none other than Christiaan Huygens, and then returned to Basel to assume, in an interesting twist in the drama, his brother's job on the faculty that became available after his brother's death. While Johann's world may have centered on math, his interests, inspired in part by Leibniz, took him beyond math and into the world of natural philosophy. As started by Galileo and furthered by Newton and Leibniz, these two worlds were continuing to become more and more intertwined.

Specifically regarding energy, the many contributions from Johann were significant. As he wrestled to reconcile the same concepts as those before him, he built upon existing concepts and proposed new ones.

One of his key contributions concerned his new interpretation of Leibniz's *vis mortua*. While Leibniz never really grasped the relationship between this concept and his *vis viva*, Johann recognized that *vis mortua* represented a stored "live force"[4]—recall that Leibniz's *vis* referred to energy but translated to force—such as exists in a stretched spring or a raised weight, that is consumed whenever *vis viva* is generated, and vice versa. This idea originated from his belief in the equality between cause and effect, although one must wonder about the depth of this belief in light of his pursuit of perpetual motion machines.

First Use of the Word "Energy" in Physics

Regarding another of his key contributions, we finally come to the first use of the word "energy" in physics. In his now famous 1717 letter to Pierre Varignon, Johann explored a mechanical system in equilibrium. As many had done before him, he created, in concept, a very small and reversible overall displacement of the system that resulted in a set of similar displacements throughout the system, with each "virtual" displacement having a "virtual velocity." Johann was the first to use the term "virtual" in this context. When considering such an event, Johann proposed that,

> *In every case of equilibrium of forces, in whatever way they are applied, and in whatever directions they act on (one) another, either mediately or immediately; the sum of the positive energies will be equal to the sum of the negative energies, taken as positive.*[5]

It's interesting that the word *energy* rose from an incorrect assumption. As Descartes and others did before him, Johann selected "velocity" as the important quantity when analyzing a system in static equilibrium, which was incorrect. The correct quantity was instead

---

[4] Yes, the term *mortua*, a Latin word for dead, is confusing in light of the concept of "live force." The concept later evolved to "potential."

[5] (Hiebert, 1981) p. 82. Johann Bernoulli personal letter (1717) to Pierre Varignon of the Paris Academy of Science. Published posthumously in Varignon's *Nouvelle mécanique ou statique* in 1725.

"displacement" in the direction of the force, thus leading to the eventually corrected term "Principle of Virtual Work" and much later to the term *mechanical work*, coined by Coriolis in 1829. So while the term velocity was incorrect, it motivated the new term "energy." Despite this issue, Johann's insight was sound and the name "energy" would eventually become a centerpiece in physics, starting with Thomas Young's use of this word in his 1807 book, *A Course of Lectures in Natural Philosophy and the Mechanical Arts*,[6] and then, as discussed in Chapter 33, with the official adoption of this word by William Thomson as he and others established the new science of thermodynamics in the early 1850s.

Daniel Bernoulli

Into this realm of thinking arrived Johann's second son, Daniel Bernoulli, "the unsung hero of energy."[7] Born in 1700 while his father was still in the Netherlands, Daniel grew up in a world of wind and water, memories of which would remain years later as he attempted to be the first to describe the basic rules of motion for these fluids. Like his father before him, Johann strongly encouraged Daniel toward a practical profession, first in commerce and then, when that (fortunately for physics) failed, in medicine. As was the case with his own father, despite the direction of such encouragement, Daniel found himself fascinated in another direction: mathematics.

In secret, Daniel's older brother, Nikolaus II, tutored Daniel in mathematics. And in secret, Daniel discovered Newton. Newton's Laws of Motion resonated deeply within Daniel, leading him to wonder whether the laws as derived for solid bodies could apply equally well to the fluids of his youth. Later, Daniel actively supported Newton, promoting his work on the continent and referring to him as the "crown prince" of physics,[8] a rather striking contrast to his father's vehemently opposite feelings toward Newton.

As Daniel's interest in mathematics became readily apparent to Johann, a pact was made between them. Johann would teach Daniel mathematics, and Daniel would enroll in medical training. In this way, Daniel received an education in both mathematics and natural philosophy from a master. A child prodigy in his own right, Daniel quickly soaked it all up. He naturally learned all about Leibniz's calculus and *vis viva*. He learned about the proposed conservation of *vis viva* and how stored live force is consumed whenever *vis viva* is generated. Within Daniel's mind, Newton and Leibniz began to intertwine, arguably the first time the two theories met within a single mind intelligent enough and open enough to unite them.

Daniel's career led him through medical school in 1716 and eventually, upon receiving an invitation from Catherine I, Empress of Russia, to St. Petersburg in 1725, to become a mathematics professor in the Imperial Academy of Sciences. The Empress sought to populate this relatively new city and academy with some of the finest minds Europe had to offer. She got two of the finest in Daniel and his friend and colleague, Leonhard Euler, whom we'll discuss shortly.

---

[6] (Smith, 1998) p. 8.
[7] (Coopersmith, 2010) p. 114.
[8] (Coopersmith, 2010) p. 120.

While in St. Petersburg from 1725 to 1732, Daniel flourished, bringing all of his education to bear on the study of motion, synthesizing Leibniz and Newton into his own thinking and culminating these efforts with his 1738 masterpiece, *Hydrodynamica*. Throughout this work, he fully embraced the concept of conservation proposed by Leibniz and furthered by his own father, but did so at a higher level, shifting the terminology from the loosely worded "conservation of live forces" to the more definitive "equality between the actual descent and potential ascent." He was the first to introduce the word "potential" to physics.[9] Daniel had a deeper intuition than both his father and Leibniz about the importance of this term, putting it at a level equal to that of "live force" and actively using it in his investigations of natural phenomena on a much wider scale than used before.[10]

Contained within Daniel Bernoulli's *Hydrodynamica* is a wealth of ideas motivated by his conservation principle. He proposed the first kinetic theory of gases in which "very minute corpuscles, which are driven hither and thither with a very rapid motion,"[11] cause such phenomena as pressure and temperature. He studied the conceptual use of a cylinder filled with such kinetic gas corpuscles to determine what happens during compression of an "elastic" gas with a moving piston, a concept to be studied many times in subsequent years and decades in light of the rise of the piston-driven engine in the industrial world. He integrated Newton's 2nd Law, incorporating his inverse square equation for gravitational force, over distance to quantify the speed of the Moon's orbit based on its free fall toward Earth from an infinite distance, one of the first such demonstrations of the connection between Newton's concept of force and Leibniz's concept of *vis viva*. This was the same approach used by Niels Bohr years later in the early days of quantum mechanics to analyze the electron's orbit about a hydrogen nucleus. He considered the efficiency of machines and the absolute limits to such efficiencies, ideas that would re-appear in the nineteenth century development of thermodynamics. He showed insight into the potential energy stored in chemical bonds by attempting to quantify the "live force which is latent in a cubic foot of coal."[12] He applied the "minimum principle" to mechanical problems. In short, he combined and then applied Newton and Leibniz in new ways with new insights.

The Bernoulli Equation for Fluid Flow

An excellent manifestation of Daniel's accomplishment in *Hydrodynamica* was his derivation of his famous equation for fluid flow, which is a story in and of itself. Working at the interface between medicine and physics, instinctively following Galileo's lead of discovering new ideas at the interface between disciplines, Daniel found himself intrigued by blood circulation.

---

[9] (Coopersmith, 2010) p. 111.

[10] The terminology here is somewhat confusing for it once again seems that the origin of a critical word we use today came from an incorrect connection in the past. "Actual descent" refers to a change in potential energy during fall, while "potential ascent" refers to kinetic energy as quantified by its ability to rise against gravity should its direction be shifted from horizontal to vertical. While Daniel's use of the word "potential" was a first in the study of motion, he used it to describe not potential energy but kinetic energy, since horizontal kinetic energy quantifies how high (vertically) a body could potentially rise against gravity and since there was no other way to measure velocity.

[11] (Coopersmith, 2010) p. 72.

[12] (Coopersmith, 2010) p. 118.

Apparently, it was customary in those days for doctors to intentionally bleed patients to relieve certain maladies. But there was no quantitative guide for such a procedure. In looking at this, Daniel wondered if there were a way to measure the pressure of the blood in the vein beforehand, so that a doctor could use the information to guide his efforts. It was known that if you punctured someone's artery, blood would spurt out. Daniel reasoned that the height of the spurt should be a direct measure of the pressure within. You can see where this is going because the height achieved by a moving body is also a measure of the body's *potential ascent* or kinetic energy. Seeking to connect concepts and theories, Daniel wondered, *could there be a relationship between kinetic energy and pressure in a contained moving fluid?*

The study of fluids was not virgin territory. Indeed, the Ancients and especially Archimedes pretty much solved the science of hydro*statics*, showing, for example, how pressure increases linearly with depth in a static fluid or how total force of a fluid equals the pressure times the area against which it acts. The Roman Empire subsequently turned these concepts into practical use, as manifested by their engineering marvels for moving large quantities of water. The deeper theoretical aspects of fluids were explored by the groundbreaking work of the Italians, starting with Leonardo da Vinci in the fifteenth and sixteenth centuries and continuing with Galileo, Evangelista Torricelli, and Domenico Guglielmini in the seventeenth century. But the problem remained. How does pressure change once a fluid shifts from *static* to *dynamic*? And why does motion change pressure to begin with? The problem remained a problem because its solution required an advanced understanding of multiple concepts such as fluid pressure, force, *vis viva*, and the conservation of living force. In other words, it remained until someone could philosophically and mathematically combine the work of Newton and Leibniz in the study of a moving fluid. And this is what Daniel did.

As eighteenth century's Galileo, Daniel instinctively attacked this problem with theory, experiment, and mathematics. First, he had to figure out a way to even measure the pressure of a flowing fluid. Building on his earlier insights, but not wanting to puncture arteries, he instead punctured the wall of a horizontal pipe and put a small hollow tube into the wall. As fluid flowed through the main pipe, water rose in the small tube to a certain height. Here Daniel discovered a means to indirectly measure the pressure of a flowing fluid. This was a tremendous breakthrough in the study of fluid dynamics and a concept used in the university laboratory today to study fluid flow. While physicians started using this concept to measure patients' blood pressures by sticking sharp glass tubes into arteries,[13] Daniel moved on and used this concept to discover that as water flows from a large-diameter pipe to a small-diameter pipe, its flow rate increases while its pressure decreases. That speed increases with decreasing pipe diameter made sense to him based on da Vinci's "Law of Continuity" that said fluid speed through a pipe increases as pipe diameter decreases in order to maintain constant mass flow. But it wasn't clear to him why pressure decreased.

Daniel then sat down and did what amounted to the first ever energy balance around a flowing incompressible fluid. Using Leibniz's infinitesimals, Newton's 2nd Law (the instantaneous acceleration, dv/dt, of a fluid is proportional to the force applied to that fluid, force being pressure times area), and the conservation of living force based on his principle of *equality between*

---

[13] You can see here the motivation in the market place for the invention of a less obtrusive measure of blood pressure.

the actual descent and potential ascent, Daniel worked the math and arrived at an equation describing the flow of an incompressible fluid. Modifying his equation with terms we're used to today (ignoring changes in elevation) brings us to the now famous Bernoulli Equation:

$$P + \tfrac{1}{2}\,\rho \times v^2 = \text{Constant} \qquad\qquad\qquad\qquad [11.1]$$

where P is fluid pressure, ρ is density, and v is flow speed.

As a fluid moves through a pipe, its total energy remains constant. But what's happening at the molecular level to cause this?[14] Bernoulli naturally didn't address this in the eighteenth century, but let's look at this based on current knowledge, starting by considering the total energy of the fluid. It's the average kinetic energy of the molecules relative to a stationary mean-velocity reference. The kinetic energy is based on the speed of the molecules, again relative to the reference, which is equal to:

$$\text{Flow energy} \propto \left(\text{Molecular Speed} \right)^2 = v_x^2 + v_{radial}^2$$

In this equation, the flow rate down the pipe corresponds to v_x and is the flow speed in Equation 11.1, while pressure corresponds to v_{radial}. As the diameter of the pipe decreases, the increasing flow speed (v_x) causes molecules moving in the flow stream to collide with greater speed into the molecules moving radially toward the pipe wall. This then deflects the molecules' flow direction from perpendicular to axial, decreasing their strike speed (v_{radial}) and so decreasing the measured pressure, which is proportional to the square of v_{radial}. Thus, with decreasing pipe diameter, v_x increases while v_{radial} and thus P decrease. Both molecular speed and thus total flow energy remain constant as the component energies trade-off with each other.

In 1732 Daniel assembled this finding together with all others into his *Hydrodynamica*, which he finally (after a long 6 years!) published in 1738. And now we return to the family drama.

The Bernoulli Family Drama

All had not been going smoothly between father and son during this time. In 1734 both were recognized as co-winners of a French Academy competition. Instead of a father's pride for his son, and a son's pride for his father, the two egos battled each other, the father pained that the son had risen to his level, the son not helping any by arrogantly embracing this fact. One year after Daniel's work was published, his father prepared a manuscript of his own work, *Hydraulics*, but dated it 1732 to make it appear that his book had been written prior to his son's. This may have been fine except for one fact. His father had included in his book analysis of the changes in pressure against the side of a vessel due to water flowing through the vessel. While Daniel could never prove so, he felt that his father plagiarized his own work, writing, "Of my entire *Hydrodynamica*, of which indeed I in truth need not credit one iota to my father, I am robbed all of a sudden, and therefore in one hour I lose the fruits of a work of ten years. All [of my father's] propositions are taken from my [own work]."[15]

---

[14] For molecular motion during fluid flow, consider this re-arrangement of the molecules' vector components: the vector is total energy, the radial component is pressure, the axial component is flow rate.

[15] (Coopersmith, 2010) p. 122.

Despite the drama, one can only think, what an amazing family, a university within themselves! The Bernoullis collectively contributed much to the progress of both mathematics and physics, including the study of motion. Ironically enough, the closeness of the family is what actually made this happen. They formed a team of intellectual talent, encouraging each other through the early stages of education, inspiring each other to learn and master and achieve. Nothing inspires so much as wanting to keep up with an older brother, wanting to make father proud, wanting to be better. To have so much talent concentrated in one single family was remarkable. The closeness served as the crucible for learning. To paraphrase Newton, each saw further because they stood on the others' shoulders. It's unfortunate that while some competition can be exciting and encouraging, other competition can be bitter and adversarial. It all depends on where the ego sits. Alas, for the Bernoulli family, the egos were strong, ultimately dissolving relations between Jakob and Johann and then again between Johann and Daniel.

Leonhard Euler

Read Euler, read Euler, he is the master of us all.
– Laplace[16]

The other key eighteenth century player in the history of energy was also the other key participant in the Bernoulli family drama: Leonhard Euler (1707–1783). A child prodigy himself, Euler enrolled in the University of Basel in 1720 at age 13 and, inspired by his father's own account of listening to the lectures of Johann Bernoulli, sought to be coached by this famous man. After solving every problem thrown at him by Johann in record time, Euler won him over and became his most famous and gifted student. In so doing, Leonhard naturally developed a relationship with Daniel. The two hit it off, and after Daniel's brother Nikolaus II's unfortunate death from tuberculosis while working with Daniel in Russia, caused in large part by the extreme Russian winter, Daniel invited Euler to join him. The two collaborated for many years in Russia.

It's hard to imagine the magnitude of the intellectual power of the Bernoulli–Euler collaboration. Their complementary strengths, Daniel helping Euler with the physics, Euler helping Daniel with the mathematics, surely provided compounded returns on their time together in the Russian Academy. It's unfortunate that Euler necessarily got entangled in the ill-will between father and son when he professionally reviewed both *Hydrodynamica* and *Hydraulics*. His public favoring of Johann's latter to Daniel's former was enough to drive Daniel to despair. "[You] diminish my inventions in a field of which I am fully the first, even the only, author."[17] And this really begs the question, after so many years of working together, how is it that Euler wasn't fully aware of and on top of Daniel's accomplishments in fluid flow? How could he not have known that the solution of fluid dynamics was truly Daniel's and not Johann's? But enough of the drama. Let's move on. Let's discuss what Euler did.

[16] Laplace quote cited in (Dunham, 1999) p. xiii.
[17] (Guillen, 1996) p. 109.

Simply put, Euler united Newton and Leibniz into a mathematized mechanics that became *the* standard at that time and ever since. Clearly he didn't accomplish this task alone. His learning the very pinnacle of the state-of-the-art in calculus from Johann (and indirectly Jakob) Bernoulli was the initial and perhaps most important step in his career as it provided him a tremendous opportunity to further the young calculus. And he took every advantage of this opportunity, moving calculus from art to rigorous science, and in so doing, rising to become the most important mathematician of the eighteenth century. But this is not why Euler is part of this story.

As with other leading mathematicians of that time, such as d'Alembert and Lagrange, Euler was also very much a physicist, having absorbed the writings of Descartes, Huygens, Leibniz, and Newton, and also having accelerated his mastery of the issues through his learning from Johann and his collaborations with Daniel. Euler clearly benefited from others. But in the end, major steps forward still typically reflect the efforts of a single individual, and Euler was this individual. He was the right person, in the right place, at the right time.

If there is a word that captures the union of mathematician and physicist, the word would apply to Euler. It was Euler's seat at the interface between calculus and mechanics that enabled him to simultaneously consider both, thus enabling one of the most remarkable parallel developments of two separate fields in history. Breakthroughs in either led to breakthroughs in the other. Euler was playing in a wide-open field of combined mathematics and physics that no one had really touched. He seized the opportunity in front of him, generating over 500 books and papers during his lifetime and about 300 more after his death, as others brought his unfinished manuscripts to light. His prolific nature continued to his end, despite the onset of blindness in 1766. Working with scribes and a phenomenal memory, Euler continued to publish at an astounding rate.

Euler brought a sharp mind and cold heart to the task of mathematizing mechanics. He used a machete to cut through the jungle of ideas and thoughts and speculations. He removed the mysterious forces, the hand-waving explanations, the meta- prefix, and founded a solid and clear structure on which to build the future. Incorporated into his work was his guiding principle, "where a change is, there must be a cause," reflecting his deep belief in the fact that something *must* be conserved, thus philosophically aligning himself with Johann Bernoulli, Leibniz, and Descartes.

The starting point for Euler's daunting effort was Newton. The *Principia* held the answer to many problems except itself. Many simply could not comprehend the value of this book, which hindered its acceptance, especially on the Continent. At root was the fact that while Newton used his calculus to discover his Laws of Motion and Universal Gravitation, he didn't explicitly include its use in the *Principia*, instead favoring the use of geometric analysis. This is reminiscent of Galileo's use of the water-drop experiment to prove his Law of Fall, while, as the story goes, having actually determined the law from a less respected but no less valid experiment involving musical cadence. Regardless of Newton's rationale, this decision slowed the *Principia*'s impact.

Euler took Newton's family of ideas and re-wrote them into the language of calculus, thus making it much easier for others to understand and appreciate Newton's accomplishment. One significant example of this concerned Newton's 2nd Law. Euler was the one who analytically formalized the 2nd law into the equation we know today, $F = d(mv)/dt = ma$. Go back and re-read Newton's 2nd Law as he wrote it, "A change in motion is proportional to the motive force impressed and takes place along the straight line in which that force is impressed." This was a proportionality based on motion and not mass. Mass had been hidden in Newton's celestial work simply because the fall of bodies in gravitational fields does not depend on mass. Euler was the one who wrote the equation in terms of differential calculus and explicitly included mass.

He translated, clarified, compared, and built on the philosophies of not only Newton and Leibniz, but also of Bernoulli, Huygens, and Descartes. He endorsed Newton's "force" as an external cause of change. He used the continuity of calculus to make mechanics continuous, an approach that lasted until Planck's discovery of the quantum forced a revision. He developed the concept of *state* and a *change of state*, concepts we continue to use to this day in thermodynamics, and proposed that all changes in state result from the action of an outside force.

As Daniel Bernoulli also did, Euler used Newton's 2nd Law to show that the integration of force over distance is quantified by the change in kinetic energy. Why was this important? Because this integration was the first demonstration of the theoretical connection between Newton's Laws of motion and Leibniz's *vis viva*. Recall that Newton never directly addressed kinetic energy in the *Principia* simply because it wasn't needed. He developed his laws primarily to explain planetary motion, and for this, momentum proved totally sufficient and kinetic energy wasn't needed. From a larger perspective, Newton simply didn't need to address the concept of energy.

Today, thanks largely to Euler's effort, we practice Newton's physics using Leibniz's calculus. Euler's many publications, and especially his 1736 *Mechanica*, became the standards on both mechanics and calculus.[18]

Bernoulli and Euler

It is notorious that this very promising pioneer work by Daniel Bernoulli was neglected by succeeding generation of physicists and chemists.

– D. S. L. Cardwell[19]

Daniel Bernoulli and Leonhard Euler, separately and together, helped break the logjam created when Newton's and Leibniz's respective works met each other in the early eighteenth century. They brought these two different paths together and so created the bridge to the nineteenth century's completion of the conservation of energy. They each published significant works and

---

[18] Before closing this section on Euler, I must mention one additional contribution of Euler's of the many to choose from, in the area of relativity, which is otherwise (way) outside the scope of this book. He brought rigor to this subject by merging Newton's theory of absolute time and space with Descartes' and Leibniz's theory of relative motion. In so doing, he 1) emphasized that velocity is an arbitrary magnitude since it's a function of frame of reference, 2) formulated the Galilean Principle of Relativity in which the fundamental laws of physics are the same in all inertial frames, 3) introduced the concept of an "observer" into relativity, and 4) raised the discussion around the concept of "simultaneity." All of these efforts established *the* standard in relativity and helped propel Einstein forward to his own theory of relativity within which the conservation of energy and momentum are contained.

[19] (Cardwell, 1971) p. 25.

became famous in and of their own right. But some of their key ideas were so advanced that it simply took time for them to disseminate and take root.

Take for example Daniel's kinetic theory of gases. This theory went largely unnoticed for many years, which is an interesting history in and of itself[20] as this theory had to be "discovered" time and time again, seemingly for the first time, by many. After Bernoulli's work (1738) came Henry Cavendish (~1787; unpublished?), John Herapath (1816), and John James Waterston (1845), each proposing a kinetic theory of gases, each working largely in isolation and each unaware of their predecessors' work. The theory became public in 1857 when Rudolf Clausius "discovered" it yet again and then became final ten years later when James Clerk Maxwell completed it, recognizing Herapath's efforts in so doing.

We read about this now and may ask, *but why didn't people quickly grasp a hold of such powerful ideas? Why was there even an issue? These ideas make so much sense!* But this is from our perspective today. Picture yourself as a scientist back in the 1700s. Many ideas were swarming around you. Old ideas, new ideas, right ideas, wrong ideas. Considering that on top of this were the confusion of terminology and language and the heat of emotional attachment, one starts to appreciate how muddy the water truly was. Our history books tend to show us only the winners, the ideas that worked, and this makes for a very misleading view. We are shown a nice linear and logical flow as opposed to the reality of the mess, the many strands of ideas that led nowhere. To truly appreciate this, to truly appreciate this reality of history, one must, as Nassim Taleb recommended, read the diaries of all those involved at that time. They capture the unadulterated thinking at a specific moment in time. They reflect the mess that was.

---

[20] (Coopersmith, 2010) p. 71–75. Bernoulli's kinetic theory of gases faced two challenges in the early 1700s: it had little or no predictive value and so couldn't be experimentally validated, and it conflicted with Newton's static model of atoms that repel one another. As discussed by (Cardwell, 1971) p. 25–26, the existence of atoms remained speculative and "the best scientific explanations would be those that dispensed with [atoms]."

> The formulation of the conservation law for mechanical energy had its scientific roots in at least three areas of theoretical mechanics: (1) The principle of conservation of mechanical work... (2) The principle of conservation of vis viva... (3) The principle of conservation of (1) and (2) taken conjointly... that in every transfer of potential energy into kinetic energy and vice versa the total energy remains unchanged. By 1750 this law of conservation of energy had been accepted for ideal mechanical systems
>
> – Erwin N. Hiebert[1]

Science entered the eighteenth century with a solid foundation in Newtonian mechanics, Leibnizian calculus, and a strong guiding belief that within such a foundation, some*thing* must be conserved based on the impossibility of perpetual motion and the equality between cause and effect. This belief naturally led scientists to ask, what exactly is this *thing* and how is it quantified? Without quantification, conservation would be impossible to prove.

Significant progress toward answering these questions had been achieved by 1750 with the quiet arrival of what would later be recognized as the conservation of mechanical energy in which the sum of kinetic and potential energies in the moving parts of an ideal and isolated mechanical system remains constant. Based on many, many observations, this "law" was quietly accepted with no defining eureka moment by the scientific community; its deeper meaning as regards energy wouldn't be known until around 100 years later.

As we now understand, two quantities were involved in the creation of this conservation law, potential energy and kinetic energy. The discovery of each arguably came about due to the presence of Earth's gravitational field.

Simple Machines Revealed Potential Energy in the Form of mgΔh (Mechanical Work)

Study of the simple machines involving subtle up and down movements of weights in this gravitational field revealed the significance of the product of weight times change in vertical height (Figure 12.1). This product evolved slowly from analysis of the lever but originally did so based on "virtual" motions. Real engineers needed simple machines to do real work involving real motions. And so weight times vertical distance shifted from "virtual" to "real" and this

---

[1] (Hiebert, 1981) p. 95.

Block by Block: The Historical and Theoretical Foundations of Thermodynamics,
Robert T. Hanlon, Oxford University Press (2020). © Robert T. Hanlon.
DOI: 10.1093/oso/9780198851547.001.0001

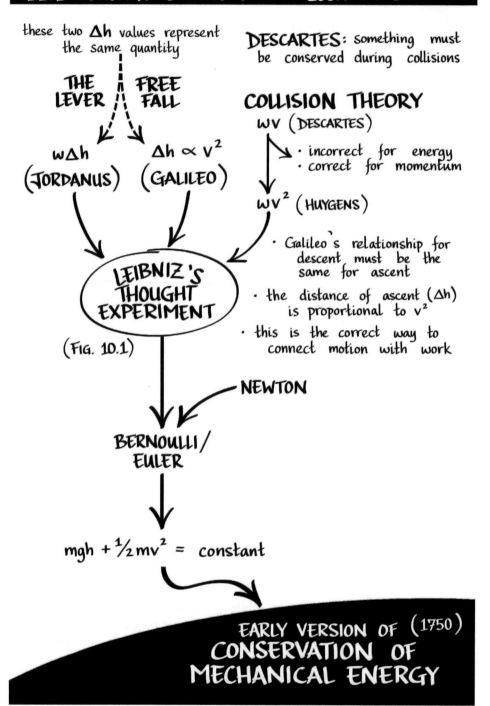

12.1   ENERGY DISCOVERY MAP ZOOM-IN ©RTH + CLS

these two **Δh** values represent the same quantity

DESCARTES: something must be conserved during collisions

THE LEVER / FREE FALL

COLLISION THEORY
WV (DESCARTES)

wΔh (JORDANUS)

Δh ∝ v² (GALILEO)

- incorrect for energy
- correct for momentum

WV² (HUYGENS)

LEIBNIZ'S THOUGHT EXPERIMENT

(FIG. 10.1)

- Galileo's relationship for descent must be the same for ascent
- the distance of ascent (Δh) is proportional to v²
- this is the correct way to connect motion with work

NEWTON

BERNOULLI / EULER

$mgh + \frac{1}{2}mv^2 = constant$

EARLY VERSION OF (1750) CONSERVATION OF MECHANICAL ENERGY

Figure 12.1 Energy discovery map zoom-in

term ultimately became *the* way to quantify the work done by a machine.[2] In such theoretically ideal machines, weights hang in balance with each other such that their subtle up-and-down movements are reversible; balance is achieved by a mechanical design that has heavier weights moving over shorter vertical distances than lighter weights. In another variation of Archimedes' quote, "Give me a lever long enough and I could move the world," a light person can hold a heavy rock off the ground with a lever when his or her lever arm is longer than that of the rock. In this equilibrium balance, zero net force and thus zero acceleration is present.

Free Fall and Ascent Revealed the Interplay between v^2 and Δh

Study of free fall in the gravitational field revealed the relationship between change in vertical height and the square of the speed of the falling body. The accepted impossibility of perpetual motion showed that this relationship worked in both directions, free fall and ascent. The value of v^2 at the bottom of a fall depended on the initial elevated height of the body. The elevated height of ascent depended on the body's initial value of v^2. The interplay between these two properties suggested conservation of some unifying property, although this interplay wasn't initially understood.

Leibniz Revealed the Logical Connection between $mg\Delta h$ and mv^2

It was Leibniz who unified the lever and free fall under the same umbrella that he called *vis* and which we now call energy, although he didn't totally realize the full extent of what he had done. His conceptual linkage of the two phenomena enabled his declaration of the importance of mv^2 (*vis viva*) in dynamics, a linkage that could not be identified with free fall alone since weight is not a factor.

In sum, analysis of the simple machines showed the significance of $mg\Delta h$; analysis of free fall showed how Δh is related to v^2; analysis of both together provided Leibniz the theoretical justification to link $mg\Delta h$ with mv^2.

When all was said and done, the critical element tying this story together was really Galileo's study of free fall and Huygens' furtherance of this study. As Hiebert stated, "Substantially all of the significant eighteenth century energy discussions fall back in one way or another upon the mechanics of Galileo and Huygens."[3] It wouldn't be until physicists such as Bernoulli and Euler started to work with Newton's mechanics using Leibniz's calculus that the connections between the free-fall quantities became apparent, that *vis viva* (mv^2) evolved to $\frac{1}{2} mv^2$, which was later called kinetic energy, and that Galileo's free-fall Δh was theoretically connected with the simple machines' Δh.

---

[2] The rate at which the work was done became known as power. So many horses can lift so much weight to such a height in such a time. Hence the origin of horsepower.

[3] (Hiebert, 1981) p. 99.

Newton and Leibniz Revealed the Fundamental Connection between Force, mgΔh, and ½ mv²

With Newton's force-based laws, Leibniz's energy-based concepts, and the arrival of calculus, physicists began to discover the relationships between the above quantities that originated from different paths. The integral

$$\int F \, dx = \text{Mechanical work} = mg\Delta h \qquad\qquad [12.1]$$

evolved over time to quantify mechanical work. While the likes of Leibniz indeed performed physics calculations based on the integration of force over distance to quantify work done, they didn't so conceptualize it.[4] This integral wasn't given its special name until Coriolis provided one in 1829 (Chapter 11). Prior to this, it was a mathematical expression and not a concept.

Revisiting the content from Chapter 5 using today's terminology, the above equation can be employed to analyze simple machines. The work done in (slowly) raising or lowering a body equals the body's weight (mg) times the distance through which it travels, Δh. Note the emphasis of slow movement. In simple machines, the movements are very slow as the system is in an equilibrium state for which no net force exists and thus no acceleration occurs. Kinetic energy plays no role in the ideal simple machine.

For free fall in a gravitational field, a net force naturally does exist. As demonstrated in Chapter 5, using Newton's 2nd Law as clarified by Euler (F = d(mv) / dt),

$$mg\Delta h = \Delta(\tfrac{1}{2} \, mv^2) \qquad\qquad [5.11]$$

where Δh is taken as a positive value quantifying the distance of fall.

As the values of mgΔh in Equations 12.1 and 5.11 are one and the same, this result crystallizes the fundamental connection between work (potential energy) and speed (kinetic energy). The increase in potential energy of a body raised in a gravitational field is exactly equal to the increase in kinetic energy of that body during its free fall back to the starting point. This result also explains why kinetic energy is quantified as being equal to one-half of Leibniz's *vis viva*. As Euler and others brought Newton's work to bear on mechanics, the resulting terms had to be mathematically consistent with each other.

Note that in both extremes of simple machines (statics) and free fall (dynamics), the mechanical work experienced by the body remains the same and is determined solely by the change in its initial and final positions—or recalling Euler's work, by the *change in state* of the body—and not by the way in which that change occurs. In other words, the body experiences the same change in potential energy independent of path. With the simple machine, potential energy exchanges with potential energy, and with free fall, potential energy exchanges with kinetic energy. Newton's force model quantifies the conserved energy in both.

Once freed from its association with gravitational force, the concept of mechanical work became a very powerful means of calculating the change in energy of a body or a system caused by any external force and became a unifying seed around which new ideas and concepts related to energy began to crystallize.

---

[4] As shared with the author in a 2017 email from Professor Kenneth Caneva.

From the Single Body to Many Particles

Looking back on this history, it's interesting that another of the stand-alone concepts from the eighteenth century, namely kinetic energy or *vis viva* (equal to twice kinetic energy), ended up <u>not</u> playing a critical role in the final steps leading to the creation of energy and its conservation, at least for two of those involved, Robert Mayer and James Joule.[5] It quietly slipped away and into the background. Clearly this concept played a critical role, serving to motivate, as it did, the invention of "potential energy" as a counterbalance to kinetic energy to ensure conservation of the sum of the two, but in the end, when the energy discovery was complete, the concept of *vis viva* was not needed. The reason was that another measurement of kinetic energy took its place.

Newton's force-based mechanics worked extremely well for the movement of celestial and terrestrial bodies that we could see, but it was initially useless for the movement of bodies we couldn't see. Accounting for such movements became the single most critical next step in the development of energy entering the nineteenth century, and the only means of doing such accounting at that time was with a thermometer. The beauty of the thermometer was that it enabled discovery of energy and its conservation without requiring one to understand the link between its measurement and the movement of those invisible bodies. That understanding would come later once early physicists began using Newtonian mechanics to model gases as a collection of colliding hard spheres.

The extent of the gulf that existed between eighteenth century mechanics and nineteenth century thermodynamics cannot be overstated. It was huge. Indeed, the discovery of energy and its conservation during the mid-1800s was an extremely complex affair. Energy itself is complicated, even to this day. There was no single discovery waiting to be found by some definitive experiment. Because energy manifests itself in many different forms, many different experiments were required. Development of a "law" demands such efforts. Indeed, this law had to be "forged."[6]

---

[5] In a 2017 email to the author, Professor Ken Caneva shared that the concept of *vis viva* was essential to Helmholtz's formulation of energy conservation. I discuss Helmholtz's contributions in Chapter 32.

[6] (Coopersmith, 2010) p. 350.

13 Heat: the missing piece to the puzzle

The Historians' Perspectives

C. B. Boyer[1] suggested that the discovery of energy and its conservation consisted of three rather complex steps:

1. The formation of the concept of energy
2. The statement that energy is neither created nor destroyed
3. A reasonable justification of (2), such as Mayer's and Joule's respective quantifications of the mechanical equivalent of heat.

I present them here to set the stage for what's to come as regards the path toward energy.

As nuanced earlier, a large stumbling block toward (1) was heat. People could easily sense heat by touch but couldn't fathom that it was part of a bigger picture. History has cleaned up this situation for us to some extent such that we can't see the mess for what it was, with ill-defined concepts and ideas cluttering the discussion. Heat wasn't really on anyone's mind when considering such concepts as "conservation" and "perpetual motion" in the eighteenth century. People read other peoples' works on this general subject but didn't understand what they were reading. Euler did a great job bringing structure to Newton's mechanics. But entering the nineteenth century, the world needed another Euler to bring structure to the world of heat.

Two theories contributed much to the clutter. These theories, widely circulated and embraced at the time, were flat-out wrong and very distracting. The first was a theory about combustion that proposed the existence of a fire-like element called "phlogiston" within combustible bodies that is released during combustion. Lavoisier did as much as anyone to bury this theory, demonstrating that combustion instead involves the chemical reaction of a portion of common air, later discovered to be oxygen, with matter. In these and other reaction experiments, Lavoisier carefully showed that matter is neither created nor destroyed and so effectively established the law of conservation of matter.

The second theory was, ironically enough, kept very much alive by the same Lavoisier who killed the first theory. People simply could not understand what heat was or how it flowed. Lavoisier proposed his version of the "caloric" theory to explain the observed heat-related phenomena, suggesting the presence of a fluid called "caloric" that is conserved and that flows from hotter bodies to colder bodies, never disappearing or forming, always present. While

---

[1] (Boyer, 1959).

Block by Block: The Historical and Theoretical Foundations of Thermodynamics,
Robert T. Hanlon, Oxford University Press (2020). © Robert T. Hanlon.
DOI: 10.1093/oso/9780198851547.001.0001

incorrect, this theory unfortunately provided sufficient explanations to remain standing for a long time, persisting as a viable theory in the scientific literature until the end of the nineteenth century when it was eventually overthrown by the mechanical theory of heat (heat–work equivalence, or work–heat equivalence; the two are the same and I use them interchangeably), the kinetic theory of gases, and the definitive discovery of the atom.

So in the midst of this mess, how was clarity achieved? How were Boyer's steps realized? There was no single path here. There were multiple paths, reflecting the multiple manifestations of energy. Some paths interconnected, while others didn't. We start by considering some of the unexplained observations made by scientists relating to heat. While many understood the basic concepts and interconnections involved in the mechanical world of simple machines and free fall, none realized that heat, or more accurately thermal energy, belonged in this same discussion. As Cardwell noted, "Work and heat belong to radically different realms of experience."[2] Heat was a mystery. Fortunately, the mystery became a strong magnet and attracted those creative minds that wanted to solve it and explain it. The observations and experiments they made became some of the key individual pieces contributing to the completion of the final puzzle.

The 1st Law of Thermodynamics – Revisited

As shared in Chapter 5, in 1850 Rudolf Clausius (1822–1888) was the first to capture the conservation of energy in an equation, which we now call the 1st Law of Thermodynamics:

$$\Delta U = Q - W \tag{5.12}$$

This seemingly simple equation is really very profound, representing as it does a tremendous amount of experience and thinking. It says that the internal energy (U) of a closed system (no mass in or out) can *only* change if heat (Q) is added *to* the system or if work (W) is done *by* the system. The equality sign represents the exact quantitative nature that a conservation law demands. As physicists, chemists, and engineers dealing with systems comprised of trillions of atoms and molecules as opposed to the single-particle world of the particle physicist, we use this equation and others built from it to analyze energy changes in any given chemical process.

The Q and W concepts provide a means to link any two physical systems together. Q entering one must depart the other; W being done *by* one must be done *to* the other. Q and W represent the processes of movement of energy between systems. They don't quantify a thing or a property of matter. They quantify the change that happens. In the cause–effect world of energy, the changes must exactly balance each other such that the change in total energy of all systems is zero. This equation captures this exactness.

(To be more technically specific regarding Q, there is no *thing* that moves between two systems during heat exchange. When two systems exchange thermal energy via conduction, it is the change in thermal energy of each system that quantifies Q. A hot system experiences a decrease in thermal energy and a cold system experiences an increase in thermal energy—of the same exact amount—when put in contact with each other such that the net change in energy is zero.)

---

[2] (Cardwell, 1989) p. 184.

Critical to note are the variables *not* included in this equation. If both Q and W equal zero and no mass enters or exits, or in other words, if the system is isolated, then ΔU equals zero and U doesn't change. If nothing external to the system interacts with the system, then the system's energy doesn't change, even if a chemical explosion were to happen inside the system. Indeed, the conservation of energy applies even for non-equilibrated states. There's no other magical "vital force" or "caloric" that can appear or disappear and so cause U to change. This is really what the 1st Law says. U only changes through Q and W.

While we'll later learn (in Chapter 31) about how Clausius arrived at this equation and what his newly created term, U, really means, I bring it up here to elucidate the different pieces that needed to be connected to complete this puzzle. We learned about the historical development of the simple machines that led to the concept of work (W) and now shift our attention to the historical development of the concept of heat (Q). I'll be referring back to this equation to keep focus on the exact interconnectedness between U, Q, and W in the context of energy.

The Thermometer

Man has sensed heat since the beginning of time. Hot and cold by touch, freezing and boiling by sight. Many figured that such effects must be related to the motions of invisible parts. When Leibniz considered inelastic collisions, he intuitively suggested that any loss of *vis viva* from two colliding large bodies is simply gained by the "parts [that] receive it." But such statements are a far cry from quantification, and per Boyer, quantification was what was needed to provide a reasonable justification to state that energy is neither created nor destroyed.

A story unto itself,[3] the development of the thermometer provided the means to finally quantify the energy associated with heat. Because fluids—gases or liquids—change volume with temperature, the fluid-based thermometer became the central physical phenomenon behind temperature quantification. Galileo arguably originated the concept around 1597 based on the expansion and contraction of air with temperature. Subsequent experiments with different materials and apparatuses to identify "best practices" ultimately led to use of mercury as the fluid, as selected and tested by Daniel Gabriel Fahrenheit (1686–1736) in 1714–21, building on the work of many others. His chosen reference points were 0° for "the most intense cold obtained artificially in a mixture of water, ice, and sal-ammoniac" and 96° for "the limit of the heat found in the blood of a healthy man."[4] Eventually the selection of reference points became better anchored to the freezing and boiling point of water, corresponding to 32° and 212° respectively on Fahrenheit's scale and 0° and 100° on the Celsius (or centigrade) scale.[5] William Thomson (Lord Kelvin) eventually determined temperature to be an absolute measure referenced to absolute zero and showed how to convert the Celsius scale to the absolute Kelvin scale, which was naturally named in his honor. We'll learn more about this history and the concept of temperature in Chapter 33, but suffice to say for now that the availability of such

---

[3] (Chang, 2007) provides a thorough overview of the development of the thermometer and the concept of temperature. Additional discussion: (Chalmers, 1952) p. 84–89. (Cardwell, 1971) pp. 17–24.

[4] (Chalmers, 1952) p. 85.

[5] Celsius' original proposal in 1742 had the reference points as 100° for freezing and 0° for boiling. More than one physicist subsequently inverted this to the scale used today.

a heat-measuring device provided a critical step toward the necessary quantification of heat, or (again) more accurately thermal energy, and also provided the means to compare experiments conducted in different locations.

*   *   *

Interestingly enough, the thermometer arrived before anyone understood what it actually measured. The physicists of the mid-eighteenth century mistakenly thought that the thermometer measured the "heat content" of a body. But, as we now know, temperature is just one aspect of "heat content" or thermal energy. The other aspect is what became known as heat capacity. A change in thermal energy is quantified by the change in temperature times heat capacity, which itself is the product of mass (m) times *specific* heat capacity (C). Based on the 1st Law of Thermodynamics, if heat (Q) is added to the system and if no work is done by the system (W = 0), such as by expanding against a piston, then the heat capacity can be calculated from the following equation:

$$\Delta U = Q = m\, C \Delta T \qquad\qquad [13.1]$$

Calorimeters were invented to measure the specific heat capacities of materials based on this understanding. Typically, as discussed in Chapter 5, a material at one temperature was put into contact with a reference material at another temperature and allowed to equilibrate. By the conservation of energy, the gain of heat by one must equal the loss of heat by the other. Water was typically used as the reference material with its specific heat capacity *arbitrarily* set to 1.0 British Thermal Unit (Btu) per pound per degree Fahrenheit. In other words, 1.0 Btu is the energy needed to raise 1 pound of water by 1 °F.

$$\Delta Q_{\text{test material}} = \Delta Q_{\text{water}}$$

$$\left(m\, C \Delta T\right)_{\text{test material}} = \left(m\, C\, \Delta T\right)_{\text{water}}$$

$$C_{\text{test material}} = (m \times 1.0 \times \Delta T)_{\text{water}} / \left(m\, \Delta T\right)_{\text{test material}} \qquad\qquad [5.7]$$

Knowing the respective mass and change in temperature of the two materials involved and assuming no phase change occurs, one can calculate the specific heat capacity of the test material (relative to water) in units of Btu/lb/F. A similar approach can also be used with isothermal events such as phase change or reaction, wherein the heat gained by the water reference equals the heat released by either phenomena. This short technical diversion was needed since the calorimeter played significant roles in both the completion of Boyer's steps toward the discovery of energy and also in the understanding of the connections between heats of reaction, free energy, and chemical reaction spontaneity. Focusing now on the former role, let's look at the development of heat capacity measurements through the experience of the man who led the way, Joseph Black.

Interlude – the Various Theories of Heat

In what follows will be discussions of the three main theories proposed to account for the phenomena associated with heat. To help keep things straight in your own mind, I outline each theory here (Figure 13.1).

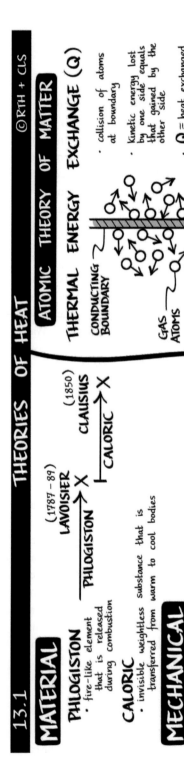

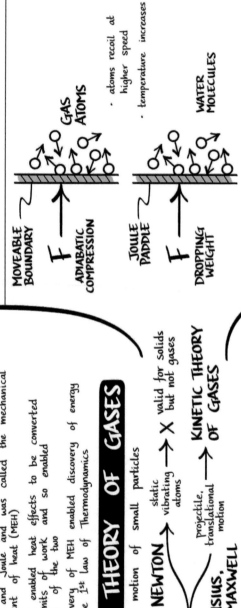

MATERIAL

PHILOGISTON $\xrightarrow{(1787-89)\ \text{LAVOISIER}}$ PHILOGISTON X $\xrightarrow{(1850)\ \text{CLAUSIUS}}$ CALORIC X

PHILOGISTON
- fire-like element that is released during combustion

CALORIC
- invisible weightless substance that is transferred from warm to cool bodies

MECHANICAL

WORK <=> HEAT EQUIVALENCE
- the exact exchange ratio was first determined by Mayer and Joule and was called the mechanical equivalent of heat (MEH)
- the MEH enabled heat effects to be converted into units of work and so enabled addition of the two
- the discovery of MEH enabled discovery of energy and the 1st law of Thermodynamics

KINETIC THEORY OF GASES

- heat as motion of small particles

NEWTON $\xrightarrow{\text{static vibrating atoms}}$ X valid for solids but not gases $\longrightarrow$ KINETIC THEORY OF GASES

GASEOUS ATOMS

CLAUSIUS, MAXWELL $\xrightarrow{\text{projectile, translational motion}}$

ATOMIC THEORY OF MATTER

THERMAL ENERGY EXCHANGE (Q)

CONDUCTING BOUNDARY

GAS ATOMS

- collision of atoms at boundary
- kinetic energy lost by one side equals that gained by the other side
- Q = heat exchanged

WORK <=> HEAT ($\Delta U = Q - W$)

MOVEABLE BOUNDARY

GAS ATOMS

ADIABATIC COMPRESSION

- atoms recoil at higher speed
- temperature increases

JOULE PADDLE

DROPPING WEIGHT

WATER MOLECULES

Figure 13.1 Theories of heat

The Material Theory of Heat: Caloric

The material theory of heat primarily dealt with the concept of caloric. Viewed as a weightless material that flowed from hot to cold, caloric was also viewed as being a conserved substance, and it was these (incorrect) assumptions on which Sadi Carnot based his analysis of the steam engine.

The Mechanical Theory of Heat: Work–Heat Equivalence

The mechanical theory of heat, also called the dynamical theory of heat,[6] addresses the theory of work–heat equivalence, also referred to as heat–work equivalence, which is the belief that work can cause heat and heat can cause work and that there's an exact relationship—the mechanical equivalent of heat (MEH)—between the conversion of one to the other. Unlike the caloric theory, the mechanical theory says that heat is not conserved; it is consumed as work is done. This was a critical distinction between the two. As will be discussed in the following chapters, Robert Mayer and James Joule, working separately, used experimental data to quantify the value of MEH (Boyer Step 3) and then went on to suggest a higher-level conservation law in which heat and work exist as two forms of a conserved quantity that would later be called energy (Boyer Steps 1 and 2). Rudolf Clausius later encapsulated these concepts in Equation 5.12, which arguably launched classical thermodynamics. While this theory acknowledged that heat is caused by the motions of atoms and ether, the theory did not assume the nature of such motion.

The Kinetic Theory of Gases

This theory assumes that the temperature of a gas is related to the kinetic energies of gaseous atoms moving through space (translational motion). The supporting mathematics relied on the mechanics developed by Newton, Leibniz, Bernoulli, and Euler. This theory provided the explanation for work–heat equivalence and also the developmental framework for statistical mechanics and statistical thermodynamics.

*　*　*

While both the mechanical theory of heat and the kinetic theory of gases fell under the umbrella of the theory of energy and its conservation, only the mechanical theory (work–heat equivalence) was required to establish classical thermodynamics, which was based solely on macroscopic properties. The link between the kinetic theory of gases (heat) and work was not initially evident as best reflected by the fact that the arguable founder of this theory, Daniel Bernoulli, was able to develop the ideal gas law in 1732 while not realizing that the temperature of a gas must rise during adiabatic compressions.[7] It was during the later rise of this theory in the early to mid1800s that the link was made clear.

---

[6] The mechanical theory and the dynamical theory of heat evolved somewhat independently over time with changing definitions and eventually merged together into synonymous words. I chose to not attempt a distinction between them and instead used Clausius' definition of the mechanical theory as described in (Clausius, 1879) as the encompassing basis for both theories.

[7] (Kuhn, 1959) pp. 355–356.

14 Joseph Black and the rise of heat capacity

The Enlightenment [was] the variously defined and unsystematic body of thought that led from the humanism of Erasmus in the 1500s to the skepticism and the questioning of authority by Voltaire and Hume in the 1700s

– David Fromkin[1]

They distinguished the northern kingdom, and though remote from the French style in gravity of mien, their radicalism brought Scotland closer in spirit to the Enlightenment than ever was England in her eighteenth-century complacency

– Charles Coulston Gillispie[2]

Marked by the embrace of human reason, the rise of the Scottish Enlightenment in the eighteenth century, encouraged by the philosophies of David Hume (1711–1776) and Adam Smith (1723–1790), led to an exciting environment of accomplishment. As a deeply religious man with a mind tuned toward logic, observation, and analytical thinking, George Martine (1700–1741) was born into this inspiring environment. While he established his career around medicine, he found himself taken by the work of Herman Boerhaave (1668–1738), his advisor at the University of Leyden, and Pieter van Musschenbroek (1692–1761) on the concept of heat and thermometry and so ended up publishing a number practical essays on this subject between 1738 and 1740. It was this excellent work that "marked the beginning of the great Scottish school of science,"[3] especially as regards the doctrines of energy and thermodynamics developed by such illustrious alumni as James Watt, William Thomson (Lord Kelvin), William Rankine, and James Clerk Maxwell.

In yet another case of a thin historical thread being critical to the development of thermodynamics, Martine's essays fortunately found their way into the hands of Joseph Black (1728–1799). As both student and professor between both Glasgow and Edinburgh Universities, and perhaps in keeping with the character of a learned Scot during that time of reason, Black seems to have been motivated, in part, by the need to explain the unexplained. In addition to Martine's work, another inspirational starting point for Black was arguably the results of an experiment carried out by William Cullen (1710–1790), a chemistry lecturer at Glasgow University who was to become Black's mentor, colleague, and close friend. Cullen observed the simple fact that, upon reducing pressure on a receiver of ether, the ether boiled while

[1] (Fromkin, 2000) p. 152.
[2] (Gillispie, 1990) p. 206.
[3] (Cardwell, 1971) p. 31.

Block by Block: The Historical and Theoretical Foundations of Thermodynamics,
Robert T. Hanlon, Oxford University Press (2020). © Robert T. Hanlon.
DOI: 10.1093/oso/9780198851547.001.0001

simultaneously and very curiously cooling. Cullen duly published this finding in 1748 without analyzing its significance. The resulting "*Why?*" generated in Black's mind by this experiment and others was all the motivation he needed. A new and critical path in science was about to begin.

Joseph Black – Separated Temperature from Heat

Galileo's breakthrough occurred when he recognized the importance of doing rather than just sitting around and thinking. So too for Black. Much thinking about heat was circulating through science at this time. The leading paradigm was that heat was caused by the presence of caloric in a substance and that "temperature" quantified the amount of this caloric; heat and temperature were essentially synonymous. Black was sitting in this caloric paradigm and could have easily attempted to manipulate it to explain Cullen's and others' findings. But something bothered him about this approach and so, taking a page from Galileo's playbook, he turned away from the hand-waving theorizing and turned toward concrete experimentation. It's not that he rejected caloric; it's just that he didn't let caloric stand in his way.

By 1765 Black had carefully conducted many experiments involving temperature equilibration between various test materials and water, and so proved that different substances have different heat capacities, a finding at odds with the caloric theory that held that total caloric is uniquely defined by temperature. In so doing, Black also successfully untangled heat into its two components, temperature and heat capacity, and unknowingly began laying the groundwork on which the concept of energy would be built. As he summarized to his students, the quantity of heat in a body corresponds to the amount of "heat substance" which it contains; its temperature, as indicated by a thermometer, corresponds to the "intensity of the heat."[4] This was a fascinating finding because Black developed the approach to quantifying heat capacity that we still use to this day ($Q = m \times C \times \Delta T$) while working under the *wrong paradigm*. As an adherent of the material theory of heat, Black likely used the terms "heat substance" to refer to caloric fluid, "heat capacity"—one of the heat-related "anachronisms"[5] we continue to live with—to refer to the ability of matter to hold or absorb this fluid, almost as if the fluid permeated the voids in the material, and heat intensity to refer to the amount of caloric per unit of weight. Regardless of the paradigm he lived in, Black's new ideas and findings were indeed significant in and of themselves. They stood on their own. Having said this, it is still very interesting that the incompatibilities of his findings with the caloric theory—if you deduced the mathematical consequences of Black's assumptions that a body contains a certain amount of caloric of certain intensity, you'll find that caloric has different properties between different bodies, which is inconsistent with the idea that caloric was supposed to be unchangeable—did not lead Black to any overt desire to overthrow the theory. To Black, the caloric theory was "the most probable of any I know."[6] This was a very telling quote and suggested how difficult it was to find a replacement theory.

---

[4] (Chalmers, 1952) p. 88.
[5] (Cardwell, 1971) p. 289.
[6] (Chalmers, 1952) p. 91.

Evaporative Cooling Does Not Make Sense in the Caloric World

We return now to Cullen's curious finding relating to ether vaporization. Black's above clarification of heat dealt with "sensible heat," the type of heat we can "sense" by touch and so measure with a thermometer and for which a change in heat correlates directly with a change in temperature. So how could Cullen's ether cool during vaporization? This would imply loss of caloric, and yet the caloric thinking proposed that evaporation was supposed to be accompanied by a small *gain* in heat.[7] The false theory of caloric created much confusion back then; its *ad hoc* explanations for specific phenomena often ran ashore when generalized to explain other phenomena. For Black, he simply couldn't reconcile the phenomena of evaporative cooling with the caloric theory and so set about to bring the spotlight of experimentation to the issue. Like Galileo, he let go of the theory and did the experiments.

The Science of Calorimetry and the Doctrines of Sensible and Latent Heats

In a set of rather clever experiments that arguably founded the science of calorimetry, Black quantified the heat associated with melting ice and evaporating water.[8] In a large room maintained at constant temperature (47 °F), he hung two glass globes, one containing ice (32 °F) and the other water (33 °F), and then measured the temperature in each over time. He found that the water heated to 40 °F in about 30 minutes and that the ice melted and then also rose to 40 °F in about 10 ½ hours. Assuming an equal rate of heat flow from the room to the globes, he concluded that the heat absorbed by the ice was 21 times the heat absorbed by the water to get to the same 40 °F final temperature. Using this same general method but naturally with a source of heat at higher temperature, Black and his assistant, James Watt—yes, he of the steam engine—were also able to quantify the relative heat absorbed by heating and boiling water. Applying simple math to these results and using modern terminology, Black measured a latent heat of melting of ice of 139 Btu/lb and a latent heat of vaporization of water of 850 Btu/lb, each reasonably close to their currently known values of 144 Btu/lb and 970 Btu/lb.

Black was again confronted by the conflict between his results and his theory. Large amounts of heat were absorbed during ice melting and water vaporization without any corresponding changes in temperature. But how could so much *material* heat flow into matter without a change in the quantitative measure (T) of its presence? (To Black, because there was no change in temperature, the heat was "concealed" or "latent".) For whatever reason, Black once again did not use this opportunity to challenge convention but instead remained adhered to the material theory of heat. It's interesting that the early conflict between the conventional theory and the findings of Cullen and others motivated Black to experiment with heat in the first

---

[7] (Chalmers, 1952) p. 90.
[8] (Chalmers, 1952) pp. 89–93.

place, while the subsequent conflict between this theory and his own new findings did not similarly motivate him to question the theory. How deep the paradigm must have been!

By introducing both specific and latent heats to the science community, Black "advanced our understanding of heat more than any other philosopher in the eighteenth century."[9] His work and meticulous experimentation opened the door to a new realm of study even though his thinking remained in the caloric world. The remarkable thing in all of this is that he kept his experiments and his theories separate. Like Joule would later do, he brought an open mind to his experimentation, following the evidence wherever it led him. This was a rather remarkable testament to Black's character.

The Dulong–Petit Law

Black's groundbreaking efforts and famous lectures—he was a better lecturer than publisher[10]—inspired many to follow. Efforts to further understand the specific and latent heat phenomena involving gases, liquids, and solids ultimately changed from mere quantification to a deep probe into the nature of matter. The shift started in 1819 when Pierre Louis Dulong (1785–1838) and Alexis Thérèse Petit (1791–1820) looked at the specific heat capacities of a number of metals and, in their own mini-eureka moment, noticed that the product of specific heat times atomic weight, a new property at that time as motivated by Dalton's work, was relatively constant.[11] Thus was born the Dulong–Petit Law stating that the specific heat *per atom* is constant for solids. (This discovery, by the way, became a strong supporting piece of evidence later on for the atomic theory of matter.) Intrigued, Dulong pursued this finding, unfortunately without Petit who had died in 1820 at the young age of 29 due to tuberculosis, and in 1829 found the same result for gases—although, as discussed below, the values were different than those for solids—thus extending the Dulong–Petit Law to include gases. And this begged, if not screamed, the question, *why?*

Understanding Heat Capacity at the Atomic Scale

Heat capacity quantifies how the temperature of a body responds to the addition of thermal energy. In considering its defining equation based on *specific* heat capacity (based on mass), $C = Q/\Delta T$, heat capacity can be viewed from two different perspectives. It quantifies the amount of heat or thermal energy needed to raise the temperature of a body by a certain amount (high heat capacity means more heat needed) or it quantifies the change in temperature caused by the addition of a certain amount of thermal energy (high heat capacity means low temperature change). Regardless, the quantity measured is dependent on the structures of the atoms and molecules comprising the body. The way that they move in response to absorbed thermal energy is what determines heat capacity.

---

[9] (Coopersmith, 2010) p. 78.
[10] (Cardwell, 1971) p. 34 and p. 40.
[11] (Chalmers, 1952) pp. 93–95, pp. 123–124. (Cardwell, 1971) pp. 137–141. (Fox, 1971) pp. 238–248.

To better understand what was happening in Dulong and Petit's studies, let's zoom in. Consider the collision of two non-interacting, monatomic gas atoms. When they collide, total energy is conserved. Moreover, since kinetic energy is the only form of energy involved, then kinetic energy is also conserved. There is no opportunity to irreversibly consume energy, such as would occur in the event of deformation, by the single atoms during collision. Because kinetic energy is conserved, the collisions are termed elastic. The kinetic energy lost by one atom is gained by the other.

Now assume you have two systems of such gases in contact with each other, separated by an infinitesimal membrane, each having a different temperature. As the atoms from one system collide at the membrane interface with atoms from the other, they experience a change in kinetic energy. In general, for each collision, the atom having the higher kinetic energy will experience a decrease in kinetic energy and the atom having the lower energy will experience an increase in energy; thus, the two system temperatures move toward each other. In this situation, because the total change in kinetic energy of each system is simply the sum of the changes of the many individual collisions, the total kinetic energy lost by one system is gained by the other.

In this scenario, we say that heat as quantified by Q flows from hot to cold. But these words don't reflect what actually happens. Remember that no *thing* actually moves from one system to the other; the concept of a "flowing heat"—yet another anachronism we have to live with—originated from a time when we thought "caloric" moved from one to the other. Unfortunately, when we later realized that there is no such thing as caloric, we never went back and updated the language to make it reflect reality. So Q represents a change in total energy for a system caused by thermal contact. In this thought experiment, Q represents the decrease in total kinetic energy of the hot system and also the increase in total kinetic energy of the cold system. The absolute values of the two changes are equal because all the contributing collisions are elastic.

What else do we know about these two systems? Because each contains an ideal, monatomic gas, the total energy of each is equal to the total number of atoms times the kinetic energy of each, which we'll assume to be the mean as represented by the $<\;>$ symbol.

Hot System (1) Total Energy: $N_1 <\tfrac{1}{2}\, m_1 v_1^2>$

Cold System (2) Total Energy: $N_2 <\tfrac{1}{2}\, m_2 v_2^2>$

Assuming the same number of atoms (N) in each system,

$$\Delta <\tfrac{1}{2}\, m_1 v_1^2> = \Delta <\tfrac{1}{2}\, m_2 v_2^2>$$

As we'll see later, the average kinetic energy of the atoms in a monatomic ideal gas is proportional to absolute temperature. Thus,

$$\Delta T_1 = \Delta T_2$$

Considering that $Q = C\,\Delta T$ and that both Q and ΔT are equal between the two systems, then one must conclude that

$$C_1 = C_2$$

What this says is the heat capacity of all ideal monatomic gases must be equal to each other, regardless of their respective mass or atomic weight. This makes sense since it's based on the fact that the temperature of any monatomic gas is dependent on kinetic energy and not solely on atomic weight of the atoms involved. At equilibrium, a mixture of light and heavy atoms will have the same average kinetic energy; they'll just have different average speeds.

Conclusion: The Heat Capacity per Atom is the Same for Monatomic Gases = 1.5 R

As we'll see in Chapter 40, the heat capacity at constant volume (C_v) of a monatomic ideal gas using either an atomic or a molar basis as the two are the same in this scenario,

$$C_{v,\,atom} = 1.5R \text{ monatomic ideal gas} \left(\text{per atom} \right)$$

Now this is a rather fascinating conclusion as it suggests that the heat capacity on a per-atom basis for helium, having an atomic weight of 4, is equal to that for argon, having an atomic weight of 40. And in fact, this is indeed the case. The heat capacity at 1 atm for each of the noble gases from helium up through radon is the same at about 1.5R or 12.5 J/mol/K. This is entirely consistent with the conservation of kinetic energy during elastic collisions of hard spheres.

C_v of Crystalline Solid = 3 R

This is fine for ideal gases, but what about for solids? Recall that the Dulong–Petit Law applies to both. They determined that the specific heat capacity for a given *solid* on a per-atom basis is also constant, just as it is for a monatomic ideal gas, except twice as high, for the following reason. In a crystalline solid, the atoms are locked in a three-dimensional structure with each individual atom jiggling around its own center point. As each atom moves further from its center point, the force for it to return increases. In a way, the atoms behave as swinging pendulums with kinetic and potential energy exchanging while total energy remains constant. Now there's something called the equipartition theory that says that the total energy contained in such a crystalline solid is divided equally between kinetic and potential energies.[12] So whereas a change in gas phase energy leads to the direct 1:1 change in temperature, a similar change in solid phase energy leads to a 50% change in kinetic energy (temperature) since the other 50% change occurs in potential energy. The specific heat capacity on a per-atom basis for a crystalline solid is thus 3R, exactly twice that for an ideal gas, independent of the size of the atoms or types of elements involved. For example, the specific heat capacity (per atom) of iron (mol. wt. 56) is 3.02R while that for gold (mol. wt. 197) is 3.05R. On a per-atom basis, it takes twice the heat to raise temperature by a fixed amount for a crystalline solid than for a monatomic gas.

$$C_{v,\,atom} = 3R \quad C_v \text{ of crystalline solid} \left(\text{per atom} \right)$$

---

[12] Technically, the energy is divided equally among all degrees of freedom present in the system. In the case of a crystalline solid for which vibration is the only motion occurring, vibration happens in three directions each having a kinetic and a potential energy component. This makes for six degrees of freedom total, equally split between kinetic and potential components.

The Dulong–Petit law is a natural outcome of the atomic theory of matter combined with the conservation of energy. For all its simplicity, this law works quite well, indicating why specific heat capacity for individual atoms is approximately bounded at the low end by that for monatomic ideal gases (1.5R) and at the high end for crystalline solids (3R). We'll discuss this some more in Chapters 40–42 on the kinetic theory of gases and statistical mechanics.

The study of specific heat capacity became a great means to use macroscopic properties to study microscopic behavior. And as we know, microscopic behavior can be described by the mechanical theory of heat only up to a point. It didn't take long for other scientists to discover this point when they noted unexplained departures from Dulong–Petit during studies at more extreme temperatures, thus raising the "why?" question all over again.

It was left to Einstein (who was this guy?!) in 1907, with further enhancements by others and especially by Debye in 1912, to make final sense of all the data. In the mechanical world, the change in energy between all possible motions—translational, vibrational, rotational—occurs as a continuum. But in the quantum mechanical world, the change is quantized and temperature dependent. A certain minimum temperature is required to enable certain energy transitions. Below that level, the transitions or motions are "frozen out"[13] meaning that as temperature decreases, certain vibrating molecules cease vibrating. When this happens, energy can no longer be stored in vibrational motion and heat capacity decreases. Einstein's (and others') theory and mathematics successfully explained the impact of such quantum effects on specific heat capacities.

We'll continue discussion of specific heat capacities later, and especially the heat capacities of gases at constant volume, C_v, and constant pressure, C_p, since the difference between them played an important role in Mayer's calculation of the mechanical equivalent of heat. Right now, though, we'll continue with the history of energy.

---

[13] (Feynman et al., 1989a) Volume I, 40–6.

15 Lavoisier and the birth of modern chemistry

Between 1790 and 1825 France produced the brightest galaxy of scientific genius that the world has witnessed to date

– D. S. L. Cardwell[1]

Around the turn of the nineteenth century, France (specifically Paris) and Britain established themselves as the two centers of science in the world, with acknowledged outposts in Switzerland, Scandinavia, and Russia (St. Petersburg). German education was not yet there, with its best scientists having moved to the likes of Paris or St. Petersburg to pursue their fortunes since Germany was not strong enough politically, economically and militaristically to support them.[2] This would come later, as Germany would rapidly improve during the nineteenth century until it was accepted as the best in the world, driven throughout by its own ongoing unification process that culminated in 1871.

The French government, believing strongly in the need for a centralized approach to intellectual study and the importance of science to the country, created an atmosphere of higher education around Paris to provide formal training and funding to scientists and engineers, including those in the military. This government directive led to the rise of the professional scientist, an ideal role for anyone wishing to pursue science for science's sake. Ah, the luxury of pure thought. This philosophy started in 1699 with the establishment of the Royal Academy of Sciences and has survived well ever since, even through the bumps encountered during the French Revolution (1789–1799) when the masses, having been inspired by the new Enlightenment principles that were sweeping the Western World, violently overthrew absolute monarchy and all the hierarchical and traditional privileging that went along with it. After the Revolution, the educational system grew even stronger under the direction of Napoleon. Some of the notable scientists and mathematicians nurtured by this centralized philosophy should sound familiar to you—Lagrange, Laplace, Lavoisier, Biot, Poisson, Regnault, Gay-Lussac, Fourier, Coulomb, Coriolis, Ampère, Legendre, Lazare, and Sadi Carnot—and many of these men were instructors of the famed École Polytechnique (founded in 1794). This government strategy was rewarded by the rise of Paris to the center stage of world science by the early 1800s.

* * *

The French, it often seems, formulate things, and the English do them
– Charles Coulston Gillispie[3]

---

[1] (Cardwell, 1971) pp. 119–120.
[2] (Cardwell, 1971) p. 32.
[3] (Gillispie, 1990) p. 205.

Block by Block: The Historical and Theoretical Foundations of Thermodynamics,
Robert T. Hanlon, Oxford University Press (2020). © Robert T. Hanlon.
DOI: 10.1093/oso/9780198851547.001.0001

Britain also rose in world science and shared center stage with Paris but traveled a different path, one not based on a formal training structure but instead one based on the individual genius, privately tinkering away in his own home or factory. Over time, such men found each other and eventually formed societies to discuss ideas and share results. With no central philosophy guiding this process, such societies formed not only in London but also throughout the country, such as in Scotland (Glasgow and Edinburgh), Cornwall, and Manchester. In the background of these activities burned the motivational forces lit by the Industrial Revolution. Having started in Britain around this time and being fueled by the presence of plenty of coal, the Industrial Revolution was fast developing momentum. During this period, the individual, driven by Britain's embrace of capitalism, commerce, and a supportive patent system, was financially inspired to apply his scientific and engineering findings to practical use. Also during this time period, the individual who was already wealthy used his money to pursue his own scientific passions. Both groups of individuals were consumed with hands-on *doing*.

While not a precise delineation between the two cultures,[4] the thinker–doer contrast between France and Britain played itself out in the development of two major breakthroughs of relevance to energy: chemistry and the engine. We'll start with chemistry by conducting a rapid-fire review of the birth of modern chemistry during the late eighteenth century. The key event during that period was the death of phlogiston, a fictitious material invented to explain the unexplained.

Joseph Priestley and Oxygen

The belief [in phlogiston] delayed the advance of knowledge and misled chemists into many entanglements

– T. W. Chalmers[5]

As of 1750, man was ignorant of both the composition of atmospheric air and the fact that it played such a critical role in chemical phenomena. Different types of "airs" had been found and isolated, but no one could put the pieces together. Remember that at this time, chemistry was still a fledgling subject, struggling to come out from behind the alchemists' confusing speculations. It would be an understatement to say that experimental evidence was lacking. It was also a time when many believed in the existence of but four fundamental forms of matter, a philosophy stemming from the ancient Greek world of fire, water, air, earth. Fire itself was viewed as a material substance that Lavoisier later named "caloric", from *calor*, Latin for heat.

---

[4] In a 2017 email to the author, historian Robert Fox suggested that Carnot (France) be considered as both thinker and doer in light of the fact that his training and background were in engineering (artillery officer). He further suggested that the distinction with regards to France might go three ways: scientists (e.g. Laplace and members of the Academy of Science), theoretically trained engineers (Carnot), and practical engineers, the real "doers" (e.g. Nicolas Clément or an engine-builder such as J.-C. Périer). Also, while the French aristocrats rather distanced themselves from practical experiments, Lavoisier, while wealthy, was not an aristocrat and wasn't afraid to get his hands dirty, and Cavendish, who was an aristocrat, probably didn't disdain experimenting either. My own added comment: Henri Victor Regnault and his meticulous experimental programs was clearly a "doer" as well.

[5] (Chalmers, 1952) p. 101.

Further confusing the scientists in the mid-1700s was the presence of an extremely deep and false paradigm. Man simply could not understand combustion. It's an easy concept for us today but back then, without really knowing anything about chemistry, people looked at fire and grasped for an explanation. As nature abhors a vacuum, an explanation was born. First proposed by Johann Joachim Becher (1635–1682) in 1667 and then furthered by Georg Ernst Stahl (1659–1734) in 1703, the phlogiston theory rose from man's mind to explain two prominent phenomena: combustion and calcination (oxidation). Stahl proposed the existence of phlogiston (phlog is Greek for flame) as an elemental substance that is driven out of substances by the application of heat. However lacking in experimental evidence this theory was, it <u>did</u> offer an explanation and thus, in the absence of anything better, survived for close to 100 years. I should have said "unfortunately survived" as its presence created much confusion and significantly delayed scientific progress. So let's follow the path of its rightful death.

In science, the doers generally precede the thinkers. So was the case in this story. Let's start with the doers in Britain and later move on to the thinkers in France. A sequence of British scientists started the downfall of phlogiston by experimentally isolating the gaseous components of air and combustion products. Joseph Black arguably took the first step in this paradigm-altering process when he heated limestone ($CaCO_3$) to form quicklime (CaO) and "fixed air" (CO_2) and later combusted carbonaceous matter to produce a material similar to his fixed air. Not enough was known then for Black to understand the details of the two chemical reactions involved: $CaCO_3 + heat \rightarrow CaO + CO_2$; $C + O_2 + heat \rightarrow CO_2$. Furthermore, his thinking was no doubt clouded by his continued belief in phlogiston. Black's work formed the basis of his M.D. thesis, published in 1754. (Once again, a tremendous thesis accomplishment, like Madame Curie's.)

Black's findings excited other Brits to study other known forms of "gas." One of his pupils, Daniel Rutherford (1749–1819), discovered nitrogen in 1772, although, like Black, he didn't realize it. By confining animals with air in closed chambers, Rutherford found that the gas (CO_2) produced by natural respiration, which he isolated via absorption into a caustic solution, was the same as Black's "fixed air" and also found that the "residue air" (N_2) was capable of killing life and extinguishing flames.

Joseph Priestley (1733–1804) followed suit by finding, but not truly discovering, oxygen in 1774. As part of his more extensive studies of air, Priestley found that, when heated, calcined mercury (HgO) released a gas (O_2) that, unlike Rutherford's nitrogen, supported both life and flame. Priestley's analysis was also clouded by phlogiston. A strong believer in the phlogiston theory until the end of his life, Priestley believed that because this released gas couldn't be burned in air (O_2/N_2), it must be *without* phlogiston and so called it dephlogisticated air. (The theories that are proposed to keep other theories alive!) Priestley's false belief system is what led others to call his achievement a finding as opposed to a true discovery.

Some have argued that Priestley's finding was preceded by Swedish chemist Carl Wilhelm Scheele (1742–1786). Scheele was the first to deduce that common air is comprised of two different substances, "fire air" (O_2) and "vitiated air" (N_2). He experimentally observed similar behavior between his "fire air" and the gas evolved from heating calcined mercury and was thus one step away from truly finding if not discovering oxygen, except that his mind too was clouded by the phlogiston paradigm. Scheele generated his results in 1773 but didn't publish them until 1777. Some question whether Priestley may have learned of Scheele's 1773 experiments and used them without acknowledgement to further his own reported findings of 1774. An interesting quote in this regards came from Dr. Eason of Manchester, who wrote to Black in 1787: "Priestley, not having anyone to steal from at present, I believe is quiet, unless it is to trouble the world with his religious nonsense."[6]

Cavendish Resolves the Water Controversy – 2 Parts Hydrogen, 1 Part Oxygen

The isolation of oxygen opened the door to a series of critical discoveries: the composition of the atmosphere (common air), the true nature of combustion, and the composition of water. It was only a matter of time until these lined-up dominoes rapidly fell. Which brings us to Cavendish.

Born into one of the richest and most well-connected families in England, Henry Cavendish (1731–1810) inherited a fortune that enabled him to pursue his passions in science without concern for income. He did much but unfortunately published little, reflecting a rather eccentric and very shy personality that strongly rejected attention and publicity. Many of his thoughts and findings in such areas as the radiant theory of heat, the gas laws, the conservation of Leibniz's *vis viva*, and electricity phenomena, remained hidden—another example of Taleb's "silent evidence" of history—until their discovery by others after his death, including the discovery of his writings on heat in 1969 (!). The increased understanding of his achievements over time elevated Cavendish to genius level. Relevant to our story here is what Cavendish *did* publish, which brings us to hydrogen and water.

First isolated in 1671 by Robert Boyle when he reacted metals with dilute acids—metal + acid → salt + H_2—hydrogen became known as an "inflammable air" because of its combustibility in common air. Seeking to understand this air better, Cavendish conducted a wonderfully elegant experiment in 1781 in which he exploded the "inflammable air" with common air. His quantitative exactness in the recipe for this experiment that lead to an almost exact two volumes of hydrogen per one volume of oxygen manifested Cavendish's keen thinking, research and planning on this subject. This ratio wasn't random. Cavendish clearly suspected water to be comprised of 2 parts hydrogen and 1 part oxygen.

A fascinating and critical observation of this experiment was that the product water showed no sign of acidity. Why was this an issue? Because many earlier similar studies produced an acidic liquid, leading to the conclusion that nitric acid is the primary combustion product of "inflammable air" and "dephlogisticated air." But what was actually happening was this. When burning hydrogen in air, if oxygen is in excess, then the high temperature generated by the combustion will cause oxygen and nitrogen to react to form nitric acid (HNO_3). By ensuring only a stoichiometric presence of oxygen, Cavendish minimized the formation of nitric acid, thus revealing pure, non-acidic water to indeed be the sole combustion product of hydrogen and oxygen, thereby resolving the "water controversy" that was swirling around this subject back then. Alas, Cavendish failed to fully understand the nature of the reactions involved as he too was led astray by his firm belief in phlogiston.

---

[6] (Chalmers, 1952) p. 106.

The British generated excellent experimental findings but couldn't put the pieces together. They sat in a fog caused by the phlogiston paradigm. The way out required a thinker, which brings us back to France.

Lavoisier – Making Sense of the Data

Almost overnight, it now seems, Lavoisier brushed away the cobwebs and displayed a picture in which the composition of water, Priestley's "dephlogisticated air," Scheele's "vitiated air," and Black's "fixed air", were to be seen fitting harmoniously and consistently into the scheme of a grand composition.

– T. W. Chalmers[7]

Recall Galileo. He was raised in the ancient Greek paradigm of motion, but then started generating and analyzing data that conflicted with this paradigm. Rather than questioning the data, he questioned the paradigm and changed the direction of science. So too was the experience of Antoine Lavoisier (1743–1794).[8] He was raised in the phlogiston paradigm, but then started seeing conflicting data. He believed in the data and so questioned the paradigm. With "boundless energy and wide-ranging curiosity,"[9] he broke with the past, quickly killed phlogiston and moved on to create the modern era of chemistry, in which phlogiston played no role.

The Death of Phlogiston

The key moment in Lavoisier's transformation was undoubtedly his receipt of a pre-print in 1775 of Priestley's publication about his finding of "dephlogisticated air." (This was another case in scientific history of questionable conduct as Lavoisier never acknowledged Priestley's work.) Ignoring Priestley's own phlogiston-based theories on the nature of the air, Lavoisier quickly grasped the truth. Combustion and calcination consisted of nothing more than the union of two bodies, specifically Priestley's "dephlogisticated air" (or Scheele's "fire air"), which Lavoisier first named *l'air vital* and later named oxygen, and either carbon (combustion) or metal (calcination). He furthered his theories upon receipt in 1783 of an early copy of Cavendish's findings and realized that these too could be fit into his evolving thesis on chemical reactions. One such reaction involved water. Water "fascinated Lavoisier;"[10] to him, it was not a simple substance but instead a union of "inflammable air" and oxygen. He proved this by running the reverse reaction. Based on a suggestion from his laboratory assistant, Pierre-Simon Laplace (1749–1827), Lavoisier passed steam through a red-hot iron gun barrel and found the presence of some "inflammable air" in the exhaust. With such solid proof in hand,

---

[7] (Chalmers, 1952) p. 109.

[8] Others of note who were raised in one paradigm and transformed to another were William Thomson (Lord Kelvin) who was raised with caloric and transformed to the mechanical theory of heat, and Planck who was raised with "I don't believe that atoms exist" Mach and transformed to the atomic theory.

[9] (Guerlac, 2008) p. 68.

[10] (Guerlac, 2008) p. 70.

Lavoisier renamed the "inflammable air" hydrogen to indicate that it generates (-gen) water (hydro-). He completed this short story by identifying Rutherford's "residual air" and Scheele's "vitiated air" as being one and the same substance, which he named "azote" and which was later changed by others to nitrogen.

Oxygen	"dephlogisticated air"; "fire air"; l'air vital
Nitrogen	"residual air"; "vitiated air"; azote
Hydrogen	"inflammable air"

Conservation of Mass

Lavoisier did not need phlogiston to explain any of these reactions. Its necessity disappeared in one fell swoop. Lavoisier publicized his findings on these reactions, including supportive stoichiometric relations between the reactants and products that would later be used to great effect in validating the atomic theory. He obtained the data by using a balance to take very accurate and highly meticulous weight measurements. Within his calculations laid a critical and very profound assumption. The change in weight of the reactants exactly equals the change in weight of the products. Matter is neither created nor destroyed; it is conserved. It was Lavoisier who gave us this conservation law, writing "nothing is created either in the operations of the laboratory, or in those of nature, and one can affirm as an axiom that, in every operation, there is an equal quantity of matter before and after the operation."[11]

As physicists, chemists, and engineers, we are indebted to Lavoisier for his perfection of the mass balance as a powerful device for understanding any process. We've since mastered a similar approach based on conservation of energy and employ both when conducting heat & material (or energy & mass) balances, the tried-and-true trademark of our professions.

Lavoisier Lays the Foundation for Modern Chemistry

If I have to describe in one word what is the prime motive which underlies a scientist's work, I would say systematization

– S. Chandrasekhar[12]

Lavoisier went on to lay a strong foundation for modern chemistry. He swept aside the arbitrary language of the alchemists, especially related to the newly discarded phlogiston theory, and replaced it with a new and highly effective structure of terms based on the new understanding. His 1787 and '89 publications on this new chemistry included replacement of the Ancient's 4 elements—fire, earth, air, water—with 33 "simple substances," of which 23 would

---

[11] (Guerlac, 2008) p. 83. Per Guerlac, p. 86. "Lavoisier deserves credit for applying [the conservation of mass] specifically to the operations of the chemist and for spelling out a law of the conservation of matter in chemical reactions."

[12] (Chandrasekhar, 1998) p. 13.

remain as correctly identified true elements and the remaining 10 would later be identified as compounds, such as silica and alumina, containing metals not on the list of 23. He later was the first to make explicit the concept of state, a continuation of Euler's work: all substances can exist in three states—solid, liquid, vapor—and the quantity of heat within a substance determines its state. All-in-all, truly remarkable accomplishments.

A Real Guinea Pig Experiment – Respiration as a Form of Combustion

For an encore, Lavoisier directed his attention to respiration. With the experimental assistance of Laplace, who devised an ice calorimeter, unknowingly repeating Black's invention, Lavoisier conducted a heat and mass balance on a living creature. He placed the subject, a small guinea pig, into an "ice bomb." After suitable cooling, suffocating, and starving, he passed a quantified amount of air into the bomb, which the rodent eagerly sucked in, and measured the response. In this way, he identified "fixed air" (CO_2) as a product of respiration and also quantified the heat of reaction based on separately measured specific heats and heat emissions (amount of ice melted within a given time).

Lavoisier's conclusions from this work? Respiration is a form of combustion in which oxygen reacts with carbon. When combustion happens, oxygen is consumed, carbon dioxide produced and heat released. And here is where the story takes an interesting turn.

The Two Competing Theories of Heat – Material (Caloric) vs. Motion

Two theories of heat were circulating in the science world at this point in time, one based on material (heat as a substance) and one based on motion. Heat-as-motion, which later became part of the mechanical theory of heat, rose in popularity during the seventeenth and early eighteenth centuries. But then opinion shifted, due in part to a lack of understanding of the interconnectedness between such concepts as heat, temperature, Newton's force, and Johann Bernoulli's energy. This confusion allowed the material theory of heat, also known as the caloric theory of heat, to rise in popularity, not necessarily based on merit or because it offered something better but simply because an explanation of heat *was* needed (man abhors a vacuum) and it offered a simpler explanation than motion. Also, at this time there was a rise in popularity in the material view of combustion created by the phlogiston theory and this certainly contributed to the rise in popularity of the material view of heat. Both theories played in a similar arena, seeking to describe both the flame and the heat resulting from combustion.

So here was Lavoisier, a genius who pulled back the veil of nature to reveal the workings of modern chemistry, and he unfortunately sided with the wrong theory. He killed phlogiston and replaced it, in a way, with "caloric," one of his listed elements that could not be decomposed. To him, it was caloric that caused changes in heat as it moved between bodies. In his mind, since caloric was an element, it could neither be created nor destroyed; it must be conserved. His earlier philosophy regarding conservation of mass likely led him to this conclusion.

Lavoisier (Unfortunately) Gave Significance to Caloric

Regardless of his reasons for doing so, Lavoisier's public elevation of caloric onto a pedestal proved a significant setback to the discovery of energy and its conservation. He didn't invent caloric. The early Greek philosophers claim credit for this. But he did keep it alive when he could have killed it, like he killed phlogiston. "It is a curious commentary on the manner in which scientific knowledge has progressed that these views [on caloric] could have been held by one who discovered the cause of combustion."[13]

And it wasn't necessarily the concept of caloric that was the issue. It was more importantly the concept of its conservation. Caloric and heat eventually merged into the same concept of a conserved quantity. And it was this that ultimately held back progress.

Lavoisier's monumental contribution to chemistry is unquestioned. He walked into a muddle and created a tight and coherent structure. He together with Laplace "put the science of heat on a satisfactory and systematic basis."[14] Yes, he unfortunately helped keep caloric and its conservation alive, but I only draw attention to this not to diminish Lavoisier, but to indicate that it *did* impact the trajectory of the history in this chapter.

Given more time, Lavoisier may have eventually transformed his paradigm on heat from material to motion. New findings from Rumford and Davy representing the first significant challenge to the caloric paradigm were but 10 years away. But alas, things didn't work out this way. On May 8, 1794, Lavoisier, age 50, was sent to the guillotine. The French Revolution's Reign of Terror caught him in its web of suspicion largely because of Lavoisier's work as a tax collector. Said Lagrange, "It took but a moment to make this head fall, but even a hundred years may not be enough to produce a like one."[15]

Now let's follow the path of the conserved caloric by returning to Britain for the beginning of another doer–thinker historical development, this time regarding the engine. While France was revolting, Britain was industrializing.

[13] (Chalmers, 1952) p. 27.
[14] (Cardwell, 1971) p. 62.
[15] (Guerlac, 2008) p. 85.

Rise of the steam engine

The discovery of energy and its conservation necessarily involved the death of phlogiston and caloric. While Lavoisier eliminated phlogiston, he promoted caloric, which unfortunately influenced the work of many scientists who followed, including Sadi Carnot who based his seminal 1824 publication, *Reflections on the Motive Power of Fire*, on the conservation of caloric. To Carnot, it was the fall of caloric from high temperature to low that generated work in a steam engine, which left one with the central problem of trying to understand where the work came from if no caloric was consumed in the process. It would take until 1850 for this problem to be solved. But we're not ready for that yet. Instead, to understand the depth of the problem and thus the achievement of its solution, we must first consider the rise of the steam engine.

*   *   *

The steam engine helped launch the Industrial Revolution, and the Industrial Revolution helped launch the steam engine. This topic makes for a great chicken-or-egg discussion that many others have addressed. I'll simply just say that each spurred the other and move on.

Britain had long relied on engines to convert the energy of wind, water, animal, and man into the motion of gears, pumps, machines, and so on. But such engines were limited by nature. Wind only blows so fast. Water only sits so high. The arrival of the steam engine changed everything, shifting the limit from nature to man's ingenuity. *How can I build a stronger boiler to withstand higher pressure steam?* The better man's designs, the more steam power could be generated. Each improvement resulted in better productivity which motivated further improvement.

Coal was the primary source of power for the steam engine. But coal cost money and so it was only a matter of time until the engineers focused on improving the efficiency of the engine to achieve more output for a given amount of coal fed to the furnace. The struggle for improved efficiency is what ultimately led to the discovery of one of the most complex concepts in science: entropy. We'll cover this concept and its history in Part IV. For now though, we'll go through a short primer on the rise of the steam engine and its connection to the concept of energy.

It Started with the Newcomen Engine

In 1712 Thomas Newcomen (1663–1729), an English ironmonger, invented the first commercially successful steam engine based on a condensing steam design that provided an approximate one atmosphere (condensing steam creates a vacuum) driving force for simple tasks such

Block by Block: The Historical and Theoretical Foundations of Thermodynamics,
Robert T. Hanlon, Oxford University Press (2020). © Robert T. Hanlon.
DOI: 10.1093/oso/9780198851547.001.0001

as pumping water out of coal mines. In 1763–75, James Watt (1736–1819), a Scottish instrument maker for none other than Joseph Black, improved upon the condensing steam concept after being asked by Black upon arrival at the University of Glasgow to repair a bench-top model of Newcomen's engine. Watt's curiosity was piqued. *What is limiting the operation of this engine? How can I make it work better?*

Watt's powerful conceptual thinking guided his subsequent tinkering efforts, resulting in breakthrough after breakthrough in the design and performance of steam engines. One of these breakthroughs, particularly relevant here was this. While Newcomen used ambient air to push the piston down into the vacuum created in the cylinder, Watt used boiler steam (atmospheric pressure) to improve thermal efficiency. In this scenario, as the steam pushed the piston to do work, it expanded and lost energy as manifested by a temperature decrease. As early as 1769, Watt imagined the possibility of this "expansive operation," an operation we continue to study to this day in thermodynamics (adiabatic expansion). Jonathan Hornblower (1753–1815), Cornish engineer and chief rival to Watt, independently arrived upon the same idea at some later date. The strong capitalistic environment that existed in Britain at that time eventually brought the two of them together in court to settle a key marketing dispute: whose engine was better? Hornblower relied upon a Cornish politician who also happened to be an Oxford-taught mathematician, Davies Gilbert (born Davies Giddy) (1767–1839), to "prove by *fluxions* [this was Newton's Britain after all] the superiority of their engine."[1] Gilbert's efforts in 1792 led to the identification of the area under the pressure/volume curve as the *work* done by an actual, expansively operated steam engine. While Daniel Bernoulli earlier identified the area under the curve, or in calculus terms, $\int P\,dV$, as a critical parameter, which went largely unnoticed, Gilbert was the first to recognize its meaning in an industrial setting.

For a constant pressure driving force, work is simply the product of the fixed pressure times the area of the piston, i.e., force, times the length of the stroke, i.e., distance. But in Watt's new system, pressure wasn't constant. It continuously dropped during the motion of the piston. So how could work be calculated for this situation? How could one integrate under the P–V curve for a real operating piston? How could one even obtain the data? There were no process control charts back then.

Into this situation stepped Watt's assistant, John Southern (1758–1815). As this was in the days before advanced process control, Southern devised an analog solution in 1796 by attaching a pencil to the spring-loaded pressure gauge and attaching a piece of paper to the moving piston. As the piston moved, the pencil traced the pressure reading, and so created the first real "indicator" diagram. In this way Watt and Southern monitored the change in pressure with the change in volume (piston location), using the data to drive their own continual improvement program, and so became the first to directly measure the famous $\int P dV$.[2] (One can quantify this area by either direct but tedious measurement or more simply by physically cutting the curve out and weighing it on a scale.) The indicator diagram would later gain fame when Émile

[1] Cited in (Cardwell, 1971), p. 79–80. Letter from Watt to Southern about progress of litigation. In an interesting aside, Watt should have used Newton's *fluents* as opposed to his *fluxions* to denote integration as opposed to differentiation.

[2] (Lewis, 2007) p. 55–57. In 1797 Joseph Black's colleague John Robinson created an approach similar to Southern's in a publication on the analysis of work done by a steam engine.

Clapeyron put it to use in analyzing the Carnot cycle. Most any introductory thermodynamics text includes this diagram.

Gilbert's theory of the indicator diagram and Southern's invented process to capture this diagram were two important forward-moving steps in the evolution of the steam engine, both inspired by the inventive thinking of Watt and Hornblower. Additional developments from Watt included the definition of steam engine efficiency as the ratio of "work done" to fuel consumed, and then the follow-up definition of work done based on the common engineering practice of weight times height, a measure that had been initiated far earlier by others. This shifted work from the "virtual" world to the real world. He further introduced the concept of power as the *rate* at which work is done. He called this term horsepower and defined 1 horsepower equal to 33,000 foot-pounds per minute based on his actual observation of horses doing work in a mill.

Watt's new terms and concepts surrounding improved engine efficiency were driven largely by money as opposed to any desire to gain a deeper theoretical understanding of the physics involved. He and his business partner Matthew Boulton (1728–1809) believed in patenting and so didn't believe in publishing, preferring to keep their inventions behind closed doors. Furthermore, when they went to the market place, they had to understand and especially had to quantify the value proposition of their technology relative to the competition. How else to price their product? The creation of the term horsepower reflected this need since their sales pitch often targeted the replacement of horses with steam to accomplish the same amount of work.

In 1800 Boulton and Watt's patents finally lapsed, thus opening the floodgates for a new generation of engineers, particularly in the highly active mining region of Cornwall and especially Richard Trevithick (1771–1833) who, enabled by his own design of improved pressurized boilers, pioneered the (safe) use of high-pressure steam. The move from atmospheric pressure to high pressure was a milestone. Equipment size shrank, equipment power jumped. Railway locomotives were born. The steam engine rose to dominance, staying there until the advent of electrical power and the internal combustion engine. We still use steam to this day, mostly in power generation.

So why did the steam engine rise solely in Britain? Financial incentive. Because a steam engine was inherently expensive to build and maintain, a strong financial incentive was needed to buy one. The incentive appeared in the mines. As England's demand for coal and metals to support the Industrial Revolution increased, any technology that could provide the means to drive down mining costs was welcomed. One significant cost concerned the presence of water in the mines. Once the mine hit water, the water needed to be pumped out to enable further mining. The conventional approach was to use horses to lift the water out. The steam engine offered a lower-cost option. But realization of this option wasn't easy. Indeed, the rise of the steam engine required entrepreneurial and inventive genius. Newcomen and later the team of Boulton and Watt had both.

As often happens, new technologies by the doers outpace the accompanying new theories by the thinkers. The development of the steam engine flourished in the absence of any fundamental understanding behind it. It would be another 40 years before it was recognized that this engine ran on *heat* rather than steam. The theoreticians had to catch up. And so we once again return to France.

Lazare Carnot and the Reversible Process

The path forward from here is a confusing one largely on account of the fact that science in the early 1800s *was* confusing. Many ideas, terms, and theories were all in play, regardless of whether they were right or wrong. Contributing to this was the presence of many manifestations of energy. Individuals were studying the individual strands, not necessarily communicating with each other, or, if they were, not necessarily understanding each other. So many differences muddied the water. Different countries, different languages, different underlying philosophies, different paradigms. Each person was approaching this new territory from a different perspective using different terms and definitions. All of these strands started converging in the early 1800s, drawn together by the unifying fact that energy is both real and conserved. As the strands converged, they collided and caused confusion, but only initially. Eventually it all became clear.

One of the key strands that emerged represented one of the most critical evolutionary paths in the history of energy and thermodynamics. As we learned earlier, Rudolf Clausius was the first to quantitatively define energy in his 1st Law of Thermodynamics. The path he took to reach both this law and his 2nd on entropy involved the synthesis of several other paths, the most important of which involved the Carnot family and Émile Clapeyron, all members of the French government university system. It was this path that contributed the most to the death of caloric, even though, ironically enough, those along the path helped keep caloric alive. While other paths toward energy conservation proved viable without relying on Carnot's work, it was this work that crystallized Clausius' thinking, so enabling him to step forward and say that it's not caloric that's conserved, but energy. And it all arguably started with Lazare Carnot.

"The Great Carnot"[3] – Referring to Father, not Son

Before Lazare Carnot (1753–1823) became a famous military commander and statesman under Napoleon, he published a 1784 study of machines, making him one more in a long list of inquisitive people who were fascinated by machines. With a background in both mathematics and practical engineering—which places him alongside his son in Robert Fox's group of theoretically trained French engineers—Lazare established a Galilean stance at the interface between the two different disciplines. Good things happen at the interface and this was no exception. Lazare's contribution wasn't just one more added to the heap. It broke open new ground.

Lazare's interest in machines together with his multidisciplinary background led him to be *the first* to bring fundamental theory into the analysis of real operation. Using Leibniz's conservation of *vis viva* as his guiding light and the concept of infinitesimal virtual motions, or as he called them, "geometric motions," as his analytical tool, Lazare contemplated the real water wheel based on "ideal" features. In so doing he pulled the concept of work from the real world of engineering into the world of physics. Both Watt and Lazare embraced the concept of work, Watt to advance steam technology and Lazare to enable improved theoretical understanding.

---

[3] (Cropper, 2001) p. 44.

This continuation of a very long effort to define work culminated in 1829 when Coriolis finally and ultimately replaced the traditional engineering definitions of work as "force times distance" and "weight times height" with the ultimate final definition of work as the *integral* of force times distance,[4] thus allowing quantification of work for a varying force such as exists in an expanding piston. (Note that for fluids, $\int Fdx$ equals $\int PdV$.)

Lazare was also the first to introduce the concept of *reversibility* into physics, defining this well-known term as any "geometric motion" wherein the "contrary motion is always possible,"[5] such as exists with a perfectly balanced lever. Embedded within this concept lay basic philosophies of static equilibrium of simple machines and the impossibility of perpetual motion. As we now know, this concept of reversibility would play a prominent role in the development of thermodynamic equilibrium.

While important in and of itself, Lazare's 1784 publication was but a prelude to the real breakthrough that followed, the birth of his first son, Sadi.

Sadi Carnot

Much as Daniel Bernoulli got a head start in life by learning from his father (and older brother), so too did Sadi Carnot (1796–1832). Sadi grew up in a family of fame and wealth. His intellectual gifts equipped him to take advantage of such privilege, especially in the form of his strong educational upbringing during the first 16 years of his life as directed by his father.

After attending the École Polytechnique and then the École de l'Artillerie et du Génie, where he received training as an artillery officer, Sadi retired at half-pay from the Army at 24 years of age, moved into a small apartment in Paris with his brother Hippolyte and commenced to live the life that each of us at one time probably wished for, his own version of Newton's *Annus Mirabilis*. While furthering his studies of physics and economics at a range of universities and while enjoying the luxury of not having to work, Sadi thought. He thought about engines, especially steam engines on account of their growing importance toward progress in France. But he didn't think about the detailed nuances of engine operation. Instead, influenced by his father's abilities to rise above such technical details, he sought a generalized approach by considering the fundamental processes at work within such a complex process.

From this generalized perspective, Sadi employed analogy to arrive at an extremely powerful insight (Figure 16.1). Just as water flows from high elevation to low when doing work in a water wheel, so too does some *thing* flow from high temperature to low when doing work in a heat engine. To Carnot, this *thing* was caloric. He felt that caloric was, in some way, transported by steam from the hot furnace to the cold condenser, and that the act of this transportation, in and of itself, in some way, produced work without any loss of the caloric, just as there is no loss of water when it falls. As Mendoza pointed out,[6] the French physicists such as Carnot and later Clapeyron simultaneously believed two conflicting theories, that heat or caloric (they were used interchangeably) is conserved in all thermal processes, *and* that heat is equivalent to

[4] (Coopersmith, 2010) p. 157.
[5] (Coopersmith, 2010) p. 157.
[6] (Carnot et al., 1988) Mendoza in Introduction, p. xiv.

16.1 CARNOT'S HEAT ENGINE AND THE CALORIC THEORY ©RTH + CLS

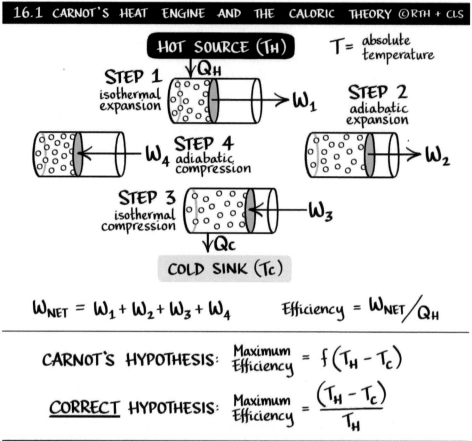

$$W_{NET} = W_1 + W_2 + W_3 + W_4 \qquad Efficiency = W_{NET}/Q_H$$

CARNOT'S HYPOTHESIS: $\dfrac{Maximum}{Efficiency} = f(T_H - T_c)$

<u>CORRECT</u> HYPOTHESIS: $\dfrac{Maximum}{Efficiency} = \dfrac{(T_H - T_c)}{T_H}$

CARNOT'S HYPOTHESIS:	
WATER WHEEL	IDEAL HEAT ENGINE
water stream Δh	caloric stream $\Delta T = T_H - T_c$
MOTOR POWER DEPENDS ON: water flowrate & Δh	heat flowrate (Q_H) & $\Delta T = T_H - T_c$
WHAT IS CONSERVED? water flow	caloric flow $(Q_H = Q_c)$

Figure 16.1 Carnot's heat engine and the caloric theory

work, meaning that it *wasn't* conserved in any thermal process producing work. It would take Joule's mechanical equivalent of heat experiments to resolve this paradox. (Figure 16.1 will be discussed in further detail in Chapter 30.)

Regardless of the mistaken assumptions embraced by Carnot, one must still sit back and say, Wow! What a monumental point in history! Sadi created a highly abstract and yet very powerful image of a heat engine involving the flow of an entity called caloric from the furnace to the condenser. His subsequent detailed analysis of the implications of this insight, including his conception of the closed-cycle operation that proved very useful in all future analyses of heat engines, all of which we'll cover in Chapter 30, became the basis for his monumental 1824 publication *Reflections on the Motive Power of Fire*.

Saved by a Thread

Monumental is an interesting word to use here because this breakthrough effort almost didn't live to see the light of day. You see, as pointed out by Mendoza, the official world of science received Sadi's publication "in utter silence."[7] With hardly any sales, it almost died a quick death. But all it takes is one thread to keep an idea alive, and that thread was Émile Clapeyron. Clapeyron discovered Sadi's work and, duly inspired, incorporated it into his own 1834 publication titled *Memoir on the Motive Power of Heat*. This work fortunately ended up in the hands of William Thomson (Lord Kelvin) and then Rudolf Clausius. We'll continue this storyline later but for now will simply acknowledge with gratitude the strength of such single threads.

Sadi Carnot unfortunately didn't live to see the eventual respectful acceptance of his work. By 1824, the party was over and Sadi was recalled to full-time military service. He retired again in 1828 to devote himself to further study, but caught scarlet fever in 1832, relocated to the country to convalesce, caught cholera while there during the epidemic, and died a quick death. He was 36 years old.

*   *   *

Figuring out the flaw in Sadi Carnot's otherwise powerful work contributed greatly to the fall of caloric and the rise of energy. But other works contributed too, specifically three that directly challenged and thus helped weaken the caloric theory. And so we now turn to them.

---

[7] (Carnot et al., 1988) Mendoza in Introduction, p. xii. Perhaps the word "utter" isn't entirely accurate as (Gillmor, 1982) suggests the existence of a larger French audience.

17 Caloric theory: beginning of its end

Although the caloric theory is apparently no longer of interest in physics, it is necessary to describe it... to show why it constituted a real obstacle to the adoption of the modern theory of energy

– Stephen Brush[1]

The rise of energy was as much about the fall of caloric as anything else. So before diving into the mid-1800s discovery of energy, we first return to the early 1800s to discuss the events leading to caloric's demise as well as to discuss why its demise was so slow in the face of such events.

The material conception of heat from which the caloric theory evolved was generally accepted until the early seventeenth century when opinion began to grow that heat was not a substance but rather some form of motion. As Francis Bacon asserted in 1620, "heat itself... is motion and nothing else."[2] Other philosophers agreed with this until the early eighteenth century when general opinion gradually shifted away from this view and back to the material theory but with one difference. Instead of regarding heat as consisting of particles, physicists during the second half of the eighteenth century developed the theory that heat was some kind of invisible, indestructible fluid. Lavoisier embraced this fluid substance as caloric, although the name arrived without a strict definition, and gave it prominence by placing it on his list of elements. It was generally thought that, among other properties and characteristics, caloric was conserved, just as any element was, and that caloric was released by friction and hammering. But no quantified proof accompanied these thoughts.

Entering the 1800s, the caloric theory pervaded science and stood as the sole hypothesis to explain much but not all of the phenomena related to heat. For those phenomena it couldn't explain, its adapting nature gave rise to *ad hoc* corrections that provided explanations and thus gave scientists a certain peace of mind that the theory was good enough. As Joseph Black shared, he accepted the caloric theory was "the most probable of any I know."

To some, however, the caloric theory just didn't make sense. In addition to the fact that, as indicated above, the theory lacked quantitative predictive power, the theory also couldn't address the following question: If caloric was indeed conserved, then how could a constant source of friction continually generate unlimited caloric in the form of heat when caloric was not unlimited? Such questions, in addition to questions addressing the fact that caloric fundamentally offered no quantitative predictive power, led to demonstrative experiments that

---

[1] (Brush, 2003a) p. 9.
[2] (Coopersmith, 2010) p. 46.

Block by Block: The Historical and Theoretical Foundations of Thermodynamics,
Robert T. Hanlon, Oxford University Press (2020). © Robert T. Hanlon.
DOI: 10.1093/oso/9780198851547.001.0001

directly challenged caloric. Complete refutation should have resulted except for one problem. There was no alternate hypothesis waiting to replace caloric. In this chapter I discuss both the attack against caloric and its unlikely survival.

Early 1800s Provides New Evidence Challenging the Caloric Theory

In the ideal world, the scientific method would have been used to generate a range of hypotheses to explain the range of heat-related phenomena and then an experimental program would have been executed to systemically challenge and eliminate each hypothesis until only one was left standing. The early 1800s was not this ideal world. The scientific method was still in its infancy and there was no coordinated effort to employ it to solve the heat problem, thus leaving one hypothesis alive but on shaky and unconvincing ground. Focus was initially placed on attacking this hypothesis and later on developing and validating an alternate hypothesis.

Rumford Bores a Cannon and so Boils Water

The story of Count Rumford (1753–1814) is another to whom I can't do justice. How can one capture in anything other than a full book the very colorful story of Benjamin Thompson? Born in Woburn, Massachusetts, Thompson fled America due to loyalist ties and then, among other activities, somehow ended up marrying (and later separating from) Lavoisier's widow and eventually became Count of the Holy Roman Empire in Bavaria. It was during this latter event that Thompson adopted the name Count of Rumford, based on the former name of Concord, Massachusetts, site of his early home. As my limited space doesn't permit a full accounting here, I focus instead on one specific area, Rumford's life-long interest in the study of heat.

The caloric theory never captured Rumford's imagination. He doubted it as early as 1778 when he observed that gun barrels were hotter when firing without the bullet. How insightful! How many others had conducted the same experiment but never made this observation? While then folks didn't understand the reason behind this phenomenon, today we do. Without having to do any work to accelerate the bullet, the hot air produced by the combustion didn't cool off and so remained hot. It's interesting that it was this concept involving the transformation of work done by an expanding gas against a resistant force into a cooling effect of the gas that was the basis for Mayer's subsequent calculation of the mechanical equivalent of heat.

Rumford continued to doubt the caloric theory when in 1789 he observed a cannon manufacturing process in the Munich arsenal. As a new brass cylinder was being bored out to form the cannon barrel, Rumford noted that the barrel and metal chips all became very hot. While he was surely not the first to notice this, he was the first to think about what it meant for he simply could not walk away from such an observation without thinking about its cause. Inspired by his inquiry, Rumford proceeded to conduct one of the most illuminating and legendary experiments in the history of physics. He set up a boring process with an intentionally dull drill bit pressed hard against the bottom of the bore of the cylinder. As the cylinder was turned by the efforts of two walking horses, the temperature of the brass cannon rose. Figuring that the best way to share this finding was to show it to people, Rumford encased the barrel with a box filled with water and then waited. Sure enough, after a while, the water began to boil, a true "ooh" and "ahhh" moment for the crowd that gathered since there was no fire there to do the boiling, making it one of the most famous public-demonstration experiments ever.

But why? Why was this such a big deal? We easily grasp the concept of friction today. Rub two objects together, hard, and then touch them afterwards. They're hot. We now know this to be true. We know that atoms interacting with each other during such contact is what hinders motion and so causes friction and so causes heat. Friction moves atoms out of their normal resting place. When they snap back, like a stretched rubber band that's released, potential energy is converted into kinetic energy. But back in the late 1700s, the phenomena called friction was very baffling, especially in the context of the caloric theory of heat since this theory couldn't explain it. Nothing was really happening in Rumford's experiment other than the rubbing together of two pieces of metal. And yet water boiled. How? To the calorists, the heat was produced by squeezing caloric out of the metal. But to Rumford, this could not be true, for how could a material like caloric be continuously squeezed from a body <u>without</u> <u>limit</u>? This simply did not make sense to him, especially in light of the belief that caloric was conserved. The only thing that did make sense to him was to state that the motion caused the heat.

From this demonstration and other supporting evidence, Rumford became convinced that heat is caused by motion—"heat is merely the *vis viva* of the constituent atoms of all material bodies"[3]—and shared his results and views in a public forum called the Royal Institution of London, which he himself founded in 1800. (I *told* you that his life was rich!) While historians acknowledge that Rumford's demonstration revealed weakness in the caloric theory, they hesitate to acknowledge beyond this.[4] Rumford proposed to jettison caloric but offered no principle to replace it with, as he really didn't have any concept of energy, and his famous experiment became famous in retrospect. But his work did inspire others, especially Humphrey Davy (1778–1829), Thomas Young (1773–1829), and later James Joule, and because of this, Rumford belongs in this history.

Davy Melts Ice by Using Friction

Davy followed in Rumford's footsteps and conducted a neat little experiment to test the relationship between heat and friction. By rubbing two ice blocks against each other, he caused them to melt and immediately leapt to the conclusion that caloric does not exist. How could friction or any other process liberate caloric from two blocks of ice? Perhaps the leap was a little too far as others rightfully pointed out that other variables could have come into play, such as the fact that pressure itself could have caused melting—by decreasing ice's melting point—and that the ambient air temperature may also have contributed. But the effect was

---

[3] (Cardwell, 1971) p. 103.
[4] (Cardwell, 1971) pp. 95–107. (Brush, 2003a) p. 10. (Fox, 1971) p. 4. Fox suggested that Rumford's work was a "red herring for the historian."

there for others to visually see and so this too contributed to a growing skepticism of caloric. Yet, as with Rumford's work, Davy's findings did not provide the quantified experimental proof needed to reject caloric, and so left the door open for the calorists to develop *ad hoc* explanations, which they did. They accepted the results of both Rumford and Davy, but were able to interpret them from their own caloric viewpoint. It is so difficult at times to let go of a paradigm. Davy publicly denied the caloric theory and some 12 years later completely accepted that "The immediate cause of the phenomena of heat then is motion."[5] He went on to become famous in and of his own right for, among other things, discovering potassium, sodium, chlorine, and other elements, and for mentoring another budding experimental genius, Michael Faraday.

Young and the Connection between Light and Radiant Heat

Thomas Young was yet another genius of wide-ranging talents, as best evidenced by the fact that, of all things, he did the ground-breaking work that enabled Jean-François Champollion to decipher the Rosetta Stone in 1822. Young traveled a different scientific path from Davy, that of light. It was Young who in 1801 devised experiments showing that light interferes with itself. The passing of light through Young's famed double-slit and the resulting repeating pattern of bright and dark zones on the wall behind beautifully demonstrated that light is indeed a wave. Young went on to suggest that heat and light were really the same thing,[6] which furthered the concept of "radiant heat" proposed by Carl Wilhelm Scheele in 1777.[7] Others were to pick up on this finding. In 1816 Augustin-Jean Fresnel (1788–1827) furthered the theoretical aspects of "light-as-wave" and in 1831 Macedonio Melloni (1798–1854) and soon thereafter James Forbes (1809–1868) experimentally verified that radiant heat and light were of the same nature. In the caloric theory, radiant heat was comprised of material particles as conceived of by Fourier in his extensive heat studies to account for heat conduction. Thus, the proposal by Young and those who followed that radiant heat is a wave challenged this theory.

While these results were powerful, they didn't settle the matter. Yes, caloric couldn't explain the wave-nature of radiant heat, but then again neither could the mechanical theory of heat. It wasn't readily apparent how to merge the two. It would take many more years to finalize this aspect of the energy story by the discovery of others that light is both wave *and* particle and has an energy equal to hv.

It should be noted that Young also proposed in print the replacement of "living force," which was the English translation of Leibniz's Latin *vis viva*, with "energy," the first use of this word in physics since Johann Bernoulli had used it in the context of "virtual work" back in 1717.[8] Although Young employed the word in the context of mechanical energy, others soon started

---

[5] (Fox, 1971) p. 119.
[6] (Coopersmith, 2010) p. 174.
[7] (Cardwell, 1971) p. 60.
[8] (Young, 1807) p. 52. "The product of the mass of a body into the square of its velocity may properly be termed its energy. This product has been called the living or ascending force, since the height of vertical ascent is in proportion to it; and some have considered it as the true measure of the quantity of motion; but although this opinion has been very universally rejected, yet the force thus estimated well deserves a distinct denomination."

using it to describe other energies relating to electricity, chemistry, radiation, and so on. It became a catch-all for that which had yet to be defined.

The Challenge of Invalidating Caloric – There was no Competing Hypothesis

Count Rumford and Humphrey Davy fought against the tide by showing, in non-rigorous, non-quantifying, but highly illuminating ways, that mechanical work can generate heat where there once was none, implying that heat can be generated and so can't be conserved. While both helped raise awareness, neither made any immediate impression on the general body of physicists. The inertia and resistance to change in the established caloric doctrine was more powerful than logic and well into the first part of the nineteenth century the caloric theory continued to command general acceptance. It was no surprise that the leading theorists of that time, and especially Sadi Carnot and William Thomson, initially believed in caloric themselves.

The fact that the caloric theory had much staying power in the face of such challenges was due to several things. First, the strongly entrenched materialistic aspect of the phlogiston theory influenced the subsequent acceptance of the materialistic aspect of caloric. Second, the presence of an otherwise weightless substance that flows from hot to cold could indeed explain a lot of general heat-related phenomena. Starting from this model, scientists often arrived at the right answer for the wrong reason. Furthermore, the theory was rather adaptable in that it was "tweaked" at times to explain away any disagreements between theory and experiment. Third, there was no viable alternate hypothesis available to replace caloric. And fourth, there was a lack of understanding at that time about the difference between force and energy, energy being the integral of force through distance, and widespread confusion around the difference between heat and temperature.

Having said this, the accumulating evidence against caloric, including Young's work, proved too much for caloric to handle and by the 1840s, physicists were ready to abandon caloric and grasp an alternate theory (finally!), namely energy. And at this point we bring Sadi Carnot back into the discussion because the critical final deathblow to caloric occurred as a result of his work. It's interesting that Carnot did not originally set out to challenge caloric. Instead, he published a monumental work that would have, in and of itself, established the beginnings of thermodynamics except for a single major flawed assumption: conservation of caloric. Clausius' seemingly simple and yet very profound change of this assumption to the conservation of energy resulted in the first step toward a complete thermodynamic discipline while simultaneously killing caloric once and for all.

Mayer and Joule – Succeeded in Killing Caloric by Ignoring It

The history of these final stages and the completion of Boyer's three steps in the seemingly sudden and rapid discovery of energy is fascinating, largely because two of the critical scientists involved, Robert Mayer and James Joule, achieved their success by simply ignoring caloric.

It's not that they didn't believe in caloric or that they didn't *not* believe in caloric. It's that they simply didn't think about caloric. Not having been raised in the academic institutions that taught caloric, they unintentionally ignored it as they pursued with sound logic and rationale their respective research paths. It was only later, when Clausius unsuccessfully tried to reconcile Carnot's and Joule's findings that caloric died and energy was born. The two theories couldn't co-exist. Either one or the other was conserved, but both couldn't be. In the end, it was really the rise of the mechanical theory of heat (work–heat equivalence) and the higher-level theory of energy that finally killed caloric.

Another interesting historical fact was this. While the mechanical theory of heat embraced heat-as-motion, this concept played no role in the work of Mayer and Joule. It wasn't necessary. It was only work–heat equivalence for which heat is *not* conserved that governed their work. This was especially true for Mayer. To him, heat simply ceased to exist as it was transformed into work. After heat as a form of energy was defined, heat as a form of motion soon followed, thus opening the door to such powerful new concepts as the kinetic theory of gases and statistical thermodynamics, both of which relied on the use of Newton's Laws of Motion to model the behavior of moving atoms.

Work ⇔ Heat

One of the most significant breakthroughs in this history of energy was the demonstration that work could be transformed into heat and vice versa by a fixed ratio called the mechanical equivalent of heat (MEH), thus supporting conservation of their sum. Many thought that as a conserved quantity, caloric could be neither created nor destroyed. But lacking rigid definitions at that time, many also thought of caloric and heat as the same thing and so thought that heat as an element could also be neither created nor destroyed. But Joule demonstrated with his very simple experiment that a falling weight could stir a bath of water and so cause its temperature to rise. What a simple experiment to prove such a monumental concept. Today this subject appears simple and obvious: the spinning paddle hits the water molecules and so increases their speed and the water's temperature; the paddle is able to continuously spin through the viscous water by the force created by the dropping weight. Yesterday it was extremely difficult. Why should foot-pounds be connected at all to degrees Fahrenheit? Whereas Rumford demonstrated the transformation of work to heat, more so than Davy, Joule quantified it with phenomenal accuracy, thus enabling him to propose a conservation theory based on this transformation.

* * *

The fact that it is energy and not heat that's conserved all at once brought tremendous clarity to science. Things were so messy and confusing at that time, as best exemplified by the fact that the French embraced *both* the conservation of heat/caloric and the ability of heat to cause work.

* * *

Recall that by 1750 the rudimentary version of the conservation of mechanical energy involving visible bodies had been generally accepted even though the concept of energy had naturally not yet been established. It was only a matter of time until the concept of mechanical work became connected with the concept of heat since many understood that friction (resulting from force through distance) leads to heating. Rumford and Davy showed this cause–effect connection, but this raised questions: *if the sum of mechanical work and vis viva is conserved, where does invisible heat fit in? How could a conservation theory embrace heat (caloric) that's also conserved, for certainly whatever terms fell under the conservation umbrella could not themselves be conserved, right?* One approach would be to demonstrate in a more quantitative way the relationship between work and heat. Neither Rumford nor Davy sought to do this; they never set out to determine the mechanical equivalent of heat but instead sought to challenge caloric and, in so doing, to demonstrate a work–heat connection. Besides, given their experimental set-ups, determination of the mechanical equivalent of heat would have been very difficult. Rumford's cannon would not have been isothermal and would have required some concept of average temperature to measure the total change in energy, while Davy's experiment didn't offer a ready means to quantify the force used to rub the ice until it melted. Mayer's and Joule's works were blessed by the fact that the work–heat effect occurred inside a well-mixed isothermal system, gas and water, respectively.

*   *   *

It's one thing to eliminate a hypothesis but it's another to validate one. The core of all efforts in this discussion involved the concept of conservation. Something had to be conserved in a world where perpetual motion was impossible. If not caloric, then what? Whatever it was, the only path to validation was Boyer's 3rd step—"reasonable justification" that energy is neither created nor destroyed and this necessarily required the quantification of the various forms of energy, starting with work and heat. If something is lost, then something must be gained, and whatever *it* was, it had to work in both forward and reverse directions. If the conversion of work into heat could be quantified by a ratio, then that same exact ratio must also be valid for the conversion of heat into work. The achievement of such quantification would require extremely accurate measurements and one of the starting points for these measurements was development of the ideal gas law.

18 The ideal gas

The historical importance of the physics of gases was not only that it provided a direct insight into the nature of heat but also that when heat was applied to gases and vapours its effects were more obvious and more easily quantifiable than when it was applied to solid and liquids. In other words, gases and vapours provided the means whereby heat could readily produce mechanical effects.

– D. S. L. Cardwell[1]

Thank goodness for the ideal gas. It lends itself so beautifully to study and education. The exactness of the relationship between its properties can't get much simpler.

$$PV = nRT \qquad\qquad [18.1]$$

But the simplicity is somewhat misleading, as many, many years were required to put this law together. Ingenuity, patience, accuracy, reliable equipment, new analytical technologies, meticulous experiments, and a thinking mind were all required to realize this tremendous accomplishment.

With this law, one can tease out aspects of nature. For example, one can, as Daniel Bernoulli first did, derive this law by assuming gas to be a billiard-ball world of colliding atoms that adhere to Newton's laws of motion. The success of this derivation supported the different but connected theories behind the assumptions, specifically the theory of energy, the kinetic theory of gas, and the atomic theory of matter. This is indeed what happened in the late 1800s.

The study of gases naturally branched out to explore other properties, such as heat capacity, and other behaviors, such as adiabatic compression and expansion. The gathering results found many applications in both theory and practice. For example, the advent of the heat engine demanded such supportive studies since the heart of the process involved the expansion of a gas against a piston to do work while simultaneously losing thermal energy as manifested by a decreasing temperature. The understanding that this process actually represented the conversion of heat into work contributed greatly to the final completion of the theory of energy. To truly understand the history of energy, we must first understand the history of the study of gases, starting with the ideal gas.

* * *

The great thing about the ideal gases is that their properties, and especially the *change* in their properties since energy is all about change, readily lend themselves to direct measurement. The

---

[1] (Cardwell, 1971) pp. 128–129.

Block by Block: The Historical and Theoretical Foundations of Thermodynamics,
Robert T. Hanlon, Oxford University Press (2020). © Robert T. Hanlon.
DOI: 10.1093/oso/9780198851547.001.0001

first to demonstrate this was Robert Boyle (1627–1691). During his travels through Europe and during his reading of Otto von Guericke's (1602–1668) work (Chapter 25) on the power of atmospheric pressure, Boyle obtained threads of knowledge regarding the behaviors of air. Upon return to his Oxford laboratories, he and his assistant, Robert Hooke (1635–1703),[2] sought to probe the subject in greater detail. As a dedicated and persistent experimentalist and researcher whose demonstrated skills led to his being selected as a founder of the Royal Society, Boyle together with Hooke studied the relationship between pressure and volume of a gas trapped in an upside-down tube immersed in a bath of mercury at constant ambient temperature. As the height of mercury was changed, so changed the volume of the gas. From this, Boyle determined in 1662 that gas pressure (as measured by the height of mercury) is inversely proportional to volume.

$$P \propto 1/V \qquad \text{Boyle's Law}\big(\text{fixed temperature and mass}\big)$$

Boyle's work, inspired in part by Guericke and motivated to its conclusion by Francis Line who challenged Boyle's earlier results,[3] contributed to the increased understanding of the behavior of air and really of any gas, all of which would be used in subsequent analyses of the steam engine. But we're not there yet. We're simply following a timeline here.

Although it took a while, others followed Boyle's lead and pursued their own studies of the relationship between any two of the variables while keeping the third constant (this inherently assumes that the number of moles, n, is fixed). Since Boyle developed a relationship for pressure and volume, temperature remained to be addressed, which presented somewhat of a conceptual challenge as one could envision zero volume and zero pressure. But zero temperature? What could this mean? And where would it be found?

While Guillaume Amontons (1663–1705) seems to have been the first to recognize the importance of actually measuring how gas volume and pressure vary with temperature, it would take another century until Joseph-Louis Gay-Lussac (1778–1850) obtained the necessary accuracy in measurements to draw conclusions. By 1802 Gay-Lussac determined that air and other gases, such as oxygen, nitrogen, hydrogen, and carbon dioxide, all expand by the same fraction when heated through the same temperature.[4] He also determined the coefficient of expansion to be 1/267, which thus meant that upon adding the constant of 267 to the temperature reading (degrees Celsius), the following relationship would hold.

$$V \propto T \qquad \text{Gay-Lussac's Law}\big(\text{volume} - \text{temperature}\big)$$

The above law is sometimes referred to as Charles' Law, especially in the English-speaking world, based on the work of Jacques Charles (1746–1823) some fifteen years earlier. But such reference was due to an English-favoring historical revision by P. G. Tait. As Cardwell commented,

---

[2] During the course of this work, Boyle and Hooke also invented their famed air pump, which they used to create a vacuum. While this device was not directly used in the development of Boyle's Law, it served a role in the early history of scientific experimentation as captured in *Leviathan and the Air-Pump* (Shapin and Schaffer, 2011).

[3] (Brush, 2003b) pp. 422–424. Brush described Francis Line (a.k.a. "Franciscus Linus") as a gadfly: an outspoken critic who challenges the ideas of geniuses, forcing them to revise and improve those ideas, resulting in new knowledge for which the genius gets the credit while the gadfly is forgotten. In this history, Line "argued that while air may have some degree of weight and elasticity, it does not have enough to account of the behavior of the mercury in a Torricelli barometer." Boyle's work was intended to answer this criticism.

[4] (Holbrow and Amato, 2011).

"the facts do not in any way justify the ascription of this law to J. A. C. Charles … [and so] we deny the claim made on behalf of Charles and shall in future refer to the law as Gay-Lussac's law."[5]

In 1807, Gay-Lussac continued his experimental studies by carrying out a series of experiments designed to determine the relationship between the specific heats of gases and their densities.[6] In the course of these studies, he found that the change of gas temperature was directly proportional to the change of pressure.

$$P \propto T \qquad \text{Gay-Lussac's Law (pressure – temperature)}$$

As the above studies involved constant mass, mass thus became the final gas property requiring study. In 1811 Amedeo Avogadro (1776–1856), building on Gay-Lussac's work, proposed his famed hypothesis,[7] writing, "the quantitative proportions of substances in compounds seem only to depend on the relative number of molecules … The first hypothesis to present itself in this connection, and apparently even the only admissible one, is the supposition that the number of integral molecules in any gas is always the same for equal volumes." Fixed volume (at the same temperature and pressure) meant fixed number of molecules. Knowing mass then enabled Avogadro to determine the relative molecular weights of various gases: the ratio of weights must equal the ratio of molecular weights. His results were comparable to Dalton's which provided validating support to his hypothesis and also provided experimentalists a new approach to determining molecular weight. The contemporary approach at that time was based on Dalton's addition of weights (referenced to hydrogen) comprising the molecule.[8] As regards this discussion, Avogadro's work told us the following relationship.

$$V \propto n \qquad \text{Avogadro's Law}$$

It was finally Émile Clapeyron in 1834 who tied together the P-V-T properties into the ideal gas law,[9,10] which, when including Avogadro's work showing that the law can be scaled by the number of molecules, became:

$$PV = nRT \qquad \text{Ideal Gas Law}$$

---

[5] (Cardwell, 1971) p. 130–131.

[6] (Crosland, 2008a).

[7] (Crosland, 2008b).

[8] (Thackray, 2008) Dalton's method for calculating molecular weights was based on his theory of mixed gases. If A reacts with B, then the weight of the product AB must equal the sum of the weights of the two reactants A and B. In this way he calculated the weights of gaseous atoms and molecules referenced against the lightest molecule, hydrogen. It was his mechanistic insight and physical model of molecules that yielded such results and helped modernize the atomic theory of matter.

[9] (Carnot et al., 1988) p. 82. In his publication, Clapeyron references his use of Mariotte's law for the P–V relationship. Boyle published his law in 1662 (Hall, 2008) while the French physicist Edme Mariotte (1620–1684) independently discovered the same law in 1679 (Mahoney, 2008), thus giving Boyle priority. Thus this law is sometimes referred to as Mariotte's law or the Boyle–Mariotte law. Two points of interest. First, this law was originally proposed by Richard Towneley (1629–1707) and Henry Power (1623–1668); Boyle confirmed it and popularized it through experimentation. Second, in his experiments, Boyle initially wanted to prove that air has elasticity like a spring. It was Francis Linus' challenge of this that motivated Boyle to conduct additional analysis which led him—"almost as an afterthought," (Reilly, 1969) p. 80—to his law of constant PV. See (Brush, 2003b) pp. 422–424.

[10] (Lewis, 2007) p. 33. Siméon-Denise Poisson was arguably the first to tie together P, V, and T in an early version of the ideal gas law in 1823.

Ironically, John Dalton (1776–1844) used the gas laws in the development of his atomic theory even though he was, being a disciple of Lavoisier, also a staunch believer in the caloric theory.[11] These two theories are so contradictory, to us at least, that one would think that upon discovery of the atomic theory, Dalton would have immediately jumped to discovering the kinetic theory of matter. But this wasn't the case. The caloric theory survived even this strong challenge, simply because Dalton found a way to use the caloric theory to explain his atomic theory. In Dalton's mind, gas was comprised of atoms but surprisingly his atoms were not the projectiles we know them to be today. Instead, they were static, fixed in space, stacked together as closely as possible, and surrounded by an atmosphere of caloric that was the source of the repulsive forces between them. (Being British, Dalton was strongly influenced by Newton's static model of gases,[12] which will be discussed further in Chapter 40.) Total gas volume was determined by the space filled by these caloric-wrapped atoms. In this model, temperature alone determined the amount of caloric surrounding the atoms. If you increase temperature, you increase caloric and thus increase volume. Such *ad hoc* explanations provided the means to arrive at the right answer for the wrong reason and so (wrongly) served to keep caloric alive for quite some time.

Theorists Attempt to Explain Heat Capacity

Since Galileo, the concept of the experiment started opening up the behavior of the natural world to man's analysis. It was this concept that led to the discovery of the ideal gas law. Its success motivated more experiments to quantify other properties and behaviors of gases, liquids and solids.

One, property of study was heat capacity. As we learned before, this property together with temperature was needed to quantify what was thought to be the "total heat" of a system. Recognizing this importance, experimentalists such as Joseph Black designed calorimetry experiments to quantify specific heat capacity by dividing a known amount of absorbed heat by the resulting temperature change ($C = Q/\Delta T$). Others including Adair Crawford (1748–1795), Gay-Lussac, and Dulong and Petit subsequently built on Black's work.

The challenge in this approach was that as heat is added to a gas, temperature isn't the only property that changes. If the container holding the gas is solid, such as for a calorimeter bomb, then volume is constant, pressure increases and the resulting heat capacity of that gas is called C_V, with the subscript "V" denoting constant volume. However, if the container is flexible and the boundary movable, such as occurs when using a balloon or inflatable bladder, then pressure is constant, volume increases, and the resulting heat capacity is called C_p. In this latter case, because the gas pushes against a movable and receding boundary, it does work and experiences a drop in temperature. The total temperature rise in the calorimeter is then <u>less</u> for constant pressure heating than for constant volume heating, meaning that C_p must be <u>greater</u> than C_V since the given amount of heat added is being divided by a smaller number. The understanding

---

[11] (Cardwell, 1971) Chapter 5.
[12] (Fox, 1971) pp. 6–8, 111.

of the significance of this difference between C_p and C_v was to become one of the critical components of Robert Mayer's calculation of the mechanical equivalent of heat (Figure 5.2).

Another approach to effecting temperature change in gas is through adiabatic compression or expansion. When gas atoms strike a receding boundary during expansion, they rebound at a lower speed and so cool. When they strike an approaching boundary during compression, they rebound at a higher speed and so heat.

As experimentalists quantified the data from the above specific heat and expansion/compression studies, theorists tried to explain them. In what should have been a clear-cut victory for the kinetic theory of gases, explaining at a molecular level exactly why heat and work are related, the caloric theory once again miraculously survived. It took some creativity, but it did indeed survive. In an utterly incorrect but effective argument, the calorists said that for a given amount of matter (atoms), the amount of caloric surrounding the atoms is governed by the total volume available. Caloric continues attaching to the atoms until all the volume is filled. This is somewhat different than Dalton's belief that temperature alone determines the amount of caloric, but stay with me here.

The importance of heat capacity measurements in the above context was very real at this time in France, so much so that in 1811 the French Institut proposed a competition to determine the most accurate experimental approach. One of the more significant mistakes in the history of thermodynamics ensued when F. Delaroche (1781–1813) and J.-E. Bérard (1789–1869) rose to the challenge of this contest by carrying out a series of investigations that seemed to work out well with the exception of one single data point mistakenly suggesting that heat capacity varied with volume.[13]

To the calorist, total caloric was constant and was split between two separate components, "latent" caloric that was chemically bound to the atoms and "sensible" caloric that filled the space between the latent spheres. In this model, the latent caloric accounted for volume; it wrapped around the atoms and expanded or contracted to fill the available volume. Also in this model, total caloric was conserved. Thus, any change in latent caloric had to be exactly compensated for by the opposing changes in sensible caloric, which itself was quantified by temperature since temperature measured the presence of sensible caloric only. So putting this all together, as gas volume decreased during compression, caloric shifted from latent to sensible, and so increased temperature. And since heat capacity was thought to quantify total latent caloric, the decreasing volume decreased latent caloric and so decreased heat capacity. This was the theory anyway.

Unfortunately for history, Delaroche and Bérard's experimental data supported this false theory. They shared their results with the utmost confidence, stating what they thought was the obvious: "Everyone knows that when air is compressed heat is disengaged [shift of latent to sensible caloric]. This phenomenon has long been explained by the change supposed to take place in its specific heat; but the explanation was founded upon mere supposition, without any direct proof. The experiments which we have carried out seem to us to remove all doubts upon the subject."[14]

[13] (Carnot et al., 1988) Read E. Mendoza's Introduction for an account of this history. Also read (Cardwell, 1971) pp. 135–137.
[14] (Cardwell, 1971) p. 136–137.

These incorrect data and conclusions, suggesting that the specific heats of gases increase with their volumes, led both Carnot and Clapeyron astray, increasing the significance and influence of the caloric theory on their and others' thought processes and calculations. As just one example of how badly off-course these conclusions could lead one, imagine increasing volume to achieve the ultimate low-density gas, i.e., a vacuum. Based on the caloric view, the resulting increase in heat capacity would be extremely high, the highest of all. As Mendoza wrote, "probably no other bad data have upset the development of thermodynamics more than these."[15]

*   *   *

The ideal gas law provided an excellent quantified relationship between readily measurable properties of gas and served to help move discussions from the abstract to the concrete. Carnot, Clapeyron, Thomson, and Clausius all relied on equation 18.1 and its inherent simplicity to help guide their specific analysis of the steam engine. The law itself was the result of extensive and accurate studies by a range of scientists and fortunately was done independent of any guiding philosophy of heat. Pressure, volume, and temperature stood on their own as directly measurable properties of nature. How scientists interpreted the results was another matter, as was reflected in the work of Delaroche and Bérard in which their caloric philosophy tainted their studies and so slowed progress.

The beauty of the ideal gas law lies not only in its simplicity but also in the power of what it shows, specifically the relationship between PV and T, for in this relationship one catches a glimpse of the relationship between work and heat which became the basis for the mechanical theory of heat and the subsequent higher-level theory of energy and its conservation. This is the direction we'll now continue to explore.

[15] (Carnot et al., 1988) Mendoza in his Introduction, p. xvi.

19 The final steps to energy and its conservation

The critical pieces to the energy puzzle were spread out on the table. The importance of weight times height and its connection with mv² in the rudimentary law of conservation of energy, Lavoisier's chemistry, Black's heat capacity, the ideal gas law, and the impossibility of perpetual motion all combined to serve as a guide. It was all there, just waiting for some seed around which all these ideas could crystallize. In Chapter 13 I shared Boyer's suggestions on the three steps required to discover energy and its conservation. In this chapter I share the triggers that had those steps be realized.

Why were triggers needed? It seems straightforward to us today how these pieces fit together. But it clearly wasn't straightforward in the early nineteenth century. There wasn't an open space in which the concept of energy could arise. There was no clean sheet of paper to create the concept. There already was a concept called caloric that could explain many of the energy-related phenomena. And what it couldn't explain, well, such things were either addressed via *ad hoc* explanations or put off to the side and ignored. The caloric paradigm was like a huge trench that people simply couldn't climb out of. Look at the work of Daniel Bernoulli. To a certain degree, he captured the essence of both the concept of energy and the kinetic theory of gases, and yet the French establishment largely ignored him. The trench was that deep.

Why am I taking you through all this? Because the search for a way out of the trench was a big deal, and it's important for us as physicists, chemists, and engineers to feel the frustration involved with this search and so better appreciate the magnitude of the challenge and the accomplishment. The caloric theory survived quite well, despite the cracks in its structure that started to form on account of the observance of phenomena that it couldn't explain. It's survival in the face of such overwhelming evidence to the contrary could be attributed largely to the fact that, as spelled out by Thomas Kuhn,[1] absent a crisis, scientists typically seek to confirm prevailing paradigms rather than invent new ones. Life (and funding) is easier, safer, and more certain this way.

So what were the eventual triggers? According to Kuhn,[2] there were three:

1. Availability of conversion processes. The arrival of processes, such as the steam engine and the electric motor, that could transform energy from one form to another.

2. Concern with engines. The rise in the study of engines, critically including the identification of the concept of work as being important. Triggers (1) and (2) alone revealed quantifiable connections between such phenomena as heat, work, chemistry, and electricity for those able to see such connections. This is where the third trigger was needed.

[1] (Kuhn, 1996).
[2] (Kuhn, 1959).

Block by Block: The Historical and Theoretical Foundations of Thermodynamics,
Robert T. Hanlon, Oxford University Press (2020). © Robert T. Hanlon.
DOI: 10.1093/oso/9780198851547.001.0001

3. Philosophy of nature. The *inner belief* that such phenomena should even be connected in the first place. As Kuhn noted, "the discoverers of energy conservation were deeply predisposed to see a single indestructible force at the root of all natural phenomena."[3] One could argue that this trigger had been in existence prior to the nineteenth century, arising as it did from a large body of work supporting such theories as the impossibility of perpetual motion, the conservation of mechanical energy, and Leibniz's conservation of *vis viva*. But I'll leave this argument for another time. All of these concepts were important, regardless of whether they were actual triggers or not.

One additional condition, not so much addressed by Kuhn, perhaps a variant of Kuhn's third trigger, was important toward finalizing energy conservation: an education outside established science. Fresh eyes were needed in the analysis of energy, eyes not tainted by the caloric theory doctrine taught in the universities, new eyes that could take in the above ideas and put them together without being hindered by the thinking of others. As Kuhn himself noted, those who are new to a given field of study are "less deeply [committed] than most of their contemporaries to the world view and rules determined by the old paradigm."[4] To this group, embracing a new paradigm that better explains the factors causing a crisis in the old paradigm is relatively easy.

But the Science of Heat was Already Complete, Right?

Before continuing, a brief comment is warranted here. An interesting aspect of this time from 1790 to 1830 was that the science of heat wasn't really all that interesting. In contrast to new and exciting activities in other areas of science, heat was already complete. Or at least Fourier suggested it was, writing in 1822 that "we have demonstrated all the principles of the theory of heat."[5] During this time, France led the scientific world, largely through the work of Laplace who used Newton's work to bring many physical phenomena into the realm of mathematics, thus establishing "physics" as a standalone science. The spread of this mathematized physics to other countries such as Britain helped foster the surge in Western science during the mid-1800s. As part of this surge, Fourier's *Analytical Theory of Heat* (1822)[6] did much to clarify the concept of heat and rates of heat transfer, bringing his powerful mathematics to "solve all the fundamental problems."[7] But Fourier was indifferent to the fundamental nature of heat, not being interested in whether the caloric or mechanical theories were correct.[8] "Primary causes are unknown to us; but are subject to simple and constant laws, which may be discovered by observation, the study of them being the object of natural philosophy... The object of our work is to set forth the mathematical laws which [heat] obeys."[9] His predictive mathematics worked and that's all that mattered. However, the science of heat at this time was far from complete.

---

[3] (Kuhn, 1959) p. 337.
[4] (Kuhn, 1996) p. 144.
[5] (Fourier, 1952) p. 174.
[6] (Fourier, 1952)
[7] (Fourier, 1952) p. 174.
[8] In a 2017 email to the author, historian Robert Fox shared how striking it was that Fourier could do what he did without taking any position on the nature of heat, which suggests that the issue of the nature of heat was becoming an irrelevance by the 1820s.
[9] (Fourier, 1952) p. 169.

20 Julius Robert Mayer

In the fields of observation chance favors only the prepared mind
– Louis Pasteur[1]

Mayer [made] the only calculation then possible of [the mechanical equivalent of heat] from published data

– Kenneth Caneva[2]

Many thought that the first calculation of the mechanical equivalent of heat was a fluke. How could a non-scientist ever be the first to arrive on top of such a profound pinnacle of achievement? His path was an enigma. His sources obscure. He had no pedigree. Seemingly out of the blue, Julius Robert Mayer (1814–1878) arrived at the mechanical equivalent of heat in 1842 and then at a more encompassing theory of energy in 1845. But Mayer was no fluke. He was a brilliant thinker who had the simple misfortune of being an outsider.

Born in Heilbronn, Germany, son of an apothecary, Mayer grew up with a fascination with machines. So strong was this fascination that at one point he attempted to build his own perpetual motion machine. His failure led to an extremely valuable experience, one that would guide his thinking in later years: one can't generate mechanical work out of nothing. The importance of this experience cannot be overemphasized, for it helped establish Kuhn's "inner belief" trigger inside Mayer's philosophy.

The Java

After obtaining his medical degree from Tübingen in 1838, Mayer jumped aboard the *Java* in 1840 as it sailed from Rotterdam to the Dutch East Indies, earning his passage as the ship's doctor. Alone during the trip with much free time on his hands, Mayer occupied himself "zealously and unremittingly with the physiology of the blood."[3] He studied hard, thought hard and noticed much. His mind was prepared for what happened next.

Upon arrival in Indonesia, some of the crew became ill, necessitating the common cure of the day, blood letting. In Mayer's own words, "In the copious bloodlettings I performed, the blood

[1] Louis Pasteur, Lecture, University of Lille (7 December 1854).
[2] (Caneva, 1993) p. 38.
[3] (Caneva, 1993) p. 7.

Block by Block: The Historical and Theoretical Foundations of Thermodynamics,
Robert T. Hanlon, Oxford University Press (2020). © Robert T. Hanlon.
DOI: 10.1093/oso/9780198851547.001.0001

let from the vein in the arm had an uncommon redness."[4] Out of all the many observations Mayer made, this one stood out. His education prepared him to notice something as "uncommon" as this. Venous blood wasn't supposed to be so red.

Many ignore the uncommon observation, the outlier, and simply let it float away. But others pounce on it, grasp it, and pull it apart, seeking the deeper underlying cause. Mayer was one such person. There was no plan here, no nice, neat path in front of him. He could have turned away to another path, but he didn't, a strong testament to his intense need to understand the cause behind the effect. *Why was the blood so red?*

Mayer was neither physicist nor scientist. He had no obvious strengths in mathematics or natural science. He was a doctor. But behind this exterior was a deep and original thinker who had an extremely logical mind that adhered to a deep belief in causality as manifested by his oft-spoken slogan, *causa aequat effectum* (cause equals effect), and as inspired by his childhood failure in building a perpetual motion machine. And with this mind Mayer went on his own journey, seeking to understand the very simple but profound observation that blood is redder in warmer climates.

The Meaning of Bright Red Blood

When lungs pull in air, oxygen latches on to the hemoglobin in the blood, turning it bright red. The blood carries the oxygen to the cells through the arteries, where it reacts with the food to generate both cell structure and heat to keep us warm. This "combustion" process yields carbon dioxide as a product which the now darker-red (oxygen-depleted) blood carries back through the veins to the lungs for release.

Based on his reading of Lavoisier, Mayer knew *some* aspects of this physiology. He knew that respiration is a form of combustion. He also knew that combustion releases heat. Using his penchant for step-by-step, cause–effect deductive reasoning, Mayer deduced that respiration must release heat and so must be what keeps animals warm. This may seem obvious to us now, but at that time many were skeptical of Lavoisier's theories and thought oxygen's role was solely to remove waste carbon from the body. Mayer didn't align with the consensus, a critical attribute for anyone seeking to create something new. Instead, he embraced Lavoisier's theories, understood their importance and instinctively held them as valid. He sensed the answer to the redder blood: in warmer climates, less heat generation is needed to keep warm, so less oxygen is combusted, resulting in a higher oxygen content in the venous blood, making it redder.

Drawing on his passion for analogies, Mayer developed a view of the human body as a working machine, perhaps inspired by the widely used organism-as-machine metaphor appearing in the German scientific literature in the early nineteenth century.[5] To him, the energy first released by the reaction between food and oxygen is "transformed partly into heat, partly into motion; both taken together naturally again give the measure of the first."[6] Thus, suggested Mayer, and this was a HUGE "thus," since heat and work vary in the daily life of an animal but

---

[4] (Caneva, 1993) p. 27.
[5] (Caneva, 1993) p. 142.
[6] (Caneva, 1993) p. 239.

must otherwise sum to the same fixed number to ensure the impossibility of perpetual motion, there must be a *quantitative* equivalence between heat and work. This was one of the first conceptual energy balances conducted around the human body.[7]

All of this thinking materialized during his voyage. Like Newton's time in the country, Bernoulli's in Russia, and Carnot's in his small Parisian apartment, Mayer's time onboard the *Java* provided him the solitude necessary for deep creative thinking. A critical outcome was his realization that in order to continue progress, he had to crossover from physiology into the world of physics. Thus did Mayer become yet another grand discoverer to be enabled by a Galilean stance at the interface between different disciplines.

Return to Heilbronn

Upon his return to Heilbronn in 1841, Mayer practiced medicine but also continued his impassioned journey into this new world of physics by reading widely and appealing for help from professors and scientists in Tübingen and Heidelberg. But having entered someone else's world as an outsider, he toiled alone, writing: "with regard to the science at hand I am almost as isolated here as on the ship."[8] He was on his own in uncharted territory, working at a time in a country that itself was relatively isolated from the world of physics. Even the text books of the day couldn't help, at times causing more confusion than clarity. How could they offer clarity? He was creating something new, something not contained in any book.

The difficulty of converting his rather abstract and novel ideas into the mathematics, equations, and terminologies of the physics community was a confusing struggle. His attempted 1841 publication was deemed not publishable. Undaunted, his second attempt in 1842 succeeded, eventually becoming a landmark paper, capturing for the first time ever the numerical value for the mechanical equivalent of heat. While he didn't lay out his calculation in this paper, Mayer did in his monumental 1845 publication in which he also put forth his theory of energy, which he called Kraft.

Mayer's Technical Journey

I can't go into all the details of Mayer's journey as this would require a full book, but I can share the highlights that marked his path, starting with his remarkable calculation. In a fascinating thought process, building from his childhood failure at perpetual motion and his original insights while on the *Java*, Mayer realized that as motion disappears, heat appears, and vice versa, writing: "in order to become heat, motion – be it a simple or a vibratory motion like light, radiant heat, etc., – must cease to be motion"[9] He reasoned that there must be an absolute causal connection between the two.

---

[7] One can sees similarities between Mayer's thought process and Lavoisier's experimental heat balance around his guinea pig.

[8] (Caneva, 1993) p. 255.

[9] (Caneva, 1993) p. 25.

Inspired by one of his friends who said, "if you can base a new experiment on your theories, then, then you've got it made,"[10] Mayer recognized his need to focus. He knew motion can cause heat, having himself warmed water by shaking it and having read about Rumford's work. But he wasn't about to launch into his own set of experiments; he was not an experimentalist, but more of a theorist. So where could he obtain the data needed to quantify the connection between motion and heat?

The Challenge of Building a Conservation Law that includes Heat

Conservation laws demand quantification, for how else can they be tested? One can readily add up energy quantities within a given classification, such as work (weight-distance), kinetic energy (mass × distance²/time²), and heat (Btu). For a simple lever, the change in weight-vertical distance for one side must exactly equal the change in weight-distance on the other. For an elastic collision, the change in the mass × distance²/time² of one body must exactly equal the change in mass × distance²/time² of the other. But what if there's a transformation from one classification to another? How do you add the units? For the conversion of gravitational potential energy into kinetic energy, we multiply weight-distance times the gravitational constant (mass-distance/weight-time²) to get mass × distance²/time², which are the units of kinetic energy. But what about heat?

The Piston

Scientists striving to build a conservation law slowly realized the need to connect the two increasingly popular measurements, mechanical work and heat, especially since the advent of the engines, one of Kuhn's triggers, used the latter to generate the former. The issue at hand manifested itself in the action of the piston. As the gas expanded to move the piston and so do work, the temperature of the gas decreased. But what was the relationship? The need to compute efficiency for pragmatic purposes initially demanded this conversion factor, but then the theorists also needed the conversion factor to determine whether energy is indeed conserved. Exact conservation demanded exact measurements and experimentation.

What was needed here was a conversion factor to equate the units of work (foot-pounds) to the units of heat (Btu's). Only by doing this could you bring work and heat together into the same units and so enable their numerical addition, a necessary requirement when assessing a conserved quantity. But how to determine this? How much change in foot-pounds results in how much change in Btu's? Leibniz did this in his famous thought experiment linking the work exerted in raising a body to the resulting kinetic energy of the released body at the bottom of its fall. But this was for a single body. Heat represents the kinetic energy of billions of bodies

---

[10] (Caneva, 1993) p. 252.

and so no direct calculation along the lines of Leibniz's approach was possible, leaving direct measurement as the only path forward.

One approach, the one used by Joule, would be to slowly lower (no kinetic energy) a weight through a known distance. The descending weight would be connected to a string tied to a rotating paddle placed in a water bath; the rotating paddle would increase the temperature of the water. Another approach, conceptually similar to what happens in a piston-in-cylinder assembly, would be use the same descending weight to compress a gas and so increase its temperature, or the reverse—raise the weight by expanding the gas which decreases its temperature, a phenomenon noticed by Rumford with the hot gun barrels (no bullet meant no work meant no cooling) and conceptually used by Mayer. In both approaches, the change in potential energy of the weight must be exactly balanced by the change in kinetic energy of the water or the gas, as quantified by the change in temperature. The mechanical equivalent of heat could only be determined based on the assumption of this work–heat equivalence according to the conservation of energy.

Mayer's Logic

Given these options, and given that he wasn't about to launch into his own set of experiments, Mayer turned toward the compressed gas approach in light of the readily available data from the early 1800s gas studies. Mayer understood the logic in the above paragraph. He knew that heat causes gas to expand, and that compression causes gas to heat. He also knew of Gay-Lussac's experiments showing no temperature change for a gas expanding into a vacuum, otherwise known as free expansion. While the contemporary caloric theory could not explain this, Mayer correctly realized that since the gas did no work during the expansion, then it wouldn't change temperature. James Joule and William Thomson would later name this "Mayer's Hypothesis" and prove its validity for ideal gases having no intermolecular interactions.

But there was one problem. No direct data had been published showing a connection between a falling weight and an increase in gas temperature. Of course this was so, because no one had yet connected the two conceptually and quantitatively, and so no one thought to do the experiment.

In one of the more brilliant historical moments in thinking, Mayer realized that the answer to this problem lay hidden in the specific heat data and analyses by the likes of Gay-Lussac, Delaroche and Bérard, and Dulong, who had published a comprehensive review of the specific heats of gases in 1829. Recall that two sets of data were available, the specific heat at constant volume (C_v) and the specific heat at constant pressure (C_p). Mayer, perhaps influenced by conversations within his newly formed physics community, correctly interpreted the difference in these values as being meaningful and critical. Heating a gas at constant pressure causes it to expand and thus do work by pushing against the surrounding environment. Since doing work causes temperature to decrease, more heat is required to raise gas temperature by a fixed number of degrees at constant pressure than at constant volume, meaning that C_p must be greater than C_v by a fixed amount having nothing to do with the heat capacity of the system.

The Math

Let's look at the mathematics here, starting with the heating of an ideal gas at constant volume. The heat capacity at constant volume, C_v, can be determined by the following:

$dU = Q - W$ ___ Ideal gas

$dU = Q_v = C_v dT$ ___ Heating ideal gas at constant volume; no work

$C_v = Q_v / dT$

Now say you changed your apparatus to allow the system to expand while maintaining constant pressure. When you heat it this time to achieve the same dT, you'd need to add more heat since some of the heat would have to be diverted to doing expansion work. How much total heat then are we talking about?

$Q_p = Q_v + PdV$ ___ Heating ideal gas at constant pressure to achieve same dT

$Q_p = C_p dT$

$C_p = Q_p / dT$

So where does this leave us?

$Q_p = Q_v + PdV$

$C_p dT = C_v dT + PdV$

$C_p dT - C_v dT = PdV$

$(C_p - C_v) = P(dV/dT)$

For an ideal gas, PV = nRT and so

$P(dV/dT) = P(nR/P) = nR$

Thus, dropping the "n" from "nR" as the C values are on "specific heat capacity" basis,

$(C_p - C_v) = R$ ___ Ideal gas

$R = f$ (the mechanical equivalent of heat [MEH])

In this calculation, the gas constant R quantifies the amount of heat required to do a certain amount of work and thus contains within it the mechanical equivalent of heat. This constant is needed to complete the ideal gas law, which shows the connection between work (PV) and heat (RT) for an ideal gas under the unifying umbrella of energy.

The Mechanical Equivalent of Heat (MEH)

Mayer attacked his calculation similarly as above. He realized that C_p is greater than C_v, the difference representing the work done in expanding that gas. Recognizing work to be the product of weight times change in vertical height, Mayer calculated this work as the product of the change in volume (he assumed that +1 degree increases volume by 1/274 based on ideal gas law) and the weight being lifted vertically (a 760 mm high column of mercury – atmospheric pressure).[11] When you do the math, and convert to British units, Mayer calculated that 1 Btu is equal to 665 foot-pounds of work, which is about 85% of the commonly accepted value. Historians later validated this approach and ascribed the error to errors in the published specific heat data.

The Community's Response

Mayer's rather abstract approach to calculating the mechanical equivalent of heat was not wholly welcomed by the physics community as it relied on the ideal gas assumption that all heat added to the gas was transformed into work as opposed to having some of it diverted to separating the gas particles. William Thomson and P.G. Tait in particular would famously challenge this assumption, as we'll see in Chapter 33. While Mayer later supported this hypothesis by pointing to Gay-Lussac's experiment showing that free expansion of a gas into a vacuum results in zero temperature change, it still cast doubt among some at the time. Also, at least some felt that Mayer's multi-step approach involving two separate measurements (C_p and C_v) plus a concluding calculation was less reliable than a clean, single, direct experiment.

Given all this, what a stroke of genius! Who would have thought that the mechanical equivalent of heat was hidden in the specific heat of gases? Others had access to the same data; only one did the calculation.

Mayer's passion continued as he sensed that his above work was but part of a larger whole. The doctor who was fascinated by machines as a child was now driving toward the heart of their operation. There must be something larger governing this equality between heat and work. Helping guide him was his stubborn attachment to simplicity—á la Occam's Razor—*nature must be simple!*—his equally stubborn attachment to cause–effect logic, his unrelenting examination of the step-by-step consequences of such causality, and his inherent originality, a necessary trait for anyone seeking to discover something new in an area where many others had already tread.

Returning to Kuhn's "Inner Belief" Trigger

As brought up earlier, also helping guide Mayer was something deeper, something connected to his deeply religious philosophy. Mayer believed, or wanted to believe, in the existence and immortality of the soul. Because of this, he strongly opposed the philosophy of materialism since this philosophy inherently excluded the possibility of the immaterial soul. Mayer proposed the concept of an "immaterial" energy as a challenge to the pre-eminence of materialism. (Note: Mayer actually used the German word *Kraft*, which translates to *force* even though his concept was more aligned with the modern concept of energy; I will continue to translate

[11] (Caneva, 1993) p. 37–38.

Kraft to energy.) Success of his concept would validate the existence of the immaterial and so validate the possibility of the soul.

Mayer developed this philosophy into a theory of energy that remains largely valid to this day, and he did this through the use of yet another powerful analogy. Believing in a certain symmetrical duality of nature, Mayer proposed as his fundamental concept of nature that the existence of both the material and immaterial is reflected in the existence of both matter and energy, respectively. Up until this point, all energy concepts, such as vital or life force and *vis viva*, were regarded as properties of matter. Mayer separated energy from matter, putting each on its own terms, and defined them as the two components of the physical world. And then, exploiting the mass–energy analogy to the extreme, he assigned similar properties to energy as then existed for matter. As he wrote in 1841, remaining with the literal translation of *Kraft*, "it is the same with the theory of forces (physics) as with the theory of material substances (chemistry); both must be based on the same principles."[12]

Note that Mayer explicitly rejected a material form of energy. The two competing theories of that day, caloric and heat-as-motion, both related heat with matter. Instead of aligning with either, Mayer chose neither. His belief in the existence of energy as separate from and independent of matter led him to create his own novel theory of an immaterial yet quantifiable energy. He gave the matter–energy duality a radical new interpretation. Furthermore, he didn't feel a need to define what energy actually was, much in the same way as Galileo and Newton didn't feel a need to define what gravity actually was. As he wrote, "I don't know what heat, what electricity, etc., is as regards its internal essence…[or] how motion passes over into heat – to demand clarification on those matters would be to demand too much of the human mind."[13] But he didn't let this void stop him. The point is that he didn't need to know the essence of energy. He just needed to know that heat was quantified by Black's equation, $Q = mC\Delta T$. While heat is indeed caused by motion, this understanding wasn't required to develop the conservation of energy. This is why Leibniz's *vis viva* wasn't needed in and of itself for the invention of energy, at least for Mayer's and Joule's respective approaches.[14] It only became critical later in the development of the kinetic theory of gases.

One of the most critical "principles" that Mayer linked from matter to energy was indestructibility. While Lavoisier conducted his work with an assumed understanding that mass is a conserved quantity, there was no explicit formulation of this conservation law in the physics and chemistry literature at that time. Mayer had, after reading Lavoisier, after contemplating the immortality of the soul, and after otherwise thinking his own thoughts, defined this law. It made sense to him and his cause–effect belief system. So in keeping with the analogy, it was an easy next step for Mayer to say that just as matter is indestructible, so too is energy. Mayer created, in essence, both conservation principles simultaneously.

Oddly enough, it took some time before Mayer could embrace the *uncreatability* of matter and energy. An internal struggle played itself out within Mayer on this subject since at that time there was a strong but fading belief in the "vital force" in which some *thing* operating

[12] (Caneva, 1993) p. 243.
[13] (Caneva, 1993) pp. 28–29.
[14] Helmholtz did use the concept of *vis viva* in his own approach to energy and its conservation as you'll read in Chapter 32.

outside the laws of physics and chemistry was at work in creating energy in living organisms and in the universe, which some thought to be a "divine organism." Mayer's breakthrough occurred when his open-mindedness led him to reject the paradigm and choose what he believed to be true. In 1842 he didn't believe that energy was conserved, only indestructible. By 1845 his thinking had evolved to a belief that energy was *both* indestructible and not able to be created.

Mayer's 1845 Publication

In his 1845 publication, which contained his fullest statement of his theory of energy, Mayer started with one of his favorite guiding slogans, "*Ex nihilo nil fit*" (from nothing, nothing is made), moved into a discussion about the equivalence between heat and work and the details behind his 1842 calculation of the mechanical equivalent of heat, and then went on to define his larger theory of energy. While he never said the single simple phrase we know today, *energy is conserved*, he did in essence say it: "In all physical and chemical processes…the given force remains a constant quantity."[15] He identified what he took to be the five principal forms of energy, namely gravity, motion, heat, magnetism/electricity, and chemical, and then listed forms that each could take, anticipating a full network of interconnections such as converting mechanical effects into electricity and magnetism, converting gravity (dropping weight) into heat, etc. Mayer didn't start with these different conversion processes in mind to create his theory; he created his theory and then looked around to locate the different conversion processes to pull into his discussion.

Mayer's Fall and Rise

In a story we've heard before, Mayer's publication was largely ignored. It didn't help that Mayer was considered somewhat quirky and unsophisticated, had difficulty converting his abstract thoughts into readily coinable words and phrases, and had "a rather poor reputation as a thinker ever since anyone began to take notice of him."[16] But still, ignored? The fact was that Mayer was an unimportant figure to many of the scientists in the mid-nineteenth century, who neither understood nor appreciated him. This reminds me of the scene in the movie, *Immortal Beloved*, in which Beethoven's nephew complains about his deaf uncle, "He does no work. All he does is scribble incomprehensible phrases. Then he bellows this stupid, childish tune at the top of his lungs….", whereupon he starts imitating Beethoven's bellowing of the main melody of the 9th symphony. "I think it's ridiculous." Mayer was a highly original genius who was never really accepted by the "in crowd," whether due to the fact that no one understood his "bellowing" or the fact that many held him in low esteem.

Fortunately, unlike Ludwig Boltzmann's similar situation years later that had no second act, this story ended rather well. After a decade of personal tragedy including a mental breakdown

---

[15] (Caneva, 1993) p. 261.
[16] (Caneva, 1993) p. 324.

and suicide attempt, caused in part by the rejection of his ideas and the scorn of his peers, all of which led to his institutionalization, Mayer re-emerged fully healed around 1860 at a time when his work was finally being recognized and acknowledged by such established scientists as Hermann von Helmholtz and Rudolf Clausius in Germany and John Tyndall (1820–1893) in England. Such support became important as the earlier neglect of his work was leading others to incorrectly suggest James Joule as the first to discover the energy principle. But the resulting cross-continent dispute, reminiscent of the Leibniz–Newton dispute, was never really between Mayer and Joule. While they never met, Mayer was full of admiration for Joule and Joule respected, perhaps with less admiration, Mayer. While they did debate their technical work in publications, there was no animosity between them. The dispute was really between their supporters, Helmholtz, Clausius, and Tyndall for Mayer and Tait, Thomson, and Rankine for Joule. I will discuss this dispute in more detail in Chapter 33.

Mayer was eventually recognized as the first to get to energy and received awards around 1870 for having done so. With such recognition came status and Mayer rose to become a hero of science amongst the German public. A memorial statue in Heilbronn marks his birthplace.

But Joule was a very close second in this race.[17] He worked totally independent of Mayer and many consider him the co-discoverer of the concept of energy. So let's now turn to him. They took very different approaches to arrive at the same point, Mayer's more theoretical and Joule's more experimental, and so in that way they complemented each other. One wonders at the possibilities had they worked together.

[17] Not to be overlooked is the accomplishment of Ludwig Colding (1815–1888), a Danish civil engineer and physicist who quantified the mechanical equivalent of heat around 1850, after Mayer and Joule.

21 James Joule

There are only very few men who have stood in a similar position and who have been urged by the love of some truth which they were confident was to be found though its form was as yet undefined to devote themselves to minute observations and patient manual and mental toil to bring their thoughts into exact accordance with things as they are

– James Clerk Maxwell on James Joule[1]

Poking at small effects you can't explain can be a way of unraveling a much bigger piece of physics

– Professor Carl Carlson, theoretical physicist[2]

Forging a law around the concept of energy and its conservation demanded many observations, many experiences, and many measurements. The long history of such has led us to two of the key discoverers of this law, J. R. Mayer, discussed previously, and now James Joule. While Mayer's rather stunning achievement was based on one "eureka" moment followed by one calculation—okay, it wasn't as fast and easy as I'm portraying, but it's amazing how far he went with so little—Joule travelled a different path involving the painstaking collection of an unassailable trove of data. Their respective approaches couldn't have been more different.

Born to a prosperous brewer near the great industrial city of Manchester, James Joule (1818–1889) enjoyed a private income that he earned by logging many hours, 9 a.m. to 6 p.m. daily, in support of the family brewery until its sale in 1854. Fortunately for science, his income afforded him a certain freedom in his after-hours, freedom he used for the pursuit of his passion, science. Despite receiving no formal education, Joule rose to become a renowned self-taught scientist, clearly benefitting from home schooling by his father through the first 16 years of his life and then private tutoring for two years by none other than John Dalton. It was through this latter connection to such a commanding figure that Joule was introduced to other distinguished members of Manchester's scientific and engineering community as embodied in their local Literature and Philosophical Society (Lit & Phil, as it was known).

While Manchester wasn't an established academic center of scientific thought, it wasn't a backwater either. Indeed, it was a city populated by men of exceptional skills, men from distant places such as Scotland, Germany, and eastern Europe, all practical, common-sense men involved in industry, business, engineering, and technology. The camaraderie offered by such a group of hard working, practical-minded men enabled Joule to thrive. At a time when

[1] (Cardwell, 1989) p. 266.
[2] Carlson quote cited in (Grant, 2013).

Block by Block: The Historical and Theoretical Foundations of Thermodynamics,
Robert T. Hanlon, Oxford University Press (2020). © Robert T. Hanlon.
DOI: 10.1093/oso/9780198851547.001.0001

textbooks and periodicals were rare, such personal relationships were critical, especially for someone not enrolled in academia.

During these formative years, Joule became well versed in the previous works involving mechanical energy before it became officially "energy," quoting, for example, from Leibniz in one of his papers, "force [energy[3]] is measured by the square of the velocity of a body or the height to which it can rise against gravity."[4] Growing up in such an industrial engine-driven environment as Manchester, he also had ready access to the real-world engineering practices involving, for example, Watt's concept of work as being force times change in vertical height. Fortunately, by *not* growing up in an academic environment, Joule was also insulated from certain scientific doctrines, such as those of Fourier, who had presumably already closed the book on heat, which could easily have inhibited his developments of the mechanical theory of heat and the conservation of energy.

A key turning point in Joule's early life was when he started to investigate the feasibility of replacing the brewery's steam engines with the newly invented electric motor. England's growing industrial economy was founded on the use of engines powered by falling water and later pressurized steam, but things changed around 1830 with the arrival of the electric motor. One can imagine the ripple of anticipation caused by this during the early days of the Industrial Revolution. Already witness to the arrival of the great locomotive steam engines, Joule also became witness to the arrival of the great electric motor.

This fabulous invention arrived as the result of a rapid sequence of historical milestones, starting with Alessandro Volta's (1745–1827) invention of the battery in 1800, Hans Christian Ørsted's (1777–1851) and André-Marie Ampère's (1775–1836) separate works in 1819–1820 showing that electricity and magnetism interact with each other, and then Michael Faraday's (1791–1867) invention of the electric motor in 1821, when he used magnets to demonstrate that electricity could produce motion and motion could produce electricity. The self-educated Faraday produced a monumental body[5] of work showing that the flow of electricity must be accompanied by chemical change, noticing the accumulation of zinc oxide on the copper electrodes and copper on the zinc electrodes, both resulting from the depletion of zinc and copper from their respective electrodes. His studies demonstrated yet another way to accomplish the raising of a weight and so afforded yet another clue to those insightful few who believed that all of these different ways of lifting a weight were fundamentally connected.

It Began with the Electric Motor

The arrival of the electric motor created much excitement throughout Europe and the United States as many thought that it offered improved performance over steam. Some even thought

---

[3] As discussed before, Leibniz's Latin term *vis viva* which would become kinetic energy unfortunately translates into *living force* in English, thus creating some confusion in light of Newton's definition of force.

[4] (Cardwell, 1989) p. 301.

[5] This body of work included Faraday's demonstration that both electricity and magnetism could impact light, a fact that James Clerk Maxwell successfully uncovered two decades later when he showed through his famous equations that light is a form of electromagnetism. Through the 1830s, Faraday demonstrated in one experiment after another how electricity, magnetism, and chemistry were all related. But it took time for these results to become disseminated.

that "perpetual motion" lay hidden deep within the electro-magnetic device, prodded by Moritz Jacobi's writings in 1835 that suggested the availability of near-infinite power from such a device. Many tried to find this pot of gold, not having yet truly understood Faraday's conclusion that it didn't exist. Electricity is generated as a result of chemical reactions; depletion of the chemicals causes flow of the electricity. The conversion of chemical energy into electrical energy was a difficult concept to comprehend, and its reality hadn't yet had time to fully settle into the minds of the public.

Along with others, Joule got caught up in this "electric euphoria."[6] His initial teenage enthusiasm for all things electric, as revealed in his joy of shocking himself and friends with his new electric devices, eventually gave way to more substantive studies. As a practical engineer seeking to improve the brewing process, Joule eagerly evaluated the switch from steam to electricity.

Joule did not set out to study energy. There was no significant, long-term goal at the beginning of his journey. Instead, at age 19 he simply and informally set out to build an improved electro-magnetic engine. Initial tinkering piqued his passion for experimentation and data. His father fed this passion by building him his own laboratory in the cellar of the family home. Over time he sensed himself brushing up against some deeper truth, moving from experiment to experiment, all the while guided by an instinct that knew which experiment needed to be done next. Whereas Mayer's compass was his rigorous cause–effect mind, Joule's was his keen experimental insight, his meticulous accuracy, and his relentless pursuit of truth driven by a deep curiosity.

The deeper truth revealed itself early. Recall Mayer's childhood failure to build a perpetual motion machine? Joule hit a similar brick wall. To him, unlimited opportunities seemed available with an electric motor until he learned in his experiments that flowing electricity always and unavoidably produced heat whose appearance corresponded to a decrease in the ability of the motor to do work. While at that time the connection between heat and waste of power was "obscure,"[7] Joule sensed that there *was* a connection, an inherent limitation to his motor. Compounding this was the fact that the electricity itself wasn't free; it only flowed when zinc was consumed. Based on his experiments involving the use of a zinc battery to drive an electro-magnetic motor to effect work, the results of which inherently included heat inefficiencies, Joule calculated that the use of high-priced zinc was very unprofitable when compared to coal combustion and so felt discouraged, writing, "I almost despair of the success of electro-magnetic attractions as an economic source of power."[8]

Joule's early experiments with electro-magnetic motors led to several laws of electrical energy, including his famed law showing that the heating effect of an electrical current is equal to the square of the current times the resistance, and further led to deeper insight into the physics of voltaic electricity. But it was his serendipitous observation that electricity generates heat that became a boon to science as it shifted his focus away from electricity and toward this "despairing" effect.

---

[6] (Coopersmith, 2010) p. 252.
[7] (Cardwell, 1989) p. 35.
[8] (Cardwell, 1989) p. 37.

The Impossibility of Perpetual Motion

Mayer and Joule each attempted the impossible early in their respective careers and each failed. The reality embedded in "perpetual motion is impossible" hit them both and became, per Kuhn, an "inner belief" trigger for each, teaching them deep down a hard but extremely valuable lesson. You can't get more out than you put in. But more of what? What exactly was it that's conserved? Caloric?

Historians[9] don't know if Joule ever accepted the caloric theory of heat at any point in his career. He probably knew about it, but whether or not he accepted it is another issue. His mentor, John Dalton, believed in it, which is rather ironic since it was Dalton's concept of the atom that would later provide the foundation for the competing heat-as-motion theory. But for Joule, there's no record. As Cardwell suggested,[10] it's actually hard to find any underlying philosophy, including religion, behind Joule's work. His only philosophy was the pursuit of truth.

Joule didn't *need* a theory regarding the nature of heat in order to develop the heat–work relationship, much as neither Galileo nor Newton needed a theory regarding the nature of gravity to develop their own respective free-fall relationships. Not being one to speculate, Joule didn't even venture in this direction until his research led him there, at which point his thinking evolved from the mechanical theory of heat based on work–heat transformation toward the kinetic theory of gases based on matter-in-motion, perhaps initially inspired in this direction by Dalton's atom and then later inspired by the work of John Herapath (1790–1868) on this topic.[11] No, instead of delving into or otherwise confronting theory in these early years, Joule focused solely on experiments and in this way eventually connected two radically different realms of experience, work and heat.

This background helps us to better comprehend Joule's work and reflects well on his tenacity to conduct research in the face of no support, for he was living in a world sitting contentedly in the caloric paradigm, disinterested in heat. Over time, Joule gradually created a new career for himself as a "gentlemanly specialist"[12] of independent means who gained expertise within a specialized branch of science and subsequent credibility within the associated gentlemanly scientific societies and institutions.

Turning Away from the Electro-Magnetic Motor toward the Magneto-Electric Generator

Joule's early experiments on the heating effect of electricity started with the electro-magnetic motor driven by a battery, for it was here that he first observed the unwanted and irritating heating effect. But where was the heating coming from? Since the caloric theory was still alive at this point, one possibility was that caloric was flowing out of the battery along with the

---

[9] (Cardwell, 1989) p. 45.

[10] (Cardwell, 1989) p. 62.

[11] (Cardwell, 1989) p. 96. Note in (Lewis, 2007) p. 78 that Joule's starting assumption in 1844 was a heat-as-motion theory based on static gas atoms. It was toward the late 1840s that he evolved his assumption to the correct theory based on the translational motion of gas atoms.

[12] (Smith, 1998) p. 55.

electricity, shifting from latent to sensible inside the coils. But why then didn't the battery feel cooler?

The presence of the battery complicated things, so Joule eliminated it. The battery generated electricity to drive an electro-magnetic motor to effect work or "lifting power."[13] But the reverse process was also possible. Work can drive the motor to generate electricity. In this scenario the motor ceases as a motor and becomes a magneto-electric generator. This was the new direction Joule turned towards and it unintentionally and fortunately brought the concept of "work" into his studies.

Hand Crank Turns → Magneto-Electric Motor Spins → Current Flows → Water Warms

Joule initially used a hand-crank to spin a coil surrounded by electromagnets (powered by Daniel cells) to generate electricity through the coil, which sat inside a water calorimeter. Again, even with this set-up, even without a direct connection to a chemical source to generate the electricity, the water still grew warmer. *Now* where was the caloric coming from? The only possible source was the work Joule was doing by turning the crank. But this didn't make sense, at least not in the caloric world where heat can't be created.

Joule's growing sense of a connection between work and heat led him to the critical turning point in this story. As he wrote, "it became an object of great interest to inquire whether a constant ratio existed between [the heat generated] and the mechanical power gained or lost."[14] Once Joule settled on converting this interest into an experimental objective, his future was set and our history was made. Not to overemphasize this, but there was no preordained theory guiding this moment in history. Joule was "feeling his way"[15] with his curiosity and instinct serving as the guides.

Weight Falls → Magneto-Electric Motor Spins → Current Flows → Water Warms

Joule knew that the previous process didn't constitute a sound experimental procedure, especially his use of a hand-crank. He instinctively understood that for his studies to stand up to the scrutiny of the scientific community he would need to introduce quantification, and the quantification had to be sound, accurate, and reproducible by others. Bringing his experimental obsession for accuracy to bear, he modified his procedures to eliminate heat loss and friction. And then he took the significant step of eliminating the hard-to-quantify hand-crank and replacing it with a falling weight, just heavy enough to slowly descend against the resistance presented by the spinning magnets, but not so heavy as to cause the weight to overwhelm the system, accelerate, and crash into the floor. He needed to make sure that the fall resulted only

[13] (Cardwell, 1989) p. 33.
[14] (Cardwell, 1989) p. 56.
[15] (Cardwell, 1989) p. 57.

in the generation of electricity and thus the water's temperature rise, and not in external deformation or heating. This set-up gave him an extraordinarily accurate means to measure the mechanical effort, since as an engineer he understood the value and significance of weight times change in vertical height. This was the point in history when the mechanical equivalent of heat was put into the units of work (weight × distance), and it was Joule who did this.

In 1843, Joule presented his results at a scientific conference in Cork, Ireland. Based on his measurements, he estimated that the fall of an 838-pound weight by one foot would raise one pound of water one degree Fahrenheit. He also stated that the heat is *generated* by the flow of electricity through the wire coil and is not transferred from another part of the circuit. His first measurement of the mechanical equivalent of heat was important in and of itself, but his latter statement was perhaps even more important, as it effectively drove a stake through the heart of the caloric theory. Heat is generated! It is not pre-existing; it is not conserved. As he wrote, "the mechanical power exerted [by the descending weights] in turning a magneto-electric machine is *converted into the heat* evolved by the passage of the currents of induction through its coils."[16]

The audience response to this monumental finding? Silence. [How many times have we heard this story before?] Even Joule admitted that the subject "did not excite much general attention."[17] A range of reasons for such indifference was possible. Most may have lacked understanding of the full meaning of Joule's findings, and admittedly it was rather confusing when electricity, magnetism, heat, and motion were all mixed together into a talk in the days before energy became a unifying concept. It probably didn't help that Joule wasn't a captivating speaker, being somewhat unassertive and shy in company. Some may have sensed that Joule was making too much of too little, as only one experimental set-up was being used to infer grand designs of nature. Some may have simply resisted the ideas in light of the caloric theory's stranglehold. Or, it may simply have been a lack of interest in heat, since anyone who was competent in mathematics knew that Fourier had already completed the science of heat, and besides, many other things were happening in science at that time. Joule's novel work in this area was thus confusing and rather unwelcome.

Direct Friction → Water Warms

Recognizing the need to obtain more supportive data, Joule next commenced studies using mechanical configurations that directly and intentionally caused friction. He heated water by forcing it through tiny capillary tubes, heated air by compression, and cooled air by expansion. It didn't matter. Regardless of the approach, when he did the math to put everything onto the same work-unit basis, the same number kept popping up: dropping an approximate 800-pound weight by one foot raises the temperature of one pound of water one degree Fahrenheit. The quantity of work he put into the process resulted in about the same quantity of heat rise in the water. "[T]he grand agents of nature are, by the Creator's fiat, indestructible; . . . wherever mechanical force is expended, an exact equivalent of heat is always obtained."[18]

---

[16] (Cardwell, 1971) p. 232–233.
[17] (Johnson, 2009) p. 100.
[18] (Coopersmith, 2010) p. 255.

Joule's approach to the gas compression/expansion studies was very telling of his work ethic. He assumed that all work done in compressing a gas is converted into heat, and he also assumed, like Mayer, that expansion in and of itself, i.e. free expansion, does not cause a change in gas temperature. Unaware of Mayer's own work, Joule proved the latter to himself, by reproducing Gay-Lussac's experiment.[19] Of course! What else would such a thorough experimentalist do?! This finding naturally contradicted the caloric theory, which proposed that free expansion should cause a shift in caloric from "sensible" to "latent" and so cause cooling. But this didn't happen. Joule validated Gay-Lussac's findings—and thus "Mayer's Hypothesis"—and so proved that temperature only changes when the expanding ideal gas does work against an *external* force. The caloric theory couldn't explain this phenomenon.

By 1845 his experiments had thoroughly convinced Joule of the falsity of caloric and the truth of the not-yet-created energy. He hadn't set out to arrive at this discovery. His insight led him there. He recognized that conservation of caloric in a heat engine couldn't explain the production of work and in fact led to the wrong conclusion that energy, which was still being developed as a concept, could be destroyed. To Joule, energy was the conserved quantity, meaning that the production of work by the heat engine is exactly balanced by the loss of heat from the steam, as quantified by the difference between heat in and heat out. Wrote Joule in 1844 as a direct critique of Carnot's and Clapeyron's respective works embracing conservation of caloric, "Believing that the power to destroy belongs to the Creator alone I affirm...that any theory which, when carried out, demands the annihilation of force [i.e., energy], is necessarily erroneous."[20] As Cardwell noted, this statement combined with Joule's experimental work "contains Joule's statement of the principle of the conservation of energy."[21]

The only way to truly prove this was through quantification. The quantification had to connect the seemingly disparate phenomena of work (also referred to as "attraction through space" which eventually became potential energy), electricity, magnetism, and light since each could cause heat to appear. Something lost, something gained. Both Joule and Mayer recognized the

[19] (Noakes, 1957) p. 251, (Cardwell, 1971) p. 133, 234. In this experiment, Joule connected two copper vessels and placed each in its own calorimeter bath. One contained pressurized dry air; the other, a vacuum. When he opened the stopcock, air passed from one to the other until the pressure equalized. The pressure vessel cooled a bit and the vacuum vessel heated a bit, but the overall effects canceled each other, leaving no net change, thus validating Gay-Lussac's findings and "Mayer's Hypothesis." For an ideal gas, internal energy is solely dependent on T; in this experiment, there was no heat (Q) and no work (W) and thus no change in internal energy, meaning that temperature remained constant. But why the slight cooling and heating in the separate vessels? One possible explanation is this: When the stopcock was opened, there was a net flux of atoms from one vessel to the other (vacuum). The atoms that would have otherwise bounced off the wall of the pressure vessel at the stopcock ended up going right through the stopcock opening into the other vessel without any change in speed. They zipped right through and started hitting the walls inside the other vessel at speeds consistent with the initial temperature. But there's a very subtle phenomena that likely happened. Not all atoms heading toward the stopcock opening got into the other vessel. Some came back. Why? Because they were traveling faster than the other atoms in front of them that were also heading toward the stopcock opening. The slower atoms got in the way. What happened as a result? Just like what happens when they hit a receding wall, these atoms recoiled at a slower speed than before the collision while the atoms that got hit from behind headed into the empty chamber at a higher speed than before the collision. Hence the pressure vessel cooled a bit and the vacuum vessel heated a bit, all due to a form of work at a molecular scale but not due to the presence of intermolecular attraction forces. But after pressure equalized, temperature, too, equalized and remained unmoved from the initial temperature of the pressure vessel.
[20] (Cardwell, 1971) p. 235.
[21] (Cardwell, 1989) p. 68.

need for this quantification, and both focused their demonstration on the readily available work–heat conversion.

Weight Falls → Paddle Spins → Water Warms

While a discovery requires but a single observation, a law requires a body of supporting data. Joule knew this. He sensed the existence of a single exact exchange rate between work and heat and realized that the best way to capture it and display it to the world was through the use of a definitive, clean, and simple experimental set-up. So, in the true spirit of Occam's Razor, he eliminated his original interest, the electric motor, and stripped down his apparatus to the bare essentials: a falling weight connected by a string to a spinning paddle inside a water-bath calorimeter.[22] The design of this simple set-up was remarkable in and of itself, as it offered a direct measure of the conversion of work into heat, not at all what Joule originally set out to do. And then he worked to make these measurements perfect. And how he worked! Over and over and over, eliminating all sources of error, all sources of heat loss, all sources of friction save the one that mattered: the paddle moving through the water.

The amazing aspect of this whole endeavor was that the actual measurement Joule sought to capture was very, very tiny. By the time he completed the main thrust of his work in 1848,[23] he was attempting to accurately measure a total increase in temperature of only about a half a degree Fahrenheit! To achieve even this small temperature rise required him, in one set of experiments, to carefully let a pair of 29-pound lead weights slowly fall 63 inches, over and over again, 20 times in all, rewinding the weights back to the starting point after each fall, all to obtain a single data point. And then starting the whole process all over again. The mean figure he reported for his first 40 tests was 0.575250 degrees Fahrenheit. Talk about significant figures. Such claimed accuracy naturally invites questions. But the fact is that he employed extremely accurate thermometers from the famed Mr. Dancer of Manchester[24] and claimed that he could, with much practice, read off with the naked eye to 1/20th the division, which translated into the detection of temperature changes as small as 1/200th of a degree![25] On top of this, he felt that his use of statistical averages allowed him to use so many significant figures in his numbers, an assumption that may not stand up to rigorous statistical analysis but one we won't argue here. A truly amazing piece of experimental work.[26]

Obsessed, Joule sought increasing accuracy until he arrived at a final number in 1849: 772 foot-pounds fall raises one pound of water one degree Fahrenheit. This published result, which was to be inscribed on his tombstone along with a wonderfully relevant and inspiring quote from the Gospel of John (9:4), "I must work the works of him that sent me, while it is day: the

---

[22] One could consider the spinning paddle as having a similar effect as a moving piston. Molecules that strike either moving structure rebound with higher energy, thus resulting in an increase in temperature. (Figure 13.1)

[23] (Chalmers, 1952) Chapter II-3.

[24] The brewers in Manchester demanded such accurate measurements to guide their brewing processes.

[25] (Chalmers, 1952) p. 37.

[26] To truly appreciate how amazing Joule's experimental work was, see in (Sibum, 1995) the challenges that Heinz Otto Sibum encountered in attempting to replicate the precision and accuracy of Joule's experiments, including, for example, the challenge of dealing with the radiative heating effect that his body had on the apparatus upon entering the room.

night cometh when no man can work," will be regarded as the historical occasion on which the mechanical equivalent of heat was fixed to within a few figures. Later work eventually brought this number up slightly to 778. We use a variation of this number (1 calorie = 4.18 Joules, one Joule being defined as a unit of energy equal to the work done by the movement of one Newton of force through one meter of distance) to convert thermal energy to the same energy units as mechanical work, such units naturally being named in Joule's honor.

Note that this result was about the same as that which Joule determined in his original design with the dynamo. Like the board game of Mouse Trap, the weights fell, the crank turned, electricity flowed, the wires heated, and water temperature rose. Each cause in this chain exactly equaled the resulting effect. Nothing was lost. You could remove pieces or switch their order, heat-to-work or work-to-heat; it didn't matter. Joule realized that each step was linked to the others by an abstract exactness such that a negative change in one reappeared as a positive change in the other.

Joule's was a lone voice operating in the wilderness outside the establishment. There was little pull at that time from a physics community that even questioned the premise that one could heat up water merely by shaking it. The caloric theory said nothing to this effect. Indeed, the conversion of abstract work into tangible heat, a fundamental aspect of the mechanical theory of heat, was an unacceptable concept to many, made all the more so by the fact that the effect was so small that only the most trained of experimentalists could detect it. Detection required someone of great skill and patience. It required Joule. He was second to none.

Thomson meets Joule

Joule is, I am sure, wrong in many of his ideas, but he seems to have discovered some facts of extreme importance as for instance, that heat is developed by the friction of fluids.
– William Thomson[27]

During an attempted 1847 presentation at Oxford, Joule, still an outsider, was asked to *not* read his paper since time was pressing but instead to simply give a brief verbal description of his experiments while displaying his paddle-wheel apparatus. During his presentation, no discussion was invited and apparently no discussion happened. Joule believed that no response at all would have happened had it not been for the fortuitous fact that a curious 23-year-old scientist named William Thomson, later to become Lord Kelvin, was in the audience. After the talk, Thomson approached the 29-year-old Joule and engaged him in discussion.[28] In yet another of the great serendipitous encounters in history, Thomson, an inside member of the physics community, discovered Joule, brought attention to his work, and, according to Joule, "rescued him from obscurity."[29] What a key moment in history this was, marking as it did the real beginning of acceptance of the mechanical theory of heat and the concept of energy. Joule and Thomson

---

[27] William Thomson in a letter to his father (July 1847) as cited in (Smith and Wise, 2009) p. 302.
[28] Another, more Hollywood version of this story had Thomson momentously rising from the audience at the end of Joule's talk with a significant question, but Thomson himself denied this. See (Cardwell, 1989) p. 83.
[29] (Smith and Wise, 2009) p. 304.

developed a life-long friendship after this meeting, which led to the two becoming one of the first and most famous experimentalist–theorist science teams in history.[30]

A short note on Thomson is warranted here, more to be provided later (Chapter 32), to help set the stage for understanding his partnership with Joule.[31] Born in Belfast, educated in Glasgow and then Cambridge, Thomson was extremely well-versed in mathematics, having mastered, for example, Fourier's *Analytical Theory of Heat* at 16 years old. His further education in Paris at Henri Victor Regnault's laboratory, where he studied the properties of steam, exposed him to the works of Carnot and Clapeyron and thus contributed to his belief in the caloric theory of heat. Wrote Thomson in 1848: "The conversion of heat (or caloric) into mechanical effect is probably impossible, certainly undiscovered."[32] This is important to understand since it explains how Thomson initially entered into his relationship with Joule as a believer in the caloric theory and underscores the size of the hurdle he needed to clear. The subsequent transformation of Thomson from the caloric to the energy paradigm was remarkable and attests to his genius. Needless to say, extracting oneself from a paradigm is extremely difficult. It requires a powerful intellect combined with a certain degree of confidence to admit, "I was wrong."

Continuing now with Joule, it is rather unfortunate that Joule's early work didn't receive the attention it deserved on its own. But the reality was that neither his personality nor his background was strong enough to earn him (deserved) credibility within the scientific establishment. He needed a champion within the establishment, someone possessing what he lacked. Fortunately, Joule found Thomson, and, one could argue, fortunately Thomson found Joule, and this made all the difference. It's unfortunate that Mayer never found his own Thomson. While he eventually did receive recognition, it was after the fact, meaning that none of his ideas were incorporated into the growing structure of physics.

Having said this, it's not as if this meeting between Thomson and Joule caused a curtain to suddenly rise and reveal Joule's grand findings to the world. Yes, Helmholtz latched onto these findings very fast, using them to support his famous 1847 paper on the conservation of energy. Clausius formalized them in 1850 into his aforementioned mathematical expression of the 1st Law of Thermodynamics that related energy, heat, and work. But on the whole, it was difficult for many to truly comprehend what Joule's data, or Mayer's work for that matter, were saying, as best reflected by the fact that it took even the bright young Thomson three years to come around to acceptance, largely due to the difficulty of his own shift from caloric to energy. Thomson acknowledged that Joule showed that work could generate heat but did not believe that Joule showed the reverse, that heat could generate work.[33] He found this reverse process

---

[30] It's interesting to consider such experimentalist–theorist relationships through history, such as that of Tycho Brahe and Johannes Kepler, although their relationship was more sequential in time than simultaneous. Kepler was Brahe's assistant from 1600 until Brahe's death in 1601, and it was afterwards that Kepler used Brahe's astronomical data to develop his laws of planetary motion.

[31] (Cardwell, 1989) p. 84.

[32] (Cardwell, 1989) p. 95.

[33] (Cardwell, 1989) p. 97. While Thomson felt that Joule had not demonstrated the conversion of heat to work, Joule felt otherwise based on his demonstrated cooling of compressed air when expanded against atmospheric pressure. It was Gustave-Adolphe Hirn (1815–1890) who delivered the first real demonstration of heat-to-work conversion by conducting an energy balance around a steam engine. See (Cardwell, 1989) pp. 221–226.

difficult to believe since it involved the consumption of heat. This was very confronting to someone who had embraced Clapeyron's principles. As his brother James wrote to him, "If some of the heat can absolutely be turned into mechanical effect, Clapeyron may be wrong."[34] Especially confusing was how to handle heat flow via Fourier's conduction, which was originally based on the conservation of heat.[35] Such flow didn't result in work. So how was this different than the heat flow that did cause work? A new theory was clearly needed to put all of these phenomena into an internally consistent framework. Joule did this by showing that heat and work were simply two sides of the same coin, two different aspects of a larger concept that later would be called energy.

Joule contributed to the rise of physics in Britain as an independent discipline and eventually—it did take some time—rose in stature in Britain's scientific community, receiving well-earned recognition and financial support for his further experimentation on the mechanical equivalent of heat. Following his above tour-de-force, he continued work on energy concepts, including his famous collaboration with Thomson between 1852 and 1862 on the Joule–Thomson effect involving the change in temperature resulting from the expansion of non-ideal gases. This work, which "marked a new departure in physics—a collaboration, sustained over a long period, by an experimentalist and a theoretician,"[36] stemmed largely from Thomson's doubt as to the full validity of "Mayer's Hypothesis," the term he and Joule gave to Mayer's assumption that there is no change in gas temperature due to expansion without work. In spite of Gay-Lussac's findings, Thomson felt that much further study was needed, reflecting an inner belief that gas atoms behaved like springs and could thus store compressive energy. The Joule–Thomson experiments[37] showed that when gases expand without doing external work, they cool, some (carbon dioxide) more so than others (air), except for hydrogen, which heats. Their work helped distinguish ideal gases from non-ideal gases, the former following Mayer's Hypothesis and the latter not, while also providing further data and theory on the properties of steam, which was needed in the ongoing efforts to understand steam engines.

Until his death in 1889, Joule continued experimenting and inventing, mostly out of his home laboratory, as he never took an academic appointment. As Cardwell observed, all in all, Joule's "was one of the most satisfactory scientific lives that can be imagined… [one] rounded off by the accomplishment… [of establishing] the first law of thermodynamics."[38]

[When Joule] overthrew the baleful giant FORCE, and firmly established, by lawful means, the beneficent rule of the rightful monarch, ENERGY! Then, and not till then, were the marvelous achievements of Sadi Carnot rendered fully available; and Science silently underwent a revolution more swift and more tremendous than ever befell a nation.

– Anonymous review of Joule's Scientific Papers (1884)[39]

[34] (Cardwell, 1989) p. 87.
[35] (Smith and Wise, 2009) p. 159.
[36] (Cardwell, 1989) p. 191.
[37] (Rowlinson, 2010)
[38] (Cardwell, 1971) p. 232.
[39] Cited in (Smith, 1998) p. 1

Mayer and Joule – Conclusion

Mayer arrived at the mechanical equivalent of heat first with Joule a close second. In the days before the instantaneous communication we're used to today, clarity as to who arrived first took time and unfortunately in this case caused acrimony, but not primarily between Mayer and Joule. Yes, they challenged each other over priority in letters to *Comptes Rendus* in 1848 in which Joule gave credit to Mayer for the idea of the mechanical equivalent of heat but claimed credit for its experimental verification. But this was a more of a professional dispute that occurs in science in comparison to the primary dispute that was actually driving it. On the one side you had Thomson, Joule's main advocate, and Peter Guthrie Tait, while on the other you had John Tyndall (1820–1893), an Irish-born physicist with an advanced degree from Germany who felt that Mayer's work had been unfairly neglected. The roots to this famed and very heated dispute ran deep and will be discussed in detail in Chapter 33.

22 The 1st Law of Thermodynamics

Between 1842 and 1847, the hypothesis of energy conservation was publicly announced by four widely scattered European scientists—Mayer, Joule, Colding, and Helmholtz—all but the last working in complete ignorance of the other.

– Thomas S. Kuhn[1]

Joule and Mayer were not the only scientists involved in the discovery of energy. Thomas Kuhn[2] identified twelve such pioneers who engaged in experiments and theories related to energy during the critical years around 1830 to 1850. Sadi Carnot (1796–1832) in 1824, Marc Séguin (1786–1875) in 1839, Karl Holtzmann (1811–1865) in 1845, and G. A. Hirn in 1854 each wrote on the convertibility of heat and work, but none recognized this as a special case of energy conservation. Between 1837 and 1844, C.F. Mohr (1806–1879), William Grove (1811–1896), Michael Faraday, and Justus von Liebig (1803–1873) each described nature as manifesting but a single "force" that could never be created or destroyed, but none provided quantification to back up such claims. It was the final four, Mayer, Joule, Ludwig Colding (1815–1888), and Hermann von Helmholtz (1821–1894), who discovered energy and its conservation between 1842 and 1847 by completing Boyer's three steps as outlined earlier in Chapter 13. Each of these four arrived at the same essential answer even though it took the subsequent efforts of others such as Thomson and Clausius to bring clarity to this answer, unifying the pioneers' findings around an appropriate set of words, concepts, and equations.

Of the four, I have covered two, Mayer and Joule, as they best exemplified the challenges involved in the discovery of energy, especially involving their respective quantification of the mechanical equivalent of heat, which served to justify (Boyer's 3rd step) the proposal (Boyer's 2nd step) that energy is neither created nor destroyed. Their histories also provide very rich and very different historical contexts, Mayer having traveled the theoretical route and Joule the experimental. Regarding select others, had he lived, Carnot may have eventually seen the larger picture as suggested by the progression of his thinking from 1824 to 1832 after his memoir and prior to death. His private notebooks from this time reveal his internal conflict. In his 1824 publication he stated that heat is transformed into work and that heat, in the form of caloric, is conserved. The fundamental incompatibility between these two theories bothered Carnot deeply and after publication led him away from the caloric theory and toward the energy-based theory of heat, as significantly revealed by his calculation found in these

---

[1] (Kuhn, 1959) p. 321.
[2] (Kuhn, 1959)

Block by Block: The Historical and Theoretical Foundations of Thermodynamics,
Robert T. Hanlon, Oxford University Press (2020). © Robert T. Hanlon.
DOI: 10.1093/oso/9780198851547.001.0001

notebooks of the mechanical equivalent of heat that wasn't too far off Joule's eventual determination.[3] Unfortunately, his shortened life didn't permit him to complete this journey. Regarding Hirn, it was he who best complemented Joule's work-to-heat experiments by attempting to measure the conversion of heat-to-work.[4] But he too didn't see the larger picture. Finally, of the remaining two works, Helmholtz's, which I cover in more detail in Chapter 32, was inspired by muscle metabolism, while Colding's was based on his 1843 experiments involving the generation of heat by friction.

Returning to Kuhn – the Triggers that led to Completion of Boyer's Three Steps

It's time now to return to Kuhn's work. In his analysis of such history, Kuhn raised an excellent question. How was it that these twelve individuals grasped some if not all of the essential features of energy within such a relatively short historical range of time? What was it that triggered the seemingly sudden convergence of ideas within scientists who weren't really interacting with each other? These twelve were indeed scattered across Britain and the European continent, working independently and in relative isolation, each studying different aspects of energy, and each grasping key aspects of the larger picture. It was not pure coincidence that all arrived at the same point at roughly the same time. So what trigger was common to all?

If a single answer had to be given, it would be the rise of the steam engine in the Western World during the late 1700s. But this wouldn't be entirely correct, as things were more complicated than this. The rise of the steam engine wasn't the trigger, per se, but it was the source of the more direct triggers that rose during the same critical years of 1830–1850.

One such trigger identified by Kuhn was the discovery of an alternative to steam, namely electricity. The conversion of coal into work using the steam engine had been well established by 1800 and was a critical component of the Industrial Revolution. But the rise of electricity spurred by Volta's invention of the battery (which consumed chemicals) in 1800 and its use to power an electric motor to generate the same work as done by the steam engine (which consumed coal) left many wondering: are these phenomena connected? The intensity of such questioning around these various processes that converted energy from one form to another (and especially to work) rose sharply in the decade after 1830 as the scientists began to unite such seemingly disparate strands of energy as mechanical, chemical, thermal, electrical,

---

[3] (Carnot, 1986) pp. 31–32.

[4] (Cardwell, 1989) pp. 221–226. Gustave-Adolphe Hirn (1815–1890) lived near Colmar and, like Joule, was home schooled and thus didn't suffer the constraints of formal discipline. Hirn was convinced that work and heat were interchangeable and sought to quantify the conversion of heat into work, something that Joule thought he had solved in 1844 when he showed that compressed air cooled as it expanded and did work, but his wasn't a strongly conclusive example to some such as William Thomson. Hirn sought a very direct measure of heat into work by conducting an energy balance around a steam engine and successfully quantifying the consumed heat to the generated work. While Joule got full credit for quantifying the conversion of work into heat, Hirn got the credit for quantifying heat into work and thus complemented Joule. His book of 1862 has been claimed as the first comprehensive textbook on the mechanical theory of heat and led to further study and development of the steam engine.

magnetic, and even radiant. Faraday's statement in 1834 captured the moment, saying that these different powers are "all connected and due to one common cause."[5]

Not only was the engine's arrival one of Kuhn's suggested triggers, but so was the theoretical analysis of its operation. Such analysis was driven largely by the huge financial reward available to any who could improve the engine's performance and efficiency. Critical to such metrics-driven improvement was the rise of strict definitions for both heat and work.

Around 1820, theoretical papers on the theory of machines and industrial mechanics began populating French literature. Such literature focused on both the energy of motion, as embodied by Leibniz's *vis viva*, mv^2, and the corresponding energy of displacement, weight times vertical displacement. Recall that these two concepts had already merged by 1750 to provide a foundation for an early form of the conservation of mechanical energy.

At this point, the energy pioneers focused predominately on work, while discussions of *vis viva* faded into the background. The reason for such focus was that the product of weight times vertical displacement offered a simple and yet very powerful measure of the most important industrial aspect of energy, the ability to move matter. This product became more rigorously defined as the integral of force times differential displacement $\int Fdx$, thus providing the crucial mathematical and conceptual link between energy and Newton's theories on motion and gravitation.

The recognition of work, after a very long journey that started in ancient Greece, as being an important quantity provided tremendous support to the theoretical analysis of engines. It established itself in the scientific community by gaining significance in the worlds of both engineering, through the efforts of such figures as Watt, and physics, through the efforts of such figures as Lazare Carnot, Coriolis, and others.

But work was only part of the story. Heat was the other part. What was needed to complete the theory of energy was establishing both work and heat as two separate forms of energy. Yes, the other forms of energy would need to be addressed and they were, but work and heat were the dominant forms and so required priority. While *vis viva* played a role in developing the conservation of mechanical energy for moving (visible) bodies and would play a later role in developing the kinetic theory of gases, it played no role in the development of the conservation of energy by Mayer and Joule,[6] even though the *vis viva* of small particles was the actual cause of heat. The theories of energy and classical thermodynamics as a whole were independent of any particular view as to the nature of heat since the relationship between work and heat in these theories was based on the mechanical theory. But still, building on Boyer's 3rd step, the need to quantify the work–heat relationship remained. A law of conservation demanded it. Lavoisier, Black, and others provided the means to quantify heat through their conceptual and experimental development of the calorimeter and this complemented the already available means to quantify work.

The rise of strong definitions for both work and heat was at the core of Kuhn's engine-analysis second trigger, the former perhaps a little more so. Since the likes of Joule and Helmholtz wrote in a way that placed the concept of work at the center of energy,[7] it became

---

[5] (Kuhn, 1959) p. 327.
[6] *Vis viva* did play a role in Helmholtz's development of energy as will be discussed in Chapter 32.
[7] (Brush, 2003a) p. 21.

so. Work evolved to become *the* means to validate the conservation theory. It became *the* standard of reference, *the* de facto rallying point for all energy studies as best manifested by the eternal questions surrounding engine efficiency, *how much fuel will I need to achieve a given amount of work with this engine?*

Recall that Kuhn's third and final trigger had to do with philosophy. Kuhn proposed that the pioneers were "deeply predisposed to see a single indestructible force at the root of all natural phenomena"[8] and further proposed that the rise of the *Naturphilosophie* movement in the early nineteenth century, being based on the belief of this single unifying principle, was the source of this disposition. However, while many of the pioneers, such as Colding, for example, may have embraced this philosophy, our two key protagonists did not. Joule's upbringing combined with his suspicion of speculation uncontrolled by experiment "made him unreceptive of such fashions as *Naturphilosophie*."[9] He showed no sign of any philosophical or religious influence in his work other than his belief in the existence of some type of structure and balance behind nature that prevented perpetual motion from occurring. As for Mayer, the role *Naturphilosophie* played in the development of his ideas was "not only minor at best, but also largely negative."[10] He was a persevering genius when it came to following a highly logical, cause–effect train of thought. There was no underlying philosophical movement behind his ideas.

So what are we left with then regarding this third trigger? It seems that philosophy was indeed at its heart, but I would argue that it had more to do with absence than presence, specifically the absence of the caloric theory in the minds of the pioneers.

Interpreting Nature without Prejudice

Alongside the rise of engines and their study as reflected in Kuhn's first two triggers was the rise of those doing the studying. What was different about this group that resulted in the occurrence of the third trigger? Because engines were not born in the laboratories and thus not bound to life inside academia, but were instead born in the real, industrial, and very practical world, they attracted a group of intellectual and curious individuals who understood the impossibility of perpetual motion but who more importantly didn't simultaneously embrace the caloric theory since this theory was largely an academic paradigm. The belief in the impossibility of perpetual motion was a critical component of this third trigger, but this belief was not the trigger in and of itself since it had been around well before 1830. It was more of a prerequisite. The crux of this trigger was the absence of caloric. One could not discover energy and its conservation without understanding the connection between work and heat, and one could not understand this connection while simultaneously holding onto the caloric theory in which heat is a conserved material.

Mayer the theorist and Joule the experimentalist may have followed very different paths in arriving at the concept of energy and its conservation, but they shared one critical thing in

---

[8] (Clagett, 1969) p. 337.
[9] (Cardwell, 1989) p. 184.
[10] (Caneva, 1993) p. 319.

common. Neither was raised under the influence of caloric. Both fought an uphill battle in getting their ideas into the public since neither was a member of the established science community. But it was just this isolation that most contributed to their breakthrough work. Since neither was raised in the caloric paradigm, each was free to interpret nature without prejudice. Consider the difficulty Thomson had in changing his paradigm from caloric to dynamic theories, even when overwhelming supportive evidence from Joule was sitting right in front of him. Mayer and Joule had deep and untainted faith in their conviction that perpetual motion is impossible, and this faith together with their perseverance in the face of a status quo of indifference if not outright resistance and hostility, and also together with their ability to think and execute—like Galileo, they weren't ones to sit idly by with their ideas—led them to the discovery of energy.

Expanding this discussion further, many of the other pioneers who helped break the hold of the caloric theory and so helped pave the road to the energy theory similarly had no connection with universities or other academic institutions.[11] As discussed more in Chapter 32, only Helmholtz worked on the inside of academia, but even he fought against the status quo, working to move German science away from the existence of special, vital forces such as *Lebenskraft* (life force) to explain natural phenomena and toward the rational deterministic science of cause–effect.[12] Most in this group were engineers, but not all as in the case of Mayer. But Mayer was clearly influenced by his own animal-as-engine viewpoint. The rise of engines brought this group forth, people with a fresh set of clear eyes who could study and experiment with engines outside the influence of established science.

Reflections on Work–Heat Equivalence

The conceptual road that the pioneers followed was extremely challenging, especially with regards to the work–heat relationship. These were such different concepts, evolving from such different histories. And yet at the molecular level they were very much linked. Divert a moving body from horizontal to vertical, much as Huygens theorized, and kinetic energy is transformed into gravitational potential energy. This is really the basis of the conservation of mechanical energy. But such simplicity is lost when dealing with billions of such bodies for which kinetic and potential are no longer able to be isolated and quantified, thus motivating the rise of such macroscopic measurements as temperature and pressure. But the pioneers didn't understand this depth at the time; they couldn't. They simply had the new concept of work and the new concept of heat. Without the fundamental molecular understanding of either, it wasn't obvious to them that these two concepts should even be related to begin with.

The faith that they were connected, arising as it did from the impossibility of perpetual motion, helped many of the pioneers endure. It wasn't necessary to know *why* there were connected. In fact, Mayer's and Joule's reliance on the work–heat conversion as their key reference toward discovering the larger concept of energy didn't depend on knowledge of the underlying

---

[11] (Cardwell, 1989) p. 272.
[12] Even with this exception though, people question whether Helmholtz discovered energy on his own or first learned of it by reading either Mayer's or Joule's work and only then developing it further.

physics behind heat. To Mayer, when heat was consumed, it simply ceased to exist and was then replaced by a "mechanical equivalent." The numbers worked out and there was no need to ask any deeper question about what was actually happening. Joule brought much more to this fundamental discussion. He realized that heat wasn't a substance like caloric but instead believed it to be a state of motion. He saw the connection between Dalton's atoms and the heat-as-motion theory as best reflected in his attempt to develop a kinetic theory of gases. But even he didn't need this understanding to discover energy. Nor did he even want to speculate about it. Mayer and Joule and the other pioneers used their own untainted instincts to journey along the path laid by nature and didn't allow man's philosophy about such false theories as caloric to divert them.

It certainly helped their efforts that the competing caloric theory was simultaneously crumbling as it failed to explain the phenomena associated with such events as Rumford's cannon and Joule's spinning paddle. Caloric's faltering left a corresponding vacuum that was rapidly filled by energy. All at once, those in established science such as Thomson and Clausius saw and embraced this new reality. In that moment, caloric died.

Energy – a New Hypothesis

Energy is an abstract concept. You can't touch it or point to it. It's a number. One can calculate the energies for a wide range of natural phenomena and then add the resulting numbers together in the belief that the sum remains constant no matter what changes occur. This concept is both powerful and complicated. It represents a paradigm, a way of thinking. So imagine the challenge scientists had in the mid-1800s in discovering such a concept, especially while they held tightly to other competing but incorrect concepts such as caloric. Once they accepted the paradigm, everything fell quickly into place. But until that moment occurred, everything remained confusing and muddled.

Acceptance of this new paradigm required faith. Mayer and Joule had a deep faith that they were on the right path in defining energy and its conservation. This faith is what pulled them through. With this paradigm, they saw a path forward in explaining those things that could not yet be explained. They didn't accept the absence of cause in the effects they studied. Yes, the caloric theory could explain many things, but not all things. And it was these discrepancies that helped drive them forward. While many may not have even understood that there was a problem that needed solving, Mayer and Joule did. In the true spirit of science, they latched onto the discrepancies between data and theory and relentlessly pursued resolution. Such very tiny discrepancies can open a door into an entirely new universe. Look at the discrepancy between Newton and Einstein regarding the amount of the bending of star light around the Sun. A very small discrepancy, a very large meaning. Look at what Joule tried to do in measuring the temperature rise of water. The calorists didn't believe that such a rise would even occur. Joule showed them that it would, by about a half a degree. A very small discrepancy, a very large meaning.

The strength of Mayer's and Joule's efforts is all the more evident when also considering that heat physics was considered done by 1830. Recall that Fourier had already said the last word on heat with his own masterpiece of work. His mathematics were superb; his understanding of the molecular physics involved was not. The French establishment, and so too the English establishment, held strong to such beliefs as well as others and so couldn't see energy. It was up to the outsiders, those insulated from such beliefs as caloric, to see the concept of energy and in so doing overthrow caloric. Mayer and Joule were both outsiders. This enabled them to grasp the concept but unfortunately thwarted them in trying to convince the physics community. In the end, Mayer and Joule needed the help of others, those strongly situated inside the community, men such as Thomson and Clausius.

Energy Moves into Academia and the 1st Law of Thermodynamics is Born

The work of the pioneers, and especially Joule, prepared the ground for the academic and professional physicists, such as Thomson, Clausius, Rankine, and Helmholtz, to build structure and mathematics around energy. As introduced earlier, Clausius captured the essence of the concept of energy in the 1st Law of Thermodynamics ($\Delta U = Q - W$) by explaining Carnot's engine in terms of energy as opposed to caloric. Energy enters the engine from the hot furnace, some is converted into work and the remainder leaves the engine through the cold condenser. The exiting energy plus the work done must equal the entering energy. In short, Clausius united Carnot and Joule by shifting the conversation from the conservation of caloric to the conservation of energy, uniting both heat and work as two different forms of a conserved quantity called energy, and then re-analyzed Carnot's work accordingly. In doing so, Clausius "[brought] order out of confusion."[13]

Furthermore, in a tremendous conceptual leap, by replacing caloric with heat, Clausius showed that it's heat (and not caloric) that flows from hot to cold and, more importantly, that hot-to-cold is the only direction heat can flow, for otherwise the impossibility of perpetual motion would be violated even though the conservation of energy wouldn't. Such thinking led Clausius to discover entropy and his own version of the 2nd Law of Thermodynamics, which will be discussed in Part IV.

Equally important to Clausius' mathematical structure was the accompanying language structure. William Thomson placed Thomas Young's *energy* center stage in the creation of the new terminology. Energy became the unifying concept for two subordinate concepts, kinetic energy and potential energy. The former was first captured in its modern sense by Gustave-Gaspard Coriolis in 1829 and the latter was first captured by William Rankine in 1853, unknowingly building on Leibniz's *vis mortua* and Bernoulli's *potential*. Hermann von Helmholtz provided the final term that glued these pieces together: conservation. Energy is a conserved quantity comprised of both kinetic and potential components.

Both Thomson and Clausius and others such as Helmholtz and Maxwell raised energy to a powerful position, one that gained comparable status with Newton's force, helping to distinguish between the two in the process. Clausius' "internal energy" opened the door to expand energy into the world of chemistry, something not readily available when starting with force,

[13] (Gibbs, 1889) p. 460.

and this evolved to become one of the major achievements of nineteenth-century science.[14] The unifying nature of energy and its conservation soon encompassed most if not all phenomena, such as heat, light, electricity, and magnetism. It even (conceptually) expanded into fields beyond science, such as economics, sociology, history, and so on.

The Kinetic Theory of Gases: Matter-in-Motion

Before moving on, let's return one more time to Leibniz's *vis viva*. As noted earlier, the discovery of energy and its conservation did not directly rely on *vis viva*. *Vis viva* and its conservation in "the parts [that] receive it" certainly helped raise awareness of such concepts as energy of motion and of the need to have some *thing* be conserved, and it was certainly needed to complete the rudimentary version of the conservation of mechanical energy around 1750. But there was no direct link between this philosophy and the breakthrough work of Mayer and Joule. Perhaps *vis viva* was implicitly in the background, as a familiar concept in the minds of any who studied mechanics, and helped form the inner belief in a conservation law.

Having said this, we must recognize that the energy story did not end when Clausius wrote $\Delta U = Q - W$. One could say that this equation represented the end of one journey and the beginning of another, one involving the link of energy concepts to the growing understanding of the atomic world. A key starting point was the ever-present question, *what is heat?*

Heat as a substance or a property of state effectively died in 1850. The false caloric theory simply couldn't survive in the face of the new data and the arrival of the new alternate hypothesis of energy. But the death of caloric left some confusion in its wake. The mechanical theory of heat based on work–heat equivalence provided the basis to move forward with classical thermodynamics as it did not depend on the nature of heat and one would have thought that it would have also provided the basis to move forward with the kinetic theory of gases as well, but there were different ways to picture heat-as-motion.

How can this be?, you might ask. Heat-as-motion certainly makes sense. We learned in school about how gaseous atoms zip all around us and the faster they move the hotter it is. But don't forget that you were raised to comprehend this. You were taught this at an early age with no competing paradigm to confuse you. In 1850, two competing visions of the heat-as-motion theory existed for gases, one involving static vibrating atoms and one involving non-static projectiles rapidly moving across space. As discussed in Chapter 40, the static-atom concept unintentionally originated with Newton when he used it as the basis for a mathematical model of the relationship between pressure and volume in a gas. He was not proposing that this model reflect reality but this message was lost when later on others latched onto his static-atom approach. The projectile model, which became known as the kinetic theory of gases, originated with Daniel Bernoulli when he made use of Leibniz's *vis viva* theory and was later more fully developed by others such as John Herapath (1790–1868), John James Waterston (1811–1883), August Karl Krönig (1822–1879), and even Joule himself with his 1848 paper on the subject (once he evolved away from the static-atom view). Of note, as of 1852, the theorist Thomson, who envisioned "atmospheres" spinning around individual atoms, had not understood or

---

[14] (Smith, 1998) pp. 302–306.

accepted the significance of experimentalist Joule's kinetic theory of gases. The success of the kinetic theory of gases, including the supportive role of *vis viva*, led to the concluding work of Clausius, Maxwell, and Boltzmann in their development of statistical mechanics and statistical thermodynamics. These subjects will also be addressed in Chapters 40–42.

Conclusion

We've reached the end of this section on the historical and theoretical foundations of energy and its conservation and also on its inclusion in the 1st Law of Thermodynamics. As physicists, chemists, and engineers, we frequently use the concepts of mass and energy conservation in the work we do. Einstein united these two in his theory of relativity in which he proposed that mass itself is a form of energy. But as we'll most likely never work with Einstein's energy tensor that captures this, we'll leave this for the physicists.

While it appears that this section is indeed concluded, it isn't. Recall that a law can't be proven and that the conservation of energy was just such a law. An overwhelming amount of evidence can validate a law, but it only takes a single data point to invalidate it. With this, let's turn to the epilogue of this story.

Epilogue: The mystery of beta decay

Some very interesting phenomenon seems to be involved in beta-ray disintegration
– Ellis and Wooster in 1925[1]

The problem of beta-ray expulsion lies outside the reach of the classical conservation principles of energy and momentum

– Bohr in 1929[2]

Rutherford's discovery of alpha-rays (He-4 nuclei) and beta-rays (electrons) between 1895 and 1898 attracted much subsequent study. By 1905 William Bragg (1862–1942) and his assistant Richard Kleeman (1875–1932) showed that alpha-rays ejected from any single emitter all travel at the same speed. Thinking that the same result should occur for the beta-rays, Otto Hahn and Lise Meitner commenced a similar study in 1907 that led to much unexpected confusion, as the beta-spectrum was more complex than first appeared. After numerous improvements to the experimental procedures, this line of study was brought to a jarring conclusion in 1914 by James Chadwick. In attempting to count the distinct spectrum of energy levels he anticipated with his improved experimental apparatus, Chadwick noted that "with the counter I can't even find the ghost of a line. There is probably some silly mistake somewhere."[3] But he made no mistake. The beta-rays gave a *continuous* spectrum.

This was an extremely disturbing finding. But why? When we look at the problem of two bodies flying away from each other due to the presence of a strong repulsive force between them, we know that both momentum and total energy are conserved. Doing the math for this two-body problem results in a single, unique speed (kinetic energy) of the alpha particle away from the nucleus emitter. Beta-decay should have followed the same logic. But it didn't, rightly casting doubt on the assumptions behind the logic.

Further confusing the issue were the results from a separate study conducted by Charles Ellis (1895–1980) and William Wooster in 1925. Using a calorimeter, they sought to measure the total energy released per individual beta-decay. Their thinking went like this. *Perhaps another decay product is emitted in addition to the beta-ray that would account for the missing energy.* If they could trap such a particle in, say, lead, together with the beta-ray, then they could capture the total energy resulting from each decay. They further figured that given the wide range of energies of the beta-rays, the total energy trapped should equal the maximum

[1] Ellis and Wooster quote cited in (Pais, 1988) p. 303.
[2] Bohr quote cited in (Pais, 1988) p. 311.
[3] (Pais, 1988) p. 159.

Block by Block: The Historical and Theoretical Foundations of Thermodynamics,
Robert T. Hanlon, Oxford University Press (2020). © Robert T. Hanlon.
DOI: 10.1093/oso/9780198851547.001.0001

energy of the beta-ray spectrum, since this ray would presumably have occurred during a specific decay in which the energy of the other undetected particle was negligible. But what they found instead was that the energy they trapped was accounted for by the beta-rays alone. Such momentous results, reflecting as they did the absence of other detectable products, were met with silence. Neither Ellis nor Wooster nor Meitner had anything of substance to say about the meaning of these results. "It was clear that a crisis was at hand."[4]

That Chadwick's and then Ellis and Wooster's results jolted the physics community is best reflected by Bohr's willingness to consider forgoing the energy law. This wasn't a total shock since it had been attempted before, back in 1924 when the difficulties generated by the arrival of wave–particle duality had some in the physics community reaching for extreme explanations involving *non*-conserved energy. Even Einstein held this view briefly. But all was put to bed in 1925 when Compton's scattering experiments demonstrated that energy and momentum are indeed conserved in individual events between particles, including photons. This finding, combined with the arrival of quantum mechanics shortly thereafter, caused doubts about energy conservation to evaporate, or so it seemed, for after considering Chadwick's results, Bohr proposed in 1930 that energy is not conserved in beta decay. This was like throwing a gauntlet in front of the physics community, not all of whom agreed with Bohr. Said Dirac, "I should prefer to keep rigorous conservation of energy at all costs."[5]

Into this situation stepped Wolfgang Pauli, who was inspired when "those vehement discussions about the continuous beta-spectrum between [Meitner] and Ellis…at once awakened [his] interest."[6] "Desperate"[7] for a solution to save the energy law, Pauli proposed in 1930 the existence of a new particle, later to be called the neutrino, later changed to anti-neutrino, in response to Chadwick's discovery of the much larger neutron in 1932. Pauli predicted that this anti-neutrino would not be easily detectable, especially not in a lead block calorimeter, but would carry-off the additional energy from beta-decay required by energy conservation. In this scenario, "the continuous beta-spectrum would then become understandable from the assumption that in beta-decay [an anti-neutrino] is emitted along with the electron, in such a way that the sum of the energies of the [anti-neutrino] and the electron is constant."[8] While the theoretical proposal of new, previously undetected particles is common today, it wasn't at all common back then. Pauli's proposal was quite courageous, as even he recognized when he wrote, "But only he who dares wins…"[9]

Enrico Fermi furthered Pauli's cause in 1934 when he wrapped the structure of quantum-field theory mathematics around beta-decay. Using simple assumptions about the random sharing of energy between the emitted electron and anti-neutrino, Fermi was able to derive an energy distribution of the electrons that agreed quite well with the beta-spectrum measurements. Through this effort, Fermi became the first to reveal knowledge about the existence of a new kind of force or interaction, eventually to be called the "weak interaction," that was responsible for beta-decay.

[4] (Weinberg, 2003) p. 146.
[5] (Pais, 1988) p. 313.
[6] (Pais, 1988) p. 314.
[7] (Pais, 1988) p. 315.
[8] (Pais, 1988) p. 315.
[9] (Pais, 1988) p. 315.

Following Fermi's work, the presence of the hypothesized anti-neutrino gradually gained favor over the non-conservation of energy and by 1936 Bohr conceded that the issue was resolved, even though the anti-neutrino still had not yet been detected. It's valuable to recall that during this same timeframe, up until 1932, the neutron also had not yet been detected. Without the neutron, attempts to explain the behavior and properties of the nucleus, including the emission of electrons (beta rays) during beta-decay, led many physicists to assume that Rutherford's nucleus was comprised of electrons and protons. Chadwick's discovery of the neutron combined with Pauli's and Fermi's theories led to a new understanding of beta-decay in which a neutron is converted into a proton, electron, and anti-neutrino. Beta-decay became seen as a means for nature to adjust the proton:neutron ratio inside a nucleus toward more favorable, low-energy states.

$$\text{Beta--decay: Neutron} \rightarrow \text{Proton} + \text{electron} + \text{anti-neutrino}$$

While Fermi himself never wrote about this subject again, his line of work continued and helped solved many quantum-mechanical problems. Milestones were reached in the late 1950s when Richard Feynman (1918–1988) and Murray Gell-Mann (1929–) formulated a more advanced theory of weak interactions, in 1960 when Sheldon Glashow (1932–) combined the electromagnetic and weak interactions, and finally in 1967 when Steven Weinberg (1933–) and Abdus Salam (1926–1996) incorporated Glashow's electroweak theory into the modern form of the Standard Model. In this model, beta-decay involves the conversion of a "down quark" into an "up quark," thus effectively converting a neutron into a proton and simultaneously emitting a W^- boson, which subsequently decays into an electron and an anti-neutrino. It's this subsequent decay that yields a wide range of electron and anti-neutrino energy combinations, each of which sums to the same constant value.

$$\text{Beta-decay: Neutron} \rightarrow \text{Proton} + W^- \text{ boson}$$

$$W^- \text{ boson} \rightarrow \text{electron} + \text{anti-neutrino}$$

By the early 1950s, the anti-neutrino postulate had been widely accepted. And yet still no one had yet seen it. The anti-neutrino is incredibly hard to find, easily passing without hindrance through matter as reflected by the fact that the absorption of one single anti-neutrino would require a light-year's thickness of lead.[10] So there was only one way out: increase the concentration of anti-neutrinos. It was Frederick Reines (1918–1998) and Clyde L. Cowan, Jr. (1919–1974) who figured this out and so set out to find such a huge source of anti-neutrinos. Their answer: a nuclear reactor. There's a huge flux of anti-neutrinos in a nuclear reactor caused by the presence of so many decaying neutrons. In 1956 they set-up their experiments at the Savannah River nuclear reactor and so detected with sufficient conviction Pauli's "foolish child."[11] They sent Pauli a telegram announcing this finding, to which Pauli responded, "Thanks for the message. Everything comes to him who knows how to wait. Pauli."[12] Pauli died shortly thereafter, in 1958.

At the end of this chapter in physics history, energy once again remained conserved.

---

[10] (Pais, 1988) p. 569.
[11] (Pais, 1988) p. 314.
[12] (Pauli, 2013) p. 19.

What's next?

Why do we as physicists, chemists, and engineers even care about energy? Simply put, we need energy to survive and make the world a better place to live in. We continually use energy to change our environment. We know that things just don't happen on their own and that there is no perpetual motion machine. When we imagine what it is we want to accomplish, we can sit down and calculate the amount of energy required. This is how we can design furnaces, boilers, and reactors.

But it's not only the concept of energy that we use in such calculations. There's another concept. Stop and think about it. It's something that we just know but perhaps don't recognize that we know because it's so obvious. Heat flows only from hot to cold. How simple this sounds, right? But how complicated the explanation and the larger implications.

The conservation of energy only tells us that energy is conserved. Nothing more. Clausius recognized this and concluded that we needed a 2nd Law of Thermodynamics, stating that heat only flows from hot to cold. But it's more complicated than this, as reflected by the fact that this law brought forth one of the most complicated concepts we encounter in our studies: entropy. Its arrival led to significant advances in thermodynamics and provided a means for us to determine what is and is not possible in all the endeavors we seek. It's now time to go to Part IV.

PART IV

Entropy and the Laws of Thermodynamics

ENTROPY

THE STEAM ENGINE

PRACTICE

THEORY

IDEAS

HERO OF ALEXANDRIA
- first successful demonstration of a steam engine

TORRICELLI (1642)
- we live at the bottom of a sea of air

GUERICKE (1657)
- demonstrated power of a vacuum

HUYGENS (1673)
- first piston-in-cylinder engine based on exploding gunpowder

PAPIN (1690)
- condensed steam in cylinder, ambient air
- drove piston down into resulting vacuum and could thus be used to generate work

NEWCOMEN (1712)
- developed Papin's concept to establish first engine capable of raising water from deep mines

WATT (1765)
- breakthrough improvements of Newcomen engine based on condensing steam to create vacuum

TREVITHICK, WOOLF, & THE CORNISH ENGINE (1800-1824)
- Trevithick shifted steam from vacuum to safe high-pressure operation. Woolf and the Cornish engineers further improved the engine to higher pressure (50 PSIG) and greater efficiency

LAZARE CARNOT (1784)
- developed theories on the water wheel engine

SADI CARNOT (1824)
- inspired by progress in Britain, brought new thinking to theories on heat engines based on analogy with water wheel
- proposed maximum work is proportional to caloric from T_H to T_C and that caloric is conserved

CLAPEYRON (1834)
- brought advanced mathematics to Carnot's analysis

THOMSON (1848)
- studied Clapeyron's paper
- published his own analysis

CLAUSIUS (1850)
- reconciled Carnot and Joule
- work is done by conversion of heat

ENERGY THEORY

JOULE / MAYER
- heat-work equivalence
- (caloric theory rejected)

1ST LAW OF THERMODYNAMICS
$$dU = \partial Q - \partial W$$

ENTROPY

TOWARDS ENTROPY

→ **THOMSON / RANKINE**
TAIT / MAXWELL
HELMHOLTZ
   (1850s)
- the "North British" raised
  energy to unifying concept

CREATION OF THERMODYNAMICS

THOMSON (1851)
- first glimpse of $\partial Q / T$
  inside Carnot's engine

THOMSON (1854)
$$\Sigma \left(\frac{\partial Q}{T} \right) = 0 \text{ in closed cycle}$$
- reversible process

CLAUSIUS (1854)
$$\int \left(\frac{\partial Q}{T} \right) = 0 \text{ in closed cycle}$$
- reversible process

CLAUSIUS (1854)
$\frac{\partial Q}{T}$ must be differential of state property

$$dS = \frac{\partial Q}{T} \qquad S = \text{entropy}$$

↘ **CLAUSIUS** (1865)
- the energy of the universe is constant
- the entropy of the universe increases
  to a maximum

↓

GIBBS (1873-78)
$$dU = TdS - PdV + \Sigma \mu_i dm_i$$
- equillibrium of heterogeneous
  substances establishes Classical
  Thermodynamics

KINETIC THEORY OF GASES & STATISTICAL MECHANICS

→ **CLAUSIUS** (1857)
- modeled gas as colliding
  non-interacting atoms
- demonstrated connection
  between T and $\frac{1}{2}mv^2$

↓

MAXWELL (1860-65)
- proposed Gaussian-based
  distribution of gaseous speeds

↓

BOLTZMANN (1872-77)
- mathematically proved that
  any given distribution of speeds
  always moves toward Maxwell's
  distribution, but later had to
  recant saying *almost* always
- an outcome of this was his
  introduction of probability into
  physics and his declaring that:

  entropy (S) quantifies the
  most probable state of a system

2ND LAW OF THERMODYNAMICS
$$S = k_B \ln W$$

ENTROPY   BOLTZMANN CONSTANT

the number of different ways
that a system of molecules can
be arranged (location, momentum)
for a given set of macroscopic
properties

24 Entropy: science (and some history)

I never understood entropy
– anonymous engineer

We can't measure it directly. We rarely use it directly. Few of us truly understand it. It has frustrated us no end. And yet...

And yet, shortly after its discovery by Rudolf Clausius in 1865, the concept of entropy became one of the founding cornerstones of classical thermodynamics. This extremely powerful concept, a concept that almost died an early death were it not for the genius of such individuals as J. Willard Gibbs, enabled a mathematically rigorous means to ensure the impossibility of perpetual motion while also providing the critical bridge from the mechanical world to the chemical world, enabling physicists, chemists, and engineers a means to quantitatively master both phase and reaction equilibria.

Yes, entropy has frustrated us since its inception, its underlying meaning elusive to our continued inquiries. But while many don't fully understand it, this hasn't stopped us from using it.

In keeping with the motivation of this book, the intent of Part IV is to raise your level of understanding of entropy and its role in the 2nd Law of Thermodynamics, especially regarding the world of chemistry. As with previous parts, this lead-in chapter to Part IV serves to introduce the fundamental science behind entropy, while subsequent chapters share the history that brought us to the discovery or creation of such fundamentals—at some point the concepts of discovery and creation blur. It is through sharing the history that I take an even deeper dive into the fundamental science involved. So let's start.

Entropy as a Consequence of Our Discrete World

The understanding of [temperature and entropy]...is dependent on the atomic constituency of matter.

– Arieh Ben-Naim[1]

To understand entropy, one must zoom into the atomic world and consider what happens with real atoms and molecules—it is because nature is discrete[2] that entropy exists.

---

[1] (Ben-Naim, 2008) p. 2.
[2] Ignoring wave–particle duality for now.

Block by Block: The Historical and Theoretical Foundations of Thermodynamics,
Robert T. Hanlon, Oxford University Press (2020). © Robert T. Hanlon.
DOI: 10.1093/oso/9780198851547.001.0001

Nature is comprised of discrete entities: atoms and molecules, photons and fundamental particles. We're taught in physics about how to apply Newton's laws of motion to analyze two colliding bodies, but we're rarely taught how to analyze a large system of such colliding bodies, as the concepts are more complicated. It's in the large system that entropy exists, which is rather fascinating since who would think that any kind of structure would exist inside such a seemingly frenetic system and who would think that such a system could be characterized by a single number called entropy?

Atoms in a Box

Put a single moving atom into an isolated box and, in the ideal Classical World, it will continue to move, bouncing off the walls, this way and that, like a billiard ball ricocheting off the sides of a pool table. Since the system is isolated (and ideal), then according to the 1st Law of Thermodynamics its energy remains constant. And since the only energy in this system is the kinetic energy of the atom, then this too remains constant. The only parameters that change are the atom's location and direction of movement, or its momentum, due to successive collisions with the box's walls, which we treat as ideal reflectors. For this system, ignoring quantum mechanical effects, knowing the atom's initial location and velocity, we can predict its motion until the end of time.

Now add another atom to this system. We can similarly model this but with the added feature of the two atoms colliding with each other. During such collisions, both kinetic energy and momentum are conserved as per Newton. With such a simple system, we can again predict the future motions of both atoms to the end of time.

Now add a third atom. The same conservation laws apply. Continuing to ignore quantum mechanics, the particles will move exactly as Newton would predict. It's just that the mathematics become a little more complicated.

Now add more atoms. Add a trillion atoms. There's no difference. The conservation laws continue to apply and all trillion atoms will move according to Newton's dictums. But while there may be no inherent physical difference as we move from the small to the large, there is a tremendous difference in how we approach the modeling. In short, we can calculate the small but not the large, for that task is nearly impossible. We've made significant progress over the years in modeling such an ideal system due to the increase in available computing power, but when we start accounting for the large numbers involved and start adding the real-world complexity of intermolecular interactions and quantum effects, we reach 'impossible' rather quickly.[3]

Fortunately for us though, there's a way around this limitation, a way based on the fundamental fact that nature is not arbitrary, although things get fuzzy at the scale of Heisenberg's uncertain world, which we can safely ignore for now. Because nature largely follows a cause–effect world at the microscopic level, certain macroscopic properties and relationships between these properties exist, which tremendously simplifies our quantitative approach to modeling nature. Instead of following trillions of colliding atoms, we can instead follow such macroscopic properties as temperature and pressure.

---

[3] (Kaznessis, 2012) see p. 5 for a discussion of this.

We saw this in Part III. In an isolated system, regardless of what's going on inside the system, if no heat enters the system and no work is done on the system, then the energy of the system remains constant. This 1st Law of Thermodynamics rises directly from the behavior of atoms at the microscopic level where energy is conserved during each collision. We can pretty quickly grasp the meaning of this law. It makes sense to us. Energy just doesn't appear or disappear. Things just don't happen. You can't get something for nothing. There's no such thing as a free lunch. As human beings, we intuitively grasp the meaning of this law as reflected by the fact that it pops up frequently during our daily conversations. We may not always apply the law correctly, but we do grasp its meaning.

The 2nd Law of Thermodynamics in which entropy is imbedded is more complicated. While this law similarly rises as a consequence of the cause–effect nature of the microscopic world, it unfortunately doesn't lend itself to such a quick grasp of meaning. But let's try.

The Appearance of Structure in Large Systems of Colliding Atoms

In the above system of a trillion atoms, for now assuming they behave as an ideal gas (no inter-actions), the atoms follow the cause–effect world according to Newton. The cause–effect is there, operating in the background, during each and every collision, beyond our limits of detection and modeling. But while all of these collisions are going on, creating a seemingly muddy pool of kinetic confusion, something quite remarkable happens. Two beautifully symmetric and smooth population distributions appear, one for location and the other for velocity.[4] Given enough time for many collisions to occur, the distribution of the atoms becomes uniform with respect to location and Gaussian—forming the famous shape of the bell curve—with respect to velocity.[5] This fascinating behavior always happens for such a large number of atoms. It is this behavior that is the basis for the highest-level definition of the 2nd Law of Thermodynamics and that explains the lower-level manifestations of this law such as heat always flows from hot to cold and perpetual motion is impossible. While we'll get to all of this later, for right now, it's important to realize that this is really it. Given enough time, atoms spread evenly with regards to location and Gaussian with regards to velocity.

So why do these two specific distributions result? Probability. There's no driving force at work other than probability. They result as a natural consequence of the probability associated with a large number of events such as collisions between atoms in the world of Avogadro-sized systems.

In short, nature moves toward its most probable state. If you flip a coin a thousand times, the most probable result is that you'll end up with 500 heads and 500 tails, even though each

---

[4] Even if you were small enough, you couldn't directly see these distributions with your eyes as you'd be one step removed. You'd have to find some magical way to measure the instantaneous location and velocity of every atom and then graph the location and velocity population distributions to see them.

[5] Technically the Gaussian distribution is based on the population distribution of one component of velocity such as v_x as opposed to speed, which is $\sqrt{(v_x^2+v_y^2+v_z^2)}$, or energy. Because these properties are all tied together, you'll often see population distributions based on one or the other or both. When I speak of a Gaussian-velocity distribution, I am referring to v_x, which has both positive and negative values centered around zero for a non-moving system.

individual flip is random. There's no driving force involved. This is simply the result of how probability works for large numbers.[6] Is it possible to observe a thousand heads for a thousand flips? Yes. Probable? No. You'll never see this in your lifetime. After many flips, the 50:50 distribution (with some very small error bars) is a near certainty. This same basic probability logic applies to colliding atoms. Are other distributions possible? Yes. Probable? No. The most probable distributions for a system of colliding atoms are uniform-location and Gaussian-velocity. The probability that each occurs is (just shy of) 100%.

Why Uniform-Location and Gaussian-Velocity Distributions Make Sense

Let's dig a little bit deeper into this discussion to get a better feel for the concepts involved. And let's do this by considering some examples, starting first with location. Envision all the trillion atoms located solely in one half of the box. You would look at this and realize that something isn't right. Nature doesn't behave this way. On the other hand, envision all the atoms spread evenly throughout the entire box. You would look at this and say, *this looks right*, just as you would if you placed a drop of dye into a glass of water and saw it eventually spread and create a uniform coloring throughout the entire glass. You would expect to see such an even distribution for location. This has been your experience based on your own real-world observations. And this is indeed how nature behaves. Atoms move in an isolated system until density is uniform throughout (assuming no external field like gravity is acting on the system). The cause is uniform probability. If all of the atoms have access to all the volume, then, given enough time, they'll eventually populate this volume without preference. As a result, out of all the possible distributions that could possibly exist, the uniform distribution prevails.

With velocity, it's a little more complicated. Imagine you have the power to 'see' the velocities—here based on both speed and direction—of all of these many atoms, and say you saw just a few really fast atoms accounting for all the energy,[7] while all of the (many) other atoms were moving really slowly. You would similarly look at this and realize that something isn't right. Nature doesn't behave this way. We just empirically know that given enough time, all of the atoms would be in motion, hitting each other, with no atom left untouched for long. Head on or glancing blow, nature has no preference. Every type of collision you could imagine occurs. If in this scenario we could 'see' the velocities of all the atoms or, better yet, if we could somehow see the distribution of velocities and how it changes over time, you would see a Gaussian distribution (assuming you're looking at a single velocity component, such as v_x) very tightly fluctuating around some average, which is in fact zero if the entire system is stationary, and

---

[6] In another example, have you ever visited a museum in which a large-scale machine called the Galton box demonstrates how balls cascading down a vertical board containing interleaved rows of pins, bouncing either left or right as they hit each pin, grows into the Normal or Gaussian distribution as a large number of such balls collect in the one-ball wide bins at the bottom? Such demonstrations are pretty cool to watch unfold and reflect how many sequential 50:50 decisions result in a Gaussian distribution.

[7] Recall that velocity converts to energy per the equation, $E = \frac{1}{2} mv^2$. When you have an ideal monatomic gas with no interactions, energy is comprised solely of the kinetic energy of the atoms, which is related to velocity by the term $\frac{1}{2} mv^2$.

every once in a while, a non-Gaussian distribution would flicker across your screen, but this would be a rare event. You'd look at this movie and think, *this looks right.* It would look like many other Gaussian distributions you've seen before such as a population distribution based on height.

Initially, even though the distribution might look right, you likely wouldn't be able to explain why. But if you thought about it some, you'd start to realize that there's a reason for this shape. When atoms collide, energy is conserved. If one gains, the other loses. Moreover, it's typically the slower one that gains and the faster one that loses, meaning that fast atoms typically don't get faster. Statistically, the faster they get, the greater the concentration of slow atoms in their immediate neighborhood and the higher the probability they'll hit a slow atom and thus be pulled back toward the average, which is either the mean kinetic energy or zero velocity for the x, y or z component, depending on which property you're studying. In the world of statistics, this "pull" is termed "regression toward the mean." There's a stronger and stronger pull toward the mean as a given atom's velocity increases further and further away from the mean.

Having said all this, it *is* possible that you could have such a highly unusual distribution of atoms inside the system—the flicker I mentioned above—in which the motion of a single atom amongst the trillion accounts for all of the energy, such as could be the case if the atom were bouncing exactly perpendicularly between two parallel walls with all the other atoms sitting off to the side with zero velocity. Is this possible? Yes. Probable? Most assuredly not. And you'd simply know this by looking at such a distribution. You'd think that it simply doesn't look right. You'd think that nature is more symmetrical than this. And you'd be correct.

The Microstate and Statistical Mechanics

At this point, you're likely wondering what this whole conversation has to do with entropy. I'm getting there but along a path that's different than typically taught. Please bear with me. I'm introducing you first to one physical model—there are others—used to guide the mathematical approach to analyzing such a system. This is a very short preview of the path taken by Ludwig Boltzmann to arrive at the probabilistic nature of entropy.

* * *

Imagine an isolated system of a trillion monatomic gas-phase atoms for which total energy (U), volume (V), and number of atoms (N) are all fixed. Further imagine that you could watch a high-speed movie of these atoms. As discussed above, you would see them evenly distributed throughout the system with small rare fluctuations. You'd see fast, you'd see slow. You'd see constant collisions. It'd be like watching a game of Pong on your television with a huge number of balls in colliding motion.

Now say that you wanted to analyze this in more detail. Taking the earlier discussion to a higher level that enables application of advanced mathematics, you could take the movie, cut out all the individual frames, take many measurements, and then for each and every frame record each atom's location and velocity (speed and direction; you would need to use two sequential frames to determine velocities). Furthermore, you could convert the raw data into population graphs, showing the number of atoms as a function of location, velocity, or energy.

In this way, each single frame of the movie becomes a single data point representing a unique way of arranging the N atoms in the given volume with fixed energy. You could even make a new movie wherein each frame consists of one of the population distributions you want to analyze. You could run the movie and watch how the distribution changes with time.

In the world of statistical mechanics, each frame of the movie is called a microstate and represents a unique distribution of the available atoms that complies with fixed U–V–N constraints. While the microstate is a man-made concept representing a significant breakthrough in our attempt to model and understand the thermodynamics of such large systems, it is also an excellent approximation of reality as it captures the real fluctuations that occur in nature. At steady state, the time average of the microstates equals the macroscopic state of the system.[8] When we measure macroscopic properties, such as temperature and pressure, the numbers aren't exactly constant because of these microstate fluctuations. We typically don't see such fluctuations, as they're easy to miss if you're not looking for them. But if you do want to see them, there is one place to look—under the microscope, for it is there that you can observe the famed Brownian motion, which is more accurately a *consequence* of the fluctuations rather than the fluctuations themselves. It was Einstein who made sense of this phenomenon by using Boltzmann's work to quantify the fluctuations and resulting motions and in so doing provided substantive evidence toward the existence of atoms.

The microstate is a *mathematical construct* that was developed to model nature. While nature doesn't know anything about microstates, I again emphasize that the above is not simply a mathematical exercise. It's a highly effective means of understanding and predicting nature, just as Einstein demonstrated. Experiments have since further proven that the models used in statistical mechanics truly do reflect how nature behaves and fascinatingly creative experiments have even made visual for us the Gaussian distribution of velocity. While so far we've only addressed the simplest of systems—monatomic, non-interacting atoms—the approach also works (with adjustments) for more complicated systems involving interactions and/or vibrating, rotating, reacting molecules. No matter how complicated the system, the population distributions based on location, velocity, and energy follow similar distributions laws. The distributions are the predictable consequence of Newtonian mechanics combined with energy conservation playing itself out over a very large number of atoms.

Mathematical Approach to Analyzing the Microstates – Balls in Buckets

Of course, it gets more complicated than this. You have a viable physical model. Now you need to bring in the mathematics. How would you go about modeling the cut-up movie?

*　*　*

If you'll recall from probability theory, there are only so many ways to arrange a given number of balls into a given number of buckets. For example, if you had two balls and two buckets, there would be four possible arrangements: both in the left bucket, both in the right bucket,

---

[8] The assumption that the time average represents the macroscopic state, or the ensemble average, is called the ergodic theorem.

and then, assuming a red and a blue ball, red in left / blue in right and then blue in left / red in right.[9] The point here is that the mathematics exists to calculate this number of possible arrangements.

Now imagine that the above buckets are not physical objects but instead conceptual ones based on location or velocity or both, with infinitesimally sized ranges for each variable (ranges are needed to create the buckets). And further imagine that instead of balls you have atoms. The same mathematics applies. You could (with much work) sit down and calculate the total number of different ways to place the atoms into the location buckets and the velocity buckets. It would get kind of complicated as you'd be working in six dimensions—three for location (x-y-z coordinates) and three for velocity (v_x-v_y-v_z vectors)—but the math would still work. In the end you would arrive at a finite number, meaning that there is a fixed number of ways to arrange a fixed number of atoms into a huge group of location–velocity buckets, all the while maintaining fixed U-V-N.

Each and every one of these arrangements aligns with a unique frame or microstate in your movie. Each microstate is thus comprised of its own unique distribution of atoms based on both location and velocity.

An interesting fact about each and every one of these microstates or movie frames or arrangements is this. Each is just as likely to occur as any other. Each is equiprobable with all the others. This sounds unusual, right? Consider that one possible frame is for all atoms to be in one half of the given volume V. It is indeed possible that this will occur. Consider that another possible frame is for all atoms to be spread evenly across the entire volume. It is equally possible that this will occur. But this doesn't make sense, does it? How can this be? The ½ volume frame looks very unusual if not highly improbable, like getting a thousand heads in a row while flipping a coin. Understanding the logic here gets you to the heart of this discussion.

If you count up all the possible movie frames or all the possible arrangements, you'd arrive at a finite number. Let's call this number J. If you were to start analyzing the different arrangements comprising J you'd realize something very quickly. Most all of the arrangements look the same. Most all of the arrangements would show a similar distribution of uniform-location and Gaussian-velocity. In fact, you could even do the mathematics to prove this. You could pose the question, which distribution accounts for the greatest number of arrangements?, and then use advanced calculus and probability-based mathematics to solve. Make "number of arrangements" the dependent y-variable, "distribution" the independent x-variable, differentiate (dy/dx), set equal to zero, and find the maximum. The answer: uniform-location and Gaussian-velocity. These two distributions account for (by far) the greatest number of arrangements. You could then quantify the total number of these arrangements and call this number W.

The final fascinating thing about this exercise is that for large numbers, W is almost exactly equal to J and for all intents and purposes the two can be assumed to be the same, an assumption which significantly simplifies subsequent mathematics. In looking at this, the reason that uniform-location and Gaussian-velocity are the most probable distributions is that they occur much, much more frequently than any other distribution. Yes, a given distribution of all atoms

---

[9] I don't address in this book the issue of distinguishable versus indistinguishable particles as it unnecessarily raises the complexity of the content beyond my objectives.

on one side has the same probability of occurring as a given distribution of uniform-location. But there are overwhelmingly more uniform-location distributions that occur, thus making them the most frequent and thus the most probable. The frequency of occurrence of all such unusual distributions—the "flickers"—is located in the value of J minus W, and this value is very small but not zero – or else Brownian motion would never exist. (Highly improbable events are not impossible.)

Nature shows no preference for how atoms and molecules distribute themselves within an isolated system so long as the U-V-N constraints are maintained. There's no "driving force" to move toward a given distribution other than pure probability. Let's look at what this means. Say you were to set up an improbable system, such as one in which all atoms were in one half of the volume or one in which all of the faster moving atoms were in one half of the volume with the slower moving atoms in the other half. You could do this by putting a sliding door into the system to separate one half from the other and then adding all atoms to one side or heating up one of the sides. In this way you'd create a pressure gradient or a temperature gradient, respectively, between the two halves. At the exact moment you removed this internal constraint by either pulling out the door for the former or by somehow enabling thermal conduction across the door for the latter, you would have created a highly improbable system. How would you know? Because just as the drop of dye in a glass of water doesn't last for long, neither does the state of these two systems. If you sat and watched what happened, you'd see the atoms rush to fill the void and you'd see the two temperatures move toward each other. In other words, just as a drop of dye moves to fill the entire glass of water, you'd see these two improbable initial states move toward the most probable final states of uniform-location and Gaussian-velocity.

Finally, the Relevance of the Number of Microstates

At this point (again) you might be asking, even more emphatically, so what? Here's where we arrive at the answer. Consider the statement: a given system of U-V-N has a *certain specific number* of ways (called microstates) that it can be arranged. This statement has profound implications because it effectively says that given U-V-N, the number of arrangements is fixed and thus this number itself is a property of the system, just like temperature, pressure, and internal energy. An unusual property, yes, but a property nonetheless.

Well, along the path taken by Boltzmann that I chose to share here, the concept of "number of microstates" is directly related to the main theme of this chapter: entropy. In fact, entropy, which we denote as S, is calculated from this number, which we denote as W, per the famous Boltzmann equation,

$$S = k_B \ln W$$

in which k_B is the famed Boltzmann's constant.

Entropy is a fundamental property of matter, a state property, and is related to the number of different ways atoms, molecules, and even photons can be arranged while maintaining fixed U-V-N. Once U-V-N are fixed, the number W also becomes fixed, just like the number of different arrangements of balls-in-buckets is fixed once you fix the numbers of both balls and buckets.

Entropy is a difficult-to-comprehend property, but it's a property nonetheless. As we'll see later, because we know from classical thermodynamics that S is dependent on U-V-N, then we also know that W is dependent on U-V-N as well and can thus establish a direct link between the atomic world and the bulk properties of matter. Boltzmann's equation is the bridge connecting the two worlds.

*   *   *

From the standpoint of classical thermodynamics, we don't *really* need this link between entropy and the number of microstates. We don't need Boltzmann's equation. In fact, we pretty much don't need any aspect of the entire preceding discussion. It's largely irrelevant to our use of entropy in the solving of problems relating to, for example, phase and reaction equilibria.

So why am I spending time telling you about it? Because this link has everything to do with explaining what entropy is. In classical thermodynamics we use entropy without really understanding why. While there's nothing wrong with that, there is something very limiting about that, something that limits our deeper understanding of what we're really doing. We end up not understanding entropy and so become almost afraid to confront it, embrace it, and use it.

The Discovery of Entropy – Clausius and the Steam Engine

One logical question to ask based on the above is, why don't we learn the deeper meaning behind entropy first, right at the beginning? And that's a good question. One reason is that we don't *need* to understand it; we can use it perfectly well to solve problems without the understanding. Another reason has to do with history. It turns out that if you start from statistical mechanics and apply some very powerful mathematics, you can arrive at the following rather familiar and famous equation

$$dS = \delta Q\,/\,T \quad \text{reversible process}$$

This equation—valid for a reversible process involving an equilibrated system, concepts we'll get to later—states that the increase in entropy of a system, regardless of the contents of that system, is equal to the heat absorbed by the system divided by absolute temperature, another non-trivial concept we'll get to later. The contents could be hydrogen gas, water, or titanium. It doesn't matter. The relationship remains the same. This equation is a *consequence* of the probabilistic nature of entropy and is buried deep down in the mathematics of statistical mechanics as just one of many equations and relationships that can be derived.

It just so happens that there's another way to discover this equation: inside a steam engine. Who would have thought? Through his deep, insightful contemplation of Sadi Carnot's work, Clausius realized, in a truly monumental step in history that broke open a huge logjam in the world of physical chemistry, that not only does entropy exist as a property but that it changes in such a way as to govern each step of Carnot's four-step engine reversible cycle: the two isothermal volume-change steps, for which the change in entropy equals the total heat absorbed or rejected divided by temperature (since temperature is constant), and the two adiabatic volume-change steps, for which there is no change in entropy since there is no heat exchange. (Figure 16.1 with further discussion in Chapter 30)

Entropy Enabled Completion of Thermodynamics

Clausius' above equation is not the sole defining equation of entropy. As noted before, it's not even *the* defining equation as that prize belongs to Boltzmann's equation. But it is *one* way and in fact *the* way that enabled completion of classical thermodynamics. Recall Clausius' separate defining equation for the 1st Law of Thermodynamics (in infinitesimal form):

$$dU = \delta Q - \delta W$$

We know from his work that heat flow into the system, Q, and work done by the system, W, do not describe properties of matter, as such, but instead describe *changes* in properties. Each quantifies a specific change inside the system. We know from Part III that the infinitesimal work (δW) done by a fluid (gas or liquid) is equal to the pressure of the system times the infinitesimal change in volume, PdV. We also know that the infinitesimal heat (δQ) added is equal to TdS. Combining these two relationships enabled conversion of the 1st Law into one based purely on properties of the system, independent of the heat / work path taken to move from one state to another.

$$dU = TdS - PdV$$

The arrival of entropy in 1865 enabled the 1st Law of Thermodynamics to be upgraded to the first fundamental equation of thermodynamics, or more accurately, the differential of the first fundamental equation. This differential equation is based solely on thermodynamics properties without reference to heat and work processes and is the source from which all other thermodynamic equations can be derived. It was this equation that served, in part, as the basis for Gibbs' monumental 300-page treatise *Equilibrium of Heterogeneous Substances*. The other part that Gibbs relied on was a fascinating characteristic of entropy that Clausius discovered.

Die Entropie der Welt strebt einem Maximum zu[10] – Clausius' Version of the 2nd Law

Clausius' entire nine-memoir analysis of Carnot's steam engine, an analysis in which the announcement of entropy was but a single part, firmly established the two laws of thermodynamics. The 1st Law as written above was based on energy and its conservation. The conservation inherent to this law said that an energy increase in one system must be compensated by an energy decrease of identical magnitude in another system. This law in and of itself had a tremendous impact on the science community. But it wasn't enough.

Carnot wrote *Reflexions* to address "whether the motive power of heat is unbounded."[11] His theoretical approach, while groundbreaking, could not be completed since it was based on the false caloric theory of heat. In correcting Carnot's work by shifting to the mechanical theory of

---

[10] (Clausius, 1867) p. 365. The entropy of the world increases to a maximum. Clausius concluded his ninth and final memoir with this bold and far-reaching statement. Gibbs used this statement as the epigraph to his own masterpiece, *On the Equilibrium of Heterogeneous Substances*.

[11] (Carnot et al., 1988) p. 5.

heat as quantified by Joule, Clausius and also Thomson were left with the residual challenge of how to quantify the maximum performance of an engine. They struggled with identifying what it was that limited performance. It was Clausius who ultimately solved this problem by first showing that the conservation of energy embedded in the 1st Law was a necessary but not sufficient condition for defining the limit. Why? Because the 1st Law did not prevent the flow of heat from cold to hot. Without this constraint, the immense thermal energy contained in the ocean on account of its mass could be used to generate an immense amount of work, a situation Thomson (Chapters 32–33) deemed impossible. To solve this problem, Clausius added a 2nd Law saying that heat can *only* flow from hot to cold, which was really conventional wisdom at that point in time, but then upgraded this to the unprecedented statement: the entropy of the universe increases to a maximum. Whether or not he meant this statement to also be valid for an isolated system is open to question and I'll address this later, in Chapter 34. But in fact this statement is valid for an isolated system—and I'll continue to refer to it as being owned by Clausius—and became one of the critical starting hypotheses for the subsequent work of Gibbs. Because of its importance, additional discussion is warranted.

Toward a Deeper Understanding of Clausius' 2nd Law

The increasing-entropy version of the 2nd Law of Thermodynamics that started with Clausius is more complicated than it sounds. Let's consider it in more depth by first creating an improbable system having internal energy gradients of pressure (mechanical), temperature (thermal), or chemical potential (chemical), a property that will be covered in Chapter 36. A way to create such a system would be to connect two separate systems having different values for any of these properties and then enabling these systems to exchange energy with each other by installing, for example, a flexible membrane, a conducting wall, or a porous wall, respectively. In this scenario, the combined system will move toward its most probable state, the gradients will dissipate, and isobaric, isothermal, and iso-potential conditions will result.

But where does entropy fit into this discussion? Well, entropy is an extensive property just like volume and mass. So take the two separate systems presented above. Assume that each is initially equilibrated (a reasonable assumption), that each has its own entropy value, and that these two entropy values can be added together to quantify the total entropy of the initial set-up prior to combination. Now at the exact moment you connect the two systems, the situation immediately changes and the combined system evolves toward one in which temperature, pressure, and chemical potential are constant throughout. When you do the math, this final evolved state, for which the energy gradients have dissipated, is the most probable state and comprises the greatest number of arrangements or microstates. When the atoms and molecules are distributed in accordance with this, equilibrium is achieved and entropy established.[12] Now here's the critical issue in all of this. This final value of entropy for the combined system

---

[12] Technically, entropy is defined at the outset of this scenario even in a state of non-equilibrium. Once you define the state of the system by defining U-V-N, you define W and thus S. Even in the non-equilibrium state, the values of U-V-N and thus S are valid. But it is only in the equilibrium state that the atoms' locations and velocity distributions match those associated with the most probable state as defined by S.

is *always* greater than or equal to the sum of the entropies of the two initial systems prior to combination.

To repeat, the final entropy of the combined equilibrated system will *always* be greater than or equal to the combined entropies of the two separate systems prior to contact. A combined system of many parts will always strive toward its most probable state, the entropy of which will always be greater than the sum of the entropies of all the components.

$$\Sigma\,S_{separate\,parts} \quad \leq \quad S_{combined\,system}$$

It was this characteristic of entropy that Gibbs employed to fully define the concept of chemical equilibria in his groundbreaking publication. I'll discuss this in greater detail later but wanted to address it now because this understanding serves to emphasize the importance of not only entropy but also its relevance to the field of equilibria.

* * *

Let's look even further into the scenario involving the contact of two separate systems, this time solely involving temperature. If you graphed the initial population distribution of all of the atoms in the combined system against their respective kinetic energies (as opposed to their respective velocities), you'd see a bimodal distribution with each peak corresponding to one or the other of the two temperatures. For the same reason an isothermal system will never naturally split into a bimodal distribution like this, an initial bimodal distribution will never exist forever. It will always (eventually) move toward a single distribution indicative of the most probable isothermal state.

Even though in the above scenario the atoms of the two systems don't mix, the system can still be viewed as a single system with regards to the fact that the energies in the two systems do "mix" via conduction. As the atoms between the two systems "talk" to each other at the conducting boundary between them by colliding with the boundary wall atoms, the hot ones slow, the cold ones speed up and the two temperatures move toward each other. When the two temperatures equal each other, no thermal energy gradient remains and the population distribution becomes a single distribution based on kinetic energy, i.e., temperature, which reflects a Gaussian distribution based on velocity.

If you were to calculate the entropy of each body before and after combination, you'd find that the total sum of the two entropies before contact is less than the total entropy of the combined and equilibrated system. This is the nature of entropy. Energy gradients of any kind reflect an improbable state. As a system moves from improbable to probable, gradients dissipate and total entropy increases. This is because the entropy *increase* of the lower energy system is greater than the entropy *decrease* of the higher energy system. This is the nature of entropy. And this is why the sum of the two changes is always positive. This is the context of Clausius' version of the 2nd Law: entropy increases to a maximum.

The Historical Path toward Entropy

The historical path leading toward our understanding of entropy is fascinating. It began with early forms of the 2nd Law, then continued with the discovery of entropy and subsequent upgrading of the 2nd Law, and finally ended with the arrival of statistical mechanics. While fascinating, to this author, this path is arguably not the ideal path by which to learn about entropy as it's difficult to learn about any subject by first learning about its consequences. It was our good fortune that Clausius remarkably discovered entropy from a very unusual direction, the steam engine, well before statistical mechanics was developed. But it was in statistical mechanics where entropy would eventually be understood.

So to this author, a challenge in understanding entropy and the 2nd Law is caused, in part, by the fact that the cart came before the horse during entropy's history.[13] The property arrived before we had any concept of what it was. For better or for worse, the structure of our education reflects this cart-before-horse history. We are educated according to history's timeline. Learning about entropy from a steam engine is not easy. But then again learning about entropy from statistical distributions isn't easy either. Many links in the logic chain separate the simple two-body collision physics of Newton from the statistical arrangement physics of Boltzmann. Each link along the way isn't that difficult to understand but the full leap from Newton to Boltzmann *is* difficult to make, especially since entropy doesn't lend itself to the visceral understanding we have for other properties, such as temperature and pressure, which we can directly sense and measure. Perhaps learning about entropy is simply an inherent challenge, no matter which way you approach it. But I prefer not to give up so easily. In fact, one of the over-riding goals for writing this book was to provide some new ideas about how to better teach the concept to students, and so here we are.

* * *

As undergraduates, we learn the classical thermodynamics of Clausius, Thomson, Gibbs, and others and while we may not understand what entropy is, we can still learn how to use it to solve problems. The statistical thermodynamics of Maxwell, Boltzmann, and Gibbs again provides insight into entropy but is taught second, typically in graduate school, if at all. So we learn how to get by without truly understanding entropy. Not helping matters is that classical thermodynamics is itself confusing. Gibbs' treatise, in which entropy played a central role, is very difficult to read. And his work competed with other works to establish the field. Many had their own version of this field: Clausius, Maxwell, Helmholtz, Duhem, Planck, Nernst, and others. Entropy was important to some, not so much to others. The atomic theory of matter came into being after completion of classical thermodynamics and so helped transform this subject into the statistical mechanics of Boltzmann. Indeed, the success of statistical mechanics helped validate its starting atomic-theory hypothesis (Chapter 4). Again, this history is indeed fascinating, but the sequence of how this all happened and how it ended up in the textbooks with different and antiquated terminologies led to even more confusion, above and beyond the technical challenge itself. Each author had his own way of explaining the material. As Feynman said, "*This is why thermodynamics is hard, because everyone uses a different approach. If we could only sit down once and decide on our variables, and stick to them, it would be fairly easy.*"[14]

* * *

---

[13] The author's feeling is that teaching entropy by following this cart-before-the-horse history creates some unnecessary challenges. The concept is indeed challenging enough to grasp without trying to figure it out while also figuring out Carnot's heat engine. Perhaps starting with the fundamental probabilistic explanation of entropy and then moving into the consequences, including that the fact that heat only flows from hot to cold, would be more effective. Discussion of engine limitations would come second in this methodology.

[14] (Feynman et al., 1989a) Volume I, pp. 44–9.

Man's incessant need to figure things out, to determine the cause of an effect, led him into a brick wall many times regarding the journey toward entropy, its meaning, and its relation to the 2nd Law of Thermodynamics. Many interpretations were given, many books were written, and many mistakes were made about this concept. And while we have finally arrived at a detailed definition of this concept, to this day, 150 years after its discovery by Rudolf Clausius, many otherwise successful graduates in physics, chemistry, and engineering still don't fully understand entropy. Part IV is my own approach to bringing clarity to the subject by sharing with you the rich, technical history that led to its discovery and use.

25 It started with the piston

The development of machines like the steam engine in the eighteenth century almost forced man to recognize the enormous power...of heat, the grand moving-agent of the universe.
– D. S. L. Cardwell[1]

Heat and Work Met for the First Time Inside the Steam Engine

In the chemical world, fluids like gases and liquids are more common than solids, which populate the mechanical world. This makes life mathematically convenient for us. Since fluids spread evenly across the surface of the system's container and since pressure is uniform inside the equilibrated fluids, the force exerted by the system on the external environment is simply the fluid pressure times the system boundary surface area. Since work is defined as integral of force times distance, then the work that a fluid does on the external environment is:

$$\text{Work}\left(\text{done by a fluid on the external environment}\right) =$$
$$\int \text{Force dx} = \int (\text{Pressure} \times \text{Area})\ \text{dx} = \int \text{Pressure dV}$$

For work to occur, the fluid pressure must be either higher or lower than the ambient environment, and the boundary between the two must naturally be able to move.[2] This is how a piston-in-cylinder operates inside a steam engine. The pressure inside the cylinder is typically higher than that of the environment, thus causing the piston to be pushed outward to effect work, like lifting a weight. The famed indicator diagram arises from this phenomenon: as the piston moves out and the volume increases, work is done as quantified by the area underneath the PV-curve.

Recall from Part III that the pressure and temperature of fluids, and especially gases, are demonstrably related, as most clearly shown in an ideal gas for which PV = nRT. You can readily "see" how they're connected through simple and direct measurements and accompanying mathematical analyses. And this makes sense since both pressure and temperature arise from the same molecular phenomenon: Bernoulli's rapid "hither and thither" motion of small particles. The only difference between them is that pressure is also dependent on the density of the

---

[1] (Cardwell, 1971) p. 292.

[2] Other work forms involving, for example, interfacial, electrical, or magnetic forces, exist but will not be treated here.

Block by Block: The Historical and Theoretical Foundations of Thermodynamics,
Robert T. Hanlon, Oxford University Press (2020). © Robert T. Hanlon.
DOI: 10.1093/oso/9780198851547.001.0001

gas. If you were to fix density by fixing both volume and number of atoms, then you would lock pressure and temperature into a specific relationship, the one named for Gay-Lussac (Chapter 18), meaning that if you changed one, you'd change the other.

Through this simple flow of logic, one can envision how heat (*thermo*) and its relation to temperature, and how work (*dynamics*) and its relation to pressure, fundamentally link under the umbrella of thermodynamics. Heat and work are both fundamentally related to the same thing, the kinetic energies of the system's atoms and molecules. It's not always as easy to envision this for non-fluid systems, especially when we deal with, for example, electrical processes in which the work function relates to pushing electrons through a wire. But the same basic principles apply, just with appropriate modifications.

As we saw in Part III, it was inside the steam engine where heat and work first met in a useful relationship. Indeed, it was the rise of the steam engine that forced the science community to confront the relationship between the two. In addition to delivering to us the concept of energy and its conservation, the resolution of this relationship also delivered to us the concepts of entropy and chemical thermodynamics.

Discovering that We Live at the Bottom of a Sea of Air

While the technical journey to entropy started with the steam engine, the historical journey started with steam itself, so that's where we'll start now.

It's interesting that in the world of science, almost any historical journey starts in antiquity.[3] It's no different with the steam engine. We go back about two thousand years to Hero of Alexandria, the famed mathematician and engineer whom we met in Chapter 9. In the first recorded instance of a steam engine, Hero neatly demonstrated steam's potential by running it into a metal sphere from which protruded two pipes, 180° apart, each extending straight out from the sphere, each taking a 90° turn at the end such that they pointed in opposite directions. As steam rushed out of these pipes, the resulting "thrust" from each spun the sphere on a rod. It was only one step from this to connecting the spinning sphere with a rope that would wrap around the sphere and so lift an object and thus do work. While we don't believe this was actually done, *Hero's engine*, as it is now called, made the point. When you make steam, you are one step away from doing work.

Other engineers in ancient Greece and Rome knew of such properties associated with steam and also the expansive power of hot air, but none made any attempt to apply this knowledge to a useful purpose. There was no need to since slaves were available to do the work required of the fortunate minority.[4] It wouldn't be until the Renaissance that the role of technology as a means to positively impact society grew in value and that practical uses for steam and air would be investigated. Even then, the early studies were mostly science-for-science's sake studies. This would change, but there's a twist in this story. The direction I'm about to go in with this history may not be the one you expect right now. Today we are used to using high

---

[3] (Seiler, 2012) p. 44. "Science was born in ancient Greece among the pre-Socratics, who were the first to look for natural explanations of the world around them."

[4] (Rolt and Allen, 1977) p. 14.

pressure steam to push against a piston or a turbine and so do work. Hero started us thinking in this direction. But the first true breakthrough in the use of steam to do work came from the exact opposite direction, one that started during the aforementioned Renaissance studies.

In 1641, Italian engineers employed by Cosimo II de' Medici constructed a suction pump to draw water up from a 50 foot deep well and were completely mystified when, regardless of what they tried, they could only draw the water up 32 feet and no more. They took this problem to Galileo (1564–1642); he too was mystified. The following year, after Galileo had died, his favorite pupil, Evangelista Torricelli (1608–1647), took up the problem, and it was he who first realized that the reason for this limit was that the Earth's atmosphere had weight, a weight equivalent to 32 feet of water. It was this weight that was pushing the water up the well toward the vacuum created by the suction pump. He later proved this theory by demonstrating that the atmosphere could only push up 28 inches of mercury, which carries the same weight as 32 feet of water due to mercury's much higher density. In addition to inventing the barometer out of this exercise, Torricelli also figured out that we live at the bottom of a sea of air.

Moving from Italy to France, in 1647, Blaise Pascal (1623–1662) proposed that if we do live under such a sea, then, moving to a higher elevation should decrease pressure just as it does beneath water. Sure enough, this proved to be so. Pascal's brother-in-law, Florin Périer, climbed the 4800-foot Puy de Dôme (France) in 1648 with a mercury barometer and found that the mercury column fell by 3½ inches.

The Power of a Vacuum

It was only a matter of time until someone concluded that something useful could be done with this new knowledge that the atmosphere had weight and could therefore exert pressure, for if a vacuum could be created, then this atmospheric pressure could be a source of power. Recall that to effect work, one needs access to a pressure differential, which creates a set of unbalanced forces and so sets bodies in motion. Typically when we think of work, and especially when we think of work caused by steam, we tend to think of systems involving high pressure that push out against the surrounding atmosphere. But the opposite can also occur.

Some of you may have seen in person or in photographs what happens in a chemical plant when a vacuum is accidentally created inside a large vessel that wasn't designed for such. The vessel is crushed by the atmosphere. One of my college professors demonstrated this for us in the class room when he put a can of water on a Bunsen burner, boiled the water to fill the can with steam, and then took the can off the flame and sealed it shut. Slowly but surely, the steam condensed and the can collapsed in on itself.[5] Sometimes we don't even realize this pressure of the atmosphere around us, all 14.7 pounds per square inch of pressure (at sea level). It took us quite awhile to even discover this. Because we live in it, we're not always aware of it.

---

[5] In a 2017 email to the author, Phil Hosken, Chairman of the Trevithick Society, shared how violent could be the use of atmospheric pressure against a vacuum as reflected by what happens when you pour a little cold water (à la Newcomen) onto the sealed hot can; it collapses with a bang! Hosken further commented that this is why the iron bound wooden beams in Newcomen's engine (Chapter 27) were so substantial. Otherwise, this collapsing force would "snap them like a carrot."

One of the great public demonstrations was based on the power offered by atmospheric pressure. It was conducted in Germany, 1657 by Otto von Guericke (1602–1686). Unaware of the earlier work of Torricelli and Pascal, Guericke himself tried to intentionally create a vacuum. Because of his interest in astronomy, he originally intended to see if by studying such a vacuum, he could mimic the vacuum he felt *must* exist in outer space, since if it were instead air and not a vacuum, he felt that the resulting friction would slow down the stars and the planets. So in this roundabout way, and after many experiments, Guericke finally devised a rather ingenious system involving the manual withdrawal of a piston from a cylinder to create a vacuum in a closed vessel. Seeking to showcase the potential power of this phenomenon, Guericke created a public demonstration, like Hero's engine before and Rumford's cannon after, for Emperor Ferdinand III and members of his government. He hooked up two teams of horses, eight on each team, to two identical metal hemispheres, called Magdeburg hemispheres for the location where they were used in demonstration, each about 22" in diameter. After placing the two hemispheres against each other to create a full sphere, Guericke used his pump to evacuate the air trapped inside. When the two teams of horses attempted to pull the sphere apart, they couldn't. The air pressure pushing the two hemispheres together was too strong for the horses to overcome. Historians rightfully feel that von Guericke may have been exaggerating, or that he was using pretty feeble horses, but this was missing the larger conclusion. Guericke too discovered that we live under a sea of air, although Torricelli got there first. But Guericke's demonstration, which he later published to great effect, helped to get this knowledge out into the public's mind.

∗　∗　∗

The question, either acknowledged or not, that then faced man was, what to do with this atmospheric pressure? What to do with the concept of a vacuum? Could the two be meshed together to create something useful, like raising water, which was what started this whole process to begin with and which continued to be of practical and commercial interest to many?

The Persistence of Denis Papin

In 1671, Denis Papin (1647–1713), a young Parisian doctor, became an assistant to Christiaan Huygens, whom we met in Chapter 10, famed member of the French Academy.[6] It was there where Papin commenced studies, drawn up by Huygens, on the increasingly popular air and vacuum. Included as part of the latter was pursuit of an interesting idea from Huygens about the use of gunpowder to create a vacuum.

Gunpowder?! Where did that come from? Some background here is clearly warranted. If you look at a revised version of the ideal gas law, P = R (n/V) T, you'll see two approaches to creating a vacuum (P < 0 psig), you can either decrease temperature or decrease density. Density can be decreased by removing gaseous atoms from the given volume. Condensation is one approach. Another is evacuation by using a pump like Guericke's. You can also decrease density by increasing volume, such as by pulling a piston out of an otherwise sealed cylinder. But there's another approach to decreasing density. In a true case study of a technology looking

---

[6] (Galloway, 1881) p. 14.

for an application, there existed during the time of Huygens great interest in assessing how gunpowder's inherent force—firing cannons, blasting mines—could be directed toward useful work. Based in part on Jean de Hautefeuille's efforts in 1678 to use gunpowder in the design of a system to pump water from the Seine to the new palace of Versailles[7] and in part on his own desire to solve the Versailles problem, Huygens began exploring ways to explode a small charge of gunpowder in a vented cylinder to raise a piston, the vents being used to ensure that the system wouldn't pressurize. Once the gunpowder exploded, Huygens closed the vents to lock-in a fixed volume (but now at lower density) and allowed the gases to cool. The resulting vacuum pulled the piston back into the cylinder, driven by atmospheric pressure. As this approach offered a means to access an in-and-out cycling pumping action, as opposed to Guericke's manual unit, Huygens is credited with constructing the first, albeit small-scale, piston-in-cylinder engine in 1673. While others may have had ideas about this concept, which would become the basis for the future steam engine, Huygens was the first to make it real. At this point, Huygens moved on to other pursuits, but Papin took the concept with him as he moved forward in his own career.

Papin's travels took him to England, where from 1675 to 1679 he worked alongside Boyle, bringing Huygens' and his own ideas about air, steam, pumps, and so on with him, and so actively cross-fertilized ideas with Boyle and also with Hooke, whom he met while there. He moved from Boyle (Oxford) to the Royal Society (London), back to France, back to England, over to Italy, back to England, and finally to Marburg, Germany in 1688. In all this time, Papin continued his studies, including those focused on creating vacuums in vessels as a means to raise water. After attempting various options, including improved designs of Huygens' gunpowder concept, Papin finally struck gold as captured in his celebrated memoir of 1690 titled, "A New Way to obtain very great Motive Powers at small cost."

In this memoir, Papin finally gave up on the gunpowder approach to a vacuum, citing its inconveniences. In its place, he proposed a new approach: "I therefore endeavoured to attain the same end in another way: and since it is a property of water, that a small quantity of it turned into vapour by heat has an elastic force like air, but upon cold supervening is resolved again into water, so that no trace of the said elastic force remains: I at once saw that machines could be constructed, in which water, by the help of a moderate heat, and at little cost, might produce that perfect vacuum which could by no means be obtained by the aid of gunpowder."[8] In short, Papin proposed the first atmospheric steam engine.

Papin demonstrated his concept by building a lab apparatus and then continued to work on it for the next fifteen years, even corresponding about it with Gottfried Leibniz at one point. But while his ideas were grand, his execution wasn't. Papin, on the cusp of greatness, simply couldn't move his concept out of the laboratory and into the real world. But his ideas, ahead of

---

[7] (Galloway, 1881) Chapter 2, pp. 17–25. "The problem of raising water from the Seine to supply the palace at Versailles caused great attention to be directed at this time to all forms of motive power available." (p. 21). In response to the request for viable options to supply the new palace of Versailles with much needed water from the Seine, Jean de Hautefeuille proposed two gunpowder schemes in 1678. One in which the exploding gunpowder pushes water up an enclosed pipe, and the other in which the hot combustion gases from exploded gunpowder are allowed to cool and so create a vacuum. Hautefeuille did not propose a piston-in-cylinder for this application. The credit for this goes to Huygens.

[8] (Galloway, 1881) p. 48.

their time as they were, on how to harness steam for useful purposes would live on and play a key role in the eventual commercialization of the steam engine.

*   *   *

Consider Papin's idea. What would happen if you put steam into a piston-in-cylinder assembly and then cooled the cylinder to condense the steam and so create a vacuum? The ambient air would drive the piston into the cylinder. This action is really no different than the opposite in which high pressure steam drives the piston out of the cylinder. Each causes the piston to move. By attaching the piston to another device, like a water pump, one could then effect work. In the end, the vacuum was the path initially chosen. The alternative simply wasn't yet ready to commercialize since contemporary vessel construction technology couldn't (safely) contain high-pressure steam. Soon, steam shifted from mere curiosity to serving as the impetus behind the Industrial Revolution. And this brings us back again to England.

26 Britain and the steam engine

Before getting into the sequence of inventive breakthroughs in steam engine technology, it's important to set the context. While ideas require little financial support to generate, either within one's mind or into one's notebook, such is not the case when moving ideas to commercial realization. In free-market capitalism, there must be a financial incentive to make this happen as the commercialization costs can be considerable. With steam engine technology, the incentive was mining, especially for coal and certain metals such as tin and copper, while with water wheel technology the incentives were milling and textile manufacturing. For each, it was critical to maximize the amount of productive work for a given input of energy. The coal required to generate the steam was expensive. The water required to turn the wheel was limited by nature; the stream only flowed so fast or fell so far. This strong financial incentive to maximize productive work spurred the technology community to focus on improvements and later the academic community to focus on understanding why some improvements worked better than others. While there were no clear boundaries between these two, most of the former happened in Britain (especially Cornwall) as part of the Industrial Revolution, while the critical steps of the latter happened in France. The English do, the French think.

Turning to Britain, we really need to take a huge step back in history to make this context even stronger. Largely as a result of being located near the equator in its very distant past, specifically during the aptly named Carboniferous period (360–290 million BC), Britain was blessed with a huge load of plant life that eventually transformed into a huge load of hard coal, thanks to the fact that the plants degraded under mud and water and in the absence of oxygen.[1] Seams of this coal remained near the Earth's surface and were thus relatively easy to access, especially in areas where running rivers carved down through the earth and exposed the seams, as the River Tyne did near Newcastle. The seams there were good, thick, and, most importantly, as they were above the water line, readily accessible.

But for a long time, Britain found the available coal more of a curiosity than a useful natural resource. The occupying Romans carved and polished jewelry out of it. At times, people burned it but the emitted fumes stank and eventually became despised by the populace, especially in the days before the invention of the chimney.

Things changed when a critical ratio increased: Britain's population relative to the availability of wood. Wood had long been the most valued of Britain's natural resources, being used as timber for structures, as fuel for home heating and cooking, and as an energy source for industries such as brewing. In a separate context, wood was also viewed as a hindrance to those

---

[1] (Freese, 2003) p. 20.

Block by Block: The Historical and Theoretical Foundations of Thermodynamics,
Robert T. Hanlon, Oxford University Press (2020). © Robert T. Hanlon.
DOI: 10.1093/oso/9780198851547.001.0001

wanting to raise sheep for the lucrative wool trade, which led to the clearing of large tracts of forests. What this meant was that over time as the population and accompanying needs climbed, in spite of such devastating events as the Bubonic Plague that wiped out close to half the population between around 1350 to 1500, and as the availability of wood fell, Britain faced an energy crisis and the need for the consumption of coal increased, slowly at first and then faster as the Industrial Revolution lifted off. The sharp rise in coal commerce was greatly facilitated by the many waterways that served to move the coal from source to use. In short, Britain became the first western nation to mine and burn coal on a large scale.

What greatly facilitated the sharp rise in coal commerce was western capitalism, a story in and of itself, the short version of which is this. For many years, the Roman Catholic Church owned and controlled much of the land, including the English lands containing coal. In a rather fascinating turn in history caused by a single event, all of this came to an end when Catherine of Aragon failed to birth a male heir to King Henry VIII. In 1534, after he decided to end his marriage and try again, and after the pope earlier refused (1527) to grant him an annulment, Henry made his famous break with Rome and declared himself head of the Church of England. Many consequences followed, but the one of interest here was the significant shift in property rights. Henry seized the church's land and sold it to the growing class of merchants, thus unleashing the power of capitalism as the merchants then aggressively pursued profits by, among other things, developing the coal mines on that land. Domestic coal use surged. By 1600, coal had become Britain's main source of both fuel and pollution, especially in London, where the smoke was thick and nasty. By 1700, coal mining grew into a large industry and Britain was mining more coal than the rest of the world combined.

The rise in coal consumption soon depleted the readily available surface coal and so forced the miners to dig deeper into the earth where poisonous and flammable gases waited along with collapsing walls, making mining one of the more, if not the most, dangerous professions. But of relevance here, water was also waiting—more so in Cornwall than, say, Wales where open-pit mining could be done high on a mountainside[2]—and the path to "deeper" was water removal. If the owners were fortunate, their mines were located near falling water or high winds where the associated energy could be tapped for water removal. But for most, either manpower or horsepower was needed to bucket-out the water. The larger mines might have up to sixty horses working round-the-clock to lift the water out. The high costs of feeding and caring for the horses made the economic incentive to find a lower-cost option very high.

So the value proposition was set. If you could offer a cheaper way to pump water out from deep mines, cheaper than horses, you would be given the job. And it was this opportunity that drew the strong interest of those seeking financial reward. Many ideas were put forth. Many ideas failed. And now we continue with the technical history, picking up where Papin left off.

To Summarize

In Britain and especially England, prior to 1700, coal could be mined near the Earth's surface and was thus relatively easy to access. But as the Industrial Revolution commenced, the

---

[2] (Hosken, 2011) p. 67.

demand for coal rose rapidly, forcing mines deeper and deeper as the supply of surface coal declined. With depth came water and any means to remove the water enabled more depth and so was eagerly sought. The advent of the steam engine was just such a means. To those who could provide a safe and reliable engine to pump out such mines without having to consume a lot of the coal to do so would go significant financial rewards. With the rules of the free market set, progress happened rapidly.

Selecting the Path from the Steam Engine to Sadi Carnot

As previously discussed in Chapter 16, the theories Sadi Carnot developed in his groundbreaking 1824 *Reflections* were inspired by the steam engines he saw coming out of Britain. While I only briefly touched on the related history of the steam engine in Chapter 16 as this was all that was needed, I delve much deeper now to bring to light the sources behind Carnot's theories. The challenge naturally is that the history of the steam engine in Britain fills volumes. As one author wrote, "The trouble with the steam engine was that its inventor was not one man, but several."[3] Furthermore, while several may have been the inventor, many more were strong contributors. Many men, many technical breakthroughs and improvements, many conflicting histories, many disputes, many unsung heroes were all involved. Given this, in the following chapters I narrow things down by telling the story along a selective timeline based on certain individuals. Should you be interested in more depth to this fascinating history, the books I reference would be excellent starting points.

---

[3] (Todd, 1967) p. 102.

The Newcomen engine

Before Newcomen came Savery's patent

The importance of the first major step was more conceptual than real. In 1698, Thomas Savery (1650–1715), a rather prolific inventor of his day,[1] patented one of the first concepts for using steam to pump water. Called *The Miner's Friend* by Savery himself, called Savery's Engine by historians, this engine's simple (no moving parts) two-step operation centered on a large metal cylinder (no piston) that opened into a vertical pipe with valves above and below the connection. In the first step, with the bottom valve open to connect the cylinder to the downward pipe situated in standing water, the boiler steam valve was opened to flow steam into the cylinder and then close again to close the cylinder, at which point an externally applied water spray was used to cool the cylinder and condense the steam. The resulting vacuum drew the water up the pipe and into the cylinder, the driving force being ambient air at atmospheric pressure. This process was limited to a maximum length of the distance between the top of the water and the cylinder; it had to be somewhat less than Torricelli's limit of 32 feet. In the second step, the down-valve was closed and the up-valve opened, enabling boiler steam to again flow into cylinder and push the water out of the cylinder and up the pipe to a height dependent on the steam's pressure.

The commercial success of Savery's invention is open to question. While some engines were indeed built and operated, the numbers were small and of uncertain success.[2] Among other difficulties of fulfilling on the design's intent, such as the fact that if the mine were deeper than 32 feet then the engine had to be located below ground where operation would have been both difficult and dangerous,[3] the boiler had to generate pressurized steam in order to push the water up the pipe. Attempts to do so were thwarted by the boiler's bad habit of blowing open its joints. "Had Savery been content to restrict his engine's operations to the drawing of water and not attempted to raise it further by steam pressure he might have been much more successful."[4]

Despite these challenges and limitations, Savery's simple, practical design did receive a patent and in the highly competitive free market, that meant a lot. Savery fought hard for this patent and was well-rewarded because it broadly covered *all* water-pumping engines based on fire and also because, although the original grant was for 14 years (1698–1712), an Act of Parliament in 1699 extended it another 21 years to 1733. It was thought that the government

---

[1] (Rolt and Allen, 1977) p. 26.
[2] (Cardwell, 1971) p. 13.
[3] (Hosken, 2011) p. 20.
[4] (Hosken, 2011) p. 21.

Block by Block: The Historical and Theoretical Foundations of Thermodynamics,
Robert T. Hanlon, Oxford University Press (2020). © Robert T. Hanlon.
DOI: 10.1093/oso/9780198851547.001.0001

passed this Act because they knew of Savery's difficulty in getting the engine to successfully work.[5] As we will see later in greater detail, while he may not have sold many engines, his patent coverage provided him an income since during this time all improvements to the "fire engine" had to go through him. Ironically, the future technical development of the steam engine followed a different path, one previously demonstrated by Guericke and further developed by Papin. Which brings us to Thomas Newcomen (1663–1729).

*   *   *

[The Newcomen engine] belongs to that small but select group of inventions that have decisively changed the course of history."

– D. S. L. Cardwell[6]

To the whole world, [Newcomen] gave the art of converting fuel into useful power

– Rolt and Allen[7]

To the small scientific world of London whose centre was the Royal Society, it seemed inconceivable that such a man could excel the brightest intellects of the day in the power of invention.

– Rolt and Allen[8]

Little is known of his personal history. He had no university degree. He was a practical tradesman, an ironmonger who made and sold tools and hardware to the miners. We can only guess how he came up with his design or even when he started building it.

While many simply couldn't come to accept that such a man could, on his own, do what he did when so many others couldn't, the fact is that Thomas Newcomen (1664–1729) did it. In 1712, Newcomen established the first truly successful engine capable of raising water from deep mines. Its arrival was sudden, seemingly from nowhere, mature-on-arrival with most all of the key problems already worked out. There it was, fully formed, entirely new to industry, "an immense advance upon anything which had gone before."[9] Yes, others such as Guericke, Savery, and Papin worked on the pieces, and yes, Newcomen was aware to some extent or the other of those pieces, although the real extent is not fully known, but, in the end, Newcomen owed little or nothing to them.[10] Newcomen had combined the basic major components developed by those before him into a mature, well-thought-out commercial engine.

This was no accident. It wasn't like Newcomen was a nobody who came from nowhere. Most certainly not. As an ironmonger from Devon, England, he frequented the mines of Devon and Cornwall and saw the challenges of using horses to remove the water. He was also familiar with metalworking and the competence levels of the craftsmen of the day. As a hard-working,

[5] (Rolt and Allen, 1977) p. 39. "Despite the claims of the *Miner's Friend* it must soon have become apparent that Savery's engine had no practical value in mining and that the purpose of the extension of the patent by Act of Parliament in order to enable him to further improve the device was not likely to be fulfilled."
[6] (Cardwell, 1971) p. 17.
[7] (Rolt and Allen, 1977) p. 11.
[8] (Rolt and Allen, 1977) p. 12.
[9] (Rolt and Allen, 1977) p. 38.
[10] (Rolt and Allen, 1977) p. 38.

thinking, and practical man with an entrepreneurial bent, Newcomen saw a problem that needed to be solved, and slowly, over at least 10 years in his little Dartmouth workshop, developed a solution. Unfortunately, historians know little of the path Newcomen took, and so we are left to wonder and, at times, conjecture.[11]

But this doesn't change the fact that he was the first to solve the problem. His solution was real. His engine was a huge hit with the coal mine operators, enabling them to raise the water from much lower depths and thus enabling them to release the horses. While it wasn't efficient, it worked and created value.

So how did it work? Newcomen based his advanced engine design on Huygens' piston-in-cylinder and on Papin's condensing (atmospheric) steam concepts, thus avoiding the dangers associated with Savery's high-pressure steam design (Figure 27.1). The implementation of the piston was a significant technical hurdle in and of itself, especially concerning the need for reliable seals. The condensation process was also a hurdle. Spraying water on the external wall of the cylinder worked but provided relatively slow condensation. So Newcomen redesigned the process to spray the water directly into the cylinder itself. The much faster condensation created a much faster vacuum, resulting in a much faster pumping action as the atmosphere pushed the piston down into the resulting vacuum inside the cylinder. The descending piston pulled down one end of a large beam set on a central pivot, thereby lifting the other end, which was connected to a rod penetrating deep into the mine shaft so that it could lift water out. The reverse motion was caused by heaviness of the mine-end of the beam, which served as a counterweight, thus pulling the piston back out and drawing steam back into the cylinder. While both processes condensed steam to create a vacuum, Savery's did so to draw water up, while Newcomen's did so to move a piston with a lifting force equal to one atmosphere of pressure times the area of the piston head. Since this force could be used to physically lift the water rather than drawing it up, it could operate on water below Savery's 32-foot limited depth—the initial maximum lift for a Newcomen engine was up nearly 130 feet.[12]

Let's return to Newcomen's idea for the direct injection of water. While this may seem obvious to us today, it wasn't so back then. In Papin's design concept, the cylinder cooled down naturally. In Savery's design, it cooled down with an external spray. Neither worked particularly well. With no science guiding him, Newcomen himself started with Savery's approach. But then serendipity struck and "Almighty God...allowed something very special to occur."[13] While operating his prototype, a leak opened up between the external cooling jacket and the interior of the cylinder. The cold water forced itself into the cylinder and immediately condensed the steam and created a vacuum. The piston slammed down into the cylinder with such force that it damaged the equipment. Such is the way discovery sometimes happens. Newcomen's subsequent incorporation of this design change was a significant advance, helping to increase the speed at which his engine operated.

---

[11] (Rolt and Allen, 1977) p. 36. "Because of the lack of solid facts about the birth of Newcomen's great invention, his biographer can do no more than choose the most likely and credible path through the maze of legend and conjecture which earlier writers have created."

[12] (Rolt and Allen, 1977) p. 104.

[13] (Rolt and Allen, 1977) p. 42. Marten Triewald's account based on Newcomen's own words.

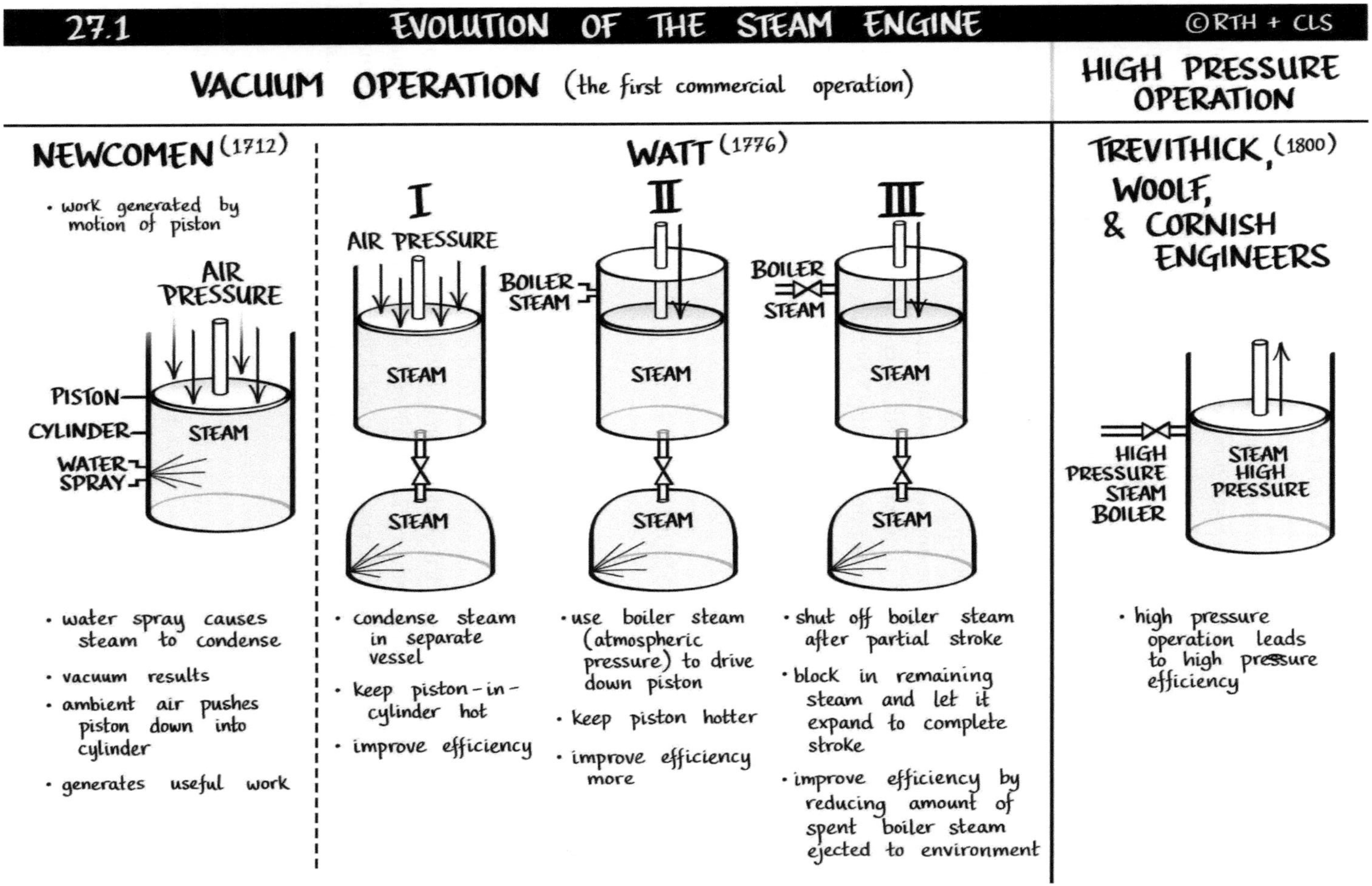

Figure 27.1 Evolution of the steam engine

That Newcomen's engine[14] was economically favorable and highly sought after by the mining companies was rather telling, because its *maximum* thermal efficiency was only about 19% based on a 100 °C heat source (steam temperature at atmospheric pressure) and a 30 °C heat sink and its actual thermal efficiency was estimated to have been around 4%.[15] These numbers meant that the engine's value to the miners had to be large to overcome the fact that it consumed a lot of coal for the amount of water it pumped. But the critical question was not absolute efficiency but relative efficiency. The only other water-removal means enabling the mining industry to dig deep were horses, whose maintenance costs were high as reflected by the fact that many mines switched over to Newcomen's engine. The safe, reliable, and relatively simple Newcomen engine won this contest and became the first practical design that employed (condensing) steam to effect work and thus "ushered in a period of British technological supremacy that was to last about 150 years."[16]

Despite the inefficiency of Newcomen's engine, it found success in the Midlands where coal was relatively inexpensive. But it was a tougher sell in the strong metal-mining region of Cornwall where coal had to be imported, making the consumption of coal rather costly. However, with the help of his connection with Joseph Hornblower, whom Newcomen met through their shared Baptist community, the Newcomen engine was also able to find success—very good success—in Cornwall. This effort signified the start of a series of Hornblower family involvements with the pioneering development of the steam engine in England.

The Hornblowers – a Short Overview

To put the Shropshire Hornblower family into quick perspective regarding their famous involvement in the rise of the steam engine, Joseph (1696? – 1762) erected the first Newcomen engine in Cornwall. His son, Jonathan (1717–1780), continued in his father's footsteps by himself becoming a well-known engine builder, while his other son, Josiah (1729–1809) was responsible for introducing the Newcomen engine to the New World, specifically at Colonel John Schuyler's copper mine in what would become New Jersey, starting the engine up in 1755. Jonathan then had two sons, Jonathan (1753–1815) and Jabez (1744–1814), who got entangled with Boulton & Watt's patents. We met this second Jonathan, whom I'll call Jonathan #2, in Chapter 16 regarding the indicator diagram. He is the focus of additional discussion in Chapter 28.

---

[14] In a 2017 email to the author, Phil Hosken shared that Newcomen's and later Watt's engines were called "fire engines" as opposed to "steam engines" since it was the fire that was known and seen to be doing the work, not the steam. Steam only became apparent when Trevithick built his "puffers."

[15] (Carr, 2013) Carr estimated an overall efficiency based on work done per coal energy supplied of less than 2%. Also, (Lewis, 2007) p. 98: "In the 1790s measurements made by Watt's friend George Lee of the heat consumed in the furnace compared with the heat recovered in the condenser of a steam engine had revealed no detectable difference; in retrospect, this was understandable, since the efficiency of engines of the day was of the order of 2 percent." Such lower efficiency numbers likely contributed to the belief that heat or "caloric" was conserved rather than consumed in a heat engine.

[16] (Cardwell, 1971) p. 15.

The Newcomen and Savery Partnership

The commercial success of the Newcomen engine was aided, interestingly enough, by Savery's patent. Newcomen realized that, even though his design was totally different from Savery's and thus very patentable on its own, he would still have to go through Savery to operate it. Savery's patent was that strong. In the end though, this turned out not to be a bad thing, for it brought Newcomen and Savery together. Historians felt that Newcomen sought a win–win partnership with Savery. In this way, he avoided the high costs of obtaining his own patent while also benefiting from Savery's very favorable patent extension. In effect, his engine was sheltered by Savery's patent. Savery's reward was a percentage of the future revenue that would be generated by Newcomen's advanced design.

Savery died in 1715, thus ending their partnership, while Newcomen died in 1729. Before the expiration of Savery's patent in 1733, at least 100 Newcomen engines were built throughout England and the European continent and at least 1,500 engines total were built during the eighteenth century, continuing to dominate the landscape, even in the face of strong competition from James Watt's new design. The fact that Newcomen, "the first great mechanical engineer,"[17] himself received no honors in his lifetime and only limited posthumous recognition can be attributed to several causes. First was the fact that his engine fell under the umbrella of Savery's patent, thus confusing people as to whose engine it really was, with some discussions or publications using the rather misleading term "Savery's Engine." Second was the fact that Watt's success later in the century became so strong and dominant that it overshadowed Newcomen's groundbreaking work. Finally was the fact that, as we saw to some extent with the history of Mayer and Joule in Part III, many simply couldn't fathom that such a practical man with no formal scientific training could possibly have done what he did when the true intellectuals didn't. "The majority ignored Newcomen and his engine completely, while the few who deigned to mention his achievement did their best to belittle it and to explain it away…Anything rather than admit that so potent a machine could be the child of Newcomen's own genius and unwearying perseverance."[18] This was a classic case where the arrogance and contempt of the elites was directed toward the practitioners.

While the low efficiency of the Newcomen design wasn't immediately apparent, especially as there was no alternative by which it could be judged and since the fuel itself was cheap, being available right there at the mine itself, it did become an issue over time. As suggested previously, one cause for the low efficiency was the fact that the steam temperature in the engine only got up to 100 °C versus the greater than 1000 °C temperature of the coal flame; this issue will be discussed in greater detail later. The other cause was the cycling of the main cylinder between hot and cold. The continual cool-down and heat-up of the single main process vessel unnecessarily consumed a lot of energy. The fact that it took a long time to address such inefficiencies reflects the patent strength of Newcomen's position. But that couldn't last forever.

---

[17] (Rolt and Allen, 1977) p. 13.
[18] (Rolt and Allen, 1977) p. 12.

The Increasing Relevance of Engine Efficiency

Mining was the most important growth industry throughout Europe in the early to mid-1700s and the Newcomen engine the most important technological innovation. Inextricably linked, each enabled growth of the other. The steam engine provided access to the coal and metals buried deep underground, and the coal, in turn, fueled the steam engine's operation, both in the mines and also in the newly forming mechanized factories. Such engines became significant sources of power, beyond what was previously available with wood, water, wind, and horses. Coal also became a critical feed stock component to the iron industry, being used to help chemically reduce the iron ore to pure iron. Steam, iron, and coal developed a mutually reinforcing relationship between themselves, which served to power England's Industrial Revolution later in the eighteenth century, providing Britain a good half-century head start in industrialization over other countries, a lead they would hold for a long time, thus earning them the name, "Workshop of the world."[19] I'm getting a little ahead of things here, but I wanted to emphasize the importance of Newcomen to the start of this revolution.

Over time, mining's favorable economics provided a strong driving-force to improve Newcomen's design. One of the main focal points concerned overall engine efficiency. With the rising cost of coal itself, its use as the fuel for the engine became a target for the cost-cutting engineers. Well rewarded were the systemic improvements by the likes of such engineers as the famed John Smeaton (1724–1792), arguably the greatest of Newcomen's successors, an early continuous improvement engineer who conducted, perhaps for the first time ever, efficiency surveys of a range of operating engines, and then identified and implemented improvements such as the use of insulation to Newcomen's design that just about doubled the engine efficiency (work done per bushel of coal consumed). But even with such improvements, Newcomen's engine was destined to be surpassed, all because of two very fundamental inefficiencies of the design: low-pressure operation and temperature cycling.

Into this context arrived James Watt.

---

[19] (Freese, 2003) p. 69.

28 James Watt

The whole thing suddenly became arranged in my mind.
– James Watt[1]

James Watt (1736–1819) (Chapter 16) was one in a long list of scientists from the famed Scottish school (Chapter 14) who contributed significantly to the science of heat, the technology of power, and ultimately to the establishment of thermodynamics.[2] For Watt, this environment within "one of the most active scientific groups in the world"[3] provided strong cultural encouragement. As a young instrument maker at Glasgow University during Joseph Black's time as distinguished professor, Watt was initially exposed, perhaps through discussions with Black, perhaps through other discussions, or perhaps through his own reading, to the concept of using steam to generate power and had dabbled some, as a hands-on inventor would, in considering ways to use pressurized steam, as opposed to vacuum-inducing, condensing steam, to develop power. But the difficulties of constructing a boiler to withstand the pressures, a difficulty that went all the way back to the Savery engine, could not be overcome. As we'll see, this obstacle would influence the course of technology development as it tainted Watt's thinking, resulting in his outright rejection of any steam engine involving steam pressures greater than atmospheric. It's not that Watt didn't attempt to work with high-pressure steam because he did, despite the danger. It's just that he couldn't design and fabricate the equipment to make it work. "He gave up strong steam because it beat him."[4]

The future of England was destined to change as a result of an assignment Watt took on during the winter of 1763–4. He was asked to repair a working, small-scale model of the Newcomen engine that was used in one of the university's natural philosophy classes. His tinkering, roll-up-the-sleeves, entrepreneurial mind went to work when he observed that the model simply wasn't doing what it should be doing based on the reported performance of the large-scale commercial unit.

---

[1] In yet another, "as legend has it" moment in history, it has been reported that Watt's sudden insight occurred while walking through Glasgow Green Park on a Sunday in May, 1765, lost in thought, struggling with how to improve the steam engine that he was working on. "The whole thing suddenly became arranged in my mind." Watt had a flash of inspiration, the idea of a separate condenser that would set the Industrial Revolution in motion. Today that inspirational spot is marked with a boulder. This story was reported in (Dickinson and Jenkins, 1927) p. 23 based on a statement Watt made to Robert Hart, an engineer in Glasgow in 1813 and reported in a publication (mentioned) in 1859. In a footnote, Dickinson and Jenkins state that they could find no record of this event in Watt's papers.

[2] (Cardwell, 1971) p. 33.

[3] (Cardwell, 1971) p. 54.

[4] From a 2017 email from Phil Hosken to the author.

Block by Block: The Historical and Theoretical Foundations of Thermodynamics,
Robert T. Hanlon, Oxford University Press (2020). © Robert T. Hanlon.
DOI: 10.1093/oso/9780198851547.001.0001

Most likely inspired by ideas of Herman Boerhaave (1668–1738) regarding the effect of scale on steam engine efficiency,[5] Watt put his finger on the root cause of this problem. The inefficient design involving the cylinder's cycling between high and low temperatures became more prominent at the smaller scale where the cylinder's surface area (and mass, as historians determined later) to volume ratio was higher. In a way, it was fortuitous that the Newcomen engine's inherent inefficiency was exaggerated by the small-scale model Watt was working on. If it hadn't been, Watt's curiosity may never have been piqued and history would have taken a different turn.[6]

Engine Efficiency – Moving the Condenser out of Newcomen's One-Cylinder Operation

At this point, it was only a matter of time until Watt identified a major breakthrough. After much contemplation and experimentation, the latter being needed to quantify heat losses associated with scale, which at that time was not trivial, Watt experienced his own eureka moment during his famed walk through Glasgow Green Park in May 1765. Legend has it that during this walk, Watt realized that he could figuratively split the single Newcomen cylinder into two; one could be kept hot at all times, the other cold. Such a seemingly simple solution, but only from our eyes today. We were raised to understand the concepts of heat, energy, and the concept of thermal efficiency, but Watt wasn't. It was still over 85 years until the concept of energy would be firmly established. Just think. Newcomen's design of 1712 lasted over 50 years until Watt's tremendous insight. Watt worked this out in his head, with just his clear thinking and his supportive experiments, and then completed a reliable and efficient design that succeeded in earning Watt the necessary business and financial backing to take his engine commercial. (Figure 27.1)

Continuous Improvement

But the cylinder-plus-separate-condenser approach was not Watt's only breakthrough. The second major breakthrough, building on his increasing understanding of thermal efficiency, regarded the fact that ambient air in Newcomen's engine was what drove the piston down into the cylinder. When the piston was pulled out by the action of the descending pump rod at the other end of the beam, steam flowed into the cylinder, after which a pipe connecting the cylinder to the condenser was opened. The resulting access of the hot steam to the cold condenser caused the steam to rush from hot to cold, where it condensed, resulting in a vacuum in both vessels. When this vacuum formed, ambient air pushed down the piston which raised the pump rod; the raising either lifted water out or, when subsequently lowered, pushed water up.[7] Watt realized that contacting ambient air with the hot piston and actually with the increasing

---

[5] (Cardwell, 1971) p. 43.
[6] (Rolt and Allen, 1977) p. 40.
[7] (Carr, 2013)

surface area of the inside of the cylinder as the piston was pushed in further and further, led to the same problem that he had just solved. Realizing that when high and low temperatures touch, thermal inefficiencies result, an insight that Carnot would latch onto some 50 years later, Watt made a design change to replace air as the driving force with atmospheric steam available from the boiler. It was at this moment that steam itself became the driving force of piston-in-cylinder work. But note that in this configuration, the steam acted on the outside of the piston and pushed the piston in. It wouldn't be until the arrival of a pressurized system (P > 0 psig) that steam would act on the inside and push the piston out.

He wasn't through. Watt realized that the sudden rush of steam from the cylinder to the condenser led to additional inefficiencies, since one could conceivably use the rush to drive a small turbine and thus do additional work. Much as allowing hot and cold to directly contact each other was inefficient, so too was allowing high pressure to directly open to low pressure. Again, this was the leading edge of thinking that would culminate in the creation of Carnot's reversible process concept and, later, entropy. Gradients in temperature, pressure, or chemical potential offer the opportunity to do work; a system in equilibrium without gradients does not. So Watt added another design change. Part way through the steam's downward push of the piston, the boiler steam supply was shut off while the blocked-in steam continued to push down the piston into the vacuum. Watt designed the timing of the shut-off such that the final steam pressure at the end of the push would just be slightly above the residual pressure in the condenser. While the power delivered in this way was less than it would have been if the supply of steam had not been cut off, this was more than compensated for by the fuel savings that resulted from not having to discard a cylinder-full of atmospheric-pressure steam at the end of each stroke. In this way, the blocked-in steam expanded adiabatically and thus marked the first known application of the "expansive principle" (1769).[8]

Watt's Early Steps toward the Founding of Thermodynamics

From all of this work and all of these improvements, Watt became, without forethought or plan but instead by following his entrepreneurial mind, one of the founders of the science of thermodynamics. A similar process would be repeated years later by James Joule during his own unplanned journey toward the discovery of energy. Watt's instinctive steps in unknowingly setting the first stones into the foundation of thermodynamics led to a multitude of consequential jigsaw pieces that Sadi Carnot would eventually put together over 50 years later. Among these pieces were:

- the early concept of a reversible process in which gradients in either temperature or pressure (and later, chemical potential) are minimized
- the creation of the "indicator diagram" by John Southern, Watt's assistant (Chapter 16)
- the recognition of the need to maximize work for a given amount of heat

[8] (Carnot, 1986) As recounted by editor and translator Robert Fox, pp. 5–7, Watt himself made little use of this principle but others such as Jonathan Hornblower did.

- the emphasis on thermal efficiency by ensuring, "during the whole time the engine is at work…[that the steam cylinder]…be kept as hot as the steam which enters it"[9]
- invention of the "expansive principle" in 1769.[10]

Even the design concept of physically creating a stand-alone condenser contributed to the foundation, since by isolating it, he created a conceptual image of the "flow" of heat from a furnace to the condenser, an image that would later inadvertently draw the attention of curious theoretical scientists, such as Carnot. Indeed, Watt's isolated condenser and the accompanying *flow of heat* concept from hot to cold became an essential feature of the heat engine, something that couldn't be seen in either Newcomen's engine (because there was no condenser) or Trevithick's (because the condenser was eliminated). Watt's engine made visually clear the use of a hot source and a cold sink.

The fact that it would take another sixty years for someone, namely Sadi Carnot, to synthesize most of these puzzle pieces together is rather amazing and reflects how advanced Watt's thinking was. It also may have reflected, in part, Watt's lack of sharing his ideas through publications. One must recall that Watt's interest in all of his efforts wasn't the establishment of the science of thermodynamics. Instead, his interest, prompted by Mathew Boulton, was financial, and when one's in business to earn money, one doesn't share publicly. One only shares what one must to receive a patent, which is where most of Watt's "publishing" ended up. Many of his insights, and frankly many of Black's insights, didn't see the light of day in their own words. (Black preferred teaching to writing.) It took others to make known the works of both Black and Watt; our indirect access to their ideas most likely slowed the advancement of science as a result.

Similar to Galileo before him, Watt's achievements were due in part to the straddling of two worlds, the theoretical world of the Scottish school and the practical world of hands-on work with the Newcomen engine. With a strong knowledge of science and a strong capacity to put ideas into practice, he brought measurable, quantifiable improvements to his heat engine and in so doing also advanced the concept of using measurement as a means to drive improvement. The improvements he made over Smeaton's improved Newcomen engine were real and significant, nearly four times more efficient than Newcomen's, as reflected by the fact that people in the mining industry, and especially in Cornwall where power and economy were equally important, purchased his technology, making Watt's associated monetary rewards real. Needless to say, Watt's accomplishments played an important role during the early years of the Industrial Revolution covering the lifetime of his patent (1775–1800).

But Watt's success wasn't immediate. For one thing, the Newcomen engine had established a strong hold on the market, even though Savery's patent had expired. Watt didn't just have to be more efficient; he had to be significantly more efficient to justify, in some if not many cases, the replacement of an existing piece of capital equipment with a new one.

[9] (Cardwell, 1971) p. 55.
[10] (Cardwell, 1971) p. 52.

The Boulton–Watt partnership

But this wasn't the only challenge Watt faced. The entire scale-up process required much backing, both financial and technical, neither of which was a trivial undertaking. An idea may be great, but it doesn't mean anything until it's successfully scaled-up from the small-scale laboratory unit to the full-scale operating engine. For Watt, that final step was a big one. The financial backing came from several, including Black himself, John Roebuck, and eventually Matthew Boulton, a businessman who ultimately formed a long-term partnership with Watt.[11] The money went largely toward getting the business up off the ground, starting with securing a patent, which was indeed expensive but which paid out many times over as it gave them 25 years of protection from its 1775 award date until 1800. The technical challenges were equally challenging, especially with the machining of the cylinder and piston to ensure a tight seal. Helping significantly in this endeavor was Watt's access to some of the best ironworkers in the world through Boulton, who owned the Soho Foundry works near Birmingham. Some of this effort was outsourced to specialists, including John Wilkinson, famed industrialist, who was given exclusive rights to use his own advanced boring machine to manufacture the cylinder for Boulton & Watt's first commercial engine. "The importance to Boulton & Watt of the timely aid of Wilkinson's boring machine can hardly be overestimated. It made the steam engine a commercial success, and was probably the first metal-working tool capable of doing large, heavy work with anything like present-day accuracy."[12]

The Boulton and Watt partnership reaped success for those 25 patent-protected years.

*   *   *

The "fire engine" continued to evolve. In the original design, the speed of the piston being pushed in (by ambient air) was not the same as the speed of the piston being pulled back out (by the beam counterweight).[13] Two separate and unequal forces were involved. To address this, Watt ingeniously invented a double-acting cylinder, connecting each end to both the steam source and the condenser sink. In this way, steam at atmospheric pressure worked to both push the piston in and then push it back out again, significantly smoothing the operation and enabling the additional use of the fire engine to provide the smooth rotation needed in other industries, such as textiles.

*   *   *

Thanks to their respective strong patent positions, Savery & Newcomen and later Boulton & Watt dominated the business of steam power throughout the 1700s and effectively thwarted the interest of others to invent improvements.[14] Regarding Boulton & Watt, the reason for their dominance was simple and best explained by the situation in Cornwall. At that time, Cornwall was one of the wealthiest mining areas in the world, especially in the mining of non-ferrous metals (copper, tin, etc.), and needed fire engines to pump out the mines. Unfortunately, coal

[11] (Hosken, 2011) p. 31. "It would be Watt's name that would go down in history for his work on the improvement of the atmospheric steam engine but it was Boulton's ability to fund the development that made this possible."

[12] (Roe, 1916) p. 3.

[13] In a 2017 email to the author, Phil Hosken shared that the speed of the atmospheric stroke was very fast, modified only by the weight of the pump rod and water. If there was no water the engine could be damaged.

[14] (Cardwell, 1971) pp. 72–73. The best manifestation of this dominance is shown in Cardwell's figure on p. 158.

fuel was unattainable in Cornwall and had to be imported by sea, mainly from South Wales.[15] There was, therefore, every incentive to develop engines of the highest efficiency and greatest reliability. The economic driving force there would eventually translate into Cornwall's manufacturing some of the finest steam engines in the world during the eighteenth century. Since the Boulton & Watt engine offered close to four times[16] the efficiency of the best Newcomen/Smeaton engine, the dominance of this design was understandable. But such dominance was not welcomed. As you can guess, Boulton & Watt had pretty much cornered the market in Cornwall. They offered far-and-away the best technology and so could demand a steep price—a percentage of the savings in coal consumption on an annual basis for 25 years (!)—while also preventing anyone else from encroaching on their patented territory, even going so far as to fight against incremental improvements to their design. Their domination won no friends in Cornwall, which leads us to return to Jonathan Hornblower.

Jonathan (#2) Hornblower

Jonathan Hornblower[17] (1753–1815), whom I referred to in Chapter 27 as Jonathan #2, son of Jonathan and grandson of Joseph, was one of those in Cornwall who fought the legal battle against Boulton & Watt. As recounted in Chapter 16, working in Cornwall, Hornblower patented a compound engine in 1781 involving a two-stage process using high-pressure steam to drive a piston down into one cylinder and then using the low-pressure exhaust steam to drive down a piston into another cylinder. Even though Watt had envisioned "expansive operation" as early as 1769, Hornblower, having independently arrived at the same idea but at a later date, embedded the concept into his patented compound engine. You can see how the lawsuit emerged from this. To be effective, Hornblower's compound engine had to operate at a reasonably high pressure, which in 1781 was hardly feasible, or at a reasonably low vacuum, which meant that he'd have to use Watt's condenser to increase utilization of the steam's condensing energy. He was somewhat trapped between the two options. "Hornblower, for all his gifts, was hamstrung [in part] by Watt's patents."[18] As discussed in Chapter 16, in the ensuing lawsuit Davies Gilbert was called to analyze the expansive principle theoretically in 1792 while John Southern gave it form in 1796 by inventing an instrument to literally capture the principle in his "indicator diagram." I won't go further into these legal issues here.[19]

Although Hornblower's lawsuit with and others' challenges to Boulton & Watt would become a moot point, the issues they dealt with set the context for what would come next. The year 1800 was fast approaching and this would be the year when Watt's patents would finally lapse. Many in Cornwall eagerly awaited this event, for it would open their doors to start building whatever engines they wanted to, being limited only by their own minds and the

[15] The iron needed in Cornwall to manufacture the engines was also imported from South Wales.
[16] (Cardwell, 1971) p. 158. Figure 12.
[17] (Cardwell, 1971) p. 78.
[18] (Cardwell, 1971) p. 79.
[19] (Todd, 1967) p. 94. "The tragedy of Hornblower was that his intellectual organization was never quite adequate for the demands made upon it by his curiosity, and by a fanatical wish to lead the attack against Watt."

availability of equipment that could withstand the new conditions their minds sought. But before that year arrived, there was a key pivot point about which the transition from Watt to Cornwall turned. It was the paradigm-shifting moment when the steam engine transformed from low to high pressure—*how to control the devil?*[20]—and it started in 1797 with Richard Trevithick (1771–1833).

[20] (Hosken, 2011) pp. 16–17. "Before Trevithick showed how this amazing gas [steam] could be controlled it appeared to have a will of its own, something that God-fearing people viewed with foreboding. In many cases the mysterious activities within a vessel and pipework designed to hold steam would frighten the public as they saw the gas escaping in an ethereal cloud. It is no coincidence that the vessels were believed to hold the Devil and that Trevithick's first steam locomotive was described by a woman in Camborne as a 'Puffing Devil.'"

29 Trevithick, Woolf, and high-pressure steam

[Trevithick's] genius left us with the control of high-pressure fluids that is still used extensively in industry and in all nuclear- and fossil-fuelled electricity power stations. However there were those, including some fellow Cornishmen, who were unable to come to terms with his outstanding achievement, and insisted upon crediting it elsewhere. While they appreciated the outcome of his work they were loath to acknowledge that it was from a man of humble beginnings who lacked their education and social standing. Trevithick may not have had grace and refinement, but his brain and determination provided the world with the technology it had been seeking for over a thousand years. The sad result is that historians and raconteurs have found it in their interests to ridicule Trevithick for his failures rather than to bless and thank him for his achievements.

– Philip Hosken[1]

It was only a matter of time until someone considered that if steam pressure helps drive the piston, then higher steam pressure should help drive the piston with more force. Recall that Watt's design depended on the creation of a vacuum by condensing steam in a vessel, known as a "hot well," separate from the piston-in-cylinder assembly. Boiler steam slightly above atmospheric pressure flowed into the other side of the piston and then pushed the piston down into the vacuum in the cylinder. This action pulled up the pump rod at the opposite end of the beam; when the vacuum inside the cylinder was broken, the heavy pump rod lowered back down and so pulled the piston back out again. The use of steam in this action, along with the use of a steam jacket around the cylinder, kept the cylinder hot during the pumping operation and thus improved thermal efficiency. But in Watt's design the steam's pressure offered no real advantage, as it was the same as that of the ambient air (one atmosphere) and actually averaged less than this as it decreased during expansive operation once the valve to the boiler was closed. Thus, purely from a pressure standpoint, steam offered less advantage than the earlier design based on (free) ambient air, which pushed down with constant atmospheric pressure throughout the stroke (but which also unfavorably cooled the cylinder). But Watt was driven by a different incentive: increased efficiency by keeping the cylinder hot. This was really why he favored the use of steam to drive the piston.

Watt, being by nature more cautious than adventurous, especially in light of early and dangerous commercial failures, and also being very singularly focused on commercial success, inspired by Boulton in this regard, was reluctant to stray into the new realm of pressurized steam. His reluctance became a tremendous barrier to progress, both his and anyone else's,

---

[1] (Hosken, 2013) From the Author's Note.

Block by Block: The Historical and Theoretical Foundations of Thermodynamics,
Robert T. Hanlon, Oxford University Press (2020). © Robert T. Hanlon.
DOI: 10.1093/oso/9780198851547.001.0001

since even a high-pressure engine would have infringed on his patent as it covered all engines that used steam.

Others were not so reluctant. Among other incentives, they viewed high pressure as a path to shrink and even eliminate vessels. With the above context set, we come to Richard Trevithick.

Cornwall – "the Cradle of the Steam Engine"

Richard Trevithick (1771–1833) was an enthusiastic, energetic, and intuitive man who grew up in the world of mining, being born to two mining parents, his father a mine captain (engineering manager) and his mother a miner's daughter, in the midst of a strong mining section of Cornwall. He experienced mining as a child, being witness to the use of fire engines to pump out the mines. But his interest lay in a different direction: he was intrigued by the idea of using steam to replace the role of the horse in transportation.[2] This would require a much smaller engine than had been used to date and the path toward "small" was high pressure. While Trevithick was not the first to consider "strong steam" in light of Watt's and Hornblower's (unsuccessful) previous efforts, he was the first to make it work, drawing on such early experiences and later ones involving, for example, his own study of high-pressure water engines powered by high local falls.[3]

At 26 years of age, a young Trevithick became engineer at Cornwall's Ding Dong Mine in 1797. There he met Edward Bull, who had started constructing engines for Boulton & Watt back in 1781. Bull's own history had already led him into commercial conflict with Boulton & Watt. Challenges from him along with others from Cornwall against Boulton & Watt's dominance led to a contentious time of many skirmishes and lawsuits, with much money and time being spent in the court room. While Boulton & Watt prevailed in most cases, it certainly didn't stop the Cornish inventors from trying, especially with the year 1800 fast approaching. It was into this environment that Trevithick arrived in 1797, a young buck of high energy and a willingness and entrepreneurial aggressiveness to break from the past and develop a whole new steam engine (hence the comment by (Harris, 1966)[4] used as the title to this section).

"Trevithick Created the Engine that was Destined to Drive the Industrial and Transport Revolutions"

Captain Trevithick informed me that the idea of the high-pressure engine occurred to him suddenly one day whilst at breakfast, and that before dinner-time he had the drawing complete, on which the first steam-carriage was constructed

– James M. Gerard[5]

---

[2] (Hosken, 2011) p. 153. "[Trevithick] saw horses as nasty, spiteful things that required to be fed even when they were not at work."

[3] (Todd, 1967) p. 79. Trevithick's water-pressure engine at Wheal Druid used a column of water brought down Carn Brea Hill.

[4] (Harris, 1966) p. 8.

[5] Gerard quote cited in (Hosken, 2011) p. 136. James M. Gerard as told to Thomas Edmunds.

While the use of high-pressure steam had been bandied about by the engineers in England and had an experimental history of failures and modest successes, its possibility took on a new light in Trevithick's eyes as he turned toward transportation. As shared by Trevithick biographer Philip Hosken,[6] prior to the above fateful breakfast, Trevithick must have been thinking for months, maybe years, about how to make high pressure work. And in a sudden moment, all of his thoughts and ideas must have crystallized into a "technology beyond the comprehension of any man, other than Trevithick." His design of a safe pressurized operating system was much more than a redesign of the piston-in-cylinder assembly; it led to the appearance of new ancillary equipment—cylindrical boiler, internal fire tube, boiler recharging pump, valve gear— together with a layout that enabled the wheels to turn.[7] Nothing had ever been built anything like it before.[8]

During this design process, as Trevithick considered ways to reduce size and eliminate vessels, he arrived at a question. *Just how critical was Watt's condenser?*

Trevithick felt that if he could use high-pressure steam to drive the piston and then release the exhausted steam to the atmosphere, thus creating the "puffs" of steam and noise that would mark his "Puffin' devil" engine design, then he could eliminate the condenser so long as the associated cost—in this scenario the vacuum would no longer contribute to the pressure driving force—wasn't high. To help with this technical question, he turned to his engineering mentor, Davies Gilbert,[9] the same who provided technical support to Hornblower in Chapter 16. Gilbert told Trevithick that the exhaust of atmospheric-pressure steam would have only a small impact on the overall engine performance. He later recalled, "I never saw a Man more delighted."[10]

This was all Trevithick needed to hear. He then set off with youthful energy and enthusiasm, confident that his design—with no condenser—together with Cornwall's fabrication community would lead to success. His confidence was justified when in 1799 he successfully commercialized the first practical high-pressure "puffer" engine in Cornwall. While his initial motivation was replacing the horse in transportation, it didn't take long until the opportunity to use Trevithick's advanced engine to replace Watt's in mining became apparent.[11]

Watt fought hard against Trevithick's move to pressurized steam, using his reputation and influence to publically challenge its use as being too dangerous to be practical, even going so

---

[6] From a 2017 email to the author from Phil Hosken, which also provided the title for this section, for which I thank him.

[7] (Hosken, 2011) p. 150. "Trevithick's high-pressure boiler, which subsequently became known as the Cornish boiler, was essential to the future development of the steam engine."

[8] (Hosken, 2011) p. 148–156. Trevithick designed a full-sized road locomotive in 1801 that included the riveted construction in wrought iron of a furnace tube with a 180° bend that was fire, water, and pressure proof.

[9] Davies Gilbert (1767–1839) was a well-educated (studied mathematics at Oxford) civil servant who played an active role in supporting both Jonathan (#2) Hornblower and Richard Trevithick. Born Davies Giddy, he changed his name in 1816 to Davies Gilbert.

[10] (Todd, 1967) p. 83.

[11] In a 2017 email to the author, Phil Hosken shared that compared to large and expensive Watt engines (600–800 tons with the granite engine house, chimney, etc.) the Trevithick puffers, at less than 5 tons, were versatile and offered a considerable reduction in capital cost. They were built in a manner that could be assembled in a variety of ways so that they could be readily adapted for uses, so extending their working lives. So high were the capital costs of Watt engines, plus the contract charges during their lives, that many operators bought Newcomen engines long after Watt finished production. A maximum of 500 Watt engines were built whilst Newcomen exceeded 2000.

far as to suggest, as claimed by Trevithick, that Trevithick "deserved hanging for bringing into use the high-pressure engine."[12] Watt clearly thought that high pressure was too dangerous to harness (and too threatening to his business);[13] Trevithick thought otherwise. With the courage of his convictions, Trevithick proceeded and succeeded. His engine's eventual demonstration of superiority over the Boulton & Watt engine led to public support and it becoming a familiar feature in the Cornish mines. As this rapid development occurred toward the end of the life of Boulton and Watt's patents, this discussion became a moot point. In 1800, Watt left Boulton, left Cornwall, and retired.[14]

The arrival of compact high-pressure steam engines for both transportation and mining, encouraged by Trevithick and many others in Cornwall, truly changed the way things happened in England and throughout the world. As Cardwell stated, "Of Trevithick we may say that he was one of the main sources of British wealth and power during the nineteenth century. Although he was rather erratic in his personal relationships, few men have deserved better of their fellow-countrymen, or indeed of the world."[15]

Arthur Woolf – Failure in London led to Success in Cornwall

"His name is associated with the improvements in the drainage of the Cornish mines"
– from Woolf's death notice in 1837[16]

In the meantime, Arthur Woolf (1766–1837) was also experimenting with higher pressures. Born and raised in Cornwall, where he likely witnessed the "fire engine" in action, Woolf eventually left for London to seek better job opportunities, landing one with Joseph Bramah and his engineering works. With the metal working skills he learned and practiced, he developed into a reputable engineer and was hired by the firm of Hornblower[17] and Maberly[18] to help with the installation of a small rotative steam engine (not the compound type developed by Jonathan #2) in the Griffin Brewery. After the 1796 installation,[19] the brewery hired Woolf on as resident engineer and he continued to work there until 1805. It was during this period that Trevithick happened to be in London with his steam carriage and Woolf called to see it.[20] It was also

---

[12] (Todd, 1967) p. 107. Recollection of Trevithick in letter to Davies Gilbert.

[13] (Hosken, 2011) p. 71–72. Watt did not stand in the way of Trevithick other than to enforce his patent...[He] did not want to be a personal problem to Trevithick. The interests of the two men barely overlapped. Boulton and Watt's denunciations of [Trevithick's] high-pressure steam were the ploys of their marketing strategy."

[14] (Hosken, 2011) p. 71–72. "Watt's position was that of an unhappy innovator...[who was] out of his depth in dealing with Cornish mine owners...He could barely wait for his patent to expire so he could leave Cornwall...and when the day of Watt's release arrived he fled and never returned."

[15] (Cardwell, 1971) p. 154. Also, per (Todd, 1967) p. 97, not helping Trevithick any was that he "had a notoriously bad head for business," and p. 109, he had no Boulton to partner with: "what Trevithick had always needed was a business partner."

[16] Cited in (Harris, 1966) p. 102. From Woolf's death notice in 1837.

[17] (Harris, 1966) p. 22. Jabez Carter Hornblower was the Hornblower of the firm.

[18] John Maberly (1770–1839)

[19] (Harris, 1966) p. 25. As the design was made in 1796 and Watt's patent wasn't to expire until 1800, Boulton & Watt commenced proceedings against Hornblower and Maberly. Hornblower lost the case.

[20] (Harris, 1966) p. 27. Woolf was five years older than Trevithick. As both grew up in small towns not too far from each other in Cornwall, there is a strong probability that they were acquainted in their youth.

during this period when Woolf, "ever alive to the improvements being effected in the engineering world,"[21] took all of his learning and experience and began investigating higher-pressure engine designs based on his own belief that he could save substantial amounts of coal consumption by operating at high pressure. This effort culminated in 1804 when he patented "an entirely new engine"[22] that combined two principles: Trevithick's high-pressure non-condensing engine with Watt's low-pressure condensing engine, no longer prevented from using a condenser by Watt's then-elapsed patents. His thinking was that the exhaust from the former could still be utilized to generate additional work in the latter. Thus was born the "Woolf Engine." While sound in principle, this double-cylinder design was complicated in practice.

It needs to be noted that Woolf's design was not "entirely new" in light of Jonathan (#2) Hornblower's two-cylinder, compound engine design from the early 1790s. Because of Bolton & Watt's resulting lawsuit against Jonathan, no one really followed up on this design until Woolf either knowingly or unknowingly revived the concept.[23]

In 1805 Woolf erected his new engine, as a personal challenge to himself to improve upon Hornblower and Maberly's engine, but when it began operation, even though it ran at higher pressure (40 psig in the boiler), it failed to deliver the required work, likely caused by the difficulty Woolf encountered when trying to find suitable materials and workmanship to construct high-pressure equipment. The silver lining in this trial was Woolf's confirmation of improved efficiency (low coal consumption for work done). Unfortunately, the continued problems Woolf faced with demonstrating his engine caused concern with the brewery management. In several trials during 1808, one attended by Trevithick, Woolf's engine failed to demonstrate clear superiority over Boulton & Watt's engine doing the same amount of work. Thus, the brewery chose Boulton & Watt, whereupon Woolf resigned, which was "perhaps the best thing that could have happened for the subsequent development of the high-pressure steam engine,"[24] for he soon partnered with Humphrey Edwards in London to improve his engine, and then in 1811 took his improvements with him back to Cornwall.

Cornwall was ready for the shot of adrenaline Woolf brought with him, because while he was away, "the Cornish engine had dwindled into a filthy jumble of a thing."[25] It wasn't his new engine design that provided that shot, for the double-cylinder approach was sufficiently complicated to "condemn"[26] it, but instead it was the engineering skill he brought with him. During the 1814 construction of his first Cornish engine at Wheal Abraham[27]—yes, he did success-

---

[21] (Harris, 1966) p. 29.

[22] (Harris, 1966) p. 36.

[23] (Cardwell, 1971) p. 153.

[24] (Harris, 1966) p. 41.

[25] (Harris, 1966) p. 50. Quote from Matthew Loam.

[26] (Harris, 1966) p. 53.

[27] (Cardwell, 1971) p. 155. By 1816 his engine demonstrated about twice the efficiency of Watt's low-pressure engine. These results spread through the industrial community. It was a big deal. Woolf's success represented a triumph for his Cornwall homeland and all who came before him. Note (Ward, 1843) p. 264, that this may not have been the first commercialization of the compound engine. In 1807 in the English Midlands, Enoch Wood (1759–1840), obtained a patent for an improved steam engine based on Trevithick's high-pressure steam and Watt's condenser. Wood applied this invention for many years to a steam-mill and to the drainage of coalmines at the Bycars near Burslem with improved productivity.

fully erect and operate his engine in Cornwall—Woolf created an effective training ground from which the subsequent leaders of Cornish engineering emerged, and he eventually rose to become one of the top engineers in Cornwall alongside Trevithick, who had also returned to Cornwall after his own failures in London. While these "rival geniuses in Cornwall"[28] didn't always get along, with their individual stubbornness clashing against each other, leading to an eventual bitter falling out, their strong advocacy of high-pressure steam helped inspire continual improvements to the steam engine design. "Although Trevithick may have been a more mercurially brilliant man than Woolf, the former lacked qualities which were necessary to bring the steam engine to that degree of perfection which it attained in the second and third decades of the nineteenth century."[29]

The Cornish Engine

While Woolf was developing his compound engine, Watt's influence was starting to wane. Once his patents elapsed in 1800, the new generation of Cornish engineers, at least some of whom Woolf helped train, rallied together, supported by strong community encouragement, to steadily improve the steam engine, building on the principles of Trevithick's design and appreciating Woolf's accomplishment, although this accomplishment was recognized as being complicated and expensive. The environment was competitive but the competition was positive. The final product became known as *the Cornish Engine*, consisting of a single cylinder, either single or dual action,[30] that combined the best of Trevithick and Watt while eliminating the complexity and associated high cost of Woolf. High pressure drove the piston from one side while a condenser created a vacuum on the other side. The use of the condenser rendered the engine virtually silent, unlike Trevithick's design, which "puffed" exhaust steam to the atmosphere. The favorable elimination of Hornblower's and Woolf's multi-cylinder approach didn't incur significant technical penalty as it was discovered that a similar economy could be obtained by working at a high rate of expansion with a single cylinder.[31] Contributing to this success was Woolf's open-minded acceptance of Trevithick's "Cornish boiler" design into the final Cornish engine.

By the mid-1840s the Cornish Engine had achieved sustainable operation around 40 psig.[32] Additional equipment modifications, such as the use of insulation, pre-heating the cold water feed with exhaust steam, and other forms of heat integration, combined with the rising pressures to increase overall efficiency. Whereas Watt's best steam engine achieved about 20 million ft-lbs of work per bushel of coal, the Cornish engine reached a top performance of about 125 million ft-lbs.

The technical success didn't rapidly translate into an economic success for the improved engines coming out of Cornwall because many mine owners already had operating engines

[28] (Harris, 1966) p. 64.
[29] (Harris, 1966) p. 64.
[30] (Hosken, 2013) When you need reciprocal motion you have to drive the piston from both sides, employ vacuum on both sides, and use the mighty flywheel to overcome the top dead center problem.
[31] (Galloway, 1881) p. 192.
[32] (Rolt and Allen, 1977)

from either Newcomen or Watt, so purchasing a new engine of superior efficiency had to offer the owners a better financial deal than continuing with the existing low-efficiency engine for which the high capital costs were already sunk. Given this, the new Cornish engines made sense where operating costs (such as cost of coal) were high, such as in Cornwall, or where there were no existing engines, such as in France and elsewhere.

Thermal Efficiency

Certain key advantages of high-pressure steam appeared immediately obvious to Trevithick with his design,[33] specifically smaller vessels and the opportunity to eliminate the condenser without undue loss in efficiency. Both were key toward creating compact, portable engines that could replace the horse.

But one advantage that was not so obvious was improved thermal efficiency. There are two types of efficiency to consider here, one being the actual efficiency as measured in the real world and the other being the maximum possible efficiency as governed by the 2nd Law of Thermodynamics. The former can be improved only up to the latter, but no further. The latter, as we'll get to shortly, can really only be improved in a steam engine by increasing steam temperature. Because temperature and pressure are connected for saturated steam, which is what the engineers were working with in Cornwall, this meant that designing engines to operate with higher-pressure steam would, in and of itself, improve thermal efficiency. For reference, when the high-pressure engines were commercialized in the early 1800s, the operating pressure eventually rose from 0 to 50 psig,[34] corresponding to a rise in steam temperature from 100 °C (19% maximum efficiency) to about 150 °C (28%).

The Cornish engineers reasonably implemented modifications to improve thermal efficiency that we still embrace to this day. But what they didn't understand at that point in history was that simply operating at higher pressure would improve thermal efficiency as well: "contemporary science was in no position to explain the superior efficiency and economy of the high-pressure steam-engine."[35] So as the engineers started operating their steam engines at continually higher pressures and noticed continually higher efficiencies—while engine efficiency had hit a plateau between 1780 and 1810 reflecting some combination of Watt's monopoly on the market together with his refusal to sanction high-pressure operation, efficiency increased steadily thereafter when the high-pressure condensing engine arrived—the cause–effect wasn't clear. Some of this increase was due to improved operations such as use of insulation, which they realized, but some was due to the higher operating pressure itself, which they didn't realize.

It's not that the engineers didn't anticipate certain efficiency benefits from operating at higher pressure, it's just that they didn't anticipate it on theoretical grounds. Trevithick intuitively felt that high-pressure engines might be inherently more economical than low-pressure

---

[33] (Cardwell, 1971) p. 84.

[34] In a 2017 email to the author, Phil Hosken shared that Trevithick recorded an engine working at 149 psig (185 °C) in the early 1800s.

[35] (Cardwell, 1971) p. 164.

ones, but his thinking was that the small losses typical of any engine would just be less of a factor at high-pressure. Anyway, recall that his original motivation behind going to high pressure had nothing to do with efficiency; it was a means to replace the horse. As for Woolf, he thought that high pressure could help reduce the consumption of coal, but this was based more on intuition than on fundamental understanding.

So the theoretical efficiency advantage of high-pressure steam was waiting to be discovered, but who had the time in the early 1800s to sit and think of such things when progress was moving so fast? And even if they did have time, how could one have even identified the true causes behind the observed effects in such a complicated and rapidly developing technology?[36]

Measurement Drives Improvement

In considering this unfolding history of thermodynamics, it is very apparent that the singular practice of measurement played a critical role. Just as Galileo had demonstrated the value of measurement in science, as opposed to just thinking, so too did the likes of Smeaton, Watt, and the Cornish engineers demonstrate the value of measurement in engineering. Whereas the pursuit of truth drove Galileo, it was free-market capitalism that largely drove the engineers since, in addition to capital cost, measured efficiency also became an important selling point as it was the only definitive way to show a potential customer that one engine was quantifiably better, i.e., consumes less coal for the same job, than another.

Indeed, the rise of the steam engine in England brought forth the measurement of thermal efficiencies for the conversion of heat (coal combustion) into work (lifting water). The measurement in and of itself brought forth a drive to innovate and improve. But at some point the measurement also brought forth, *had* to bring forth since man's curiosity couldn't tolerate otherwise, the question, how efficient can the steam engine possibly be?

The Role of the Water Wheel in This Story

During the late 1700s, Watt's steam engine and its support of the mining industry was not the only show in town. Also playing a major role in the Industrial Revolution was the water wheel and its support of the other industries, such as milling and textiles. Each involved the transformation of energy into work but each evolved along different paths, which were fortuitously brought together through the sharing of ideas, especially regarding efficiency, between the engineers like Trevithick who worked on both.

The textile industry required a steady uninterrupted source of energy to spin quality threads from a range of animal and vegetable fibers. The early steam engines were simply too clunky to provide such a source of energy, especially relative to the alternative of smoothly flowing or falling water. Although horses walked smoothly in circles and did provide certain benefits, especially once all the best river sites were occupied, as the horses could be located anywhere,

---

[36] Also note that the low efficiency of the early steam engines made data gathering rather difficult to quantify a difference between two large numbers (heat absorbed and heat rejected) since the relative work generated was low.

they simply weren't competitive due in part to the high costs associating with caring for them. But their long and reliable use in the industrial world did give them everlasting fame in the fact that "horsepower" became *the* unit of measure of power (rate at which work is done).

Since the textile mills required either running streams or falling water or both, they were massively decentralized in order to tap into each and every supportive water spot along the large streams and rivers throughout the textile counties of England. Because of this, they were also very rural, not becoming centralized until the steam engine technology improved to replace the water wheel. A key driving force in all of this was the power shortage resulting from strong economic growth combined with the fact that there were only so many flowing and falling streams to tap.

Note that the water wheel design was applied more to running streams with some degree of fall than to larger falls themselves. For falls exceeding thirty to forty feet, the water wheel became too large[37] and the so-called "column-of-water" engine was created, which is the type currently used in the large hydraulic plants tied to dams. The limits to these column-of-water engines were the materials of construction requirements in light of the very high pressures at the bottom of the column where high-pressure water was directed through a turbine to generate power. Recall that Trevithick acquired some experience in high-pressure operation as a result of working on such engines. Conceptually, calculations to determine the maximum work and power available from either the water wheel or the column-of-water were both based on the same conservation of mechanical energy ideas that were generally accepted around 1750. How much total kinetic (velocity[2]) plus potential (height) energy does each design provide? This represented the maximum work that could be done.

Given this background, you can see the strong economic incentive for the improved design of high-efficiency water wheels. The stream was what it was. Its flow rate was what it was; you couldn't control it. You were left to the whims of nature, especially during the summer months when streams would dry up. Similarly, the waterfall was what it was, its height also fixed; you couldn't control this either. Given these fixed resources, the only thing you *could* control was the equipment design. The better the design, the more work you could get at your location on the stream.

As was the case for the steam engine, the financial incentives to improve the water wheel design from rule-of-thumb craft to advanced technology were significant. A major step in this journey occurred in 1704 when Antoine Parent (1666–1716) carried out "the first sophisticated dynamic analysis of the vertical water wheel."[38] He considered both the volume and velocity of a flowing river in his calculations, embracing Leibniz's and Huygens' mechanics of *vis viva* (mv^2)—he was French, after all—as opposed to Newton's rival mechanics of force, which were less directly applicable. No falls were involved as most supportive experimental studies were done near either Paris or London, which were regions of low-fall, high-volume rivers.

Parent's analysis broke the ground for many to follow, such as Lazare Carnot. Frequent were the efficiency comparisons between "overshot" and "undershot" wheels, the former being used more for falls and the latter more for flowing streams without falls. In the end, scientists and

[37] It is noted, however, that a 72-foot wheel was built in the Isle of Man in 1854.
[38] (Reynolds, 1983) p. 205.

engineers began to realize the importance of eliminating energy waste in all of its forms, including, for example, sources of friction and sudden changes in flow when water entered and then exited the wheel. Ideally, 1) the wheel would be moving at the speed of the water flowing into it so as to eliminate the turbulent impact of the water against moving buckets and blades, and 2) the wheel would be moving at a slow-enough rotation such that the water would be released gently into the river so as to minimize the loss of kinetic energy (implying the need for large buckets).

In the end, scientists and engineers recognized that the efficiency of any water engine, no matter how advanced, couldn't be greater than 100%. No matter how efficient the water wheel, the total accomplished work couldn't be greater than the total kinetic and potential energy available in the moving stream relative to the reference conditions (height, flow) downstream of the engine. As discussed earlier, this fact was largely known by 1750 when the early forms of the conservation of mechanical energy became generally accepted.

The Origin of Reversibility

Out of the many discussions involved with water wheel efficiency arose the early concepts about "reversibility." In 1752, Antoine Déparcieux (1703–1768), famed French mathematician, considered a perfect water wheel driving an identical perfect wheel *in reverse* and stated that the upper limit of efficiency must be 1. This argument aligned with the overall conclusion on the conservation of mechanical energy and the impossibility of perpetual motion. In this argument, building on the Leibnizian approach, the available input energy (kinetic + potential) was set equal to maximum output energy. The flowing/falling water could only, under ideal conditions, raise an equivalent amount of water to an equivalent energy height but no more. The fact that the units for each form of energy—kinetic, potential, work—were the same meant that the concept of mechanical energy conservation could be applied and that a maximum efficiency of 100% could be identified. (Note: the kinetic energy of flowing water could be calculated similarly as that of a solid object.)

The water wheel and steam engine evolved side-by-side while addressing different needs during the Industrial Revolution. The water wheel provided direct and smooth rotational motion to the textile plants at relatively low cost, especially since the "fuel" was effectively free, and generally out-performed the steam engine until well into the nineteenth century.[39] The rise of the steam engine reflected the fact that, unlike the water wheel, it wasn't tied down to any geographic location; it could be placed pretty much anywhere, such as at a mine, or in a centralized location like a city where water power was not available, or even on a carriage. Also, the power output of a coal-fired steam engine could be as large as one could build the engine, whereas that of the water engine was limited by the flow and height of the water available.

This symbiotic evolution saw the sharing of real-world improvement ideas between the two power technologies, as well as early formulations of theoretical understanding. As the water wheel evolved earlier than the steam engine, such ideas flowed more from the former to the latter.

---

[39] (Cardwell, 1971) p. 70.

Recall that all such idea exchanges, both practical and theoretical, occurred during a time years before the concept of energy and its conservation had been defined (around 1850). So while the engineers and scientists may have understood the concept of water wheel efficiency based on the mechanical energy associated with moving bodies (solid or liquid) and gravitational potential energy, these concepts were not sufficient to analyze the efficiencies of steam engines in which other forms of energy, such as thermal energy (kinetic energy of invisible bodies) and chemical potential energy (coal combustion), were involved.

Sadi Carnot

When the Napoleonic Wars finally ended shortly after the Battle of Waterloo in the summer of 1815 and freedom of movement between the involved countries returned, the curtain surrounding England was pulled back and the Europeans found themselves staring at an industrialized nation comprised of iron foundries, heavy machine-tools, steam engines, textile mills, canals, and mines. All of a sudden, to the outside world, a new England emerged, the "workshop of the world."[40] What a surprise and shock this must have been to the Continent. For all the groundbreaking efforts of the French State to establish a strong, central presence in science and engineering in Paris and for all of the genius they attracted to this presence, nothing like what happened in England had happened in France. While French technology did indeed advance in certain areas during the England–France warring period of 1793 to 1815, many of the French visitors to this recently evolved England concluded that "the backwardness of most French technology was beyond question," and furthermore that "nowhere was the disparity between the two countries more glaring than in the realm of power technology."[41]

As one could imagine, all aspects of England's great industrialization, especially the steam engine, aroused much interest in France. The last data point of reference for France prior to 1815 was during a brief moment of peace in 1802–1803 in the form of Watt's engine. It was when the curtain closed around England shortly thereafter that such men as Trevithick, Woolf, and the Cornish engineers caused the technology to take off as best manifested by the commercialization of high-pressure steam operation.

While many, and especially Trevithick, contributed to the rapid rise of steam engine technology in England in the early 1800s, it was Woolf's high-pressure two-cylinder engine that "won the admiration of the French as soon as peace was restored."[42] There's an interesting story behind this event as chance played a key role in Woolf's rise in France. Right around the transition year of 1815, the reputation of Woolf's engine reached a peak in England, for it was only then, after years of unsuccessful trials, that Woolf successfully demonstrated in Cornwall his engine and its superiority to Watt's engine. But this success was short-lived. As discussed previously, it wasn't long thereafter that the British mining engineers soured on the Woolf engines, as they were expensive and complicated systems whose operation could not be sustained. The

---

[40] (Cardwell, 1971) p. 156.
[41] (Carnot, 1986) p. 3.
[42] (Carnot, 1986) p. 4.

engineers soon turned to the simpler Cornish Engine (one cylinder) as it was of similar performance efficiency and far more reliable and easier to maintain.

But during this brief moment of Woolf's technical success, his former business partner, Humphrey Edwards, used his entrepreneurial skills to actively import the Woolf engine into France. In a country where coal was scarce, any engine offering reduced coal consumption would dominate. Indeed, Woolf's engine and its improved efficiency relative to Watt's, the benchmark of the day, achieved a "spectacular" success in France.[43]

So what's the relevance of this little story? Well, it was the Woolf engine that was regarded as the most successful engine in France at the time that Sadi Carnot sat down in his Parisian apartment to write *Réflexions sur la Puissance du Feu* (*Reflections on the Motive Power of Fire*), hereafter referred to as *Reflections*.

Conclusion

To conclude this history of the early development of the steam engine, or more accurately, the history that led to Sadi Carnot, by the end of the eighteenth century and into the early nineteenth century, the rise of engines to transform the energy associated with either coal or falling water into useful work brought attention to the efficiency of such transformations. Coal cost real money; falling water was limited; efficiency became critical. This attention melded with the separate realization, based on a separate history, that perpetual motion is impossible. Because the various water engine technologies were governed solely by mechanical considerations of the kinetic energy of flowing water, the potential energy of falling water, and the useful work that could be generated by each, man deduced that the efficiency of conversion could not exceed 100%. The conservation of mechanical energy (1750) dictated this limit.

But with the steam engine there was no readily available means to quantify efficiency. The mechanical world dealt with moving visible bodies, including water. The thermal world dealt with moving invisible bodies, specifically those associated with kinetic energies of these bodies as well as the chemical interactions and bonding between them, and these concepts were naturally not established science at that time. The key issue of efficiency may have been understood, as reflected by the continuous improvement of the steam engines' commercial operation, but the fundamental science of thermodynamics working in the background wasn't. This science would lead to the 2nd Law of Thermodynamics and ultimately dictate the maximum thermal efficiency possible, a limit that could not be exceeded even if all the continuous improvement ideas were successfully implemented. The engineers and scientists didn't know how to calculate the maximum work possible for a given bushel of coal, nor did they fundamentally realize why there should even be a maximum to begin with.

The rise of the steam engine in Britain, which took "heat" out of the laboratories and into the real world, together with Joseph Black's and his successors' establishment of the science to quantify heat, combined to play a critical role in the early history of thermodynamics. In the

---

[43] (Carnot, 1986) p. 8. Also, in a 2017 email to the author, Phil Hosken shared that the absence of steam engines in France at the time reflected Watt's reluctance to see his engines abroad due to the likelihood of their being copied outside the protection of the English patent. Thus, this market was left open to others.

beginning, thermodynamics, as the name coined by Thomson[44] suggests, dealt primarily with the transformation between the newly commercialized heat of the steam engine and the work it generated; over time, it came to encompass the transformations of all forms of energy and especially the limits to those transformations, including the limits involved with phase and reaction equilibria. Because of the financial success achieved with the energy conversion technologies, it didn't take long until thermodynamic theory became of strong interest.

The strength of England's "doing" played perfectly into the strength of France's "thinking." The time was right for theory to explain and ultimately lead practice. So many pieces to this complicated jigsaw puzzle were available, just waiting for someone to come along and identify the relevant pieces, cast aside the confusing red herrings, and ultimately complete the puzzle.

As the chemists slowly lost interest in the science of heat,[45] largely due to their difficulty in understanding what heat actually was, especially since the caloric theory wasn't providing the necessary answers, the physicists stepped forward to take on these experimental studies, starting with the study of the behavior of gases and vapors as motivated in large part by the rise of the steam engine and its intriguing piston-in-cylinder process. This is where the early days of thermodynamics played out.

But it wouldn't be the physicists who would make the first major step. Yes, their studies helped provide the basis for the step. But the step itself would be left to an engineer. So now we again return to Sadi Carnot and his book, *Reflections on the Motive Power of Fire*, which was arguably the most important book published in the history of thermodynamics, the one that started it all, inspiring as it did the groundbreaking work of William Rankine, William Thomson, and Rudolf Clausius around 1850. It's interesting that Carnot's sole publication (at but 28 years of age) would become so influential, just as was the case for Copernicus, clearly showing that quantity is not a prerequisite for the impact one can have in life.

---

[44] (Cardwell, 1971) p. 243.
[45] (Fox, 1971) p. 279.

30 Sadi Carnot

> *Notwithstanding the work of all kinds done by steam engines…their theory is very little understood, and the attempts to improve them are still directed almost by chance*
>
> – Sadi Carnot[1]

As manifested in his famed book, 28-year-old Sadi Carnot was clearly impressed by what he saw come out of Britain, paying homage to the efforts of Savery, Newcomen, Smeaton, "the famous Watt," Woolf, and Trevithick, among others. He saw what they had accomplished, saw the technology gap they had created between Britain and France and realized the importance of increasing the level of understanding of these engines for the benefit of France if not mankind.

Raised and educated by an inquiring father who himself studied the efficiency of mechanical engines, Carnot couldn't help but look at the steam engine and wonder how it worked, and especially how it was limited. "*The question has often been raised whether the motive power of heat is unbounded.*"[2] While Carnot's question, as indicated, was not original, his approach to answering it was.

Historical Context

> *Out on the edge you see all the kinds of things you can't see from the center*
>
> – Kurt Vonnegut[3]

To help set some historical context, consider first that even though Carnot was educated in the prestigious École Polytechnique, he led most of his adult years outside academia as an officer amongst his fellow engineers within the French military. This situation was somewhat similar to Galileo's. Indeed, there are several similarities between Carnot's scientific journey and Galileo's that are interesting to reflect on, including this issue of working at interfaces, as they provide at least some insight into the conditions that support paradigm-shifting breakthroughs in science. Carnot worked at the interface between engineering and academia. Galileo worked at the interface between craftsmanship and academia. The exposure each had to a world apart from academia helped enable them to approach problems differently, if not uniquely.

---

[1] (Carnot et al., 1988) p. 5.
[2] (Carnot et al., 1988) p. 5.
[3] (Vonnegut, 2009) p. 84.

Block by Block: The Historical and Theoretical Foundations of Thermodynamics,
Robert T. Hanlon, Oxford University Press (2020). © Robert T. Hanlon.
DOI: 10.1093/oso/9780198851547.001.0001

This is not meant to minimize the role academia played in the development of each. As a student in Pisa and teacher in Padua, Galileo clearly had a foot in the academic world. For Carnot, his academic bearing started with the combined benefits of his father's strong technical influence and his own subsequent public education and then continued with the classes he was taking at the universities during his famous post-military existence in Paris. Of special importance regarding the latter was the relationship and friendship he developed with Nicolas Clément, an industrialist and chemist turned professor who gave evening chemistry lectures for working men. Clément was developing a theory of heat engines with supportive experimental studies at just the time when Carnot was becoming interested in the problem, and he too was interested in knowing the maximum work that could be obtained from a steam engine for a given amount of fuel. Clément's lectures, and especially his 1819 publication with his father-in-law, Charles-Bernard Desormes, on the theory of steam engines clearly influenced Carnot's general approach to this problem in addition to influencing some of the specific details involved, such as adiabatic expansion of steam in a cylinder. While Carnot acknowledged Clément very well in one very long footnote,[4] the depth of this influence would not really be known until 1968 when personal correspondence between the two were subsequently found.[5]

Just as neither Galileo nor Newton created his respective masterpiece in a vacuum, neither did Carnot. Yes, each worked alone and perhaps achieved their critical creative breakthroughs by sitting in solitude and thinking quietly, but each also had influences. As revealed in his *Reflections on the Motive Power of Fire* (1824), Carnot was very well read in the contemporary literature of physics and engineering and so constructed his idealized engine based on the theories commonly accepted at that time. Furthermore, he had two strong technical mentors, namely his father, Lazare, and Clément. Through the latter, he was connected to the academic world, at least to some extent, although it needs to be pointed out that Carnot himself was not of the academic world. His personal relationship with Clément was more based on science than academia. As discussed by Fox, "both Sadi and Clément were solitary men whose writings earned them little recognition within the scientific establishment."[6]

In the end, while recognizing the role of influence behind most all great works, and recognizing that many of the questions Carnot sought to address already existed in France's technical world, *Reflections* was "highly original work…and the way in which existing ideas and elements were united and molded to their new purpose was truly creative."[7] A similar statement could be applied to both *Discourses* and the *Principia*.

Thermodynamic Context

To establish additional context before proceeding, let's consider what we know to be true. By 1750, an early version of the conservation of mechanical energy arrived without much if any

---

[4] (Carnot et al., 1988) p. 50.
[5] (Carnot, 1986) p. 49.
[6] (Carnot, 1986) p. 18.
[7] (Carnot, 1986) p. 21.

fanfare and stated that the sum of what we now call kinetic and potential energies in the moving parts of an ideal and isolated mechanical system remains constant. So, as a given body freely falls, its gravitational potential energy (mgh) transforms into kinetic energy ($1/2\ mv^2$), such that the sum of the two always remains constant. Similarly, when two bodies elastically collide (with no change in potential energy), the kinetic energy of each may change, but their sum doesn't.

The transformation of heat into work is simply an extension of the conservation of mechanical energy into the atomic world. When atoms bounce off a moving piston, causing the piston to move out and ultimately push a weight up against gravity, the average kinetic energy of the atoms (temperature) decreases while work is generated; heat is transformed into work. The challenge here is that because we can't directly measure what's going on in this world, we must rely on bulk properties to quantify it.

Thermo (heat) – dynamics (work) naturally encompasses more than just heat and work. It really encompasses the transformations of all forms of energy—kinetic and potential, mechanical, thermal, chemical, electrical—into other forms of energy along with the limits on these transformations.

* * *

Now let's add some of these pieces together to create an understanding of an operating steam engine. The chemical energy of coal is converted into thermal energy in the furnace, which is then converted into mechanical energy when it vaporizes water contained in the boiler. The resulting high-pressure steam pushes against a piston to do work at the expense of internal energy. Upon completion of its expansion, the low-energy steam is exhausted, condensed, and pumped back to the boiler to start its journey all over again.

The overall efficiency of this process, and really of any heat engine process, is the ratio of work generated to fuel consumed (Equation 30.1). As both quantities can be put into the same energy units, overall efficiency is based on an energy balance and thus cannot exceed 1.0. You can't get more energy out than you put in. The last term in Equation 30.1 is called thermal efficiency and is defined as the ratio of work generated to heat absorbed by the heat engine (Equation 30.2).

$$\begin{matrix} \text{Overall efficiency} \\ \text{for the conversion} \\ \text{of coal to Work} \end{matrix} = \frac{\text{Heat into engine}}{\begin{matrix}\text{Heat generated by}\\ \text{burning coal}\end{matrix}} \times \frac{\text{Work done by engine}}{\text{Heat into engine}} \qquad [30.1]$$

In addition to the aforementioned limit on overall efficiency, a limit also exists for thermal efficiency. This limit, shown in Equation 30.3, is the subject of this chapter, for it was Sadi Carnot's work that led to its discovery.

$$\text{Thermal efficiency of heat engine} = \frac{\text{Work done by engine}}{\text{Heat into engine}} \leq 1.0 \qquad [30.2]$$

$$\text{Maximum thermal efficiency of heat engine} = (T_H - T_C)/T_H \qquad [30.3]$$

In this equation, T_H and T_C represent the temperatures of the heat source and the cold sink, respectively. The derivation of this equation and the meaning of these terms will be discussed in more detail shortly.

*   *   *

When Carnot looked at the steam engine, the question for him was not whether such a machine worked, because it did. It was a real machine, burning real coal, doing real work, positively impacting real economies. It wasn't a theory. Rather the question for Carnot and for many others was, *how good could it be?* Theory wasn't needed to commercialize the steam engine but it would be needed to determine its maximum performance and so guide technology development.

The empiricist engineers weren't waiting on theory to guide them. The water engine taught them the value of eliminating anything that decreased work output, including friction (that slowed the wheel and generated useless heat in the gear mechanisms), leaks (that had water spilling out of the wheel and freely falling without doing any work), and even high-velocity exit streams (since they represented excess *vis viva* that could have otherwise been converted into useful work). They applied similar logic to the heat engines by eliminating friction in the moving mechanical parts and by using insulation to eliminate leaks that had heat "spilling" out of the engine via conduction or radiation without doing any work. They eliminated additional waste by using the hot furnace exhaust to pre-heat the cool water entering the hot boiler.

Yet, in the midst of all of these continual improvements, there was one variable for which the empiricists had no guide: steam temperature. In the early 1800s, a typical coal furnace operated at well over 1000 °C, far higher than the 150 °C temperature of the saturated steam (4 atm gauge) in the best of engines. The intellectually curious—I'm not exactly sure who other than Carnot they were but am guessing that they did exist—couldn't help but look at this temperature difference as a form of some kind of loss. For those few, there was no way to even approach quantification of this loss, nor its inclusion into the broader quantification of efficiency itself.

The water engine didn't really provide guidance here; its efficiency was determined as a dimensionless number based on conservation of mechanical energy. But as regards Carnot, since the larger concept of the conservation of energy, which necessarily includes thermal energy, wouldn't arrive for another 25 years, there was no corresponding equation available for the steam engine. Work produced relative to bushels of coal consumed was clearly not dimensionless. Without this, there was no way to determine how good the steam engine could be. Furthermore, there was no way to determine whether such a limit even existed.

Clearly, a theory that encompassed all of the aforementioned issues was lacking.

The Stage Was Set for Carnot's Inquiry

The theories, insights, hypotheses, concerns, calculations, and discussions in *Reflections* are many, to the point of being confusing at times. In part, this was due to the fact that *Reflections* <u>was</u> confusing at a most fundamental level, since Carnot wrote it while working under the false paradigm involving the caloric theory. This didn't invalidate the entirety of

Reflections, just certain parts. But, all the same, it made the work rather difficult to interpret. When Clausius eventually modified and corrected Carnot's work by eliminating caloric and replacing it with energy, *Reflections* became much easier to understand, its strength became evident to the science community, and its conclusions spurred the creation of classical thermodynamics.

To facilitate your understanding of my own attempt to interpret *Reflections*, I feel it's best to tell you the ending first. To this end, a short overview of Carnot's achievements in *Reflections* is offered here, an overview greatly aided by the translations and excellent in-depth commentaries of Eric Mendoza[8] and Robert Fox,[9] Fox's separate book on the caloric theory,[10] and finally additional and very valuable communications with Dr. Fox, for which I am deeply appreciative.

∗   ∗   ∗

The theory governing the maximum value for thermal efficiency was non-existent during Carnot's time and remains a challenge for many to understand today. How does one squeeze every bit of work out the heat entering an engine? Seeking insight by analogy with the water wheel, one would want to eliminate all "leaks," whatever that meant in the thermal world, and generate an effluent stream that would match the properties of the ambient environment and so not be capable of doing any further work. And you'd want the process to be reversible such that it would exist in a near-equilibrium state with infinitesimal driving forces such that an infinitesimal change in conditions would flip the driving forces and send the engine in the reverse direction. In the thermal world, reversing a steam engine results in the refrigeration process, which will be covered later.

In this chapter we'll go into the details involved in this discussion but for now will state that it was Sadi Carnot who, focusing primarily on thermal efficiency, devised the ideal process to convert heat into the maximum work possible. To do this required the creation of new ideas and concepts, which would later become core to the rising science of thermodynamics. These ideas and concepts included

- The importance of temperature in governing maximum engine efficiency
- Maximum work
- The closed-cycle process as a means to analyze the performance of a steam engine (or really, any heat engine)
- The first, very faint glimmer of entropy
- The first, very faint glimmer of the famed Clapeyron and Clausius–Clapeyron equations.

While I cover each of these points, I feel it important to start by describing the caloric paradigm that Carnot embraced and that prevented him from completing his inquiry and in fact left him in a state of frustration, which I'll discuss more at the end of this chapter.

---

[8] (Carnot et al., 1988)
[9] (Carnot, 1986)
[10] (Fox, 1971)

Revisiting the Caloric Theory

To further facilitate your understanding of this discussion, here is a brief review of the caloric theory as manifested in a short rulebook[11] that Carnot used. Some of the rules may be different than those described before but such was the nature of this theory; it was adapted on an "as needed" basis at times to explain that which it couldn't otherwise explain.

- Caloric is a weightless substance
- Caloric flows from hot to cold
- Caloric is conserved – it can be neither created nor destroyed
- Caloric is comprised of two forms: sensible and latent
- The amount of sensible, or free, caloric present is quantified by a thermometer
- The amount of latent, or chemically bound, caloric increases or decreases with available volume. If volume increases, such as during "expansive operation", latent caloric increases. Since total caloric is conserved, sensible caloric must decrease, and this is why temperature decreases.

The fictitious caloric had many properties that would later and rightfully be assigned to heat. Indeed, to Carnot, the two were the same. As he wrote, "we employ these two expressions indifferently."[12] The critical conflict inherent to this statement became apparent only later since the property of caloric that could not be assigned to heat concerned conservation. Sadi's caloric was conserved; Clausius' heat was consumed.

One final comment is needed. I wrote out Carnot's key ideas and concepts above in a way to make them flow in a logical, linear order. However, *Reflections* wasn't written as such, which furthered the confusion somewhat. While I stuck with Carnot's sequencing, please refer back to the above list as needed for clarity. So with the above in mind, let's now dive into *Reflections* to see how Carnot arrived at these achievements.

> Reflections on the Motive Power of Fire

Where would one even start in attempting to bring theory to such an intricate and complicated piece of moving machinery?

With an engineer's eye and a father's embedded lesson[13] on the value of generalizing, Carnot took a step back and sought a less complicated, more idealized view of the commercial operation, a timeless best practice that's integral to any sound education. He stripped away the complexity involving the multitude of vessels, pipes, valves, and moving beams, and brought focus to the simple fact that coal and cooling water enter the engine, and work and slightly warmer cooling water exit. The coal is burned in a furnace to produce steam to drive a piston to do work. The spent

[11] (Fox, 1971) This book provides an excellent overview of the caloric theory.
[12] (Carnot et al., 1988) p. 9.
[13] (Carnot, 1986) p. 13. Lazare's attempt to establish a general science of machines exerted such a clear influence on Sadi that it deserves special mention. Sadi sought to create a general theory of heat engines, reduced to their essentials and with all practical considerations laid aside, just as Lazare did for the mechanical machines.

steam is exhausted from the cylinder, condensed by the cooling water and sent back to the furnace boiler to complete its journey. He then zeroed in on the machine's core essence, the expansion of a gas against a piston in a cylinder. In this generalized approach, while Carnot made no reference to a particular engine, clearly he was inspired by Woolf's newly arrived pressurized steam engine, even though he was apparently unaware of this engine's practical shortcomings.[14]

He took yet another step back by shifting focus from the specific engine based on steam to the more general heat engine based on gas (as the working substance). This was a remarkably simplifying generalization as steam engines of that time operated at saturated conditions (superheat technology wouldn't arrive commercially until around 1900) necessitating analysis of evaporation and condensation processes in the days before vapor–liquid equilibrium theories were even available. Analysis of heat engines based on gas required none of this and so provided Carnot with an easier path forward.

While the simplifications did indeed facilitate analysis, their simplicity was a mirage, for Carnot's next steps dove deep into the development of concepts new to the world of engineering if not physics, concepts that had no precedent. Over the previous one hundred plus years of commercially operating steam engines, no one other than possibly his main influence, Clément, seems to have approached this technology in such a powerfully insightful way as Carnot did. It was, by definition, a hard task; had it been easy, it would have already been done. Said one of Carnot's biographers, Robert Fox, the seeming simplicity of Carnot's book was a "snare."[15] While Carnot intended *Reflections* to be a popular book for a general audience, it was anything but.[16] Carnot wrote as an engineer in the French manner, with an interest in the theory governing the performance of an *ideal* engine, this as opposed to the rougher British writing approach.[17] It took the strength of such giants as William Rankine, William Thomson, and Rudolf Clausius to grasp what Carnot was getting at and to use this as their starting point for the creation of classical thermodynamics.

Carnot's steps were guided by the generally accepted laws regarding the conservation of mechanical energy (1750) and the impossibility of perpetual motion in the mechanical world.[18] He knew and understood the limits inherent to these mechanical laws and felt strongly that they must also apply to the world of heat and electricity as well,[19] highlighting this last point by making special reference to Alessandro Volta and his voltaic pile (the first electrical battery). While some had regarded the pile as a potential source of perpetual motion, Carnot discounted this possibility, as James Joule would do later, writing that "the apparatus has always exhibited sensible deteriorations when its action has been sustained for a time with any energy."[20] He furthered his argument by suggesting that even the chemical world must have limits, as reflected in his question, "is it possible to conceive the phenomena of heat and electricity as due to anything else than some kind of motion of the body, and as such should they not be

---

[14] (Carnot, 1986) p. 7–8.
[15] (Carnot, 1986) p. 2.
[16] (Carnot et al., 1988) p. x.
[17] As shared by historian Robert Fox with author in a 2017 email.
[18] (Carnot et al., 1988) p. 12.
[19] (Carnot et al., 1988) p. 12.
[20] (Carnot et al., 1988) p. 12.

subjected to the general laws of mechanics?"[21] It's fascinating how close Carnot appeared to be regarding the mechanical theory of heat and the kinetic theory of gases in such a statement, and yet equally fascinating to realize how far away from such theories he actually was.

Regarding his view of the steam engine, Carnot realized that the work had to come from some source, that some *thing* must be depleted to cause the observed effect, and that whatever that *thing* was, the cause–effect relationship between its depletion and work generated must ultimately result in a maximum limit of the latter. He knew that the combustion of coal generated steam that drove the engine that did the work. But he needed a fundamental cause behind the work, something that could explain its presence and, more importantly, its limit. Yes, the main cause was the coal, and commercial engineers recognized this by constantly measuring and monitoring their engine's efficiency by the metric of work done per bushel of coal consumed. But there was no theoretical underpinning to this ratio; it wasn't dimensionless. This was the basic task in front of Carnot, the one that he faced head-on.

To truly understand what Carnot did next in his approach to this tremendous challenge, you need to first enter his mind, to see the world as he saw it, which was through the caloric paradigm.

The Caloric Paradigm

In France in the early 1800s, the caloric theory remained prominent in the textbook literature due largely to the advocacy of such influential figures as Laplace, lab assistant to the arguable founder of the modern caloric theory, Lavoisier. As a student in the French system, Carnot was raised in this paradigm. His belief became so strong that he wrote in *Reflections*, with specific reference to the conservation of caloric, "To deny it would be to overthrow the whole theory of heat."[22] It's interesting that William Thomson would later utter a similar thought in writing that if heat is indeed not conserved, as Joule proclaimed, it would "overturn the whole theory of heat."[23] As an interesting aside, the caloric theory started to fade by the 1820s, for reasons I'll address later, when Carnot took to writing *Reflections*, but unfortunately, Carnot, not traveling in the mainstream academic circles, wasn't completely aware of this and retained the caloric theory longer than others.[24]

It is rather interesting to note that both Galileo and Carnot started out in a deep paradigm of science—as William Thomson would do later—and struggled to climb out of this depth toward a new paradigm, and each was enabled by their working at different interfaces. Galileo really operated at four different interfaces, two philosophical ones, Aristotle's physics and modern physics, reflecting the paradigm shift he led but didn't complete, and also two professional ones, academia and craftsmanship, which helped him to see things differently and so helped enable the paradigm shift. Carnot also operated at four different interfaces, two philosophical ones, caloric and energy, reflecting the paradigm shift he led but didn't complete, and

[21] (Carnot et al., 1988) p. 12.
[22] (Carnot et al., 1988) p. 19.
[23] (Cardwell, 1971) p. 244.
[24] (Fox, 1971) p. 275.

also two professional ones, academia and military engineering, which also helped him to see things differently and so helped enable his start toward a paradigm shift.

Being able to cause one's own paradigm shift is a testament to one's genius. Challenging the status quo in which one was raised, as Galileo and Carnot had done, is indeed a formidable task. It takes a courageous genius to let go of one paradigm and grab hold of, or, better yet, create another. The fact that neither quite made it doesn't detract at all from their accomplishments. Perhaps the depth in which they initially lived in their respective paradigms made their accomplishments in (almost) climbing out all the more remarkable.

Carnot was a scientist at heart, in the truest sense of the word, "conspicuously reflective and relentlessly logical in his arguments,"[25] as opposed to blindly following conventional wisdom. Throughout *Reflections*, he was a "cautious"[26] calorist, willingly challenging this theory and his own belief in it, acknowledging, for example, that "many experimental facts appear almost inexplicable in the present state of this theory."[27] But, in the time and place of Lavoisier and Laplace, it was all that he had to work with. While caloric was failing, nothing else was available to replace it. As we'll see, it was only later, toward the end of his life after publishing *Reflections*, that he would break free from caloric and come so very close to discovering energy.

Perhaps, in hindsight, if I may speculate, it was a good thing that Carnot, however conflicted he may have been, still believed in caloric enough to see him through to completion. His belief provided him with at least some structure, however wrong, to move forward. If he had lost this belief, and if he didn't have any replacement, then this could have stopped him from even attempting *Reflections*.

*   *   *

To Carnot, the critical challenge was that there was no obvious *thing* that was lost or consumed in the steam engine to balance the generation of work, at least none that anyone had identified before.[28] Yes, coal was consumed, but as energy had yet to be discovered as a concept, it wasn't clear how to account for this. Rather than fighting this battle, he simplified things even further by ignoring the generation of heat by coal combustion, choosing instead to focus on what happens once that heat was available. In other words, he focused on thermal efficiency and not overall efficiency (more discussion on this in Figure 30.2).

In his approach to analyzing the transfer of heat into the steam engine, he latched onto what he knew: the caloric theory. This theory gained traction in the 1770s when Lavoisier's caloric theory was applied to a gas, which, at that time, was envisioned as being comprised of static atoms as Newton had originally assumed.[29] Weightless caloric moved through the static atoms as a conserved quantity that was portioned between latent and sensible forms, with shifts between the two accounting for all heat-related phenomena, such as changes in temperature.

---

[25] From Robert Fox in a 2013 email to the author.

[26] (Carnot, 1986) p. 26.

[27] (Carnot et al., 1988) p. 19.

[28] Recall from Chapter 27 that the low thermal efficiencies of the operating steam engines around 1800 provided little indication that heat was consumed in the engine.

[29] Newton's assumption of a gas structure comprised of static atoms to model Boyle's Law, which he never intended to represent a real physical model, ended up hindering the path toward the kinetic theory of gases as much as any other assumption.

Of special importance to Carnot was the application of this theory to analyzing gas expansion and compression. According to the theory, for example, during adiabatic expansions ("expansive operation"), increasing gas volume forces a sensible-to-latent shift, thus decreasing temperature.[30] This manifested the reason for caloric's unfortunate staying power. It sometimes gave the right answer for the wrong fundamental reason. In this scenario, temperature decrease and work generation aren't connected in any theoretical way, almost as if the temperature change were some "oh, by the way" occurrence of little importance as opposed to a quantification of energy loss.

Carnot's Hypothesis: Thermal Efficiency = $f(T_H - T_C)$

With this background, Carnot proposed a new hypothesis about caloric's behavior, specifically the existence of a relationship between the flow of caloric and the production of work in a heat engine. In one of the more noted uses of analogy in the history of science, reaching back to his father's work on water engines, Carnot proposed something entirely new to the world of physics:[31]

> The motive power of a waterfall depends on its height and on the quantity of liquid; the motive power of heat depends also on the quantity of caloric used, and on what may be termed, on what in fact we will call, the height of its fall,* that is to say, the difference of temperature of the bodies between which the exchange of caloric is made. In the waterfall the motive power is exactly proportional to the difference of level between the higher and lower reservoirs. In the fall of caloric, the motive power undoubtedly increases with the difference of temperature between the warm and the cold bodies; but we do not know whether it is proportional to this difference... It is a question we propose to examine.
>
> * The matter here dealt with being entirely new, we are obliged to employ expressions not in use as yet, and which perhaps are less clear than is desirable.

In this analogy, as caloric falls from high temperature to low, a concept that was consistent with the everyday observation that heat flows from hot to cold, total caloric is conserved while latent caloric increases to fill the larger expanded volume, thus leaving sensible caloric and hence temperature to decrease. Carnot didn't spell it out like this,[32] but it does follow from a physical model based on caloric (refer back to Figure 16.1).

When trying to understand the logic behind this rather significant hypothesis, historians are stymied. They don't really know the thinking that went on behind Carnot's bold leap.[33] Carnot didn't spend time explaining. His hypothesis was not the result of logical build-up of arguments based on a detailed microscopic analysis. Instead, it was the result of a macroscopic intuition derived from his creative comparison of two engines running on different sources of power mixed in together with influences from his father, Clément, publications,

[30] (Kuhn, 1958) This paper contains a good discussion around the caloric theory of gases and its staying power.
[31] (Carnot et al., 1988) p. 15.
[32] Carnot likely didn't spell it out because the cause–effect logic involved with the caloric theory is difficult to follow since the theory itself is wrong.
[33] (Carnot, 1986) p. 119.

and text books. Carnot did not say how his hypothesis worked. He simply proposed that it must be so.

The core logic in Carnot's hypothesis was naturally wrong, as it embraced the fictitious caloric. Building on his water engine analogy, Carnot envisioned[34] that the caloric flowed out of the furnace and somehow incorporated itself into the steam, and then flowed into the piston-in-cylinder assembly to *"perform some function"*[35] [my italics], and from there into the condenser, where the cold water took "possession"[36] of it. To Carnot, the steam was only there as a means to transport the caloric and the production of work was due "not to an actual consumption of caloric, but to its transportation from a warm body to a cold body."[37] Furthermore, the maximum possible work was related in some way to the magnitude of the temperature difference between the heat source and the heat sink, presumably quantifying the amount of shift that occurred between latent and sensible caloric. It is very confusing in reading this to see exactly how Carnot proposed that the transfer of caloric from high temperature to low resulted in work. He had to, in effect, wave his hands at this, using such nebulous statements as "perform some function." He had no other way to fight this battle of inquiry.

But given this, Carnot's hypothesis was largely correct and motivated his inquiry toward the right direction. As we'll shortly see, the maximum possible thermal efficiency for the steam engine is indeed related to the temperature difference between the furnace and the condenser. Indeed, *Reflections* was largely correct, despite the incorrect starting point.

In the end, Carnot created, as one of his biographers Robert Fox put it, a "flawed masterpiece."[38] But perhaps it was just this flaw that enhanced the influence of this body of work, for it was due to this flaw that Carnot inadvertently put his finger squarely on one of the most critical conflicts confronting mid-nineteenth century science. No one had seen the flaw in such sharp detail prior to Clausius' analysis. Initially, the strength of Carnot's argument strongly persuaded Clausius. But then again so did the strength of Joule's argument. Clausius' struggle to reconcile both led him to see the flaw in the diamond. It took an expert to see this. Clausius was that expert. Once recognized, the flaw attracted his attention, for it stood out like a sore thumb. Realizing what he then needed to do, in one fell swoop, Clausius eliminated caloric from the world of science and re-interpreted *Reflections* based on a conserved quantity that would later be called energy, with heat and work being two different forms. Thus was laid the initial foundation of classical thermodynamics.

But I'm getting ahead of myself. Let's look now into even more detail behind the journey that Carnot took.

> Engine efficiency and temperature

---

[34] (Carnot et al., 1988) pp. 6–7.
[35] (Carnot et al., 1988) p. 7.
[36] (Carnot et al., 1988) p. 7.
[37] (Carnot et al., 1988) p. 7.
[38] (Carnot, 1986) p. 43.

Carnot was original, to the point of being mystifying, in his insistence that the savings achieved with a high-pressure engine should be attributed to the temperature of the steam rather than to its pressure.

– Robert Fox[39]

Utterly without precedent and dense with implications
– David Lindley[40]

It was out of Carnot's flawed analogy that his correct focus on temperature emerged as *the* critical variable when analyzing the performance of the steam engine. At a time when everyone else was focused on pressure, Carnot proposed temperature. The central role of temperature in the emerging field of thermodynamics arguably started in *Reflections*. Up until this time, there was nothing really special about temperature. It was simply a property of matter. If anything, temperature was somewhat less special than its companion, pressure, especially when studying the steam engine. Carnot changed that.

But why temperature? Why not stay with the mainstream thinking and consider the fall of caloric from high to low *pressure*? Unfortunately, Carnot didn't clarify his logic. *Reflections* was his sole publication and didn't include a lot of supportive logic behind such ideas. Historians indeed have been challenged by Carnot's thinking, leaving many to speculate. Allow me to join this group.

As a true engineer, Carnot realized the need to bring quantification to his work. But what variables did he really have to work with? There were only two to consider, pressure and temperature, as each was easy to measure. The former was favored, for at the heart of the steam engine operated the piston-in-cylinder, and this was all about pressure. While not fully understanding the underlying theory, the British engineers of the day such as Trevithick brought a common-sense focus on pressure as being *the* critical variable. It was a practical fact that higher pressure was better, so long as the equipment could handle it. In addition to decreasing the size of the equipment, it also manifestly increased engine efficiency although as discussed before the cause–effect connection wasn't fully understood at the time. French theorists of the day attributed the increased efficiency of Woolf's engine to both its higher operating pressure and its effective use of Watt's expansive principle, but these theories were still incomplete and had many loose ends.[41]

When Carnot came along and joined such theoretical discussions, which were popular in Paris at the time of his work, he chose to follow a different path, his own path. He insisted that the critical variable in determining the potential savings achieved with a high-pressure engine was temperature and not pressure.[42] This idea and others would have seemed "perversely cumbersome"[43] to other engineers of the day, being so out of the mainstream as to have most likely contributed significantly to *Reflections'* low readership since "few engineers would have bothered even to follow the argument."[44]

---

[39] (Carnot, 1986) p. 11.
[40] (Lindley, 2004) p. 69.
[41] (Carnot, 1986) p. 10–11.
[42] (Carnot, 1986) p. 11.
[43] (Carnot, 1986) p. 25.
[44] (Carnot, 1986) p. 25.

Carnot felt that the flow of caloric from furnace to condenser was critical toward assessing the efficiency of the steam engine. These two vessels were very visibly prominent in the engine's operation, the condenser being made more so by Watt's isolation of it. Recall that Woolf maintained the condenser in his design and that this lay-out was what Carnot most likely experienced in Paris. When considering the engine from the furnace-condenser perspective, the only simple important measurement common to both was temperature and not pressure. The engineers were focused on ensuring a hot furnace and a cold condenser. And so my speculation is that it was this thought process by the engineer in Carnot that led him to temperature as opposed to pressure. It aligned better with his concept of caloric flow from hot to cold.

An additional inspiration for Carnot may have been this. As addressed earlier, by being a believer in caloric, Carnot recognized that temperature was a means to quantify shifts within the total caloric between its "sensible" (free) and "latent" (bound) components, temperature being an indicator of the former and not the latter. Again, Carnot did not get into a discussion about his logic here, but perhaps he considered temperature as a means to quantify this shift, while considering pressure less relevant. In the end, scientists would eventually validate Carnot's decision to focus on temperature on theoretical grounds.[45]

* * *

It wasn't only the identification of temperature that made Carnot's hypothesis so novel and profound. It was also his proposal that work was related in some unknown way to the distance of fall in temperature between the hot source (furnace) and the cold sink (condenser); the greater the fall, the greater the work. There was absolutely no precedent for this hypothesis. His pursuit of this inquiry led him to consider the criteria to maximize work for a given amount of heat input, from which sprung new ideas that would later inspire Thomson and Clausius in the development of classical thermodynamics. While it had yet to be proven that a proposed (unknown) relationship existed between maximum work, or really heat-engine efficiency, and temperature difference, it became a great starting point for Carnot.

Maximum Work

What is the sense of the word maximum?
– Sadi Carnot[46]

We remember the rather confronting concept of "maximum work" from homework problems and test questions in thermodynamics. For better or for worse, we have Carnot to thank for this. It all started in *Reflections* as a logical progression of his water engine analogy.

Recall that Carnot wanted to quantify the maximum efficiency possible for a heat engine and to do this he had to determine the maximum amount of work that could be generated from a given "fall" of caloric. In his mind, just as maximum work occurred in a water engine when all of the water flow was productive, with no leaks from the process or any residual *vis*

[45] (Coopersmith, 2010) p. 186. Temperature is the more fundamental thermodynamic property because every body has a temperature while only gases have pressures. Also, temperature is the property used in the definition of entropy as: $dS = dQ/T$.
[46] (Carnot et al., 1988) p. 12.

viva energy present in the stream upon exiting the wheel, maximum work occurred in a steam engine when all of the caloric flow was productive. While the meaning of this wasn't as clear as with the water engine, Carnot's pursuit of the meaning (this was new territory after all) led to the early development of the reversible process. To Carnot, the necessary condition for maximum work was that no unproductive falls of caloric were allowed anywhere in the process. This meant that no temperature gaps between the engine and the environment could exist: the working substance temperature had to equal that of the high-temperature heat source on the one end and that of the cold-temperature sink on the other. To Carnot, any bodies that make contact in a steam engine must "differ as little as possible in temperature,"[47] for in his mind, if there were a finite difference in temperature between two bodies, caloric would flow without work being done. It also meant that no "leaks" of caloric from the process to the ambient environment were allowed, meaning, for example, that the engine had to be thoroughly insulated and that no direct contact (thus leading to conduction) could exist between the heat source and the cold sink.

In considering the achievement of maximum work, Carnot considered the path taken by the caloric during its flow and especially the shifting equilibrium between caloric's sensible and latent forms along this path. The furnace started by serving to first "destroy" this equilibrium and causing a latent-to-sensible shift (increase in temperature). Nature then sought to reestablish this equilibrium once the caloric entered the steam. Carnot proposed two routes for this re-establishment. The first involved the aforementioned heat conduction. To Carnot, this route represented an unproductive "fall" of caloric, one that should be minimized, and the best way to do this was to ensure that only minimal temperature differences exist when two bodies are brought in contact with each other.

The second route for re-establishing caloric's equilibrium involved a change in volume. When a gas expands, it does work and so decreases in temperature. Total caloric is conserved, latent caloric increases to fill the larger expanded volume, and sensible caloric decreases, thus dropping temperature.

Carnot saw that of the two possible routes, only the second resulted in work. Thus, he wanted all of the caloric to "fall" via this route and none to fall via the first. "The necessary condition of the maximum is, then, that in the bodies employed to realize the motive power of heat there should not occur any change of temperature which may not be due to a change of volume. Reciprocally, every time that this condition is fulfilled the maximum will be attained."[48] Temperature is only allowed to decrease in Carnot's ideal engine when the working substance volume increases while doing work.

* * *

Yes, the logic here is somewhat confusing as we catch but glimpses of reality set against a false background. Again, he was largely correct, but for the wrong reason. So let's pause here to consider Carnot's thoughts against the reality of what we know today.

It's important to distinguish between the two separate concepts involved in Carnot's above discussion of thermal efficiency. The first involves any process by which heat is lost from an engine. There's naturally the conductive and radiant heat transfer (caused by a difference in

---

[47] (Carnot et al., 1988) p. 13.
[48] (Carnot et al., 1988) p. 13.

temperatures) out of the engine and into the environment due to lack of insulation. There's also the lost capability to do work associated with the mixing of two process streams of different temperature. The losses associated with these phenomena were the common-sense truth at that time and Carnot was correct in pointing them out. Cornish engineers observed increases in thermal efficiencies when they applied thermal insulation to the equipment and also when they used hot furnace exhaust to pre-heat the cold make-up water entering the hot boiler.[49] In general, any time that two bodies of different temperature make contact, including times when process streams are exhausted to the environment at temperatures higher than ambient, the opportunity to generate work by hooking them up to, say, a Carnot engine is lost.

The second concept involved in Carnot's thermal efficiency discussion is a little more complicated. Consider the temperature difference between the furnace and the steam, and then again between the steam and the condenser. Clearly, some difference must exist or else the respective heat transfer rates would be zero and the engine wouldn't run. Carnot recognized the need for some difference, but not too much, stating that the difference "may be supposed as slight as we please."[50] We know today that we want to indeed minimize these two temperature differences, but not because of Carnot's belief that they resulted in the non-productive flow of caloric, but instead because we want to stretch the steam cycle to its maximum limits. In other words, these temperature differences don't represent a loss of heat from the engine but instead represent a loss of opportunity for what the engine *could* do given the respective temperatures available at the furnace and the condenser. The closer the steam temperature approaches the furnace temperature on the hot end of the cycle and the condenser temperature on the cold end, then the further apart they are and the more work can be done. We'll return to this later.

In sum, Carnot's goal of minimizing temperature differences at all points within the heat engine is associated with two separate and distinct principles: minimize heat loss (eliminate friction; use insulation; don't mix streams of different temperature; don't exhaust "hot" streams) and maximize temperature spread of the working fluid. He believed that achieving this goal and so ensuring a "reversible process" would maximize work, which indeed is correct.

Building on this discussion further, Carnot's deep belief in the concept that a maximum amount of work fundamentally exists for a given amount of heat is more profound that it sounds. Why? Well, because in following the above discussion, the limit is set by the range of temperatures available between the furnace and the condenser; a greater amount of work isn't possible since the working fluid temperature can't be stretched beyond this range. Since heat can only flow from high temperature to low, the steam temperature can never go higher than the furnace nor lower than the cooling water. You can see how such thinking led Carnot to focus on temperature and also how such thinking eventually led to the early formulations of the 2nd Law of Thermodynamics.

> The Closed Cycle

---

[49] (Cardwell, 1971) pp. 153–157.
[50] (Carnot et al., 1988) p. 13.

[Carnot] invented the closed cycle of operations
– E. Mendoza[51]

To summarize where we are now, Carnot hypothesized, with no supporting evidence, how the performance of a heat engine is limited based on analogy with a water engine. Caloric flows from the furnace into the steam and then from the steam into the condenser. During this journey, caloric's temperature drops from the furnace to the condenser and this (somehow) produces work. So long as it's solely this passage that the caloric travels through, which is achieved by minimizing all temperature differences along the way, then the work produced is the maximum it can possibly be. Again, the goal here is not necessarily to follow the technical logic, but to attempt to see the world that Carnot saw while he was writing.

Embedded in this logic was Carnot's correct focus on the way temperature changes a body's volume. This was a key point in considering an engine that continuously converts heat to work. Some volume changes are small, some large. Some are positive (heating water to vapor), some negative (heating ice to water). Feynman illustrated this issue by envisioning a rubber-band engine in which a heat lamp was used to alternately heat and cool the rubber, causing it to contract and relax, respectively,[52] which then spun a wheel and did work. The unusual negative volume change in the rubber upon heating resulted from a shift in intermolecular forces; with increasing temperature, more rotational configurations become accessible which leads to a shortening of the end-to-end distances and thus contraction. Whether the change is negative or positive, it doesn't matter. It's the fact that it's not zero that matters. Thermo (heat) – dynamics (work) is about the conversion from one to the other through this physical mechanism.

To prove his hypothesis, Carnot realized that he next had to address the physics around the "ensemble"[53] of individual steps comprising his macroscopic hypothesis. He needed to address the details relating to the piston's operation and the different ways it could be used to produce work. By doing this, he hoped to develop equations to quantify the overall thermal efficiency of the engine, or more specifically, the maximum possible efficiency that the engine could attain, and then assess whether his hypothesis was correct or not. Was the maximum efficiency some function of the difference in temperatures between the furnace and the condenser, or not? To facilitate our understanding of this, we need to first review the key historical-technical steps in the evolution of the steam engine.

* * *

Recall that the rise of the steam engine began with Newcomen. He flowed steam into a cylinder fitted with a piston and then sprayed cool water into this space to condense the steam, pull a vacuum, and thus enable the outside atmosphere to push the piston down into the cylinder and do work.

Watt modified Newcomen's concept by moving the steam-condensing step to an external condenser, thus enabling higher efficiencies since this allowed the piston-in-cylinder system to remain at high temperature. He kept the piston-in-cylinder even hotter (even more thermal efficiency) by switching the external driving force from ambient air to hot atmospheric boiler

---

[51] (Carnot et al., 1988) p. x.
[52] (Feynman et al., 1989a) Volume I, p. 44-2.
[53] (Carnot et al., 1988) p. 21.

steam and by adding a steam jacket around the cylinder.[54] He next employed a dual-pump action (smoother operation) wherein both sides of the piston had access to both boiler steam and condenser cooling.

Watt's next improvement addressed the inefficiency of condensing atmospheric-pressure steam to create the vacuum. When the hot cylinder was opened to the cold condenser (vacuum), the sudden pressure differential resulted in a rush of steam from the former to the latter. Watt correctly viewed this as a source of wasted potential to do incremental work. Steam at atmospheric pressure can do work when a vacuum is available; why waste it by condensing it? Hence was born his "expansive" principle whereby part-way through each stroke, the boiler steam supply was shut off and the steam allowed to further expand (adiabatically) while losing pressure such that at the end of the stroke, the pressure was just above the condenser vacuum pressure. The resulting increase in thermal efficiency more than compensated for the decreased power of the full stroke.[55] Hornblower, Trevithick, and Woolf came along later, took advantage of improvements in vessel fabrication and improved Watt's basic concepts by employing higher-pressure steam.

*   *   *

One feature was common to all of these designs. Water flowed around each in a *closed cycle*, meaning that water was neither lost nor gained during the cycle. It was boiled to steam by the furnace, did work in the piston, was condensed and then pumped back to the boiler to start its cycle all over again. The entire cycle was closed, with no inlet or outlet. It was admittedly hard to recognize the different steps involved in this closed cycle. It was much easier for the engineers to focus solely on the expansion of the steam in the piston to do work, since it was this step that inspired many to undertake experimental studies of gas behavior as a step toward bringing theoretical understanding to the efficiency of steam engines. But it was this step that was also just one aspect of a larger multi-step process. Others missed this insight. Carnot didn't. He was one of the first to truly recognize the closed cycle for what it was and to bring analytical inquiry to it.

Out of his analysis, Carnot was able to create a means to account for all of the relevant steps in the closed cycle by designing a hypothetical heat engine based on a working substance that remained <u>inside</u> the piston-in-cylinder system at all times during its operation. As part of his simplification process to cut through all the many moving parts of the engine to get to its heart, Carnot eliminated steam flow into and out of the cylinder, and created the pumping action of expansion and contraction by conceptually moving the piston-in-cylinder system back and forth between heat source and cold sink, respectively. In this way, Carnot transformed the complicated real steam engine into a simplified ideal heat engine, enabling the theoretical analysis that followed. While it may seem obvious in hindsight, as most great ideas usually are, it's quite amazing to realize how significant this transformation was. We still learn about the Carnot cycle through this idealization. It was an amazingly simple and rather monumental step in the history of engine analysis.

---

[54] Another inherent thermal inefficiency to the piston-in-cylinder is that while it was indeed desirable to keep the cylinder hot, the cooling of the gas during expansion lead to temperature differences between the gas and the cylinder wall.

[55] (Carnot, 1986) p. 7.

But what working substance would Carnot use in this engine? He could use water, as was being done in the commercial steam engines, employing the hot and cold sources to evaporate and condense as needed. In fact, he initially started down this path in *Reflections* but soon stopped. Why? He didn't say, but I'm assuming that it was because the theory and mathematics were very complicated. During all four steps, liquid water and vapor steam remained in equilibrium with each other, even when the process shifted from high temperature and pressure to low temperature and pressure, the analysis of which would later lead to the rise of the Clapeyron and Clausius–Clapeyron equations. Furthermore, during the adiabatic volume change steps, did incremental water vaporize to steam or did incremental steam condense to water?[56] In short, attempting to model the changes in such an equilibrated liquid–vapor system likely proved both theoretically and mathematically challenging to Carnot, which contributed to his choosing to operate his ideal engine with a non-condensing ideal gas. When combined with his closed-cycle approach, this decision tremendously simplified and clarified the analysis. It didn't matter that no such engine existed; it didn't matter that such an engine would be difficult to even operate due to the difficulty of achieving reasonable heat transfer rates into and out of the cylinder during the expansion and compression strokes;[57] it's practicality and speed/power were irrelevant; it was an ideal engine and all that mattered was that the approach and analysis worked.

Carnot next had to construct the steps of his ideal-gas closed cycle. As helped by the lessons of and discussions with Clément and also by his own readings, Carnot developed a modified approach of Watt's two-step power stroke, one that would, he hypothesized, generate the maximum amount of work from a given amount of heat based on his sole guiding principle of temperature difference minimization. He assumed no heat loss from the engine to the external environment and further assumed that gas temperature reached (just barely) that of the furnace and also of the condenser. But how would he operate the expansion–compression cycling of the enclosed gas to minimize temperature differences between furnace and the gas and then between the gas and the condenser?

Carnot was well aware of the fact that a gas cools as it expands while doing work.[58] He figured that he could use the furnace to keep the gas at a constant temperature, just below that of the furnace, throughout its initial expansion step (Step 1), and then conceptually remove the cylinder from the heat source to allow the subsequent adiabatic expansion, i.e., Watt's expansive step, to proceed (Step 2), during which additional work was done while both pressure and temperature decreased. These two steps are best understood when considering their trajectories on the PV-diagram in Figure 30.1. Such diagrams were first started but kept secret by Watts

---

[56] (Pupin, 1898) p. 88. As an interesting historical fact, some researchers in the 1800s reported measuring negative values for the specific heat of saturated steam. What they didn't realize was that when saturated liquid–vapor (water) expands adiabatically (Step 2 of Carnot's cycle), temperature decreases (in consequence of external work done by the expansion) more rapidly than the density decreases, resulting in condensation. In order to prevent this condensation, it would be necessary to supply heat during the expansion. During adiabatic compression, on the other hand, the vapor would become superheated.

[57] This engine was commercialized by Robert Stirling in 1816 and found some limited use.

[58] (Carnot et al., 1988) p. 16. Sadi Carnot: "The change of temperature occasioned in the gas by the change of volume may be regarded as one of the most important facts of physics…and at the same time as one of the most difficult to illustrate, and to measure by decisive experiments." Recall that in the caloric theory, expansion leads to a sensible-to-latent caloric shift.

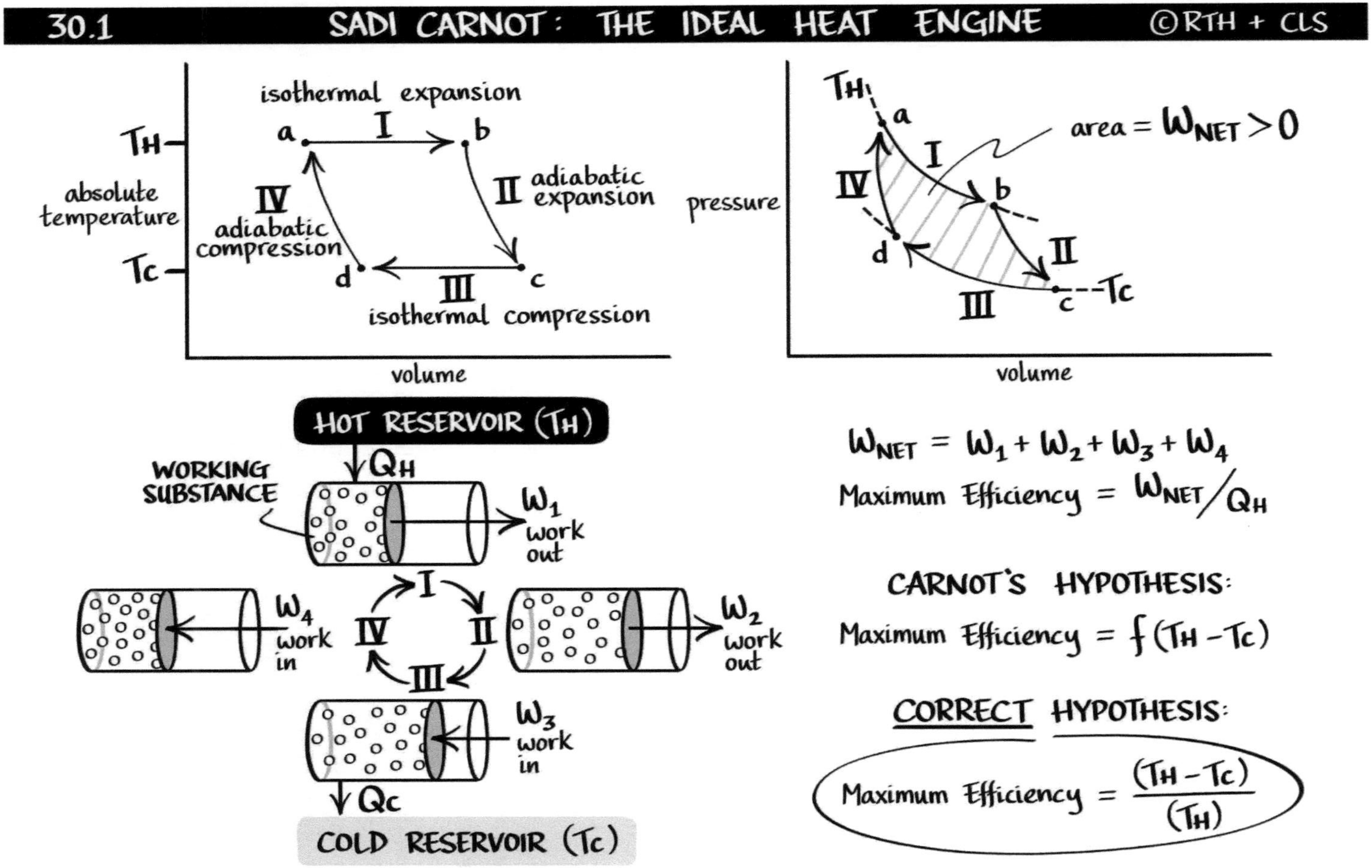

Figure 30.1 Sadi Carnot: The ideal heat engine

and Southern in their highly guarded indicator diagram quantification for work done by a moving piston and were then more fully developed by Davies Gilbert and later independently constructed for public consumption by Clapeyron to aid in his published analysis of Carnot's work, and this approach continues to aid us today. Both steps complied with Carnot's principle of minimizing temperature differences and ensuring that temperature changes result solely from volume changes, i.e., work generation. The work-generating adiabatic expansion provided inherent "productive" cooling of the gas from the furnace to the condenser without the need for unproductive conduction-based cooling.

In the end, Carnot's process generated work from the combination of these two sequential expansion steps, isothermal followed by adiabatic. So we're done right? Well, no. With only these two steps, you wouldn't be done. You would generate work once, and then that'd be it. This is where one of the "snares" Fox referenced lay in wait.

As inferred earlier but highlighted more now, Carnot's breakthrough in solving this puzzle happened when, from his larger perspective, he considered the fact that a continuous process is, well, continuous. This sounds silly, but for Carnot it was a critically important insight. A heat engine is a continuous process, not a one-time event. Once pressurized gas expands to its furthest point to do work, then what? For a continuous process, this step has to happen again, and again, and again.

It's easy to look at Step 1 of Carnot's cycle and call it a starting point. But this is somewhat misleading because you're starting with a pressurized gas, and such a gas isn't just sitting around in the environment, free, waiting to do work. It has to be prepared, and the energy needed to do this must be accounted for. All continuous heat engines that convert heat to work must account for the work required to close the cycle.

In Carnot's system, after the two-step expansion, the gas needs to be recompressed to its starting point, to close the cycle, so to speak, so that it can expand again and continue to do work, and this preparation itself requires work. In other words, some amount of work is required to do a larger amount of work on a continuous basis. So long as the latter is greater than the former, net work is positive. Carnot explained this considering all of the steps in the complete closed cycle.[59]

*   *   *

So how did Carnot close the cycle? How did he compress the gas back to its exact original starting position? Well, he could have simply reversed the expansion steps, starting with adiabatic compression followed by isothermal compression. But there would have been just one problem with this approach. No net work would have been generated. The work generated in the two-step expansion would have been exactly cancelled by the identical amount of work required in the two-step compression, thus rendering this engine a simple demonstration of heat-to-work and work-to-heat conversions.

Carnot realized that he couldn't go back the way he came. He needed a different path, one that required *less* work. This historical moment of realization arguably commenced our

---

[59] While the closed cycle was also present in the steam engine, it was somewhat hidden since the vaporization of the steam occurred in a vessel (the boiler) quite removed from the piston-in-cylinder assembly and the main work input occurred when pumping the condensate (vacuum) into the boiler (4 atm).

distinction between the path-dependent nature of heat and work and the path-independent nature of state properties.

Carnot's insight was that taking the gas at the end of its final expansion and compressing it isothermally and then adiabatically would require less work than the work generated during the expansion.[60] Why? Because, as shown in Figure 30.1, the isothermal compression occurs at lower pressure and work is proportional to pressure ($W = \int P dV$). By continually cooling the gas during its compression, using the condenser to maintain isothermal conditions just barely above the condenser temperature, thus fulfilling Carnot's temperature minimization guideline, pressure remains low until at just the right moment the cylinder is removed from thermal contact with the condenser and the remaining compression occurs adiabatically, thus raising both pressure and temperature back exactly to the original starting point, thereby completing the full cycle.

> The Real Physics Involved

Let's take a look at our current understanding of Carnot's closed-cycle operation, while using Figure 30.1 as reference, to understand the physics behind the four-step cycle. We'll make things easy by assuming an ideal gas. As we'll see later, it doesn't matter what the working substance is. Note that Carnot himself did not do these calculations. I'm doing them here to serve as a reference of what it was that Carnot did versus what the correct answer actually was.

In the first step, the gas expands ($V_2 > V_1$) isothermally at the temperature of the furnace (T_H) and does work. Since for an ideal gas in this isothermal process, $PV = RT = $ constant, then,

Step 1: Work done by ideal gas during isothermal expansion[61]

$$Work_1 = \int P \, dV = R\int \left(T/V\right) dV = RT_H \ln\left(V_2/V_1\right)$$

Similarly, the work done *by* the gas during isothermal compression in Step 3, which is negative since work is actually being done *on* the gas to reduce its volume ($V_4 < V_3$),[62] is

Step 3: work done on ideal gas during isothermal compression

$$Work_3 = \int P \, dV = R\int \left(T/V\right) dV = RT_C \ln\left(V_4/V_3\right)$$

---

[60] You can't obtain additional work by expansion to a temperature lower than the cold sink, because when you start your isothermal re-compression, there's no colder sink to maintain isothermal conditions since heat doesn't flow from colder to cold. So you have to do an adiabatic compression right back the way you came from; net work equals zero for going below cold sink.

[61] It's an important point to note here, a point not isolated by either Carnot or Clapeyron, that the inherent assumption in this and all other work-related calculations is that the pressure inside the cylinder is just infinitesimally higher or lower than that outside in the environment, depending on whether or not work is being done by the gas or on the gas, respectively. In effect, this is the same as saying that there isn't a finite gradient in pressure between the system and the environment because such a gradient would result in the presence of a net force and hence irreversible acceleration of the piston, causing it to slam into the cylinder support, leading to unwanted conversion of kinetic energy to waste heat. One can imagine the piston "reversibly" pushing up a number of small weights. As its pressure declines, small weights would be removed onto a platform at that height of the piston. The summation of the successively positioned small weights times their respective increasing heights is then the total work done.

[62] Clausius was a stickler with nomenclature, thank goodness. In his nomenclature for the 1st Law, $dU = \delta Q - \delta W$, the W refers to work done *by* the system. If work is done *on* the system, then the W in this equation is taken as the negative of this work. Internal energy increases when heat is added to or work is done on the system.

While I don't want to use space here to show the mathematics, which are available for the interested reader in most basic textbooks, because of the way Carnot designed his process, the work done by the ideal gas in Step 2 during adiabatic expansion and on the gas in Step 4 during adiabatic compression are the same and so cancel each other. After making these simplifications including the use of this relationship, $V_3/V_4 = V_2/V_1$,[63] whose proof is also available in a basic textbook, the net work done during Carnot's cycle using an ideal gas is the sum of the two isothermal steps.

$$\text{Work}_{net} = \text{Work}_1 + \text{Work}_3$$
$$= RT_H \ln(V_2/V_1) - RT_C \ln(V_2/V_1)$$

To calculate overall thermal efficiency, we now need to quantify the total heat required, Q_{in}, to do this work. Because internal energy (U) is a function of temperature only for an ideal gas—this property of the ideal gas played a critical role in Robert Mayer's work and will be discussed further in later chapters—there's no change in internal energy during isothermal expansion, and so the heat added (Q) is equal to the work done (W).

$$Q_{in} = \text{Work}_1 = RT_H \ln(V_2/V_1)$$

We can now calculate the maximum thermal efficiency for Carnot's cycle, which returns us to Equation 30.3.

Maximum Thermal efficiency for Carnot cycle
$$= \text{Work}_{net}/Q_{in}$$
$$= (T_H \ln(V_2/V_1) - T_C \ln(V_2/V_1))/T_H \ln(V_2/V_1)$$
$$= (T_H - T_C)/T_H \qquad\qquad [30.3]$$

Because T_H is greater than T_C, the maximum efficiency is positive and less than one. This result is based on the fact that more work is generated by the expansion of a given gas when it's hot (higher pressure) than when it's cold (lower pressure) and thus less work is needed to compress a given gas when its cold (lower pressure) rather than hot.[64]

Later I'll show how this equation evolves even more fundamentally from the concept of entropy and thus also show how it applies not only to an ideal gas but to any and all working substances. For right now, it provides us with a theoretical understanding of how the maximum thermal efficiency of Carnot's cycle is indeed governed by the furnace and the condenser, as Carnot hypothesized but wasn't able to mathematically prove as he assumed that heat is conserved and not consumed.

It's important to consider different interpretations of this equation. First and foremost, this equation sets the upper limit on how much work can possibly be generated on a continuous basis from a given amount of heat. It can't be used to determine the actual efficiency of a commercially operating heat engine itself; that is dependent on the operating conditions and

---

[63] (Planck and Ogg, 1990) p. 66.

[64] As we'll see later with the Thomson brothers' ice engine, it's not always the case that the dominant work generation step aligns with T_H. With ice-water as the working substance, the work step aligns with T_C for it's during this step that water freezes, expands, and does work.

procedures. But it can be used to tell you whether you did your calculations correctly. If the actual efficiency you calculate is larger than the maximum, then you did your calculations incorrectly. Work won't always reach the maximum, but it'll never exceed it.

Now let's consider how to make the maximum theoretical efficiency higher. There are two options. You could increase the temperature of the furnace—here it's inherently assumed that an increase in furnace temperature means an increase in working substance temperature as the two values are assumed equal in a reversible engine—but in doing so, the question you would face is whether the materials of construction and fabrication techniques you would use to build your engine would be sufficient to ensure safe operation of a gas contained at such higher furnace temperatures. At temperatures around, say, 500 °C, stainless steel starts to creep.

You could also decrease the condenser temperature. The concern here isn't the same as with the furnace temperature. With the condenser temperature, it's not whether you have the metallurgy and fabrication to withstand the cold temperatures; it's the fact that cold temperatures aren't readily accessible. We're pretty much stuck with ambient temperature, which is about 10 °C in ambient water and air. To go lower than ambient would itself require work; it would require refrigeration. Yes, we have access to a much lower ambient temperature (3 K) in outer space, but this raises all sorts of other issues. Bottom line: it takes energy to move matter away from ambient temperature and pressure. So ambient conditions become our reference point.

Assuming that the gas temperature is limited on the high end by mechanical constraints to 500 °C and that it can closely approach the condenser temperature (10 °C) on the low end given enough time (large surface area heat exchangers), then the maximum thermal efficiency for a heat engine is about 63%. Trying to achieve 100% with a heat engine is pretty much impossible.[65]

What Does the Thermal Efficiency of a Heat Engine Really Mean?

What does this last sentence of the previous paragraph mean? It does not mean that it's impossible to transform heat into work at 100% efficiency. Indeed, in Step 1 of Carnot's cycle with the ideal gas, the heat entering (Q) is directly converted (100%) into work (W). But this isn't what Equation 30.3 is based on. This is an important distinction. This equation says that it's impossible to convert the *continuous* flow of heat into work with 100% efficiency. It's this continuous conversion that's limited, simply because some of the work you generate in Step 1 is needed to re-invest back into the process to close the cycle.

To delve just a bit deeper into this, realize that while the above equation is based on the continuous conversion of heat into work, heat itself is <u>not</u> energy. Heat quantifies the *change* in

[65] (Keenan, 1970) See Chapter XII, pp. 175–200, for good discussion on heat engine cycles. The typical coal combustion product is around 1900 °C while the best material possible can operate at around 500 °C. A large part of the reason that the best modern heat engines can hardly attain efficiencies in excess of 30% is attributed to the large gap between the temperature of the hot source and the maximum operating temperature of the working fluid.

energy when thermal energy is exchanged between bodies. There is no absolute temperature assigned to heat (change in thermal energy) as manifested by the way that heat is quantified:

Change in energy due to thermal exchange = "heat" = Q = mCdT

Taking this one step further, consider that you have a hot gas stream resulting from, say, coal combustion (Figure 30.2). Assume for now that you have a Carnot heat engine that can continuously transform the thermal energy contained in the stream into work. In the ideal (maximum work) approach, an infinitesimal amount of heat is transferred to the heat engine at T_H. Again, the heat quantifies the change in energy of the hot gas stream. The heat engine converts the heat to work with an efficiency of $(T_H - T_C)/T_H$. Now what? Well, you still have a hot gas stream with just a little less energy and a little lower temperature. You could then have this stream transfer another infinitesimal amount of heat to another Carnot heat engine to generate more work with a somewhat lower efficiency since the exchange temperature is just a little bit lower than the initial T_H. You could line up an infinite series of Carnot engines to continue to generate increments of work from the gas stream at ever decreasing exchange temperatures until a final exchange temperature of T_C is reached at which point no further work could be generated. The series of Carnot heat engines operating in this way provides a mathematical construct to calculate the maximum amount of work that could be generated from the hot gas stream based on thermal energy exchange alone. This is a different answer than that quantifying the maximum amount of work that could be generated from one engine.

When you do the mathematics to quantify the maximum amount of work that could be generated from a continuous flowing stream—the calculation inherently assumes use of only reversible processes such as, for example, the aforementioned series of Carnot heat engines and reversible turbines that can generate shaft work—the answer is:

$$\text{Maximum work} = -(\Delta H_{rxn} - T\Delta S_{rxn}) \qquad [30.4]$$

where the Δ quantifies the change in the state properties of enthalpy (H) and entropy (S) between the initial and final states and the final temperature is equal to ambient conditions (T_C). We'll discuss these terms in more detail in Chapters 37 and 39 but for right now I wanted to show you where this discussion ultimately goes.[66]

*   *   *

In Carnot's ideal world, the only "fall" allowed for caloric is that which is productive. In our ideal world, the only contact allowed between bodies is when there's an infinitesimal temperature difference between them, for if there were a finite difference, then additional incremental work by a Carnot engine could be accomplished.

Carnot invented the optimal means by which to continuously transform thermal energy to work. His engine converts the energy of a hot stream at low pressure (which can't do any work "as is") into a hot stream at high pressure (by heating a closed system containing a movable boundary) that can then do work. He ingeniously combined isothermal and adiabatic volume change steps into a single continuous process that generates work from a heat source while eliminating all temperature differences between contacting bodies. His reversible process is really an equilibrium process, just like a teeter-totter. A little more weight on one side or

---

[66] For a more in-depth discussion, see (Tester and Modell, 1997) Chapter 14.

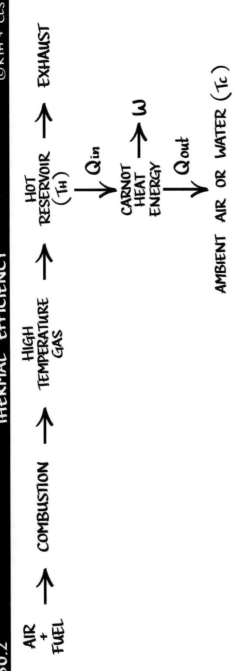

Figure 30.2 Thermal efficiency

the other, and his process can shift from heat-to-work to work-to-heat, which is the refrigeration process and will be discussed shortly. This invention was truly remarkable and has withstood the test of time.

Carnot Did the Best He Could with What He Had

In *Reflections*, Carnot didn't provide a rigorous mathematical approach to constructing his cycle. His brother, Hippolyte, indicated that this may have been intentional, suggesting that Sadi wished to write a readable text for the general public about the potential of steam engines.[67] While this may have been part of the reason, I don't believe it was the whole reason. Clearly, if Carnot knew how to do the math, he would have. He proposed, based on his intuition, an unknown relationship between the maximum thermal efficiency of the steam engine and the difference in temperature between the furnace and the condenser, and he also constructed a closed-cycle, multi-step operation to achieve maximum work. The next logical step would have been to link the two together by writing out the caloric equivalent of an energy balance for each step and then combining the equations in a similar method to the one shown above to determine the form of the unknown relationship. Surely, if he could have done this, his passion on this subject would have had him do it. And it's not that he didn't try, because he did, including a multitude of calculations throughout *Reflections* dealing with physical property relationships and such, but not the critical fundamental equations involving some aspect of a caloric balance (as Clapeyron would later do) or an energy balance (as Clausius would do) around each step. Why not? It wasn't that Carnot didn't know what he needed to do. It's that he didn't have the right tools to do it with. It's time again to revisit the culprit behind this.

*   *   *

In the early 1800s, Carnot was working in a scientific community struggling to understand gas behavior. We take it for granted today the existence of the conservation of energy and the kinetic theory of gases and can readily apply these theories to the study of gas expansion and compression. But in 1824 such theories were still far off into the future, leaving as the only theories available the caloric theory of gas combined with the static-atom theory of gas structure. This is why Carnot saw the steam as nothing more than a substance needed to carry the caloric, nothing more.

While progress by the science community on this front was happening, it was quite slow, hampered by the "extraordinary degree of ignorance about adiabatic heating in France during the eighteenth century."[68] As of 1819, in spite of the experimental progress of such eminent French experimentalists as Gay-Lussac, Clément and Desormes, and Delaroche and Bérard, there were no published measurements quantifying temperature rise in a gas due to compression: "the heat capacity of existing thermometers was too large, and they responded too slowly to permit measurement by straightforward techniques."[69]

---

[67] (Carnot et al., 1988) Mendoza's Introduction, p. x.
[68] (Fox, 1971) p. 49.
[69] (Kuhn, 1958) p. 137.

Further revealing the community's ignorance of adiabatic processes in particular was the fact that attempts to calculate the related temperature changes were essentially abandoned since the processes were assumed to be isothermal.[70] Why? Because it made the mathematics much easier. But this assumption was clearly a mistake, as indicated by Antoine Vène,[71] a military engineer and Polytechnique graduate (four years before Carnot), who strongly endorsed the need to address the drop in temperature during Watt's expansive operation, after the steam cut-off, since the failure to do so led to a tendency to significantly overestimate the work produced during this step. Vène also pointed out that no attempt to study the cooling, either experimentally or theoretically, had been successful. The fact that Vène wrote this in 1837 only serves to further emphasize the challenges Carnot faced.

The bottom line was that science was in a muddle on this subject. The physical model and related calculations Carnot needed in order to answer his rather challenging questions were not available. Consider the questions: *How much heat is needed to maintain the temperature during isothermal expansion?, How much does temperature drop during adiabatic expansion?, How does one calculate the heat requirement when considering a saturated vapor (liquid water vaporizing and expanding in the cylinder) as opposed to an ideal gas?* To this day, such questions confront students in the classroom, even during open-book tests. Carnot understood the importance of the challenge, writing "The change of temperature occasioned in the gas by the change of volume may be regarded as one of the most important facts of physics, because of the numerous consequences which it entails, and at the same time as one of the most difficult to illustrate, and to measure by decisive experiments."[72] Given the difficulty of the calculations as well as the lack of experimentally validated theories and equations to work with, Carnot still moved forward, doing the best that he could with the caloric theory, proposing and hypothesizing new ideas as needed, while admitting at times, with reference to how caloric was involved in such expansion and compression calculations, "we do not know what laws [the caloric] follows... Experiment has taught us nothing on this subject."[73]

That Carnot moved forward at all in spite of the above reflects that he had passion, intelligence, new ideas, and at least some good information to work with. For example, he did recognize at the time of his writing, at least directionally, the observed adiabatic effects of compression and expansion on temperature, just not the theory behind them. As he wrote, "When a gaseous fluid is rapidly compressed, its temperature rises. It falls, on the contrary, when it is rapidly dilated. This is one of the facts best demonstrated by experiment. We will take it for the basis of our demonstration."[74] He cited the work of Gay-Lussac and Welter, and also that of Laplace on the speed of sound, in which air was assumed to heat during sudden compression in sound waves. The fact that the underlying cause of these temperature changes and their associated mathematical formulations wouldn't truly be understood until the conservation of energy (and eventually the kinetic theory of gases) had arrived some twenty-five

---

[70] (Fox, 1971) p. 181–182.
[71] (Carnot, 1986) p. 38.
[72] (Carnot et al., 1988) p. 16.
[73] (Carnot et al., 1988) p. 16.
[74] (Carnot et al., 1988) p. 15.

years later didn't hinder Carnot from moving forward. He had some of the key experimental results and a theory, however false, to start with, and that was enough.

Entropy – A Preview

It's amazing how significant Carnot's invention of the closed-cycle analysis really was. He was the first to recognize the need for this approach. While some of the most talented minds of the day had begun to take an interest in the theoretical problems raised by the new steam engines, all except Carnot "had considered the work done in the expansion stroke *only* [my italics] and not the work done in a complete cycle of operation."[75] Carnot understood the need to consider the full cycle when analyzing the heat engine.

As shown in the mathematics above, the closed-cycle analysis leads directly to the relationship between thermal efficiency and temperature, something that Carnot believed must exist but was uncertain as to its form. While Carnot himself never got to this finding, the logic behind it led the way to Clausius' successful re-interpretation. But this was just one of several important findings that rose from Carnot's logic, the combined total of which would lead to the 2nd Law of Thermodynamics and heat–work energy limits. And this is where our challenge in understanding Carnot lies.

Certain findings from the analysis of Carnot's closed-cycle process, findings made by Carnot himself and by others at a later date, arise from the fact that a state property called entropy exists. This property manifests itself in different ways, which I'll get into later. Unfortunately, the historical timeline placed the manifestations before the discovery of the property in 1865. Had it been the other way around, the ability of entropy to unify the different aspects of Carnot's cycle may have brought more clarity to classical thermodynamics. But then again, perhaps not, since entropy itself is quite difficult to understand and wouldn't be understood until long after Clausius discovered it.

Carnot's extension of the impossibility of perpetual motion into the realm of heat-to-work conversions led Clausius in 1850 to identify two principles:[76]

1. Heat–work equivalence: In all cases in which work is produced by the agency of heat, a quantity of heat is consumed which is proportional to the work done (and vice versa)

2. A transfer of heat from a hotter to a colder body always occurs in those cases in which work is done by heat, and in which also the condition is fulfilled that the working substance is in the same state at the end as at the beginning of the operation.

As will be discussed in later chapters, these two principles later evolved into the two laws of thermodynamics based on energy (1850) and entropy (1865), respectively. Clausius finished what Carnot started.

I raise this point now because I'm about to delve into the key interpretations as laid out by Carnot. I feel that it's important to share Carnot's thinking here as it importantly and significantly influenced Clausius. I realize that it may be confusing at times to follow but will attempt

[75] (Fox, 1971) p. 179.
[76] (Carnot et al., 1988) p. 112 and 133.

where I can to shed light on the physics behind the interpretation to help facilitate your own understanding.

Carnot's Logic for Why His Engine Was the Best

Carnot stated that the closed-cycle heat engine that he created was second to none. To him, his engine generated the maximum amount of work possible since the entire fall of caloric from high to low temperature occurred while work was generated, with no caloric being wasted via non-volume-generating conduction.

This rather bold leap was based on Carnot's deep belief in the impossibility of perpetual motion. He intuited that the conservation of mechanical energy should also apply to the unseen world. He just couldn't prove this to be true; the presence of caloric in his unseen world prevented him from doing so. While Carnot's caloric-influenced idea to maximize work by ensuring no caloric "fall" without work was based on a false premise, it led to the correct answer as it minimized temperature differences in his closed-cycle engine. Minimizing such differences eliminated heat loss from the engine while stretching the steam cycle to the full extent possible as defined by the temperatures of the furnace and the condenser. As a fictitious concept, caloric naturally had nothing to do with this reality.

The crux of Carnot's main argument in support of his "second-to-none" hypothesis—it was still an unproven hypothesis at this time—was that his ideal heat engine cycle was, in fact, reversible.[77] Recall Déparcieux's concept, which Carnot very likely had read about, of a reversible water wheel in which all changes were done over infinitesimal gradients. Carnot applied a similar logic to his engine. By minimizing all temperature differences throughout his cycle, each point in the cycle could be easily reversed with an infinitesimally small change in temperature.[78] Thus, the entire cycle could be reversed in this way, which brings us to the refrigeration cycle.

The Refrigeration Cycle

In the refrigeration cycle, or the air conditioning cycle, both specific applications of the general heat pump cycle (Figure 30.3), starting at the same initial point (a) in Carnot's work-generating cycle as before, with a compressed gas, instead of isothermally expanding as per Step 1, the gas could adiabatically expand as per the reverse of Step 4. Then, once temperature reached the condenser temperature, the gas could isothermally expand as per the reverse of Step 3. The

---

[77] (Lewis and Randall, 1923) p. 112. "Such an ideal process, which we will call reversible, is one in which all friction, electrical resistance, or other such sources of dissipation are eliminated. It is to be regarded as a limit of actually realizable processes." Generally speaking, a reversible process is one in which the process exists in near-equilibrium with only an infinitesimal energy gradient(s) to drive the process in one direction or the other. In such a process, no energy is lost from the system, meaning that with a change in direction of the gradient(s), the process would operate in reverse with no need of additional energy.

[78] Throughout his cycle, it was understood although not stated that the force caused by the pressure of the gas was exactly balanced by an equal and opposite force, such as could be generated by manipulating weights on top of the piston.

30.3   CARNOT: THE IDEAL HEAT ENGINE IN REVERSE (HEAT PUMP)   ©RTH + CLS

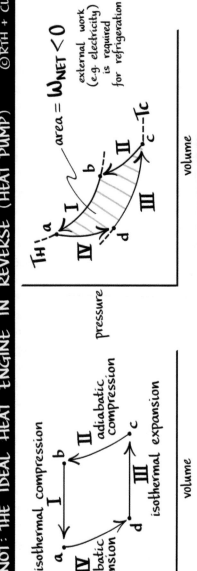

isothermal compression

II adiabatic compression

IV adiabatic expansion

III isothermal expansion

volume

T_H ♨ absolute temperature

❆ T_c

(in the case of air conditioning)

♨ T_H
- outdoor summer air cools the working fluid during compression (heat flows from working fluid into outdoor air)

❆ T_c
- indoor air is cooled as it is passed over the cold coils containing the working fluid (heat flows from family room into cold fluid)

area = $W_{NET} < 0$

external work (e.g. electricity) is required for refrigeration

T_H : a   I   b
II
IV   d
III
c ⋯ T_c

pressure

volume

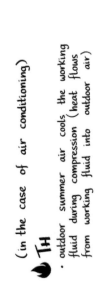

HOT RESERVOIR (T_H)   ← Q_{out}

WORKING SUBSTANCE

work in

work out

I
II
IV
III

work in

work out

COLD RESERVOIR (T_c)   ← Q_{in}

if the cold reservoir is a family room, then this heat pump performs as an air conditioner

Figure 30.3 Carnot: The ideal heat engine in reverse (heat pump)

presence of the condenser would prevent the gas temperature from decreasing further; the condenser in this scenario ends up "heating" the gas instead of cooling it; conversely, the gas ends up cooling the fluid in the condenser. At the end of this expansion, external work would be needed to adiabatically compress the gas back up to the hot temperature source and then isothermally compress the gas back to its starting point. In this last step, the heat source actually ends up "cooling" the gas instead of heating it in order to prevent its temperature from rising during the compression. As you have already realized, this reverse cycle is the refrigeration cycle. The net work is negative, meaning that work must be done on the gas system (by the environment). Contemporary refrigeration cycles depend on electrical work—hence high summer electrical bills—to operate the compression pumps, and also rely on materials that can be condensed, just like in the steam engine. High-pressure liquid is vaporized via expansion to low pressure; the cool vapor cools the water or air that is circulated through the living quarters. The low-pressure vapor is then compressed and condensed via cooling by passing the pressurized gas through an outdoor coil. In this scenario, while the outdoors is hotter than the indoors, it's the outdoor temperature that actually cools the hot fluid during compression.

Carnot stated that the work done to drive the reverse cycle must *exactly* equal the work generated by the original cycle since caloric is not lost in either cycle. Furthermore, out of this logic, he hypothesized that no other cycle, involving, say, a different working substance, could possibly be more efficient than his ideal cycle. His cycle produces the maximum amount of work in which all steps are reversible with no flow of caloric wasted (and no accumulation or depletion of caloric in the working substance itself). No other process could do better because if it could, then perpetual motion would result. Here's Carnot's next level of logic on why this is so.

In Carnot's mind, if there were a more efficient process available than his, one that could produce more work for a given "fall" of caloric from high to low temperature, then this process could be connected to his work-generating process, with the work output from his serving as work input to the other operating in the reverse direction. (Bear with me here; it does get confusing at times.) In this scenario, if Carnot used a given amount of caloric in *his* cycle to generate a certain amount of work, then this work could be used to drive the reverse operation of the "more efficient" cycle, which would result in a higher flow of caloric from cold to hot than Carnot initially used to flow from hot to cold. In the end, the net work done would be zero, but the net flow of caloric would be from cold to hot. This would set up perpetual motion since this flow of caloric would occur from nothing (remember, net work equal to zero) and so could be used to drive Carnot's heat engine to generate unbounded work. This result would also run counter to the common everyday experience that heat only flows from hot to cold. So this was Carnot's basis for stating that his engine was the best, second to none, the governing hypothesis being that perpetual motion is impossible. In reality, the real basis was entropy. Because all thermal energy exchange occurs with only infinitesimal temperature gradients, entropy change for Carnot's reversible process is zero. We'll get more into this later.

Nature of Working Substance is Irrelevant

As mentioned earlier, another concept that entropy made impossible was that any working substance could be better than any other. This was such an amazing (and correct) conclusion of Carnot's. How could anyone then have imagined such a thing to be true? To this day, it's difficult to grasp that the working substance is irrelevant to the engine's maximum performance. The decision on what substance to use is purely practical and not theoretical; while some fluids may be ideal for an engine, meaning that they don't break down when spanning the temperature range and don't otherwise damage equipment, others may not be. Whether ideal gas, non-ideal gas, liquid–vapor systems, solids, liquids, or Feynman's rubber band, it doesn't matter. None performs any better or worse in the Carnot cycle. In his ideal engine, all fluids work, both ideal and non-ideal. So long as Carnot's minimum-temperature-difference criteria are met for the complete cycle, furnace and condenser temperatures alone determine the maximum thermal efficiency, nothing else. In Carnot's words, as long as the working substance is employed in the most advantageous manner possible, there is no "useless re-establishment of equilibrium...in the caloric,"[79] meaning that work is maximized. In today's words, it's because the addition of an infinitesimal amount of heat (δQ) to a body at a given temperature (T) increases entropy by $dS = \delta Q/T$, *regardless of the composition of the body*, that the nature of the working substance is irrelevant. What a truly remarkable property of matter.

Carnot's conclusion on the irrelevance of the working substance was especially remarkable given that it was "totally at odds with contemporary opinion."[80] While some of Carnot's contemporaries thought that air was better than steam, and others felt that steam was better than air, the one solution that no one other than Carnot considered was that neither was better than the other. This conclusion was so unexpectedly different and unusual that it too may have contributed to the low readership of and interest in Carnot's work. Why would someone care about such theoretical abstractness when more pressing practical matters were at hand, such as welding techniques?[81]

To be clear, the fact that the working substance is irrelevant to a heat engine's maximum theoretical performance doesn't mean that it doesn't impact the process design and operation. Being an engineer, Carnot recognized this by readily highlighting the practical difficulties of working with certain substances that appeared "ill fitted"[82] for use in a heat engine, as they could result in unwieldy large pressures or volumes or even corrosive and unsafe environments depending on their physical property relationships, such as saturated pressure for a given furnace temperature. While any substance would perform as well as any other, in theory, some were clearly more practical than others. Carnot's favorite substances were gases and vapors, since they "are the means really adapted to the development of the motive power of heat."[83] Even with gases, Carnot was very practically minded, stating that working at high pressures could be costly since you would need strong vessels to contain the pressure but also large vessels to contain the fully expanded gas. Thus, instead of building one large and thick vessel, he was a proponent of building several vessels as part of a multi-stage engine such as Woolf's for which the initial high-pressure vessels would be small and thick while the later low-pressure vessels would be large and thin. The net effect he was targeting was lower total vessel cost. Carnot's conclusion about the theoretical irrelevance of the working substance remained very

---

[79] (Carnot et al., 1988) p. 18.
[80] (Carnot, 1986) p. 24.
[81] (Carnot, 1986) p. 24–25.
[82] (Carnot et al., 1988) p. 47.
[83] (Carnot et al., 1988) p. 48.

powerful since it told engineers what *not* to do; it told them to not use improved theoretical efficiency as a criterion for seeking other working substances.

The Clapeyron and Clausius–Clapeyron Equations

Imagine instead of an ideal gas that a liquid–vapor system is used in Carnot's cycle. Imagine next that heat enters the cylinder and causes the working substance to isothermally expand (due to vaporization) to do work, as per Step 1 of Carnot's cycle. Since the liquid–vapor system remains throughout (not all of the liquid evaporates), then pressure also remains constant during the isothermal expansion. The huge expansion upon vaporization is what does the work.

Next remove the cylinder from the furnace and allow it to expand further (Step 2), cooling the system as it does. Now imagine Step 3 when the liquid–vapor system is isothermally compressed at a lower temperature and pressure than in Step 1. The condenser maintains the isothermal conditions. Because the pressure is lower, less work is required to do this than the work generated in Step 1. Then remove the cylinder from the condenser and compress the working substance further, adiabatically such that the temperature and pressure rise back to the initial starting point. During all four steps, the working substance continues to be in vapor–liquid equilibrium the entire time.

The complete cycle involves equations based on equilibrium temperature and pressure (P^{sat}) for the working substance along with the substance's heat of vaporization. The heat input (during Step 1) is equal to the heat (enthalpy) of vaporization times the total mass vaporized.[84] The __net__ work done in the closed cycle is equal to the increase in volume upon vaporization (of that mass) times the difference in saturated vapor pressure between the two isotherms. The fascinating thing is that when you write out these equations to quantify the maximum thermal efficiency of this system, the answer you get *must* be identical to the one you got for an ideal gas. This is one amazing implication of Carnot's conclusion that the working substance is irrelevant. When you equate the maximum thermal efficiencies for vapor–liquid and ideal gas, you arrive at an equation relating the change in saturated vapor pressure with temperature (dP^{sat}/dT) to the heat of vaporization. This is the famed Clapeyron equation.[85]

$$dP^{sat} / dT = \Delta H_{vap} / \left(T \, \Delta V_{vap} \right) \qquad [30.5]$$

Carnot really deserved his name on this equation as well. It was he who started the inquiry into this situation when he led off *Reflections* with a physical description of such a liquid–vapor system based on water, which simply reflected the reality of the steam engine at that time in which the steam used in Step 1 was always saturated; superheat wouldn't arrive commercially

---

[84] $dU = \delta Q - PdV = \delta Q - dPV$ (for isobaric process). Thus, $dU + dPV = d(U+PV) = dH = \delta Q$ = enthalpy of vaporization. The use of enthalpy accounts for the isobaric work done.

[85] This equation is also sometimes referred to as the Clausius–Clapeyron equation but in truth, the Clausius–Clapeyron equation is of a different form, being derived from the Clapeyron equation by assuming the vapor to be an ideal gas: $d \ln Psat / d (1/T) = \Delta Hvap / R$.

until around 1900.[86] Carnot began *Reflections* with this system, but then shifted to the ideal gas system later on as it was conceptually easier to handle. But Clapeyron picked up the thread, brought rigorous mathematics to it based on the caloric theory, and then Clausius revised the results based on the energy theory. All-in-all, the evolution of this equation was a fascinating application of Carnot's work to describe real-world phenomena. It was Clausius' subsequent validation of this equation (among others) with published experimental data that provided him the larger validation he needed to convert Carnot's findings from hypothesis (assumption) to law, what he called his 1st and 2nd Principles and what later became the 1st and 2nd Laws of Thermodynamics.

Reflections Wrap-up

> *We will show how far we are from having realized ... all the motive power of combustibles*
> – Sadi Carnot[87]

Carnot began his book with the stated goal of determining the maximum efficiency that could be obtained for a given amount of heat. His mind was indeed focused on the larger issue. Throughout his book, he raised the right questions and answered them the best he could using his own ideas together with the theoretical understanding and experimental data he had. Although his belief in caloric ended up limiting his analysis, he was intelligent enough at the end of *Reflections* to return his focus to the key issue at hand: overall efficiency and not just thermal efficiency. As Bernoulli had hypothesized about the work contained in a given weight of coal, so too did Carnot, using the new ideas he brought to the world.

Carnot took all that he had, ideas, theories, and data, especially the heat released during the combustion of coal based on calorimeter data, and estimated that only 1/20th of the maximum work potential in coal was being utilized by contemporary steam engines. Yes, others had noted the probable existence of such a large gap. But still, Carnot's head-on calculation of it using his ideas was a great conclusion to *Reflections*.

*   *   *

Despite his incorrect belief in the presence of caloric during this inquiry, Carnot's thinking remains largely valid to this day. His scientific approach and logic were sound. His identification of certain "best practices" was true. The conclusions around the working substance's irrelevance and the closed-cycle approach to analyzing thermal efficiency were valid. The main issue that wasn't valid, that flawed his book, was caloric. Because of this, the correct mathematical analysis couldn't be completed. Work was generated while heat (in the form of caloric) was conserved, a mistaken assumption that Clausius would later correct. Carnot's lack of tools to solve this problem left him somewhat stranded, without a means to quantify thermal efficiency, although as just discussed he made an excellent attempt at the end of his book to do just this starting from coal.

*   *   *

---

[86] In 1898 Dr. Schmidt started the commercial application of superheaters to railway locomotives, the first superheated engines being placed in service on the Prussian State Railway.

[87] (Carnot et al., 1988) p. 57.

There. We made it through Carnot's work, all the terms, concepts, and logic.

So given this rather monumental effort of thinking, how was Carnot's book received by the public? It wasn't. Apparently hardly anyone bought it.[88] While Carnot's working at the "margins"[89] of French science may have helped keep his mind open to new ways of thinking, it certainly didn't help generate a large readership. Also not helping in this regard was the fact that his ideas may have been so different as to not have been interesting to those seeking practical engineering guidance. Indeed, his ideas "were of a kind that invited neglect."[90] Given this, why then was Carnot's work so monumental? Why did William Thomson proclaim it to be one of the greatest scientific works ever written?[91] To answer this, we take a step back to gain a larger view of his work and his impact.

Sadi Carnot – A True Scientist

When a hypothesis no longer suffices to explain phenomena, it should be abandoned. This is the case with the hypothesis which regards caloric as matter, as a subtle fluid.

– Sadi Carnot[92]

Generally speaking, when one is living on (seemingly) solid ground inside a given paradigm, life is comfortable. All is known; all is certain. But when the foundations of that paradigm start to crack, the comfort fades. This is especially so when one realizes the need to let go of the existing paradigm but without having a new paradigm ready to grab onto, like a trapeze artist letting go of the swing, high above the ground with no other swing waiting.

Such discomfort and uncertainty are felt by any scientist thinking about making a break from the past. Galileo most likely felt this as he let go of Aristotle but had not yet seen the modern era of physics waiting, the one that Newton would create. Similarly, Carnot felt the discomfort as he started to let go of the caloric theory but had not yet seen energy, entropy, and the rise of classical thermodynamics. His discomfort was perhaps not due to fear but rather due to immense frustration. He was well on his way to making this transition from one paradigm to the next when his unfortunate and untimely death arrived. He never made the full transition, unknowingly leaving this task to Joule, Rankine,[93] Thomson, and Clausius, whose subsequent work arguably brought meaning and posthumous fame to Carnot. At the time they discovered him, Carnot's ideas had not yet been validated.

Galileo, Carnot, and also Thomson each started their scientific lives firmly and deeply planted in a false paradigm and then accomplished remarkable progress during their respective attempts to climb out. The advances each made reflect the genius each possessed. They opened the door for others to pass through. Steven Hawking once said, "Galileo, perhaps more than any other single person, was responsible for the birth of modern science."[94] The same

[88] (Carnot et al., 1988) p. xi.
[89] (Carnot, 1986) p. 43.
[90] (Carnot, 1986) p. 25.
[91] (Carnot, 1986) p. 1.
[92] (Carnot et al., 1988) p. 68.
[93] (Smith and Wise, 2009) p. 327. "Thomson's interaction with Rankine in 1850 was crucial."
[94] (Hawking, 1988) p. 179.

could be said of Carnot and his role in the birth of classical thermodynamics. Unfortunately for Carnot, this wasn't realized by anyone, including himself, before his death.

Carnot's was a very honest scientific inquiry, an honest quest for truth, not fame, and a strong open-minded willingness to challenge his own assumptions. While this philosophy made its presence known in the different parts of *Reflections* where Carnot openly questioned the caloric theory, it even more profoundly showed itself in Carnot's private notes[95] that were originally unpublished and thus almost permanently categorized as yet another case of Taleb's "silent evidence" had his brother, Hippolyte, not found them and made them public in 1878 long after Sadi's death.

In these notes, which were interestingly enough written about the same time as *Reflections*, one learns that at some moment in time before his death, Carnot finally rejected the caloric theory. It is rather striking to read his refutation, which was based on a list of experimental facts that the theory couldn't explain, including, for example, Rumford's work and especially Gay-Lussac's experiment showing no change in temperature for the free expansion of a gas into a vacuum. The caloric theory simply couldn't handle this phenomenon since, according to the theory, such a change should have noticeably decreased temperature. Recall that during adiabatic expansion, temperature was predicted to drop due to the sensible-to-latent shift. But Gay-Lussac demonstrated that this didn't happen. While the caloric theory could explain many phenomena, it couldn't explain this one, a very demonstrable one, among others. As Carnot firmly stated, "When a hypothesis no longer suffices to explain phenomena, it should be abandoned. This is the case with the hypothesis which regards caloric as matter, as a subtle fluid."[96]

One can almost sense the frustration and despair Carnot experienced at this moment of writing, which read as if he were ready to fall on his sword in defeat. It reminded me of the similar experience of Anthony Hopkins' character in *The Remains of the Day* (1993), who served as a butler to English nobility during WWII, and suddenly realized at the end, when he discovered his master's sympathetic ties to the Nazi cause, that what he thought was so, wasn't. Reality came down hard on what he believed and left him questioning what he had accomplished. Reality must have come down hard on Carnot. Disheartening, to say the least.

The truly frustrating aspect of all this for Carnot must have been the fact that while he had apparently let go of one paradigm, he didn't have another to grab onto. No second trapeze swing was waiting for him. Yes, he seemed to be moving quickly toward the concept of energy as manifested by his private notes where he made an attempt to quantify the mechanical equivalent of heat, where he proposed new experiments to explore in better detail the theory of heat, including some that were later conducted by Joule, and where he even spoke of the logical progression of the conservation of mechanical energy into the world of molecules and heat.

So what stopped him? Why didn't he, or why *couldn't* he readily see the connection between heat and motion, just as Bernoulli did earlier in his approach to a kinetic theory of gases? Well, there's a lot of guesswork to this, but a reasonable attempt can be constructed, once again by

---

[95] These notes are included in (Carnot et al., 1988).
[96] (Carnot et al., 1988) p. 68.

trying to see the world through Carnot's eyes, which were somewhat clouded by the Newtonian view of the structure of gases.

While Carnot and others could conceivably eliminate caloric and alternatively account for heat with the *vis viva* of gas molecules, they couldn't envision the molecules themselves as freely moving. Instead, they viewed them as static vibrating entities. This was Newton's unintentional doing. When he proposed[97] in the *Principia* that Boyle's Law (PV = constant) could be predicted by assuming an inverse-square law of repulsion between two adjacent gas atoms, Newton assumed that the gas atoms were static for the sake of mathematical ease and readily admitted that while the question of whether or not the gas atoms were indeed static had not been answered and was a matter for future discussions, as was the question of what exactly the cause of the repulsive force was, the important thing was that the mathematics worked. Future Newtonians ignored his caveat, grasped onto the "static-atom" theory, and eventually incorporated the presence of a subtle fluid called caloric between the static atoms that could account for the repulsive force between the atoms and also could account for all heat-related phenomena.

As of the early 1800s, both the caloric theory and the static-atom theory remained strong in the minds of many, combined as they were into the caloric theory of gases, establishing in the process an even deeper paradigm about the nature of matter, one that had the French physicists effectively ignoring Bernoulli's kinetic model of his "hither and thither" gas atoms. When this combined theory started to fall apart, such as in failing to explain Gay-Lussac's findings, it wasn't the entire theory that was thrown overboard but only one aspect of it, caloric. The static theory remained standing. So, as Mendoza pointed out,[98] while Carnot indicated in his private notes that he accepted the macroscopic equivalence of heat and work, he couldn't see how this equivalence rose from the microscopic mechanics, largely because his mind was stuck on static gas atoms, and he couldn't understand, Newton couldn't understand, no one could understand, how this structure was possible. No one could explain how the gas atoms remained apart from each other in a static arrangement, especially once the caloric theory was jettisoned. Again, one gets the sense that Carnot must have felt stranded in no-man's land.

* * *

Now one didn't need to believe in the atomic theory of matter to arrive at the concept of energy and the equivalence between heat and work. Mayer certainly didn't. He believed in the existence of energy as separate from and independent of matter, but this was irrelevant since he rightly felt he didn't really need to define the essence of energy in order to conclude that it was a "constant" quantity, comprising both heat and work. One didn't need to know why, just as one didn't need to know why Newton's gravity did what it did. The math worked.

It seems, though, that Carnot needed to have a valid physical model to work from before he could truly embrace energy. Without it, he couldn't make the jump, despite having all the evidence he needed. His questions reveal the barrier that he couldn't overcome, strongly suggesting that he simply couldn't see how the mechanical theories could work in a gas of static atoms. "How can one imagine the forces acting on the molecules, if these are never in contact with one another, if each one of them is completely isolated? To postulate a subtle fluid in between

[97] (Carnot, 1986) pp. 6–8.
[98] (Carnot et al., 1988) pp. xviii–xix.

would only postpone the difficulty, because this fluid would necessarily be composed of molecules…Is heat the result of a vibratory motion of molecules?"[99]

A great speculative question for historians to answer is this.[100] What would have happened if Carnot had lived? By the end of his life, he had rejected the caloric theory. In fact, he had started this process while he was in the midst of writing *Reflections*, having made corrective alterations to his manuscript prior to publication.[101] But given this starting point, that of him living, what would he have done? He was still very far away from discovering such concepts as energy and entropy. He had laid out experiments similar to Joule's. But even with Joule's findings, would he have completed his paradigm shift from caloric to energy as William Thomson would eventually do? If so, would he have revised *Reflections* and created the same piece of work that Clausius did when he reconciled Carnot's original work with Joule's findings? Or would he have been stuck after rejecting caloric, not really knowing where to turn, left to sit in doubt about his whole approach to the matter of steam engine efficiency. As Mendoza proposed, by the time *Reflections* was finished, Carnot began to suspect the whole idea of cycle of operations.[102]

* * *

While Carnot, in the end, was not able to prove his hypothesis, the techniques and approaches he used in confronting the problem became extremely valuable to those who eventually did. He blazed the path that others followed in creating the foundation of classical thermodynamics. While he himself did not found this field, as Hawking said of Galileo, he was responsible for its birth.

In the end, the water wheel engine analogy that launched Carnot's work pointed him in the right direction,[103] even if it was for the wrong reason. He set up the correct framework and logic for the closed-cycle engine, but couldn't solve them. This is where caloric truly failed him. In the end, the caloric theory didn't necessarily invalidate *Reflections*, it just "flawed" it.

* * *

While many may not have read *Reflections*, all it took was one. Thank goodness for Émile Clapeyron.[104] It was he who kept Carnot alive by discovering *Reflections* and feeling compelled to clarify the work through the use of mathematics and the very helpful PV-indicator diagrams which made understanding Carnot's multi-step cycle much easier. His 1834 publication (*Memoir on the Motive Power of Heat*, translated to English in 1837) retained Carnot's assumptions, especially the existence of caloric and its conservation, and also his main conclusions. While this publication, as for Carnot's original, didn't draw a large audience, it did give a new look to Carnot, one that eventually found its way to William Thomson, among others.

---

[99] (Carnot et al., 1988) p. 62.

[100] An equally great speculative question is, what would have happened if there hadn't been a Sadi Carnot? How would entropy have been discovered?

[101] (Carnot et al., 1988) p. 46.

[102] (Carnot et al., 1988) p. 46.

[103] (Lewis, 2007) p. 55. "It is intriguing that the development of thermodynamics should owe so much to the water-wheel as well as to the steam engine."

[104] (Lewis, 2007) p. 59. Lewis speculates that Carnot could possibly have met Clapeyron in 1832, prior to Carnot's death.

It's funny how things happen. Consider this very thin thread that connected Carnot to Clapeyron and then to Thomson and eventually to Clausius. It could have easily broken at some point along the way. Thankfully it didn't. But it's still a fascinating guessing game to wonder what would have happened if it had. How would entropy have been discovered without this thread? Perhaps an even more fascinating game is to wonder what discoveries *haven't* yet been made because of broken threads we don't even know about?

31 Rudolf Clausius

The establishment of thermodynamics during the middle years of the nineteenth century – one of the great creative epochs of science – was so rapid and extensive that a whole new branch of knowledge was suddenly created, where previously there had been the old science of heat plus the substantially empirical technology of heat-engines.

– D. S. L. Cardwell[1]

The revival of Carnot in the mid-1840s was perfectly though inadvertently timed to collide with the emergence of Joule. The basis of each body of work represented two fundamentally incompatible views of the conversion of heat into work. But which one was right?

Different aspects of this question were addressed in a group of publications ranging from the mid-1840s to the early 1850s by a small group of scientists. One of the first to confront the question head-on was Joule himself. In his 1845 publication on work–heat equivalence, Joule wrote:

> The principles I have adopted lead to a theory of the steam-engine very different from the one generally received, but at the same time much more accordant with the facts. It is the opinion of many philosophers that the mechanical power of the steam-engine arises simply from the passage of heat from a hot to a cold body, no heat being necessarily lost during the transfer. This view has been adopted by Mr. E. Clapeyron [who is in agreement with] Mr. Carnot...I conceive that this theory, however ingenious, is opposed to the recognized principles of philosophy. Believing that the power to destroy belongs to the Creator alone...any theory which, when carried out, demands the annihilation of force [i.e., energy], is necessarily erroneous. The principles...which I have advanced in this paper are free from this difficulty. From them we may infer that the steam, while expanding in the cylinder, loses heat in quantity exactly proportional to the mechanical force which it communicates by means of the piston...Supposing no loss of heat by radiation, the theory here advanced demands that the heat given out in the condenser shall be less than that communicated to the boiler from the furnace, in exact proportion to the equivalent of mechanical power developed.[2]

It's remarkable to me how advanced and clear Joule was in his thinking. Had he wrapped his theory in mathematics, he most certainly would have gained at least some of the attention and fame that Clausius would later receive for completing this task.

---

[1] (Cardwell, 1971) p. 277.
[2] (Joule, 1845)

Block by Block: The Historical and Theoretical Foundations of Thermodynamics,
Robert T. Hanlon, Oxford University Press (2020). © Robert T. Hanlon.
DOI: 10.1093/oso/9780198851547.001.0001

So, who else was involved? There was Karl von Holtzmann who, in 1845, published his analysis of Clapeyron's work that included a calculation of the mechanical equivalent of heat.[3] Interestingly enough, though, and as Clausius pointed out, Holtzmann did this calculation even though he "tacitly assumes that the quantity of heat is constant."[4] Then there was Hermann von Helmholtz, who, in his famous 1847 publication *On the Conservation of Force*, which brought the concept of "conservation" into the thermodynamic lexicon, referenced Holtzmann, Clapeyron, and Joule, but didn't appear to grasp the importance of applying Joule's work–heat equivalence to the operation of Clapeyron's steam engine. Then you had William Thomson, who published an account of Carnot's theory in 1849 that embraced the conservation of heat even though, again, interestingly enough, he was well aware of Joule's work, the two having met in 1847. Additionally, you had William Rankine who published his own ideas in 1850 based on his reading of Clapeyron in 1837 while a student at Edinburgh. Apparently, Rankine had developed his fundamental theories on heat–work conversions in 1842 but, for lack of data, had to wait until Regnault's data were available in 1849 to complete his work in 1850.[5] Finally, we arrive at Rudolf Clausius. As touched on in Chapter 22, Clausius, inspired by Thomson's 1849 publication and identifying the need to correct its underlying premise, came out with his own famed publication in 1850.

Not to be forgotten were the experimentalists, namely Joule, with his experimental determination and quantification of work–heat equivalence, and, just as importantly, Victor Regnault (1810–1878), whose experimental data on the properties of saturated steam proved critical. Without the collective data of both, the theorists would have been left without a reality check. Just as Johann Kepler had Tycho Brahe, the founding theorists of thermodynamics had Joule and Regnault, Joule for his singular focus on determining the mechanical equivalent of heat nearly a dozen different ways[6] and Regnault for determining with a high degree of accuracy properties of saturated steam across a wide range of temperatures and pressures, a particularly important contribution in light of the fact that in one of Carnot's two ideal steam engines based on liquid–vapor, steam is always present at saturated conditions. The other ideal engine was based on an ideal gas. As Cardwell pointed out, "Without Regnault's experimental data [and Joule's, I might add] thermodynamics would have been impossible, just as Kepler's laws could not have been established without Brahe's extensive and accurate ephemerides."[7]

The flood of ideas, some right, some wrong, some right for the wrong reason, released during this time period was tremendous, and from this flood rose the foundations of classical thermodynamics and statistical mechanics. The conversion of these ideas into the written word has provided great source material for the historians, especially as concerns priorities to certain of the ideas. For the sake of brevity, however, I choose not to get into the detailed nuances of the thinking of each of these aforementioned great men, but choose instead to focus primarily on the path taken by Clausius, as it was at the end of this path that the concept of

[3] Clapeyron's 1834 publication was translated into English in 1837 and into German in 1843.
[4] (Carnot et al., 1988) p. 111.
[5] (Hutchison, 1981)
[6] (Cardwell, 1989) p. 208.
[7] (Cardwell, 1971) p. 240. Furthering this quote: "Regnault and Brahe, together with Kelvin and Kepler, exemplify the truth of T. H. Huxley's dictum that the advance of science depends on two contrasting types of men: the one bold, speculative and synthetic, the other positive, critical and analytic. It is worth remarking, too, that apart from his experimental gifts Regnault, like Brahe before him, was very conservative in his scientific views."

entropy materialized in 1865. Having chosen this path does not mean to imply that Clausius worked alone, for he certainly benefited from the ideas of others, such as Joule and especially Thomson. After all, it was Thomson's 1849 publication with its intriguing question—*what happens to the mechanical effect which is lost when heat flows from a hot to a cold body not by way of a heat engine but by simple conduction?*—that strongly drew Clausius' attention. The back and forth, yin–yang publications of Thomson and Clausius over the subsequent years are central to the history of thermodynamics.[8] But in the end, Clausius was really the first to pull it all together and so I focus on his path, which began with Thomson.

> William Thomson

[For Thomson, Reflections was] one of the greatest scientific works ever written.
> – Robert Fox[9]

In 1845, while spending four-and-a-half months expanding his education in the laboratory of experimental physicist Henri Victor Regnault in Paris, Thomson, all of 21 years old and newly graduated from Cambridge, chanced upon Clapeyron's publication and its reference to Carnot. Unfortunately, Carnot's *Reflections* was a "bibliographical rarity"[10] in Paris when Thomson tried in vain to track it down by stopping at every book shop he could think of. While some of the booksellers had heard of Lazare and Hippolyte, who was to later rise to fame in the French political world, none had heard of Sadi.[11]

When he couldn't find Carnot's referenced work, which he wouldn't find until 1848 and which must have cast at least some doubt on its validity during the time between his reading Clapeyron and finding Carnot,[12] Thomson trusted his judgment—perhaps it was his youth that kept him open-minded—that there was something of importance to Carnot's line of approach and so pursued a course of study in this direction. When he finally read and absorbed Carnot's work, he published *An Account of Carnot's Theory* in 1849 and it was this publication that then inspired Clausius toward his own 1850 *On the Motive Power of Heat* in which Clausius stated that he hadn't been able to get a copy of Carnot's original article and only knew of his work through the works of Thomson, Clapeyron, and Holtzmann.

> Rudolf Clausius

---

[8] The relationship between Thomson and Clausius served science well as it afforded the opportunity for healthy Darwinian competition between their differing views. They weren't "yes men" to each other; instead, they challenged each other and we're the better for it. Each read the other's work and built on what they read with their own ideas. Thomson's analysis of Carnot and Clapeyron in his works of 1848 and 1849 caught the attention of Clausius. Clausius' revision of Carnot with work–heat equivalence caught the attention of Thomson. It would go back and forth like this in the years to come. They both struggled with conduction and a true understanding of heat; each was slow to accept the concept of energy. Clausius focused in, Thomson focused out. The scientific debate between them was healthy and hugely beneficial to science. It's unfortunate that priority claims, as spurred on by Tait, had to get involved.

[9] (Carnot, 1986) p. 1.

[10] (Carnot, 1986) p. 40.

[11] (Thompson, 1910) Vol. 1, pp. 132–3.

[12] (Carnot, 1986) p. 40.

When Clausius sat down to study the works of Carnot and Joule, he had to reconcile the very fundamental difference between them. This wasn't at all easy. While the supporting evidence behind Carnot's caloric remained elusive, and while caloric's star power had already started to fade, the fact was that the caloric paradigm ran very deep in the minds of many and so held a tight grasp on progress. The shift from Carnot's caloric to Joule's work–heat equivalence should have been straightforward, but it wasn't, largely because Joule's correct ideas were still very new. Frankly, to many it wasn't totally clear what Mayer and Joule were even talking about in the mid-1840s.

Not helping the matter was that Mayer's approach, although entirely correct and ground-breaking, was somewhat complicated, involving as it did the indirect determination of the mechanical equivalent of heat through calculations involving two separate measurements of ideal gas heat capacities (C_p and C_v). Fortunately, Joule's work was much more direct. The weight fell, the paddle spun, the water's temperature rose. People could see it, in effect, with their own eyes.

The problem, though, was that while Joule's experiments were visually real, the central explanation behind them wasn't. Energy was, and continues to be, an intellectually challenging concept, especially when compared against the easy-to-visualize stream of flowing caloric, however wrong this vision was. The fact was that the shift from caloric to energy represented a "tremendous leap in abstraction."[13]

So while one may reflect, *"But Joule's experiments were clear. What more was needed?"*, it wasn't that easy. Consider, for example, that while Joule demonstrated the conversion of work (falling weight) into heat (increasing water temperature), he didn't show the opposite, the conversion of heat into work, or at least he didn't do this is in a way that was clear to others.[14] But why was this distinction important? Because the steam engine was all about the conversion of heat into work, not the other way around. This was the issue that confronted such intelligent men as Thomson. Recall from Chapter 21 that Thomson wrote in 1848, "The conversion of heat (or caloric) into mechanical effect is probably impossible, certainly undiscovered."[15] This was a remarkable statement from one who was intimately familiar with Joule's research that demonstrated the conversion of mechanical effect (work) into heat. Needless to say, achieving clarity in this situation was not trivial. To Joule, work to heat and heat to work were the same. To Thomson, they weren't.

Furthering the challenge of reconciling Carnot and Joule was that both approaches were clearly sound. However wrong the basis, Carnot's work was mostly and brilliantly correct. But how could both be sound and yet both be so fundamentally different? Surely, either one or the other must be correct, but not both. Thomson, who was raised in and strongly influenced by the caloric theory, recognized this. In reviewing Carnot's work, he referenced Joule's work showing that heat is *not* a conserved quantity and was deeply bothered. "If we abandon this principle [the conservation of heat], we meet with…an entire reconstruction of the theory of heat from its foundation."[16] He couldn't easily let go, even in the face of such powerful data. But

---

[13] As shared in a 2013 email to the author from Professor Hans von Baeyer.

[14] As pointed out in Chapter 21, Joule felt that he had demonstrated heat to work conversion based on his demonstrated cooling of compressed air when expanded against atmospheric pressure.

[15] (Cardwell, 1989) p. 95.

[16] (Carnot et al., 1988) p. 111.

eventually he did let go and became part of the team doing the reconstruction effort that must have seemed so daunting when he wrote this sentence.

Fortunately, it was at this time in 1850 that Clausius, all of twenty-eight years old, the same age as Carnot was when he wrote *Reflections*, published his famed paper that finally brought much needed clarity to this situation, sweeping out the cobwebs, so to speak, and effectively killing caloric in the process. In looking at the discrepancy between Carnot and Joule, and perhaps gently chiding Thomson along the way, Clausius stated, "we should not be daunted by these difficulties."[17] He simply cut the Gordian knot by switching the situation from "either/or" to "and." One didn't need to reject one or the other. One could keep both, but just not as they originally stood, or more accurately, not as Carnot originally stood, for this is where the critical flaw lay. Clausius saw the truth in Joule's finding of work–heat equivalence. As shared in the previous chapter, he also saw the truth in Carnot's work: heat must flow from hot to cold whenever work is done in a cyclic process. It was Clausius' keen mind that identified the single flaw that needed to be dealt with: Carnot's belief in caloric and specifically his belief that caloric (heat) is conserved. By decisively uniting Carnot and Joule through casting aside the belief that heat was conserved, Clausius arguably launched the age of thermodynamics with his 1850 publication and the two "consequences" he extracted from Carnot's *Reflections*, forerunners to the two laws of thermodynamics. As Gibbs would later claim, this work of Clausius' brought "order out of confusion"[18] and marked "an epoch in the history of physics."[19]

The Challenge of Creating a New Paradigm

Before moving on, a note is warranted regarding Clausius' writing style. As a stickler for exact definitions and terminology, Clausius was quite rigorous with his writing, which helped for the sake of brevity and intentionality. The words he chose and later created meant very specific concepts. Unfortunately, as it would also be for Gibbs, his writing approach for such a complicated subject didn't always result in clarity for the reader. If you didn't understand it the first time around, there was no second time. Clausius didn't spend much additional time clarifying certain of his statements, perhaps because such statements were already clear to him, thus leaving the reader to have to work rather hard to understand or guess at his logic, but also perhaps because he himself struggled to conceptually unite Carnot and Joule. One can only imagine how often Clausius may have deeply questioned the highly abstract concept of energy. In such a mindset, the challenge of untangling the concepts and terminologies associated with the dying caloric from those that would soon be associated with energy was quite large. In fact, this task was never completed. Consider that we continue to talk about a "flow" of heat and the different varieties of heat: latent, sensible, liberated, and free. All of these and more, like heat capacity and heat balance, as opposed to energy balance, are relics of the caloric theory. These terms confuse us, largely when we attempt to interpret their literal meaning in today's world of

---

[17] (Carnot et al., 1988) p. 111.
[18] (Gibbs, 1889) p. 460.
[19] (Gibbs, 1889) p. 459.

energy. Clausius certainly faced similar issues. The point here is simply to convey that the job of creating a new paradigm with strictly defined words was an immense undertaking.

The Critical Question Confronting Clausius: What is Heat?

At the core of his theoretical journey, Clausius had to confront the question, *what is heat?* The concept of heat still existed—the work of Joseph Black remained valid. It's just that it wasn't conserved. Nor was it a thing or a substance. So it needed a new definition, a new look, a new feel.

How Clausius actually viewed heat in 1850, especially the underlying fundamental explanation of what heat was, I'm not sure. He may have had ideas as revealed in a subsequent 1857 publication,[20] the first in a powerful sequence of works establishing the kinetic theory of gases, in which he mentioned that he had already formed a "distinct conception" of the nature of heat-as-motion prior to his 1850 publication. But in 1850 he didn't want to share these ideas as he sought a more generalized approach, one not dependent on (nor potentially jeopardized by) any assumption regarding the nature of heat itself. This was the same logic used by Joseph Black and Mayer, neither of whom speculated about what heat was, and even Newton, who didn't speculate about what gravity was. Such definitions weren't needed for each to build their respective arguments. In fact, venturing into such territory with conjecture could have unnecessarily jeopardized their work, as it ultimately did for Rankine, whose contributions to thermodynamics, while valid and significant, didn't yield him deserved fame largely because he rested his theories on the rather nebulous proposal that heat phenomena resulted from spinning atoms. "It was this metaphysical assumption, together with a certain obscurity of style, that deprived Rankine of the full measure of credit that would otherwise be his as one of the two founders of the modern theory of thermodynamics."[21]

But while Clausius intentionally kept discussion of heat-as-motion out of this 1850 paper, you can read between the lines his tacit support for this hypothesis. For example, he wrote, "facts have lately become known which support the view, that heat is not a substance, but consists in a motion of the least parts of bodies. If this view is correct, it is admissible to apply to heat the general mechanical principle that a motion may be transformed into work, and in such a manner that the loss of *vis viva* is proportional to the work accomplished."[22] He was careful not to claim whether or not he believed such a view, but his writing strongly suggests that he did.

Interestingly, given his rising belief in heat-as-motion, Clausius never flatly rejected caloric. In fact, he never even used the word, but instead relied on the word "heat." For Clausius, it was all about the rejection of the conservation of heat. Such differences and nuances may appear subtle, but in trying to reconstruct how someone viewed nature, such differences are significant.

---

[20] (Clausius, 2003a) p. 111.
[21] (Cardwell, 1971) p. 254.
[22] (Carnot et al., 1988) p. 110.

It's possible that for Clausius there was no caloric to reject but just a concept of heat that needed to be re-defined. I'm really not sure as historians don't know much about Clausius' thinking apart from his papers; no extensive secondary literature on Clausius the person exists.[23] At times, I read in his words that while he may have rejected the conservation of heat, he may still have retained certain concepts originating from the caloric theory. In referring to Clausius' view (post-1850) that the work accomplished in an engine cycle is caused by the fall of heat, Daub wrote, "Undoubtedly, Clausius was still under the influence of the caloric theory."[24]

Clausius' Two Principles

As mentioned in Chapter 30, in his 1850 publication, seeking simplicity and generalization, Clausius identified two "consequences" that became "principles"—predecessors to the two laws of thermodynamics—after his analysis of Carnot's work:

1. Heat–work equivalence
2. Heat must flow from hot to cold whenever work is done in a cyclic process.

Because Clausius arrived at these principles after testing their consequences against available data, the two principles started off, in effect, as two hypotheses in Clausius' mind. Thus, his publication itself became a great manifestation of how to use the scientific method to test and validate hypotheses. His sound and very logical approach is what made this publication strong and influential and thus worthy of spending time exploring, as we'll do now.

> Clausius' 1st Hypothesis:
> Heat–work Equivalence

Clausius tested the heat–work equivalence hypothesis by first challenging Carnot's central assumption that heat is a conserved quantity. Wrote Clausius, "I am not aware…that it has been sufficiently proved by experiment that no loss of heat occurs when work is done."[25] "On the contrary, [it may] be asserted with more correctness…to be…highly probable"[26] that heat is *not* lost but instead consumed when work is done. Here he references, among others, Joule's falling-weight experiments in which heat is produced by mechanical work. He then proposed to test the hypothesis that heat and work are connected by assessing the consequences of such a hypothesis, "since it is only in this way that we can obtain the means wherewith to confirm or to disprove it."[27] This was the scientific method at work, and it was the weapon that Clausius chose to use in this battle of ideas. Writing about one such consequence, "It is common to

[23] (Daub, 1970a) p. 311.
[24] (Daub, 1970a) p. 305–6.
[25] (Carnot et al., 1988) p. 110.
[26] (Carnot et al., 1988) p. 110.
[27] (Carnot et al., 1988) p. 112.

speak of the *total heat* of bodies, especially gases and vapors, by which term is understood the sum of the free and latent heat, and to assume that this is a quantity dependent only on the actual condition of the body considered…"[28] Here he was using the word "heat" in the sense of caloric. Clapeyron had (incorrectly) treated heat/caloric as just such a property of state and performed calculations involving the total differentiation of heat based on its pressure and volume dependence: $Q = f(P,V)$. In a rather monumental historical step, Clausius broke from the past by saying that this assumption "is no longer admissible if our principle is adopted." In other words, since per the 1st Principle heat isn't conserved, then heat can't be a property of state.[29]

So while Carnot saw the fictional caloric flowing from the boiler to the condenser, with work somehow being generated by this "fall" from high temperature to low due to some caloric shift from sensible to latent with total caloric being conserved, Clausius said no, it doesn't happen like this. Heat—again, Clausius never mentioned the word caloric—isn't conserved. It enters the steam from the boiler; some is then converted into work while the remainder is rejected to the environment. Further, as Joule showed, a direct relationship exists between the conversion of the heat and the generation of work; they're quantitatively linked by the mechanical equivalent of heat that puts heat and work into the same units so that they can be mathematically combined. As we understand today, it is the generation of work itself that *simultaneously* results in a decrease in energy and temperature; atoms rebound from the outward moving piston at slower speeds. Whereas Carnot's thought process at times reflected some disconnect in time between these two events,[30] Clausius correctly saw the two as occurring simultaneously.

Heat-work Equivalence Provides New Approach to Analyzing the Heat Engine

With this logic, Clausius employed Joule's equivalence of heat and work to take a new look at the heat engine. Recall that Carnot hypothesized (through analogy) that the amount of work generated by the flow of caloric from hot to cold temperature was proportional to some unknown function of the height of the fall $(T_H - T_C)$. In Carnot's mind, total caloric didn't

[28] (Carnot et al., 1988) p. 112.
[29] You can start at a given point on the PV-indicator diagram and move around the diagram by doing varying amounts of heat and work (to effect changes in P and V), but when you arrive back at the starting point, the net change in heat and the net amount of work done in the closed-cycle will only be zero if you follow the *exact* same path back as you followed out. You can add heat to move P and V, and then remove the same amount of heat and return to the original P and V. Likewise, you can do work to move P and V, and then have work be done to return to the original P and V. In these two scenarios, the net change in heat entering and leaving the body is indeed zero, and the net change in work done by and on the body is also zero and so one could argue that (especially) heat and work are properties of the body. But this logic falls apart when you start combining both heat and work during the cycle, as Carnot showed, i.e., as you start choosing a different path to return that you used to go out. Here the net changes in heat and work are non-zero. This is how work can be generated from heat. This is what Carnot showed, although he wasn't able to see the equivalence between work and heat on account of his belief that heat was conserved.
[30] In Carnot's writing the decrease in temperature seemed to occur almost as an afterthought once expansive work was done; he made no direct connection between work done and temperature decrease. To him, the temperature decrease was simply caused by the shift in caloric from sensible to latent.

change, just the distribution of caloric between sensible and latent forms. So long as caloric was never lost from the process due to, say, conduction or radiation, it was conserved. But with his 1st Principle in hand, Clausius saw something different. By conducting what amounted to an overall energy balance of the engine and ignoring what happened inside the engine as this didn't change, Clausius saw that a certain amount of heat entered the engine, a portion was consumed in the generation of work and the remainder was rejected.[31] Work doesn't just magically appear—there really was no rationale for the generation of work in Carnot's explanation—but instead is only generated at the same time as heat is consumed. In other words, the total work generated is directly proportional to the total heat consumed. The below equations capture this difference between Carnot and Clausius. Their simplicity belies the complexity of the concepts involved.

Carnot: Heat is conserved

$$Q_{in} = Q_{out}$$

Work $= f(T_H - T_C)$ and results from caloric shift from sensible to latent

Clausius: Heat is consumed

$$Q_{in} - Q_{out} = A \times Work \qquad\qquad [31.1]$$

$$A = \text{heat consumed/work done} = 1/(MEH)^{32} \qquad\qquad [31.2]$$

where MEH = the mechanical equivalent of heat = 1/A

Clausius said that the existence of his 1st Principle must be valid, largely on account of Joule's meticulously gathered data. He referenced the fact that Joule quantified the value of 1/A from multiple work–heat experiments, such as heat produced from electricity, work produced when atmospheric air is heated, and heat produced by falling-weight / paddle-wheel experiments using either water or mercury. To Clausius, the agreement of all of these 1/A's measured from totally different experimental concepts left "no further doubt of the correctness of the fundamental principle of the equivalence of heat and work."[33]

More Consequences of Heat–Work Equivalence – Again Asking, Where Does the Heat Go?

Assuming heat–work equivalence, Clausius then asked, as he would continue to ask time and time again, if heat can be consumed, *where does it go?* Such a seemingly simple but quite

---

[31] The concept of heat entering and leaving ("rejection") a body is naturally misleading. The gas needs to be cooled during compression, so yes, its heat is "rejected" but in reality the collisions between the gas and the cold sink at the boundary cause the gas atoms to decrease in kinetic energy, and thus the gas to cool, as quantified by the term, Q. Q quantifies the change in energy resulting from the exchange of thermal energy. During this exchange, no "substance" is lost from the gas. Thus, the word "rejection" itself doesn't help clarify what's actually happening.

[32] Typically, the mechanical equivalent of heat (MEH) is not included in thermodynamic equations as the units of heat (Q) and work (W) are both based on energy. I included "A" here as it re-appears in this chapter as an important aspect of Clausius' work.

[33] (Carnot et al., 1988) p. 152.

insightful question. In Clausius' new world, if heat disappears in one place, it must re-appear in another, either as work done or heat generated (one body heating another). These were the only two possibilities. In a significant simplification, he viewed all thermal phenomena in terms of heat and work.

But this wasn't enough. Clausius needed to go deeper. As Joule showed, heat can be consumed to do work, but what *kind* of work? Embracing his "We should not be daunted!" philosophy, Clausius confronted this question.

The work associated with the expansion and contraction of a gas in a piston was reasonably well known and generally accepted at the time. It was quantified by the area under Clapeyron's PV-indicator diagram ($\int PdV$). But this only quantified the amount of work done in relation to the environment, such as in the lifting of a weight. There was another type of work that needed to be considered, specifically the work done in relation to the body itself, even more specifically the work associated with moving the interacting atoms apart from each other. Although Clausius admitted no atoms in this discussion, he acknowledged that matter existed and required work to separate itself from other matter as quantified by the integration of the attractive force over the distance moved.

Clausius accounted for this type of, as he called it, "internal work"[34] when he considered the vaporization of a liquid. He realized that work is needed to push a vapor molecule of water away from the bulk liquid in order to overcome the forces of attraction, and that this work consumes thermal energy, or heat. It's the same thing as lifting a weight in a gravitational field. Each involves application of a force on a body through a distance in a direction opposing an attractive force, either gravitational or chemical. Work is done on the body and the body's potential energy increases such that total energy remains constant. The rules of the mechanical world apply largely intact to the chemical world.

Summarizing his assessment, Clausius concluded that when heat (Q) enters a body—here, his focus was clearly influenced by the first step in Carnot's cycle in which the working substance receives heat from a hot source—causing changes in, say, volume and temperature, it can do any or all of the following: increase temperature (this is related to Black's sensible or free heat), do internal work (this is related to Black's latent heat), or do external work (W). This made four terms in all to account for. The Q and W terms were well known at the time. The challenge was with the two terms describing the internal conditions of the body itself, free heat and internal work. The former was philosophically tied to Joseph Black's sensible heat and had been captured quite well by Black's studies relating thermal heat addition and heat capacity to a body's change in temperature. The latter was philosophically tied to Black's latent heat but still represented a whole new and somewhat nebulous concept and so needed its own definition.

The Arrival of Internal Energy (U) and the 1st Law of Thermodynamics

In yet another historical cutting-the-Gordian-knot moment, Clausius reasoned that he didn't need to isolate free heat and internal work. It wasn't necessary. He didn't need to know what

_____

[34] (Carnot et al., 1988) p. 122.

was happening inside the proverbial "black box" body of interest. He just needed to account for its total internal change and this could be done by a simple equation.[35]

$$\text{Heat Entering a Body} = [\,(\text{change free heat in body}) + A \times (\text{do internal work within body})\,] + A \times (\text{do external work on the environment})$$

Heat Entering a Body = Change in Internal State of Body + A × External Work

$$Q = \Delta U + A \times W \tag{31.3}$$

$1/A$ = mechanical equivalent of heat

Clausius' 1st Principle of heat–work equivalence validated the additive nature of the internal heat and work terms comprising U. How these terms were distributed was, at least in the above form of his 1st Principle, irrelevant to Clausius. He just cared about the sum of the two. As I share in Chapter 34, he would later attempt to distinguish between them by introducing a term called *disgregation*.

From this thought process, Clausius created a property, U, as "an arbitrary function of V and T."[36] While left unnamed at the time, which was uncharacteristic of Clausius,[37] this term later became referenced as *internal energy*, one of the core properties we use in thermodynamics. To Clausius, the change in U results from "the free heat that has entered[38] and the heat consumed in doing internal work" and, furthermore, "has the properties which are commonly assigned to total heat."[39] The historical use of the symbol Q by Clapeyron,[40] among others, to bring mathematics to caloric-related concepts, however vain such attempts were, as reflected by the fact that Q could not be strictly and mathematically defined, had been incorrectly and confusingly thought of as being associated with two entirely different entities, the thermal heat entering a body and the total heat contained within the body. It was this latter association that incorrectly led to the consideration of heat as a property of the body, one uniquely determined by pressure and volume, as reflected in Clapeyron's differential equations for Q. Clausius clarified this situation, limiting Q to a single definition only, namely the quantity of thermal energy entering a body, and then creating a separate term, U, to account for a certain total quantity internal to the body, not "total heat" but what would become "total energy" or simply internal energy, a quantity comprised of both free heat (kinetic energy) and internal work (potential energy). Here, I'm using energy terms that would arrive later to explain things in ways that we

---

[35] I modified the equation Clausius actually used, which was dQ = dU + dW, to this equation, Q = ΔU + W, to reflect that neither Q nor W are differentiable, as we now know these terms to represent quantities of energy exchanged between systems as opposed to properties of the systems themselves.

[36] (Carnot et al., 1988) p. 121.

[37] (Daub, 1970a) p. 304–305. According to Daub, the reason why Clausius didn't name U was because the heat in a body H (sensible) and the internal work J (latent) were his fundamental concepts. Thomson, on the other hand, in developing his mechanical (or dynamical) theory of heat in 1851 coined the name *intrinsic energy*, which later evolved to *internal energy*, "since it represented the total mechanical work that might be theoretically obtained from a substance."

[38] This is another interesting statement of Clausius' that again reflected remnants of caloric, "free heat" being different than heat itself, seeming like the sensible heat of the caloric theory, only detectable by temperature.

[39] (Carnot et al., 1988) p. 122.

[40] (Jensen, 2010) "Clapeyron most likely selected the letter Q to emphasize that he was dealing with the *quantity* of heat rather than with its *intensity* or temperature, for which he used an uppercase letter T."

understand today. As Clausius defined it, U was not based on any assumed molecular model, a fact rendering it with all the more significance, one that all could embrace, not just those who agreed with any given molecular assumption.

Embedded in this equation is the 1st Law of Thermodynamics based on the conservation of energy. In a body isolated from its environment, for which both Q and W are equal to zero, there can be no net change in internal energy.[41] Also embedded is work–heat equivalence, showing how to use the mechanical equivalent of heat (MEH) to unite the two. In this paper, however, Clausius did not draw attention to this sense of conservation, even though his equation said as much. In fact, Clausius wouldn't embrace the larger umbrella concept of energy until years later.[42]

In a final observation about this development of his 1st Principle, it's rather telling that Clausius started his equation by writing "Q =" as opposed to how we're currently taught the 1st Law of Thermodynamics, in which we write the identical equation, but starting with "ΔU =." A small difference and yet one that revealed an insight into Clausius' thinking, focused as it was on a key question regarding Carnot's first step, *where does heat go once it enters a body?*

Using the 1st Law to Study the Properties of Saturated Steam

With new equations based on heat–work equivalence in hand, Clausius could begin quantifying the cause–effect phenomena of the material world and so further validate the 1st Principle. Of particular relevance at that time was naturally the properties of the dominant working fluid: saturated steam.

Recall in Watt's expansive operation that saturated steam from the boiler entered the cylinder and pushed the piston down into the cylinder vacuum. In this mode of operation, the steam's pressure at the end of the stroke remained at that of the boiler and so when this piston-volume of saturated steam was exhausted to the vacuum condenser, energy was wasted since this pressurized steam could have otherwise been used to generate more work. Watt fully understood this and so creatively improved the process by identifying a point during the stroke at which the boiler feed line would be shut off, thus isolating the cylinder from the boiler, leaving the steam to expand adiabatically while experiencing a decrease in both temperature and pressure. Ideally, the piston stroke would end when the pressure was just slightly above vacuum.

In this new and improved mode of operation, which became incorporated into Carnot's ideal cycle as Step 2 (adiabatic expansion), the interesting question for many was, did the steam remain saturated at the end of the adiabatic stroke? What an interesting problem to solve! Using his 1st Principle of work–heat equivalence within Carnot's cycle, Clausius concluded, after a detailed series of calculations that I won't go into here, that a certain amount of vapor condenses during the adiabatic expansion stroke, while during adiabatic compression,

---

[41] Expanding this concept to the universe, itself an isolated body (or so we believe!), Clausius later stated that the energy of the universe is constant.

[42] (Clausius, 1867) Sixth Memoir – Appendix (1864), p. 251. "The definition I have given [U], being for general purposes too long to serve as the name of the quantity, several more convenient ones have been proposed... The term *energy* employed by Thomson appears to me to be very appropriate... I have no hesitation, therefore, in adopting, for the quantity U, the expression *energy of the body.*"

the temperature rise is great enough to cause the saturated steam to become superheated. Wrote Clausius, "It must be admitted that this result is exactly opposed to the common view [involving conservation of heat]; yet I do not believe that it is contradicted by any experimental fact," whereupon he referenced the work of Pambour[43] in which the temperature of steam exhausted to the atmosphere at the end of the expansion stroke matches that of saturated steam, which would be consistent with such steam appearing as white puffs, indicating condensation—steam is otherwise invisible. What a great extension of his theory—and a great validation of the 1st Principle—to help resolve a long-standing question that didn't readily lend itself to direct measurement.[44]

The usefulness of this analysis became more than an academic question when new and improved steamengines were designed.[45] While the Carnot cycle is indeed the most efficient of all heat engine cycles, it doesn't lend itself to practical commercialization. The rate of heat transfer across the cylinder walls required to maintain isothermal conditions during work generation in Step 1 is simply too high for any reasonable amount of power generation (work per time). It is for this reason that an alternative cycle evolved in which continual evaporation of water takes place in Step 1, thus meeting Carnot's isothermal criteria. Water continually enters a boiler, is evaporated, and then drives a turbine to do work. The complete cycle involved here constitutes what is known as the Rankine cycle. So why bring this up here? Because during the expansive work of the saturated steam in the turbine, some degree of condensation occurs, as predicted by Clausius, thus necessitating special equipment design to handle its presence and mitigate its damage.

> Clausius' 2nd Hypothesis:
> Heat Must Flow from Hot to Cold
> Whenever Work is Done in a Cyclic Process

Now we move on to the more fascinating second part of Clausius' 1850 publication. Not that the first part wasn't fascinating, as it laid out a clear validation of Joule's theory of the equivalence of heat and work along with some very helpful extensions of meaning and application. But it was the second part that truly broke open new ground and started us toward entropy.

Broadly speaking, Clausius evaluated the 2nd hypothesis by focusing on Carnot's claim inherent to this hypothesis that the nature of the working substance was irrelevant to the efficiency of the ideal heat engine. Clausius believed that if he could prove this were true, using his already-validated 1st Principle to help in doing so, then it must follow that the 2nd Hypothesis must also be valid, thereby converting it into the 2nd Principle, with both principles being one step shy of what Rankine would eventually turn into "law." Let's now dive into the details.

* * *

---

[43] (Carnot et al., 1988) p. 127.

[44] Adiabatic expansion (isentropic) of saturated steam to generate work sufficiently decreases energy and so causes condensation; the exhaust steam thus contains some amount of condensate. Adiabatic expansion of saturated steam involving no work, such as when exhausting pressurized steam through a relief valve, will superheat the steam if below the inflection point on the Mollier diagram (around 600 psig).

[45] (Smith and Van Ness, 1975) pp. 490–492.

Clausius saw the truth in Carnot's "flawed masterpiece." "[Heat] always shows a tendency to equalize temperature differences and therefore to pass from hotter to colder bodies."[46] At its core, after all the convoluted discussion about both forward and reverse heat engines that Carnot created and Clausius furthered, the 2nd Law as applied to thermal energy exchange is really about this statement. More generally, at its ultimate core as presented in Chapter 24, which Clausius wouldn't discover until later, the 2nd Law is all about entropy and the fact the entropy of a combined (equilibrated) system is always greater than or equal to the sum of the entropies of the individual (equilibrated) system components prior to combination.

$$\Sigma\, S_{individual\ equilibrated\ parts} \leq S_{combined\ equilibrated\ system} \qquad [31.4]$$

But if we remain solely with the concept of temperature equilibration via thermal energy exchange, then the 2nd Law is about the above truth that Clausius saw in Carnot's work. From this truth arose the key hypotheses that Carnot developed, that the maximum efficiency for the continuous conversion of heat into work in a cyclical process is equivalent for all working substances—"The motive power of heat is independent of the agents employed to realize it"[47]—and is an unknown function of temperature. Recall that Carnot didn't start with the concept of equalization of temperature differences. Instead, he started with the proposition, based purely on intuition, that the maximum amount of work was determined by the height of caloric's fall from high to low temperature, and that the working substance was but a means to carry the caloric and had no bearing on the work generation. Again embracing the scientific method, Clausius sought to use his already-validated 1st Principle to test Carnot's hypothesis by its consequences.

As he did in establishing his 1st Principle, Clausius started this effort by writing out what amounted to heat–work energy balances for each step of the complete Carnot cycle, the details of which I don't address here but which are available in most introductory texts on thermodynamics. Then, instead of considering the relation between work done and heat *consumed*, which he had earlier used to quantify the mechanical equivalent of heat, A, he considered the relation between work done and the heat *absorbed* from the boiler, which he used to quantify engine efficiency. Per Carnot's hypothesis, he set this efficiency equal to an unknown function of temperature, dT/C(T). The inclusion of dT represents the fact that Clausius simplified his analysis, as Clapeyron had done, by considering infinitesimal changes. Clausius' use of 1/C(T) as the unknown function rather than just plain C(T) was a consequence of Clapeyron's mathematics.

Maximum Efficiency of an Infinitesimal Heat Engine =

$$\left(\text{Maximum Work Done}\right)/\left(\text{Heat In from Boiler}\right) = \delta W_{max}/\delta Q_{in} = dT/C(T) \qquad [31.5][48]$$

To solve this equation, Clausius re-interpreted the individual steps in Carnot's cycle, re-doing Clapeyron's mathematics in the process but accounting for the work–heat equivalence while

[46] (Carnot et al., 1988) p. 134.
[47] (Carnot et al., 1988) p. 20.
[48] The use of δ in front of both W and Q is often used to indicate infinitesimal values for as opposed to mathematical differentiation of these terms.

eliminating the idea that heat is conserved. He then solved the equation for both an ideal gas and a saturated vapor (water), choosing these two because they were "…easily submitted to calculation…and [also] the most interesting."[49]

Looking first at the case of the ideal gas, he used his 1st Principle to write out the appropriate terms in the above equation, doing so with infinitesimal changes that enabled use of exact differential equations, drawing on Clapeyron's ideal-gas relationship as created from the combined laws of Mariotte[50] and Gay-Lussac, and found that the function, C (T), was equal to the product of one over the mechanical equivalent of heat, A, and absolute temperature, T, which validated Carnot's hypothesis that the maximum heat engine efficiency is dependent solely on temperature, at least for an ideal gas.[51]

Max Work Done / Heat In = $dT / (A \times T)$ - - - for an infinitesimal heat engine using ideal gas

To arrive at this conclusion, Clausius used a critical assumption—which he called his "subsidiary hypothesis"[52]—that would draw much critical attention from Thomson later on. Recall that Mayer assumed that all heat added to an expanding (ideal) gas is transformed into external work as opposed to having some of it diverted to separating the gas particles (internal work). In Clausius' new understanding, this meant that in Step 1 of Carnot's cycle, the work generated (W) equaled the heat added (Q) since there was no change in internal energy (U) as the *vis viva* remained constant (due to isothermal conditions) and no internal work was done. While the experiments of Gay-Lussac and later Joule validated this hypothesis by demonstrating no change in temperature upon expansion of an ideal gas into a vacuum, a consensus hadn't yet been reached by 1850. In fact, Thomson initially felt that this assumption, which he called "Mayer's Hypothesis,"[53] was unfounded, and it was his challenge of it that led him to launch the extensive and famed Joule–Thomson experiments (see Chapter 33).

To complete the test of Carnot's hypothesis, Clausius then wrote out the corresponding equations to calculate the heat engine efficiency for a liquid–vapor system and, in the critical final move, set this efficiency to the same function, $dT / (A \times T)$, as for the ideal gas. According to Carnot, since the maximum work generated was not dependent on the nature of the working substance, this resulting equation for the liquid–vapor engine should be just as valid as that for the ideal gas.

Clausius' final equation looked like this:

$$r = [A \times (a+t)](s-\sigma)\, dp / dt$$

for which

r = heat (enthalpy) of vaporization of the liquid to the vapor
s = volume of a unit weight of saturated vapor
σ = volume of a unit weight of liquid at the same temperature as the vapor

---

[49] (Carnot et al., 1988) p. 115.
[50] As noted in Chapter 18, Boyle (1662) arrived at this law prior to Mariotte (1679).
[51] Of note is that Clausius, following Clapeyron's lead, used an infinitesimal approach to his mathematics. The equation here is for small increments of work and heat caused by moving a small difference (dT) from T_H to T_C. When fully integrated from T_H to T_C, the equation reads as the familiar: Maximum Work Done / Heat In = $(T_H - T_C) / T_H$.
[52] (Carnot et al., 1988) p. 128.
[53] (Smith, 1998) p. 75.

p = pressure

t = temperature (the symbol t was used for degrees Celsius; the value of *a* was needed to eventually convert to absolute temperature, T, in Kelvin)

1/A = the mechanical equivalent of heat (MEH)

Note that with some re-arrangement and updating of symbols, this equation becomes

$$dP^{sat} / dT = \Delta H^{vap} / (T \Delta V^{vap})$$

[31.6]

As introduced in Chapter 30, this is the famed Clapeyron equation quantifying how saturation pressure changes with temperature for a given substance based on its heat of vaporization and volume expansion properties. It's based on a Carnot cycle using liquid–vapor wherein the Step 1 expansion occurs at a given P^{sat}-T combination, each being a unique function of the other, and the Step 3 compression occurs at infinitesimally lower P^{sat}-T combination.

Clausius tested the validity of his "r =" equation using Renault's extensive liquid–vapor data for water and the resulting value for 1/A that he calculated (772 foot-pounds) was very close to the experimental results (780–838) he cited for Joule. What a fantastic eureka moment this must have been for Clausius. "Such an agreement between results which are obtained from entirely different principles cannot be accidental; it rather serves as a powerful confirmation of the two principles and the first subsidiary hypothesis annexed to them."[54] To Clausius, his 1st Principle stating work–heat equivalence was validated by Joule's meticulously obtained data, while his 2nd Principle, when combined with the "Mayer's (first subsidiary) hypothesis," stating that maximum heat engine efficiency depends solely on temperature and not working substance, was validated by the fact that the mechanical equivalent of heat determined by this principle showed very good agreement with those values directly measured by Joule.

Two very different thermodynamic journeys, one taken by Carnot–Clapeyron–Clausius and the other by Joule, arrived at the same end point. To have demonstrated with theory and experimental data that the maximum efficiency of ideal gas and liquid–vapor Carnot heat engines were identical to each other and that the mechanical equivalent of heat derived from them was in close agreement to Joule's separately determined value was beyond any logical and linear progression of thought. Carnot's instinct was justified. As Clausius would write, "The agreement…leaves really no further doubt of the correctness of the fundamental principle of the equivalence of heat and work; and confirms in a similar way the correctness of Carnot's principle," which was Clausius' 2nd Principle. I'll discuss these results in more detail in later chapters.

Clausius Took the First Major Step toward Thermodynamics

By reconciling Carnot with Joule in this 1850 publication, Clausius arguably launched classical thermodynamics, although one could also argue that it was Thomson's validation of Clausius with his 1851 publication *On the Dynamic Theory of Heat* that really launched things as it led

---

[54] (Carnot et al., 1988) p. 138.

to a larger North British movement to establish thermodynamics as a new field of study and also to a larger English–German groundswell of acceptance. Regardless, it was Clausius who took the first major step. But, as with most all major steps in science, it's important to realize that, as with Newton, Clausius too stood on the shoulders of others to make it. There's a reason why one of his pinnacle results is called the Clapeyron Equation. Clausius wouldn't have arrived at this or any of his other conclusions had the ground not already been prepared first and foremost by Carnot, for it did all start with him, and subsequently by Clapeyron and Thomson, among others. In fact, the main equations that Clausius arrived at were nearly identical to those published by Carnot and Clapeyron. Nearly. There was one critical difference: his incorporation of Joule's heat–work equivalence.

It's interesting that Carnot's work and critical hypotheses weren't really impacted by his incorrect belief in the conservation of heat. His main hypotheses and conclusions remained valid. Even the equations rising from his work as formalized by Clapeyron were essentially correct, hence the similarity between Clausius and Clapeyron. It's just that when it came time to complete the mathematics, his answer was wrong. The energy balance didn't close, largely because energy, as a concept that would unite both heat and work, had not yet been discovered. It was here that the (incorrect) belief in the conservation of heat caused Carnot, Clapeyron, and even Thomson to become stuck. They couldn't get the numbers to work out in a way that made sense. And this is why Clausius' correction enabled him to break through the barrier that stopped the others, to complete the job that the others had started, and to blaze a path that ultimately led to the discovery of entropy. Some say that it was Carnot who discovered the 2nd Law with his statement linking work to the movement of heat from hot to cold, but the gap between this statement and the real 2nd Law was huge. He may have gotten the words largely correct, but the understanding behind the words was far removed from a true understanding of nature.

While acknowledging the shoulders he stood on, it was Clausius alone who solved the problem of Carnot and Joule. What Clausius did was a big deal, and it's what made his publication a true seminal work, one that left him with priority based on publication date. Yes, Rankine published on the same subject prior to Clausius, but his work was not as encompassing. He too inserted Joule's heat–work equivalence into Carnot's engine but did not drive the mathematics to the final conclusion achieved by Clausius. Thomson's acknowledgement of Clausius' accomplishment was the final word on this, "The whole theory of the motive power of heat is founded on the two following propositions, due respectively to Joule, and to Carnot and Clausius,"[55] whereupon he noted the first proposition to be work–heat equivalence and the second to be that a reversible heat engine produces the maximum amount of work, which he later modified to his own proposal that, "It is impossible…to derive mechanical effect from any portion of matter by cooling it below the temperature of the coldest of the surrounding objects."[56] We'll pick this up again in the chapters that follow.

Additionally, it was only through the incorporation of Joule's findings into his analysis of Carnot that Clausius arrived at the first mathematical statement of the 1st Law of Thermodynamics, although he didn't spend time in this 1850 publication raising the flag about this.

[55] (Thomson, 1851) Paragraph #9.
[56] (Thomson, 1851) Paragraph #12.

The science at the time was in such a significant state of flux, caused by the arrival of the concept of energy and its collision with Carnot, that it must have felt for many like being on a small boat in a rough sea. Clausius, with a steady hand on the tiller, chose to move forward slowly but very surely, keeping a low but strong profile, slowly building classical thermodynamics' foundation.

32 William Thomson (Lord Kelvin)

[Thomson] existed as scientist and technologist, academic and entrepreneur, a philosopher and a practical man rolled into one... [He] had been described in his lifetime as Britain's and perhaps the world's greatest scientist.

– David Lindley[1]

The rapid back-and-forth progress achieved by William Thomson and Rudolf Clausius in the early 1850s was remarkable, especially since the two never met. It was Thomson who initially helped set the course Clausius followed. It was then Clausius who surpassed Thomson and inspired him to catch up. It was then Thomson who indeed did catch up and perhaps surpassed Clausius by becoming the leading force behind the rise of energy as a unifying theory in the thermodynamic movement in North Britain which saw the establishment of the thermodynamic fundamentals, structure, and terminology that we learn and use to this day. Then it was Clausius who launched the concept of entropy. All told, their collective efforts opened the door to an entirely new realm of science based on energy and entropy. Each fed off the other, which makes this a good segue from one to the other. We briefly met Thomson in the previous chapter. Now it's time to meet him again, this time in more detail.

* * *

This chapter needs a bit of a roadmap as several different themes are addressed under one umbrella named William Thomson. The challenge involved with presenting Thomson reflects the fact that there were two distinct periods in his scientific life, the caloric period and the energy period. No other personality in this book better reflects the challenge of undergoing a paradigm shift. The depth in which Thomson embraced the caloric paradigm needs to be understood to fully appreciate the genius he expressed in lifting himself out and fully embracing the new paradigm based on energy.

The first half of the chapter deals with Thomson's caloric years and two significant challenges he faced in trying to reconcile his belief in caloric with the new emerging theories. Challenge #1 involved Carnot's hypothesis that the nature of the working substance is irrelevant. If this were indeed so, then a system based on solid–liquid (ice–water) should perform exactly the same as one based on liquid–vapor or even gas alone. Working with his brother James, they resolved this challenge in a fabulous way: experimental proof. Unfortunately, this victory led to their continued embrace of the caloric theory. Challenge #2 was much more conceptually difficult. Neither William nor James could understand how the caloric theory could explain the difference

---

[1] (Lindley, 2004) pp. 4–5.

Block by Block: The Historical and Theoretical Foundations of Thermodynamics,
Robert T. Hanlon, Oxford University Press (2020). © Robert T. Hanlon.
DOI: 10.1093/oso/9780198851547.001.0001

between the "fall" in temperature in Carnot's engine and that which occurs in the phenomenon of conduction. This challenge was much more difficult than it sounds to you right now.

The second half of the chapter deals with Thomson's energy years. They were years of intense break-out activity as he took his full embrace of energy and raced ahead with it, benefiting from a phenomenal group of like-minded "North British" scientists with whom he surrounded himself, a situation standing in sharp contrast to Clausius' own isolated approach. We'll meet two of these scientists, William Rankine and Hermann von Helmholtz, who, like Joule, was not "North British" but became adopted by that community. We'll also meet the origins of the two laws of thermodynamics and the specific confusion over what the 2nd Law was all about. And finally, we'll meet a rather interesting result of Thomson's new energy paradigm: How would *you* go about calculating the age of the Earth?

> William Thomson
>
> –
>
> The Caloric Years

As we turn our attention from Clausius to Thomson, we see the difference in approach taken by these two giants. As a theoretical physicist, Clausius initially worked in his Berlin office[2] largely on his own, not involving himself in experiments, preferring instead the reading and thinking of others' work while creating his own ideas and publications. He followed the beat of his own drum, pursuing his own interests while, at times, showing low to zero interest in the works of others, most notably in later years those of Gibbs and Boltzmann.[3] He had a distinctive approach to his own work, characterized by "an excellent grasp of the fundamental facts and equations relevant to the phenomena, a microscopic model to account for them, and an attempt to correlate the two with mathematics."[4]

Thomson, on the other hand, was about as diverse in his studies and as well connected as any in the world of science. As was the case with Galileo, Thomson succeeded by operating at the interface between fields. He dove deep into fundamental physics, including theory, mathematics, and experiment, while continually looking upwards toward practical applications. He existed as "scientist and technologist, academic and entrepreneur, a philosopher and a practical man rolled into one."[5] His involvement with such a range of interests naturally led to his involvement with a range of fellow scientists and engineers. Indeed, being highly energized by a zest for science as manifested by others' descriptions of him—"never idle...always got his work done...a lifelong habit of incessant activity"[6] and "worked every working hour at science"[7]—Thomson actively surrounded himself with a wonderful network of prominent scientists, such as Regnault, Joule, Rankine, and later, Peter Guthrie Tait, Hermann von Helmholtz, and James Clerk Maxwell, among others. One could argue that he sat at the center of this

---

[2] (Smith, 1998) p. 97. Clausius was appointed both as physics teacher in the Berlin Artillery and Engineering School and as Privatdozent at Berlin University in 1850. His research focused almost wholly upon heat.

[3] (Daub, 1970a) p. 310.

[4] (Daub, 1970a) p. 303.

[5] (Lindley, 2004) p. 4.

[6] (Lindley, 2004) p. 20.

[7] (Lindley, 2004) p. 42.

diverse collective. The places he went and the things he did, ranging from his early life of being born in Ireland, schooled in Scotland, further educated at Cambridge, where he became a star student, to his time as a young researcher learning experimental techniques under Regnault in Paris,[8] and to his first job as a young professor at Glasgow,[9] where he founded experimental physics by building his own laboratory in the mold of Regnault's own laboratory and where he also was able to immediately try out his new ideas in front of the bright young minds in his classroom, all of these opened doors of opportunity to him, which he seized and nurtured, rising over time in stature and acclaim until he earned the title Lord Kelvin in 1892. Even his family played a prominent role in this network. Both his father, Dr. James Thomson (1786–1849), and then his brother, James (1822–1892), an established engineer and academic physicist in his own right, influenced William's development as a natural philosopher, the latter's communications and collaborations being especially important.[10]

Enhancing the strength of this relationship network was the fact that Thomson eagerly and prolifically contributed many of his own ideas to the ongoing discussions. Out of his full engagement with such a group of highly talented individuals, Thomson, working with Tait, amalgamated the voices, including his own and including Clausius', into the written structure of classical thermodynamics that we are taught to this day. It was Thomson around whom this work was done. He was the central figure.

To understand this more deeply, we need to delve more into Thomson's history around heat, a history that I'll tell in a somewhat linear path of time.

Thomson's Early Education was Influenced by French Thought – Heat is Conserved

[Thomson was convinced] that the whole science of heat rested on what since 1783 had been the basic axiom of conservation

– D. S. L. Cardwell[11]

Thomson's educational upbringing regarding heat was strongly influenced by the French, and to the French, heat was conserved. When Lavoisier added caloric to his list of thirty-three elements, heat implicitly became subject to the conservation of matter. To the later students of Lavoisier's new chemistry, the conservation of heat simply became a given. By immersing himself in French science, starting with an enthusiastic devouring of Fourier's work within two weeks as a sixteen-year-old, and continuing with his time spent studying in France, Thomson came to accept this belief. His subsequent deep affection for Carnot's work was but a continuation of this devotion to French thought. In all of this time, he never had to commit to a theory of heat, nor be challenged by a competing theory. It was a theory given to him, one he never really had to choose. Heat was conserved, and that's just the way it was.

---

[8] (Smith and Wise, 2009) p. 105.

[9] (Cardwell, 1989) p. 103.

[10] (Smith and Wise, 2009) p. 283. "The role that James Thomson played in the accomplishments of his younger brother has not been adequately appreciated."

[11] (Cardwell, 1971) p. 244. 1783 was the year Lavoisier published "Réflexions sur le phlogistique."

In short, Thomson never had his beliefs challenged; thus, Thomson himself never had to challenge his own beliefs.

So given this, one can better understand Thomson's reaction to Joule coming along in June of 1847 and saying that heat is not conserved. To Thomson, if this were indeed true, it would "overturn the whole theory of heat"[12] and necessitate a re-interpretation of all existing data within a new framework, a task too overwhelming to consider. How ironic then that this is exactly what happened, with Thomson leading the way.

It's perhaps telling that when both Thomson and Clausius heard of Joule's results and conclusions, Clausius, who wasn't raised under such a strong French influence, readily accepted them and melded them with Carnot's, while Thomson became stuck, not knowing which way to turn. Indeed, Thomson's confrontation with Joule's findings was a significant moment in history, leading him, as it did, to a "soul-searching examination of the grounds of his scientific beliefs."[13] Said his biographer, Silvanus Thompson, "The apparent conflict took possession of (Thomson's) mind and dominated his thoughts."[14]

The fact that Thomson successfully pulled himself out of this situation and shifted his paradigm the way that he did was truly amazing. To let go of something that one so deeply believes in and to then grab onto an entirely different belief requires a strong open-minded intellect and a devotion to the ultimate spirit of science, the seeking of truth. Above all, Thomson sought truth.

The Thomson Brothers Discover Carnot

The Thomson brothers were primed to be excited by Carnot. Both were interested in how to get more work out of a given engine. Thus, Carnot's new ideas about heat-to-work conversions and the use of water wheel analogies to analyze the steam engine spoke right into their listening.

The desire to improve engine efficiency was a common theme in England, where such improvement could be readily translated into economic reward. This was especially true in Scotland where the Presbyterians embraced the belief that sin and idleness were wasteful, a belief that was well embodied in James Watt who lived and breathed the desire to reduce waste in everything even to the point of carrying around a "Waste Book" to note and correct all sources of waste in his daily life.[15] As educational descendants of Watt and this Scottish culture, James and William also sought to reduce waste, especially in engines. Their problem was that there was no theoretical underpinning available to guide their initial efforts.

This situation changed in 1844 when James wrote an interesting note to William that included two intriguing statements.[16] The first concerned the possibility of operating steam engines *without fuel* [my italics] "by using over again the heat which is thrown out in the hot water from the condenser." This statement demonstrated the Thomsons' fascination with

12 (Cardwell, 1971) p. 244.
13 (Smith and Wise, 2009) p. 342.
14 (Cardwell, 1971) p. 244.
15 (Smith, 1998) p. 33.
16 (Smith, 1998) p. 42.

heat-to-work conversions and especially the fundamental theory behind their limitations, which was clearly lacking. The second was the statement, "I shall have to enter on the subject of the paper you mentioned to me." This was the first known evidence of when the Thomsons learned of Carnot. While neither Carnot nor Clapeyron was referenced directly in this note, the language that James used to discuss the ideas with William, specifically the fall in temperature and its relation to work being done, reflected a certain familiarity with Carnot's ideas. James likely learned of Carnot through his reading of the 1837 translation of Clapeyron that had appeared in England in Taylor's *Scientific Memoirs*. He may have also caught glimpses of Carnot's (either Lazare or Sadi or both) ideas in the teachings of Lewis Gordon (1815–1876), his professor at Glasgow who had traveled the continent during the 1830s and, among other things, learned of the science of engines and machinery in France.[17]

In 1845 William read Clapeyron while working with Regnault in Paris and by 1848 had finally gotten hold of Carnot's original paper from Lewis Gordon.[18] His total enthusiasm for Carnot inspired his 1849 publication, *An Account of Carnot's Theory of the Motive Power of Heat*[19] in which he embraced Carnot's work which was based on the "ordinarily-received and almost universally-acknowledged"[20] caloric theory. It was in this publication that Thomson launched the word and concept, *thermo-dynamic*, while embracing Thomas Young's *energy*, which he embedded in the statement, "Nothing can be lost in the operations of nature – no energy can be destroyed."[21] We'll discuss this latter term in more detail shortly, but its use did reflect the strong belief of most everyone in some concept of conservation, although what it exactly was that was conserved wasn't known; Thomson's use of the term preceded its correct definition. While Thomson followed much the same path created by Carnot and Clapeyron, his open-minded approach was evident. Pointing to Joule's results that heat can indeed be generated, a phenomenon at odds with caloric, William wrote that Carnot's caloric approach "may ultimately require to be reconstructed upon another foundation, when our experimental data are more complete."[22]

One may wonder how Thomson could proceed with publishing this paper after having learned of Joule's results in 1847. The two were fundamentally at odds with each other. Again, he was sitting right on the edge of a huge paradigm shift in science at that time, so it's easy to look back on such questions from the comfort of knowing the right answer. But it's still worth looking at this. First, Thomson did not agree with all of Joule's results. Recall his note to his father, "Joule was wrong in many of his ideas, but he seems to have discovered some facts of extreme importance, as for instance that heat is developed by the friction of fluids in motion."[23] This may have diminished the magnitude of the conflict in his mind. But another consideration is simply that even if he were to reject caloric, he didn't have any better theory to replace

---

[17] (Smith and Wise, 2009) p. 289.
[18] (Thompson, 1910) Vol. 1, p. 133.
[19] (W. Thomson, 1849)
[20] (W. Thomson, 1849)
[21] (Smith, 1998) p. 77.
[22] (W. Thomson, 1849)
[23] (Smith, 1998) p. 79.

it with,[24] as represented by his writing, "I shall refer to Carnot's fundamental principle, in all that follows, as if its truth were thoroughly established."[25]

Regardless, in the end, Carnot's hypotheses were strong and powerful to Thomson, leaving him with no compelling reason to jettison caloric. Indeed, he felt that Carnot's work was for the most part sound, as the following historical example shows.

Challenge #1 – the Ice Engine

In a fascinating demonstration of the power of pure thinking, the Thomson brothers looked at the Carnot engine and inquired, what if the working substance consisted of solid–liquid instead of liquid–vapor? Specifically, they contemplated a Carnot engine operating with an ice–water mixture, thinking that the engine's efficiency with this working substance must equal that with the established water–steam system (Figure 32.1). That's what Carnot hypothesized when he proposed working substance to be irrelevant. The phase shift for either causes a change in volume that, in turn, does work. Boiling water to steam in Step 1 of Carnot's cycle results in a huge change in volume, so here it's quite obvious what's going on. But how about for ice–water? When water freezes, it expands. This was common knowledge to anyone who attempted to freeze sealed bottles of water only to witness the resulting breakage of the seal. So it must be this change in volume that does the work. But exactly how would this fit inside Carnot's engine? This is the issue that confronted the Thomsons.

At that time, it was another common knowledge, or so many thought, that water froze at a single temperature of 0 °C. There weren't any data to prove otherwise. To the Thomson brothers, however, this meant that if the freezing of water to ice (expansion) and the thawing of ice back to water (contraction) both occurred at one and the same temperature, then the pumping action of this cycle could yield a net amount of work while operating at a single temperature. This result, if true, would indicate a fatal flaw in Carnot's proposal since work would be generated in the absence of a temperature difference across which caloric could fall. You'd be generating work without a flow of heat and thus achieving perpetual motion.

While William struggled with this puzzle, reflecting that his grasp of this new subject was evolving in "fits and starts,"[26] James saw the way out. As he was to later explain, to avoid such perpetual motion, "it occurred to me that…the freezing point [of water] becomes lower as the pressure…is increased."[27] Wow! What a remarkable insight, all based on pure reasoning. Not only did he correctly predict that the freezing point of water must change as pressure is increased, he also correctly predicted the direction of the change.

The question now is, why? Why must the freezing point change with temperature? The basic classical thermodynamic answer is as follows.[28] In a cyclic process in which work is done, heat must always be added to the system during the high-temperature isothermal step. It doesn't

---

[24] (Cardwell, 1971) p. 242.

[25] (W. Thomson, 1849)

[26] (Lindley, 2004) p. 102.

[27] (Smith and Wise, 2009) p. 298.

[28] The physics answer is that when water freezes, the resulting solid ice structure fills a larger volume due to hydrogen bonding. Thus, removing heat effects work.

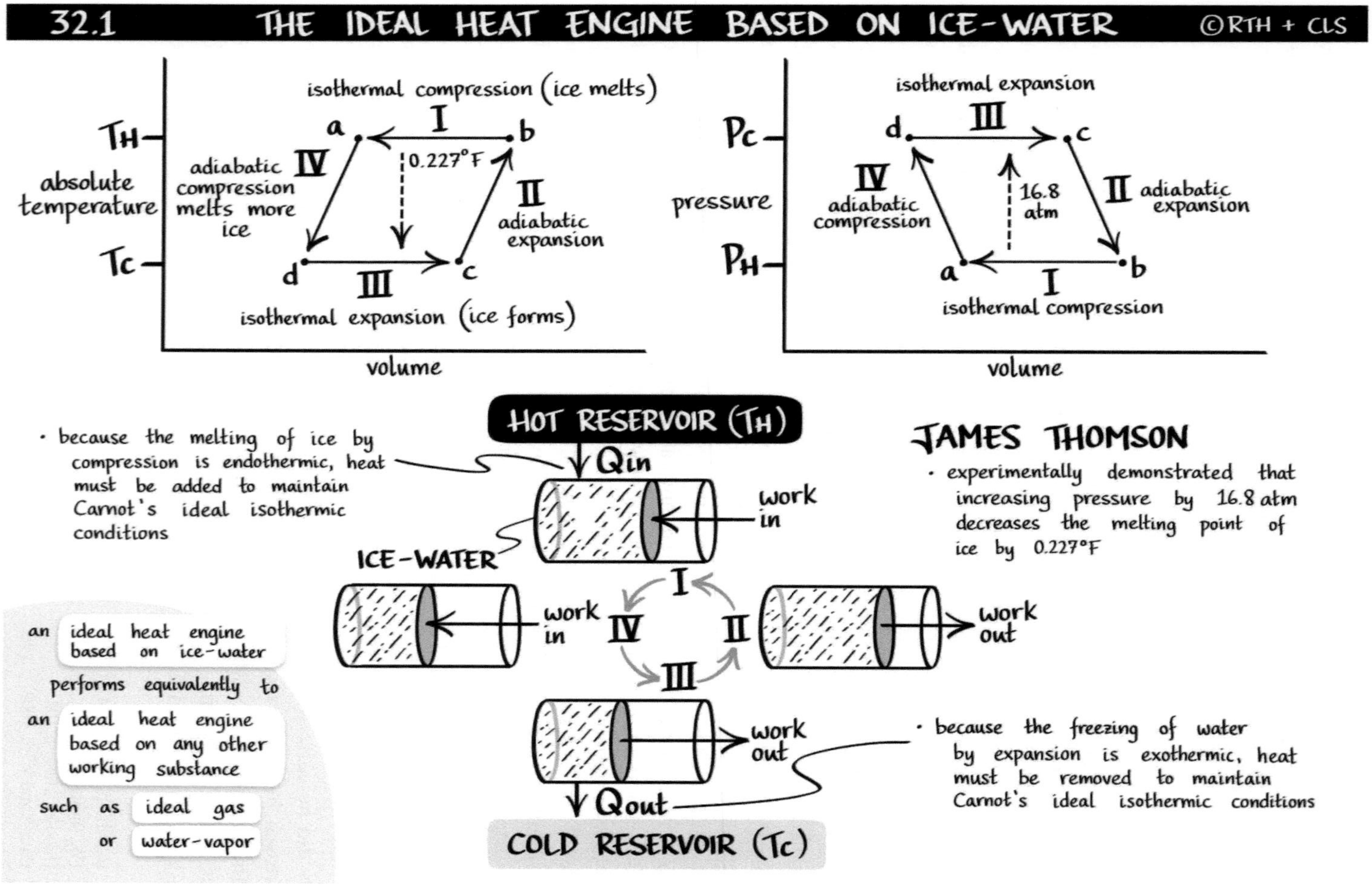

THE IDEAL HEAT ENGINE BASED ON ICE-WATER ©RTH + CLS

isothermal compression (ice melts)

T_H — absolute temperature — T_c

I — 0.227°F
IV adiabatic compression melts more ice
II adiabatic expansion
III
isothermal expansion (ice forms)

volume

isothermal expansion

P_c — pressure — P_H

III — 16.8 atm
IV adiabatic compression
II adiabatic expansion
I isothermal compression

volume

JAMES THOMSON

- experimentally demonstrated that increasing pressure by 16.8 atm decreases the melting point of ice by 0.227°F

HOT RESERVOIR (T_H) Q_{in}

work in

ICE-WATER

work in IV I work in II III work out

work out

COLD RESERVOIR (T_c) Q_{out}

- because the melting of ice by compression is endothermic, heat must be added to maintain Carnot's ideal isothermic conditions

- because the freezing of water by expansion is exothermic, heat must be removed to maintain Carnot's ideal isothermic conditions

an ideal heat engine based on ice-water performs equivalently to an ideal heat engine based on any other working substance such as ideal gas or water-vapor

Figure 32.1 The ideal heat engine based on ice-water

matter whether this step is high pressure or low, expansion or compression, generating work or requiring that work be added. It just matters that for the generation of work, heat enters at the high temperature. The reason is due to the fact that 1) Q_{in} must be greater than Q_{out} since, per the 1st Law, work generated equals $Q_{in} - Q_{out}$, and 2) per the 2nd Law which we'll get to, $Q_{in} / T_H = Q_{out} / T_C$. When you do the math, T_H must be greater than T_C and so Q_{in} must occur at the high temperature.

Although Clausius didn't lay it out as such, one consequence of these combined laws is his 2nd Principle: Heat must flow from hot to cold whenever work is done in a cyclic process.

Now it just so happens that while the above logic is valid and true, it is not true that the primary work generation step always corresponds to the step where heat flows into the engine. For the ideal gas and the water–steam engines, this does occur. Add heat, increase volume, do work. However, for the ice–water engine, because of the unique properties of water, this is not so. It's just the opposite. Remove heat from water, turn water into ice, increase volume, do work. The high-pressure work generation step occurs at low temperature, just the opposite of the gas engine. The laws of thermodynamics still apply. It's just the temperature–pressure relationship between the phase equilibrium that has changed.

It was out of this thought process that James reasoned that high-pressure, water-to-ice transformation must occur at a lower temperature than low-pressure ice-to-water. In other words, James reasoned that, if Carnot were indeed correct in his thinking, the freezing of water to cause expansive work must occur by using the low-temperature reservoir (T_c) to withdraw heat. Since this is the work-producing step, it must further occur at high pressure. Thus, the freezing point (T_c) of water at high pressure must be lower than the melting point (T_h) of water at low pressure and so increasing pressure must decrease the temperature needed to melt ice.[29]

It's rather interesting that William didn't realize this as quickly. William, who had thoroughly immersed himself in Carnot and Clapeyron, couldn't easily make the leap from liquid–vapor to solid–liquid. He couldn't grasp the meaning behind, "working substance is irrelevant," or the implications of Carnot's hypothesis.

In one of those beautiful moments that occurs infrequently but momentously in science, as when Eddington quantified the bending of star light around the Sun which validated Einstein's prediction (predicting a first-time-ever, never-before-seen phenomenon) based on his Theory of General Relativity, James' prediction[30] of an aspect of nature that was never before seen was validated by experiment. Based on Carnot's hypotheses, James predicted that the freezing point of water should fall by 0.227 °F when pressure is raised to 16.8 atmospheres. William then tested this in the lab, announcing his experimental results in January, 1850.[31] He measured a decrease in freezing point of 0.232 °F. As William wrote, it was a "…highly favorable comparison." I would have used a different adjective: phenomenal.

*　*　*

What a great piece of work this was by the Thomson brothers. A new and experimentally validated theory on the nature of matter arrived in our midst. But here's what happened next. As Cardwell writes, due to the positive results of this theory-experiment, "…Carnot's theory

---

[29] One way to melt ice at 0 °C is to put pressure on it, which is why ice skates easily glide over ice.
[30] (J. Thomson, 1849) Note in (Smith and Wise, 2009) p. 298, footnote 35 that William claimed credit for the theory.
[31] (Thomson, 1850)

was not merely saved, it had received something like a triumphant vindication."[32] And it was indeed a vindication, but only for a *part* of Carnot's theory. Carnot proposed that when net work is done by his cyclical process, heat must flow from hot to cold. The Thomsons demonstrated that this is indeed true, even when the working substance is something as unusual as ice–water. But to take this as a victory for Carnot's full theory was misleading. The conflict between the conservation of heat and Joule's findings was nowhere needed in this specific investigation; hence, the Thomson brothers' analysis shed no light on the situation, other than a somewhat misleading light of validation.

Prelude to Challenge #2 – Switching Paradigms

> *Thomson's dramatic shift from a broad-minded uncertainty about [heat–work] interconversion in 1849 to this uncompromising assertion of irreversible dissipation in 1851 signals a transformation in both the style and content of his science.*
>
> – Crosbie Smith[33]

The benefit of William's sitting at the center of a network of such highly skilled scientists was his exposure to a multitude of ideas and results, especially those that weren't aligned with his own. The fact that he was open to considering them, however slowly that occurred, reflected well on his abilities as a true scientist who sought natural truth. Three voices in particular swayed him in the end: James Joule, William Rankine, and Rudolf Clausius. The direct communications Thomson had with Joule and Rankine clearly sharpened his thinking, left him with new ideas, and led him down the path to the mechanical (or dynamical) theory of heat. The indirect communications Thomson had with Clausius through Rankine were equally if not more critical.

It was Joule who, with the results of his definitive experiments in hand, strongly and persistently encouraged William toward work–heat equivalence. Such persistence was necessary since both William and James initially felt resistant to Joule's ideas. Recall William's note to his father, "Joule was wrong in many of his ideas."[34] And then we had James to William in 1847, "[some of Joule's views] have a slight tendency to unsettle one's mind as to the accuracy of Clapeyron's principles."[35] I can only imagine how frustrating this must have been for Joule, to see something but not have others see the same thing.

While Joule was working on William from one side, Rankine was doing likewise from the other. In February, 1850, Rankine published *On the Mechanical Action of Heat*. He had started developing this paper in 1842 after reading Clapeyron's paper, and so was ahead of the Thomsons on the timeline, but didn't finish until he had Regnault's experimental data in hand. It was J.D. Forbes, Rankine's professor at Edinburgh, who sent Rankine's paper to William in Glasgow for critical review, thus commencing the Thomson–Rankine relationship. Being positively influenced by Rankine's vision of heat as motion, even though he didn't buy into

[32] (Cardwell, 1971) pp. 243–4.
[33] (Smith and Wise, 2009) p. 317.
[34] (Smith, 1998) p. 79.
[35] (Smith and Wise, 2009) p. 307.

Rankine's specific mechanical model based on his vortex theory, William "was soon prepared to accept a *general* mechanical theory of heat, namely that heat was *vis viva* of some kind."[36]

But it wasn't only in this way that Rankine influenced Thomson, for it was Rankine who first introduced Thomson to the key elements in Clausius' work. The timeline here is rather interesting as it was Thomson who first brought Clausius to Rankine's attention. You see, when Thomson became aware of Clausius' publication that came out in April of 1850, he informed Rankine about it while he himself didn't truly absorb it. Rankine then delved into the paper and interpreted it for Thomson in a September letter, along with some less-than-enthusiastic comments that did nothing to encourage Thomson to delve into the paper himself.[37] Even as late as October 1850, Thomson had not fully assimilated Clausius' ideas, writing Joule, "I have not yet been able to make myself fully acquainted with this [Clausius'] paper."[38] The discussion around this timeline of events has interested the science historians, as some question how much of Thomson's work was truly original as opposed to an extension of Clausius. But the point is that Thomson's knowledge of Clausius' work through Rankine was surely relevant, if not influential, as it showed Thomson a way out of his dilemma. While it was Joule who initially helped move Thomson toward the mechanical theory of heat, it was this added effort of Rankine, both directly and indirectly through Clausius, that was "crucial"[39] to Thomson's dramatic metamorphosis.

The fact that it took the combined efforts of these three giants to sway Thomson from the conservation of heat to Joule's work–heat equivalence is telling. There's no simple explanation here, at least as recorded by history, especially since Thomson didn't spend much time writing about his views on the fundamental nature of heat. Perhaps, in the end, it was the fact that, as David Lindley noted, "Fourier made Thomson."[40] What I mean by this is that the influence of Fourier's theory of heat transfer on Thomson was so strong as to blind Thomson to seeing what heat truly was, which provides us a great segue to conduction.

Challenge #2 – Conduction

It's instructive to understand the depth of Thomson's conflict circa 1850 so as to better appreciate the height he had to scale to get out. We treat such concepts as heat so matter-of-factly today, and yet they and their origins are quite profound and abstract. To gain a deeper and more beneficial mastery of the subject, we need to do yet another deep dive.

At the crux of Thomson's conflict was a rather fascinating quandary. While William could accept Joule's concept of work-to-heat, he couldn't accept the reverse concept of heat-to-work, which is, after all, what the heat engine was all about. Unfortunately, while Thomson clearly understood Joule's demonstration of the former (drop weight, raise temperature) such was not the case with the latter. While Joule felt he had demonstrated heat-to-work when he demonstrated the expansive cooling of compressed air, the conclusion wasn't evident to Thomson.

---

[36] (Smith and Wise, 2009) p. 320.
[37] (Smith and Wise, 2009) p. 326.
[38] (Smith and Wise, 2009) p. 327.
[39] (Smith and Wise, 2009) p. 327.
[40] (Lindley, 2004) p. 36.

One would have thought that the cooling effect of reversible adiabatic expansion (Step 2 in Carnot's cycle) would have fulfilled this need, but it didn't. While such a progression of thought may not have been a significant issue for Joule, since he readily accepted that work-to-heat was the same as heat-to-work, it was for Thomson.

So why, exactly, was the concept of heat-to-work such a significant stumbling block? The answer, or at least part of the answer, lies in the phenomenon of conduction (Figure 32.2).

To Fourier, Thomson's guiding light, the conduction of heat was something that could be mathematically modeled without the need to rely on first principles for validation. The success of the math *was* the validation. Fourier didn't need to identify the *why* behind the math, just as Newton didn't need to explain the *why* behind his math when he proposed his law of Universal Gravitation. The math worked and that's all that mattered.

Thomson's own approach largely emulated Fourier's as he didn't spend much time writing on the true nature of heat. But this approach faltered when confronted by the concept of work. Fourier didn't address work in his theory of heat. At his time, the two weren't connected. And there was no reason for Thomson to question this exclusion prior to the arrival of Carnot and Joule into his life. But after their arrival, everything changed, as it forced Thomson to figure out how heat and work fit together.

The single main stumbling block regarding conduction was this. With Fourier's conduction, heat flowed down a metal bar from high temperature to low. At first, there was nothing really complicated about this, but once Carnot's conceptual heat engine arrived, things changed. Carnot said that the fall of heat from high temperature to low causes work to be done while heat or caloric was conserved. The work resulted from a sensible-to-latent shift within the caloric, and a corresponding volume increase, although Carnot never flat-out stated the cause–effect involved and didn't get into the details how exactly this all worked. It was simply an integral part of the caloric theory. Since William was raised within this French view of heat, he didn't have the wherewithal to solve this problem and neither did James. He couldn't explain conduction. In his mind, during conduction, as heat flowed from the hot reservoir to the cold reservoir, its temperature dropped, which must have meant that there was a shift in the otherwise conserved caloric from sensible to latent. But where was the "effect" that this should have produced? Where did the work end up? Did it somehow attach itself to the heat and simply get swept away into the cold reservoir? The confusion around this phenomenon was epitomized in William's writing:

> When 'thermal agency' is thus spent in conducting heat through a solid, what becomes of the mechanical effect which it might produce? Nothing can be lost in the operations of nature—no energy can be destroyed. What effect then is produced in place of the mechanical effect which is lost? We may hope that the very perplexing question in the theory of heat, by which we are at present arrested, will, before long, be cleared up.[41]

Carnot acknowledged that conduction of heat from hot to cold resulted in lost work potential but didn't provide a reason. He basically said that if heat short-circuited and flowed directly from the hot to the cold reservoir, then the total work produced would be less than the maximum possible.

---

[41] Thomson quote cited in (Smith, 1998) p. 94.

CALORIC THEORY

CALORIC FLOWS DOWN
THE METAL ROD

METAL ROD

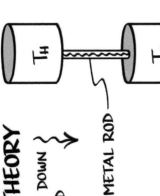

- in the caloric paradigm, the material caloric is conserved as it flows down the metal rod from hot to cold during conduction

- this transfer should generate work in theory

  But where does the work end up?

  Why does caloric lose temperature as it flows?

  What becomes of the caloric during conduction?

ENERGY THEORY

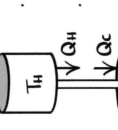

$Q_H = \,^-Q_c$

$Q_H/T_H = \Delta S_H$

$Q_c/T_c = \Delta S_c$

Q_H is negative relative to T_H

Q_c is positive relative to T_c

- in the energy paradigm, nothing flows from hot to cold

- collisions at the atomic scale result in energy loss from the hot reservoir (Q_H) and energy gain by the cold reservoir (Q_c)

- the two temperatures move towards each other and total entropy increases

$$\Delta S_{TOTAL} = \frac{Q_H}{T_H} + \frac{Q_c}{T_c} = Q_H \left(\frac{1}{T_H} - \frac{1}{T_c} \right) > 0$$

negative value

Figure 32.2 The challenge of understanding conduction in the caloric paradigm

Thomson initially took Carnot's theory at face value, never really confronting how exactly the heat-to-work conversion happened. Yes, as cited earlier, he did have some reservations, but without an alternative path forward, he went ahead and analyzed Carnot "as is." It was his meeting with Joule that became the straw that broke the camel's back. If Carnot's work bothered Thomson, certainly Joule's work bothered him ten times as much since Joule effectively said that the production of work had nothing to do with a change in quality of heat but instead was due to the consumption of heat. Neither Carnot nor especially Joule fit into any puzzle Thomson had been forming around heat.

The intriguing concept of a wasted mechanical effect, i.e., work, that couldn't be found drew Thomson's full attention. He lived his life to be prepared for this dilemma. If the work generated by conduction was lost to the cold reservoir, could it be recovered as work again? Recall James' aforementioned rather stunning and telling inquiry about the possibility of operating steam engines *without fuel* [my italics] "by using over again the heat which is thrown out in the hot water from the condenser." William now took up this inquiry for himself.

The answer, as we now know, was no. It couldn't be recovered. It was not an entity that remained as being recoverable. It dissipated and was lost to man. This is what Thomson finally realized, even if he didn't know exactly why this was so. And my language here is inaccurate and inappropriate. There is no "it" regarding heat. It's not a thing. And the more we talk about a heat that flows and dissipates, the more confusing things become when we try to truly understand it.

William's conquering of this quandary was pivotal. This was perhaps one of the singular key moments in his evolution as a scientist, for it was this thinking that gave rise to his concept of "dissipation," which became formalized if not immortalized, as we'll see, in his own version of the 2nd Law of Thermodynamics.

He went through the fire and came out the other side. Because he wrestled with this very fundamental problem, William's arrival at the eventual solution became deeply powerful to him from both science and religious perspectives. Learning about the latter is important to understanding the direction he took the former. But before going there, I want to take a short detour to discuss the challenges of conduction.

The Science of Conduction – There is No Such Thing as "Heat Flow"

Do you think that conduction is easy to grasp? Think again. Even after William Thomson's breakthrough to the mechanical theory of heat in 1851, did he truly understand conduction? I'm not sure. A note from James Clerk Maxwell to William in 1855, years after the launch of energy, is very telling. Wrote Maxwell, "By the way do you profess to account for what becomes of the *vis viva* of heat when it passes through a conductor from hot to cold? You must either modify Fourier's laws or give up your theory, at least so it seems to me."[42] Such questioning revealed a continuing large gap in knowledge regarding energy in the years after its launch in 1850–1.

---

[42] (Smith, 1998) p. 247.

What was it about conduction that was so difficult to grasp? The answer, as Maxwell's question suggests, was that heat was still thought of as a flowing entity. Even today, many are confused about this because we still use the words, heat flow. The fact is that there is no flow of heat. There is no entity called heat.

Heat is nothing more than the *change in state* of a system that results from the exchange of thermal energy across a stationary boundary. The change results from the direct collisions between atoms. For example, in the first step of Carnot's cycle, molecules from a hot source (combustion products) smash into the external metal wall of a cylinder and cause the cylinder's metal atoms to vibrate faster around their fixed positions, thereby striking the gas molecules inside the cylinder with higher kinetic energy. As the gas molecules move faster, both temperature and pressure increase and the piston gets pushed out to accomplish work. In analyzing this, we say that heat "leaves" the hot source and "enters" the piston. But this isn't correct. In reality, the energy of the hot source decreases while the energy of the gas inside the cylinder increases by exactly the same amount, as the conservation of energy is applied to the collisions of the zillions of atoms and molecules involved. The *changes* are quantified by a term called "heat", as opposed to similar changes that result when the boundary is moving rather than stationary, which we call "work." They're both caused by the same phenomenon: the collision of atoms at interfaces. In this discussion, it doesn't confuse us too much to refer to heat as leaving or entering. We're used to this terminology. But when we turn to the process of conduction, confusion appears.

Put a metal bar between a hot source and a cold sink (Figure 32.2). The bar provides a means for heat to figuratively drain out of the hot and into the cold, bypassing the engine, thus resulting in a decrease in energy of the former and an increase in energy of the latter. The magnitude of each change we quantify as "heat." But then we say that heat "leaves" the hot, "flows" down the bar and "enters" the cold. Here we meet the source of our confusion, because nothing, in fact, flows. In the world comprised of a flowing and conserved heat, before energy was discovered and the atomic theory of matter became known, it was difficult to grasp conduction in the mid-1800s.

Challenge yourself on what you really know about conduction. When we look at the steady-state conduction of heat down a metal bar from high temperature to low, we assume that the heat leaving the high equals the heat entering the low since there's no accumulation (no change in temperature of the bar over time). We refer to this heat as a flow from a high to a low "elevation" which we call temperature. The analogy here, created by Carnot and enthusiastically embraced by the Thomsons, who had separately shown keen interest in water wheel power generation, was indeed the water wheel in which high elevation water flowed into the water wheel buckets, turning the shaft as they fell to lower elevation. It was easy back then to understand this from an "$mg\Delta h$" argument. But when you look at conduction, the analogy breaks down. You're left asking, what happens when heat flows from high temperature to low? Something must be happening, right? Temperature decreases. But does this mean that energy also decreases? But this can't be since the energy leaving the hot must equal the energy entering the cold. And this is the exact paradigm that Thomson was stuck in, all because of this mental model of a "flowing heat."

Again, our use of such out-of-date terms and concepts really makes our lives difficult. There is no flow. There is no temperature of heat. There is no entity called heat. It's an incorrect mechanical model that, while sufficient to guide Fourier's math, was insufficient to guide Thomson's understanding of heat and work together in the same context. He couldn't understand how a flowing heat could change its temperature and cause work in Carnot's engine but cause nothing in conduction. How confusing this must have been for both brothers. Life was fine when only dealing with conduction. But with the conflicting arrival of Carnot's heat engine and Joule's experiments, life was no longer fine.

Even though William may not have fully understood conduction at the time of Maxwell's note, which may have contributed to his struggle with the draft manuscript of his 1851 *Dynamical Theory*,[43] the need to bring new and exact definitions to the new world of energy became clear to him. When it did, Thomson eventually and carefully changed his use of the word "heat," ceasing to speak of the heat *in* a body, "referring only to heat absorbed or emitted with respect to the surroundings, and of 'thermal motions' inside."[44] But even with this new definition, I'm not sure that Thomson understood the distinction between heat and energy.

And to answer the Thomson brothers' question, if such a short-circuit existed to cause thermal energy to exchange between the hot reservoir and the cold reservoir, the lost opportunity to do work indeed ends up in the cold reservoir in the form of an increase in thermal kinetic energy. But you'll likely never see this. Why? Because the mass of the cold reservoir is typically quite large, leaving the resulting temperature rise infinitesimal.

The 1st Law Revisited – a Closer Look at Heat and Work

Consider again Clausius' creation of internal energy, U, and its lead placement into the 1st Law of Thermodynamics.

$$\Delta U = Q - W$$

This law seems so simple and yet it says so much, even though Clausius may not have originally realized this as he wouldn't embrace the concept of energy until years later. What this equation says is that the energy of a closed system can *only* change if either Q or W is non-zero. If both are zero, then ΔU is zero, meaning that no matter what is happening inside the system, such as, for example, a chemical reaction, the total energy of the system does not change. It's conserved. This became a much higher-level statement than work–heat equivalence, embodying as it did the conservation of energy.

The fact that heat and work are both included as additive terms in the 1st Law shows how fundamentally connected they are. Consider the barrier separating the closed system from its environment. The atoms and molecules comprising the system continually collide with the atoms comprising the barrier, which then continually collide with the atoms comprising the external environment, which can also be considered as its own closed system. If the temperature of the external environment is different than that of the closed system, then the collision of the atoms between systems will result in a *change* in internal energy of both systems, which we call "heat" and quantify by Q. The faster atoms will slow down, the slower atoms will speed

[43] (Smith, 1998) p. 101.
[44] (Smith and Wise, 2009) p. 338.

up, and the temperature of the two will approach each other. But this result occurs without anything actually flowing between the two systems.

Likewise, if the pressure between the two systems is different and the barrier separating the two is movable, then the barrier will indeed move until the pressures approach each other and the resulting change in energy is called "work" which we quantify by W. It's the atom-collision impacts involved with both heat and work that cause change in internal energy as captured by the 1st Law. Again, this result occurs without anything actually flowing between the two systems. The concept of heat–work equivalence as expressed in Clausius' 1st Law simply says that both Q and W represent and quantify the two different ways that the energy of a closed system can change. Either way, whether it's Q or W, it's all about the atomic collisions between the closed system and the boundary.

The challenge sometimes in applying the 1st Law, or the 2nd Law for that matter, arises when trying to figure out what the system actually is. In many homework problems in thermodynamics, this isn't an easy task, as many who have attempted such problems can attest to. Even something as seemingly easy as Joule's experiment requires some thought. One could argue that the closed system involved was comprised of both the dropping weight and the agitated water. Since this system was closed with no exchanges of either work or heat energy with the *external* environment, then the total energy inside the system had to remain constant, thus resulting in Joule's conclusion that work done (W) equaled heat generated (Q). It's easy after-the-fact to see this. But not so easy to one sitting in the classroom struggling to analyze this problem.

The 1st Law of Thermodynamics can be used to explain all that William Thomson sought to explain, especially regarding Carnot's heat engine and Fourier's conduction. The challenge, as we will see, was to consider these explanations and realize that they were not yet complete. The story wasn't over. A second law would be required.

The 2nd Law Struggles to Emerge

Before continuing, I need to take step back here and consider what was happening during this time period in the larger context.

Why was it so challenging to arrive at the 2nd Law of Thermodynamics? Well, one reason was that entropy itself was a challenging concept to both discover and then comprehend. But there was another reason and it was this. From the 2nd Law based on the probabilistic nature of entropy, one can deduce many consequences, almost as if this ultimate form sits at the top of a tree and all of the branches below represent the consequences. The challenge that resulted was that scientists discovered the multiple consequences first, hence the multiple different versions of the 2nd Law that would appear, and had a difficult time seeing the larger tree that they were a part of.

Fleshing this discussion out further, consider the entire concept of entropy as the full tree (Figure 32.3). At the top of the tree is the statistical mechanical explanation of entropy as defined by Boltzmann. The entropy of a system quantifies the maximum number of different ways that the entities comprising a given system can be configured for a given set of fixed macroscopic properties. Now consider that all the many consequences of entropy sit down in

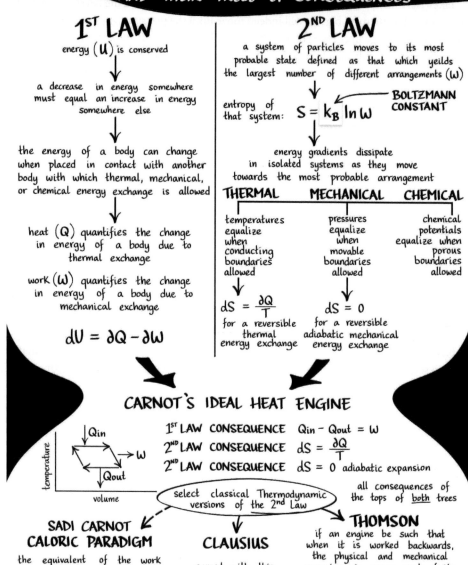

32.3 THE TWO LAWS OF THERMODYNAMICS AND THEIR TREES OF CONSEQUENCES ©RTH + CLS

1ST LAW

energy (U) is conserved

a decrease in energy somewhere must equal an increase in energy somewhere else

the energy of a body can change when placed in contact with another body with which thermal, mechanical, or chemical energy exchange is allowed

heat (Q) quantifies the change in energy of a body due to thermal exchange

work (W) quantifies the change in energy of a body due to mechanical exchange

$$dU = \partial Q - \partial W$$

2ND LAW

a system of particles moves to its most probable state defined as that which yeilds the largest number of different arrangements (W)

entropy of that system: $S = k_B \ln W$ — BOLTZMANN CONSTANT

energy gradients dissipate in isolated systems as they move towards the most probable arrangement

THERMAL   **MECHANICAL**   **CHEMICAL**

temperatures equalize when conducting boundaries allowed

pressures equalize when movable boundaries allowed

chemical potentials equalize when porous boundaries allowed

$dS = \dfrac{\partial Q}{T}$ $dS = 0$

for a reversible thermal energy exchange

for a reversible adiabatic mechanical energy exchange

CARNOT'S IDEAL HEAT ENGINE

1ST LAW CONSEQUENCE $Q_{in} - Q_{out} = W$

2ND LAW CONSEQUENCE $dS = \dfrac{\partial Q}{T}$

2ND LAW CONSEQUENCE $dS = 0$ adiabatic expansion

all consequences of the tops of both trees

select classical Thermodynamic versions of the 2nd Law

SADI CARNOT CALORIC PARADIGM

the equivalent of the work done by heat is found in the mere transfer of heat from a hotter to a cooler body

+

while the quantity of heat remains undiminished

CLAUSIUS

agreed with this, called it Carnot's 2nd principle

corrected this based on new theory of energy

THOMSON

if an engine be such that when it is worked backwards, the physical and mechanical agencies in every part of its motions are all reversed, it produces as much mechanical effect as can be produced by any thermodynamic engine, with the same temperatures of course and refrigerator from a given quantity of heat

plus many additional versions of the 2ND Law based in part on Carnot's engine

Figure 32.3 The two laws of thermodynamics and their trees of consequences

the branches of the tree. The sum of the entropies of separate systems is less than or equal to the entropy of the combined systems once equilibrated. And then to lower branches. The temperatures of two bodies that can exchange thermal energy move toward each other. The pressures of two bodies that share a movable boundary move toward each other. The chemical potentials of a given molecule in two systems that share a porous boundary move toward each other (as the molecule moves from one system to the other). Now down to even lower branches, heat flows from hot to cold.

There's another tree right next to the one we're on, one based on the 1st Law of Thermodynamics. The concept of energy and its conservation sits at the top of the tree and down in the branches sits work–heat equivalence (Clausius' 1st Principle) stating that the trade-off between thermal energy and mechanical work must be one-for-one.

Then further down in the world of consequences the two trees intertwine and you arrive at Carnot's intriguing heat engine in which the maximum work done cannot exceed the thermal energy consumed. In this ideal machine—no heat loss due to conduction, temperature range of the working substance is equivalent to the temperature range of the hot source and cold sink—the work done (W) is equal to the difference in the loss of thermal energy from the heat source (Q_{in}) and the gain in thermal energy of the cold sink (Q_{out}). $W = Q_{in} - Q_{out}$. And not only that, but W is the maximum that it can be, which means that Q_{out} is the minimum it can be. The quantity Q_{out} occurs as a natural consequence of Carnot's cycle; heat must be removed from the working substance when using work to bring it back to its starting point. As we'll discuss in future chapters, the absolute minimum that this can be is determined by the temperature ratio of the hot and cold reservoirs, because $Q_{in}/T_H (= \Delta S_H)$ must equal $Q_{out}/T_C (= \Delta S_C)$, which means that $Q_{out} = Q_{in}(T_C/T_H)$. The temperature terms reflect the temperatures of the working substance, which in Carnot's ideal engine equal the temperatures of the reservoirs. The fact that heat can only flow from hot to cold means that these temperatures cannot exceed the range set by the reservoirs. And this is what ultimately sets the theoretical "best" the engine can perform. The working substance can only do what the reservoirs provide in terms of heat flow and temperatures, no more. The combination of the 1st and 2nd Laws dictates this.

Given all these different consequences, one can understand the challenge faced by the early physicists and sciences in discovering the 2nd Law. They started down in the branches, each seeing a different consequence, all consequences being some part, either separate or combined, of two trees, each leading toward the top where entropy and energy are defined. Clausius latched onto the branch that said heat must flow from hot to cold whenever work is done in a cyclic process (2nd Principle). Thomson would latch onto the branch that said when $T_H = T_C$, no work is possible. Both were caught down in the branches. Neither would truly be able to rise to the top, although Clausius would come very close.

A Final Note on Thomson's Consequence Branch of the Entropy Tree

Just a note here regarding the reference to Thomson's branch, which we'll get to in more detail shortly. In Carnot's engine, is it possible for the working substance to achieve higher temperatures than the hot reservoir or colder temperatures than the cold reservoir? Absolutely yes. Just over-compress during the adiabatic Step 4 to achieve the former, and over-expand during adiabatic Step 2 to achieve the latter. But does this provide a means to generate more work? No. Because when you start the subsequent respective isothermal steps, you need to be at the temperature of the reservoir, and the only way to get there is by undoing the "over" steps, thus exactly negating whatever effect the "over" step caused. For example, if you over-expand in adiabatic Step 2 and generate a lower temperature than the cold reservoir, you need to raise the temperature prior to Step 3 isothermal compression or else you'll create a temperature gradient and thus lose efficiency. The way to bring the temperature back up would be to do an adiabatic compression. But this would necessarily follow the identical path of the adiabatic over-expansion and thus yield nothing. The extra work generated by the over-expansion would be exactly cancelled by the extra work required by the subsequent compression.

William Rankine

We have now arrived at the point in history where Thomson embraced the concept of energy and thus aligned with Clausius. The resulting confluence of these two great and complementary minds is really what launched classical thermodynamics. We'll see shortly how Thomson led this launch after Clausius' 1850 breakthrough and in so doing developed a whole new realm in physics built around energy, taking much more of a leadership role in creating this than Clausius did, even though the work of Clausius really inspired its formation. But before moving on with this story, I pause here to consider two significant historical questions with the story up until now, the first being, why Thomson and not Rankine, and the second, what role did Clausius play in Thomson's breakthrough?

Starting with the first, William Rankine (1820–1872) was born in Edinburgh and soon became yet another of those precocious scientists who somehow managed to learn higher mathematics at an early age, having been so inspired by reading Newton's *Principia* by the age of fourteen, a very humbling thing to hear for most of us. In 1836, Rankine entered the University of Edinburgh, where he studied natural philosophy under Professor James David Forbes. Despite showing much promise, limited financial resources forced him to leave without a degree in 1838 to start a professional career in engineering. Rankine complemented the development of his practical side with an ongoing academic pursuit of natural philosophy. For example, as mentioned earlier, in 1842 he started to develop his own theory of heat, had to stop due to lack of experimental data, but eventually completed the task in 1850 after he obtained Regnault's data.

Back at the University of Edinburgh, Professor Forbes, as Secretary of the Royal Society of Edinburgh (RSE), soon became a vital player in this story as he started actively seeking papers relating to the new theories around heat and the limitations of steam engines. Rankine, a former student of Forbes, provided one such paper. In February, 1850, Rankine shared his work with the RSE when he read *On the Centrifugal Theory of Elasticity, as Applied to Gases and Vapors*. In preparation for this reading, Forbes had sent his counterpart in Glasgow, none other than William Thomson, a draft of Rankine's paper for review, thus commencing the famed Thomson–Rankine relationship. Upon receiving his own invitation, Thomson, also in turn, shared his work with the RSE in March 1851 when he read *On the Dynamical Theory of Heat*.

Rankine would eventually enter into a series of ongoing scientific dialogues with Thomson, a situation that received a major boost when the geographic gap was closed in 1855 upon Rankine's relocation to Glasgow to succeed Lewis Gordon as Chair of Engineering. Rankine remained at Glasgow, teaching and publishing a classic series of engineering treatises, until his death in 1872, when he was succeeded by none other than James Thomson. Small world.

Rankine was clearly in front of Thomson in developing the mechanical theory of heat. But unfortunately, he too strongly tied his otherwise valid theory to his rather complex "hypothesis of molecular vortices" and then wrapped it all up in a language that wasn't easy to penetrate. The vortex theory was based on an incorrect mechanical model involving static atoms surrounded by elastic spinning atmospheres whose *vis viva* of revolutions defined the atom's quantity of heat. The interesting thing is that since the model inherently linked the *vis viva* ($\frac{1}{2}$ mv^2) that originates from a centrifugal force analysis to heat, the answers derived from it were essentially correct. This was yet another case in which a wrong model resulted in correct answers. So close, and yet so far. The model wasn't too far-fetched, relying as it did on the work of others, including the eminent Sir Humphry Davy. And people didn't know it was wrong. But people couldn't completely understand it, and that was the real problem. Rankine's personal style of writing was itself confusing, which only compounded the situation. Many, including Thomson, found it difficult to understand Rankine in this regard although they did value his thinking, his ideas, and his work in general. So while Rankine was very influential during this time, especially with regards to Thomson, his influence faded afterwards, while Thomson's rose. We unfortunately rarely speak of Rankine today in the history of Thermodynamics, even though he wrote one of the first recognized treatises on the subject. Had he kept his model behind the scenes, as Clausius did with his own thermodynamics—"We shall not consider here the kind of motion which can be conceived of as taking place within bodies"[45]—and had he been more clear with his writing, then his place in history would have likely risen. He properly associated heat with the *vis viva* motion "of the small parts" and arrived at the correct mechanical answers. It's just that his picture of the small parts was wrong. Again, he seemed constrained by the static-atom model going back to Newton.

What Role did Clausius Play in Thomson's Eventual Paradigm Shift?

Now to address my second question. What role did Clausius really play? While the influences of Joule and Rankine on moving Thomson from caloric to energy were direct, not so the influence of Clausius, who never met Thomson. While Thomson knew that Clausius reconciled Joule with Carnot by embracing work–heat equivalence, he didn't know the details of how and so went about taking his own path to this reconciliation and eventually to his own development of the 2nd Law of Thermodynamics.

Thomson's mind was moving so fast during this time period that it's difficult to determine the extent to which Clausius really did influence him. It was certainly greater than zero percent, but also certainly less than one hundred. Clausius' 1850 publication clearly preceded

---

[45] (Carnot et al., 1988) p. 112.

Thomson's 1851 publication, but while the latter paralleled much of the former, Thomson claimed that his own interpretation of Carnot's work was done prior to reading Clausius.[46] So even though he certainly gave credit to Clausius for publishing first, writing "It is with no wish to claim priority that I make these statements, as the merit of first establishing the proposition upon correct principles is entirely due to Clausius," he saw fit to add a caveat, "I may be allowed to add, that I have given the demonstration exactly as it occurred to me before I knew that Clausius had either enunciated or demonstrated the proposition."[47] It wasn't only his pride talking; he did see things differently, especially in one key area. Whereas Clausius initially saw equilibrium, Thomson saw irreversibility.

It's quite difficult to lay down an exact timeline of Thomson's caloric-to-energy transition and Clausius' role in the matter. The reason is that the transition didn't occur as a single eureka moment for Thomson. Instead, as Crosbie commented, the transition occurred more slowly over time, as "tentatively, ever so cautiously, Thomson had begun to toy with the mechanical theory of heat. How far he actually progressed in reconciling it with Carnot's criterion before reading Clausius, we can only surmise."[48]

Based on his writings, both published and unpublished, his correspondences with Joule and Rankine, and the notes obtained from students attending his classroom lectures, the transition seems to have occurred at some point in early to mid-1850.[49] It's as if the scales were (almost) removed from his eyes, as he quickly realized the need to re-interpret Carnot based on Joule's work–heat equivalence. This led to his famed series of papers between 1851 and 1855 *On the Dynamical Theory of Heat*, which Crosbie Smith labeled "the most creative phase in the development of Thomson's thermodynamics"[50] and a "critical watershed" in Thomson's career.[51] He was finally convinced of Joule's arguments. Heat is not a substance!

The Value of Learning History

Why so much time spent on this? Well, our current education is based on this story, but rarely do we have time to learn the history. It's a remarkable story about how the giants of our field wrestled with the deep, fundamental concepts. Knowing the theory and history together provides us a much firmer grounding in the bedrock fundamentals, enabling us to approach new problems that we haven't seen before. The more we learn about the fundamental concepts from different angles, where mistakes were made and successes achieved, the more we learn of the path taken, with its many false turns and dead ends, the more we'll be able to solve these problems. And it's not the mathematics that I'm referring to here. More than half of the challenge in thermodynamics is in setting up the problem. As Feynman said, "You can't do physics just by plugging in the formulas...you have to have a certain *feeling* for the real situation."[52]

[46] (Thomson, 1851)
[47] (Thomson, 1851)
[48] (Smith and Wise, 2009) p. 325.
[49] (Smith and Wise, 2009) pp. 323–324.
[50] (Smith and Wise, 2009) p. 327.
[51] (Smith and Wise, 2009) p. 334.
[52] (Feynman et al., 2013) p. 56.

Ask a student to explain conduction. Can they? Chances are, no. But I'm guessing that after learning this history of Thomson and Clausius wrestling with conduction, the answer to this question would change to yes.

> William Thomson
> –
> The Energy Years
> –
> The Two Laws

The science of thermodynamics is based on two laws, the first of which states the fact of the mutual convertibility of heat and mechanical energy, while the second shows to what extent the mutual conversion of those two forms of energy takes place under given circumstances. In the course of the last few years the first law has been completely "popularized;" it has been amply explained in books and lectures, composed in a clear and captivating style, and illustrated by examples at once familiar and interesting, so as to make it easily understood by those who do not make science a professional pursuit.

The second law, on the other hand, although it is not less important than the first, and although it has been recognized as a scientific principle for nearly as long a time, has been much neglected by the authors of popular (as distinguished from elementary) works; and the consequence is that most of those who depend altogether on such works for their scientific information remain in ignorance, not only of the second law, but of the fact that there is a second law; and knowing the first law only, imagine that they know the whole principles of thermodynamics. The latter is the worst evil of the two: "a little learning" is not "a dangerous thing" in itself, but becomes so when its possessor is ignorant of its littleness.

– William Rankine (1867)[53]

Once Thomson finally embraced the mechanical theory of heat, he quickly brought his full scientific genius to bear on Carnot's work, an effort that culminated in his March, 1851 reading at the RSE of *On the Dynamical Theory of Heat*, which was a catch-all title capturing what would become a series of publications on this topic in the early 1850s. In these publications, Thomson did his own reconciliation of Carnot and Joule and arrived at essentially the same place as Clausius had in 1850. There are many similarities between the two authors' findings that I won't go into here. But what I do want to address are the two main themes having to do with the two laws of thermodynamics.

The rise of these two laws was not pre-ordained. Instead, they evolved as a natural progression of Clausius' interpretation of Carnot in which he wisely separated the two issues at hand, namely Joule's heat–work equivalence, which replaced Carnot's conservation of heat, and Carnot's theory of conversion limitations, calling them his first and second principles. Clausius' thinking was that his two principles were needed to constrain the maximum amount of work from Carnot's engine. Thomson would build on both, calling them his two propositions, and Rankine helped conclude these statements, raising their names to the status of law.

---

[53] (Rankine et al., 1881) p. 432.

The Two Laws

Just as Newton reduced a large volume of experimental data and observations to the very concise three laws of motion, so too did Clausius and Thomson likewise reduce a large volume of experimental data and observations to the two laws of thermodynamics, to which others would later add a third (S = 0 at 0 K) based on Nernst's heat theorem, which I'll cover later. It's really quite remarkable how these laws could explain so much with so little.

So let's look at these two concepts, starting with the first. As discussed in the previous chapter, the equivalence of heat and work was established by Mayer and Joule and then recognized by Clausius and later by Thomson when they used this finding to unlock the mystery of Carnot. It's important to realize that heat–work equivalence was a necessary step on the path to the 1st Law, but it wasn't the 1st Law itself. Indeed, the 1st Law arguably came into its own when, to little fanfare, Clausius used heat–work equivalence to combine heat (Q) and work (W) under the umbrella of energy (U) in his famous equation (in infinitesimal form),[54]

$$\delta Q = dU + A \times \delta W \quad \text{Clausius, 1850}$$

in which "A" is the heat equivalent for a the unit of work. Clausius referred to U as a property of state but did not further refer to it as "energy." It was the scientists of Scotland who pulled U into their new terminology by labeling it *intrinsic energy* (later changed to *internal energy*).

It wasn't just the introduction of energy that made the 1st Law of Thermodynamics significant. It was also the implied conservation in Clausius' equation. If δQ and δW are both zero, then so too is dU. If a closed system receives no heat and does no work, in other words if it's isolated, its energy will not change, regardless of what happens inside the system. Such thinking eventually led Clausius to make the larger statement that, assuming the universe is a closed system, then the energy of the universe is constant.

In the end, as suggested by Rankine's quote above, even though the concept of energy was rather abstract, the 1st Law of Thermodynamics made visceral sense to the scientists of the day, especially once caloric disappeared from the discussion. The 1st Law's eventual and rapid acceptance was rather straightforward as recounted in the previous chapter.

One can't overstate the impact of this law on science. It provided a rallying point around which a whole new world of energy-based physics would form, one that would eventually reveal to us the inner workings of matter and the chemical world. Mayer's and Joule's independent discovery of heat–work equivalence was a critical and absolutely necessary step, but it was really Clausius' and then Thomson's efforts that wrapped this discovery into the larger and more encompassing theory of energy and its conservation.

While the 1st Law brought welcome clarity, the 2nd Law brought confusion, as reflected by the many different versions used to describe it. The confusion came about more due to the fate of history than due to the science involved. The 2nd Law was indeed embedded in Carnot's work, but only in part, as it was buried deep down in the branches of a logic tree that had the statistical mechanical entropy sitting on top with all of its different manifestations spread out below. Who would have thought that the term $\delta Q/T$ actually meant something, especially

---

[54] (Carnot et al., 1988) p. 121.

something that was entirely different from Joseph Black's seemingly similar but actually <u>very</u> different term, $\delta Q/dT$, for heat capacity?

The 2nd Law from Different Perspectives – Down in the Branches of the Entropy Tree

Carnot, Clausius, and Thomson all struggled to discover the fundamental cause limiting the generation of work in a heat engine. They sensed its presence, but just couldn't distinguish it. They were clearly on the right path but had necessarily started so far down in the tree's branches, well below $\delta Q/T$ and into the inner workings of the steam engine, that it was extremely complicated to sort things out. The historical statements we read from this time challenge our abilities to understand. But I'm not sure what other history would have eventually led to the top of tree. A great book for the historians to write would be one that speculates on what would have happened had there been no Carnot.

Let's go now to the history and walk through the actual statements that led to the 2nd Law. You can see for yourself the confusion the different statements caused. As noted by Helge Kragh, "Discussions of the second law…appear in as many forms as there are writers."[55]

Wrote Carnot,

The production of motive power is then due in steam engines not to an actual consumption of caloric, but to its transportation from a warm body to a cold body.[56]

Wherever there is a difference of temperature, motive power can be produced.[57]

The maximum of motive power resulting from the employment of steam is also the maximum of motive power realizable by any means whatever.[58]

The necessary condition of the maximum is…that in bodies employed to realize the motive power of heat there should not occur any change of temperature [such as conduction – author's words] which may not be due to a change of volume.[59]

The motive power of heat is independent of the agents employed to realize it; its quantity is fixed solely by the temperature of the bodies between which is effected, finally, the transfer of caloric.[60]

As for Clausius, he didn't really add anything new to this 2nd Law discussion in 1850. He simply read the above statements, made mention of his own thoughts, such as the observation that heat "always shows a tendency to equalize temperature differences and therefore to pass from hotter to colder bodies,"[61] validated Carnot by proving through his detailed calculations that a heat engine running on an ideal gas performs the same as one running on liquid–vapor, which in and of itself was an extremely powerful validation, and then finally settled on his 2nd principle summary:

[55] (Kragh and Weininger, 1996) p. 111.
[56] (Carnot et al., 1988) p. 7.
[57] (Carnot et al., 1988) p. 9.
[58] (Carnot et al., 1988) p. 12.
[59] (Carnot et al., 1988) p. 13.
[60] (Carnot et al., 1988) p. 20.
[61] (Carnot et al., 1988) p. 134.

The equivalent of the work done by heat is found in the mere transfer of heat from a hotter to a colder body.[62]

Regarding Thomson's role in this matter, he interpreted things just a little bit differently than Clausius, writing,[63]

Proposition II (Carnot and Clausius) – If an engine be such that, when it is worked backwards, the physical and mechanical agencies in every part of its motions are all reversed, it produces as much mechanical effect as can be produced by any thermodynamic engine, with the same temperatures of source and refrigerator, from a given quantity of heat.

to which he added,

The demonstration of the second proposition is founded on the following axiom: It is impossible, by means of inanimate material agency, to derive mechanical effect from any portion of matter by cooling it below the temperature of the coldest of the surrounding objects.[64]

along with a footnote,

If this axiom be denied for all temperatures, it would have to be admitted that a self-acting machine might be set to work and produce mechanical effect by cooling the sea or Earth, with no limit but the total loss of heat from the Earth and sea, or, in reality, from the whole material world.[65]

and a final comment on Clausius,

The following is the axiom on which Clausius' demonstration is founded: It is impossible for a self-acting machine, unaided by any external agency, to convey heat from one body to another at a higher temperature. It is easily shown, that, although this and the axiom I have used are different in form, either is a consequence of the other.[66]

Are you confused yet? If so, don't worry. All of these different statements combined with the lengthy discussions by Carnot, Clausius, and Thomson, among others, about forward and reverse engines make for very heavy reading. The thing to remember is that, when all is said and done, all of these discussions reduce to the same top of the tree, the 2nd Law of Thermodynamics based on the statistical mechanical definition of entropy in which isolated systems move to their most probable state. In fact, it was most likely this same confusion, this inability to reduce all of these statements to a simple equation as was done with the 1st Law, that gnawed at Clausius and motivated him to discover entropy in 1865. Clausius believed in the quantifiable simplicity of nature and at that point in time, the different renditions of the 2nd Law were anything but. So don't feel alone in your confusion. There was a reason the 2nd Law was largely neglected by the contemporary authors of popular works, as noted by Rankine.

Let's look at Thomson's above four statements. Acknowledging the priority of Carnot and Clausius in the first, he stated that the reversible heat engine indeed generates the maximum amount of work. This was not new. Then he drew attention to temperature and its role in

[62] (Carnot et al., 1988) p. 132.
[63] (Thomson, 1851) Paragraph #9.
[64] (Thomson, 1851) Paragraph #12.
[65] (Thomson, 1851) Footnote 5.
[66] (Thomson, 1851) Paragraph #14.

generating work in the next two statements, proposing that once matter reached the coldest temperature of its surroundings, no further work could be generated from it, even if its energy were as large as that of the ocean. Again, nothing new, as this was Carnot's key premise when he said that the amount of work possible was related to the "height of the fall" of temperature; no height, no work. In the fourth statement he cited Clausius' axiom that heat can only flow from hot to cold. Again, this was common knowledge at that time.

These statements in and of themselves were not new to the world of science. Indeed, it was really Clausius who really brought these ideas into the public discussion. But it was what Thomson did next that was new, for he wasn't content to stop here and instead probed the implications, or, as he called them, the "remarkable consequences"[67] which follow from these statements.

A Brief Return to Conduction

While Thomson's admission, "It is easily shown that…either is a consequence of the other,"[68] suggests that both his and Clausius' statements regarding what would become the 2nd Law were, in effect, the same, the fact remains that the way they were worded was different, reflecting two different approaches to Carnot. Yes, it may have been Thomson's ego that wanted to distinguish his own words from those of Clausius,[69] but there was something more going on here, and it goes back to the main stumbling block that confronted Thomson when he attempted to reconcile Carnot and Joule. Conduction.

For Clausius, the issue of conduction was straightforward. He followed Carnot's lead. In conduction, heat flowed from hot to cold, never from cold to hot. Heat lost to conduction was heat that could have otherwise been used to generate work. To realize the maximum work for a given amount of heat, "Only temperature changes due to volume changes are permissible."[70] In other words, conduction was not permitted, for it sucked the heat out of the hot source by Q as opposed to W. Not being as confronted by conduction as Thomson was, Clausius recognized this and simply assumed in the ideal process that there was no conduction. To him, conduction was simply a transfer of heat without mechanical effect. In his analysis of Carnot, he really didn't spend any time on this issue, choosing instead to focus on the ideal *reversible* engine for which conduction, caused by temperature gradients between touching bodies, played no role. In other words, it's not that Clausius understood conduction, it's that he just ignored it for many years.

For Thomson, though, the issue of conduction delved into something much more profound, something regarding the inherent *irreversibility* of nature. As he wrote, when heat is conducted from hot to cold, it is "irrecoverably lost to man, and therefore 'wasted,' although not annihilated."[71] Thomson realized that a given amount of heat could either flow through Carnot's engine to do work or flow directly to the cold sink via conduction. With the latter, the heat flow would be

---

[67] (Thomson, 1852)
[68] (Thomson, 1851) Paragraph #14.
[69] (Lindley, 2004) p. 107.
[70] (Carnot et al., 1988) p. 13.
[71] (Thomson, 1851) Paragraph #22.

constant as no work was done, but upon entry into the cold sink, the temperature difference between the flowing heat and the cold sink would be lost, which meant, according to Carnot, that the generation of work from this heat would be impossible. Hence, the opportunity to generate work from this heat would be "wasted" even though the heat itself would not be consumed. This is what Thomson was thinking when he wrote the above. Again, I'm not sure exactly what Thomson thought when he spoke of "heat flow" in this regard.

To Thomson, the decrease in temperature caused by conduction was caused by the irreversible diffusion of heat. "When heat is diffused by conduction, there is a dissipation of mechanical energy, and perfect restoration is impossible."[72] His youthful dream of recovering the heat from the cold sink was gone. Bringing his religion into his science, he continued, "As it is most certain that Creative Power alone can either call into existence or annihilate mechanical energy, the 'waste' referred to cannot be annihilation, but must be some transformation of energy."[73]

From a theoretical standpoint, I'm still not convinced that Thomson, or really anyone, including Clausius, thoroughly understood conduction in 1851, as reflected by Maxwell's aforementioned note to him and also by the above statements and others, in which Thomson refers to heat as passing from one body to another at a lower temperature. Reading all of these statements suggests that heat and energy remained some kind of entity to Thomson—even if not a material entity—that could flow and diffuse and transform. I don't know exactly what Thomson was thinking regarding the fundamental nature of heat, especially since he didn't expound on this in his writings, and I don't venture into this area to cast judgment, but merely to help interpret his writings.

Thomson's Paradigm Shift

But given this, given that Thomson may not have fully understood the physics involved in conduction, he did fully grasp the implications. Once heat got to the cold sink, it could no longer do any work. Why? Because there was no longer a driving force. This is what he finally recognized by combining Carnot with Fourier. Just as water could no longer do work once it reached sea level, heat could no longer do work once it reached the temperature of the environment.

It was this moment of realization for Thomson that finally broke the logjam in his thinking. He had long seen loss in nature, going all the way back to his earlier years when he and James would wonder about the losses in water wheel power plants, but now he finally identified what the loss represented. It wasn't a loss of energy. It was the loss of energy's ability to generate work due to its dissipation. The measure of the loss was in temperature.

In 1851, with his revised interpretation of Carnot in hand, Thomson was very well positioned to tie this all together by revealing the larger meaning behind Fourier's equations. A spatial driving force dT/dx results in the decrease in temperature (dT) over time (dt) due to, for example, the conductive flow of heat, and thus results in the lost ability to do work in a *thermo-dynamic* engine over time.

---

[72] (Thomson, 1852)
[73] (Thomson, 1852)

This transition of Thomson to the mechanical theory of heat was significant for both Thomson and history. It wasn't just that Thomson saw things in a new light, it was that this new concept resonated very deeply inside him as it aligned with his religious beliefs. The best way to understand this is by considering why Thomson intentionally brought the word *dissipation* into his science. This choice was not casual but quite intentional.

Energy Dissipation and Heat Death

Nature contains within itself the rudiments of decay

– Rev. Thomas Chalmers[74] (c.1830s)

The moment in 1517 when Martin Luther nailed his *Ninety-Five Theses* on the front door of the Catholic church in Wittenberg, Germany arguably marked the point when Christianity's rapid spread throughout Western Civilization was met by man's reflexive resistance to being told what to do by other men. Luther's action launched the Protestant Reformation and attempted to establish an equilibrium dividing line between the two opposing forces. This movement made its way over to Scotland and led to the rise of Presbyterianism in the mid-sixteenth century, which would play a central role in Scottish culture thereafter.

On a separate historical timeline, Scotland's union with England in 1707, being one primarily regarding governance, resulted in an un-mixing of the populace, as it was primarily the politicians and aristocrats who moved to London, and those associated with Scottish law, religion, medicine, and academia who stayed behind. The resulting concentration of lawyers, clergy, professors, doctors, scientists, and engineers formed a highly intelligent middle-class and ultimately led to the Scottish Enlightenment that emphasized reason, individualism, and the advancement of knowledge through the scientific method.[75] The formation of this critical mass of talent also led to the formation of a strong infrastructure of universities, marine engineering works, and scientific societies that promoted congregation, discussion, and the debate over ideas. The presence in Scotland of such talented figures as Joseph Black, James Watt, James and William Thomson, William Rankine, and James Clerk Maxwell, in addition to Adam Smith and David Hume, was no coincidence. The Scottish culture grew them.

I bring up this history because the achievements of these Scottish thinkers and doers was influenced to a degree by the fact they often mixed some of their Presbyterian religion into their Enlightened science, using their religious beliefs to help guide their interpretation of nature. Regarding our main protagonist, William Thomson was raised in just such a blended household run by an intellectual father (mathematics professor at Glasgow University) who was faithful to his church. And this was where things get interesting for us, because one of the close friends of the Thomson family was Thomas Chalmers (1780–1847), a powerful preacher and influential leader of that church. To Chalmers, God's world was not the one described by the followers of Laplace in which everything was perfectly balanced, with periodic but

---

[74] Chalmers quote cited in (Smith, 1998) p. 15.

[75] (Fromkin, 2000) p. 152. "One thing that [the Reformation and the Enlightenment] had in common is that both, rejecting authority, emphasized the freedom of the individual—freedom of conscience, freedom to shape one's own faith, freedom of thought and inquiry."

non-permanent changes, but one that was eternally changing and changing in but one direction, down. The evidence of this design was all around, in the volcanoes, hurricanes, and floods, and in decay, death, and disease.

In his role as a religious leader, Chalmers endeavored to apply this philosophy toward Glasgow life by preaching of a God who provided man with gifts, both physical and spiritual, finite in quantity, located in nature.[76] Man, with the possession of free will, could choose to either refuse or accept these gifts. The former was a sin, the latter a virtue. Chalmers preached that it was up to man and his free will—a strong Enlightenment theme—to make the most of these gifts by engaging in useful work while avoiding the waste of these gifts through idleness, all the while recognizing the designed presence of natural decay operating in the background. Chalmers sought to wake the city of Glasgow to this vision.

Around this same time period, the leaders of Glasgow University were searching for someone who embraced this view of doing useful work with God's gifts. Their natural philosophy department lacked strong leadership and they wanted to hire a new professor to help transform this department into one that again rose to the high standards set earlier by James Watt. They needed fresh blood, new ideas, and high energy. But they also needed a man of faith. It was William who best met these requirements. Wrote Chalmers to James Thomson (senior) in 1847 after William's arrival: "may your son be guided by the wisdom from above to the clear and influential discernment of this precious truth – that a sound faith and a sound [natural] philosophy are one."[77] Needless to say, William's appointment was no ordinary appointment. Expectations of the new professor were high.

With this background in mind, let's consider again the years of 1851 and after, when Thomson took his beliefs and his scientific findings and created his own path toward the 2nd Law. He needed a word to capture what happened when heat entered the cold sink. Yes, heat *diffused* in this situation. But he wanted a stronger word, one that would appeal to and be accepted by his fellow Presbyterians.[78] He found it in *dissipation* and so in 1852 named his doctrine "The Universal Dissipation of Mechanical Energy." Dissipation was indeed a word that was commonly used in the Scottish culture, carrying with it the connotation of moral and financial waste, and perhaps it was this use, and his dad's use in particular, that had William select this term on which to build his own version of the 2nd Law. Wrote his father to William in 1842, "[a Glasgow student and son of a Unitarian minister had been found] to have been attending the theater and other amusements from night to night – to have been indulging in habits of dissipation."[79] The student's acts of dissipation described the waste of what God had provided, clearly not a behavior aligned with Scottish culture.

Now can you see the parallel construct between Chalmers and Thomson? Man could either do work or be wastefully idle. Heat could either do work (via Carnot) or be wastefully conducted (via Fourier). Thomson was predisposed by his religious upbringing to see this choice between industriousness and idleness, and to also realize that, regardless of which choice was made, all energy would eventually dissipate itself into the cold sink, just as all men would

---

[76] (Smith, 1998) pp. 17–22.
[77] (Smith, 1998) p. 25
[78] (Smith, 1998) p. 124.
[79] (Smith, 1998) p. 124.

eventually die. This was how God designed nature. This was what Thomson was raised to believe. God gave us the gift of heat. Thomson's quest to understand the cause of how such a gift could best be used and not wasted became almost spiritual. The answer found its way into his science.

Age of Earth Calculations

Thomson was raised on Fourier's mathematics and so believed in the directional flow of energy in the form of heat across space over time regardless of how it was used. Fourier's mathematics quantified this for heat. He also believed that man could either divert this flow for useful work, thereby consuming the heat, or not. The energy flow remained the same either way; energy was conserved. But once this heat ultimately and irreversibly dissipated into the "coldest of the surrounding objects," its ability to do work became "lost" to man. To Thomson, only God could reverse the direction of this dissipative flow. As he wrote, "I believe the tendency in the material world is for motion to become diffused, and that as a whole the reverse of concentration is gradually going on."[80]

Thomson's discovery of Carnot through Clapeyron in 1847 didn't initiate such spiritual thinking. It started prior to this. For example, in his inaugural dissertation of 1844, William blended Fourier's conduction with Chalmers' decay to quantify the age of the Earth.[81] As he published years later, he estimated the outward heat flux from Earth based on the experimentally determined one degree Fahrenheit rise in temperature per every 50 feet of depth from the Earth's surface. Based on an assumed estimated range of the Earth's initial temperature, he then proposed the age of the Earth to be somewhere between 20 and 400 million years.[82] He also proposed that the directional cooling of the Earth meant that it had a beginning and would most certainly have an end.

In an interesting aside, this was not an insignificant estimate, as it ran directly into the heated debate over the theory of evolution. About this time, as captured in his 1859 *On the Origin of Species,* Charles Darwin (1809–1882) had proposed a much higher value for the age of the Earth than Thomson's based on his estimate that a longer time was required for random variation and natural selection to evolve life. The conflict between these estimates had significant ramifications in the battle being waged at the time between a rising generation of scientists, represented by T. H. Huxley (1825–1895), who sought to establish a science free from religious doctrine,[83] and the scientists from Scotland who saw Darwin's theory of evolution as a direct challenge to their own religious beliefs in God's direct creation of man. The latter group set about using their quantified science as a means to undermine Darwin's heretical theory and was initially effective in doing so, as it caused Darwin to pause. Wrote Darwin, "Thomson's views of the recent age of the world have been for some time one of my sorest

[80] (Smith, 1998) p. 111.
[81] Thomson also made the first attempt to calculate on the basis of physical theory the age of the Sun and the end of its life. See (Kragh, 2016).
[82] (Thomson, 1864)
[83] (Smith, 1998) p. 7.

troubles."[84] But, as we now know, Darwin was right and Thomson very wrong. The fault in Thomson's calculation was that he failed to account for two facts: 1) the heat generated by radioactive decay in the Earth's interior, something no one knew of at that time, and 2) our planet is not completely solid, which significantly changes the calculations. His estimates vastly under predicted the age of the Earth, which is now known to be over 4 billion years.[85]

Thus, the groundwork for Thomson's analysis of the cooling of Earth via heat dissipation was laid prior to 1851. It was common knowledge then that heat flowed from hot to cold. Even though Thomson took a tremendous leap from heat engine to cosmological scales, it was really not a leap in conceptual thinking.

The reason why things changed so dramatically in 1851 was that Thomson finally saw *why* heat flowed from hot to cold. Dissipation didn't just reside in Fourier's heat transfer problems; it lay at the heart of nature. The energy that God created dissipated as a naturally occurring process over time in a one-way direction, from high to low concentration, never to be reversed by man, only to be reversed by the intervention of God. It all made sense to him. The pieces finally fit together. His questions were answered, his doubts cured, his future work set. His classic 1852 publication *On a Universal Tendency in Nature to the Dissipation of Mechanical Energy* captured these ideas and provided a more substantive argument for his Proposition II, which itself was a prelude to the 2nd Law.

When energy dissipates, temperature decreases, taking the ability to do work along with it. The absolute directionality would take the world from beginning to end, as the Creator had designed. When Thomson considered the extension of time to infinity, he saw death. The Sun would turn off; the Earth would freeze. He captured this finality in his private notes, referencing Psalm 102:25–7, "They shall perish, but thou shalt endure."[86] Talk about a downer. Helmholtz would further these ideas by arguing for the universe's ultimate demise in his coined phrase "heat death." All matter would equilibrate to the same cold temperature, everywhere. All work would cease. All would die. If this doesn't depress you enough, let me use his exact words. "Within a finite period of time past the Earth must have been, and within a finite period of time to come the Earth must again be, unfit for the habitation of man as at present constituted, unless operations have been, or are to be performed, which are impossible under the laws to which the known operations going on at present are subject."[87]

[84] (Burchfield, 1974) p. 316.
[85] (England, 2007) In a classic application of Fourier's heat transfer mathematics—this would make for a great homework problem for a graduate level heat transfer course—Lord Kelvin assumed the Earth to have solidified with a uniform temperature of 7000 °F at time = 0 except for the surface temperature of 32 °F, which established a temperature gradient at the Earth's surface. The resulting conductive heat flux caused by the gradient was assumed to radiate out to space. The temperature gradient at the Earth's surface was then quantified as a function of time and the age of the Earth then quantified as the time required for the gradient to decrease to the known gradient of 1 °F rise per 50 feet depth. By assuming a solid Earth, Kelvin arrived at an age of Earth much lower than an estimate made by John Perry, an Irish engineer and mathematician, which was based on a liquid core. The higher heat transfer (convection) resulting from a liquid interior meant that the Earth's interior near the surface would cool much slower, resulting in a higher age estimate. In the end, Perry's assumption was shown to be true and accounts for the much more realistic estimates, more so than can be accounted for by the radioactive heat generation argument. See also (Perry, 1895), (Carslaw and Jaeger, 1986) pp. 85–87, and (Livio, 2013) Chapter 4, "How Old is the Earth?"
[86] (Smith, 1998) p. 111.
[87] (Thomson, 1852)

As Thomson reflectively summarized in 1874,[88] perfect reversibility fails in the natural world due to the presence of such phenomena as friction, conduction, generation of heat by electrical currents, the absorption of radiant heat and light, and so on. All such phenomena, common throughout the natural world, combined with the "all-pervading law of the conservation of energy", would lead to the conclusive downgrading of energy throughout the universe with time. Thomson awakened to the non-ideal world around him in which the reversible is the ideal and the irreversible the real.

Concluding Thoughts on Thomson

Clausius and Thomson, among others like Joule and Rankine, were slowly circling around the 2nd Law by generating such statements as, "heat only flows from hot to cold" and "when heat flows from hot to cold, it loses its ability to do work." The problem here was that such statements didn't offer any means of quantification but only a means to arrive at a yes/no answer to the question, *is this process possible?* We're going to eventually continue this story line with Clausius because it was this lack of quantification that motivated him toward the discovery of entropy. But we're not finished with Thomson yet. Following his transformative year of 1851, his industrious career would take off in a range of directions: experimental, theoretical, and commercial. He would receive accolades, culminating in the ultimate accolade of being given the title of Lord Kelvin in 1892 based on the sum of his contributions to Britain.

Regarding our sole focus here, some additional discussion of Thomson's contributions to classical thermodynamics is clearly warranted. After 1851, Thomson aligned himself to the new physics of energy and started to turn his attention to creating the structure of classical thermodynamics based on Carnot while also working with Joule to experimentally test "Mayer's Hypothesis," which led to the famed collaborative study of Joule–Thomson expansion.

Later, I'll spend some more time discussing the irreversible behavior of nature, but right now I want to return to the reversible "ideal" world from which entropy and thermodynamic equilibrium would emerge, for this is Thomson's mastery delivered.

* * *

The year 1851 was pivotal for William Thomson. The struggles he endured in arriving at his mechanical theory of heat were rewarded with his landmark publications and the golden opportunity he seized to be the first to structure a newly unified physics based on his dual principles of energy conservation and energy dissipation. He couldn't have been better positioned for the task. He was about as well connected as any in the powerful Glasgow–Edinburgh Scottish school science community. He was conversant in the most fundamental of technical issues involved. And he had a deep and intelligent passion to see this through. Whereas Clausius would turn inward at this point to probe the inner workings of matter, Thomson would turn outward to lead the group from Scotland, which included James Joule as an honorary member, in the creation of the terms, structures, and textbooks that would define classical thermodynamics. As we've seen, developing the appropriate language around the technical

---

[88] (Thomson, 1874)

concepts was absolutely critical toward achieving acceptance in the scientific community. Great ideas without great language don't get far.

Thomson indeed had many great ideas. But before he could use them to re-write physics, he would need one additional idea around which to crystallize them. This idea, or really, this set of ideas, would come from someone else.

Hermann von Helmholtz

He exceeds...all the scientific greats I know personally, in sharpness, clarity, and quickness of mind, so that at times I felt dull-witted beside him.

– Hermann von Helmholtz on William Thomson[89]

[Thomson] had an exceptional ability to sort and clarify, to resolve confusion and contradiction, and many of the standard elements of classical thermodynamics trace back to his definitions and arguments. On the other hand, at crucial points he needed a prompt from someone else

– David Lindley[90]

Of the several notable events on William Thomson's scientific timeline, the one that most helped guide his structuring of an energy-based physics was arguably his reading of Hermann von Helmholtz's *Über die Erhaltung der Kraft* (1847) in January, 1852. As the prompting by Clausius led to Thomson's *Dynamical Theory of Heat*, so this prompting by Helmholtz led to Thomson's *Dissipation of Mechanical Energy*. While he may not have excelled in the creation of the ideas or even in the acceptance of the ideas as manifested by his slow acceptance of Joule's early work (on the kinetic theory of gases), he certainly excelled in their development. Once he saw the light, he was able to bring "a profound sense of logical and mathematical rigor"[91] to their development.[92]

Before delving deeper into Helmholtz's consequential 1847 paper, let's first take a step back.

As discussed in Part III, entering the nineteenth century, science was on a sound footing with Newtonian mechanics and Leibniz's conservation philosophy based on the impossibility of perpetual motion. By 1750, the conservation of mechanical energy had quietly arrived, in which the sum of the kinetic and potential energies in the moving parts of an ideal and isolated mechanical system remained constant. The change in potential energy in this case was primarily if not solely gravitational in nature as reflected by the experimentally determined mathematical expression mgh used in conjunction with its counterpart $\frac{1}{2}mv^2$ or *vis viva*. When elevation was not an issue, such as for the case of colliding balls on a billiard table, then this law reduced to the conservation of *vis viva* alone.

The fact that mgh is actually derivable from Newton's 2nd Law of Motion wasn't obvious to all. Recall from Chapter 5 that, starting with Newton's 2nd Law of Motion, F = ma, the following equation for a body in free fall can be derived

$$\Delta\left(1/2\ mv^2\right) = -\Delta\left(mgh\right)$$

---

[89] Helmholtz quote cited in (Lindley, 2004) p. 113.

[90] (Lindley, 2004) p. 111.

[91] (Lindley, 2004) Lindley on Thomson's 1852 writing, p. 106.

[92] (Cardwell, 1989) p. 137.

Helmholtz's Rational Approach to a Science based on Cause–Effect

Physiology offered the battleground for the fight over explaining animal heat in terms of the principle of the impossibility of perpetual motion and the consequent refusal to admit vital forces. The acceptance or rejection of vital forces in explaining the origins of animal heat constituted one of the fundamental problems of mid-century German physiology.

– Fabio Bevilacqua[93]

With the above as context, let's now look at Helmholtz. Being quite well-read, Helmholtz (1821–1894) was familiar with the above theories and was motivated to take them to a higher level, recognizing the existence of forces other than gravity and also the need to account for heat. Being raised in the German educational system, Helmholtz was also familiar with and aggressively rejecting of the non-scientific theories proposing that living matter depended on a special vital force, *Lebenskraft*,[94] as he felt that such theories were only created by scientists to explain things that they couldn't otherwise explain, such as the source of the mysterious animal heat. In these situations, rather than using the scientific method to probe nature to understand such phenomena, many would simply invent fictitious forces to fill the gap. Helmholtz was motivated to eliminate such theories and to install a more rational approach to science, one based on a cause–effect mechanical foundation in which such phenomena as animal heat[95] had to result from the depletion of something else, such as the oxidative conversion of organic matter, thus ensuring the impossibility of perpetual motion.

Following in Bernoulli's and Euler's footsteps, Helmholtz united the philosophies of Newton and Leibniz and directed his focus not on the planets where gravity reigned but on the atoms where other forces came into play. He did this by generalizing the conservation of mechanical energy to include a force term, Ø, in place of the gravitation force

$$\Delta\left(1/2\,mv^2\right)=\int \emptyset\,dr$$

and made the intensity of this force dependent on the distance separating the particles. Helmholtz named Ø *Spannkraft*, which translates to tension and which we now call force, based on Newton's definition. Ø thus represented the <u>net</u> force acting on mass m due to the interactions of m with all the surrounding bodies in the environment. The change in the kinetic energy of m is equal to the integral of this net force through the distance dr, which is equal to the change in potential energy of m. As we now know, this captured the classic trade-off between kinetic and potential energies in the conservation of mechanical energy. Helmholtz readily extended this concept to a system of many bodies, thus capturing the conservation of

---

[93] (Bevilacqua, 1993) p. 298.

[94] (Smith, 1998) p. 74.

[95] (Bevilacqua, 1993) p. 303. "In 1847, while still writing the *Erhaltung*…Helmholtz tried to link the problem of animal heat to that of the mechanical force produced by muscle action. Seeking to demonstrate that heat is produced in the muscle itself [much like Joule sought to demonstrate that heat is produced in the electric resistor itself], he devised a very sensitive thermocouple which…could detect differences of temperature in the range of 1000th of a degree centigrade. His thorough experiments on twitching frogs' legs showed that heat is generated directly in the muscle tissue, that its origins are due to chemical processes, and that heat production in the nerves is negligible. He had disposed of vital force on empirical grounds."

vis viva along with the more general conservation of mechanical energy, and hence took one of the first steps toward the mathematization of energy and the underlying interplay between force and motion, between PE and KE, between h and v^2, and between work and heat, all taking place under an umbrella of a conserved energy. The balance of this interplay is what led Helmholtz to propose the *Erhaltung* (conservation) between the sum of these terms on both sides of the equation.

With this equation as his starting point, Helmholtz then turned to using it to interpret a range of natural phenomena, such as electricity, magnetism, and heat, and in so doing "boldly"[96] moved physics from the conservation of mechanical energy, which originated in 1750 from the world of gravity, to the more powerful and encompassing conservation of energy, which brought forth the concept of potential energy as a means to quantify all forces. This indeed was a large step.

Regarding his analysis of heat, first recall that around 1700 Leibniz introduced an early categorization of two different forms of *vis*, later to be re-named as "energy" by Thomson, namely *vis viva* (living force) and *vis mortua* (dead force), along with the rudiments of a conservation theory involving these two quantities. Moreover, he also extended his conservation theory of *vis viva* for elastic collisions to the analysis of inelastic collisions and concluded that any loss of *vis viva* from two colliding large bodies is gained by "the parts [that] receive it." Helmholtz built on these concepts by using his equation to envision atoms as possessing *both* motion and the forces that cause motion, and then updated Leibniz's ideas, suggesting two categories of heat, *free heat* (*vis viva* of atoms) and *latent heat* (forces between atoms). Yes, others had discussed such concepts before, but Helmholtz brought advanced mathematics to the discussion and so helped crack open the door to the new world of energy.

Helmholtz was well positioned in 1847 to open the door even further to the world of heat-energy, having by then read Joule and Clapeyron, but stopped short of doing so, most likely because he wasn't clear in his own mind about the fundamental nature of heat. This lack of clarity manifested itself in his incomplete, largely qualitative, and rather "cautious"[97] analysis of heat. Unfortunately, he wrote his memoir prior to Clausius' decisive reconciliation of Carnot and Joule in 1850, as he surely would have benefited from this. While he desired to move away from the *Lebenskraft* theories, it wasn't clear that he relegated the caloric theory itself to this category. He seemed to straddle the decision-fence between heat as material and heat as motion. At one point he wrote, with reference to the work of Carnot and Clapeyron, "The material theory of heat must necessarily assume the quantity of caloric to be constant; it can therefore develop mechanical forces only by its effort to expand itself."[98] Such a statement conveys that the material theory of heat was still alive as a viable option in his mind, suggesting as it did that, consistent with the caloric theory, the underlying phenomenon during the expansion of a gas to do work was the shift in caloric from sensible to latent forms, which is what led to a decrease in temperature. But then, at another point, he considered that a range of forces—electrostatic, galvanic, electrodynamic—could be the cause for an increase in molecular *vis*

[96] (Bevilacqua, 1993) p. 314.
[97] (Bevilacqua, 1993) p. 324.
[98] (Helmholtz, 1853) p. 131.

viva and thus an increase in heat. Despite such a cautious straddling-the-fence approach, however, his outline, while imperfect, was valid and strong.

This was quite a remarkable paper, a "masterpiece"[99] to some, pulling as it did from an extensive range of sources to organize a theoretical basis for what would become the theory of energy. Again, Helmholtz's motivation was to move German science away from the fiction of *Lebenskraft* toward the rational deterministic science of cause–effect. In this he succeeded, all at the age of twenty-six, but not in Germany. The German physics community didn't see the "masterpiece" in Helmholtz's work but instead saw the lack of original experimental data and a heavy reliance on the work of others. But while this indeed was a paper that pulled from many, it wasn't simply a regurgitation of what others had done. It offered a sophisticated theory to *explain* what others had done, and thus marked Helmholtz's emergence as a theoretical physicist. It's just that those in Germany didn't see it that way. Fortunately, this was a moot point since, regarding the rise of energy, Germany wasn't where the action was.

* * *

Timing is everything. When William Thomson received the English version of Helmholtz's essay in 1852, which was, interestingly enough as we'll see, translated by none other than John Tyndall, he was as prepared as anyone to fully absorb its meaning and significance. Upon reading it, his thoughts crystallized. Being a recent convert to the mechanical theory of heat and Clausius' Carnot–Joule reconciliation, Thomson saw the essence of Helmholtz's theories and simply ignored Helmholtz's hesitant approach to the nature of heat. Using *Erhaltung* as a template, he set about structuring the science of energy. Thomson's enthusiastic recognition of *Erhaltung*, which had originally been received in silence, led to a magnificent revival for Helmholtz. As Helmholtz himself noted in 1853, "*Erhaltung der Kraft* was better known here [in England] than in Germany."[100]

Adapting Helmholtz's ideas to his own needs,[101] Thomson published *On a Universal Tendency in Nature to the Dissipation of Mechanical Energy* in April, 1852, in which he identified energy as being *the* entity that was conserved in nature, and then, in a much fuller release of his full mathematical power, as co-author with Tait, published *Treatise of Natural Philosophy* (1867), in which he re-worked Helmholtz's mathematics from the world of material points to the world of continuum mechanics,[102] benefiting strongly from his earlier mastery of Fourier's use of the same. In this treatise, Thomson embraced the view of Newton's force as being a gradient in "potential," a term created in 1840 when Gauss shortened Green's use of the term "potential function." According to this view, 1) force equals the difference in potential between two equipotential surfaces divided by the distance between them (d PE / dr), and 2) for each force in nature, a potential function must exist. Thomson's use of such "dissipation" mathematics would later find great utilization in the study of the energy associated with electromagnetic fields and would also arguably mark the introduction of direction to thermodynamics.

---

[99] (Bevilacqua, 1993) p. 333. Fabio Bevilacqua commenting on Helmholtz's *Über die Erhaltung der Kraft*.
[100] (Smith, 1998) p. 127.
[101] (Smith, 1998) p. 138.
[102] (Smith, 1998) p. 206-208.

33 The creation of thermodynamics

> *In the years after 1850 a developed, established science with extensive theoretical structures and satisfactory experimental verification had to be re-established on a new basis; that of the axiom of the conservation of energy. This was the first time in history that such a thing had happened. It pre-dates the famous re-establishments that followed the quantum and relativity theories by something like fifty years. Thus the assertion made by popular writers of a generation ago that the calm certainty of science was not disturbed between the times of Newton and the first papers of Planck and Einstein is seen to be, in historical terms, a dangerously misleading half-truth.*
>
> – D. S. L. Cardwell[1]

> *From the early 1850s the Glasgow professor of natural philosophy, William Thomson... and his ally in engineering science, Macquorn Rankine... began replacing an older language of mechanics with terms such as 'actual' and 'potential energy'. In the same period, Rankine constructed a new 'science of thermodynamics' by which engineers could evaluate the imperfections of heat engines of all conceivable varieties. Within a very few years, Thomson and Rankine had been joined by like-minded scientific reformers, most notably the Scottish natural philosophers James Clerk Maxwell... and Peter Guthrie Tait. As individuals, as partners... and as an informal group with strong links to the [British science]... these 'North British' physicists and engineers were primarily responsible for the construction of the 'science of energy.'*
>
> – Crosbie Smith[2]

While others such as Mayer, Joule, and Clausius contributed to the rise of this new science of energy, it was largely Thomson and Rankine and later Tait and Maxwell who built the foundational structure. These men of Scotland recognized the need to develop a new structure and a new language for energy, using words and phrases that were clear, concise, and unambiguous, while eliminating those that weren't, and so combined their common organizational talents with their respective skills in theory, experiment, and practice in a way that made the end product very powerful and lasting.

The first step in this process was to declare a name for the overarching theme. They chose Helmholtz's *Kraft* but then had to determine how best to translate it into the English language. This proved difficult as the word itself translated into English as "force" even though Helmholtz's use of it applied to the motion produced by force, which Leibniz called *vis viva*. *Kraft* unfortunately had different meanings associated with it, none of which matched Newton's definition

[1] (Cardwell, 1971) p. 291.
[2] (Smith, 1998) p. 1.

Block by Block: The Historical and Theoretical Foundations of Thermodynamics,
Robert T. Hanlon, Oxford University Press (2020). © Robert T. Hanlon.
DOI: 10.1093/oso/9780198851547.001.0001

of force. A new word with a single clear meaning was needed to replace it, one that could encompass the concepts of both *vis viva* and work, among others.

In the end, Thomson selected the word *energy*.[3] From a practical standpoint, he knew that energy had long been used in everyday life and so recognized that his strategic selection of this familiar word would appeal to a wide audience. While Thomas Young arguably first used this term in public,[4] it was William Thomson who dusted it off and sent it up the flagpole, while simultaneously and quietly lowering Newton's force. Energy became a property of state, thankfully replacing heat in so doing. It became *the thing* that was conserved and thus the fundamental reason why perpetual motion was impossible. Under its umbrella, all quantifiable forms of energy, including heat, work (gravity), and also, per Helmholtz, electrical, magnetic, and radiant forms, could transform between themselves such that when summed together, they equaled the same number. Newton's mechanics and Helmholtz's *Kraft* merged to become Thomson's *energy* and the resulting energy-based physics soon became one of the most important fields in science.

It is quite important to repeat again that once Thomson established the concept of energy, he ceased speaking of the heat *in* a body and instead spoke of heat as being absorbed or emitted with respect to the surrounding.[5] This transition was significant. The replacement of heat with energy as a property of matter helped to finally break the hold of heat on the progress of science and thermodynamics. Energy was conserved; heat was not. Even Clausius, who helped lead the charge toward energy with his creation of U, held onto the concept of the heat contained in a body until years later in 1864.[6] Until then, and perhaps even after, he viewed heat as the *vis viva* contained within the body as manifested in his Sixth Memoir (1856) as reflected by his use of the phrases, the "heat present in the body," the "heat contained in bodies" and even the "passage of heat," however this was supposed to happen.

In this collective effort, Thomson, Clausius, and later Tait and Gibbs are rightfully acknowledged for remaining firm in their philosophical conviction to keep the concept unattached to any particular form of molecular motion. They recognized that the motion of the "parts" of matter comprised the *vis viva* component of energy but steered clear of defining or even speculating what that motion was. The amazing thing about classical thermodynamics was that it didn't matter.[7] This small group understood this and didn't try to force it, realizing that any attempt to speculate would jeopardize their entire work, as unfortunately was the case for Rankine. Thomson had reprimanded Rankine for linking his theories to his vortex atom. He heeded his own advice,[8] as did Clausius, who wrote, "I have taken especial care to

---

[3] Thomson initially started with the term "mechanical energy" which was later shortened.

[4] Johann Bernoulli used the term in a letter in 1717 (Chapter 11). Per (Harper, 2007): "ENERGY: From Greek energeia 'activity, action, operation,' ...Used by Aristotle with a sense of 'actuality, reality, existence' (opposed to 'potential') but this was misunderstood in Late Latin and afterward as 'force of expression,' as the power which calls up realistic mental pictures. Broader meaning of 'power' in English is first recorded 1660s. Scientific use is from 1807."

[5] (Smith and Wise, 2009) p. 338.

[6] (Clausius, 1867) Clausius, Sixth Memoir – Appendix (1864), p. 251.

[7] Today we are taught the so-called "black box" approach to classical thermodynamics, as perhaps first captured by Tait: "We are quite ignorant of the condition of energy in bodies general. We know how much goes in, and how much comes out, and we know whether at entrance or exit it is in the form of heat or work. But that is all," cited in (Klein, 1969) p. 141. Both Tait and Maxwell recommended staying away from what happens *inside bodies*.

[8] Interestingly enough, Thomson would return to the vortex theory in later years.

base the development of the equations which enter into the mechanical theory of heat upon certain general axioms, and not upon particular views regarding the molecular constitution of bodies, and accordingly I should be inclined to regard my treatment of the subject as the more appropriate one [in reference to Rankine's speculation about molecular vortices.]"[9]

Beneath this concept of energy, in addition to giving us the term *thermo-dynamic*, which he had introduced in 1848, Thomson also created two categories of energy in his *Dynamical Theory* (1851) and *Dissipation* (1852) papers by building on the two-category template used by Leibniz and Helmholtz and hinted at by Clausius through his introduction of the variable U as a means to account for both free and latent heat. Building on the concepts involved in the mechanical world, the first he called *dynamical* to encompass motion and the second *statical* to encompass the cause of the motion. A raised body is static but still possesses energy that can be converted to motion when dropped. This conceptual step in and of itself was significant, as recognized by Thomson himself, who "proudly believed this to be 'the first division of Energy into two kinds.' "[10] Admittedly though, it did parallel what others had done before.

Thomson also created the concept of a standard reference state, writing in *Dynamical Theory*, "The total mechanical energy of a body might be defined as the mechanical value of all the effect it would produce...if it were cooled to the utmost..." [11] Embedded in the background of this statement was Thomson's conceptual use of the Carnot heat engine as a means to quantify the absolute amount of work, or "mechanical effect," a body could produce given its temperature when the surrounding environment was at the "utmost" coldest temperature. Recognizing that such a calculation could not be done due to our "ignorance regarding perfect cold," Thomson showed that this could be handled by invoking the concept of "reference state" and focusing on relative energy as opposed to absolute energy, an especially important distinction since thermo-dynamics was more concerned with changes than with absolutes. The reference state would be based on whatever properties of matter were required to define the state, such as temperature and pressure.

Without raising too much attention in his *Dynamical Theory* paper, perhaps because he didn't want to highlight his initial mistaken approach to Carnot, Thomson also corrected his absolute temperature scale by adjusting his original arguments to his new understanding of Carnot's work. As discussed later in this chapter, Thomson's revised temperature scale corresponded to the temperature defined by the ideal gas law and measured by an ideal gas thermometer, which fortunately had been the apparatus already being used by the experimentalists. So while nothing new from a practical perspective really resulted from this, much from a theoretical perspective did, the most significant being that whereas Thomson's initial scale descended to minus infinity, his revised scale descended to zero, which became the

[9] (Clausius, 1867) Clausius, Seventh Memoir (1863), p. 273–274. Also, (Klein, 1969) p. 135. Clausius refrained from molecular guesswork even though he claims to have thought about the kinetic theory of gases prior to 1850.

[10] (Smith and Wise, 2009) p. 347.

[11] (Thomson, 1851) Part V, p. 222–223, Paragraph #82. "The total mechanical energy of a body might be defined as the mechanical value of all the effect it would produce...if it were cooled to the utmost...but in our present state of ignorance regarding perfect cold, and the nature of molecular forces, we cannot determine this "total mechanical energy" for any portion of matter. Hence it is convenient to choose a certain state as standard for the body under consideration, and to use the unqualified term, mechanical energy, with reference to this standard state, so that the "mechanical energy of a body in a given state" will denote the mechanical value of the effects the body would produce in passing from the state in which it is given, to the standard state."

famed *absolute zero*. It was Thomson who helped lead us to an absolute temperature scale on which zero was placed, never to be reached. As Tom Shachtman wrote, "The man who would do the most to complete the conquest of cold by reaching an understanding of heat was William Thomson."[12]

Further building this linguistic edifice, Rankine took Helmholtz's word, *Erhaltung*, and used it in his *On the General Law of the Transformation of Energy* (1853)[13] to establish the phrase *conservation of energy,* and, in a move supported by both Joule and Thomson, to also establish the terms *actual energy* and *potential energy*; Thomson relinquished *dynamical* and *statical*. As an interesting aside, while Thomson was intimately familiar with the term "potential," it was Rankine who first selected it for use as a category for energy. Thomson's ready acceptance showed that it was a fitting choice and aligned with his mathematical inclinations relating force to the gradient of potential energy. Later, in 1862, Thomson and Tait changed *actual* to *kinetic,* even though Rankine felt this was too restrictive of a term.[14]

Rankine then took the lead in packaging these concepts together by publishing *Principles of Thermodynamics*, the first recognized treatise on this subject, as a chapter in his *Manual of the Steam Engine and Other Prime Movers* (1859). In this chapter, Rankine further coined *The First and Second Laws of Thermodynamics* and stated that these laws captured the relationship between heat and motive power without being dependent on any assumed model of nature,[15] a move intended to counter the aforementioned arguments of some against his use of the vortex model of the atom in his thought process.

These packaging efforts by Rankine were huge. He spent years preparing these publications, ensuring their applicability to the practical design of heat-to-work engines as opposed to their becoming some set of abstract concepts of no use to anyone other than a theoretician. He as well as Thomson and Tait wanted their efforts to be read by, used by, and solidified in both the public and scientific communities.

In reading these publications, you can see why Crosbie Smith suggested they bordered on legal documents.[16] Rankine's wording and structure were very disciplined, designed to leave no room for misunderstanding, working to ensure everyone was aligned in approach to a whole new science. The legal feel, as manifested so clearly, for example, in his use of "Law," reflected at least to some degree the influence of the social culture he lived in, seeing as his family and friends were lawyers, businessmen, and inventors. It was as if he and Thomson sought to lay claim to the intellectual property of this new territory, which perhaps they did as a means of building their own credibility in the scientific community.

The team from Scotland, which started with Thomson and Rankine and then continued with, among others, Tait and later Maxwell, and which also included as honorary members Joule and later Helmholtz, who eventually met and became friends with Thomson in 1855, moved the process of structuring the science of energy from a dispersed effort to a coordinated approach, from the individual to the team. Their collective effort was greatly facilitated by close

---

[12] (Shachtman, 1999) p. 95.
[13] (Rankine et al., 1881) p. 203–208. Rankine, *On the General law of the Transformation of Energy*, read before the Philosophical Society of Glasgow, January 5, 1853, and published in the Proceedings of that Society, Vol. III, No. V.
[14] (Smith, 1998) p. 140.
[15] (Smith, 1998) p. 165.
[16] (Smith, 1998) p. 139.

proximity in that they could continually try out different ideas, words, concepts, and models with each other, often face to face. Their effort was further facilitated by their ties to academia which allowed immediate and direct use of the classroom as a proving ground. Such a unique situation generated an ongoing process of validation, making the final resulting structure extremely tight, strong, and enduring. Their desire for clarity and simplicity led to a major housecleaning. The old was swept out and the new brought in. Unfortunately, some relics of caloric, such as sensible heat, latent heat, and heat flow, survived, but this is a minor point.

All in all, this was a remarkably successful effort. Had it not occurred, thermodynamics could have become a tangled mess.

Temperature

The rise of energy lifted the stature of another property of matter, temperature. Prior to the rise of thermodynamics around 1850, temperature did play an important role in science but it was not center stage. Instead, it was situated somewhat behind heat. Recall that in the caloric theory, temperature was vaguely connected to how heat (or caloric) distributed itself between latent and sensible forms. According to that theory, during adiabatic expansion, for example, total heat remained constant while the latent fraction increased to fill the available volume. The resulting shift from sensible to latent caused the decrease in temperature, almost as an after-thought. In this theory, heat and temperature were confusingly intertwined.

While Joseph Black's work helped to separate heat from temperature, but more from an experimental than theoretical perspective as his results were generally interpreted from the standpoint of the caloric theory, and later while Carnot did indeed emphasize the importance of temperature in the operating efficiency of heat engines, it wasn't really until the advent of energy in the 1850s that temperature came out from behind heat to stand on its own, alongside energy, as a separate and distinct property of state. Heat rightly lost its status as a state property and instead became correctly identified as a measure of the change in energy due to thermal exchange between a body and its environment. The "temperature of heat" became a concept of the past.

The side-by-side existence of energy and temperature as two independent properties of matter became challenging to understand but nonetheless powerfully enabling. The challenge centered around the fact that the total energy of a body was dependent on temperature, but not solely so. The two were not the same. Energy was an extensive property, meaning that it was dependent on mass. More mass meant more energy. Temperature, on the other hand, was an intensive property. More mass didn't change temperature. Moreover, energy was comprised of both kinetic and potential forms, while temperature specifically quantified the kinetic energy of atoms in motion. But the seeming narrowness of temperature's meaning belied its importance, because it was temperature alone that determined the direction of heat flow between bodies and thus determined whether or not work could be generated in a Carnot heat engine. As Thomson pointed out, the ocean may have a huge amount of energy on account of its mass, but it can't be a source of work generation since its temperature is not greater than "the temperature of the coldest of the surrounding objects." As helped along by Thomson, the rise of temperature contributed to the strong foundation on which the Laws of Thermodynamics were placed.

The History of Temperature

This brings us to the need for a short primer on the history of temperature and why what Thomson did was so important. While I can't do full justice to this history here, I can provide the key highlights. For more complete details, I encourage you to read Hasok Chang's *Inventing Temperature*.[17]

People had long sought a means to quantify the physical sense of heat, a task more difficult than attempting the same for the other so-called primitive properties, especially length and volume. Length could be measured against a readily available reference, such as the length of your hand or arm. Volume would then be a multiple of these lengths or some other common standard that arose, such as a barrel. An added plus with these measures was that one could visually estimate each. Weight was also direct, although less so visually as it's hard to quantify density by sight. It too required a readily available reference, such as a set of stones or metal objects of varying size and a scale. But one could also physically "feel" the weight of a body and this helped, especially when also estimating density. Time? Well, books have been written about this variable and its quantification via falling sand, swinging pendulums, or spinning Earth, but I don't have the space here to handle time, nor is it particularly relevant in the world of classical thermodynamics. But the concept of temperature was somewhat more indirect and complicated. Yes, one could sense temperature through touch. Something is hot, cold, or room temperature. But how would one bring quantification to this?

The development of a suitable method to measure temperature started when scientists realized that matter changes volume when exposed to heat or cold and then realized that they could use this behavior to their advantage; if you know volume, you know temperature. So they set about designing equipment to very accurately and reproducibly measure the volume of matter, typically air or mercury, and then referenced this equipment using two well-known phenomena that occurred exactly the same everywhere in the world, namely the freezing and boiling of water (assuming pure water at same elevation). On the Celsius scale, these two reference points were assigned to zero and one hundred degrees, respectively, and the space between was evenly divided into one hundred (Centigrade) "degree" markings.

This was all very good and afforded scientists across different continents the opportunity to conduct research and compare results, all the while knowing that the temperature of a body measured by a calibrated thermometer in, say, France would be identical to that measured in England under the same conditions. Temperature tied together the world's research programs.

But there was a problem with this approach, and William Thomson captured it well:

> *Although we have thus a strict principle for constructing a definite system for the estimation of temperature…we cannot consider that we have arrived at an absolute scale…[W]e can only regard, in strictness, the scale actually adopted as an arbitrary series of numbered points of reference sufficiently close for the requirements of practical thermometry.*[18]

To Thomson, temperature was sitting off by itself in an arbitrarily created world, with no connection to the real world. His above concept of "absolute scale" sought to make this

---

[17] (Chang, 2007)
[18] (Thomson, 1848)

connection. Important to note here is that Thomson's use of "absolute" had nothing to do with our current concept of "absolute temperature" which is referenced against an absolute zero. In this thermodynamic context, these two references to "absolute" meant separate things. The latter was based on Guillaume Amonton's idea of a zero point in temperature that exists at the end of extrapolating the pressure–temperature data of air to zero pressure, at which point zero heat would remain and so absolute zero temperature would be reached. Thomson's concept of absolute rose out of his desire to make a "degree" of temperature actually mean something. The two concepts of absolute temperature would eventually unite, but not yet.

Thomson and the First Glimpse of δQ/T

Carnot provided the original thermodynamic basis for a theoretical relation between the three defining parameters of a heat engine, namely heat, work, and some unknown function of temperature, which became known as Carnot's function. Clapeyron brought mathematics to Carnot's ideas based on an assumed infinitesimal heat engine and arrived at the following relationship (Chapter 31).

$$\text{Heat engine efficiency}(\text{infinitesimal}) = \text{Work Out / Heat In} = dT / C(T) \qquad [31.5]$$

This equation proposed that the maximum amount of work that could be generated from a given amount of heat is inversely proportional to Carnot's function, C (T).

In his 1848 paper, *On an Absolute Thermometric Scale Founded on Carnot's Theory of the Motive Power of Heat*, Thomson latched onto the concept inherent to this equation to define an absolute temperature based on his proposal that the fall of heat through one degree of temperature would yield the same amount of work regardless of the starting point, which would indeed be the case if C(T) in the above equation were simply a constant, independent of temperature. But as we learned before, such was not the case and Thomson's proposal resulted in a minimum absolute temperature of minus infinity.[19] It was Joule who, after reading Thomson's article on absolute temperature, urged him to reformulate his idea, but this time on the basis of heat–work equivalence. It was also Joule who, based on intuition, suggested to Thomson that Carnot's function, C, was equivalent to temperature itself. While Joule spoke in one of Thomson's ears, Clausius spoke in the other with his 1850 publication that contained a revised Carnot theory based on heat–work equivalence along with a more definitive quantitative proof relative to Joule's that C was indeed equal to temperature, even though Thomson would not give him due credit for this on account of the fact that Clausius relied on "Mayer's Hypothesis" to arrive at this result.[20] These two voices combined to finally change Thomson's mind and so opened the door to his conversion to the dynamical (or mechanical) theory of heat in 1851 and then, in 1854, to a reformulation of his theory of absolute temperature,[21] one for which he established a zero point at minus 273 °C, which later became known as absolute zero on the Kelvin scale.

---

[19] (Lindley, 2004) p. 109.
[20] (Thomson, 1856)
[21] (Thomson, 1851) Paragraph 97 footnote in *Part VI. Thermoelectric Currents*, read May 1, 1854.

If we insert C (T) = T in the above equation, based on Joule's "conjecture," as Thomson put it, and on Clausius' more quantitative proof,

$$\delta W / \delta Q = dT / T \,^{22} \tag{33.1}$$

Cymbals crash! Horns blare! We are finally here. Finally! It was out of the re-arrangement of this equation that a whole new variable, $\delta Q/T$, appeared in the world of science,

$$\delta W/dT = \delta Q/T \tag{33.2}$$

In his 1854 article, Thomson discovered this result and then converted it from the infinitesimal to the finite by creating a conceptual system of many infinitesimal engines, all inter-connected such that the heat rejected by one became the heat driving another. In this mathematical construct, the only heat entering the system was from the hot reservoir, Q_{in} at T_H, and the only heat leaving the system was to the cold reservoir, Q_{out} at T_C. All other in-and-out heat flows remained inside the box while the sum of all the work ended up outside the box as work done in the external environment. From this analysis, Thomson arrived at

$$Q_{in} / Q_{out} = T_H / T_C \tag{33.3}^{23}$$

It was Carnot who first hypothesized that a unique relation must exist between heat, work, and temperature, and it was Thomson who took this to a higher level by proposing that, in concept, if one had access to a reversible Carnot heat engine, one could use the above equation to quantify temperature by considering the ratio of the heat entering to the heat leaving. The amount of work generated wouldn't even be needed in this calculation since it is uniquely determined by the difference between these two heats ($Q_{in} - Q_{out}$). It was Thomson who, perhaps inspired by his deep experience in experimental research regarding temperature while in Regnault's Paris laboratory, brought forth a new concept of temperature based on this rather ingenious use of Carnot's theory. He was all of 24 years old at the time.

Thomson's proposal provided a theoretical underpinning to temperature, which was a major accomplishment by itself. But while this looked fine in concept, it was rather impractical as no such ideal Carnot device was readily available. So, being very capable of taking theory to practice, Thomson used his results to guide his famed experiments with Joule on the cooling effect of expansion on gases. Out of this excellent piece of theoretical and experimental work,[24] he concluded that absolute temperature as used in the Carnot equation ($\delta W/\delta Q = dT/T$) could be quantified by other measurable properties such as gas volume, pressure, and heat capacity, and shown to be equal to that quantified by a thermometer based on use of a permanent or ideal

---

[22] This equation quantifies the efficiency of an ideal infinitesimal heat engine. It's the integration of this equation that provides the efficiency equation for a finite heat engine, $W/Q = (T_H - T_C) / T_H$. This may seem confusing given the term $\int(1/T)dT$ but note in Clapeyron's mathematics [(Carnot et al., 1988) p. 84] that the variable T actually refers to the temperature of the hot reservoir, thus making the infinitesimal equation, $dW/dQ = dT / T_H$. Integration over T, from T_H to T_C, leads to the finite engine equation.

[23] (Thomson, 1851) Paragraph 97 footnote in *Part VI. Thermoelectric Currents*, read May 1, 1854. "for the action of a perfect thermo-dynamic engine...the heat used is to the heat rejected in the proportion of the temperature of the source to the temperature of the refrigerator...and [is] now adopted as the definition of temperature." See also (Chang, 2007) pp. 182–6.

[24] (Chang, 2007) pp. 186–197.

gas. Up until this point, temperature, labeled with a small "t," was recorded in degrees Celsius. But by adding a constant, equal to approximately 273, to this temperature, thereby converting temperature to an "absolute" basis, denoted by large "T" and (eventually) "degrees Kelvin," the ideal gas law became valid. In the end, it was the ideal gas law (T = PV/R) combined with gas property relationships initiated by Carnot that provided the means to create a theoretical definition of absolute temperature.

The use of ideal gas temperature as the primary reference ended up uniting the two separate concepts of "absolute," namely Thomson's in which temperature was linked to other measurable parameters, such as heat flow in a Carnot engine or pressure and volume in the ideal gas equation, and Amonton's in which temperature goes to zero as either pressure or volume or both go to zero. As noted by Chang, "It was characteristic of Thomson's work to insist on an abstract mathematical formulation of a problem, but then to find ways of linking the abstract formulation back to concrete empirical situations, and his work in thermometry was no exception."[25]

Also as noted by Chang,[26] looking back, the use of Carnot's theory of heat engines wasn't the best starting point for establishing the theoretical basis for temperature. Yes, it helped Thomson take the first step in this direction, but it was rather restrictive in its being tied to the heat flows around an ideal heat engine. Not that Thomson understood such at the time, but the ultimate point on which to establish this basis would be provided by the kinetic theory of gases which was still years away. We'll get to this later.

Capturing the New Science of Thermodynamics in the Written Word

Despite Rankine's Herculean effort in preparing his aforementioned 1859 treatise, *Manual of the Steam Engine*, which included a chapter *Principles of Thermodynamics*—and I don't use the Hercules reference lightly as this was, after all, a whole new science being written with no real predecessor to build on—some still had reservations about his final product, especially regarding his assumptions about his vortex hypothesis. Some were also concerned with the manual's lack of clarity[27] and strong industrial focus. To professors, such concerns made Rankine's text unsuitable for the classroom. But they were equally disenchanted with the alternative, namely Clausius' work, which they felt was too theoretical. Even Thomson's extensive writing on the subject, which wouldn't be compiled until after 1880 in his *Mathematical and Physical Papers*, wasn't in a form suitable for teaching. The only thing left was for someone to step forward to create a new text, which brings us to Peter Guthrie Tait.

P. G. Tait

After becoming Chair of Natural Philosophy at Edinburgh in 1860, succeeding J. D. Forbes and beating out James Clerk Maxwell in the process, Professor Tait (1831–1901) searched in vain

[25] (Chang, 2007) p. 174.
[26] (Chang, 2007) p. 218.
[27] (Smith, 1998) p. 165.

for a textbook suitable to the task of developing a stronger framework for his physics course. He didn't like what was available, writing "both Clausius and Rankine are about as obscure in their [thermodynamic] writings as anyone can well be."[28] So with all the "zeal of a fresh convert"[29] to this new science, Tait finally decided to take matters into his own hands and approached the publisher Macmillan with a proposal for a text. Shortly thereafter, Thomson, who was too busy with other projects to take the lead on this but was more than willing to collaborate, joined the project for his own classroom needs in Glasgow and the two commenced their collaboration in late 1861. Their plan was quite ambitious as they sought to, in effect, modernize Newton's *Principia* by 1) eliminating the Latin and writing it in the common language of the public, and 2) centering it around energy as opposed to force. The resulting *Treatise on Natural Philosophy*, eventually published as a single volume in 1867, falling short of its original multi-volume goal, indeed established and defined the science of energy and so helped launch the field of physics.

The slow pace at which the *Treatise* progressed, due largely to Thomson's overfilled schedule, motivated parallel writing efforts by Tait and Maxwell, which culminated in their own texts, *Sketch of Thermodynamics* (1868) and *Theory of Heat* (1871), respectively.

The combined efforts of Rankine, Thomson, Tait, and Maxwell were amazing. They built a whole new science from scratch. And specifically regarding Thomson and Tait, there is no questioning their contribution to science through these endeavors. It was truly groundbreaking.

Unfortunately, there's another aspect of this otherwise wonderful story that needs to be told, for it influenced the way that students continue to learn the theoretical and historical foundation of thermodynamics. It's not a pleasant story. Simply put, while their science was truly groundbreaking, their history, especially as written by Thomson and Tait, was misleading, most likely with intent, and it was this history that infected the later textbooks.

Revisionist History

To understand how Thomson and Tait approached the historical aspects of thermodynamics in their *Treatise*, it's important to consider the range of motives governing their writing. The most direct motive naturally was their immediate need for a new textbook and the anticipated acclaim and financial reward for being the first to write it.[30] Both were entirely reasonable expectations. But there were other additional motives at work, not only for Thomson and Tait but also for those who rose against them.

One such additional motive that drove Thomson and Tait was, simply put, nationalism. They wanted to claim energy as a product of Britain, willfully ignoring the others involved in its discovery. With their flag so planted in this new frontier, they would have the scientific authority to control the territory and use it as they saw fit, almost like it was a form of intellectual property that needed to be patented and protected. The upgrade of Carnot's principles to Thomson's propositions to Rankine's laws may reflect some of this legal thinking. It was this

---

[28] (Smith, 1998) p. 166.
[29] (Smith, 1998) p. 177.
[30] (Smith, 1998) p. 195.

possessiveness where the drama started. To secure a "British pedigree for the doctrine,"[31] Tait and Thomson creatively worked Newton's physics into a whole new light, one based on energy instead of force, and one that Newton himself may not have even recognized. This in and of itself was a very beneficial technical achievement and if they had stopped here, all would have been good. But they then added their own take on the history behind the science and so bookended *Treatise* with Newton as opposed to Leibniz on one end and with Joule as opposed to Mayer on the other. In so doing, Thomson and Tait rewrote history. Maybe they didn't think they'd get called on it. But they were. Which brings us to John Tyndall.

John Tyndall

Pursuing a strong interest in science, John Tyndall (1820–1893) traveled from his birthplace in Ireland to Marburg, Germany, in 1848, where he earned his PhD. By the early 1850s he had established connections with many leaders of German science, including a good friendship with Clausius, and then, after he returned to Britain in 1851, he similarly established many connections with the leaders of British science largely as a result of becoming a member of the Fellow of the Royal Society (1852) and a Professor of Natural Philosophy at the Royal Institution of London (1853). Needless to say, Tyndall was rather well placed at the center of science and thus also well placed to challenge Thomson and Tait. He was also well motivated.

It all went down on a Friday evening in June, 1862,[32] when Tyndall delivered a public talk at the Royal Institution in London titled "On Force"[33] in which he declared Mayer a "man of genius", the discoverer of energy and Thomson as someone who had merely applied his "admirable mathematical powers to the development of theory." Wow. Talk about lighting a fuse. Tyndall didn't tip-toe into this situation. He didn't hold back. And it's not that he ignored Joule. He praised him by drawing attention to his experimental validation of the mechanical equivalent of heat. It's just that he gave priority to Mayer and started the process of righting a wrong and re-balancing history.[34] It was thanks to Tyndall, who became acquainted with Mayer's great work during his time studying in Germany, work that indeed preceded Joule's— although Joule's experimental work directly demonstrating work–heat equivalence was second to none—that Mayer's star rose to sit alongside Joule's.

So what was the real motivation driving Tyndall here? Many had thought it to be altruism, believing Tyndall to be a man who felt deep sympathy for Mayer's plight and so, out of a sense of fairness, stepped forward to give Mayer his just due. While this sounds good, and while this may have provided some degree of motive, it simply wasn't the full story.

Instead, again according to Daub,[35] the true story had to do with Tyndall's antagonistic view of Thomson, one that had its beginnings in 1850 at the very beginning of Tyndall's career when he had read a paper to the British Association for the Advancement of Science. During his talk,

---

[31] (Smith, 1998) p. 180, p. 197.

[32] While prior to the publishing of some of Thomson's and Tait's key papers, Tyndall already knew enough about Thomson and Tait's collective direction to prepare for the aggressive talk that he gave.

[33] (Smith, 1998) p. 180.

[34] (Cardwell, 1989) p. 207.

[35] (Daub, 1975)

Thomson repeatedly challenged Tyndall's ideas, so much so that Tyndall recorded the event in his journal as a verbal "Hand to hand fight."[36] Tyndall never quite got over this. The die was cast. It was only a matter of patiently waiting until the opportunity for payback arrived.

In this light, a more careful reading of his 1862 talk shows that Tyndall brought Mayer into the debate not for altruistic reasons to right the wrong done to Mayer but to question Thomson's originality for his theory of falling meteors as the origin of the Sun's heat.[37] He bore no ill will toward Joule. His target was Thomson, and specifically Thomson's claim to originality on this subject, when in fact Mayer had first proposed the falling-meteor theory. Even as this debate moved forward and Tait had entered the fray, Tyndall's focus remained Thomson. Wrote Tyndall, "The fact really is that if it could be shown that Prof. Thomson had…been aware of what Mayer had done, he would be at the present moment in as unenviable a position as could possibly be occupied by a scientific man."[38] One could easily read this as an indirect accusation that Thomson plagiarized Mayer's ideas. Daub concluded his paper by saying that the view that Tyndall's original motive in citing Mayer was to benefit Mayer sounded noble but was a myth, one largely generated by Tyndall himself. "Tyndall's supposed altruism merely masked a deeper antagonism."[39]

There was yet another motive driving Tyndall, one less personal and more professional. As discussed by Smith,[40] Tyndall was part of Huxley's rising generation of scientists who were working to establish a "scientific naturalism" that had no room for religion. Their weapons? Darwin's Theory of Evolution (1859), which Thomson was attempting to thwart with his age-of-the-Earth calculations and the new science of energy. Being well versed in the science of energy from both the German and British perspectives and realizing its most certain future prestige, Tyndall endeavored to use it, much as Huxley was using Darwin's Theory of Evolution, to promote a science that stood on its own, without the outside influence of God. Ironically enough, Tyndall bolstered his argument by seizing on the belief of Helmholtz, Thomson's friend by that time, in scientific determinism, which was more aligned with the cause–effect nature of the conservation of energy than it was with the Scottish belief in an intervening God, human free will, and a science in harmony with Christianity. But Tyndall had a major obstacle to overcome. Energy was a science controlled by those same Scots. Tyndall's solution to getting energy out of Scotland was to first break it away from Joule, who was an honorary member of Scottish academia. Tyndall's talk was strategically designed to do just this.

So let's move on now to Thomson and Tait's response to Tyndall's talk. In addition to Thomson and Tyndall's earlier interaction in 1850, there was another less direct interaction in 1860 when Tyndall published an article on the motion of glaciers.[41] From his office in Glasgow, Forbes took offense at this article since he felt that glacial motion was his personal domain of research. Since Thomson was friends with Forbes, and since Tait succeeded Forbes, Forbes'

[36] (Daub, 1975)
[37] (Kragh, 2016)
[38] (Daub, 1975)
[39] (Daub, 1975)
[40] (Smith, 1998) p. 171. In a 2017 email to the author, historian Helge Kragh shared that Tyndall was a leading exponent of "scientific naturalism" and in this regard antagonistic to Christian scientists such as Thomson and Maxwell, which undoubtedly was a factor in the Tyndall–Thomson debate.
[41] (Cardwell, 1989) p. 206.

dislike of Tyndall for the perceived slight was thus passed on to both Thomson and Tait. These interactions were motivation enough for a strong response. But there was more. Tyndall's 1862 talk directly challenged Thomson's professional credibility and indirectly, through his own strong credentials and status, threatened Tait's ambition to become the sole authority on the new science of energy. Tyndall's initial challenge of Thomson expanded into a challenge of Tait, at least in Tait's mind. You also had the fact that Tyndall's aforementioned drive to remove religion from science ran in the opposite direction from Thomson and Tait's belief in a Christian-friendly science. The final icing on the cake? Tait's character was just as combative as Tyndall's. Add it all together, and you've got not just a rebuttal, but a counter attack.

That same year, but after Tyndall's talk, Thomson and Tait published an article about the "ONE GREAT LAW of Physical Science, known as the *conservation of energy*"[42] in the Christian-friendly *Good Words*, a popular journal of the time. This article, which served as a prelude for their *Treatise on Natural Philosophy*, expanded what would become known as the Thomson–Tyndall Controversy into the public sphere by challenging the technical competency of both Tyndall, who was unnamed in the article, and Mayer, who was referred to as the "German physician."[43] According to this article, it was an all-British team who discovered energy, from Locke who had proposed heat as motion, to Davy who had proved heat as motion, to Rumford who had arrived at an approximate quantification of the mechanical equivalent of heat, and finally to Joule who stood as "The founder of the modern dynamical theory of heat, an extension immensely beyond anything previously surmised."[44]

The gloves pretty much came off at this point. The resulting arguments that were carried out in the *Philosophical Magazine* are there for historians to interpret. The battle spread, mostly due to Tait who sought to secure credit for Thomson and his energy dissipation theory[45] and who also privately expressed a desire to "render [Tyndall] an object of scorn to every man of any sense of honour."[46] Tait attacked Tyndall by attacking Mayer. In his *Sketch of Thermodynamics* (1868), Tait worked to fill in certain gaps in his teaching textbook not covered by *Treatise*, including the history of thermodynamics, in which he furthered the British cause, and especially Joule's, while lessening Germany's role, calling into question Mayer's capability in experimental physics and then pretty much neglecting the contributions of Clausius. Adding fuel to the fire, Tait further claimed that it was William Thomson "who first *correctly* adapted Carnot's magnificently original methods to the true Theory of Heat." Tait pulled Maxwell into the dispute by asking him to review *Sketch*, which is how Maxwell first learned about Clausius' 1865 concept of entropy. But Tait's mistaken comprehension of entropy found its way into Maxwell's own textbook *Theory of Heat* (1871) due to Maxwell's unfortunate trust of Tait, which ultimately brought Clausius, who had been keeping his peace until then, into the battle.[47] Barely mentioned in Maxwell's first edition, poorly mentioned in Tait's *Sketch*, and sensing that Britain was trying

---

[42] (Smith, 1998) p. 184.

[43] (Smith, 1998) p. 186.

[44] (Smith, 1998) p. 186.

[45] (Daub, 1970b) p. 330. "Quite obviously, Thomson was king for Tait, or as T. H. Huxley put it, 'Tait worships him [Thomson] with the fidelity of a large dog – which noble beast he much resembles in other ways.'"

[46] (Smith, 1998) p. 188.

[47] Maxwell corrected himself regarding the technical content in his second edition in 1872 and then in the historical content in his fourth edition in 1876. He made these corrections out of respect for Clausius' achievements.

to claim ownership for his work,[48] Clausius objected by writing into *Phil. Mag.* which led to an open debate with Tait.[49] Tait's unreasonable style in this forum led Clausius to ultimately cease engagement with a succinct closing sentence: "the tone in which Mr. Tait has written renders it impossible for me to continue the discussion." Returning to Maxwell, realizing he had been unfair, Maxwell corrected subsequent editions of his book by including Clausius. More importantly, upon reading Gibbs' work[50] of 1873, he realized and corrected his mistake in discussing entropy, whereupon he wrote to Tait, with gentlemanly understatement, "When you wrote the *Sketch* your knowledge of Clausius was somewhat defective."[51] Maxwell would subsequently become an enthusiast and advocate of Gibbs' work. It was arguably Maxwell who introduced Gibbs to Britain and Europe.

This was such a convoluted history of intertwined motivations. In the end, the famed debates on the priority rights of Joule vs. Mayer really had nothing to do with Joule and Mayer. They had to do with men of ambition fighting over ideas in the public sphere, each group "competing intensely for a monopoly on 'truth'".[52] Thomson and Tait fought to protect their credibility, their intellectual property territory, their country, and their God. Tyndall fought to (rightly) establish Mayer as being the first to articulate the concept of energy and its conservation, and in so doing attacked Thomson, both professionally and personally, while simultaneously promoting a deterministic science in which God and man's free will played no role. Tyndall had no quarrel with Joule. It's just that he really didn't like Thomson.

Yes, there were some legitimate technical issues involved in all of these arguments, especially regarding Mayer's Hypothesis. Both Mayer and later Clausius relied on this historic hypothesis in the development of their respective works. This assumption, later proven correct, was reasonable at that time, and its use certainly helped to break open new ground while also motivating Thomson and Joule to conduct their famed Joule–Thomson experiments to study the effect of irreversible expansion on temperature in a flowing gas, which also enabled Thomson to finalize his development of a theoretical basis for absolute temperature as discussed earlier. The full and open debate over assumptions such as Mayer's was all an accepted part of the scientific process and led to the Darwinian survival of only the best ideas.

And yes, there was a significant difference in style and philosophy between Britain and Germany, the former being more practical and industrial, believing that fundamental concepts should be capable of direct measurement, and the latter being more theoretical. Such different approaches to the concepts of energy and to a lesser extent entropy certainly contributed to the debate, while also contributing to the strength of the final product. Again, this is all part of the scientific process.

But much of this legendary dispute, especially coming from Thomson and Tait, seemed to originate less from a concern for science and more from a concern of personal, religious, and national dominance. Daub cited Tait as being an Anglophile and a Germanophobe, as well manifested by Tait's unreasonable public responses to Clausius.[53] This explains much. The

---

[48] (Daub, 1970b) p. 322.
[49] (Klein, 1969) p. 132.
[50] (Gibbs, 1993) See specifically the long note on p. 51–52.
[51] (Smith, 1998) p. 257.
[52] (Smith, 1998) p. 171.
[53] (Daub, 1970b) p. 322.

existence of the human condition involving pride, jealousy, anger, and even tribal instincts played itself out in this dispute. Understanding at least some of this history helps us to gain perspective on the content for many of our textbooks, for this content was largely generated by one side of this dispute, namely the Scots and especially Thomson and Tait. Knowing this helps us to better understand the absence and perhaps willful exclusion of such worthy figures as Mayer and especially Clausius despite their significant contributions. These weren't nice, neat, linear times, either technically or emotionally. Indeed, a revolution in science had occurred. And it still wasn't done.

The New Thermodynamics Still Lacking a Definitive 2nd Law

If we return to the period of time following Clausius' 1850 revision of Carnot's theory, we find the rise of a thermodynamics based on energy that was well positioned to enable great strides in science and engineering. As discussed at length in Part III, the 1st Law of Thermodynamics involving the conservation of energy was itself a tremendous step forward, and it has remained a constant, unchanging cornerstone of progress ever since, offering a means to interpret phenomena related to both many small colliding particles and massive objects. And even though the 1st Law did indeed involve a concept that was quite abstract, people could still gain a reasonable understanding of what it meant because they could see energy all around them in the activities of everyday life.

But the 1st Law wasn't sufficient to completely constrain Carnot's engine. More than an energy balance was required. Unfortunately, defining the second constraint would prove much more difficult than defining the first. From its outset, we wrestled with this constraint, struggling hard to capture it in words and symbols. Wrote Maxwell of Rankine's own attempt, "When we come to Rankine's Second Law of Thermodynamics we find that...its actual meaning is inscrutable."[54] And here Rankine was one of the thermodynamic founders. But no one else could readily clarify this situation either.

Defining the second constraint was rather simple in hindsight: heat can only flow from hot to cold. The temperature range of the working substance in Carnot's engine could not go above that of the hot reservoir nor below that of the cold. But it would take the arrival of the theory of entropy to explain this.

While Clausius' attempt to clarify the second constraint led him to entropy, the path he took was circuitous and has confused us ever since, as it got tangled up in Carnot's cause–effect hypothesis that the fall of caloric results in the generation of work. The resulting confusion slowed the embrace of entropy by the scientific community and general public and negatively influenced the subsequent attempts to teach this concept in the classroom. Fortunately, it didn't similarly slow Gibbs. His ready comprehension of the concept enabled his subsequent and highly effective use of it to unlock phase and reaction equilibrium.

Carnot's engine was a challenging mystery to nineteenth-century scientists, a puzzle to be solved. Its design was pure genius. Without it, I'm not sure how else thermodynamics as we know it today would have appeared. But its understandability was pure struggle, largely

---

[54] (Smith, 1998) p. 165.

because the caloric theory permeated Carnot's work. The arrival of the conservation of energy and Clausius' use of it to unlock Carnot proved a major step forward and, as Gibbs put it, helped "bring order out of confusion."[55] This law, which originated separately from Carnot's work, rightly replaced caloric with energy and so put a constraint on the engine's performance by stating that the amount of work done cannot be greater than the difference between the incoming and outgoing flows of heat. But why wasn't this sufficient? Why was there a need for another constraint? It certainly wasn't obvious.

Both Clausius and Thomson ventured into this inquiry, with Clausius being first in 1850 by focusing on one of Carnot's key but false hypotheses, specifically that the motive power generated by a steam engine is due to the "transportation [of caloric] from a warm body to a cold body."[56] Clausius called this constraint Carnot's 2nd Principle and modified it to his own version, substituting heat for caloric in so doing, resulting in the following statements regarding the closed-cycle, reversible process: "The equivalent of the work done by heat is found in the mere transfer of heat from a hotter to a colder body."[57] To help distinguish this statement from others made by others, I'll continue to refer to this statement and variations thereof by Clausius as Carnot's 2nd Principle. Thomson followed in 1851 with his own version: "It is impossible [in Carnot's engine]…to derive mechanical effect from any portion of matter by cooling it below the temperature of the coldest of the surrounding objects."[58]

Returning to Clausius, interestingly enough his journey along the path laid out by Carnot wasn't what led him to entropy. Instead, a separate path not really connected with Carnot's 2nd Principle led him to the fact that dS equals δQ/T for a reversible process. But Carnot's path did lead Clausius to propose without justification that entropy increases in the universe. This proposal was wrongly based on the conflation of heat flow or transfer per Carnot's 2nd Principle with the very different concept of heat flow per the phenomenon of conduction (via direct contact).

The physical and engineering science behind all of this together with the history involved makes for a fascinating story and leaves one wondering, *How did they ever successfully emerge on the other side of all this?* Let's find out.

---

[55] (Gibbs, 1889) A tribute-obituary Gibbs wrote on Clausius.
[56] (Carnot et al., 1988) p. 7.
[57] (Carnot et al., 1988) p. 132.
[58] (Thomson, 1851) Paragraph #12. His footnote to this statement reads: "If this axiom be denied for all temperature, it would have to be admitted that a self-acting machine might be set to work and produce mechanical effect by cooling the sea or earth, with no limit but the total loss of heat from the earth and sea, or, in reality, from the whole material world."

34 Clausius and the road to entropy

I especially endeavored to bring the second fundamental theorem, which is much more difficult to understand than the first, to its simplest and at the same time most general form, and to prove the necessary truth itself.

– Rudolf Clausius[1]

We now enter the final stages of man's journey toward the 2nd Law of Thermodynamics, a journey arguably started by Carnot and continued by a host of others, including Clapeyron, Holtzmann, Joule, Rankine, Thomson, and Clausius. As we'll see, it was indeed Clausius who took the final steps in this journey, and so I now bring him back to center stage.

The Power of Isolation

I have written about the power of isolation in working out theories in one's own mind and feel strongly about the benefits of this practice; the isolation enables uninterrupted focus on threads of ideas and the resultant recognition of any elusive connections between them. But it's also true that for this to work, one must first become fully conversant with the ideas of others. One can really only sit in isolation and let new ideas germinate when one has a starting point consisting of a mixture of others' ideas. You see this being true when reading the histories of such figures as Newton, Bernoulli, Carnot, Clausius, and, as we'll see later, Gibbs. Each was influenced by others, but each then worked in isolation to create new ideas from these influences. Their ideas didn't just arise out of nothing.

Although not much is known about Clausius' life, it appears that he was indeed one who, being very well read, had many influences, but was also one who generated his breakthrough ideas by working largely alone. Yes, he did interact with other scientists, such as John Tyndall who first visited in 1851 when Clausius was but 28 years old and Tyndall had just entered the scene. But for the most part, Clausius worked in solitude and left behind hardly any personal correspondence with other scientists and hardly any personal notes other than those found in the margins of the papers and books comprising his library. He left behind no autobiography and provided no content for historians to study. The secondary literature on Clausius' life is sparse.[2] The fact is that we know little of Clausius and his inner thought process, especially as

---

[1] (Clausius, 1867) Ninth Memoir (1865), p. 327
[2] (Daub, 1970a)

Block by Block: The Historical and Theoretical Foundations of Thermodynamics,
Robert T. Hanlon, Oxford University Press (2020). © Robert T. Hanlon.
DOI: 10.1093/oso/9780198851547.001.0001

compared against what we know of Thomson's rather public life inside the highly interactive physics of Scotland. But with this downside came an upside, for it was likely just this solitude that afforded Clausius the opportunity to delve deep into his own thinking, bringing in the ideas of others, yes, but largely doing the final stage of thinking on his own. Perhaps this was how he was able to successfully interpret Carnot and discovery entropy, to list just two of his many other vital accomplishments in physics.

Yes, Clausius was pretty much alone and out front in confronting the hidden depths of Carnot. It wasn't that others weren't circling around some of the same issues. Rankine and Thomson clearly were and clearly influenced Clausius' thinking, all three being spurred on by the recent discovery of the conservation of energy. But in the end, it was Clausius alone who arrived at the critical end result.

A Brief Recap of How We Got Here

Before diving into Clausius' work, a brief recap is needed regarding one of the most defining moments in the history of entropy: the arrival of δQ/T out of the mathematical analysis of Carnot's heat engine. When Clapeyron converted Carnot's theories into formal mathematics, he arrived at an equation that quantified for the maximum efficiency of an infinitesimal heat engine,

$$\delta W / \delta Q = dT / C(T) \tag{31.5}$$

which contained a variable, C, later known as Carnot's function, that was hypothesized to be some unknown function of temperature. Even though Clapeyron's mathematics were wrongly based on the conservation of caloric, the above equation remained standing when Clausius revised Carnot's assumptions in 1850 to account for heat–work equivalence and then determined $C(T)$ to equal absolute temperature.

$$\delta W / \delta Q = dT / T \tag{33.1}$$

Now the above two equations weren't as explicit as this. Each involved complicated relationships for the δW and δQ terms, making it difficult to see the larger picture, specifically that the maximum work possible from a unit of heat (1 degree temperature change) in an infinitesimal heat engine is inversely proportional to absolute temperature.

As previously discussed, Thomson too arrived at this relationship when he adopted the use of absolute temperature for $C(T)$ based on both "Joule's conjecture" as he called it and also on Clausius' own work, although Thomson didn't admit as much. Thomson then easily rearranged this equation to

$$\delta W / dT = \delta Q / T \tag{33.2}$$

which is when the famed δQ/T arrived. It was sitting there, deep inside Carnot's theory, simply waiting for someone to find it.

Rankine, too, had arrived at a similar term based on his use of geometric diagrams of energy as inspired by Watt's indicator diagram.[3] His version became known as Rankine's "thermodynamic

---

[3] (Rankine et al., 1881) p. 352. Read before the Royal Society of London on January 19, 1854, and published in the *Philosophical Transactions* for 1854.

function," which he assigned the variable F, but as Brush noted, his theory became so entangled with his hypothesis of molecular vortices that he never received much credit for this or other of his contributions to thermodynamics.[4,5]

Thomson then took the logical step of converting Clapeyron's engine from infinitesimal to finite through an interesting integration technique in which he assumed the presence of many, many infinitesimal engines, all interconnected such that the rejected heat of one became the entering heat of another. All of these engines were wrapped inside a large box, with only two heat streams penetrating the walls, one from the source, one to the sink. Out of this analysis, "in a moment of penetrating insight,"[6] Thomson arrived at what was, as he himself stated, "the mathematical expression of the second fundamental law of the dynamical theory of heat"[7] in which the summation of all the infinitesimal heats entering *and* leaving each infinitesimal engine divided by the temperature of its entry and exit, respectively, equals zero.

$$\Sigma\,(\delta Q\,/\,T)=0$$

For the entire finite system, the fully integrated result reduced to a rearranged version of [33.3]:

$$Q_{in}\,/\,T_H + Q_{out}\,/\,T_C = 0$$

This equation can also be derived by integrating [33.1] from T_H to T_C[8] to arrive at

$$W\,/\,Q_{in} = \left(T_H - T_C\right)/\,T_H \tag{34.1}$$

The quantity $T_H - T_C$ is indeed the critical parameter in the efficiency calculation, just like Carnot thought, *but* it must be referenced against T_H. It's the percentage of absolute "fall" that matters.

Since $W = Q_{in} - Q_{out}$, [34.1] and [33.3] are identical. These are the equations we're taught in the university to calculate the theoretical maximum efficiency of a thermo-dynamic engine that converts heat into work. Clausius also identified [34.3] in his Fifth Memoir (1856) and recognized that William Thomson and Rankine had already discovered it, but based, in part, on his own work.

Regardless of the influences, by 1854 the famed $\delta Q/T$ term had been isolated, and Thomson had clearly stated that this term lay at the heart of the 2nd Law of Thermodynamics. But nothing seemed to happen afterwards, as if no one really knew what to do with this mathematical term now that they had it nor how to combine it with Clausius' written form of Carnot's 2nd Principle.

---

[4] Rankine's thermodynamic function, F, appeared to be identical to $\delta Q/T$ but upon closer scrutiny showed a key difference. Using his terms, he defined F as being equal to dH/Q. His term, dH, for heat received was the same as the δQ term in [33.2], but his term, Q, was not the same as absolute temperature. He defined his Q as the "total sensible or actual heat present in the body". As Brush noted in (Brush, 1986b) p. 578, "the relation between Q and absolute temperature was obscured because at this time Rankine was trying to maintain a distinction between the 'absolute zero of gaseous tension' and the 'point of absolute cold'".

[5] (Brush, 1986b) p. 583, footnote 57. Note that it was Rankine who converted in name Carnot's 2nd Principle to the 2nd Law of Thermodynamics.

[6] (Cardwell, 1971) p. 258.

[7] (Thomson, 1851) Part VI. Thermoelectric Currents, Paragraph #101. Read in 1854.

[8] To repeat, Clapeyron's mathematics suggested that Carnot's function, C, is really a constant, T_H. The integration then is solely over the term, dT, from T_H to T_C.

Why Two Laws are Needed to Govern the Maximum Performance of Carnot's Engine

Two laws are needed to constrain the maximum amount of work from Carnot's engine. The 1st Law provides a means to ensure that the energy balance around the engine is closed. Assuming no net change in the engine itself over time, then the work generated must equal the difference between the heat absorbed and the heat rejected. This is one of the fundamental methods for conducting an energy balance around a continuous process that is taught in introductory engineering classes. It's this law that prevents perpetual motion—the generation of work from nothing. Consider one of the most fundamental equations used for the conservation of mass and energy: accumulation = In − Out = zero at steady-state. What goes in, must come out.

The 2nd Law (in the form of: heat can only flow from hot to cold) provides a means to ensure that the temperature range of the working substance does not exceed that of either reservoir. Without this constraint, the heat engine would no longer require the generation of heat from, say, coal combustion. It could simply generate work from the energy contained in a near-by lake and this just doesn't happen. If you don't have access to a heat source with a temperature *different* from that of the environment, either hotter or colder—in other words, if the v^2 of the heat source is the same as the v^2 of the environment—then you don't have the means to raise a weight relative to that environment.

The fact that heat doesn't flow from cold to hot is due to temperature and not energy. If such a flow were to happen, the 1st Law wouldn't be broken since the energy balance could still be closed. But the 2nd Law would be broken. And this is where it's important to highlight the key distinction that exists between energy and temperature. Aligned with the 1st Law, energy is an extensive property that depends on temperature, yes, but also on other properties such as mass and heat capacity. Aligned with the 2nd Law, temperature is an intensive property related to v^2, which corresponds to the possibility of work generation as spelled out in the conservation of mechanical energy. It is temperature and not energy that determines whether thermal energy can be exchanged between two bodies. It is temperature that determines whether or not work can be generated from a *thermo-dynamic* engine. It is temperature that is most associated with entropy. It is temperature that places theoretical limits on the efficiency of Carnot's heat engine.

Even though everyone knew that heat could only flow from hot to cold, the realization that this was the core of the 2nd Law itself wasn't readily apparent at the outset, thus leading to confusion as evidenced by the various forms of this law that arose. The recognition of this phenomenon wouldn't be embraced until after Clausius completed his own work.

What to Do with δQ/T?

Unfinished coming out of Carnot's pamphlet, the 2nd Law of Thermodynamics started forming around two different ideas in 1850–4, the first appearing in words when Clausius sought to interpret Carnot's 2nd Principle, the second appearing in mathematics when the term δQ/T started to appear in the works of Rankine, Thomson, and Clausius. While these three scientists were all involved with developing the 2nd Law, it was only Clausius who was able to explain the second idea and (almost) combine it with the first. This combined task would take him until 1865.

It's interesting that Clausius was the only one of the three willing, able, and available to follow this through to the end. I haven't seen any records of anyone else pursuing this issue. While both Rankine and Thomson isolated δQ/T, neither pursued the determination of its significance. Perhaps this was because they didn't recognize it as a problem waiting to be solved. Maybe Clausius simply saw what others didn't and had the time to patiently sit and contemplate. And regarding Carnot's 2nd Principle, no one other than Clausius seemed interested.

It's also possible that Rankine's attachment to his vortex model may have limited him in this regard, although it didn't seem as if much limited Rankine. He may have simply chosen to pursue work on his textbooks and especially the application of thermodynamics to the practical design and operation of engines. As for Thomson, perhaps it was his strength in idea development as opposed to idea generation that had him either not recognize or otherwise not become engaged in this challenge. Or perhaps it was the fact that he was being pulled in so many directions in pursuit of so many endeavors, both scientific and business. The "pulling" wasn't without its merits. Repeating some from Chapter 32, while William Thomson is the name I use in this book, as it was William Thomson who was so involved in the early days of thermodynamics, it was this same William Thomson who went on to great fame in Britain, contributing much to society, especially including his theoretical and practical assistance in the laying down of the first trans-Atlantic communications cable, and who eventually was awarded in 1892 a new name, Lord Kelvin.[9] The rise of his remarkable career may have left William Thomson little time to probe the deeper meaning of δQ/T.

Who really knows? The above is all speculation. All we do know is that while Rankine and Thomson, among others, certainly made contributions toward the formulation of the 2nd Law and positively influenced Clausius' thinking in so doing, it was Clausius alone who got the result and deserves the credit for doing so. Yes, Clausius struggled to understand Carnot's 2nd Principle, δQ/T, and entropy, but no one else even recognized these as things to struggle with.

> Rudolf Clausius

Heat can never pass from a colder to a warmer body without some other change, connected therewith, occurring at the same time.

– Rudolf Clausius[10] (Modified version of Carnot's 2nd Principle)

Why was Clausius driven to understand and unite the two initial forms of the 2nd Law while others weren't? What was his motivation? Simply put, regarding his own above written version of Carnot's 2nd Principle, he saw a law that was confusing, based on words only, without equations, lacking in clarity. People couldn't really do anything with this law other than to say whether or not something was possible. It provided a means to answer a question with either yes or no. Nothing else. Regarding the term δQ/T, he felt that it reflected some deeper property

---

[9] (Lindley, 2004) This book provides an excellent history of Thomson's legacy, which led to his being crowned Lord.

[10] (Clausius, 1867) Fourth Memoir (1854), p. 117.

of matter, but what, he didn't know. And it was the unknowing surrounding both of these forms that drove him forward as best understood by considering these two quotes bookending the key timeline involved:

> Fourth Memoir, 1854: "We cannot recognize [in this law] with sufficient clearness, the real nature of the theorem, and its connexion with the first fundamental theorem"[11]
> Ninth Memoir, 1865: "In my former Memoirs on the Mechanical Theory of Heat, my chief object was to secure a firm basis for the theory, and I especially endeavoured to bring the second fundamental theorem, which is much more difficult to understand than the first, to its simplest and at the same time most general form, and to prove the necessary truth thereof."[12]

The untidiness of this situation must have deeply bothered Clausius and his innermost scientific philosophy of simplicity and clarity and left him with a strong desire to finish the job. He simply couldn't leave the 2nd Law just sitting there, in two forms, messy and near useless. So he sought to resolve this situation by probing the molecular behavior involved, driven by the belief that a simple resolution existed because nature, at its core, is simple.

The challenge involved was significant. The 2nd Law was indeed a struggle from the start on account of Carnot's belief in the caloric theory. One can only wonder how much further he would have gotten had he lived to overcome this situation. But even with the replacement of caloric with energy and the subsequent revision of Carnot's assumptions and mathematics, Thomson, Rankine, Clausius, and all others continued to struggle, both in arriving at a single coherent statement of the law—think back to Tait's and Maxwell's comments on Rankine's attempt—and in understanding the physical meaning behind the statement. Rankine, Thomson, and Clausius appeared close in 1854 but this was in fact misleading. They may have been close with their respective mathematics but not with their respective interpretations. There's a good reason why it took many more years to complete this journey. Clausius almost got there in 1865. But it wasn't until Boltzmann's work years later that the underlying physical meaning behind the 2nd Law would become clear and quantified.

The fascinating thing here is that the form of the 2nd Law needed in the mid-1800s, at least from a thermal energy perspective, was what everyone already knew. Heat only flows from hot to cold. Granted, the supporting mathematics and broader law based on entropy had yet to be discovered. But still, this simple statement was really all that was needed at the time for it ensured that the isothermal steps in Carnot's cycle wouldn't exceed the range offered by the high and low temperature reservoirs.

Clausius laid out the path he took toward entropy in his famed series of nine memoirs titled *The Mechanical Theory of Heat*. Before reviewing this collective work, I thought it important to first review in more detail what is and is not really happening inside a Carnot heat engine and while doing so, to clarify the confusion and frustration that is commonly experienced in reading about such concepts as heat engines, reversible closed-cycle processes, and entropy. To this day, many textbooks aren't sufficiently clear about this subject. In fact, as we'll see, Clausius himself was not totally clear in his own mind at the completion of these memoirs, which only furthered the confusion of an already confusing subject. I felt it important here to seek an

---

[11] (Clausius, 1867) p. 111.
[12] (Clausius, 1867) p. 327.

accurate, simple, and readable explanation of this engine while simultaneously sharing some of the words, terminologies, and concepts that led to our aforementioned confusion and frustration.

Revisiting Carnot's Engine

Let's start with this fact. The performance constraints limiting Carnot's engine can be understood by two laws: 1) the conservation of energy (work–heat equivalence), and 2) heat only flows from hot to cold. This is it. Now, there are consequences of these laws. One is that when an engine runs forward and produces work from heat, heat flows into the engine from the hot reservoir and out to the cold. This is what Clausius' version of Carnot's 2nd Principle states. Another is the reverse, when the engine receives work (that high electrical bill) to effect cooling (refrigeration cycle), heat flows into the engine from the cold and out to the hot. Just as simple 0s and 1s combine to create a complex computer program, these two laws combine to create a complex Carnot heat engine. But the complexity lies in the consequences of the laws as they play out in the engine, not in the laws themselves. Starting with this high-level overview, let's now go deeper inside this engine (Figures 30.1, 32.1, 32.3).

In the Carnot four-step cycle, heat enters the engine from the hot reservoir, a net amount of work is done, and residual heat exits into the cold reservoir. So long as the temperature of the working substance in the engine maintains itself at that of the reservoir during isothermal expansion (Step 1) and then at that of the reservoir during isothermal compression (Step 3), and so long as all of the consumed heat ($Q_{in} - Q_{out}$) is converted into productive work (W), then the net work generated is the maximum possible. Also, because W is a positive value, Q_{in} must be greater than Q_{out}.

"But wait!" you say, "How does this reconcile with Carnot's 2nd Principle? Why is it that heat must <u>always</u> enter from the hot reservoir when useful work is being done?" This is somewhat more complicated, but again, it's a consequence of the two constraining laws combined with the way Carnot constructed his engine. There's a reason why Q_{in} is always associated with T_H.

Notice in the above that I didn't specify which of the two reservoirs is responsible for the work-producing step. We generally think that it's the hot reservoir that causes work. For example, in Step 1 of Carnot's heat engine operating with, say, an ideal gas (Figure 30.1), heat enters from the hot reservoir to expand the gas and so do work. But don't forget that the working substance is irrelevant. Ice–water works just as well as an ideal gas. But with ice–water (Figure 32.1), the heat entering from the hot reservoir melts ice while work is done on the system to shrink volume. So in this scenario, the work-producing step isn't caused by the hot reservoir. Instead, it's caused by the cold reservoir. Recall James Thomson's ice engine: it's the cold reservoir that causes work. When heat exits into the cold reservoir, water freezes, expands, and does work.

As manifested in both Figure 30.1 and Figure 32.1, regardless of which reservoir is responsible for producing work, heat only flows into the system from the hot reservoir. Since heat can only flow from hot to cold, then the temperature of the hot reservoir must always be slightly greater than the temperature of the working substance. Since this heat is required to maintain isothermal conditions, if it were removed the temperature of the working substance would have to fall away from that of the hot reservoir and move toward that of the cold reservoir. This is exactly what happens during Steps 2 and 4, the adiabatic steps. A reservoir is removed and temperature moves from high to low or from low to high, as opposed to moving from high to higher or from low to lower.

Looking at this from an entropy perspective, since we know that Q_{in} must be greater than Q_{out}, then we also know that T_{in} must be greater than T_{out} since we know, $Q_{in}/T_{in} = Q_{in}/T_{out}$. Thus "in" must align with the hot reservoir.

Carnot's 2nd Principle is correct because of the combined consequences of the 1st (work–heat equivalence) and 2nd (heat only flows from hot to cold) Laws acting inside a particular engine. The working substance temperature never exceeds the range established by the hot and cold reservoir temperatures; the 2nd Law prevents temperatures from moving away from each other. When the hot reservoir is no longer engaged, the working substance doesn't get hotter. When the cold reservoir is no longer engaged, the working substance doesn't get cooler. The adiabatic steps move the temperature of the working substance from one reservoir toward the other, never away.

Given all of this, you can understand the confusion surrounding Carnot's central premise, "The equivalent of the work done by heat is found in the mere transfer of heat from a hotter to a colder body." This statement originated from Carnot's false cause–effect belief that the fall of caloric causes work to be done. Clausius, among others, continued with this theme as noted previously, but modified the cause–effect logic to: if work is done by heat, then heat must be transferred from hot to cold. But even this modified statement remained confusing.

The problem inherent to Carnot's original and Clausius' modified 2nd Principle was that it incorrectly portrayed heat as a *thing* that could be transferred. Carnot envisioned a "flowing caloric" and this misleading visualization unfortunately lived on long after caloric itself faded away. To this day, we speak of "heat flow," "transfer of heat," and even "fall of heat." While Clausius rightly revised Carnot by stating, "Heat is not a substance,"[13] his writing suggested that he wasn't entirely certain what heat actually was. He referred to heat as a "thermal element", "a motion,"[14] a "sensible heat"[15] present in a body that is "carried" or "transferred" from hot to cold. Furthermore, he and others treated heat, at times, as if it were a property of state. He wrote of the "heat content of a body" and "temperature of heat" and employed calculus to differentiate heat (δQ) with respect to, for example, pressure and volume, or pressure and temperature, just as Clapeyron had done while working under the influence of caloric, even though heat can't be differentiated in this way since it's not a property.[16] Clausius readily

---

[13] (Clausius, 1867) Seventh Memoir (1863), p. 268.

[14] (Clausius, 1867) Seventh Memoir (1863), p. 268.

[15] (Clausius, 1867) Sixth Memoir (1862), p. 255.

[16] (Cardwell, 1971) p. 246. As Cardwell emphasized, the use of Q in such equations did not mean that Q is a function of P, V, and U (the old heat doctrine). This may have been technically legal, as discussed by Clausius, but it was confusing, or why else draw attention to its use. The equation $\delta Q = dU + PdV$ does not mean that Q is a function of p, v, and U (the old total heat doctrine). To me, Planck's rectification of this situation was the best as he suggested that the equation be written $Q = dU + PdV$. Also, in (Planck and Ogg, 1990) p. 57, Planck certainly helped bring clarity to this confusion by pointing out that heat (Q) quantifies the difference between two numbers and that such terms as "δQ" are inappropriate in this situation since the differential of a specific number is not the same thing as the very small difference between two specific numbers. Q should stand on its own as a measured difference in internal energy

acknowledged this caveat, explaining his correct logic for being able to still differentiate δQ as a means to calculate the amount of heat required to effect certain changes in state properties like temperature and pressure. While technically correct, the approach was still complicated and confusing. Today, we understand that the appropriate mathematics should be based on internal energy, U, which Clausius himself created.

So while Clausius clearly stated that heat was not a substance, I'm not really sure that he viewed heat, which quantifies a change in energy, as a separate and distinct concept from energy itself. At times, he seemed to use heat interchangeably in both contexts, stating that heat is a measure of the *vis viva* motion of the atoms comprising a body of matter, thus equating heat with the kinetic energy component of internal energy. Such thinking might explain why he didn't philosophically embrace the concept of energy until 1864 in the Appendix to his Sixth Memoir of 1862 when he wrote, "the term *energy* employed by Thomson appears to me to be very appropriate."[17] Energy was still a new and abstract concept at that time. As Cropper noted, "It is an impressive measure of the subtlety of the energy concept – and of Thomson's insight – that Clausius was not willing to accept Thomson's energy theory for almost fifteen years."[18]

In the end, I am not certain how immune Clausius really was to the relics of caloric. As noted earlier by Daub, some of Clausius' writing suggested that he was "still under the influence of the caloric theory."[19] Unfortunately, also as noted earlier, Clausius left behind no notes, no letters, really no anything for historians to sift through in order to better understand his off-the-record thought process on this matter.

To Summarize: What Heat is and is not

To summarize based on what we now know, there is no *thing* called heat. It is not a property. Strictly speaking, heat is a number quantifying the *change* in internal energy of matter caused by the thermal exchange between colliding atoms at the interface between the body and the environment. Heat does not flow from one body to the other; it's not an absolute number; it's a number quantifying a *change*. We use the phrase "heat flow" to simplify the discussion, and I'll continue to use it here as it does make things easier. But still, it's important to realize that there is no thing that flows.

This misleading concept of a "flowing heat" used by Carnot and continued by Clausius had implications when interpreting the 2nd Law. Again, Carnot proposed that the flow of heat from hot to cold in a continuous, closed-cycle process causes work to be done. When Clausius revised Carnot, he rightly changed the basis for his mathematics from the conservation of caloric to heat–work equivalence but unfortunately held onto Carnot's central tenet, that work results from a flowing heat. When Clausius conflated this phenomenon with the separate and

---

caused by the exchange of thermal energy between two bodies. Q is a difference, not a thing. A body doesn't have Q. See also (Lewis and Randall, 1923) p. 54 for further discussion. In this book the author has chosen to remain with δQ to denote a very small change in Q.

[17] (Clausius, 1867) Sixth Memoir (1862) p. 251.
[18] (Cropper, 1986) p. 1069.
[19] (Daub, 1970a) p. 305.

entirely different phenomenon of heat conduction, confusion resulted. That these two state-ments are indeed different is not readily apparent.

Carnot's concept of a flowing heat was very misleading as it referred to the net impact of the two separate isothermal steps in the engine. Clausius' subsequent attempt to seek a generalized theorem from this physical model was an attempt that went too far. As I'll discuss in more detail shortly, he admirably tried to conceptually pull apart the engine, categorize the pieces, and then put them back together again into a simpler and more comprehensive set of theorems. He created new terms to help, some like *disgregation* and *ergonal* which never caught on, but others like entropy that did. Again, all admirable efforts. But one of the unfortunate and unforeseen results of these efforts was confusion, for both Clausius and his readership, espe-cially when he linked this flowing heat model with the separate model involving the direct conduction of heat. While they might sound the same, using the similar concept of "flowing heat," they are not the same. While Clausius acknowledged, at least to some extent, the differ-ence between the two by stating that Carnot's concept of heat flow involves "connected transmissions,"[20] he still combined the two into the same bucket when creating his theory of transformations. He had no justification for doing this. One of his versions of Carnot's 2nd Principle was, "Heat can never pass from a colder to a warmer body without some other change, connected therewith, occurring at the same time." But with direct contact, even with man's intervention, heat can never pass from cold to warm. Period. It's statistically impossible. And this brings me to a final point.

Both Thomson and Clausius struggled with conduction (Figure 32.2). This process pre-sented a very fundamental and very challenging problem in the early development of thermo-dynamics. What really happens when heat flows from hot to cold? If you picture this as a material (caloric) flowing down a metal rod from hot to cold, then the total energy stays the same—based on conservation of energy—and yet the temperature decreases. But how can this be? This was a very difficult question to answer in the mid-1800s. To Clausius, conduction represented the transformation of heat from high temperature to low. But what did "trans-formation" really mean here? I discussed this subject before, especially Thomson's long strug-gle with it, and how such statements as this implied a mistaken belief in heat as a moving entity or as a property of matter. Again, I'm not sure that even Clausius had this figured out by the time of his concluding Ninth Memoir of 1865. Even the term "conduction" itself is misleading, implying as it does motion through a medium. I'll continue with this discussion later.

*   *   *

Before going on, it's interesting that conduction inside a Carnot heat engine occurs when thermal energy is exchanged between the reservoirs and the working substance. Heat is con-ducted across the metal boundary. Unwanted conduction also occurs when the hot reservoir is directly connected with the cold reservoir, such as would be the case if a metal bar were used to connect the two. While the phenomena of conduction along the metal bar would be identi-cal to that across the boundary, it would be a much stronger visual to imagine. One could imagine heat flowing through a metal bar more readily than across a metal boundary. And, I believe, this image of piped conduction and the flow it suggested led to an incorrect physical

[20] (Clausius, 1867) Fourth Memoir (1854) p. 118.

model and so confused those trying to understand the concepts of heat, conduction, and non-ideal engines.

Clausius' Nine Memoirs

We are now finally ready to consider Clausius' series of nine memoirs on *The Mechanical Theory of Heat* in which he laid out the path he took toward entropy. The specific focus here will be on the Fourth through the Ninth memoirs as I covered the groundbreaking First Memoir of 1850 in detail previously and the Second and Third aren't needed for this discussion. Rather than summarizing these memoirs one by one along the timeline of publications, I approach things from a different angle with a more collective summary based on the following three key themes:

1. Clausius' discovery of entropy in 1865
2. The reason for the delay
3. The Ninth Memoir's final sentence.

While the discovery of entropy (1) was naturally a critical event, my discussion of how this came about is quite fast, for it wasn't the discovery of entropy that was so fascinating in this story, but instead it was the discovery of a critical behavior of entropy, namely that the entropy of a combined equilibrated system is greater than or equal to the sum of the entropies of the separate equilibrated components.

$$\Sigma\ S_{\text{separate equilibrated parts}} \leq S_{\text{combined equilibrated system}}$$

While Clausius didn't arrive at this second discovery, he was heading there as revealed in his persistent attempts to understand what entropy meant. The challenge involved was greater than that involved with (1), hence the lengthier discussion in (2) and (3).

(1)
Clausius' Discovery of Entropy in 1865

When he applied his thinking on the mechanical theory of heat to Carnot's heat engine, Clausius arrived at the same $\delta Q/T$ term (Fourth Memoir, 1854) as had Rankine and Thomson. While I don't want to get into the details of his work as they are rather abstract, especially when compared with Thomson's approach, I do want to point out that he had actually arrived at the founding equation, namely $\delta W/\delta Q = dT/T$, behind this term in his First Memoir of 1850 once he successfully determined Carnot's function to equal absolute temperature, T. But Clausius apparently didn't recognize the importance of $\delta Q/T$ in that memoir as he didn't explicitly isolate the term, thus unintentionally leaving this opportunity for Rankine and Thomson to seize.[21]

---

[21] Clausius published his Fourth Memoir in December 1854. Thomson read his paper in May 1854 but it wasn't published until later. Rankine read his paper in January 1854 but it too wasn't published until later.

Recall that Thomson built an argument based on his own isolation of δQ/T to demonstrate that the sum of this term for all heat entering and leaving a system comprised of an infinite number of infinitesimal Carnot engines must equal zero. He then recognized the importance of his finding by claiming it to be the mathematical expression of the 2nd Law. In his Fourth Memoir, Clausius ended up not with Thomson's use of summations (Σ) but instead with the following, more fundamental, calculus-based equation describing heat flow into and out of Carnot's reversible, closed-cycle engine, regardless of the number of steps and associated reservoirs.

$$\int \delta Q / T = 0 - \text{reversible, closed-cycle heat engine}$$

In an interesting aside, while both Thomson and Clausius almost simultaneously ended up with the same basic equation, I don't believe that either knew what the other had done. Thomson read his paper on May 1, 1854,[22] but his follow-up article wasn't published until 1856. Clausius published his article in December, 1854, after Thomson's reading but prior to Thomson's publishing. So it is highly unlikely that Clausius, who was not in a geographically favorable position to attend Thomson's talk, could have been aware of Thomson's work, a conclusion consistent with the fact that Clausius did not reference Thomson in his Fourth Memoir combined with the fact that Clausius was typically quite rigorous in following science protocol.[23] The fact that they were both on the right path at the same time simply reflected their common interest in the newly revised theories of Carnot. It would have been fascinating had Clausius written an autobiography containing his retrospective view of this history, especially highlighting the motivating instances of important insights such as those described here. Unfortunately, he didn't.

While Thomson moved on to other endeavors at this point, Clausius paused, his attention being caught by something he saw in the above equation that others didn't, specifically the presence of a new property of state.

I need to take a brief step back here to establish a context for this next discussion.

As Discussed Previously, the Properties of Matter

Roughly speaking, prior to 1800 scientists typically characterized matter using certain fundamental or "primitive" properties such as volume, pressure, and temperature that lent themselves to reasonably direct measurement. Length could be measured against everyday references like the human foot while volume became a simple multiple of length. Pressure was a little more complicated. Devices were invented for its direct measurement and one could conceptually grasp the meaning of the measurement by looking at a gauge on a closed container and watching the container explode (from a safe distance) when the gauge got too high. Temperature too could be directly measured, although the deeper meaning of the resulting measurement

---

[22] While Thomson was influenced by Rankine, the concept of summing δQ/T wasn't present in Rankine's work.

[23] (Clausius, 1867) Fifth Memoir (1856), p. 161 (footnote). Clausius was very correct in concepts of priority, acknowledging, for example, work published by Rankine prior to his own publications, writing, "By the earlier publication of [Rankine's] memoir I lost, of course, all claim to priority with respect to this part of my investigation."

wouldn't become readily apparent until the development of the kinetic theory of gas. Our physical senses of touch and sight helped guide the rise of such measurements. We could see, feel, or otherwise sense these primitive properties.

Sometime after 1800, more complex properties of matter started to appear in the literature. For example, back in the days when caloric and heat were conceptually muddled together, scientists such as Clapeyron attempted to quantify the amount of heat contained in a body. While there was no analytical device available to directly measure such properties, their proposed existence logically led one to believe that they could be quantified as some unknown mathematical function of those properties that could be directly measured. The underlying belief here was that once a certain number of properties were fixed, then the "state" of the body along with all of its other properties, no matter how complicated they might be, were also fixed, with each such property being able to be placed as a single point on a graph. (I will discuss this more in Chapter 36 in the context of Gibbs' phase rule and "degrees of freedom.") So for the example of heat, its proposed existence led to such mathematical terms as Q (T,V) in scientists' writings. And while this specific term was inappropriate since heat was not a state property, its use opened the door to considering the existence of real state properties and their own associated terms.

The first such property was U and its arrival in 1850 was a powerful moment in the history of physics. While he may not have realized it at the time, especially as reflected by the fact that he didn't draw much attention to it and by the fact that he would later be challenged to accept the new concept of energy, Clausius was very much on the cutting-edge of science with this discovery. But in the context of what I'm sharing here, it wasn't the discovery of U itself that was relevant. It was the dawning realization in Clausius' mind of what a state property actually was, as described above.

Prior to Carnot, the importance of this concept was small and the need to distinguish it not great. However, once Clausius immersed himself in Carnot's work, this situation changed. He soon realized that it wasn't just the primitive properties such as pressure and volume that returned to the starting point upon closure of the cycle. It was *all* properties. Thus, the integral of *any* state property around Carnot's reversible closed cycle must equal zero. Clausius demonstrated his newfound understanding in his Fourth Memoir when he drew attention to the fact that the integral of U around the complete Carnot cycle must equal zero. He would demonstrate this understanding again when he interpreted δQ/T.

The Logic behind Clausius' Discovery of Entropy

It's no wonder that it was Clausius who first figured out δQ/T. He was the first to realize that because ∮δQ/T equals zero for a reversible, closed-cycle process, as highlighted in his 1854 Fourth Memoir, then δQ/T must, in turn, represent a fundamental aspect of nature, a newly discovered state property that doesn't experience a net change upon return to its starting point. He defined this property in his Ninth Memoir (1865) as

$$dS = \delta Q / T \tag{34.2}$$

although this was not technically a definition but rather a limited means of quantification.

In a nod to the Ancients, he combined *-trope* (Greek for transform) and *en-* (for in) to create the term *entropy*, which also created a nice alliterative sound when placed next to energy. This decision was based on his belief that entropy captured heat's ability to be *transformed* into other forms of energy, such as work; we'll discuss this in more detail later. Clausius realized that such naming was a big deal, not to be taken lightly. Being a scientist's scientist, he followed a disciplined and rigorous path in his work, understood the significance of what he had discovered and marked this by his deliberate naming. The interesting point here is that he hadn't done likewise in 1850 when he discovered U and unintentionally allowed Thomson and Rankine to name it. Again, perhaps he simply hadn't realized U's significance at the time.

Once the above proof was done, the proof itself was no longer needed. Entropy as a state property stood on its own, independent of the Carnot cycle, its existence discovered, its broader definition still to be defined. In a sense, the Carnot cycle served the role of enabling the discovery of entropy and once entropy was discovered, the enabling logic was no longer needed. Entropy was a state property that could be placed as a point on a graph, just as the primitive properties or U could be. Such concepts as heat and work, on the other hand, could not exist as a single point on a graph; instead, they served to quantify the *difference* between two states and not a given state itself.

Clausius didn't stop with his newly discovered property. He took another monumental step forward by reasoning that because the value of δQ is the same for both the 1st Law and his entropy concept, then the following must be valid:

1st Law: $dU = \delta Q - \delta W = \delta Q - P\,dV$

$dS = \delta Q / T$

Combined:[24] $dU = T\,dS - P\,dV$ [34.3]

This combined equation, which includes the assumption that work is due solely to a moving boundary, quantifies how state properties change in relation to each other, regardless of what is happening in the external environment to change them. As you'll see in the next chapter, Gibbs would take this equation and work wonders.

More Discussions about Clausius' Logic

Before proceeding further, believing in the importance of understanding the logic and definitions at work here, I thought it appropriate to pause here to discuss some additional and relevant aspects of Clausius' discovery based on what we know now and that Clausius didn't realize then.

The integral of any state property of an internally equilibrated body (maximum entropy) around any reversible, closed-cycle process must equal zero by definition. It must be since the changing body returns to its exact equilibrated starting point to close the cycle. Thus,

---

[24] (Clausius, 1867) Ninth Memoir (1865), Appendix p. 366. In the 1866 Appendix to his Ninth Memoir, Clausius wrote the equation $TdS = dU + \delta W$ from his standpoint of *"where does the heat go?"* For fluids, it was well understood that δW equaled PdV, but Clausius never explicitly included this as such.

$$\int d(\text{state property}) = 0 \quad \text{valid for reversible, closed–cycle process} \qquad [34.4]$$

It just so happens that for such a process, the only means by which entropy can change is heat transfer and these changes can be quantified by the term $\delta Q/T$. This term applies to all four steps in Carnot's cycle, capturing as it does the fact that no change in entropy occurs during the adiabatic expansion and compression steps (Steps 2 and 4) since heat transfer is zero. Based on this, the following equation applies:

$$\int \delta Q/T = 0 \quad \text{valid for reversible, closed–cycle process} \qquad [34.5]$$

Clausius' path to entropy occurred in the reverse direction. First he discovered [34.5] and then reasoned that [34.4] must be so and thus that $\delta Q/T$ must equal d (state property). Thomson also discovered [34.5] but didn't pursue it.

Carnot's Cycle was Unknowingly Designed to Reveal Entropy

It's all quite amazing, when you really think about it, how the properties of entropy lined up perfectly with Carnot's four-step cycle in such a way to allow entropy itself to be discovered. What are these properties?

1. When a body experiences an infinitesimal change in internal energy due to thermal exchange with the surrounding environment, the change in its entropy is exactly equal to $\delta Q/T$.

2. When a body undergoes reversible adiabatic expansion or compression, exchanging internal energy for external work, the net change in entropy is exactly zero. We know that entropy increases with increasing volume and we know that entropy decreases with decreasing temperature, but who would have thought that these two effects would exactly cancel each other during adiabatic expansion?

3. And a property not embedded in Carnot's cycle since his cycle was reversible: When two bodies (initially at equilibrium) exchange thermal energy to achieve a new equilibrated state, then the entropy of this combined state is always equal to or greater than the sum of the entropies of the two initial bodies. To use Thomson's term, the "dissipation" of temperature gradients is what results in an increase in entropy since the positive entropy change of the colder body is always greater than the negative entropy change of the hotter body.

The molecular understanding behind these three fascinating properties of entropy is rather complicated and will be discussed in the context of statistical mechanics in Chapter 42. But the way they played out in Carnot's engine—and somewhat separately in heat conduction—thankfully attracted the attention of Clausius. The rest, as they say, is history.

It's important to emphasize that even when a process is irreversible, the change in entropy still lends itself to calculation. It's just that the change cannot necessarily be calculated using $\delta Q/T$. As the first to discover entropy, Clausius was also the first to show how to handle these situations. He said that because entropy is a state property, it is a function solely of that state and not on the history of that state. If a body moves from one state to another, then if you know the change in the defining properties between the two states, which may in and of itself be difficult to determine and require some insightful assumptions, then you can calculate the change in any other property, such as entropy. It doesn't matter how the body moves from one state to the other. Specifically, Clausius said that since one has the above two equations to work with, namely dS = δQ/T for reversible heat transfer and dS = 0 for reversible adiabatic volume change, then one can conceptually use these two process steps—like the two knobs on an Etch-A-Sketch—to move anywhere on the PV-indicator diagram, or any other property diagram, and thus calculate the change in entropy change between any two states, regardless of how that change between them actually occurred. For example, this approach can be used to analyze free expansion even though this process is clearly irreversible. The entropy change for this irreversible process is equal to the difference in entropy between the initial and final states; the path taken is irrelevant since entropy is a state property. One could conceivably design a reversible process consisting of isothermal and adiabatic steps to identify a reversible path from initial to final state that would lend itself to ΔS quantification using thermodynamic equations.

A final comment regarding the equation dS = δQ/T. This equation simply says that if a body absorbs heat due to thermal contact with another body, then the change in entropy of this body is equal to the heat received divided by the temperature of the body. It doesn't matter if the heat transfer process itself is reversible or not; the temperature difference between the two bodies exchanging the heat is irrelevant in this calculation.[25] This is an important point to remember. So long as the only change happening to the body is caused by heat entering or leaving, then this equation is valid. For reversible adiabatic volume change, δQ equals zero and the change in entropy also equals zero. For either of these two situations, if any irreversible change occurs at the same time, such as free expansion or chemical reaction, then the equation dS = δQ/T is no longer valid since other factors also contribute to entropy change. As noted in the previous paragraph, in such situations one must find another way to calculate the resulting entropy change using hypothetical ideal reversible processes.

By combining the two concepts of closed-cycle analysis and the existence of state properties of matter, Clausius was able to discover entropy as the state property at the heart of the 2nd Law. In effect, Carnot set up his physical model in such a way that energy and entropy were unintentionally sitting there waiting to be discovered. Carnot created a puzzle that Clausius solved. But note that while it was the analysis of Carnot's work that led to the discovery of both internal energy (U) and entropy (S), these properties, once established, and others that would be derived from them afterwards, would stand apart from Carnot, existing as an inherent part of nature with no dependence on the processes that led to their discovery.

*   *   *

Clausius announced entropy in his concluding Ninth Memoir (1865). What was fascinating about this was that he really identified the key basis for the announcement in his Fourth Memoir of 1854. In reading this memoir, you can sense that he felt the presence of a new property but for some reason stopped short of drawing attention to it. In fact, even when he later did announce this new discovery, he really didn't draw much attention to it. It was a rather quiet announcement. The question is, why?

---

[25] One caveat is this. While the heat absorbed should be readily determinable, the temperature at which this occurs may not be, especially if the heat transfer rate is high enough to create internal temperature gradients.

My simple answer is this. Clausius couldn't figure out the physical meaning behind $\delta Q/T$ and thus dS. He was on somewhat shaky ground and knew it. As a man of caution, he didn't want to recklessly announce a new property that he didn't understand. Thankfully, rather than being stopped, he was inspired. The gap motivated him. His deep desire to "seek for the precise physical cause, of which this theorem is a consequence"[26] kept him going. He possibly realized that he alone was in the position to solve this problem and excited by the challenging opportunity. This then brings us to (2).

(2)
The Reason for the Delay

In 1854, Clausius had two pieces of information sitting in front of him. The first was his own version(s) of Carnot's 2nd Principle,[27]

> *In all cases where a quantity of heat is converted into work… another quantity of heat must necessarily be transferred from a warmer to a colder body.*
> *Heat can never pass from a colder to a warmer body without some other change, connected therewith, occurring at the same time.*[28]

The second concerned the term $\delta Q/T$ that fell out of Clapeyron's mathematical treatment of Carnot, specifically regarding the fact that the integral of this term for a reversible, closed-cycle process equals zero. To Clausius, that these two pieces were connected was apparent since both stemmed from Carnot's work, but *how* and *why* they were connected wasn't. This was the larger challenge confronting him.

Clausius was troubled by the first piece of information as reflected by the fact that he kept returning to it and modifying it, never being quite settled in his own mind that he had arrived at a final stopping point. In his world of exactness, the nuanced wording of this statement and its lack of quantification as reflected by the absence of a supportive equation surely frustrated him as he couldn't really do anything with it. Such was not the case with the second piece of information, for $\delta Q/T$ was the first step toward bringing a new and quantified approach to Carnot's work. It was the presence of this mathematical exactness that had Clausius pour his energy into understanding $\delta Q/T$ first, likely thinking that the achievement of this goal would then enable a more definitive statement of Carnot's 2nd Principle.

Clausius' philosophical path toward understanding $\delta Q/T$ was aptly captured by Cardwell: "Science in fact is a matter of asking the right questions."[29] To Clausius, there was one "right" question that served to guide him throughout his nine memoirs, *What happens to the heat?* This line of questioning was an entirely logical starting point as it was this heat that was the basis for all efficiency calculations. Steam engine owners spent great amounts of money in buying the coal to generate this heat. They wanted to know what they were getting in return.

---

[26] (Clausius, 1867) Sixth Memoir (1862), p. 219.
[27] (Clausius, 1867) Fourth Memoir (1854), p. 117.
[28] (Clausius, 1867) Fourth Memoir (1854), p. 116–7. Clausius modified this phrase in 1864 to, "Heat cannot by itself pass from a colder to a warmer body."
[29] (Cardwell, 1971) p. 241.

Guided by this question in his First Memoir (1850), Clausius asked, *what happens to the heat when it enters a body?* This inquiry ultimately led to the discovery of internal energy, U. Guided by this same question in later memoirs, Clausius asked, *what happens to the heat when it enters an engine?* Here, the inquiry was more challenging and led him to continually pound away at Carnot's engine, driving himself into "extremely tedious analysis"[30] of a very abstract theoretical world, all because the answer was extremely elusive to grab onto.

Reading Clausius' memoirs is not for the faint of heart. He wrote with extreme brevity, which is fine if you can follow his thinking, but not fine if you have any confusion about what he was saying. He didn't always expound on the logic behind his arguments. Not helping matters was the presence of the aforementioned relics of caloric thinking that were inadvertently sprinkled throughout his writing and which, when combined with some degree of residual confusion that stirred inside him about the differences between caloric, heat, temperature, and energy, only served to make matters more confusing.

Did his own confusion delay his announcement of entropy? To a certain extent, probably yes. Did it hurt his achievements? No, it only made them more monumental, for it suggested that Clausius himself was caught deep in the weeds, deep in the details, and was trying valiantly to find his way out, eventually (almost) doing so in spite of whatever internal struggle and confusion he was dealing with.

The Concept of "Equivalent Transformations"

Recall that when he asked, *what happens to heat when it enters a body?*, in his First Memoir (1850), Clausius asserted that heat could be *transformed* (a word he used often in reference to heat) into internal heat (*vis viva*, "sensible heat," "free heat"), internal work (*vis mortua*, "latent heat," "bound heat"), or external work. This was the starting point for much of his subsequent study.

$$\delta Q = d\left(vis\ viva \right) + d\left(internal\ work \right) + \delta W \qquad [34.6]$$

Based on the 1st Law, Clausius understood that the change between any of the above forms had to be exactly balanced by an equal and opposite change in another form. He called such transformations *equivalent transformations*.

Two points of interest here. First, note his placement of δQ at the beginning of his 1st Law equation, as opposed to our current use of dU. It shows how his mind was working. He was thinking about heat and not energy. Second, note that even after he embraced energy, he continued to refer to both Q and *vis viva* as heat, as if the two were interchangeable concepts, suggesting an incomplete distinction between the two. In this context, the more technically accurate term for *vis viva* would be the kinetic energy component of internal energy.

Since he didn't have a means to isolate and quantify "internal work," he combined it with its counterpart, "*vis viva*," to form U, which became internal energy.

$$\delta Q = dU + \delta W$$

---

[30] (Daub, 1970a) p. 304.

When δQ/T appeared in 1854, Clausius pursued a similar strategy to understand its meaning and to also hopefully discover in the process a bridge from this term to Carnot's 2nd Principle. It took him until his Sixth Memoir (1862) to share his results. He again proposed that heat could be transformed into different parts according to [34.6] but then changed the way he combined terms by writing

$$\delta Q = d\left(\textit{vis viva}\right) + T\, dZ \qquad [34.7]$$

and thus introducing a new variable, Z, that he called *disgregation*. Clausius proposed that the product of disgregation and temperature quantifies the sum of both internal and external work. The delay in sharing this, which likely caused the delay in announcing entropy as the two were linked, was likely due to Clausius' perceived breach of normal conduct because, as he noted, until this point he had "wished to avoid everything that was hypothetical"[31] by relying on the fact that with a cyclical process, there was no need to consider changes in the working substance since it always returned to its original position and so experienced no net change. But in his quest to understand δQ/T, he needed to relax his caution. Desperate times called for desperate measures.

The replacement of U with Z in the 1st Law equation provided Clausius a better means to interpret δQ/T from a molecular basis. While U quantified a mixed bag of molecular effects involving both kinetic (*vis viva*) and potential (internal work) energies, Z isolated the two by pulling internal work out of U and combining it with external work. In this way, Clausius created a means to separate molecular motion from molecular "arrangement," as he called it, his thinking being that disgregation—a quantity "fully determined when the arrangement of its constituent particles is given"[32]—provided a means to quantify the work required to achieve a certain degree of separation between molecules. It is important to note that this was, in effect, a classical thermodynamic approach to accounting for arrangement using the concept of work (force times distance). Later use of equations of state involving force functions and then statistical mechanics involving a probability-based interpretation of entropy would provide more structure to this initiative, one not requiring a separate term like disgregation.

Clausius then divided both sides of this equation by temperature, T, in an attempt to understand the two molecular-based components of δQ/T, and arrived at

$$\delta Q / T = d\left(\textit{vis viva}\right) / T + dZ \qquad [34.8]$$

but try as he might, this path proved to be a dead-end, and by his Ninth and final Memoir, Clausius seemed to accept this as a problem that he simply couldn't solve. Yes, he did suggest that the above analysis based on his new concept of disgregation helped interpret the physical meaning of δQ/T and thus entropy, but I doubt it was to his own satisfaction as best reflected by the fact that when he summarized all of his work on the mechanical theory of heat in 1879, he spent no time discussing disgregation except for one passing reference in a rebuttal to Tait's objections to his work.[33]

[31] (Clausius, 1867) Sixth Memoir (1862), p. 216.
[32] (Clausius, 1867) Sixth Memoir (1862), p. 226.
[33] (Clausius, 1879) p. 360–362. Also, (Klein, 1969) pp. 139–140. "Clausius recognized that this result of his on the true specific heat was 'considerably at variance with the ideas hitherto generally entertained of the heat contained in bodies,' and that his new assumptions would hardly command universal assent. That was why he was so careful to pre-

Consider that when Clausius created U, he was able to reasonably grasp the physical meaning behind it since he based the creation on two comprising properties that he grasped quite well, the *vis viva* of the moving atoms and the work needed to separate them. In other words, he first learned the pieces and then created the whole. With entropy, however, it was the other way around. He discovered the whole and then tried to figure out the pieces. He first discovered S and then attempted to understand it by dissecting $\delta Q/T$, which ended up being a dead-end road. Boltzmann (Chapter 42) took a different road and arrived at the right answer.

The problem here, which Clausius couldn't have known at the time, was that attempting to use the physical insight gleaned from a 1st Law comprised of such concepts as *vis viva* and work to interpret the physical meaning of the 2nd Law was simply not possible as the latter represented an entirely different concept based on the laws of probability as embedded in statistical mechanics; the underlying theories involved hadn't yet arrived to science.

* * *

In 1865, Clausius introduced entropy in his final Ninth Memoir. He likely had been thinking about its existence way back in 1854 when he first wrote $\int \delta Q/T = 0$ but his need to first understand the physical meaning behind it before publishing held him back. Even then, he published without full understanding. He had taken his ideas regarding the mechanical theory of heat as far as he could and realized that it was time to close this chapter in his life. His Ninth Memoir offered him a means to do this.

Clausius wrote his Ninth Memoir to "render a service to physicists and mechanicians"[34] who wished to use thermodynamics in their own work. It was a great compilation of his work. He wrote out most all the differential equations that we use today in thermodynamics to quantify how state properties vary with respect to other state properties. His approach touched on a wide range of thermodynamic phenomena, such as free expansion, non-homogeneity, and phase change, and offered many ideas, thoughts, and leads for future work, as later picked up by Gibbs.

Realizing that his Ninth would be his final memoir, at least with regards to the mechanical theory of heat, Clausius must have felt the need for closure. He brought the physics along as far as he could, even if he was not fully satisfied with where he stood, which sounds rather similar to Carnot's situation in which he too published with uncertainty as manifested in his posthumous manuscripts. Clausius' attempt to seek a deeper understanding of $\delta Q/T$ didn't really pan out. Disgregation offered no answer. And yet he still felt ready to conclude, to summarize his fifteen years of work on this subject, to close this chapter in a way that did justice to what he achieved and perhaps also to point the way toward future research. It was in this context that he quietly introduced entropy. The introduction of this property in and of itself would have made a fitting end to his story. And if he had stopped there, his work would have remained powerful and groundbreaking. But he didn't stop there.

Instead, in a turn of historical significance and scientific confusion, he ended his Ninth Memoir with a final sentence that was a doozy, one that arguably was the single cause for the

---

vent this work from interfering with the acceptance of his better-founded results. Thus, when Clausius reworked his papers into a systematic exposition on thermodynamics for the second edition of his book in 1876, he carefully omitted any mention of disgregation, although he promised to treat it in a third volume, which he did not live to complete."

[34] (Clausius, 1867) Ninth Memoir (1865), p. 327.

main scientific and philosophical impact of this publication. In a sudden and unexpected flash for the reader, accompanied by a most minimal lead-in discussion, Clausius made a bold leap from Carnot to the cosmos.

From Carnot to Cosmos

Clausius began his Ninth Memoir by stating his two fundamental theorems of the Mechanical Theory of Heat, namely the equivalence of heat and work and the equivalence of transformations, the latter of which we'll get to shortly, but then ended his Ninth by stating that the "the fundamental laws of the universe which correspond to these fundamental theorems" are:

1. The energy of the universe is constant. (Die Energie der Welt ist constant.)
2. The entropy of the universe tends to a maximum. (Die Entropie der Welt strebt einem Maximum zu.)

We'll get to that rather large conceptual leap embedded in the second and final sentence in a moment. But first, let's look at these final two statements together since they reflected a change of major proportions. Clausius and his need for simplicity, generality, and symmetry likely sought a grand way to bring his work to a close. His shift from Carnot and the mechanical theory of heat to the fundamental laws of matter led him to effectively declare that the science of matter was based on two fundamental state properties, energy and entropy, that held much greater significance than the primitive properties on which they were based. Gibbs recognized the rightness of this navigational direction set by Clausius and based his own work on these properties as he famously made clear by his verbatim use of these two statements, in the original German, as a lead-in to his groundbreaking paper, *On the Equilibrium of Heterogeneous Substances* (1875–78). (One could argue that Clausius provided Gibbs his roadmap.)

It Wasn't Just Entropy that was Important, but the Declaration of an <u>Increasing</u> Entropy

But it was more than the declaration of energy and entropy that made this memoir so revolutionary. It was the declaration of an *increasing* entropy in the universe that garnered the most interest and later fame for Clausius. This must rank as one of the largest conceptual leaps in the history of science.

By his Ninth Memoir, Clausius felt he had reached the peak of a mountain that only he saw, although I doubt he felt he had truly made it. But his final two sentences were, to me, his attempt to declare the peak reached. The first statement had been uttered before. The second statement, on the other hand, was brand new to the field of thermodynamics, a whole new concept, one not only introducing a new property of matter but one also introducing a unique behavior of that property. This was what was so revolutionary about this specific memoir. It represented a whole new take on the 2nd Law, one quite different from Carnot's 2nd Principle, as noted by Kragh: "To equate the second law of thermodynamics with the law of entropy increase, as is often done in modern textbooks, is not only conceptually wrong but also historically misleading."[35] Clausius' one short sentence launched a whole new field by upgrading the 2nd Law from Carnot's 2nd Principle, and other versions provided by the North British contingent, to a quantifiable entropy that truly brought the concept of direction into thermodynamics. But where did this sentence come from?

By 1865, others had started to consider the universe as fair game in the analysis of whether or not the laws of thermodynamics remained valid. As Clausius started publishing his theories on this subject, other scientists, rightly so, started to test their validity since not all accepted Carnot's 2nd Principle. This was the scientific method at work. Propose a hypothesis, then test it. Does it survive? As a true professional, Clausius respected this process and responded to each challenge with a good, open spirit. For example, he devoted much of his Seventh Memoir to successfully fighting back an interesting challenge by Gustave-Adolphe Hirn, who proposed a process that could transfer heat from cold to hot with no other compensating process, in apparent contradiction to the 2nd Law.[36]

Then in his Eighth Memoir, he fought back against an even more interesting challenge, this time from Rankine,[37] which was originally directed at Thomson's theory of dissipation but which was easily redirected at Clausius' version of the 2nd Law. Considering the potential total reflection of radiation at the edge of the universe and its reconstitution at a focal point, Rankine proposed that in this scenario, heat could flow from cold to hot with no compensating action, again in apparent contradiction to the 2nd Law. Clausius showed in much detail and long analysis how this wouldn't be so. While such magnification of radiation might change the absolute magnitudes of the quantities of heat which two bodies radiate to each other, it cannot alter the ratio of these magnitudes, and it's this ratio that drives the temperature of the two bodies toward each other. More heat will radiate from hot to cold than from cold to hot, even in the presence of magnification, for if radiation can pass from one body to another, then it must also be able to pass back. There is no selective Maxwell's demon sitting on the path, allowing one and not the other. In this way did Clausius refute Rankine's challenge.

While such challenges provided motive for Clausius to extend his work beyond Carnot's engine, it was really Thomson's work that provided the defining motive. In his Eighth Memoir, Clausius summarized the key aspects of Thomson's work on the universal tendency for energy to dissipate such that different temperatures move toward and not away from each other and used this as his main argument to refute Rankine. Clausius' work with entropy would ultimately provide the means to explain this. In this context, with precedent having already been established by both Rankine and Thomson regarding the universe, it wasn't a total surprise that Clausius looked to the universe in response, especially regarding his succinct capturing of the 1st Law of Thermodynamics, this time in terms of his newly adopted concept of energy as opposed to work–heat equivalence, which should be noted had previously been so captured by, for example, Rankine.[38] In fact, it was likely these communications, and others involving, for

[35] (Kragh and Weininger, 1996) p. 92.
[36] (Clausius, 1867) Seventh Memoir (1863) p. 280.
[37] (Rankine et al., 1881) p. 200–202.
[38] (Rankine et al., 1881) p. 344. Rankine wrote (in his 1854 paper), "if this were not so, the cycle of operations would alter the amount of energy in the universe, which is impossible."

example, Reech as discussed by Daub,[39] that shifted Clausius' thinking from Carnot to the universe, from the cycle to the single state, and, most importantly, from the reversible to the irreversible.

So the extension of the 1st Law to the universe was not unreasonable; it had already been started by others. But entropy, and especially the concept of an increasing entropy? This was different, and I contend that Clausius really had no theoretical basis or supportive evidence for doing this. It was a statement based on invalid assumptions. But it was also a very bold and intuitive statement from an otherwise cautious scientist that was essentially, albeit not entirely, correct. And this brings us to (3).

(3)

The Ninth Memoir's Final Sentence

Carnot's ideal engine operates as an equilibrium system, perfectly reversible, with nothing more than an infinitesimal temperature gradient between the reservoirs and the working substance driving it. In this world, the equal sign rules. $Q_{in} = W + Q_{out}$, $\Sigma\, Q/T = 0$, $\int \delta Q/T = 0$, $dS = \delta Q/T$, and so on. This conceptual construct was both valid and groundbreaking, leading as it did to the discovery of the rules governing the capability of this engine as laid out in the 1st and 2nd Laws of Thermodynamics. While this engine generates the maximum amount of work, it does so only because the driving forces are infinitesimally small—no gap between temperatures of working substances and reservoirs; no difference between force pushing piston out and force from environment pushing piston in[40]—thus leaving the rate of work (power) near zero.

In the real world, however, power rules. In the real world, it isn't the amount of work but the amount of work *per time* that matters. The means to increase power is to increase gradients, which unfortunately comes with a cost since gradients decrease the overall efficiency of the engine. Temperature gradients required to rapidly move heat into and out of the engine decrease the temperature span of the working substance below the maximum offered by the hot and cold reservoirs. Pressure gradients required to overcome friction (poor lubrication) generate wasted heat in the engine machinery, which is ultimately dissipated into the environment. Unwanted temperature gradients between the engine machinery and the environment combined with insufficient insulation lead to additional waste heat dissipation to the environment. In the real world, such non-idealities decrease the efficiency of the engine below the maximum possible.

Many practicing engineers understood this all the way back to the early days of the steam engine. Similar "lost work" concepts existed for waterpower engines as well. "Irreversible" became the term used to describe processes that contained efficiency-wasting gradients. Such concepts had long been circulating in the literature. But prior to Clausius, no one had been able to figure out how to apply mathematics to such an irreversible world and especially how to approach quantifying the trade-off between power and efficiency.

---

[39] (Daub, 1978)

[40] The ideal Carnot heat engine sits in perfect equilibrium, infinitesimal work, infinitesimally slow, easily reversible.

Clausius' decision to tackle the mathematics of irreversible thermodynamics was yet another critical step in his journey. Clearly, there was a practical motivation to take this step since this issue lay at the core of economic optimization between engine efficiency and rate. But this step offered another benefit, to Clausius in particular, for it provided a means for him to unite the two different forms of the 2nd Law, specifically δQ/T and Carnot's 2nd Principle.

The Rise of Irreversibility in Thermodynamics

What happens inside an irreversible engine? How would you even begin to quantify this? Clausius approached this issue by first revisiting his own version of Carnot's 2nd Principle from the perspective of "transformations":

> Transformation #1: In all cases where a quantity of heat is converted into work,
> Transformation #2: ...another quantity of heat must necessarily be transferred from a warmer to a colder body.[41]

Seeking to link this statement with δQ/T, he then took his question, *What happens to the heat?*, shifted its focus from the body to the engine and proposed that heat could be *transformed* into work and also that heat could be *transformed* from heat at high temperature into heat at low temperature. The extension of logic from the earlier one-step processes involving a body of matter that revealed such concepts as internal energy (U) and disgregation (Z) to the multi-step processes involved in an engine brought with it a much higher level of abstraction that made his Fourth and subsequent memoirs very challenging to read. It was also not entirely valid since, as we'll discuss in more detail later in this chapter, the only transformations occurring in Carnot's engine are heat–work equivalence and the conduction of heat, which were aptly covered by the 1st Law (dU = δQ − δW) and the 2nd Law as regards the directional flow of heat from hot to cold, respectively; Transformation #1 and #2 above were consequences of these laws. Further challenging the validity of this logic was what the word "transform" actually meant to begin with. At times, Clausius referred to the fact that heat disappeared when it was transformed, as if it had existed as an entity beforehand. Again, at this point in time, he had not yet embraced energy as being at the center of this discussion. Energy could change between different forms, such as kinetic and potential. Heat, however, could not, as heat itself *was the change* and not a property.

Having said all this, such very creative and rather ingenious extension of logic indeed provided the path for Clausius to link δQ/T with Carnot's 2nd Principle. He did this by putting the above two transformations comprising Carnot's 2nd Principle into quantities that could be combined, employing δQ/T as the basis of quantification, understanding fully well that this term had separately been shown to provide the theoretical foundation for the development of work, heat, and temperature relationships. To separate the cycle into the two transformations, he creatively converted Carnot's typical cycle from four steps into six (Figure 34.1). Instead of the original two sequential expansion steps, isothermal (Step 1) followed by adiabatic (Step 2), he employed four expansion steps, isothermal, adiabatic, isothermal, adiabatic, and then set

---

[41] (Clausius, 1867) Fourth Memoir (1854), p. 116.

CLAUSIUS'S FOURTH MEMOIR (1854)

seeking to better understand the two possible transformations of heat:

TRANSFORMATION #1: heat to work

TRANSFORMATION #2: heat from high temperature to low

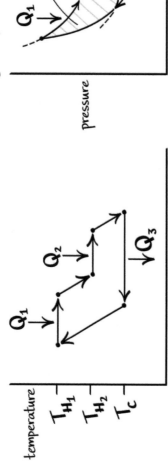

strategically designed process such that: $Q_2 = {}^-Q_3$

T#1: transformation of heat to work occurs for Q_1 as quantified by: $\dfrac{Q_1}{T_{H_1}}$

T#2: transformation of heat from high temperature to low occurs for Q_2 as quantified by: $Q_2\left(\dfrac{1}{T_{H_2}} - \dfrac{1}{T_c}\right)$

$$\begin{array}{c} \textbf{TOTAL} \\ \textbf{TRANSFORMATION} \\ \textbf{OF HEAT} \end{array} = T\#1 + T\#2 = \frac{Q_1}{T_{H_1}} + Q_2\left(\frac{1}{T_{H_2}} - \frac{1}{T_c}\right)$$

Figure 34.1 Clausius' attempt to understand the transformation of heat

the operating conditions such that the heat entering the second isothermal step from a second hot reservoir (operating at lower temperature than the first) equaled the heat rejected to the cold reservoir during the isothermal compression step, which was kept the same as in the original Carnot engine. He then used the $\delta Q/T$ transformation term to bring quantification to all three transformations, ignoring the adiabatic steps as they did not involve heat.

Heat transformations in Clausius' six-step Carnot cycle

- involving three heat reservoirs: two hot (T_1, T_2), one cold (T_3)

Isothermal Step One (expansion): $\delta Q_1 / T_1$

Isothermal Step Two (expansion): $\delta Q_2 / T_2$

Isothermal Step Three (compression): $\delta Q_3 / T_3 = -\delta Q_2 / T_3$

By designing this process such that the absolute values of heat involved in the second and third isothermal steps were equal, Clausius was able to isolate the second half of Carnot's 2nd Principle (Transformation #2) and thereby propose the following transformation values:

Transformation #1 Heat to Work $\delta Q_1 / T_1$

Transformation #2 Heat from high temperature to low $\delta Q_2 \left(1/T_2 - 1/T_3 \right)$

$$T\#1 + T\#2 = \delta Q_1 / T_1 + \delta Q_2 \left(1/T_2 - 1/T_3 \right) = 0$$

Recall that in a reversible, closed-cycle engine, the sum of all $\delta Q/T$ values must equal zero. Clausius took this thinking to a higher level by grouping the $\delta Q/T$ values into these two categories of transformation to align with Carnot's 2nd Principle: heat to work and heat from high temperature to low. This then became the basis for Clausius' theory of the equivalence of transformations and the quantifying terms became *equivalence values*.

Clausius then went to an even higher level of abstraction—requiring multiple readings to follow. Writing in an after-the-fact note to his Sixth Memoir (1862), Clausius proposed,

> "In nature…there is a general tendency to transformations of a definite direction. I have taken this tendency as the normal one, and called the transformations which occur in accordance with this tendency, positive, and those which occur in opposition to this tendency, negative."[42]

To Clausius, nature moves in a certain direction and this direction is positive and spontaneous; thus, the mathematical term quantifying this transformation must also be positive. A positive transformation, such as conduction from hot to cold, can occur on its own. A negative transformation, on the other hand, such as the transfer of heat from cold to hot, cannot occur on its own, for this would contradict Carnot's 2nd Principle. It must occur together with, or in his terms, be compensated by, a positive transformation of equal value but opposite sign. Conduction, on the other hand, can happen without compensation.[43] Note that this entire logical construct was a hypothesis, especially Clausius' choice of signs, which was an entirely arbitrary choice, likely based on the emotional feeling that what happens in nature must be positive.

---

[42] (Clausius, 1867) Sixth Memoir (1862), p. 248 (footnote added 1864).
[43] (Clausius, 1867) Sixth Memoir (1862), p. 248 (footnote added 1864).

The importance of following this logic, which admittedly is very difficult to do, is not to embrace the theory but to understand Clausius' thinking. The theory itself didn't live beyond Clausius. His approach brought unnecessary complication to the development of entropy. We don't study the *equivalence of transformations* today. Neither do we study other concepts that Clausius created, such as ergal[44] and disgregation. I certainly don't pass judgment in writing such statements. I recognize and applaud Clausius' need to attack the 2nd Law from different directions, continually probing until he achieved his breakthrough. But it is only by reading such publications, however difficult that may be, that we are enabled to understand the origin of his final sentence.

Let's look where this logic unintentionally led him. Look carefully at the quantity $\delta Q \, (1/T_H - 1/T_C)$ he created to quantify the transformation of heat from high temperature to low in Carnot's engine. Does it look familiar? Indeed it does. By following his creative path for the Carnot engine, Clausius unknowingly and fortuitously stumbled upon a critical equation in the world of thermodynamics that describes the total change in entropy when two bodies exchange (irreversibly) thermal energy. Note that in this term, δQ is negative (heat leaving the hot reservoir) and $(1/T_H - 1/T_C)$ is also negative ($T_H > T_C$), thus making the product of the two positive. That this term must be positive as a law of nature based on the theory of entropy, Clausius didn't realize at the time. We'll come back to this discussion shortly (Figure 32.2).

With these two terms in hand, Clausius knew he next needed to assign signs to enable addition. There were four transformations in total, the two spelled out in Carnot's 2nd Principle and their reverse. But which signs for which transformations? Which were the naturally occurring, positive transformations? This issue would consume him for years.

The best way to understand the logical path Clausius traveled is through the following table that I prepared based on his conceptual arguments. For each of the two transformations, which direction is positive and which is negative? Is heat to work positive or negative? Also, for each of the two, what's the equivalent transformation? For example, if heat to work is positive, what's the corresponding negative transformation involving heat transfer? As stated earlier, he initially felt that the sign for each transformation could be made arbitrary, so long as the reverse was the negative of the forward. He didn't see a scientific basis to choose positive or negative; it was more philosophical. Only later would he validate his choice by observing that heat conduction indeed yields a positive value using his theory.

Transformation	Heat Transformation	Equivalence Value	Positive Transformation	Negative Transformation
#1	Heat ⇌ work	$\delta Q/T$	?	?
#2	Heat transfer from warm to cold	$\delta Q \, (1/T_H - 1/T_C)$?		?

As Clausius considered this situation, he realized that the sign on the value of δQ was really what he had to select, for in so doing, the sign on the transformation itself would become fixed. Temperature wasn't an issue as it was always positive.

---

[44] (Clausius 2003c)

The problem with δQ was that Clausius had no basis to select a sign convention (or at least didn't realize he had one in the 1st Law that he himself wrote). This was a challenge for him throughout his memoirs and one he didn't fully conquer until the wrap-up publication in 1865. As dictated by the conservation of energy and the 1st Law of Thermodynamics, when thermal energy (Q) is exchanged between two bodies, 1) the two changes in energy are identical in number and opposite in sign, and 2) the signs themselves are not arbitrary. When heat enters a body, the value of Q *for that body* must be positive since its dU is positive. Likewise the value of Q for the other body must be negative since its dU is negative. Recall per the 1st Law (as created by Clausius), dU = δQ (absent work).

This is all easy enough for us now, but in the mid-1800s such wasn't the case. Folks were still coming out from behind the influence of the caloric theory and the new replacement concept of energy was still rather abstract for many, including, at least to some extent I suspect, Clausius, as was discussed previously.

Clausius' challenge with the sign on Q was best reflected by the fact that he started with one convention in his Fourth Memoir and then changed to the correct convention in his Ninth, writing, "[In my Fourth Memoir] a thermal element given up by a changing body to a reservoir of heat is reckoned positive, an element withdrawn from a reservoir is reckoned negative...In the present memoir, however, a quantity of heat absorbed by a changing body is positive, and a quantity given off by it is negative."[45] In reading this one senses that he viewed the reservoir and the working substance differently and the heat that passed between them as an entity that existed on its own, with its own identity and sign, depending on which direction it was moving in.

With no sound rationale available for selecting the sign convention for Q, Clausius turned to the transformation values. Selecting either would determine the other.

As brought up earlier, Clausius believed that nature allowed certain transformations and built up his logic accordingly.[46] For example, nature favored the flow of heat from hot to cold. This was common knowledge. He felt that such natural tendencies were positive and should have positive signs attached to them, with the reverse transformation process, e.g., heat flowing from cold to hot, being negative with negative signs to suggest that nature simply didn't behave that way. This decision to place positive signs on nature's tendencies was initially arbitrary. But still, a decision had to be made. So how did Clausius make it?

The fact was that none of the four transformations he was considering in his conceptual table had any natural "tendency" and the only heat-related phenomenon that did, namely conduction, wasn't a part of the table. But this didn't stop Clausius. He identified a rather creative way to use conduction as a way to make his decision on signs. Unfortunately, this decision involved an unjustified conflation of two separate phenomena, the conduction of heat from hot to cold, and the transformation of heat from high temperature to low.

And So We Return Again to Conduction

Conduction was the only natural "tendency" that Clausius knew to be true. "Heat everywhere manifests a tendency to equalize differences of temperature, and therefore to pass...from

[45] (Clausius, 1867) Ninth Memoir (1865), p. 329 (footnote).
[46] (Clausius, 1867) Fourth Memoir (1854), p. 123.

warmer to colder bodies. Without further explanation, therefore, the truth of the principle will be granted."[47] But direct conduction wasn't involved in Carnot's 2nd Principle as written by Clausius because for the reversible process, by definition, temperature gradients don't exist. For whatever reason, he analyzed conduction as if it were a separate process, uninvolved in the operation of the engine, when in fact it was one of the central albeit hidden features constraining the engine's work generation. Conduction, and specifically the fact that heat could only flow from hot to cold was really at the core of Carnot's 2nd Principle and the later 2nd Law of Thermodynamics. Clausius was unable to isolate this argument as such. In fact, I'm not sure that he saw the relevance of the conduction between the reservoirs and the working substance, noting in a rather telling statement that the finite gradients existing between them were not accounted for in his mathematical approach to irreversibility.

Clausius then proposed that the flow of heat from hot to cold during conduction was the same phenomenon as the transformation of heat from high temperature to low in Carnot's engine. He made a guess, perhaps based on an intuitive feel, that since the conduction of heat from hot to cold happens naturally, it must be a positive transformation, and since the transformation of heat from high temperature to low seems quite similar, at least in words, then it too must happen naturally and so must be positive. And this was likely how Clausius arrived at the logic he used to complete the table; no other logic was presented. And through this logic, the logic for putting a sign on Q was done and he shared these results in his Ninth Memoir as mentioned above.

So why wasn't this valid? Well, let's look at conduction. This phenomenon occurs between two bodies when they are put into direct contact. In this situation, heat can *only* flow from hot to cold. Now let's look at Carnot's engine. Here, heat flows in during one step and out during another. Two disconnected steps are involved, which Clausius rightly recognized as "connected transmissions,"[48] for which direct conduction plays no role, at least from the standpoint of Carnot's 2nd Principle. Whether or not heat flows into the engine from the hot or the cold reservoir depends on whether the engine is running forward (generating work) or backward (generating cooling). Since the engine is reversible, it's easy to switch back and forth between these two, hot to cold, cold to hot. This is the origin of Clausius' modified version of Carnot's 2nd Principle, "Heat can never pass from a colder to a warmer body without some other change, connected therewith, occurring at the same time." The "other change", in this context, referred to work being done on the system to drive the refrigeration engine. In conduction involving direct contact between two bodies, however, heat can never flow from cold to hot. Period.

Completing Clausius' Equivalent Transformation Table

With this background, here's the summary of how Clausius set about filling out his table. He knew the natural conduction of heat from hot to cold to be true. So this must be positive.

---

[47] (Clausius, 1867) Fourth Memoir (1854), p. 118.
[48] (Clausius, 1867) Fourth Memoir (1854), note p. 118.

Transformation	Heat Transformation	Equivalence Value	Positive Transformation	Negative Transformation
#1	Work $\rightleftharpoons$ Heat	$\delta Q/T$	?	?
#2	Heat transfer from one temperature to another	$\delta Q\,(1/T_H - 1/T_C)$	Direct *conduction* of heat from high to low temperature is natural and thus positive	?

He then proposed that direct conduction and the transformation of heat from high temperature to low could be treated as the same type of transformation.

Transformation	Heat Transformation	Equivalence Value	Positive Transformation	Negative Transformation
#1	Work $\rightleftharpoons$ Heat	$\delta Q/T$	?	?
#2	Heat transfer from one temperature to another	$\delta Q\,(1/T_H - 1/T_C)$	Heat transfer from a warmer to a colder body is the same "type" as direct conduction	?

With this part of the box filled out, he then revisited Carnot's 2nd Principle: Heat transforms into work, and heat is transferred from warmer to a colder body. He already solved the latter half of this principle, and could thus now solve the first. Because the heat transfer from warm to cold body is positive, the conversion of heat into work must be negative because the sum of all "equivalent transformation" values involved must equal zero.

Transformation	Heat Transformation	Equivalence Value	Positive Transformation	Negative Transformation
#1	Work $\rightleftharpoons$ Heat	$\delta Q/T$	?	Heat to work
#2	Heat transfer from one temperature to another	$\delta Q\,(1/T_H - 1/T_C)$	Heat transfer from a warmer to a colder body	?

Once these two boxes were filled in, the other two boxes were automatically determined and so completed the table.

Transformation	Heat Transformation	Equivalence Value	Positive Transformation	Negative Transformation
#1	Work $\rightleftharpoons$ Heat	$\delta Q/T$	Work to heat	Heat to work
#2	Heat transfer from one temperature to another	$\delta Q\,(1/T_H - 1/T_C)$	Heat transfer from a warmer to a colder body	Heat transfer from a colder to a warmer body

Here's how Clausius summed up his proposal, as written in his Fourth Memoir: "we shall consider the conversion of work into heat and, therefore, the passage of heat from a higher to a lower temperature as positive transformations."[49] In an after-the-fact footnote to this statement, written in 1864, Clausius said that the reason for this choice of signs would become apparent in his later memoirs. I believe that what happened was that Clausius' initial arbitrary choice, perhaps guided by intuition, was only later able to be justified, at least to him, once he better understood conduction, although again, I don't believe that this justification was sound.

It was with the completion of this table—which Clausius didn't write out but which I created based on his statements—that Clausius brought quantification to Thomson's concept of dissipation. Certain phenomena happen naturally, others don't. This was a tremendous step forward, but the logic used was not sound. Yes, heat must flow from hot to cold during conduction; this is nature's spontaneous tendency as captured in the theory of entropy. But to use this fact as the basis to then say that the net transfer of heat from a hot to a cold body in a multi-step heat conversion process is also positive and also happens spontaneously doesn't follow. And neither does Clausius' suggestion in his Eighth Memoir[50] that work-to-heat happens spontaneously in nature while heat-to-work doesn't. I never really understood such statements when I've heard them spoken by others or read them in books and now realize their possible origin.

With the completion of this table, we can now finally look at Clausius' final logic regarding the irreversible process. Building on the above, Clausius proposed in his Fourth Memoir that an irreversible process comprised of multiple transformations, both positive and negative, must ultimately show, once all equivalence values are added together, a net positive transformation.[51] He reasoned that since the negative transformations can only occur alongside their equivalent and cancelling positive transformations, so as to not contradict Carnot's 2nd Principle by allowing heat at low temperature to be transformed into heat at high temperature with no compensating process, only uncompensated and thus positive transformations could remain in the math. The irreversible process therefore must contain an excess of positive transformations; the sum of the equivalence values therefore must be greater than zero. The irreversible process occurs as a result of nature's tendency and must be positive. Got it? Don't worry if you don't. As I said before, reading this is not easy.

Applying his Concepts to Analysis of an Irreversible Heat Engine

One source of confusion caused by Clausius was when he used his theory of equivalence of transformations to analyze the irreversible, closed-cycle engine and to propose that $\int \delta Q/T < 0$ for such an engine. Starting back in 1854, Clausius became very focused on the equation $\int \delta Q/T = 0$ for the reversible engine. As originally developed, the δQ and T terms applied to the heat flow for the reservoirs only, not the working substance.[52] But because the infinitesimal

[49] (Clausius, 1867) Fourth Memoir (1854), p. 123.
[50] (Clausius, 1867) Eighth Memoir (1863), p. 290.
[51] (Clausius, 1867) Fourth Memoir (1854), p. 133.
[52] Remember that the working substance was irrelevant in a Carnot heat engine because of the way the problem was set up. The working substance must operate between the two temperatures. The fact that the temperature range

engine was reversible, the $\delta Q/T$ values for the reservoirs were identical to those for the working substance (and opposite in sign), thus meaning that the equation itself was also valid for the working substance. The signs on Q didn't matter because zero is zero. This also meant that his not considering both the reservoirs and the working substance together couldn't be seen.

Hot reservoir:	$-Q_H / T_H$
Working Substance:	Q_H / T_H
Working Substance:	$-Q_C / T_C$
Cold reservoir:	Q_C / T_C
$\int \delta Q/T_{reservoirs} =$	$-Q_H / T_H + Q_C / T_C = 0$
$\int \delta Q / T_{working\ substance} =$	$Q_H / T_H - Q_C / T_C = 0$

But when Clausius started to consider irreversible engines, the above no longer applied. Based on his theory of equivalence of transformations, the sums of the transformation values could no longer equal zero due to the imbalance between the positive and negative transformations caused by the irreversibility. I don't want to go through his entire logic here because of its high degree of abstraction, but do want share that he originally proposed that $\int \delta Q/T \geq 0$ for the working substance in the irreversible case in his Fourth Memoir and then changed this to $\int \delta Q/T \leq 0$ in his Ninth Memoir.[53] While he acknowledged that this difference was caused by his change in approach to affixing a sign on Q, which made sense in its way, the overall logic guiding this discussion was not evident. After multiple readings, I still could not see the single, individual steps of logic he took to arrive at such inequalities.

Now, this is not to say that such inequalities don't exist. But it's critical to carefully explain exactly what you're applying the mathematics to, which Clausius did not do. For example, continuing with the above, one could arrive at the conclusion that $\int \delta Q/T \leq 0$ for the working substance in an irreversible engine by considering the heat flowing into and out of the working substance in a two-reservoir, closed-cycle Carnot engine for which the following applies:

$$Q_H / T_H + Q_C / T_C = \text{Total entropy change of the working substance}$$

For a reversible engine, these two terms cancel each other (remember, $Q_C < 0$ for the working substance); there is no change in entropy. But for an irreversible engine, things change. Q_H, T_H and T_C all stay the same, assuming that the temperatures between reservoirs and working substance remain infinitesimal, but Q_c *must* increase. The reason is conservation of energy. As the engine goes from reversible to irreversible, the amount of work it generates decreases. Since $Q_H = Q_C + W$ by the 1st Law, then decreasing W must result in an increasing Q_C, thus

---

was assumed at the outset made the nature of the working substance irrelevant. Whether or not the working substance could truly operate between the two temperatures in the real world, without, for example, decomposing or damaging the mechanical equipment, was not part of the discussion.

[53] (Clausius, 1867) Ninth Memoir (1865), p. 329.

rendering the above sum of $\delta Q/T$ terms negative. (In the extreme, as W goes to zero, Q_C goes to Q_H—pure irreversible conduction.) The critical thing to recognize here is that Q_C represents *all* heat flow to the environment, not just that passing from the working substance into the cold reservoir during Step 3 – isothermal compression. It must include all of the lost heat from such irreversible steps as conduction, friction, or accelerating pistons, etc., all of which ends up as heat somewhere in the engine, which ultimately ends up being dissipated into the environment. This total lost heat represents the lost opportunity to convert Q_H into work, or as the thermodynamicists would call it, "lost work."

The confusion created by Clausius' signs on Q, his equivalence values, and his various $\int \delta Q/T$ analyses led to confusion in the reading public, especially when attempting to interpret his new concept of entropy based on these concepts. Many after-the-fact clarification notes were included throughout his compilation of all of his Memoirs,[54] likely reflecting his need to address this confusion and to update his own evolved thinking. Included in this reading public were some notable scientists, such as Tait, who misinterpreted Clausius, perhaps intentionally so to suit his own purposes as discussed by Smith,[55] by stating in his own textbook, "The entropy of the universe tends continually to zero." Unfortunately, Maxwell, trusting Tait, put this into the first edition of his book, *The Theory of Heat*. It was Gibbs who had correctly read Clausius and so in a footnote in his second paper[56] corrected Maxwell and Tait. Maxwell later revised his book based on Gibbs' work.

I certainly don't want to minimize the groundbreaking work that Clausius did in addressing the irreversible case. He intentionally confronted this complex concept out of his desire for completeness. No one else involved at that time was exploring irreversibility as deeply as Clausius. Thomson, for example, despite the fact that he was the one who introduced the concept of energy dissipation into thermodynamics, never applied this concept to $\int \delta Q/T$. While the conclusions drawn by Clausius around this work may not have been perfect, the conversation he started around this subject of irreversibility strongly influenced those who followed and led to a better understanding of entropy itself.

With this background, we now turn to the final step for Clausius when he proposed an increasing entropy in the universe.

The Entropy of the Universe Increases

> *[Clausius] alone of the founders of thermodynamics fully explored and developed the irreversible case.*
>
> – E. E. Daub[57]

It was Clausius who brought inequalities into the mathematics of thermodynamic irreversibility and so brought quantifiable direction into thermodynamics. Neither Thomson nor Rankine

---

[54] (Clausius, 1867)
[55] (Smith, 1998) p. 257.
[56] (Gibbs, 1993) p. 51–52.
[57] (Daub, 1978) p. 346.

did this to an appreciable extent, although Clausius was likely influenced by their work.[58] It was Clausius alone who recognized that any comprehensive study of thermodynamics had to include the presence of gradients (and their inherent inequalities) to accurately account for the irreversibility involved in such processes as, for example, conduction, free expansion, and friction.

His entire sign construct to handle irreversibility was built, rightly or wrongly, the issue not being relevant as we're solely interested in following Clausius' thought process, on top of the conduction of heat. Based on intuition, Clausius initially proposed that the (disconnected) flow of heat from the hot to the cold reservoir in Carnot's engine was favored by nature and so had a positive sign. He then used this to anchor his sign convention for the different types of heat transformations, which in turn anchored his sign convention of Q. Later, he validated in his own mind this intuition by demonstrating with his *equivalence of transformations* mathematics that a similar phenomenon, direct conduction, was indeed positive. But this validation process itself was not valid.

Conduction's role in the early days of thermodynamics was critical. While Clausius didn't explicitly say so, I believe that the logic he used to guide him to his famed closing sentence hinged on conduction. Numerous irreversible processes were available for him to work with, such as free expansion, friction, and chemical reaction, but the process that was most central to the heart of thermodynamics and the discovery of increasing entropy was conduction.

Unfortunately, conduction was a phenomenon that didn't lend itself to easy comprehension. Both Thomson and Clausius wrestled with it. Students continue to wrestle with it today. How can heat flow from hot to cold without losing energy? The heat that goes in one end of a metal rod must come out the other, right? Conservation of energy tells us this. But how can the temperature of the heat decrease as it moves down the rod while energy remains the same? Such questions come from a misunderstanding about the concepts of energy, heat, and temperature. Such questions, I believe, confounded both Thomson and Clausius, among many others.

As I shared before, when conducting heat from hot to cold, there's no flow of anything. The only flow is conceptual in nature. What happens is that energy decreases in the former and increases in the latter with the corresponding changes in entropy quantified as Q_H/T_H and Q_C/T_C. In these terms, $Q_H < 0$, $Q_C = - Q_H$, and $T_H > T_C$. Thus, $dS_H + dS_C = dS_{tot} > 0$. The theory of entropy states that this must be so. This would have been an unusual conclusion to have made back then. Consider from an energy standpoint for the same scenario that $dU_H + dU_C = dU_{tot} = 0$. While energy was still new and abstract, this would have made more sense in the spirit of the conservation philosophy than the equivalent entropy balance that led to an inequality. And perhaps it was such a challenge that led to Clausius (somewhat surprisingly) not considering analyzing conduction in the context of an overall change in entropy.

Clausius demonstrated this math in his Sixth Memoir (1862)[59] when considering heat equilibration inside a given (non-isothermal) body. He showed that the change in internal energy is negative for the hot and positive for the cold, and that because T_H must, by definition, be greater than T_C, that the sum of the Q/T terms for this phenomena must be positive. But it was rather telling in this example that Clausius intentionally focused on internal

[58] (Daub, 1978)
[59] (Clausius, 1867) Sixth Memoir (1862), p. 246.

equilibration—involving *the vis viva* component of internal energy—as a means to avoid having to deal with heat transfer (δQ) between the body and a reservoir of heat, suggesting he didn't understand the true relation between energy and heat. Only later in his Seventh Memoir[60] did he start discussing heat transfer between bodies, which is really what conduction and the need for δQ is all about. In this memoir, Clausius stated, with regards to the spontaneous flow of heat between bodies of varying temperatures, that the sum of the transformation values (Q/T) for the bodies increases or in the limit remains unchanged; the sum never decreases.[61] While he didn't realize this, Clausius was describing how the theory of entropy accounts for the changes caused by the conduction of heat.

In yet another telling moment, for an unknown reason, after Clausius announced entropy, he never used it to interpret his earlier work, such as conduction. Also, his focus on irreversibility remained primarily based on $\delta Q/T$ rather than dS; perhaps in his mind the link between the two for reversible processes was no longer valid for irreversible processes—I don't know. He never stated the sum of the entropies for two bodies increases ($dS_{total} > 0$) due to conduction. He didn't seem to understand entropy as a two-body phenomenon. He came close but never got all the way there, perhaps reflecting the fact that he himself was not clear with all that was involved in this issue. The pieces were there. He just didn't understand how to put them together. This was left to others such as Maxwell[62] and Gibbs to explain.

Building further on this discussion, Clausius didn't seem to fully grasp the distinctions between the reservoirs and the working substance. He treated the former as if they had no properties relevant to the situation other than temperature.[63] He continually focused on either the multiple reservoirs or the changing body, but never both together and never the work lost due to the temperature difference between them. He understood this last point, but acknowledged that his mathematics did not account for the gradient.[64] It's as if the reservoir was something other than a body of matter itself.

* * *

The convergence of both Clausius' correct assumption—and this was an assumption—for the sign on Q and his related proof that conduction is indeed a positive transformation gave him confidence that his chosen path was valid: nature spontaneously moves in a positive direction. He didn't realize that he had stumbled upon the essence of entropy, as the entire conduction argument was based on this essence as captured by his transformation term, $Q_H (1/T_H - 1/T_C)$, which was set up by his analysis of Carnot's 2nd Principle, which we now know to represent $dS_H + dS_C = dS_{tot} > 0$.

So where did this leave things?

In his Eighth Memoir, when Clausius sought to refute Rankine's challenge of the 2nd Law, he turned his focus toward the transformations occurring in the universe—the universe is filled with many transformations—using as a key reference Thomson's energy dissipation theory—heat flows in such a way to eliminate temperature differences. Clausius had previously reasoned in his Seventh Memoir that the sum of the transformation values involved in such

[60] (Clausius, 1867) Seventh Memoir (1863), p. 289.
[61] (Clausius, 1867) Seventh Memoir (1863), p. 289.
[62] (Maxwell and Pesic, 2001) Based on Maxwell's Ninth Edition (1888), p. 162–165.
[63] (Clausius, 1867) Sixth Memoir (1862), p. 217.
[64] (Clausius, 1867) Fifth Memoir (1856), p. 164.

dissipation is greater than or equal to zero based on his analysis of conduction. He further reasoned in his Ninth Memoir that since entropy quantified the total transformational content of a given body, and since the universe could be treated as one such body, however non-homogeneous it was, then the entropy of the universe must increase to a limiting value as it moved from non-homogeneous to homogeneous. Hence the final, intuitive, and rather bold statement that concluded his Ninth and final Memoir on mechanical theory of heat, a statement that was mostly true on the face of it, but a statement nonetheless that was built on arbitrary assumptions.

Consider Clausius' leading paragraph to his famed final statement and note my bold italics:

> The second fundamental theorem, in the form which I have given to it, asserts that all transformations occurring in nature may take place in a certain direction, ***which I have assumed as positive*** ... The application of this theorem to the Universe leads to a conclusion to which W. Thomson first drew attention, and of which I have spoken in the 8th Memoir ... The entire condition of the universe must always continue to change in [a positive] direction.[65]

Clausius' concept of direction was based on his assumptions that nature moves in a positive direction. He validated this assumption with conduction, but had to assume in his Ninth Memoir[66] a sign convention for Q to make this so. His understanding that, in fact, the sign on Q is not arbitrary but is in fact founded on the 1st Law, may have become stronger by his Ninth Memoir. Again, I don't know. But it is interesting that it was in this memoir that he finally wrote the 1st Law of Thermodynamics with energy in front: $dU = \delta Q - \delta W$.[67] This too was telling. The Ninth Memoir became a means for Clausius to reconcile energy and heat and so conclude his efforts.

But What Happens to the Entropy of an Isolated Body?

Clausius' hypothesis that the change in entropy of the universe is positive held no substantive backing. It was an intuition of his that happened to be correct, but in a limited way, based largely on assumptions involving the conduction of heat. We don't really know what's happening in the universe, but we do know that when two bodies meet and exchange energy due to the presence of gradients, the entropy of the equilibrated combined system is greater than the sum of the entropies of the two initial bodies. It was this thinking that Gibbs and others picked up on and further developed, extending this beyond thermal gradients to include all forms of energy gradients.

It's interesting to me that Clausius himself did not really grasp the application of entropy to a given isolated body. To him, the universe was comprised of a series of ongoing transformations, the net of which, quantified by a change in entropy, was positive. But the relevance of this logic to a closed system comprised of gradients didn't seem to attract his attention. In fact, once stated, it didn't appear that Clausius saw any further use for entropy. But fortunately, Gibbs did, to great effect. Gibbs realized that one could consider a given non-equilibrated system to be

---

[65] (Clausius, 1867) Ninth Memoir (1865), p. 364.
[66] (Clausius, 1867) Ninth Memoir (1865), p. 329.
[67] (Clausius, 1867) Ninth Memoir (1865), p. 354.

comprised of component parts, each equilibrated internally and thus characterized by its own state properties such as U and S but none equilibrated with the others. He also realized based on Clausius' work and also Thomson's that that transformations and dissipations would eventually cause the gradients—mechanical, thermal, chemical—to disappear or (per Thomson) "dissipate" and that the final fully equilibrated system would have an entropy greater than or equal to the sum of the components' initial entropies. Gibbs deeply understood the implications of what Clausius discovered much more so than Clausius himself did.

Am I being too harsh here by saying that Clausius ended his Ninth Memoir based on combined assumption and intuition as opposed to a sound, structured logic? Perhaps. Again, I certainly don't mean to diminish the masterpiece of work that it was, as it clearly provided Gibbs the path forward on his own monumental work. But evidence in support of this is provided by Clausius himself. When he summarized all of his work on thermodynamics into his comprehensive *The Mechanical Theory of Heat (1879)*,[68] he included no mention of a natural tendency for entropy to increase.

* * *

In a world embracing conservation, entropy must have come as a shock. Mass is conserved. Energy is conserved. But entropy? It's not conserved. The entropy of the equilibrated sum is greater than or equal to the combined entropies of the initial parts. What an unusual property! This is what gave Gibbs a powerful direction toward equilibrium.

* * *

> Clausius – Epilogue

[Clausius] knew exactly how to interpret and rebuild Carnot's message, and then to express his own conclusions so they could be used by another genius, Gibbs. The grandest theories make their own vital contributions and then inspire the creation of other great theories. Clausius' achievement was of this very rare kind.

– William Cropper[69]

Clausius may have the dubious distinction of being the most forgotten major nineteenth century scientist.

– William Cropper[70]

Clausius did not invent entropy. Instead, entropy along with its unique characteristics was sitting there in nature, waiting to be discovered. Across fifteen persistent years, Clausius eventually came upon entropy, took note and brought it to the attention of others. In so doing, he became one of the first theoretical physicists, filling a critical need in those early days of thermodynamics as captured by Klein, who wrote, "What was lacking was not experiment but analysis."[71] It was analysis that Clausius supplied.

[68] (Clausius, 1879) Clausius wrote this in 1875, Browne translated in 1879.
[69] (Cropper, 1986) p. 1073.
[70] (Cropper, 1986) p. 1073.
[71] (Klein, 1969) p. 131.

The fact that Clausius' Ninth Memoir would eventually launch a revolution in the science of chemistry was not immediately apparent. Entropy arrived to little fanfare. Tait made a mistake, intentional or not, in interpreting it. Thomson never used it and only rarely referred to it, preferring instead to speak in terms of his dissipation theory.[72] Even Clausius himself paid little attention to it, likely due to his own uncertainty about what it meant. Moreover, he took little interest in the later developments of the thermodynamics he helped create as manifested, for example, by the fact that he apparently never reciprocated to correspondences sent to him by Gibbs whose work was so strongly dependent on his own.[73] I would love to know the reasons for these actions, or lack thereof. I would love to know what Clausius was really thinking about regarding his discovery. If only he had kept a notebook of his thoughts as Carnot had done.

Having said this, Clausius' discovery did eventually take hold. Tait, Thomson, and Maxwell all noticed it, although they did not initially do justice to it, which became the origin of the conflict discussed earlier. But once Gibbs, and later others, such as Helmholtz, entered the picture, clarity soon arrived. It's as if Gibbs rescued this burning ember started by Clausius, blew on it, added wood, and started a fire. The ember was that close to burning out.

Gibbs relied on the works of both Clausius and Thomson to build his theories. Each offered a different and valuable view of energy. To Thomson, nature dissipates energy such that gradients disappear over time. To Clausius, the disappearance of these gradients results in a quantifiable increase in total entropy. To both, energy remains constant throughout. Gibbs united both views into his own theories. (One could argue that Thomson introduced direction into thermodynamics while Clausius quantified that direction.)

Gibbs wouldn't be alone in finding a powerful use for entropy. Others followed. Recall that one of the most powerful results of Clausius' discovery was that it united the 1st and 2nd Laws of Thermodynamics into a single equation of state: $dU = TdS - PdV$. This result unleashed the intellectual might of giants such as Gibbs, Maxwell, and Helmholtz toward the development of a solid theoretical foundation for physical chemistry. While entropy was discovered in the world of physics, it eventually found itself more at home in the world of chemistry.

Before we move on to Gibbs and the rise of thermodynamic functions, I wanted to step back and pay final homage to Clausius. He may not have been on solid ground when announcing his theory of an increasing entropy and he may not have known what this theory meant or how to use it. Maybe like Carnot, he arrived at a conclusion that he wasn't totally satisfied with. Who really knows? Regardless, his relentless pursuit to understand Carnot's 2nd Principle and to resolve it to a "simpler, general and more mathematical form" is what led him to his discovery. His solitary journey lasted fifteen years. His results caught Gibbs' attention. Gibbs took the baton and took off. Clausius was done.

But Clausius wasn't done with science. Despite his lack of follow-up to his Ninth Memoir, Clausius' career in science continued, especially in two notable areas, the first involving the kinetic theory of gases and the second electrodynamics. For the former, he wrote a seminal paper in 1857, which we'll get to in Chapter 40, and for the latter, starting in 1875, he sought to develop a theory for electrodynamics, again seeking "the simplest and therefore most probable

---

[72] (Kragh and Weininger, 1996) p. 93.
[73] (Klein, 1969) p. 128.

form."[74] We shouldn't forget that in the background Clausius had a consuming life to live as well. His career in teaching led him to Berlin (1850), Zurich (1855), Würzburg (1867), and finally Bonn (1869). "His burning patriotism"[75] for a rising Germany found him volunteering for the Franco-Prussian war in 1870–71, during which he served in the ambulance corps and was wounded in battle. In 1875, while still recovering from and dealing with his war wound, he lost his wife of sixteen years,[76] bore the responsibility of raising his six children, remarried in 1886, fathered another child, and finally died in 1888. Needless to say, during these years, the occurrence of life left him little time for theoretical physics.

Gibbs' Famed Obituary on Clausius

I should not expect to do justice to the subject, but I might do something.
– J. Willard Gibbs to J. P. Cooke, 1889, in response to request
to write a notice on the work of Clausius[77]

In 1889, Gibbs was asked to write an obituary for Clausius, a man he never met but a man whose footsteps he followed in. Recognizing that this would be a "very delicate matter… [since the]…hot-headed partisans"[78] involved in the infamous Tait–Tyndall disputes were still alive, Gibbs bravely accepted the invitation and, understanding that Clausius had been unfairly treated by these "partisans," set out to right a wrong. That he did. In "one of the most remarkable obituary notices in scientific literature,"[79] Gibbs stood up for Clausius.[80] He praised the "epoch" of Clausius' First Memoir in 1850 and also his subsequent work on the mechanical theory of heat. Seeking to be just, he also used this opportunity to acknowledge Thomson as an equal contributor, writing, "Clausius was rivaled, perhaps surpassed, in activity and versatility by Sir William Thomson."[81] But in the end, the obituary was all about Clausius and the primacy of his place in thermodynamics. Gibbs left no doubt about his opinion in the infamous disputes. He had previously and prominently included Clausius in his own work, by listing entropy in his first paper (1873) as one of the fundamental quantities of thermodynamics alongside energy and then by quoting Clausius' famous final two sentences as the opening epigraph for his famous third paper on heterogeneous equilibria (1877–8). Gibbs' alignment with Clausius was further revealed by the fact that his epigraph was written in the original language,[82] an act suggesting a need to also rebalance history between Britain and Germany.

The quote I enjoyed the best in Gibbs' tribute was this. Gibbs made reference to a statement Regnault once made about the fact that he was developing experimental data to help answer the question, which I paraphrase here, how much work can be generated from a given quantity

[74] (Daub, 1970a) p. 309.
[75] Clausius' brother Robert in St. Andrews.
[76] Clausius married Adelheid Rimpam on November 19, 1859.
[77] Gibbs quote cited in (Rukeyser, 1988) p. 300.
[78] (Klein, 1969) p. 128.
[79] Quote from J. G. Crowther (1937) in (Klein, 1969) p. 129.
[80] (Gibbs, 1889)
[81] (Gibbs, 1889) p. 460.
[82] (Klein, 1969) p. 134.

of heat? Gibbs cites the work done by Helmholtz, Joule, William and James Thomson, and Rankine in addressing this question and then wrote, referring to the time period of confusion prior to Clausius' groundbreaking 1850 First Memoir, "Truth and error were in a confusing state of mixture. Neither in France, nor in Germany, nor in Great Britain, can we find the answer to the question quoted from Regnault. The case was worse than this, for wrong answers were confidently urged by the highest authorities. That question was completely answered, on its theoretical side, in the memoir of Clausius, and the science of thermodynamics came into existence."[83] It doesn't get much better than that.

As the provider of the pivotal bridge from Carnot to Gibbs, Clausius' ideas live with us today, but unfortunately his name doesn't, other than in the Clausius–Clapeyron equation, which represents but a small fraction of his total work. He was indeed one of the most vital leaders in the launch of thermodynamics, cautious in some regards, bold in others. In his struggle to understand heat, he was willing to try new ideas and invent new terminologies, taking professional risks in doing so, all the while deeply respecting the process of science. Some of his ideas didn't work and some he was ridiculed for. But some changed the course of history. He pulled us away from the mystery of Carnot[84] and toward the thermodynamic properties of matter. His life was well captured in the final sentence of Gibbs' tribute:

> But such work as that of Clausius is not measured by counting titles or pages. His true monument lies not on the shelves of libraries, but in the thoughts of men, and in the history of more than one science.[85]

---

[83] (Gibbs, 1889) p. 459.

[84] Unfortunately, we still learn our thermodynamics through Clausius' analysis of the Carnot cycle, which is confusing as this analysis sits down in the branches of entropy's tree of consequences. Is it time to re-work how we teach entropy?

[85] (Gibbs, 1889) p. 465.

35   J. Willard Gibbs

In the world of classical thermodynamics, we are concerned with the macroscopic properties of matter and how they change in various processes. We know that whatever atomic interactions happen in the microscopic world of the system of interest, they obey the fundamental conservation laws, especially the one in which we're most interested right now, the conservation of energy. If a certain amount of energy disappears here, it must re-appear somewhere else in the exact same amount. Not the same approximate amount. The same exact amount. This law is applicable to individual atoms and thus to systems comprised of billions of these atoms and all of their collisions, interactions, and reactions. The system follows this law because each and every one of the billions of atoms follows this law. Nothing is lost in scale-up. This discovery became embedded inside the 1st Law of Thermodynamics. Internal energy is conserved and constant, only changing in response to external influences such as heat and work.

We also know that at the core of this macroscopic world exists a fundamental law: nature resides in her most probable state. There is no driving force involved here. It is purely probabilistic. Not the probability involved with Heisenberg Uncertainty but the probability of large numbers. A system of atoms and molecules consistently spreads itself over space and energy in a very repeatable way, the same way every time. This tendency of nature to follow probabilistic laws is neatly summarized in the concept of entropy. If you leave a system alone, it will move toward its most probable state and reside there forever. This is the law of large numbers in action. This discovery became the retroactive core basis of the 2nd Law of Thermodynamics.

That the two laws of thermodynamics could be tied together in a single equation—$dU = TdS - PdV$—is simply amazing. Clausius created the 1st Law of Thermodynamics for a system: $dU = \delta Q - \delta W$. The conversion of δW to PdV was not a huge step, for it had been discovered well prior. It works for fluids (pressure is equal throughout the system), and, since much of chemistry deals with liquids and gases, it was an acceptable substitution to make. The conversion of δQ, on the other hand, to TdS was a huge step and still boggles the mind to this day. That such a property as entropy exists is fascinating in and of itself. That such a property changes <u>exactly</u> as the ratio of the heat absorbed divided by absolute temperature ($\delta Q/T$) is beyond fascinating, not only from a fundamental physics point of view—*why exactly does entropy change in this way?*—but also from the viewpoint of enabling Clausius to create an equation based solely on the properties of nature: $dU = TdS - PdV$. The absolute icing on the cake, as we'll soon see, was that the structure of the equation differentiates the extensive but not the intensive properties, and that the non-differentiated intensive properties T and P also serve as the criteria for thermal and mechanical equilibrium.

Block by Block: The Historical and Theoretical Foundations of Thermodynamics,
Robert T. Hanlon, Oxford University Press (2020). © Robert T. Hanlon.
DOI: 10.1093/oso/9780198851547.001.0001

The equation, $dU = TdS - PdV$, which arguably became the starting point for classical thermodynamics, and many others that were derived from it all demonstrate that nature is not random. At its core, the physical world is comprised of atoms that move and behave in ways that we can codify as law. These movements and behaviors result in macroscopic properties that we can directly measure (T, P, V, N, m_i, x_i) and those we can't, such as energy and entropy. While each property quantifies something different than the others, they are not totally independent. If you tell me the temperature, pressure, and number of moles of an ideal gas, I'll tell you the volume. The properties are interconnected because they are measuring the same system of moving atoms, just from different angles. It's as if each property were a lever on a large mechanical device, with each lever connected to all the other levers inside the device. If you move one lever, another lever moves in response. In nature, if you move one property, another property moves in response. You cannot simply change the temperature of an ideal gas without changing some other property. The ideal gas equation spells this out. This equation and all the many others tell you how the levers are connected inside the machine called nature.

The fact that the properties of matter are connected is a prelude to our next section on J. Willard Gibbs. It was Gibbs who not only told us this but who also showed us these exact connections. In the true spirit of the scientific method, Gibbs deduced the consequences of Clausius' induced hypothesis, $dU = TdS - PdV$, by bringing an advanced level of calculus together with a relentlessly deductive mind to bear on the subject. In one fell swoop, he put the capstone on the development of classical thermodynamics, creating the field of physical chemistry in so doing.

> J. Willard Gibbs

It is an inference naturally suggested by the general increase of entropy which accompanies the changes occurring in any isolated material system that when the entropy of the system has reached a maximum, the system will be in a state of equilibrium. Although this principle has by no means escaped the attention of physicists, its importance does not appear to have been duly appreciated. Little has been done to develop the principle as a foundation for the general theory of thermodynamic equilibrium.

– J. Willard Gibbs, 1878[1]

[Gibbs is] doubtless the greatest genius produced by the United States of America.

–Ernest Cohen, Dutch physical chemist[2]

[Such men as Gibbs ask] for nothing better than to be left alone...allowed to do his work in peace and quiet...free from daily routine and material needs.

– Professor Ernest W. Brown, Yale University, 1927[3]

Gibbs' scientific work was Gibbs.

– Lynde Phelps Wheeler[4]

---

[1] (Gibbs, 1993) p. 354.
[2] Cohen quote cited in (Daub, 1976) p. 747.
[3] (Pupin, 1927) Comment from Professor Brown in his Introduction to Michael Pupin.
[4] (Wheeler, 1998) p. vii.

The creative loneliness of the impelled spirit.

– Muriel Rukeyser[5]

There are no such adventures like intellectual ones.

– Henry James[6]

Many breakthroughs in science occurred when a person first gathered knowledge and then went off to think, alone, seeking connections between threads of ideas where others hadn't. History is littered with examples of such a sequential creative process—first gather, then think. Newton, Bernoulli, Carnot, arguably Clausius. But the epitome was Gibbs.

J. Willard Gibbs (1839–1903) certainly "gathered" while growing up. He was born in New Haven, Connecticut, and experienced an educational upbringing deeply intertwined with Yale University on account of his father having been a professor of sacred literature there from 1826 to 1861. Gibbs himself received his undergraduate (1858) and Ph.D. (1863) degrees[7] in engineering from Yale and served as tutor there for three years following graduation. This base provided him a tremendous springboard to dive into the world of European science and math. Between 1866 and 1869, he spent one year each at the universities in Paris, Berlin, and Heidelberg, attending lectures in mathematics and physics, reading widely in both fields. He returned to New Haven in 1869, never left, never married, lived with his sisters in the house where he had grown up, taught at Yale—and thought. Alone.

To those around him, he seemed to live quietly. But beneath the surface, his mind was on fire.

He Found Problems to Solve in the Papers He Read

We must resign ourselves to ignorance about the philosophical ideas that no doubt presided over the birth of physical theories in Gibbs' mind.

– Pierre Duhem, 1907[8]

Historians have asked, *what caused Gibbs to turn his thoughts to thermodynamics?* He didn't leave many clues behind. No journal, no memoir—absences perhaps encouraged by his renowned modesty. He did not attend lectures on this subject while in Europe, and no stimulus came from his colleagues at Yale.[9] It seems pretty clear that what caught Gibbs' attention was what he was reading. The scientists of the early 1870s generated many publications on different aspects of the rising field of thermodynamics, such as those related to the public debates between Tait and Clausius over 2nd Law priority and those related to the concept of heat as put forth by Maxwell and summarized in his *Theory of Heat* (1871 – 1st Edition). There was also the experimental work of, for example, Thomas Andrews with carbon dioxide and his discovery of

---

[5] (Rukeyser, 1988) p. 438.
[6] (James, 1901) p. 214.
[7] (Wheeler, 1998) p. 32. Gibbs' 1863 doctorate was the second in science and the first in engineering to be given in the United States.
[8] Duhem quote cited in (Deltete, 2011) p. 94.
[9] (Klein, 1983) p. 153.

the critical point. The many ideas, data, and unresolved issues swirling around Gibbs likely captured his attention. He had the powerful combination of the physicist's drive to understand the physical world and the mathematical tools to bring structure to this understanding. He also had the ability to focus with sustained intensity. He was ideally situated. As he advised one of his students, "one good use to which anybody might put a superior training in pure mathematics was to the study of the problems set us by nature."[10] The problems Gibbs attacked were sitting right there in the open literature.

So Gibbs thought. Alone. He sought a structure for describing nature that had not yet been formed. He was Sadi Carnot in the Paris apartment all over again. Just him, others' theories, others' experimental data, and his strong deep desire to make sense of it all. Between 1873 and 1878 Gibbs published three papers that altered the course of chemical thermodynamics and helped launch the field of physical chemistry. No one at Yale or even in the United States really knew what he was doing. His papers appeared in an obscure, little-read journal—*Transactions of the Connecticut Academy*. If not for his pro-active forwarding of his papers to a wide list of scientists in Europe, and especially to James Clerk Maxwell, his work may have remained hidden for a very long time.

Gibbs Built on Clausius

So what was Gibbs' breakthrough? What *was* in those papers? The amazing thing is that for all of the daunting challenges Gibbs provided to those who read his papers, his papers' content was but a logical extension of Clausius' work. Yes, a logical extension greatly aided by a genius mind. But still, if you understood Clausius, you understood Gibbs. I'm not saying that understanding either was easy, just that the former flowed rather smoothly into the latter. Indeed, even though the two never met, and even though Clausius never responded to any of the reprints Gibbs mailed to him, Gibbs chose Clausius to build on. Out of all the acclaimed scientists that Europe and Britain had to offer, Gibbs chose Clausius. The structure that resulted has withstood the tests of time, unaltered.

Gibbs' First Two Papers

There's opportunity at interfaces. If you want to find a fertile field of study, find an interface. Gibbs certainly did. In writing his Ph.D. thesis at Yale, Gibbs became proficient in the use of mathematical graphing techniques for scientific inquiry. In his travels and studies in Europe and his own study back in New Haven, Gibbs also became proficient in additional mathematics as well as in contemporary thermodynamics. Gibbs saw his opportunity at the interface between mathematics and thermodynamics, using the former to explain and bring new insights to the latter. Needless to say, for Gibbs, this field was indeed fertile.

The use of graphical techniques in thermodynamics arguably began when Watts created the indicator diagram to quantify the efficiency of his steam engine. This classic PV diagram,

[10] (Klein, 1983) p. 155.

directly linked as it was to the quantity of work done, became the favored approach for many years thereafter and was highly effective for elucidating the steps in Carnot's cycle. But things changed upon Gibbs' discovery of Clausius in the early 1870s. Whether it occurred as a single eureka moment or a more gradual realization, I don't know. But there had to be a moment when Gibbs looked at the result of Clausius' Ninth Memoir (1865–66),[11]

$$dU = T\,dS - P\,dV \qquad\qquad [34.3]$$

and realized that an opportunity was sitting there in front of him. As a young professor with an interest in finding his own niche of contribution, he pursued this equation, following unbreakable threads of cause–effect logic. *Given that this equation is true, what must be the consequences?* Deductive logic based on an induced hypothesis at its finest.

When Clausius developed this equation, his primary focus was on the mechanical theory of heat and its role in the relationships between heat, work, and internal energy. Entropy arrived toward the end of his research, and he certainly brought attention to it, but it wasn't central to his efforts. His equation, published in 1866, sat relatively idle, largely because people, including Clausius himself, had a difficult time grasping what entropy was all about. But where others saw confusion, Gibbs saw a new beginning. He realized that regardless of what entropy actually meant, its sheer existence and features were what were important, for they enabled the creation of the above equation based purely on the thermodynamic properties of matter as governed by the 1st and 2nd Laws of Thermodynamics. It was a true equation of state, or more accurately, a derivative of such an equation, and embodied no reference to path dependency. And it was in this equation that Gibbs realized the existence of its parent equation based on the relationship U(S,V). In this way energy and entropy, two new and abstract properties, became central to Gibbs' effort to shift focus from Carnot's cycle to the properties of matter.

Moving Beyond the PV Indicator Diagram: U(S,V)

In effect, Gibbs' realization occurred by working backwards. With his mathematically trained mind, Gibbs saw Clausius' fundamental equation as the result of a mathematical process that started with a parent equation, U(S,V), whose total differential

$$dU = (\partial U / \partial S)_v\,dS + (\partial U / \partial V)_S\,dV$$

led him to conclude, by comparison with [34.3], that

$$T = (\partial U / \partial S)_v$$

$$P = -(\partial U / \partial V)_S$$

and thereby provided him an exact thermodynamic definition for temperature and pressure. Gibbs proposed that since the U(S,V) relationship for a closed system led to Clausius, and

---

[11] (Clausius, 1867) Ninth Memoir Appendix (1866), p. 366. Clausius wrote the equation TdS = dU + dW from his standpoint of, *where does the heat go?* For fluids, it was well understood that dW equaled PdV, but Clausius never explicitly included this as such in his fundamental equation.

since Clausius is valid, then the function U(S,V) must exist. He wasn't concerned with the exact nature of the function or what it looked like; that was for experiment to determine. He was solely concerned that mathematical logic proved its existence.

As Gibbs continued to work his way through this logic, he drew on his background in graphical techniques to better understand the implications of these results and also to clarify and explain his understanding to others. Within U(S,V) he saw a new set of thermodynamic coordinates on which to graph physical properties. He started with a two-dimensional graph based on S-V (first paper – 1873) and then a three-dimensional graph based on U-S-V (second paper – 1873) wherein the slopes of the resulting curves carried exact physical meaning as shown above for temperature and pressure.[12]

*   *   *

Gibbs' graphical approach in his first two papers might sound easy in concept to us today (or not), but at that point in history it wasn't. It was new territory, where but a few had explored.[13] James Thomson was the first (1871) to arrive with his three-dimensional graphical analysis of thermodynamic properties based on P-V-T coordinates, and Gibbs was a close second (1873). T. Belpaire (1872) and Macfarlane Gray (1876) independently developed ideas around the T-S diagram.[14] Gibbs' work took these approaches to a higher level. He was aware of Thomson's efforts and may have even been inspired by them. But he also saw in them the opportunity to contribute an improvement, writing that P-V-T coordinates offer less complete knowledge about a system than U-S-V coordinates, citing the fact that while the former can be determined by the latter via differentiation, the latter cannot be determined by the former.[15] An additional benefit to Gibbs' approach was that the chosen coordinates were all extensive properties, meaning that they lent themselves to simple addition, a fact which would come into play when he began considering mixtures of phases in equilibrium. Consider that on a T-P property map, a complete phase transition is shown as a single point on a given equilibrium line, whereas on the U-S-V map, the extensive property coordinates enable the entire phase transition to be shown.

*   *   *

It seems likely that Gibbs' struggle to figure out how to create such graphs certainly opened his thinking to the very core issues involved with equilibrium and the thermodynamic properties of matter and deeply motivated him to completely understand the full range of implications emerging from this unique approach based on a single equation. While I naturally can't do justice to his full body of work on thermodynamics here, I can share key highlights so that you will gain an understanding of the origin of much of the thermodynamics we study today.

---

[12] One difficulty in reading Gibbs' first two papers is that he worked in a three-dimensional world that was very hard to understand based on his written descriptions and two-dimensional drawings. He knew what he was talking about, but close study is needed by the reader to arrive at the same understanding. Some graphs are easy to read and the brain clicks on their meaning immediately. Not so the Gibbs' graphs. I write this only to emphasize that if you experience challenges in interpreting such graphs, don't worry, because they aren't necessary to understanding the underlying principles.

[13] (Klein, 1983) p. 156–7.

[14] (Crowther, 1937) p. 262.

[15] (Gibbs, 1993) p. 34.

3D Graphics

Gibbs viewed the U(S,V) relationship as a single, curved surface in three-dimensional space, with the tangential planar slopes of the curved surface again reflecting changes in T-P. Every point on this surface complied with the U(S,V) equation and thus with Clausius' original equation [34.3]. This equation is only valid once entropy is at a maximum, which occurs when all internal energy gradients have dissipated. For this reason, Gibbs referred to this curved plane as a "surface of dissipated energy." (Figure 35.1) Both sides of this surface had relevance regarding what is and is not possible. For each U-V point, assuming a single phase only for now, there exists but one S point for an equilibrated system. As this point is a maximum, then all S points *larger* than this for the given U-V point represent an impossible state, while all S points *smaller* than this represent a possible state with the caveat that all such points represent *non-equilibrated* states for which entropy is not yet at a maximum. They are legitimate states but contain energy gradients of some form or another making the combined entropy value less than maximum. In sum, the final structure of the graph is a solid filling up the (possible) non-equilibrated portion of the U-S-V coordinate system, the surface of which quantifies the equilibrated state for which Clausius' equation applies, and then emptiness signifying an impossible world.

The depiction of thermodynamic properties offered by this approach deeply inspired Maxwell, as best reflected by the fact that in his later edition of *Theory of Heat* Maxwell devoted a whole section to Gibbs' work, and also by the fact that, in the midst of all the other many activities he was involved with, he devoted his own hands-on time to sculpting a three-dimensional clay model—"a concrete image of sympathy between two great and subtle minds"[16]—of the U-S-V relationship for water, from which Maxwell prepared three plaster casts and sent one to Gibbs as a gift, keeping the other two for himself at Cambridge University.

Gibbs was particularly interested in two aspects of his own graph, the equilibrated surface itself and the distance between this surface and the points inside the non-equilibrated solid structure. As will be discussed later, this distance quantifies the energy available to do useful work during the equilibration process as the point moves from the solid interior to the surface.

Early Steps toward Phase Equilibrium of a Single Compound

[My second paper] contains, I believe, the first solution of a problem of considerable importance.
– Gibbs in letter (1891) to Ostwald[17]

While such three-dimensional graphs must initially have been challenging for Gibbs to visualize and comprehend for a single phase, they certainly became even more so once multiple phases were considered. How could they be put onto the same graph? How could situations involving equilibrium between phases be represented? How much he struggled with this

---

[16] (Rukeyser, 1988) p. 203.
[17] Gibbs quote cited in (Daub, 1976) p. 749.

GIBBS (1873): THE SURFACE OF DISSIPATED ENERGY (constant V)

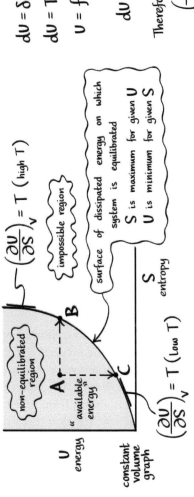

$$dU = \delta Q - \delta w \quad \text{CLAUSIUS} \,(1850)$$

$$dU = TdS - PdV \quad \text{CLAUSIUS} \,(1865)$$

$$U = f(S,V) \quad \text{GIBBS} \,(1873)$$

$$dU = \left(\frac{\partial U}{\partial S}\right)_V dS + \left(\frac{\partial U}{\partial V}\right)_S dV$$

Therefore,

$$\left(\frac{\partial U}{\partial S}\right)_V = T \qquad \left(\frac{\partial U}{\partial V}\right)_S = -P$$

AC distance (ΔU) quantifies the maximum work that can be generated from dissipating A's energy gradients

Gibbs referred to AC as "available energy"

We now call this change in "free energy" or change in "Gibbs free energy"

surface of dissipated energy on which system is equilibrated
S is maximum for given U
U is minimum for given S

$$\sum S_{parts} = S_{total} \leq S^{max}$$

A = a system that has internal energy gradients and is thus not equilibrated. may be composed of parts for which

AB = if left isolated **A** would move at constant energy and increasing entropy towards **B**

AC = if external work could be reversibly extracted, **A** would move at constant entropy and decreasing energy towards **C**

Figure 35.1   Gibbs and the surface of dissipated energy

challenge, we don't know. We only know that he resolved it, and in so doing brought a much clearer definition of the concept of equilibrium to thermodynamics.

This was such a critical moment for Gibbs. The term "equilibrium" had long been in use, but its exact definition in the world of thermodynamics and chemistry was not yet clear. Remember that this was 1873, just eight years after the arrival of the very abstract concept of entropy.

In his first two papers, Gibbs acknowledged without challenge the common scientific wisdom that if two phases of a single substance exist at the same temperature and pressure then they are in equilibrium with each other, using the logic that if each exists at those conditions on its own, then nothing would change when they were put into direct contact with each other. As he put it, they would "co-exist." The identification of such co-existing phases could be achieved through the use of his graphical techniques by evaluating the tangential-plane slopes, based on dU/dS and dU/dV, of all points on the three-dimensional U-S-V surface of the single substance. When the slopes of the tangential planes of two different U-S-V points equaled each other, then the two planes were not only parallel to each other but must be one and the same; a single plane must touch the two points. Gibbs proved this by showing that given Clausius' equation, and given the fact that temperature and pressure are equal, then the following equality must hold between the two equilibrated states that are also at equilibrium with each other:

$$(U_2 - U_1) = T(S_2 - S_1) - P(V_2 - V_1) \qquad [35.1]$$

Note that the fortuitous beauty of T-P being constant at equilibrium is that Equation 34.3 can be converted from differentials into deltas and can thus quantify finite differences between equilibrated states.

The straight line represented in Equation 35.1 is contained within the single tangential plane and connects the two endpoints in equilibrium with each other. In other words, the plane touches the surface of dissipated energy at these two points, reflecting the fact that temperature and pressure are equal between the two phases at equilibrium. Furthermore, because all three property coordinates involved in this equation are extensive, they are additive, meaning that the total system values for U-S-V are the mass-weighted-sums of the corresponding values for each equilibrated state. So during the phase transition, the properties of the components remain the same, while the system properties move along the line as the relative masses of the components change. Gibbs' embrace of this whole-equals-the-sum-of-the-parts concept would continue to play a critical role in his subsequent work.

Another characteristic of the line connecting the two points in equilibrium with each other is that the line is "derived," as Gibbs put it, by combining the properties of the two phases involved. But underneath this "derived" line remain the "primitive" U-S-V lines for the two separate phases. These single-phase lines diverge from the derived two-phase line at the point of contact with the plane in whichever direction change is leading them, whether it's via heat exchange (Q), work (W), or a combination of both. One way to look at this is to consider the heating of water at atmospheric pressure. Under ideal conditions, it's possible to heat water above 100 °C. If you were to drop some boiling chips into the superheated water, it would explode into a sudden fury of boiling. But if you didn't otherwise disturb the system, the water's temperature would rise above 100 °C and the system would remain precariously stable, following its own U-S-V curve. It would thus enter into what is called a *metastable* phase. Continued heating would lead to a critical point where the phase would become *unstable*, rendering further heating of this phase impossible.

So what's the difference between the "derived" two-phase line and each of the "primitive" lines? Entropy. For the same U-V values, entropy is higher on the "derived" line than on the corresponding "primitive" line. This is the physical reason for the single phase to split into two to begin with, to move toward the higher entropy (most probable) state. The weighted sum of the entropies of the two phases that form upon the establishment of equilibrium via boiling exceeds the entropy of the single metastable phase.

Gibbs' findings in the above work solved a problem having a long history. Others had known that in graphing the P-V coordinates for an isotherm, points of abrupt change occurred at phase transitions wherein pressure stopped changing while volume continued to change. Thomson called these points a "practical breach of continuity."[18] Up until Gibbs' publication in 1873, the problem of where to draw the horizontal constant-pressure line had not been solved; Maxwell would do so in 1875 based on a different set of thermodynamic arguments. Noted Klein,[19] what was lacking in their efforts was the use of entropy. The use of Clausius' equation based on entropy enabled Gibbs to solve this problem "in his usual brief and elegant manner," since his resulting equation [35.1] defined the end points of the line. It was this problem to which Gibbs referred in the leading quote of this section. The properties of the two co-existing phases had to fulfill the equality shown in Clausius' equation.

Entropy Maximum leads to Energy Minimum

Gibbs did not live inside a world separated from reality while developing his theories. Indeed, he was well aware of a wide range of chemical phenomena that didn't fit into nice, neat theories of the ideal world. He strove to ensure that his theories addressed such work, regardless of the challenge. He didn't allow incompleteness into his work. He provided others' experimental evidence where needed to explain or clarify (he did no experiments himself) and additionally set up the framework for others to design, execute, and interpret their own experimental studies into such phenomena. He provided the map to guide the experimentalists; he "laid down laws for problems yet unproposed."[20]

Gibbs also brought discussion to the definition of equilibrium and the irreversible approach to equilibrium by emphasizing another critical concept of Clausius. The entropy of an isolated system increases to a maximum. While Clausius appreciated the concept of an increasing entropy on a cosmic scale, he didn't appear to appreciate it or the implications of it on a more local scale. But Gibbs did, to great effect. As Gibbs considered the large topic of equilibrium and stability, he returned to this single concept time and time again. To Gibbs, stable equilibrium meant that isolated-system entropy had reached its maximum value, meaning that any perturbations of the system at that point would have to result in a decrease in entropy, which is impossible. Thus, for stable equilibrium, the mathematical formulation for defining maximum entropy was

[18] (Klein, 1983) pp. 156–7.
[19] (Klein, 1983)
[20] (Rukeyser, 1988) p. 240.

$dS = 0$ Condition for stable equilibrium at constant U

$d^2S \leq 0$ Condition for maximum S at constant U

But Gibbs also ingeniously formulated this in a different way based on energy (U). In an isolated and equilibrated system, energy is constant and entropy at a maximum. But one could also consider a "closed" equilibrated system as well, which unlike the isolated system allows heat interactions with the environment. As heat exchange is the only reversible approach to changing both energy and entropy, the net effect is that such a system would exist at maximum entropy and minimum energy, at which point:

$dU = 0$ Condition for stable equilibrium at constant S

$d^2U \geq 0$ Condition for minimum U at constant S

Gibbs showed that these two stability criteria are mathematically equivalent[21] and that the use of the energy-based criterion provides more flexibility in solving thermodynamic problems, especially those involving the concept of "maximum work" and "free energy." More on this later.

Thus, to test whether or not a system is stable, Gibbs proposed that the above mathematical equations and inequalities be used, not only for stable versus unstable systems but also for considering the metastable state for which both the first- and second-order differentials equal zero.

* * *

At some point as Gibbs was developing his graphical approach toward thermodynamic properties, likely that point at which he had to account for variations in mass and composition, he realized that he had to move on to a new approach. The graphs became hard if not impossible to draw, and even harder for the reader to understand, as the level of complexity and dimensionality rose upon his inquiry into the nature of equilibrium between multiple substances and multiple phases. As Gibbs' mind followed this path of increasing complexity, he let go of his graphs and re-directed his efforts to pure mathematics. It really was his only option. This shift in direction will be more thoroughly covered in the next chapter.

Thought Experiments Involving Isolated Systems: $(dG)_{T,P} \leq 0$ for Spontaneous Change

It was in his second paper that Gibbs first started employing the "thought experiment" as a means to understand equilibrium. Building on Clausius' own approach—exploring phase equilibria in his Ninth Memoir, Clausius wrote, "Conceive then a certain quantity of matter to be enclosed in a vessel of a given volume, and one portion of it to be in the first and the other in the second

---

[21] (Gibbs, 1993) pp. 56–62. Gibbs showed the equivalence of his entropy maximization and energy minimization criteria by stating, "it is always possible to increase both the energy and the entropy of the system, or to decrease both together, viz., by imparting heat to any part of the system or by taking it away." Variations in entropy at constant energy can be transformed into variations in energy at constant entropy through the exchange of thermal energy with the environment; both energy and entropy move together with such exchange.

state of aggregation"[22]—Gibbs imagined a rigid-walled, fully-insulated, closed vessel, referred to hereafter as an *isolated system*, for which total volume, energy, and mass did not change. For such a system, as he would later state in the epigraph to his famed third paper, Gibbs relied upon Clausius for guidance.

> *Die Energie der Welt ist constant.*
> *Die Entropie der Welt strebt einem Maximum zu.*

Substitute his imagined isolated vessel for *Welt* and the epigraph remains valid, something Gibbs realized but Clausius possibly didn't.

Recall that both concepts expressed in the epigraph are built into Clausius' original fundamental Equation 34.3 describing a fluid system, regardless of the material comprising the system, in its equilibrated state.

If a closed and equilibrated system receives no heat and does no work, then it is *isolated* and its energy remains constant ($dU = 0$). This is the conservation of energy—the first half of Gibb's epigraph. But the fact that energy is constant does not mean that nothing is happening. Indeed, should any energy gradients exist within the system, be they mechanical (P), thermal (T), or chemical (μ – this will be introduced later) in nature, they will dissipate over time, increasing the total system entropy to a maximum—the second half of Gibb's epigraph—at which point, and only at which point, Equation 34.3 for the *closed* equilibrated system, for which both heat and work can occur, becomes valid. For Gibbs, this seemingly simple framework, which he felt had not yet been "duly appreciated," became the starting point for his work.

Gibbs explored variations of this general thought experiment as a means of addressing certain problems he sought to solve. One variation appeared in his second paper. He imagined placing the body (comprised of a single substance but not necessarily of a single phase) inside the isolated vessel alongside a medium much larger in mass that served the purpose of maintaining both the body and the medium at a fixed temperature and pressure via the existence of a flexible and conducting boundary between the two. These constraints were not random but reflected Gibbs' recognition of real-world chemical operations that were controlled to fixed temperature and pressure set-points. "These problems may be modified so as to make them approach more nearly the economical problems which actually present themselves."[23]

The medium itself was internally equilibrated, uniform throughout, and its entropy at its maximum value. The body, on the other hand, was not equilibrated, not uniform throughout, and its entropy not at its maximum. Gibbs left unstated what was actually going on inside this body, only saying that it was not in a state of thermodynamic equilibrium. It didn't matter in this thought experiment. It only mattered that the body's entropy was <u>not</u> at its maximum, meaning that energy gradients, of whatever type, were present inside the body.

Gibbs then started writing out equations for both the body and the medium to show how they were conceptually connected by the fact that both resided inside his isolated vessel and by

---

[22] (Clausius, 1867) Ninth Memoir (1865), p. 348.
[23] (Gibbs, 1993) p. 53.

the fact that extensive properties are additive. Heat flow and volume change were allowed during the equilibration process between the two to maintain T and P constant. So what do the resulting equations look like? With some modifications of Gibbs' original equations, they look like this:

$$dU_{body} + dU_{medium} = 0 \quad \text{the total system energy inside the isolated vessel remains constant}$$

$$dV_{body} + dV_{medium} = 0 \quad \text{the total system volume inside the isolated vessel remains constant}$$

$$dS_{body} + dS_{medium} \geq 0 \quad \text{the total system entropy inside the isolated vessel increases}$$
$$\text{to a maximum as the body's internal gradients dissipate}$$

What else do we know? Well, the medium is fully equilibrated and remains so throughout the body's equilibration process, as all heat-flow and volume-change interactions between the two are assumed to be reversible, thus rendering valid the following equality:

$$dU_{medium} = T \, dS_{medium} - P \, dV_{medium}$$

Making substitutions and flipping signs,

$$dU_{body} + P \, dV_{body} - T \, dS_{body} \leq 0$$

Because both T and P are constant, fortuitously so in light of the simplified mathematics and of the equality of these properties between equilibrium phases, Gibbs then simplified this to

$$d(U + PV - TS)_{T,P} \leq 0 \qquad \text{for the body} \left(\text{not the medium, not the system}\right) \qquad [35.2]$$

This evolution was such a fascinating progression of thought by Gibbs. In his search to see how the inequality caused by increasing entropy played out, he isolated certain established properties of matter into a composite property of matter, wholly new to the world of thermodynamics. When temperature and pressure are held constant, this new property, U + PV − TS, which would later be named the famed Gibbs energy (G), must decrease to a minimum, a fact that Gibbs pointed out in a footnote in his second paper, almost as if it were a moment of reflection—*look at this interesting thing I found*—prior to publishing. One must wonder what he first thought when he arrived at this point. Little did he know that years later we'd be basing many of our calculations for equilibrium on this concept as well as calculations for "maximum work" and change in "free energy," which is why G is sometimes referred to as both Gibbs Energy and Gibbs Free Energy. From here on out, I will use the former.

    Note in this exercise that if the body itself were instead isolated and no medium were present, the body's energy would remain constant, its entropy would increase to a maximum, and its temperature and/or pressure would change from the initial conditions. Then through some combination of reversible heat exchange (Q) and work (W), the body could be returned to its initial T-P values. The resulting entropy at the referenced T-P would then be maximum for these new conditions but naturally different than the maximum entropy reached for the original isolated body.

Discussion of Select Results from Gibbs' First Two Papers

It's rather telling that Gibbs shared the above result with little fanfare. Some of Gibbs' former students described him as being a man of "retiring disposition"[24] and "judicious silences,"[25] "devoid...of the slightest desire to exalt himself."[26] It's almost comical the way in which Gibbs shared significant discoveries like this with the utmost brevity and with hardly any discussion of their significance or his role in the matter. As Klein wrote, "Gibbs used hardly any more words [than necessary to state his findings], as though he too were simply reminding his readers of familiar, widely known truths, and were writing them down only to establish the notation of his paper."[27] Even his use of the equation $dU = TdS - PdV$ right there in his first paper was done without discussion about the fact that this single equation was a novel concept at that point in time. His statements occurred as if they were "conventional wisdom" when he wrote them, but they weren't. His entire approach was novel. To Gibbs, it was obvious. But to the reader, not so.

This finding of Gibbs energy minimization was quite remarkable, a logical consequence of Clausius, but one that wasn't evident prior to Gibbs' thought experiment. In an isolated system comprised of a body plus a medium that serves to maintain the body at constant temperature and pressure, once equilibrium is reached, all energy gradients dissipated and total entropy maximized, this function reaches a minimum value, meaning that the value of $(dG)_{T,P}$ for the body equals zero. Once a system has equilibrated at a given temperature and pressure, entropy is maximized and thus Gibbs energy minimized. The system sits on Gibbs' surface of dissipated energy. If such a system were then perturbed at constant temperature and pressure, such as through small fluctuations of heat or work, then the equality $dU = TdS - PdV$ would still hold for the body, meaning that G would not change. G would continue to sit at a minimum.

There's a second valuable concept to consider with Gibbs energy. Recall Equation 35.1. During phase transition, say from liquid to vapor, temperature and pressure remain constant. If one were to assume 100% liquid as State 1 and 100% vapor as State 2, then [35.1] tells us that

$$(U_V - U_L) = T(S_V - S_L) - P(V_V - V_L) \qquad \text{constant } T - P$$

$$U_V + PV_V - TS_V = U_L + PV_L - TS_L$$

$$G_V = G_L$$

Thus, if two phases of the same compound—we're still in the single-compound world of Gibbs here—exist in equilibrium, then in addition to temperature and pressure equality between the two, the sum of $U + PV - TS$ and thus Gibbs Energy must also be equal between the two. As you add heat to boil water and allow reversible expansion to maintain pressure while performing work (Step 1 of Carnot's cycle), the equation $dU = TdS - PdV$ holds throughout the entire

---

[24] (Gibbs, 1993) p. xxiii. From H. A. Bumstead's introductory biographical sketch of Gibbs.
[25] (Rukeyser, 1988) p. 194.
[26] (Gibbs, 1993) p. xxv. From H. A. Bumstead's introductory biographical sketch of Gibbs.
[27] (Klein, 1983) p. 148.

process. Thus, the value of G for the beginning state of water must equal the value of G for the end state of vapor; equilibrium is maintained throughout.

From the above discussion you can detect the fundamental origin of the Clapeyron equation. Since G is equal between two phases at equilibrium for a fixed temperature and pressure, then for an infinitesimal change in temperature and pressure for which the two phases remain in equilibrium, the following relationship must be valid, here using liquid–vapor equilibrium as an example.

$$G^L = G^V \text{ for } T_1 \text{ and } P_1$$
$$G^L = G^V \text{ for } T_2 \text{ and } P_2$$

Thus, for the infinitesimal change from T_1-P_1 to T_2-P_2, the following must also be valid

$$dG^L = dG^V$$

Let's do some simple calculus:

$$G = U + PV - TS \qquad \text{for any given system}$$
$$dG = dU + PdV + VdP - TdS - SdT$$

Since $dU = TdS - PdV$, then

$$dG = VdP - SdT$$

While we used G earlier to analyze <u>constant</u> T-P processes for which dG = 0, we can also use G to analyze <u>varying</u> T-P processes. This provides a means to assess how equilibrium changes with changing T-P conditions. Here's how we do this: Since the above equation applies to any given system, then

$$dG^L = V^L dP - S^L dT$$
$$dG^V = V^V dP - S^V dT$$

Since $dG^L = dG^V$, then

$$(V^V - V^L)dP = (S^V - S^L)dT$$
$$dP/dT = (S^V - S^L)/(V^V - V^L)$$

which is the rigorous form of the Clapeyron equation. This can be simplified assuming $V^V \gg V^L$ to

$$dP/dT = \left[\Delta Hvap/T\right]/V^V$$

* * *

Returning to the discussion around Gibbs Energy minimization, you might argue that S and thus G aren't even defined in the above inequality for the body [35.2] until equilibrium is reached, and you would be correct. But that's why it is written as an inequality. Throughout his writings, Gibbs embraced the whole-equals-the-sum-of-the-parts philosophy by stating that the extensive properties of a given heterogeneous body are equal to the sum of the extensive properties of the body's homogeneous parts. So calculating the total entropy of a non-equilibrated body by summing the entropies of the equilibrated parts was entirely justified. Over time, as the parts equilibrate between themselves, the total system entropy increases to a maximum, at which point equilibrium is achieved and Clausius' fundamental equation for the body, for the medium and for the total system becomes valid.

$$\Sigma\, S_{parts} = S_{total} \le S^{max}{}_{total\ at\ equilibrium}$$

Such discussions will become important when we consider the concept of "maximum work" or changes in "free energy" because the starting point for such discussions is a body that is not internally equilibrated. It's the equilibration process itself that provides the means to generate work. In such scenarios, employing composite properties that include a value for S that is based on the sum of the parts of the non-equilibrated whole is valuable. But remember that it's only for fully equilibrated systems, for which S has achieved its maximum value and G its minimum value, that the fundamental Equation 34.3 (dU = TdS – PdV) is valid.

36 Gibbs' third paper

[Gibbs' "Equilibrium of Heterogeneous Substances"] is unquestionably among the greatest and most enduring monuments of the wonderful scientific activity of the nineteenth century.

– H. A. Bumstead[1]

Like Sir Isaac Newton's Principia, this work of Willard Gibbs stands out in the history of man's intellectual progress as an imperishable monument to the power of abstract thought and logical reasoning.

– Lynde Phelps Wheeler[2]

The greatest effort of sustained abstract thinking in the history of America
– Lawrence Henderson on Gibbs' third paper[3]

I myself had come to the conclusion that the fault was that [my third paper] was too long. I do not think that I had any sense of the value of time, of my own or others, when I wrote it.

– J. Willard Gibbs[4]

At the conclusion of his second paper, Gibbs highlighted the fact that up until then, all of his work had been concerned solely with a system that was comprised of a single component, or, as he worded it, "homogeneous in substance."[5] But Gibbs was well aware that a complete theory of thermodynamics would have to account for "the motions of diffusion and chemical or molecular changes"[6] and would thus have to handle systems comprised of multiple or "heterogeneous" substances as well. Thus he devoted his third and most famous paper to doing just that, as reflected in the title, *Equilibrium of Heterogeneous Substances*.

The task of addressing such a complicated situation was indeed difficult if not overwhelming. No one had yet thoroughly solved this problem as entropy had only just arrived and was central to the solution. Some had attempted the solution and achieved a certain degree of success. August Horstmann (1842–1929), for example, was the first (1873 – in October, prior to Gibbs in December) to investigate chemical equilibria based on entropy maximization, and Francois Massieu (1832–1896) had introduced (1869) thermodynamic composite properties

---

[1] (Gibbs, 1993) p. xii. From H. A. Bumstead's introductory biographical sketch of Gibbs.

[2] (Wheeler, 1998) p. 71.

[3] Henderson quote cited in (Rukeyser, 1988) p. 393. Lawrence Henderson, M.D., was one of the leading biochemists of the early twentieth century.

[4] Gibbs quote cited in (Rukeyser, 1988) p. 264.

[5] (Gibbs, 1993) p. 54.

[6] (Gibbs, 1993) p. 59.

Block by Block: The Historical and Theoretical Foundations of Thermodynamics,
Robert T. Hanlon, Oxford University Press (2020). © Robert T. Hanlon.
DOI: 10.1093/oso/9780198851547.001.0001

involving entropy prior to Gibbs.[7] But Gibbs took the solution to a much higher level. While he solved this problem in his second paper for a specific scenario involving a single substance and multiple phases, he sought to employ in his third paper a more generalized thought-experiment scenario involving multiple species and multiple phases to create a more generalized solution. Gibbs again imagined an isolated system but this time comprised not of one component and one medium, as he had in his second paper, but of multiple components and no medium. He then assumed this entire system to be in its equilibrium state—entropy already maximized—and asked, *what does it look like?*

While the elimination of the medium eliminated the convenience of assuming constant T-P during irreversible change, it did enable a more generalized approach for a system already at equilibrium and thus at fixed T-P, a significant goal of Gibbs in his third paper.

It's rather interesting to see just how far his efforts to generalize went in this paper. With confidence, he left no stone unturned. He wasn't intimidated in the least to cover everything. Take for example his exclusive focus on fluids in his first two papers, which offered the means to convert work done by a body into a function of pressure. He expanded the reach of his third paper by including solids and surfaces for which constant pressure throughout the system did not apply. He further expanded his reach to most all other physical phenomena of interest as well, such as osmotic pressure, capillary effects, gravity, and electrochemical cells. Gibbs' third paper was no paper; at ~300 pages and 700 equations it was a full-blown tome that laid the groundwork for those who followed. With this single paper, he made a "clean sweep"[8] of the subject with such a "degree of perfection that in fifty years almost nothing has been added."[9] All deduced from a single equation. Amazing.

Chemical Potential

The discovery of the chemical potential was the Northwest Passage of this science; it was the link between classical thermodynamics and contemporary physical chemistry and electrochemistry.
– Muriel Rukeyser[10]

If to any homogeneous mass we suppose an infinitesimal quantity of any substance to be added, the mass remaining homogeneous and its entropy and volume remaining unchanged, the increase of the energy of the mass divided by the quantity of the substance added is the "potential" for that substance in the mass considered.
– J. Willard Gibbs[11]

With his new generalized scenario, Gibbs returned to Clausius' fundamental equation. Again, this equation applied to all bodies, regardless of their nature, as captured by Gibbs: "the validity of the general equation is independent of the uniformity or diversity in respect to state of the

[7] (Crowther, 1937) p. 271–2.
[8] (Rukeyser, 1988) p. 233, quoting Larmor.
[9] (Rukeyser, 1988) p. 232, quoting Henry Adams.
[10] (Rukeyser, 1988) p. 237.
[11] (Gibbs, 1993) p. 93.

different portions of the body."[12] The equation holds for any closed system (no mass in or out) that can exchange thermal energy (Q) and work (W) with the environment. It's valid for all fluids, from the simple ideal gas to the complex multi-phase, multi-species, multi-reaction mixture. The equation doesn't care; the nature of the working substance in Carnot's engine is irrelevant. So let's re-write this equation to more clearly reflect this, using "t" to reflect Gibbs' total system.

$$dU^t = T\,dS^t - P\,dV^t \quad \text{closed system} \left(\text{heat and work exchange allowed}\right) \qquad [36.1]$$

Now in the event you have multiple phases within the total system, one would think that this same equation would apply for each phase, such as written here for a given phase, α,

$$dU^\alpha = T\,dS^\alpha - P\,dV^\alpha$$

but this would be wrong, for while the total system is closed ($dm^t = 0$), the comprising parts are open. The phases touch each other, thus permitting movement of chemical species between them in addition to the occurrence of heat and work.

Recall that in his first two papers Gibbs stated that equilibrium between phases exists once the thermal and mechanical gradients between them become absent, or as Thomson worded it, once the gradients dissipate, or as we put it today, once the temperatures and pressures of the phases become equal. But with a changing composition caused by chemical movement, the new question he faced was, *what is the equivalent concept for chemical equilibria and how would he mathematically describe it?* How would he account for the impact of variable mass on internal energy?

It was from this line of inquiry that Gibbs generated one of his major breakthroughs. He said that the way to convert Clausius' equation from a closed system to an open system was to add a term as follows:

$$dU^\alpha = TdS^\alpha - PdV^\alpha + \sum \mu_i dm_i^{\alpha} \qquad [36.2]$$

for i = 1 to n chemical components in the α phase

Thus, per Gibbs, for any given "homogeneous mass" such as the α phase

$$\mu_i = (\partial U / \partial m_i)_{S,V,m \neq i} \quad \text{for a given phase}$$

wherein i refers to each of the n chemical components present in the given phase and m the corresponding mass and wherein this open-system equation would be valid for each phase in the total closed-system vessel. With this step, Gibbs introduced a new physical property to thermodynamics, μ, which he called *chemical potential*—here using the term coined by Rankine—to capture the sense of an energy gradient caused by chemical composition. In essence, he didn't really "discover" this term as much as he created it, and once created, the floodgates opened. Gibbs introduced this equation to the world of thermodynamics on the ninth page of his third paper and then used it as a key source for the remaining ~300 pages.

---

[12] (Gibbs, 1993) p. 35.

Chemical Potential (μ) Created to Enable Analysis of Equilibrium

Before continuing, let's take a closer look at Gibbs' new equation. In keeping with the structure used by Clausius, Gibbs' added term μdm comprised two properties, one intensive and the other extensive, and as a result, the mathematics employed defined the property, specifically μ since mass was already defined, and not the other way around. Of historical note, recall that temperature and pressure were defined well *prior* to their inclusion in this equation. People could relate to T-P; they stood on their own, separate from this equation. Yes, the definition of each became formalized in this new world of thermodynamics, but their origins remained understandable. People had a visceral feel for what they meant. Not so with chemical potential. Prior to this equation, chemical potential didn't even exist. There was no device or meter to measure this property. It was created to solve a problem. As Callen wrote, we have "an intuitive response to the concepts of temperatures and pressure, which is lacking, at least to some degree, in the case of the chemical potential."[13]

As created by Gibbs, chemical potential (μ) was defined by the mathematics to equal $(\partial U / \partial m_i)$ for constant entropy, volume, and mass of all chemical species other than i in a given phase or "homogeneous mass." Similar to temperature and pressure, it is an intensive property and is quantified as the change in internal energy that results from adding an infinitesimal amount of chemical i to the system. But realize that it's actually more than this. It's the change that occurs when adding i but without changing S and V in the process. As the addition of i likely changes both S and V, μ then quantifies the *net* change in internal energy upon addition of i to the system once reversible application of heat and work are done to return the system to its original S and V starting point. It's this definition of chemical potential that Gibbs embedded in his mathematics. It's the change in energy with composition, all other things being equal.[14] The nomenclature had to be clear on this.

The conjugate pairing of intensive (zero-order in mass) and extensive (first-order in mass) properties in Gibbs' open-system equation was a natural and very fortuitous consequence of Clausius' original formulation for it enabled, as we'll see, the ready application of the multi-variable calculus methods of Euler, Legendre, and others to transform and manipulate the thermodynamic equations to suit the needs of physicists, chemists, and engineers. The wonderful work of many famed mathematicians found a new home in Gibbs' thermodynamics, a topic we'll return to later.

* * *

When Gibbs wrote out all of his open-system equations, one for each phase, along with the associated subscript and superscript nomenclature to account for both the multiple species and the multiple phases and also to account for the properties being held constant, the complexity of the mathematics jumped—perhaps not so much the technical complexity but the complexity involved in simply trying to read and comprehend the text. While the nomenclature was critical to capture the exactness required, it did make the reading of his work a challenge,

---

[13] (Callen, 1960) p. 48.

[14] This is the reason why chemical potential can be negative. For some chemical species, the system's energy must decrease as mass is added so that entropy remains constant. See (Cook and Dickerson, 1995) for further discussion.

especially for the chemists who weren't used to this world of mathematics and especially to its rules.

Gibbs took these equations, added them together, again using his whole-equals-the-sum-of-the-parts thinking, and thereby created the equation for the heterogeneous whole. Note that Gibbs applied the word "heterogeneous" to account for both multiple substances and multiple phases, which made reading confusing at times given the brevity of his context. To these equations he added constraints to acknowledge that the total system, in his scenario, was at equilibrium, meaning that temperature and pressure were assumed equal for all phases comprising the system and that *total* system mass, energy, volume, and entropy were constant since the total system was isolated with no mass in or out.

> Highlighting Three Developments in Gibbs' Third Paper

Development One: Chemical Potential (Continued)

Out of this new thought experiment arose numerous developments, three of which are shared here to illustrate the breadth and depth of Gibbs' accomplishments. The first regards the rise of *chemical potential*, considered Gibb's "greatest discovery,"[15] as a defining new parameter in chemical equilibrium. Although Gibbs did not explicitly show this, when all of the fundamental equations [Equation 36.2], one for each phase, were added together, again being made easy by the fact that temperature and pressure were the same for all, one arrived at the following

$$dU^t = TdS^t - PdV^t + \sum \mu_i dm_i \qquad [36.3]$$

for i = 1 to n components; all phases

But if the total system were closed (heat and work allowed, but not change in mass) and equilibrated (entropy at its maximum value), then Equation 36.1 applies, which then means that the following must be true,

$$0 = \sum \mu_i dm_i \quad \text{for } i = 1 \text{ to n components; all phases}$$

and the only way this could be so, given that $\sum dm_i$ across all phases also equals 0 on account of conservation of mass, is if the chemical potential of each species were equal across all phases

$$\mu_i = \text{constant in all phases at chemical equilibrium}$$

for all i = 1 to n components

It was in using this flow of logic that Gibbs identified or, more accurately, created the critical third criterion for equilibrium. In the world of phase equilibrium for a single substance, the only requirements for equilibrium were equal pressure and temperature to ensure the absence

---

[15] (Rukeyser, 1988) p. 232.

of mechanical and thermal gradients between the phases.[16] But as you add species, the additional requirement became equal chemical potential to ensure the additional absence of *chemical* gradients between each of the species in all of the phases.

To Gibbs, chemical potential quantified the "active tendency" of a chemical species to move from one phase to another. When this property is the same throughout a system, there is no tendency to move, no better place to be, no chemical gradient.[17] Total system entropy is at its maximum and equilibrium exists. And this wasn't all. Recall that Gibbs sought to account for both diffusion *and* reaction. He thus further extended his mathematics ($\sum \mu_i dm_i = 0$) to include reactions and concluded that the sum of the chemical potentials of each species involved in a reaction in a given phase times its mass-based stoichiometric value must equal zero. Thus, for example, for the equilibrium reaction $A + B \rightleftharpoons C$,

$$m_A \mu_A + m_B \mu_B = \left(m_A + m_B\right)\mu_C \quad \text{for chemical reaction at equililbrium in a given phase}$$

Wrote Donnan, "For the first time in this history of science, the method of Gibbs enabled the equation of chemical equilibrium in a homogeneous system to be expressed in an exact and yet perfectly general form."[18]

Development Two: Gibbs' Phase Rule

A second key development from Gibbs' work was the discovery of what has become known as the Gibbs Phase Rule. While the rule is just that, meaning there's no way to prove its existence, its derivation is still of interest as it's a manifestation of the structure that Gibbs built and, in fact, helped serve to validate the structure itself.

From the moment when Gibbs first proposed the existence of U(S,V) as the parent of Clausius' equation and the use of graphical techniques to visualize the relationship between U, S, and V, he assumed the existence of a definitive U-S-V equation. Furthermore, while he acknowledged the need for experiments to determine the actual equation itself, he also acknowledged that such an effort was *not* required to prove the equation's existence. He showed how this single equation could be differentiated to yield all "thermal, mechanical, and chemical properties, so far as *active tendencies* are concerned." Because of this, he called this U-S-V equation a *fundamental equation*. First derivatives of the U-S-V equation, which gives us Clausius' Equation 36.1, yield formal thermodynamic definitions of temperature, pressure, and chemical potential (for when variable mass is included in the equation, indicating an open system), while second derivatives yield such properties as the coefficient of thermal expansion, isothermal compressibility, and the specific heats at constant pressure and at constant volume.

---

[16] The condition of maximum entropy is an additional equilibrium requirement, although the maximum may be a local maximum, reflective of a metastable state. Even in this state, Gibbs' equation is valid so long as the state remains at its metastable value.

[17] Just because there's no gradient naturally doesn't mean there's no movement. Equilibrium doesn't mean zero motion, it just means zero *net* motion. Remember that entropy is a statistical phenomenon. All atoms and molecules have full freedom to move until they achieve their most statistically probable distribution, at which point they're still moving but not in a way to change the macroscopic state of the system.

[18] (Donnan, 1925) p. 463.

The Gibbs–Duhem relation expresses additional relationships between the first derivatives, as we'll shortly see, while the Maxwell relation expresses additional relationships between the second derivatives.

Before moving on, I choose at this point to shift discussion from extensive to intensive properties. Division by mass converts the former to the latter while leaving the fundamental equations valid. The discussion that follows will be based on intensive properties unless otherwise noted.

When Gibbs began studying the means to visualize the U-S-V relationship, his three-dimensional equilibrium surface of "dissipated energy" implicitly reflected the fact that if you defined any two of the properties involved, the third property became fixed. You could go to his graph and see this to be so by picking any two properties, finding the corresponding U-S-V point on the surface, and then determining the third property. You could also naturally do this by simply solving the equation for the third point but Gibbs was emphasizing the graphical approach initially to help further understanding.

Gibbs' use of two properties to fix the equilibrated state of a body was entirely consistent with the history of thermodynamics until that point. Based on hundreds of years of experimental observation, scientists developed the postulate that once you fix two properties for a given body of matter, you fix all other properties for that body.[19] So according to this postulate, the thermodynamic state of a body has two degrees of freedom (DOF). Additionally defining the mass of the body would complete the full characterization of the body, but again, we're only focused on intensive properties right now. The two most often used intensive properties prior to Gibbs were temperature and pressure as these were "primitive" and easily measured. The ideal gas law best reflected this. Once you fix temperature and pressure, you fix volume (for a given mass). But then Gibbs showed that there's nothing special about temperature and pressure. In fact, the DOF value of two would work for other properties, such as any two of his U-S-V coordinates.

Determination of DOF became more complicated once a second phase was added to the system. Historical experience showed that the temperature at which a pure liquid boiled varied with pressure, thus suggesting that a DOF value of one would suffice to describe a two-phase system. But there was no theory otherwise available that led to that answer or that led to the fact that the same answer results for other two-phase systems. How many considered that a T-P relationship tying the two together must exist for ice-water prior to James Thomson's experiment?

In his third paper, Gibbs took this discussion to a higher level. He found that while adding a second phase decreased DOF, adding additional substances increased DOF. Specifically, assuming a single phase for now, if n substances or chemical species are included in the system, then beyond the two properties needed for the single-substance system, n−1 *additional* properties, typically mass (m_i) or mole (x_i) fractions, are required to define the multi-substance system. Note here that you only need to define n−1 mass/mole fractions as the final fraction is determined by the fact that all must naturally sum to 1.0. Thus, with the presence of additional substances, one needs a total of 2 + (n−1) properties to define the system and all properties associated with the single-phase system. Again, Gibbs assumed in his discussion the existence

---

[19] (Tester and Modell, 1997) pp. 15–16.

of the appropriate *fundamental equation*—U(S,V,x$_i$) or U(S,V,m$_i$)—from which all of these properties, now including chemical potential, could be determined for any given fixed point or state.

It was the addition of phases to his addition of substances that really forced Gibbs to shift from graphs to mathematics. It was the only way to handle the resulting hyperspace coordinates. In what was arguably the single largest step he took in his third paper, a step that seems so simple to us now but which at that time had not yet been clarified, as equilibrium itself was still an ill-defined concept, he looked at the seeming complexity of the phase equilibrium mathematics and tremendously simplified it all by realizing that each phase had its own fundamental equation and that each fundamental equation was tied to all the other fundamental equations by equilibrium criteria. It wasn't until Gibbs brought his high-powered mathematics and physical insights to bear on DOF calculations that all became clear in the maze of the completely generalized system of multiple components and multiple phases. The answer fell out of his work.

Gibbs showed that one needed 2 + (n−1) intensive properties to completely define *each* phase in the system. At this point, one might then think that if you had π phases, then you would need π times [2 + (n−1)] properties to define the whole system. But this was not the case. Recall in the world of mathematics that if you have X equations quantifying relations between Y variables, then you only have to determine Y − X of the variables in order to determine the rest. Applying this to the case of heterogeneous equilibria, Gibbs' overlay of his equilibria criteria onto the mathematics reduced the DOF. Specifically, he showed that the condition of equilibria added 2+n equations—representing T-P-μ_i equality between the phases—to the mathematics for each added phase *above* the first phase, (π − 1) in number, thus making the total number of properties needed to fix the system equal to

$$DOF = \pi[2+(n-1)]-(2+n)(\pi-1)=2+n-\pi \qquad [36.4]$$

And finally we arrive at the famed Gibbs Phase Rule. Remembering the derivation involved, this rule quantifies the number of independent *intensive* properties needed to define the thermodynamic state of a body. Once this number of properties has been fixed, all other properties are fixed.

What this rule does *not* say is which intensive variables to use to define the system. Not all combinations work. For example, for a two-component, liquid-vapor azeotrope, two different sets of liquid–vapor compositions can exist for a fixed temperature and pressure depending on which side of the azeotrope the system is on. But if you switch from T-P to T-x$_i$, then this successfully defines the system.[20] This example serves to show that not all intensive variable combinations work in Gibbs' Phase Rule.

In the larger picture, the Phase Rule says that once you're at equilibrium, you can't independently change variables. The equilibrium places constraints on such changes, figuratively tying the variables together. Now there's nothing to say that you can't independently change the variables on a system at equilibrium. It's just that when you do this, you'll break equilibrium. You can raise the temperature above 100 °C on a water–vapor equilibrium system at atmospheric pressure but only one phase will remain. Variables are tied together with equilibrium; change one, you have to change the other to remain in equilibrium. Taking the above example,

---

[20] (Tester and Modell, 1997) pp. 645–6.

you can maintain the two-phase system of water–vapor above 100 °C if you also increase pressure—by an amount quantified by the Clapeyron equation.

It's important to note here that not all intensive properties are equal between different phases at equilibrium. Gibbs identified the three properties critical for defining equilibrium: pressure, temperature, and chemical potential. Other intensive properties, such as density, mole fraction, or even viscosity, are typically quite different between phases. While I have nothing further to say in this regard, this division of intensive properties into two categories, one relevant to equilibrium and the other not, is taken up by Griffiths and Wheeler,[21] among others.

The Gibbs Phase Rule looks seemingly simple but the simplicity belies the deeper meaning. The Phase Rule, and really, all of Gibbs' work, provided us the necessary theoretical structure to understand equilibrium in thermodynamics. While each step along the way is relatively easy to understand on its own (Gibbs' writing didn't always ensure this), the combined steps and especially their interconnectedness are challenging to fully grasp.

In keeping with his "retiring disposition," Gibbs didn't draw undue attention to his Phase Rule, nor did he spend time explaining it with illustrations. He simply stated it as a fact—when you have so many species and so many phases, you only need to define so many properties to define the whole system—and then moved on to other issues. Little did he know that his four-page derivation would give rise to the "The Phase Rule" as one of the core fundamental teachings in thermodynamics. Little did he know that with the help of his Phase Rule, others "created industries and saved countries,"[22] a statement that may seem a little over the top until you read the history described in Chapter 38.

Development Three: The Rise of Composite Properties

The peculiar multiplicity of formulation and reformulation of the basic thermodynamic formalism is responsible for the apparent complexity of a subject which in its naked form is quite simple.

– Herbert B. Callen[23]

We conclude this look at Gibbs' third paper with a third representative development, this one concerning his talented use of multivariable calculus to fully explore the opportunity offered by composite properties (Figure 36.1). Gibbs' mastery of calculus allowed him to easily dance through this discussion, while unfortunately leaving others, including myself, struggling to keep up. But with repetitive reading, it is possible to gain understanding.

In reading his second paper, we see Gibbs' creation of perhaps the first composite property which we now call Gibbs Energy (G), defined as U + PV − TS. In reading his papers, one realizes that while this derivation of G may have been an interesting topic for him in his second paper, the philosophy behind the derivation became a dominant feature in his third, for the beauty of such composite properties was that they could be created in such a way as to provide

---

[21] (Griffiths and Wheeler, 1970)
[22] (Crowther, 1937) p. 280.
[23] (Callen, 1960) p. 85.

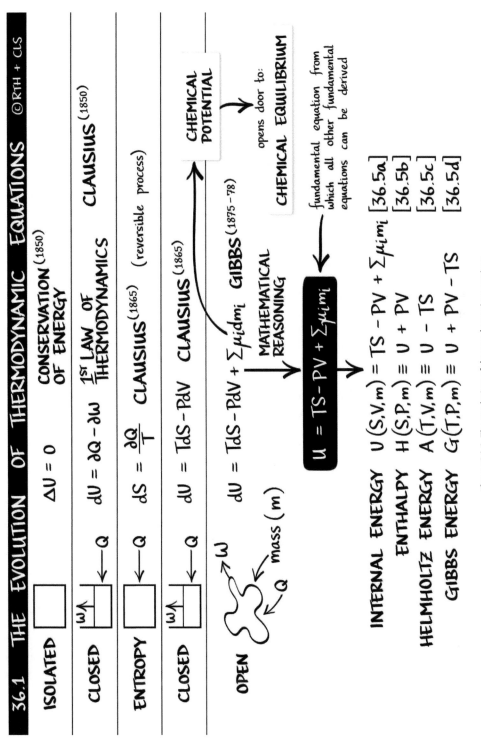

Figure 36.1 The evolution of thermodynamic equations

easier paths to mathematical solutions, with an array of different composite properties available for an array of different problems. As each of these composite properties provided the same thermodynamic information as all of the others, just in different forms, the final selection of which composite property to employ to solve a given problem was made based purely on convenience.

It was the arrival of energy, entropy, and chemical potential that enabled Gibbs to shift thermodynamics from primitive properties to derived properties, which complicated our understanding but expanded our capabilities. Even though none of these new properties could be directly measured (or even physically understood), they built a mathematical bridge from problem to solution. A problem based on such primitive properties as temperature, pressure, and composition could be transformed and solved in terms of composite properties and result in a final answer also based on the primitive properties along with the corresponding energy requirements of work and heat needed to effect the change from problem to solution. In the end, the only things physicists, chemists, and engineers really care about are the primitive properties, since these can be both measured and controlled, and the energy requirements, since these are directly tied to process design and economics.

Gibbs' progression of mathematical thought followed two separate and independent paths, one based on pure mathematics and the other on pure science. He eventually brought the two together.

Pure Mathematics

The first path was based on pure mathematics involving a function X(a,b,c) for which all of the variables are homogeneous to the first degree. For this situation, one of Euler's many theorems states that[24]

$$X = \left(dX/da\right)_{b,c} a + \left(dX/db\right)_{a,c} b + \left(dX/dc\right)_{a,b} c$$

Now let's give these variables new definitions: specifically let's make X(a,b,c) equal U(S,V,m) and let's also recognize that U, S, V, and m are all homogeneous to the first degree, meaning that, being extensive properties, they are first-order in mass (yes, I've switched back to treating these properties in their extensive form for the sake of this discussion). Then given that U(S,V,m) exists, the following must be true

$$U = \left(dU/dS\right)_{V,m} S + \left(dU/dV\right)_{S,m} V + \left(dU/dm\right)_{S,V} m$$

Pure Science

The second path based on pure science embraced the 1st and 2nd Laws of Thermodynamics in the original fundamental equation of Clausius and Gibbs for an open (heat, work, mass exchange allowed), single-phase system as follows.

---

[24] In (Gibbs, 1993), Gibbs did not explicitly reference Euler but instead on p. 87 stated that Clausius' fundamental equation (modified for variable mass) could be integrated by "supposing the quantity of the substance considered to vary from zero to any finite value, its nature and state remaining unchanged."

$$dU = TdS - PdV + \Sigma\mu_i dm_i \quad \text{for } i = 1 \text{ to } n; \text{ open, single-phase system} \qquad [36.2]$$

$$T = \left(\partial U / \partial S\right)_{V,m}$$

$$P = \left(\partial U / \partial V\right)_{S,m}$$

$$\mu_i = \left(\partial U / \partial m_i\right)_{S,V,mj\neq i}$$

Combining Math and Science

Gibbs then brought these two separate paths together and logically concluded that the following equation applies to each phase

$$U(S, V, m) = TS - PV + \Sigma\mu_i m_i \text{ for } i = 1 \text{ to } n\dots \text{ for each phase}$$

with the complete differential of this equation being

$$dU = TdS + SdT - PdV - VdP + \Sigma\mu_i dm_i + \Sigma m_i d\mu_i$$

In light of his differentiated fundamental equation [36.2], this meant that for each phase

$$0 = SdT - VdP + \Sigma m_i d\mu_i \qquad [36.4]$$

We'll come back to this final equation—the famed Gibbs–Duhem equation—later as it quantifies how equilibrium criteria are inter-related.

In summary, Gibbs brought the rules of calculus to the world of thermodynamics. Starting with the original differential [36.1], he showed that the parent equation must be a fundamental equation. His logic was that if a simple linear equation with certain characteristics leads to an equation based on science, then the original equation itself along with all possible variations of this equation must be valid. It was this logic that really opened the doors for Gibbs to pursue further applications of calculus. Indeed, thermodynamics itself is governed by the rules of calculus.[25] The properties of matter change in accordance with these rules as applied to the original fundamental equation—or to be technically accurate, the original differentials of the fundamental equation—of Clausius and Gibbs. Once you work your way through such mathematical proofs, you realize how truly fortuitous—U,S,V,m all first-order ; T,P,μ all zero-order—the structure of the original equation was.

Embracing his mastery of calculus even further, Gibbs then walked through other derivations to develop other coordinate systems, specifically using Legendre's work to demonstrate how to move from S-V-m to S-P-m and from T-V-m to T-P-m. In short, when you take the first derivative of Gibb's original fundamental equation, you generate the intensive properties T-P-μ. Each of these derivatives can be calculated for each point on the U(S,V) surface and then

---

[25] These steps and the mathematical logic involved are not trivial. The formal mathematics are available in many textbooks such as (Callen, 1960) and (Tester and Modell, 1997).

formed into a linear equation with a slope equal to the corresponding T-P-μ property (based on the fixed parameters involved in the differentiation) and an axis-intercept calculated based on the slope and the specific point of interest on the U(S,V) surface. If you do this for each point, you end up with a series of linear equations, one series for each property of interest, and these series can then be used to recreate the original U-S-V-m curve. Using this process, you can transform the fundamental equation from one set of coordinates to another. The details involved go beyond the scope of this discussion but are shared here to again demonstrate how the work of Legendre, Euler, and other mathematicians found post-mortem applications in the world where the properties of matter are tied together by the 1st and 2nd Laws of Thermodynamics.

For the original U-S-V-m system, the slope of the first derivative of interest $(dU/dS)_{V,m}$ is temperature. Thus, for this situation, S can be transformed to T. For the other situations, V could be transformed to P, and S-V to T-P. These partial Legendre transforms together with Euler's theorem can be used to generate the following fundamental equations and corresponding symbols:[26]

$$U(S,V,m) = TS - PV + \sum \mu_i m_i \quad \text{Internal Energy} \tag{36.5a}$$

$$H(S,P,m) \equiv U + PV \quad \text{Enthalpy} \tag{36.5b}$$

$$A(T,V,m) \equiv U - TS \quad \text{Helmholtz Energy}^{[27]} \tag{36.5c}$$

$$G(T,P,m) \equiv U + PV - TS \quad \text{Gibbs Energy} \tag{36.5d}$$

Upon further differentiations and select substitutions involving the complete differential of U(S,V,m) for an open system:

$$dH = TdS + VdP + \sum \mu_i dm_i$$
$$dA = -SdT - PdV + \sum \mu_i dm_i$$
$$dG = -SdT + VdP + \sum \mu_i dm_i$$

Note that it's possible to combine any properties to create new properties. The challenge is to do so in a way that helps solve real problems.

One More Demonstration of the Power of Calculus in Thermodynamics

Of note here is that even though, for example, H is stated to be a function of S-P-m, the property of pressure doesn't appear in the equation as a stand-alone variable (but does importantly

---

[26] Gibbs used other symbols; the ones listed are those in common use today. See, for example, all such functions in Table 5.2, p. 151 of (Tester and Modell, 1997).

[27] (Lindauer, 1962) Helmholtz likely used A for this property to reference Arbeit, German for work, but others suggested that A referred to *affinity* in light of its connection to this theory.

appear in the subscripted conditions). The reason for this goes back to Euler's theorem and the fact that pressure is zero-order in mass. Given

$$H = H(S,P,m)$$

and given that

H, S, and m are all first-order in mass while P is zero-order

Then Euler says:

$$H = (dH/dS)_{P,m} S + (dH/dm)_{S,P} m$$

We know separately that for an open system

$$dU = TdS - PdV + \Sigma\mu_i dm_i \text{ open system}$$

which can be re-written as

$$dU + PdV = d(U + PV) - VdP = dH - VdP = TdS + \Sigma\mu_i dm_i$$
$$dH = TdS + VdP + \Sigma\mu_i dm_i$$

From this we see that

$$(dH/dS)_{P,m} = T$$
$$(dH/dm)_{S,P} = \mu_i$$

which makes the following a valid fundamental equation

$$H = TS + \Sigma\mu_i m_i$$

As shown here, and this can be shown for the other composite properties as well, this fundamental equation for H does not explicitly include P even though H(S,P,m). However, note that this fundamental equation of H is equivalent to the previous fundamental equation, H = U + PV, which does include P and which just goes to further demonstrate how interconnected all of the properties are.

Gibbs very thoroughly and carefully laid out arguments showing that each one of the above equations (and more) based on composite properties provides exactly the same amount of thermodynamic information as any of the others. Each, upon differentiation, yields all the mechanical, thermal, and chemical (so far as "active tendencies" are concerned) properties of matter. Thus each is a fundamental equation in and of its own right. And each is recognized today by the scientifically accepted name assigned to it as shown above.

Gibbs showed exactly how the thermodynamic properties are connected. The composite properties he created were all based on the strict rules of calculus, which meant that not all possible relations, equations, and composite properties that could be created from the properties of matter have relevance as being "fundamental." I note this in light of the fact that others during that time period proposed the existence of other composite properties. For example, as mentioned before, James Thomson proposed the use of P-V-T coordinates, which was fine in and of itself, but, as Gibbs himself pointed out, this approach did not provide all the information that a fundamental equation provided.[28] Just as not all intensive properties work with the Phase Rule, not all composite combinations work to form a fundamental equation. Each must be mathematically linked to the original fundamental equation.[29]

Note that Gibbs' approach was based on using the energy function U(S,V) as his starting point. He could just as easily have used the entropy function S(U,V) that results by re-arrangement of coordinates such as can be visualized by rotating Gibbs' three-dimensional U-S-V graph 90° to provide one based on S-U-V. The equilibrium conditions would change from S maximization to U minimization but as discussed earlier, both are mathematically identical. Thus, both starting points are equally valid and the selection of which to use is again based on whichever is more convenient for solving the problem at hand. The S-U-V approach was the one taken by M.F. Massieu[30] whose work in 1869 predated Gibbs'. Gibbs rightly acknowledged Massieu's work while also noting that his own approach was more comprehensive as Massieu did not address the issue of variable mass.

Using his calculus to look at these thermodynamic mathematics from yet another direction, Gibbs showed that the intensive properties used as the criteria for defining equilibrium could be written in terms of the extensive properties from which they were determined. Thus

$$T = T(S,V,m)$$
$$P = P(S,V,m)$$
$$\mu = \mu(S,V,m)$$

Each one of these intensive-property equations (for a given homogeneous, single-phase mass) is called an *equation of state* (EOS) and is distinctly different from its parent fundamental equation in that the latter contains all the thermodynamic information about a given body while the former does not, as referenced earlier in regard to James Thomson's work with P-V-T coordinates. It would take all of the equations of state to re-create the fundamental equation and this makes sense. When you're differentiating the fundamental equation to determine a given equation of state, you're using partial derivatives. You're differentiating, for example, U with regards to S to determine the function for T while keeping both V and m constant. Thus, you're losing all information about how U changes with V and m in the process.

Because the above intensive properties are equal to the first derivative of the original homogeneous first-order equation and because they themselves are homogeneous zero-order, then it can be proven, once again using Euler, that a relationship exists between them such that once n+1 of the n+2 intensive properties are fixed for a single-phase system, then the final property is also fixed. This relationship, which later became known as the Gibbs–Duhem equation—Duhem independently arrived at this equation—was introduced earlier and is simply repeated here using the more rigorous argument.[31]

---

[28] (Gibbs, 1993) p. 34.

[29] For a good discussion on the limits of constructing such composite properties see (Dill and Bromberg, 2011) p. 139.

[30] (Gibbs, 1993) p. 86.

[31] One might argue at this point that having learned earlier about the Phase Rule calculation, that this equation adds another constraint to the DOF calculation. How can the equilibrium criteria between phases be based on T, P, and μ_i if specifying all of these results in an overly defined system per the Gibbs–Duhem equation? So for a simple system

$$\sum m_i d\mu_i = VdP - SdT \qquad\qquad\qquad [36.4]$$

The above equations and the calculus behind them provide the tools to tailor the equations toward the problem at hand. The effort needed to manipulate the equations has more to do with skills in multivariable calculus than with the need to understand the physical problem. In fact, much of thermodynamics is like this. When confronting any problem, students are taught to first use physics and a physical understanding to frame the problem and to then use calculus to solve it. With the advent of computers, the latter has become much easier almost to the point where knowledge of the calculus is no longer needed. But even with computers, the need to understand the physical model still remains, as it always will. Understanding the physical phenomena involved will always be the true challenge. What is the system we're considering? What is the question we're trying to answer? What is physically happening? Computers will not answer these questions. Again, as Richard Feynman wrote, "You can't do physics just by plugging in the formulas… you have to have a certain *feeling* for the real situation."[32]

Deduction and the Scientific Method

It is universally recognized that… publication [of Gibbs' work] was an event of the first importance in the history of chemistry.

– H. A. Bumstead[33]

The power of rigorous deductive logic in the hands of a mathematician of insight and imagination has always been one of the greatest aids in man's effort to understand that mysterious universe in which he lives.

– F. G. Donnan[34]

Here's a good question: What's the difference between deduction and induction in science? Deduction consists of following a logical progression of thought from a given starting point (Figure 36.2). One moves along a path of pure cause–effect logic to reach an array of conclusions based on that given starting point. The conclusions inferred from the starting point cannot be false if the starting point is true (and no mistakes are made). Induction, on the other hand, is the deductive process in reverse. Induction makes generalizations based on a number of specific observations such as those taken during experiments. While the scientific method is based on both processes, the first step must be induction since induction is primary and deduction depends on it.[35] From many observations, many hypotheses are proposed. Sometimes

---

of n components having n+1 degrees of freedom, typically chosen as T,P,x_i, could one not choose T, P, μ_i instead? And the answer would still be yes, but you'd still end up with n+1 degrees of freedom. Just as x_i has a constraint built into it—the sum of x_i equals zero—μ_i also has a constraint built into it per the Gibbs–Duhem equation. Whether one uses x_i or μ_i, one still arrives at a DOF value of n+1.

[32] (Feynman et al., 2013) p. 56.

[33] (Gibbs, 1993) p. xv. From H. A. Bumstead's introductory biographical sketch of Gibbs.

[34] (Donnan, 1925) p. 483.

[35] In a 2017 email to the author, David Harriman shared that induction is primary, because deduction depends upon it. A valid deduction based on false premises does not lead to any knowledge. Also, induction is both harder and less understood.

36.2 INDUCTION, DEDUCTION, AND THE SCIENTIFIC METHOD

INDUCTION

experiences

data

observations

> HYPOTHESIS
to explain it all

EXAMPLE:

RUDOLF CLAUSIUS
after considering the work of both
Sadi Carnot and James Joule, induced

$$dU = \partial Q - \partial W = TdS - PdV$$

DEDUCTION

J. WILLARD GIBBS
deduced 300+ pages of consequences
from Clausius's induced hypothesis

HYPOTHESIS

consequences that
have not yet
been observed

THE SCIENTIFIC METHOD - STRONG INFERENCE*

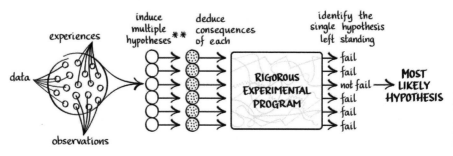

experiences

data

observations

induce
multiple
hypotheses**

deduce
consequences
of each

RIGOROUS
EXPERIMENTAL
PROGRAM

identify the
single hypothesis
left standing

→ fail
→ fail
→ not fail → **MOST LIKELY HYPOTHESIS**
→ fail
→ fail
→ fail

*PLATT, JOHN "Strong Inference" (1964)

**ensure that scientists don't become wedded to any one hypothesis
"many people get married to their ideas all the way to the grave"
(Taleb, 2005) p.240

©RTH + CLS

Figure 36.2 Induction, deduction, and the scientific method

only one is proposed, but in the world of *strong inference*,[36] many hypotheses are proposed to ensure that scientists don't become wedded to any one hypothesis. This process of hypotheses generation from experimental data is induction. Each hypothesis is then assumed to be true and the deduced consequences of this assumption are tested. If by logical sequence the proposed hypothesis doesn't lead to the specific observations, then the hypothesis is assumed to be false. As Richard Feynman famously stated, "It doesn't matter how beautiful your theory is, it doesn't matter how smart you are. If it doesn't agree with experiment, it's wrong." Only the hypothesis that cannot be proven false, the last one standing, is taken to be the correct hypothesis. Even then, this does not constitute a proof. A hypothesis is only taken to be correct if it offers a means to be tested and if those tests can't prove it incorrect.

Why this discussion now? Gibbs' first three papers were all based on deductive reasoning based on Clausius' initially induced starting point. Gibbs assumed one single equation to be true, Clausius' fundamental equation (in its derivative form) based on the two laws of thermodynamics, and then deduced all of the cause–effect consequences from this assumption using very tight logic and rigorous adherence to the exclusion of assumptions, leaving no weak links in his deductive chain. Thus, by logical reasoning, he knew that what he wrote must be true so long as he didn't make any logical mistakes. He didn't. In holding back from allowing any assumptions into his chain of logic, he contributed to physical chemistry more greatly than he might have with any more "reckless gesture."[37] With this approach, Gibbs "[handed] down the law."[38] He made no experiments; he induced no generalizations. He required no interactions with others, no discussions around, "*What do you think the results of this experiment means?*" He was working in the opposite direction.

His self-confidence in this process was remarkable. Some called him "diffident."[39] This clearly was not the case. Yes, he was humble and shy from the standpoint of not wanting recognition. But he certainly wasn't diffident. He certainly didn't lack self-confidence. He knew the shot was going to go into the basket. He didn't have to wait and see. He was heading back to the other end of the basketball court right after the ball left his hands. He was that sure. He did not spend time arguing for his own priority on many aspects of the science involved. He knew that he had established truth and that was all that mattered to him. Many others would later "re-discover" Gibbs by arriving at findings that he had already found, learning only later in a "Kilroy was here" moment that Gibbs had already completed that subject as well.

Gibbs Revealed the Central Theories of Classical Thermodynamics

Completeness is the only simplicity, even though the language be more complicated. It is complicated in order to represent complexity in the most direct way

– Muriel Rukeyser[40]

---

[36] (Platt, 1964)
[37] (Rukeyser, 1988) p. 322.
[38] (Rukeyser, 1988) p. 366.
[39] (Ostwald, 2016) p. 180. Ostwald: "Willard Gibbs was an excessively modest and diffident scholar."
[40] (Rukeyser, 1988) p. 199.

In reading Gibbs' work, it is truly amazing to see how much he accomplished with so little to start with. He took a single equation from Clausius that incorporated a new property of matter along with the characteristic behavior of that new property and then brought intense focus on defining with pure logic the implications of what he had. From seeming simplicity, he created great complexity, but complexity of the kind that was needed. The subject itself was hard and the complexity surrounding it before Gibbs was confusing. Gibbs saw through the confusion to the true fundamental underpinnings of the subject and explained what was happening. He did this with utmost brevity, challenging himself as he shared in 1881, "One of the principal objects of theoretical research in any department of knowledge is to find the point of view from which the subject appears in its greatest simplicity."[41] And while the end result was challenging to comprehend, there really was no alternative. It was a difficult subject to begin with. Years of experimental data laid waiting for a central theory to tie it all together. Without such guidance, scientists could have moved off in a multitude of different directions, creating one theory for each direction. Confusion could have reigned. Noted Professor Brown of Yale, "Before Gibbs' work came along chemistry had gotten to the point where it seemed there was nothing more to do than merely work out details. Gibbs changed the whole face of that by his work."[42] Chemistry was indeed more empirical than theoretical prior to Gibbs and maybe this atmosphere led many to believe that there wasn't even a problem to be solved or an opportunity to be seized. Regardless, absent a binding theory, things could have become very messy at this point. But thankfully, in Gibbs' hands, the central theory was revealed. And thankfully this was done in a single publication, his third, the only one needed. The textbook to explain the interconnectedness of it all from a central core theory arrived fully complete.[43] What a wonderful moment in the history of physical chemistry.

Gibbs' theory of heterogeneous equilibria provided the structure to organize, analyze, and interpret data and ultimately to explain what had previously been without explanation. His work was the capstone on the long development of classical thermodynamics and provided a new starting point and direction for those who followed.

What was even more amazing about his work was that it was all done without connection to any atomic theory of matter. This wasn't happenstance. It was by design. In an era when "much of chemistry was firmly based on the assumption of atoms and molecules,"[44] Gibbs took great effort to keep such assumptions out of his work. As he wrote in his *Elementary Principles in Statistical Mechanics* (1902): "Certainly, one is building on an insecure foundation, who rests his work on hypotheses concerning the constitution of matter."[45] It's not that he wasn't following the advances as certain sections of his third paper clearly showed. It's just that he realized he could conduct the core of his work without it, just as Clausius had done, and also realized that any attempt to marry the two risked what befell Rankine.

---

[41] (Wheeler, 1998) pp. 88–89.

[42] (Pupin, 1927)

[43] (Gibbs, 1993) p. xxiv [H. A. Bumstead's introductory biographical sketch of Gibbs]. The fact that the third paper arrived as a complete package, with no subsequent editing or additions required, reflected on Gibbs' tendency to never present his work to the public until he was satisfied with the logical structure on which it was based. The unfortunate consequence of this, as noted by one of his former students, was that his students were, in the words of H. A. Bumstead, "deprived of the advantage of seeing his great structures in process of building."

[44] (Kragh, 1993) p. 405.

[45] (Gibbs, 2008) p. x.

The progress classical thermodynamics achieved without reliance on a theory of matter was indeed amazing, as well captured by Albert Einstein:

A theory is the more impressive the greater the simplicity of its premises is, the more different kinds of things it relates, and the more extended is its area of applicability. Therefore the deep impression that classical thermodynamics made upon me. It is the only physical theory of universal content concerning which I am convinced that, within the framework of applicability of its basic concepts, it will never be overthrown.[46]

but also limiting, as well captured by McC. Lewis:

The conclusions we shall arrive at on the basis of thermodynamics are thus independent of any molecular hypothesis we may have formed in respect of the process. This, it will be seen, is in many ways a great advantage. It means that the conclusions of thermodynamics are quite general, and will remain true even if our views regarding the actual mechanism of the process considered from a molecular standpoint were to undergo radical change. Of course, it will be seen as the converse of this that thermodynamical reasoning and conclusions, although true in general, do not tell us anything of the mechanism involved in given process. This is naturally a considerable drawback, for advances in theoretical treatments seem to be most easily made along mechanical lines of thought, i.e., with the aid of molecular hypotheses.[47]

---

[46] Einstein quote cited in (Klein, 1967).

[47] Lewis quote cited in (Kragh and Weininger, 1996) p. 128. William Cudmore McCullagh Lewis (1885–1956).

37 Practical applications and Gibbs energy (G)

Classical thermodynamics doesn't tell you *how* to design a process to achieve a certain goal. But it does tell you whether or not any process you do design is even possible. With the 1st Law providing boundaries on energy requirements and the 2nd Law providing additional boundaries based on the fact that all systems have a natural tendency to move toward the most probable state, the combined laws define the ultimate boundary between possible and impossible. Your design can't go beyond Gibbs' surface of dissipated energy.

Taking another perspective, one could argue that any process one can envision is indeed possible so long as one has sufficient energy to make it happen. From this perspective, one could include the concept of external work in the thermodynamic equations. This approach is often used by physicists, chemists, and engineers when they address the question, *how much work can I generate from this process?*, or, *how much work will be required to make this process happen?* The former case is a "maximum work" calculation and the latter a "minimum work" calculation. Both lend themselves to solution using classical thermodynamics.

Including the concept of external work in one's calculations enables design of an *ideal* process that exists at equilibrium for which the overall change in entropy of the process is zero (infinitely slow process) and thus the amount of work is either maximum (generation) or minimum (consumption). Such calculations are so powerful in the conclusions they enable, telling one whether some concept or grand idea is economically feasible or even possible, that I wanted to use this chapter to draw your attention to the fundamentals behind the calculations, fundamentals based on entropy and arguably developed best by Gibbs, as initially shared in Chapter 35 and more fully developed here.

Maximum Work, Free Energy, Available Energy

Classical thermodynamics generally deals with equilibrated systems for which entropy is at a maximum. But for many industrial systems such as power generation, the initial system isn't equilibrated, for if it were, no power generation would be possible. For example, when we burn coal to generate steam to do work, the initial mixture of air and coal is clearly not equilibrated. So how do we approach such problems?

The ability to produce useful work starts with an energy gradient of some kind. A waterfall – mechanical. A steam engine – thermal. An internal combustion engine – chemical. The nature of the gradient doesn't matter; instead what matters is the fact that the initial system is not at its most probable state, that the system will naturally move toward that most probable state

Block by Block: The Historical and Theoretical Foundations of Thermodynamics,
Robert T. Hanlon, Oxford University Press (2020). © Robert T. Hanlon.
DOI: 10.1093/oso/9780198851547.001.0001

(and so dissipate the energy gradient and increase entropy) and that this natural tendency can be taken advantage of to do useful work. As Callen stated, "The propensity of physical systems to increase their entropy can be channeled to deliver useful work."[1]

Energy gradients in such forms as, for example, mechanical (P), thermal (T), chemical (μ), can exist either internal to a given system or between the system and the surroundings or both.[2] The problem to solve in any of these situations is this: how to calculate the maximum amount of work that can be generated by dissipating the gradients? According to Gibbs' energy minimization theory, the maximum amount of energy that can be removed from a system occurs when fully *reversible* processes are employed, thus leaving system entropy unchanged (Figure 35.1). As the infinitesimal energy gradient dissipates, entropy wants to increase. However, because external work is reversibly generated and causes system energy to decrease, entropy also wants to decrease. The two actions—gradient dissipation and external work production—affect system entropy in opposing directions and cancel each other out, thereby leaving system entropy constant.

The practical challenge in the above theory is this. How does one go about calculating the maximum work based on such an ideal but non-defined process? It sounds rather complicated, doesn't it? Fortunately, Gibbs provided us a way out, an approach that doesn't even depend on the process chosen but instead depends only on the initial and final states.

* * *

Consider an isolated but non-equilibrated system, one not on Gibbs' surface of dissipated energy, such as a system comprised of two chemicals that can react to create a contained explosion. How could work be produced from this reaction? One option would be to simply have the system react inside the vessel. In this scenario, system energy would remain constant while entropy would increase to a maximum, thus placing the system on the surface of dissipated energy. But just because it's on the surface doesn't mean it can't still do work. The surface simply says that the system itself is internally equilibrated but it doesn't say that the system is equilibrated with the external environment. In fact, when reactions are allowed to occur in an isolated vessel, temperature and pressure will typically change thus resulting in a T-P gradient between the system and the environment. This affords the opportunity to use some combination of reversible processes such as adiabatic expansion in a turbine or a series of Carnot heat engines to generate work while bringing the system temperature and pressure back to ambient conditions.

A second option—equally effective since it's only the end state that matters—would be to place a connection between this otherwise isolated initial system and an external device that could generate work. For example, in an electrochemical cell the entire set-up could be considered as the system and the voltage generated between the two electrodes could be used to operate a motor to lift a weight. In this scenario, the chemical potential energy stored inside the energy gradients is transformed into the mechanical potential energy of external work. Again, design details aren't needed here. We're simply interested in thermodynamics limits, which don't depend on the path taken. With the electrochemical cell, according to the 1st Law, as the external weight is lifted, the total internal energy of the isolated system must decrease by exactly the

---

[1] (Callen, 1985) p. 103.
[2] (Tester and Modell, 1997) See Section 14 for their approach to accounting for both.

same amount. If this were all done reversibly and if no heat were exchanged with the external environment, then the entropy of the enclosed system would remain constant. So in this scenario, as the gradient dissipates, energy is removed from the system to maintain constant entropy while driving energy down to a minimum value. The end state of the system (no gradients remaining) falls onto the surface of dissipated energy at a point having the same entropy as the initial system and thus at the lowest energy corresponding to that entropy. System energy is minimized at constant entropy.

As discussed by Gibbs in his 2nd paper, with the second option, since there's no PV-work done by the system (constant volume vessel) and since entropy remains constant (reversible processes, no heat exchange with external environment) then

$$\text{Maximum Work} = -d\left(U^{body}\right)_{S,V}$$

As the internal gradients dissipate and external work is done, the energy of the enclosed system decreases to a minimum at which point no more energy gradients remain and the constant value of entropy then becomes a maximum for the corresponding energy minimum. Remember that Gibbs demonstrated the mathematical equivalence between this and the other scenario in which entropy is maximized while energy remains constant.

While you could do these calculations for the second option, the answer you get might not be as useful as you'd like. The reason? Reducing energy at constant entropy does not imply constant temperature and pressure. The values of T-P for the internal system would no longer be the same as the initial values. As discussed previously, in the real world, reactions are typically carried out at constant T-P, such as in a heated bath in the lab open to the environment or inside a large well-controlled chemical reactor. So the answer above would not align with the reality of your process and you would need to make a correction, one that would bring the state of the system back to the initial T-P starting point. Now you could do this by the same procedure outlined above by using ideal reversible processes to move the body from one set of T-P conditions to another, calculate the net work required (or generated) and use this value to adjust the total work generated from gradient dissipation. Or you could use a simpler approach by putting a constant temperature–pressure medium inside the system and touching the body such that thermal (Q) and mechanical (W) energy could be reversibly exchanged. In this way, while the body + medium combined entropy would remain constant along with the combined temperature and pressure, the body's entropy could change. But because the change would be reversible, the change in entropy of the body would be countered by the change in entropy of the medium, such that total internal entropy would remain constant. This brings us to the solution offered by Gibbs.

The Use of Gibbs Energy (G) to Quantify Maximum Work for Constant Temperature and Pressure

Look again at the following equation from Chapter 35, modified by the addition of a term for Gibbs energy, for the body of interest (closed system: heat and work exchange allowed; not mass)

$$\left(dG\right)_{T,P} = d\left(U + PV - TS\right)_{T,P} \leq 0 \qquad \text{for the body} \qquad [35.2]$$

Gibbs derived this equation for a given body by assuming the total system to be comprised of the body of interest plus a large medium to maintain constant temperature and pressure. This equation was written as an inequality to show that as total system entropy increases toward its maximum value and as the body exchanges both mechanical and thermal energy with the medium to maintain constant T-P conditions, then the above quantity decreases to a minimum at which point $(dG)_{T,P}$ for the body equals zero.

When the total system reaches equilibrium, Equation 35.2 becomes the equality $(dG)_{T,P} = 0$. Consider though that there's another way to turn Equation 35.2 into an equality and that is to create an equilibrium balance between the initial non-equilibrated body and an external weight, sort of like balancing a lever by putting a weight on one side to balance a weight on the other. When Clausius first developed his fundamental equation, the work effect was typically based on the volume-change of the system. But naturally other types of external work are also possible, such as those that can be generated from electrochemical cells. Again, the process design is irrelevant to this discussion. Thus, a more general equation for a system could be written:

$$dU = TdS - \sum \left(\text{External Work} \right)$$

Any work that is done external to the system must necessarily remove an equivalent amount of energy from the system as dictated by the 1st Law. Furthermore, for a reversible process (no loss of heat to the environment), the external work must necessarily be a maximum since TdS (of the total body + medium system) equals zero, meaning that all of the decrease in energy goes to external work and nowhere else. So given this, let's rearrange the equation one more time by splitting External Work into its component parts

$$dU = TdS - PdV - \text{Maximum Work}$$

This can be re-written as follows:

$$\text{Maximum Work} = -\left(dU + PdV - TdS \right)_{T,P} = -\left(dG \right)_{T,P}$$

As mentioned earlier, the caveat here is that both body and medium are present inside the system. What this means is the entropy for either can reversibly change since the entropy of the other would have to change by exactly the same amount in the opposite direction. It's the entropy of the total system that remains constant in this conceptual thought experiment to quantify the maximum work potential.

At times the symbol Δ is used instead of the differential since Maximum Work is defined as the finite difference in Gibbs energy for constant T-P between the initial and final states of the body. Since $\int dx = \Delta x$, then

$$\text{Maximum Work} = -\left(\Delta H - T\Delta S \right)_{T,P} = -\Delta G_{T,P} \left(\text{based on body properties only} \right) \qquad [37.1]$$

In addition to aligning with reality, the maintenance of both temperature and pressure also helps facilitate the mathematics as it converts the analysis from path to state dependency: the total amount of external work possible is fixed by the two states. Here one can see how the structure of Clausius' equation involving temperature and pressure was truly fortuitous.

The *change* in energy resulting from the process is often referred to as the "available energy" or the change in free energy; the terminology and exact definitions have been somewhat confusing throughout history. It's the *difference* between two energy states that quantifies the energy stored in the internal gradients and thus represents the "maximum work" that can possibly be generated by reversibly dissipating these gradients. The energy value of each point is called either Gibbs energy or Helmholtz energy depending on the conditions assumed. Neither Gibbs nor Helmholtz originated the idea of such calculations; it's just that they brought the new property of entropy into their solutions.

To reiterate what I stated earlier, while the overall change in system entropy for this thought experiment is zero, the TdS term must still reside inside Equation 37.1 as it applies to the body only. This term is needed to quantify the exchange of thermal energy between the body and the medium required to maintain constant temperature conditions. Because this exchange is reversible and at constant temperature, the change in entropy of the body is equal to the negative of that of the medium such that their sum is zero. Thus, in this set-up, it's possible that the entropy change of the reaction at constant T-P could be negative and yet that the reaction would still proceed and generate work, something that would never happen if the body were alone in the isolated vessel. The heat supplied by the medium can play a critical role in ensuring that a negative-entropy reaction at constant temperature and pressure proceeds. More will be discussed on this important point in Chapter 39.

So with Equation 37.1 we arrive again at Equation 35.3 but with one critical difference. While Gibbs used Equation 35.3 to assess the stability of thermodynamic equilibrium, he used the concepts contained in Equation 37.1 to assess the maximum work a body could generate during equilibration.[3] The earlier relationship was written as an inequality to show the direction in which G moves during the proposed equilibration process. Equation 37.1 was written as an equality to quantify the amount of external work that can be generated by a given process, which can also be viewed as the amount of external work required to achieve a balanced equilibrium. Let's discuss this in more detail now by considering in more depth what this equation means.

Interpreting $(\Delta G)_{T,P}$ as regards Chemical Reaction Spontaneity

Let's start with the easy analysis first and consider this in light of Gibbs' graphics (Figure 35.1). If $(\Delta G^{body})_{T,P}$ equals zero, the distance between the initial state and Gibb's surface of dissipated energy is zero, meaning that the initial state is already on this surface and internally equilibrated (maximum entropy; no internal energy gradients) and thus has zero capacity to generate useful work. Furthermore, there may be another point on the dissipated surface having the same temperature and pressure (using Gibbs' conceptual three-dimensional surface concept— the same tangent plane touches both points) for which the value of Gibbs energy is the same, meaning that G^{body} is the same for the two points. These two states could co-exist with each other. In other words, the two states would be in equilibrium with each other. But again, all of this is discussed in the context of these states being on the dissipated energy surface and thus offering no potential to generate work based on internal gradients. (Naturally if a gradient exists between the body and the environment, work generation is indeed possible.)

---

[3] (Gibbs, 1993) pp. 49–54.

Now consider that we're looking at the reaction of A to B. If we calculate $(\Delta G)_{T,P}$ for this reaction and find it to be equal to zero, meaning that no work could be generated from this reaction, then this doesn't mean that nothing happens, that the reaction doesn't go. This calculation is based on 100% conversion of reactants to products. So $(\Delta G)_{T,P} = 0$ simply means that there's no thermodynamic driving force for *complete* conversion. In such situations, a more accurate tracking of the value of Gibbs energy for the entire system, one accounting for the extent of reaction and the presence of mixtures, could show that this value is minimized for partial conversion conditions. For example, if $(\Delta G)_{T,P} = 0$ for 100% conversion of A to B, then G itself would be minimum for a 50:50 mixture of the two, for it is this mixture that represents the most probable state and thus maximizes entropy.

What if $(\Delta G^{body})_{T,P}$ is negative for the reaction? This means that external work is possible, which in turn means that changes in the body will occur on their own, spontaneously, even if as mentioned above the entropy of the body itself decreases. Entropy need only increase in an isolated system. Whatever internal energy gradients exist, regardless of type, will dissipate naturally and the state of the system will move from the non-equilibrated, solid region of the Gibbs graph to the equilibrated surface. The value of external work as quantified by the value of $(\Delta G^{body})_{T,P}$ is the maximum possible and is, in effect, what would be required to balance the system, prevent change, and establish equilibrium. The design of such a system is not relevant. But in concept, if one could connect a work-generating device to the system and load it with a "maximum work" weight, then the entire set-up of system plus external weight would be in equilibrium. This can be made to happen in an electrochemical cell as will be discussed in more detail in Chapter 39.

Finally, what if $(\Delta G^{body})_{T,P}$ is positive? Does this mean that the proposed change won't happen? No. It just means that it won't happen spontaneously. No external work could be generated from the proposed change and, in fact, external work would have to be done *on* the system to make the proposed change happen. The maximum work would be a negative number, meaning that it would then become the minimum work required to make the process happen.

The entire situation encompassing all three scenarios is best observed in an electrochemical cell (T-P constant). A spontaneous reaction for which $(\Delta G^{body})_{T,P} < 0$ can be stopped by creating a voltage between the electrodes in the opposite direction of the reaction, equal in absolute magnitude to the change in Gibbs energy. In this situation, both the reaction and the flow of electricity would stop. You could slightly increase or decrease the voltage to make the reaction go in one direction or the other. Thus, you could make a ΔG-positive reaction occur by creating a gradient where there hadn't been one before. In this case, the voltage would conceptually move the system *away* from the surface of dissipated energy. The electrochemical cell will be discussed in more detail, again in Chapter 39.

Summary

The change in Gibbs energy quantifies the amount of external energy needed to balance the energy-releasing potential offered by a body containing internal energy gradients. In the mechanical world, it's the counter weight that balances a system and, in effect, quantifies the potential energy stored in the system. If you remove this weight, the system accelerates, kinetic energy results, collisions happen, heat is generated, and ultimately is lost to the environment. For example, in an electrochemical cell, you can operate the cell to generate electricity, but without a "counter balance," the electricity flow itself generates heat, which is lost to the environment. A counter voltage would slow the electrical current to the point where the system would be operating at near reversible conditions with a corresponding reduction in lost heat effects. However, most commercial processes require high rates to be profitable. And so arises yet again the classic conflict between rate and thermodynamic efficiency. This is a subject for a separate book.

While I don't want to go further with this discussion of chemical equilibrium as many textbooks already address this subject, I will say this. The same concepts involved in determining Gibbs energy are used in the study and interpretation of chemical equilibrium. Some new terms were created to facilitate these studies since scientists found that working with Gibbs' chemical potential—which is tied to Gibbs energy by the equation, $\mu_i = (\partial G/\partial m_i)_{T, P, m \neq i}$—led to certain problems with the mathematics, such as, for example, the awkward result of the chemical potential for a gas going to negative infinity as pressure goes to zero. Gilbert Lewis proposed a solution to this problem[4] by suggesting a new mathematical term called *fugacity* (*f*) to replace chemical potential as a means to quantify "escaping tendency" (Lewis) or "active tendency" (Gibbs) of a chemical species. This concept later evolved into the creation of a related term called *activity* (*a*) to quantify fugacity relative to a reference fugacity. As summarized in Lewis and Randall's highly influential textbook *Thermodynamics* (1923),[5] both of these concepts have proven to be very powerful in enabling easier mathematics for solving thermodynamic problems and are used today in computer models to solve mass and energy balances around processes involving equilibrium. But when all is said and done, such calculations ultimately still, over one hundred years later, rely on two of the key ideas presented by Gibbs as he considered the concept of equilibrium: Gibbs energy minimization and chemical potential equalization. Both approaches provide the same answer since the occurrence of either results in the other. The question of which approach to use is based largely on computer efficiency for the given problem.

---

[4] (Lewis, 1901)
[5] (Lewis and Randall, 1923)

38 Dissemination of Gibbs' work

It is a great pity that many cultivators of the science of thermodynamics since the time of Gibbs have not gone back to the fountainhead and closely correlated their results with this.

– F. G. Donnan[1]

The exceptional man is always in advance of his generation. Willard Gibbs was not alone in failing at first to attract the attention of the scientific world. He simply could not be understood.

– Michael Pupin[2]

Newton, his closest friend, had been asked whether he followed what Gibbs was saying, at a lecture; he had answered; "I saw how the second equation was derived from the first, and the third from the second; after that I was lost."

– Muriel Rukeyser[3]

As monumental as Gibbs' work was, it took some time for it to disseminate into the scientific community. Gibbs brought clarity, he really did. It's just that it was and continues to be hard to see. It's hard to get through his writings; his use of his own symbols didn't help. Each step along the way is relatively easy to understand but the totality of it all, and especially the interconnectedness of it all, is difficult to grasp. Bringing clarity to the complex can still leave things complex. Sometimes they just are.

It's rather interesting to imagine what it must have been like back in the late 1800s to first read Gibbs' three papers on thermodynamics and especially his third. Scientists were still grappling with entropy and so trying to follow the logic of its implications was a significant stretch.

Perhaps in the same way that Clapeyron saved Carnot, Maxwell saved Gibbs, well, maybe not saved, as Gibbs' work would certainly have gotten out sooner or later, especially with the help of his students, but certainly accelerated the dissemination of Gibbs throughout England and Europe. Simply put, it takes a genius to recognize a genius. Maxwell was the genius who first recognized Gibbs. He read Gibbs and was inspired, perhaps because, as with Gibbs, he too had a penchant for a geometric approach to thermodynamics. He put his own stamp of approval on Gibbs.

Maxwell also took to Gibbs' style of "strong, finely focused principles, capable of being developed in a thousand ways."[4] In a way, Gibbs became an independent third party in the

---

[1] (Donnan, 1925) p. 466.
[2] (Pupin, 1927)
[3] (Rukeyser, 1988) p. 329.
[4] (Rukeyser, 1988) p. 251.

Block by Block: The Historical and Theoretical Foundations of Thermodynamics,
Robert T. Hanlon, Oxford University Press (2020). © Robert T. Hanlon.
DOI: 10.1093/oso/9780198851547.001.0001

world of thermodynamics, a new voice that attracted Maxwell (and the Scottish school) as Maxwell had strong reservations about the second party, namely the abstractly theoretical Germans. To Maxwell, "[Gibbs] has more sense than any German."[5] Wrote Francis Everitt, "Gibbs had much in common with Maxwell. Like Maxwell he would choose some large topic…give it prolonged study, and then write a paper or book drawing all together in a grand synthesis. Among British physicists of the 1800s Gibbs was often referred to as the Maxwell of America."[6]

Maxwell's support and validation helped, at least to some extent, the scientific community comprehend Gibbs. It was a difficult enough subject to begin with, and while Gibbs handled it with exactness and brevity, this approach was both blessing and curse. The content Gibbs covered could have easily filled ten times the volume in lesser hands; "he was prolific as nature, in the limits he allowed himself."[7] So it was a blessing that he covered so much with such brevity. But there is such a thing as too much brevity. Gibbs chose each and every word very carefully. He wrote exactly what he needed to write to convey his points but no more. He chose very specific words and placed them into very specific sentences, using only as many words as exactly required. Gibbs stated things one way, his way. And this would have been fine if the reader knew exactly what Gibbs meant. But if not, and if a single step in the logic chain was not understood, then the reader had to start again, and again, and again, until such a time that things finally clicked. As Einstein wrote in reference of Gibbs' later work on statistical mechanics, "[Gibbs] is hard to read and the main points have to be read between the lines."[8] It was because of such hardship that Maxwell's "translation" of Gibbs into the simpler English of his own famed *The Theory of Heat* had such a positive impact. While Gibbs may have exceeded Maxwell in rigor, Maxwell clearly exceeded Gibbs in the transfer of knowledge.[9] Ever the teacher, Maxwell approached writing and speaking with a different philosophy than Gibbs as reflected in his writing: "there is no more powerful method for introducing knowledge into the mind than that of presenting it in as many different ways as we can."[10] It was Maxwell who first helped "in liberating the beautiful work of Gibbs from Gibbs' writing."[11]

The two complemented each other very well; together they certainly would have accelerated the development of thermodynamics. Unfortunately, this wasn't to be, for in 1879 Maxwell died of abdominal cancer at the rather young age of forty-eight. The contact between the two was broken, resulting in "one of the most tragic wrongs of waste in the history of science."[12] Noted a member of the Connecticut Academy, "Only one man ever lived who could understand Gibbs' papers. That was Maxwell, and now he is dead."[13]

---

[5] (Smith, 1998) p. 263.
[6] (Everitt, 1984) p. 120.
[7] (Rukeyser, 1988) p. 436.
[8] (Klein, 1983) p. 143.
[9] (Crowther, 1937) p. 269.
[10] (Rukeyser, 1988) p. 209.
[11] (Rukeyser, 1988) p. 351.
[12] (Rukeyser, 1988) p. 251.
[13] (Rukeyser, 1988) Quote from unidentified member of the Connecticut Academy as included in Rukeyser, p. 251.

Path 1: Gibbs → Maxwell → Pupin → Helmholtz → van't Hoff → community

Gibbs' work spread throughout Europe along two separate paths. As already mentioned, the first started with Maxwell. It was his book that indirectly led Gibbs' work to Helmholtz along an interesting path taken by Michael Pupin (1858–1935).[14] As a doctoral student in experimental physics at the University of Berlin under Helmholtz, Pupin was familiar with Maxwell's *Theory of Heat* and thus had read of Gibbs. When Pupin realized that Helmholtz was interested in the same types of problems as Gibbs, he undertook to learn more deeply about Gibbs' work and so obtained the original publications from the *Connecticut Transactions*. He "studied, studied, and studied them" until he finally understood them. He then shared what he learned with Helmholtz along with the rather brash statement, "Everything you have done and everything the physical chemists are doing today, is apparently all in Gibbs." He even wrote in his thesis, "This whole theory of physical chemistry of today is contained in Gibbs, and the science of the physical chemistry was made in the State of Connecticut and not in Germany [as was supposed at the time]." In a testament to his sense of fairness, Helmholtz approved this. Pupin would latter go on to his own success as a famed physicist and physical chemist.

It should be noted that while Helmholtz did unknowingly cover ground already trodden by Gibbs, his work was more influential to chemists. Expanding on his work in 1847 on energy conservation, Helmholtz published a very influential memoir in 1882 on chemical thermodynamics. As proposed by Kragh,[15] it was through this memoir and subsequent elaboration by van't Hoff that chemists discovered thermodynamics.

Path 2: Gibbs → van der Waals → Roozeboom → Community

The second path of dissemination throughout Europe began with another early admirer of Gibbs, Johannes Diderik van der Waals (1837–1923), a Dutch theoretical physicist. It was really this pathway that ended up validating Gibbs and putting him on the world stage. As recounted by Daub,[16] the relevant timeline started in 1884 when Rakhuis Roozeboom (1854–1907), a Dutch chemist, began his experimental research on the complex phase equilibria of the hydrates of water with different elements and compounds, including HBr. The complexity of such studies was enhanced by the fact that water could form multiple hydrates with any such given species. Thus, such systems provided a tremendously rich and complex world for the study of multiple species and multiple phases—solid, liquid, and vapor. For each of his systems, Roozeboom carefully mapped out the T-P-x_i data for the liquid and multiple solid phases involved and then used the data to create fascinating phase diagrams. For the water plus HBr system, he arrived at diagrams that he couldn't fully explain, which is when he turned to van der Waals for help. Van der Waals saw this problem as an opportunity to apply Gibbs' work and so developed a set of thermodynamic equations based on the phase rule and also on the

---

[14] (Pupin, 1927)
[15] (Kragh, 1993) p. 417.
[16] (Daub, 1976)

Clausius–Clapeyron equation and ultimately solved the puzzle, providing in so doing insight into the existence of a new hydrate ($HBr*H_2O$), a discovery that wouldn't have been possible without the guiding theoretical work of Gibbs. In his publications on this success, Roozeboom proposed that chemists adopt Gibbs' phase rule as the structure for classifying complex dissociation equilibria. Thus were the Dutch among the first to experimentally validate Gibbs and establish his theories in chemistry.

While Gibbs spent little time writing about the application of this work to industry, others did the job for him, with Roozeboom being one of the early adopters. For example, prior to Gibbs, there had been little to no theoretical understanding of metal alloys. Roozeboom used Gibbs' work to interpret the properties of steel as a two-component system and in so doing turned metallurgy into a modern science. It wasn't long before such successes inspired others in other fields. Soon Gibbs' work became the means by which to bring theoretical understanding to other industrial chemistries. Wrote Tammann, "No abstract work has had such decisive influence on the development of basic industries as Gibbs' memoir on heterogeneous equilibria."[17]

Francis Arthur Freeth – Gibbs' Phase Rule in Practice

It was hard for me to comprehend the impact of Gibbs' Phase Rule on industry. As mentioned before, I had originally thought that the accolades were somewhat over the top. But then I started to read more about this history and found it fascinating, finally understanding the significance of Gibbs in this area, just one of many.

A good appreciation of this history is provided by the experience of Francis Arthur Freeth (1884–1970).[18] Born in England and raised in a family of strong military tradition, Freeth received his education at Liverpool University where he found his passion in physical chemistry, receiving his M.Sc. under F. G. Donnan (1870–1956) in 1906. Donnan had worked with Ostwald and Jacobus Henricus van't Hoff (1852–1911) and believed that research in the new field of physical chemistry, especially including the work of Gibbs, was critical to understanding chemical theory. This belief strongly influenced Freeth. The two became friends.

Freeth's professional work began in 1907 at Brunner Mond & Co and within a year he rose to Chief Chemist at the age of 26. The company's chemistry focused on the targeted precipitation of salts, such as soda ash (sodium carbonate) from aqueous solutions during which both water content and temperature played key roles. In 1909 the Brunner Mond leadership asked Freeth to begin investigating the process requirements to produce ammonium nitrate by double decomposition of sodium nitrate and ammonium sulfate. Putting his education under Donnan to practice, Freeth took a remarkable journey, starting with a literature search. While there wasn't much, what he did find was extremely valuable. It was a paper by Dr. F. A. H. Schreinemakers (1864–1945), another member of the Dutch school of chemists involved with phase theory. The paper was in Dutch; Freeth was undaunted. He taught himself the language, read the paper, discovered the world of Dutch chemistry and soon became an established authority on the phase rule, whereupon he soon became teacher himself.

---

[17] (Crowther, 1937) Reference to G. Tammann's writing, p. 279.
[18] (Allen, 1976)

With a much deeper background in phase theory, Freeth then put his mind to immediate use. Brunner Mond had a huge collection of data on many different chemical processes involving inorganic salts. He brought theory to the data and began to make sense of it all. You would think this would be relatively easy if you had only seen simple phase diagrams involving salt precipitation. But in reviewing the phase diagrams of many of these systems, you would soon realize how extremely challenging it was to make sense out of the data, as well illustrated in *The principles of the phase theory* by Douglas A. Clibbens,[19] himself a student of Freeth. One seemingly simple example captures it all. In Figure 6 (p. 48) of that book is shown a phase diagram for a representative salt hydrate. Starting with an unsaturated solution of specific concentration at a specific temperature, water is evaporated until precipitation starts, increasing the number of phases from one to two. The sequence of events with continued evaporation is as follows:

	Unsaturated solution	1 phase
	Saturated solution/solid	2 phases
	Solid	1 phase
Evaporation	Solid/saturated solution	2 phases – saturated solution is of different composition than previous saturated solution
	Unsaturated solution	1 phase
	Saturday solution/solid	2 phases
	Solid	1 phase

As Clibbens wrote, "This remarkable behavior…would be very difficult to explain without the aid of the phase theory." It was tedious work creating phase diagrams from such data. The sequence above was for one temperature. As temperature is adjusted, a new sequence results. Without Gibbs' theoretical guidance on how to create the phase diagram, experimentalists would have been wandering in the wilderness, finding interesting things here and there, but not seeing how they all fit together.

From an industrial point of view, one can see the value of such knowledge. How much to evaporate, at what temperature, how to wash, how to purify? The process design would be guided by the phase diagram.

To continue this history, the critical time of World War I took Freeth to the battle front in France in 1915, thus continuing the military tradition of his family. But his stay at the front was short-lived. A crisis was brewing back on the home front, one requiring his expertise in phase theory. At the start of the war, England had completely underestimated the quantities of high explosives required. By 1915 the situation was urgent. Freeth was asked by the government to revisit his earlier work with the various nitrate-salts used in explosives. The achievement of purity in such systems is challenging, as captured by Freeth in reference to his research in double-decomposition: "You have a fair number of double salts in the system and you have to dodge your concentrations in order that these are not crystallized."[20] Also challenging was the timeline on which Freeth had to produce results. With the war still raging, the need for

---

[19] (Clibbens, 1920)
[20] (Allen, 1976) p. 109.

ammunition became paramount. As the Director General of Explosives Supply, Lord Moulton, told him, "My dear Freeth, do you realize the safety of England depends on this?"[21]

In the end, Freeth delivered. Under his guidance, the production of ammonium nitrate increased significantly during 1917–18. Such success plus other successful war-work requiring his skills in phase theory led Freeth to a successful career in chemistry following the war, a career in which he became a strong advocate for using "a great deal of reading and reasoning"[22] in research. Highlighting in 1920 the value of phase theory to English research, Freeth wrote in the introduction to Clibbens' book,

> Singularly little attention has been paid to [Heterogeneous Equilibrium] in this country…Far too much attention is concentrated on "curves" to the exclusion of the vital principle that every point in every [phase] diagram, whether on a curve or not, has a definite meaning. The fact that all the common forms of equilibria can be easily deduced by a system of graphical thermodynamics is not generally realized.

With this history now shared, you will better appreciate, as I now do, why Gibbs' work helped "save countries."

Translating Gibbs

Given the publicity started by Helmholtz and the Dutch, it was only a matter of time before others sought to translate Gibbs' third paper, Ostwald being the first (German, 1892) followed by Henry Louis Le Chatelier (1850–1936) (French, 1899). The fact that such men spent such time on translating this paper reflects the significance it held for them. Such translations weren't easy as even someone of Ostwald's stature faced a challenge with Gibbs. Convinced by Roozeboom's success that translation was a worthy endeavor, Ostwald, a Russian-German chemist, "soon found that the only way to study Gibbs' papers was to translate word by word because the text was already so terse that no abbreviated summary of the content was possible."[23] In his introduction to the translation, Ostwald wrote, "The author [Gibbs] has not had the time to add notes and comments; the translator has not had the courage." With open honesty, Ostwald admitted in his autobiography that while he had not been able to understand all of the mathematics in Gibbs' work, he believed that not one flaw—either in logic, mathematics, or in scientific assumptions—had yet been found there.[24]

Together with Roozeboom's publications, these translations helped cement Gibbs' status in Europe and also generated many discussions around thermodynamic theory, including ones around the phase rule.[25] Wrote Donnan, "Perhaps in the whole history of science no other linear equation of equal simplicity has had a wider significance and influence."[26] Particularly intense discussions arose around this rule, not regarding its validity but, interestingly enough,

[21] (Allen, 1976) p. 109.
[22] (Allen, 1976) p. 111.
[23] (Daub, 1976) p. 748.
[24] (Daub, 1976) p. 749.
[25] (Daub, 1976) p. 747–751.
[26] (Donnan, 1925) p. 461.

its derivation. This stumped many. Duhem was one who challenged the derivation. He admitted that Gibbs dealt with this problem in a few lines, but he promised to do it "with all possible rigor" – which resulted in a 42-page proof. Imagine how long Gibbs' entire body of work would have been in the hands of others.

While Gibbs' work started to move fast through the scientific community, it would take time for it to work its way into the textbooks. As wrote Arthur Wightman (1979), "Apart from its impact on Maxwell, [Gibbs' first two papers] had very little influence on late nineteenth century textbooks. The notion of 'fundamental equation' and the simple expression it gives for the laws of thermodynamics…only became available with the publication of 'neo-Gibbsian' textbooks and monographs in the mid-twentieth century."[27] But the point is that Gibbs eventually did secure a strong and almost unshakeable position in textbooks where his work was woven throughout.

As a fitting end to this section on Gibbs' work, I share this great tribute from Pupin in his speech at Yale.[28]

> It is given to very few men to create a new science. Some men of genius make a discovery, or make an invention, but it is very rarely given to man to create a new science. Newton created one, modern dynamics; Maxwell created one, electric magnetic theory; Gibbs created one, the modern science of physical chemistry. It is, therefore, no exaggeration to say that Gibbs belongs to the class of men in which we have Newton and Maxwell. No other American university can boast of having produced a scientist of that order of nature.

---

[27] (Klein, 1983) p. 159.
[28] (Pupin, 1927)

39 The 2nd Law, entropy, and the chemists

As a young man I tried to read thermodynamics, but I always came up against entropy as a brick wall that stopped any further progress.

– James Swinburne (1904)[1]

The second law of thermodynamics, which is known also as the law of the dissipation or degradation of energy, or the law of the increase of entropy, was developed almost simultaneously with the first law through the fundamental work of Carnot, Clausius and Kelvin. But it met with a different fate, for it seemed in no recognizable way to accord with existing thought and prejudice. The various laws of conservation had been foreshadowed long before their acceptance into the body of scientific thought. The second law came as a new thing, alien to traditional thought, with far-reaching implications in general cosmology. Because the second law seemed alien to the intuition, and even abhorrent to the philosophy of the times, many attempts were made to find exceptions to this law, and thus to disprove its universal validity. But such attempts have served rather to convince the incredulous, and to establish the second law of thermodynamics as one of the foundations of modern science.

– Lewis and Randall (1923)[2]

O.k. So when I suggested that Gibbs' work moved fast through the scientific community, that wasn't entirely accurate. His work met with resistance for several reasons, starting with the high-intensity mathematics, which most chemists weren't prepared to handle, and also with the fact that the 2nd Law of Thermodynamics was still confusing to many. As aptly put by Kragh, the 2nd Law appeared "in as many forms as there are writers."[3] Recall the different versions proposed by Carnot, Joule, Thomson, Rankine, and Clausius. While many of these forms were "primitive"[4] to the ultimate form involving entropy, this wasn't known then. And even if it had been known, it wouldn't necessarily have helped. Which brings us to the last reason for the resistance. The addition of entropy to the 2nd Law only made the confusing more confusing. Most if not all didn't even know what entropy was and to place such a "ghostly quantity"[5] at the center of a body of work did not help in gaining the work's acceptance, especially in the chemist community. In a way, this unwelcomed and ungrasped outsider was forced upon them. Even though Gibbs embraced entropy as a means to understand chemistry,

[1] Swinburne cited in (Kragh and Weininger, 1996) p. 121.
[2] (Lewis and Randall, 1923) p. 110.
[3] (Kragh and Weininger, 1996) p. 111.
[4] (Kragh and Weininger, 1996) Hinshelwood quote, p. 129.
[5] (Berry, 1913) p. ix.

Block by Block: The Historical and Theoretical Foundations of Thermodynamics,
Robert T. Hanlon, Oxford University Press (2020). © Robert T. Hanlon.
DOI: 10.1093/oso/9780198851547.001.0001

and indeed was arguably the first to give entropy prominent status in this field, he himself was not a chemist. He along with Helmholtz, a self-proclaimed "physicist among chemists,"[6] were both outsiders. Initially, his and later Helmholtz's theories became situated between physics and chemistry in the new field of physical chemistry, which wasn't a bad place to be as it afforded the physical chemists a certain comfort and freedom in a whole new field. But such a situation could only last so long. Soon the chemists realized the opportunities offered by physics and eventually adopted physical chemistry into their own now-larger world.

A Most Challenging Chapter

This chapter was one of the more difficult to write. To help you navigate the challenges, I lay out here a brief overview of what you're about to read, and it really all starts with the concept of maximum work, which I presented in the previous chapter.

As chemistry met economics, the concept of maximum work became a critical focal point. What's the maximum amount of work that this process can possibly generate? What's the minimum amount of energy required to make this process operate? And for both questions, do the answers make the process economically favorable? The financial need to understand how to approach these questions helped drive the scientists toward a deeper understanding of the physics involved.

Early thinking on maximum work theories led to the thermal theory of affinity, which basically said that maximum work is equal to the heat released of a chemical reaction (ΔH_{rxn}). But cracks in this foundation formed when the theory couldn't explain why the endothermic reaction is possible. With this backdrop, I proceed to dig into the calorimeter, for it's this device that serves to quantify ΔH_{rxn}.[7] Why dig here? Because how often does one take a step back and question what the calorimeter truly measures? This discussion then leads into my interpretation of the physical meaning of each of the terms comprising Gibbs' maximum work equation,

$$W_{max} = -(\Delta G_{rxn})_{T,P} = -(\Delta H_{rxn} - T\Delta S_{rxn})_{T,P} \qquad [39.1]$$

which then, in turn, leads into a discussion about the electrochemical cell as a great vehicle for manifesting each of these meanings. It was Gibbs' handling of this situation with the practitioners of commercial electrochemical operation that brought clarity to these meanings. I end the technical discussion with the Gibbs–Helmholtz relation, which quantifies how Gibbs' maximum work equation (and thus equilibrium) varies with temperature and then end the entire chapter with some concluding thoughts. So without further ado, it's time to learn about the thermal theory of affinity.

> Thermal Theory of Affinity

---

[6] (Kragh, 1993) p. 194.
[7] When a reaction takes place in a constant T-P calorimeter, heat is either removed or added to maintain isothermal conditions and the system expands or contracts to maintain constant pressure conditions. According to the 1st Law, $\Delta U = Q - W = Q - P\Delta V$. And thus $\Delta(U+PV)_{rxn} = \Delta H_{rxn} = Q$, which is the thermal energy removed from or added to the system.

Nothing contributed more to changing the whole complexion and theoretical priorities of physical chemistry than the recognition that the course of spontaneous processes is governed by both the first and second laws of thermodynamics, and not just the first, as the Thomsen–Berthelot principle implies.

– Erwin N. Hiebert[8]

One way for a newcomer to win over a community is to solve a problem upon arrival. And that's what entropy did. To understand this particular story, one must first go back to the early 1800s and the thermal theory of affinity.

The Thomsen–Berthelot Principle

Chemistry had developed through the monumental efforts of Lavoisier and others to the point where there existed basic understanding of what happened when, for example, you added X to Y. Piles and piles of data were available along with many different theories and laws attempting to explain them. And for the most part, it all fit together and worked rather well. Indeed, the field of chemistry was quite strong heading into the late 1800s. But beneath this seeming strength was an unstable foundation. Chemistry embodied no fundamental unifying structure. Instead, it was more empirical than theoretical and more likely to address the "what" and the "how" rather than the much harder "why."

In the mid-1800s, the beginning of the "why" was broached in the development of thermo-chemistry by Danish chemist Julius Thomsen (1826–1909) in 1854 and later French chemist Marcellin Berthelot (1827–1907) in 1864. In what became known as the Thomsen–Berthelot principle,[9] they proposed that the key criterion for determining which reactions happen and which don't was heat release. If the reaction was exothermic, then it was spontaneous; the more heat released, the more likely to happen. As evolved from the 1st Law of Thermodynamics, this criterion proposed that a strong attractive force holds chemical species together and that the strength of this force is quantified—via direct measurement by a calorimeter—as the heat released during its formation. If A and B react to form C and release heat as a result, the quantity of heat reflects the force or the affinity holding C together. It is this force that would be required to break C back apart into A and B. As more chemists entered this discussion, the principle and criterion acquired a more accurate phrase, *the thermal theory of affinity*, and became an attractive theory as it provided the means to link chemical phenomena to the mechanical theory of heat and ultimately Newtonian mechanics. But while this theory indeed explained many data, it remained a "vague concept"[10] with no theoretical underpinning. And by the early 1880s, when it was confronted with data it couldn't explain, the theory eventually fell apart. What data? The endothermic reaction. According to *affinity*, such a reaction shouldn't happen. Yet it did, and at some point this couldn't be ignored. Which brings us back to Gibbs and entropy.

---

[8] (Hiebert, 1982) p. 101.

[9] (Kragh, 1984) Berthelot first stated the principle, which was also known as the "principle of maximum work," in 1864.

[10] (Kragh and Weininger, 1996) p. 94.

THE 2ND LAW, ENTROPY, AND THE CHEMISTS

Many Wondered, Why Do We Need Entropy?

Gibbs' and later Helmholtz's groundbreaking work[11] on the concepts of "maximum work"[12] and "free energy," discussed in Chapter 37, brought clarity to reaction spontaneity. The predictive power offered by the combined two laws of thermodynamics won out over the first law alone. Assuming constant temperature and pressure, the equation dG = dH − TdS says that the endothermic reaction (dH > 0) can be spontaneous (dG < 0) so long as TdS > dH. But the story doesn't end here for this was more of a victory for thermodynamics than for entropy.

In the reversible world, entropy as a property of matter is not a necessary component of thermodynamics. Not to oversimplify, but to some, all that was needed to solve real problems was the work of Clausius before he declared dS = δQ/T. His nine-memoir analysis of energy, heat, and work, which included the behavior of the quantity of δQ/T as revealed in his analysis of Carnot's heat engine, worked quite well to address thermodynamic problems. If one didn't need entropy, why go through the pain?

Given the pain involved, one can see why two camps developed, those who embraced the pain, such as the mathematical physicists (Gibbs, Helmholtz, Duhem, and Planck) and those who avoided it, such as the chemists (van't Hoff, Walther Nernst, and even Gilbert Lewis). But it's not that this latter group lacked the intelligence to understand entropy. It's that they didn't see the need to emphasize entropy as the "main concept"[13] in classical thermodynamics, especially since it caused so much confusion. This group, arguably led by van't Hoff, believed in a more pragmatic approach to thermodynamics, a simpler approach that was more accessible and appealing to the chemists and their experiments. "If one could derive all chemically relevant equations by using Carnot cycles and works of chemical affinity, why introduce the perplexing concept of entropy?"[14] The strength of their avoidance spoke most loudly in their silence on the subject, as best reflected in van't Hoff's massive and influential textbook in physical chemistry of 1898–1900, which contained not one word on entropy.[15]

With this historical and theoretical background, we're now ready to learn about the problem that entropy solved, a feat that ultimately had the chemist community embrace entropy.

$$\boxed{\text{The Calorimeter and } \Delta H_{rxn}}$$

I undertook the writing of this book to delve more deeply into both the history and the physics behind the equations. This objective challenged me to clarify the physical meaning behind the concept of the "maximum work" calculations. I feel it worthwhile to take a step back to consider this concept in greater detail as it offers an even deeper understanding of Gibbs' work and, more generally, thermodynamics.

---

[11] August Horstmann's (1842–1929) work is also noted but it's not clear if either Gibbs or Helmholtz built on Horstmann. See (Kragh, 1993) p. 418.

[12] (Lindauer, 1962) Berthelot used the phrase "principle of maximum work" to describe this correlation, which was somewhat of a misnomer as he didn't actually link this correlation to determining maximum work.

[13] (Kragh, 2001) p. 201.

[14] (Kragh, 2001) p. 199. Kragh commenting on van't Hoff's logical thought process.

[15] (Kragh and Weininger, 1996) p. 102.

Let's start this discussion by first discussing the term ΔH_{rxn} for it was this term that lay at the heart of the Thomsen–Berthelot principle. What happens when a chemical reaction occurs? Typically the orbiting electrons rearrange themselves (Phenomenon 1 listed below) by bond breaking (reactants) and bond forming (products) so as to move the system into the most probable state. Since the electron structure in atoms and molecules contributes to the (chemical) potential energy of the system, this rearrangement causes a change in potential energy, and when done in a closed system (absent the generation of external work), a corresponding change in the opposite direction in kinetic energy. When the closed system is allowed to exchange thermal energy with a constant-temperature bath, such as what occurs in a calorimeter, if the reaction drives temperature to increase, then thermal energy is removed from the system and the reaction is termed exothermic. Conversely if the reaction drives temperature to decrease, thermal energy is added to the system and the reaction is termed endothermic. Heat removal or addition is quantified as ΔH_{rxn}. Thus ΔH_{rxn} quantifies the transformation of potential energy into kinetic energy; a change in kinetic energy results in a change in temperature, which then drives heat exchange with the constant-temperature bath. This value, negative when heat is removed, quantifies the amount of thermal energy removed or added to the system. Since some calorimeters are constant pressure and so allow volume to vary, such changes are embedded in the ΔH_{rxn} term since $H = U + PV$.

Thomsen and Berthelot's principle stated that ΔH_{rxn} quantifies the maximum amount of work a reaction could generate. Moreover, it also stated that if ΔH_{rxn} was negative (exothermic; reflecting the removal of heat), then maximum work is positive and the reaction was called spontaneous. It would happen naturally, without the need for external energy. While the theory worked quite well for most data, it didn't work well for all data. Sometimes all it takes is a single data point to eliminate a theory. In this case the damaging data point was the aforementioned endothermic reaction. Given this principle, how was it possible that an endothermic reaction (measured by a positive ΔH_{rxn}) could happen spontaneously? This the principle couldn't explain.

* * *

To address this question, let's break down what happens during a chemical reaction. It's easy to consider that the thermal energy generated by a reaction is caused solely by the rearrangement of the chemical bonds. But this view is incomplete, for there's another phenomena that can also contribute to the generation of thermal energy in going from reactants to products.

- Phenomenon 1 (P1): chemical bond rearrangement (introduced earlier)
- Phenomenon 2 (P2): changes in intermolecular forces.

In a calorimeter, since both phenomena involve changes in potential energies, both contribute to changes in kinetic energy and thus temperature. While we may generally believe that the heat released during a reaction is caused by P1, we can't ignore P2 in our thinking. But what does this really mean? Well, let's look at the calorimeter. This apparatus quantifies the total amount of energy that is transformed from potential to kinetic. This can be seen by looking at the following rearranged Gibbs' equation.

Net Heat Removed in Calorimeter = Phenomenon 1 + Phenomenon 2

$$\Delta H_{rxn} \quad = \quad \Delta G_{rxn} \quad + \quad T\Delta S_{rxn}$$

where the "rxn" subscript refers to the quantities determined by "products" minus "reactants."

The value of $\Delta G_{T,P}$ quantifies the maximum amount of work that can be generated by the dissipation of energy gradients at constant temperature-pressure for given process. For a chemical reaction, the maximum work is tied up in the chemical bonds (P1), nowhere else. When a reaction occurs in a calorimeter, this potential energy transforms to kinetic energy and contributes to the heat removed. But it isn't the only contributor, because some amount of energy (P2) is needed to establish the system of products at the constant temperature-pressure conditions. And it's the sum of both that results in the net thermal energy exchanged with the medium (ΔH_{rxn}) in the calorimeter.

The Meaning of $T\Delta S_{rxn}$

So what does $T\Delta S_{rxn}$ represent? Let's start by considering this. The entropy of a system quantifies the amount of thermal energy needed to bring the system from absolute zero to a given temperature. More accurately, it's the sum of the thermal energy increments along this path divided by the temperature of each increment: $S = \int \delta Q/T$. More specifically, δQ encompasses the total thermal energy added to a system of atoms and molecules to raise their collective temperature from absolute zero to the given temperature and includes not only the energy required to put the particles in motion (and so increase temperature) but also the energy required for them to overcome any attractive forces, such as phase change, along the way. Thus, the difference in entropy between reactants and products (ΔS_{rxn}) times temperature (T) quantifies the difference in thermal energy required to establish each. $T\Delta S_{rxn}$ thus quantifies the difference in energies required to establish the system of reactants and to establish the system of products, the system being here defined by the atomic location and momentum of each atom, which are the two sources of entropy in Boltzmann's statistical mechanics.

Recall that in his efforts to answer the question, *where does the heat go?* Clausius created the concept of disgregation and proposed Equation 34.7

$$dQ = d\left(vis\ viva\right) + T\,dZ \qquad [34.7]$$

with the product of temperature and the change in disgregation (dZ) quantifying the sum of the changes in both internal and external work. In this way he attempted to quantify the degree to which the molecules of a body are separated from each other, with TdZ quantifying the change in this separation or the change in the molecules' configuration. This "internal work" approach evolved into the latter statistical mechanic methods of Maxwell and Boltzmann.

So the value of $T\Delta S_{rxn}$ is conceptually similar in concept to what Clausius was getting at with disgregation. $T\Delta S_{rxn}$ represents the amount of thermal energy needed to establish the configuration of the products relative to that of the reactants for a given T-P and it's quantified by the thermal energy reversibly exchanged with the medium to maintain constant temperature. Regarding signs:

$T\Delta S < 0$ (thermal energy removed from body and absorbed by medium)

A decrease in entropy at constant temperature in Phenomenon 2 occurs when the attractive forces between the atoms in the products is greater than those in the reactants. The increased presence of such forces places more constraints on the locations of the atoms comprising the system

(Chapter 24). The increased presence of these forces causes atoms to accelerate toward each other, which results in the release of thermal energy, thus necessitating the removal of energy to maintain constant temperature.

$T\Delta S > 0$ (thermal energy removed from medium and absorbed by body)

An increase in entropy at constant temperature in Phenomenon 2 occurs when the attractive forces between the atoms in the products is less than those in the reactants. The decreased presence of such forces places less constraints on the locations of the atoms comprising the system. This decrease removes the acceleration forces on the atoms and so decreases their kinetic energies, resulting in a cooling effect. Heat must be added to the system to maintain temperature.

Thus, the TS term quantifies the energy contained in the structure of the system itself. It's "bound" in the system, as Helmholtz would say but perhaps not for the same physics-based reasoning I just went through here. The change in thermal energy associated with the change in this term ($T\Delta S_{rxn}$) is a component of ΔH_{rxn} as measured by a calorimeter that cannot be used to do useful work.

ΔG_{rxn} Alone Determines Reaction Spontaneity

Recognize that when all is said and done, whether or not a reaction occurs spontaneously at constant temperature and pressure is not determined by either the entropy change ($T\Delta S_{rxn}$) or by the enthalpy change (ΔH_{rxn}) of reaction. Instead, it's solely determined by the value of ΔG_{rxn}. Yes, $T\Delta S$ and ΔH data are used to calculate ΔG_{rxn}, but only because both can be directly calculated or measured whereas this is only possible for ΔG in an electrochemical cell. For regular reactions, ΔG_{rxn} must be indirectly quantified as $\Delta H_{rxn} - T\Delta S_{rxn}$. But this doesn't change the fact that ΔG_{rxn} alone determines spontaneity. Negative ΔG_{rxn} means that the reaction will occur spontaneously, that the system will move on its own toward Gibbs' surface of dissipated energy and that work will be generated. Prior to Gibbs and as shared earlier, attempts to address reaction spontaneity didn't focus on ΔG_{rxn}, as Gibbs hadn't created it yet, but instead focused on ΔH_{rxn}. Knowing what we know now, let's put this into perspective in a way often shared in textbooks but from a different viewpoint, starting with ΔG_{rxn} in the left-hand column instead of ΔH_{rxn} (Table 39.1).

So now the questions must be asked. What is it about ΔG that makes it the sole determinant of reaction spontaneity and why don't we just directly measure this? Here's the answer. While the calorimeter provides very useful information, it simply wasn't designed to separate out the two phenomena P1 and P2 contributing to the result, largely because no one realized that two phenomena were even involved in the energy balance, thus leaving the scientists stranded when attempting to explain the spontaneous endothermic reaction. And this now brings us to the rest of the story regarding the problem that entropy solved. The problem came about with the advent of the electrochemical cell. No one seemed to understand why the different baths containing the chemical reactants were giving off heat.

Table 39.1 Using ΔG_{rxn} to understand chemical reaction spontaneity. The signs in the table represent negative (–) and positive (+) quantities.

Reaction	ΔG_{rxn}	$T\Delta S_{rxn}$	ΔH_{rxn}	Comments
	measured by electrochemical cell	measured by electrochemical cell	measured by calorimeter	$\Delta G_{rxn} = \Delta H_{rxn} - T\Delta S_{rxn}$
Negative ΔG_{rxn} **Means that the reaction is spontaneous**	–	+ increase in entropy pulls heat from medium	– exothermic	
Does not require external energy to make happen and can generate external energy such as work	–	– decrease in entropy rejects heat to medium	– exothermic	It is possible for a reaction to be spontaneous at constant T-P when entropy change of reaction is negative so long as $\Delta H_{rxn} < T\Delta S_{rxn}$
	–	+	+ endothermic	It is possible for a reaction to be spontaneous at constant T-P when endothermic so long as $\Delta H_{rxn} < T\Delta S_{rxn}$
Not spontaneous **requires external energy to make happen**	+			$\Delta H_{rxn} > T\Delta S_{rxn}$

The Electrochemical Cell Directly Measures dG$_{rxn}$

The maximum work potential of a chemical reaction lives in P1. It's all about the electrons. Let's take a step back and explain this. Building on Chapter 24, and as you'll see in subsequent chapters, statistical thermodynamics is all about the arrangement of the physical particles contained in a given system. For a fixed set of particles and a fixed set of conditions, such as U-V-N, there are a finite number of ways to arrange the particles in the given volume to yield the macroscopic U-V-N properties. This finite number (W) corresponds to the entropy of the system by the Boltzmann equation, $S = k_B \ln (W)$, where k_B is the Boltzmann constant. Typically, the particles considered in a thermodynamic system are the atoms and molecules. But from a very rigorous standpoint, the particles should really be the electrons and the nuclei. How many different ways can these two types of particles be arranged to yield U-V-N?

While calculations can be done based on the quantum mechanics to quantify W based on the distribution of electrons and nuclei, they can also be done based on the distribution of the reactants and products themselves. We can measure a range of properties for the atoms and molecules involved, such as heats of reaction, bond energies, electron energies (via spectroscopy), heat capacities, and so on, and then use these properties to model the different possible distributions of atoms and molecules using a statistical mechanics model, and so determine the distribution of atoms and molecules that results in the greatest number of arrangements and thus the largest value for W and S.

When reactants are brought together such that the electrons rearrange themselves to preferentially fill the lowest energy states (most probable distribution), the shift of electrons from high to low potential energy results in the shift of atom and molecular velocities from low to high kinetic energy, which causes temperature to increase. This is what happens in a calorimeter. Alternatively, this electron shift from high to low potential energy can be used to raise a weight (by driving an electric motor) from low to high potential energy. This is what happens in an electrochemical cell.

So with this technical discussion behind us, let's now discuss the calorimeter versus the electrochemical cell. Each device can measure the energy released during a reaction but the energies they measure are fundamentally different.

In the electrochemical cell, you separate the reactants, attach a wire between them and allow the electrons to flow from high electrode potential to low as they seek to arrange themselves in the lowest-energy distribution available, generating work in an electric motor while so doing, with the difference between the potentials quantifying the amount of useful work possible. This difference is a direct measure of ΔG_{rxn}. So while a calorimeter directly measures ΔH_{rxn}, the electrochemical cell directly measures ΔG_{rxn}. But what about $T\Delta S_{rxn}$? Where does this come into play in the electrochemical cell? The answer: the heating or cooling requirements of the constant temperature bath. In essence, the electrochemical cell employs a selective membrane that allows only electrons to pass through, thus isolating the ΔG_{rxn} effect and enabling the maximum work to be generated. This can't be done in the calorimeter since the $T\Delta S_{rxn}$ effects are mixed in with the ΔG_{rxn} effects. You now see why ΔH_{rxn} doesn't represent the maximum work possible as those who espoused the *theory of affinity* once believed. Only one of the two phenomena comprising ΔH_{rxn}, specifically ΔG_{rxn}, is available to do work.

It's important to realize that the entropy change of the reaction itself might be positive or negative in this constant T-P system. Remember, the entire system in this scenario, which mirrors one of Gibbs' thought experiments, is not just the body of interest; it's the body of interest *plus* the large constant T-P medium such as a constant temperature bath. It's the combined body-plus-medium that remains at constant entropy—since there's no loss to the environment—while energy is reversibly depleted from the system to generate external work. Whether the body itself gains or loses entropy is dependent on the reaction involved. It is possible for the reaction to proceed spontaneously in this situation while the entropy of the body decreases.

The Inadvertent Contribution of the Electrochemical Cell to Thermodynamics

The design of the electrochemical cell inadvertently enabled scientists with an improved means of thermodynamic study as it effectively separated ΔG ("free energy") from $T\Delta S$, two quantities that were otherwise tangled in the calorimeter's ΔH_{rxn}. It was the arrival of entropy and its ability to bring quantification to irreversible processes that made this happen. And it was Gibbs who showed us how.

History of Gibbs' Influence on Electrochemistry Theory

There is a rather interesting history involved in this discussion that is important because it affords an even deeper understanding of the issues involved in entropy's acceptance by the chemists.

Gibbs devoted nineteen pages in his third paper to the thermodynamic theory of electro-chemical cells and cited the fact that the heating and cooling requirements to maintain isother-mal conditions are "frequently neglected"[16] in the analysis of these cells. He argued that such effects should be captured as the value of $T\Delta S$ and that the resulting equation could be used to quantify equilibrium conditions of a "perfect electro-chemical apparatus" for which the volt-age difference between the two electrodes is perfectly countered by an external voltage such that the current is negligible and easily reversed. In this scenario, the entire system is perfectly balanced, thus validating the equal signs in:

$$\text{Maximum Work} = -\Delta G_{rxn} = -\left(\Delta H_{rxn} - T\Delta S_{rxn}\right) \quad \text{chemical reaction at constant T-P} \quad [39.1]$$

The historical significance[17] of Gibbs' theory as applied to electrochemical cells is that, as previously discussed, prior to his work many chemists were influenced by the *theory of affin-ity* in which it was proposed that maximum work as quantified by the voltage difference in an electrochemical cell is equal to the change in energy (ΔH_{rxn}) of the reaction alone. Lord Kelvin (William Thomson) himself suggested this to be so. It was Gibbs' work that demon-strated the need to also account for entropy effects. In 1887 he shared this understanding in two famed letters to Sir Oliver Lodge, Secretary of the Electrolysis Committee of the British Association for the Advancement of Science. This theoretical contribution "profoundly affected the development of physical chemistry [and was] a potent factor in raising the industrial processes of [electrochemistry] from the realm of empiricism to that of an art governed by law, and thus [helped enable] the expansion of the electrochemical industry to its present vast proportions."[18]

---

[16] (Gibbs, 1993) p. 339.
[17] (Crowther, 1937) pp. 281–282.
[18] (Wheeler, 1998) p. 80.

The Gibbs–Helmholtz Equation – the Impact of Temperature on ΔG and thus on Equilibrium

Gibbs created his eponymous property as a means to solve problems involving equilibrium, such as those involving chemical reactions and the electrochemical cell. For both, the issue of temperature sensitivity became important. How does one adjust the temperature of an iso-thermal process to improve performance? The answer lay in what became known as the Gibbs–Helmholtz equation.

Let's look at the two examples above, starting with the electrochemical cell. In this example, the maximum work is governed by the value of ΔG_{rxn}. If this value is less than zero, useful work can be generated, and if greater than zero, then energy (such as electricity) needs to be pumped into the system to drive the reaction. Either way, the value of ΔG_{rxn} is important to the electro-chemist. And one way that the chemist can change ΔG_{rxn} to suit process needs is by varying temperature.

The fundamental question we're really asking here is this. When temperature is varied, how does the most probable state vary? Where do the electrons go? If we delved into the world of quantum chemistry and statistical thermodynamics involving use of the Boltzmann distribu-tion and so on, we could calculate this. Alternatively, in the world of classical thermodynamics, we could conduct specific experiments to individually quantify heat release and entropy and thus derive an appropriate equation based on theoretical concepts to show how ΔG_{rxn} calcu-lated from ΔH_{rxn} and ΔS_{rxn} data varies with temperature. Typically, the modelers use such data to improve their efforts and so improve their predictive powers, which becomes extremely valuable in those cases where experiments can't readily be done.

* * *

The first person to truly address how ΔG varies with temperature was Helmholtz (not Gibbs) and it was entropy that enabled him. To paraphrase their combined derivations, and focusing here solely on Gibbs' G and not Helmholtz's A, recall that

$$G = H - TS$$

Thus, first seeking to quantify how G itself and not ΔG varies with temperature (while keeping pressure constant)

$$(\partial G/\partial T)_P = (\partial H/\partial T)_P - T(\partial S/\partial T)_P - S$$

As can be shown via calculus, the second and third terms cancel each other, thus leaving

$$(\partial G/\partial T)_P = -S = (G - H)/T$$

So for a given pressure, the slope of G with respect to T is equal to the negative entropy at that temperature. Since each term is a property, this equation can be applied to both the system of reactants and the system of products. Furthermore, the two equations can be combined by subtracting reactants from products. Assuming that ΔG is based on the difference in G between the reactants and products at the same temperature and pressure, the value of $(\partial \Delta G/\partial T)_P$ comes out to this.

$$(\partial \Delta G/\partial T)_P = -\Delta S = (\Delta G - \Delta H)/T \text{ for constant P}$$ [39.1]

There are alternate forms of this equation such as one developed by creating another term, G/T, and differentiating it like this

$$d(G/T) = 1/T\, dG + G\, d(1/T)$$

From Chapter 36 we learned that

$$dG = -SdT + VdP$$

Thus,

$$d(G/T) = 1/T\,(-SdT + VdP) + (H - TS)\, d(1/T)$$

With some calculus maneuvers and again subtracting reactants from products

$$d(\Delta G/T)/dT = -\Delta H/T^2 \quad \text{for constant P} \tag{39.2}$$

which leaves one with the original Gibbs–Helmholtz equation.[19]

Applying Equation [39.1] to the world of electrochemistry and setting ΔG equal to the difference in electrode potentials in an electrochemical cell, which we'll call EP for now,

$$d(EP)/dT = -(EP - \Delta H)/T$$

So if one were to measure EP for an electrochemical cell at a range of temperatures, one could calculate the heat of reaction, ΔH_{rxn}, from the data using the above equation, an approach "far more accurate than any of the calorimetric values."[20] With this value, one could then fully predict how the electrode potential will vary over a range of temperatures.

As not all reactions lend themselves to the electrochemical cell, one could alternatively use experimentally accessible enthalpy data from the calorimeter in either equation to predict how ΔG varies with temperature. But then why is this important? This brings us to the second example, chemical reaction equilibrium.

* * *

Using more advanced thermodynamics than I care to go into here, at a given temperature for an equilibrated reaction, one can calculate an equilibrium constant, K, based on a ratio of concentrations of products over reactants. There are some additional complications involved but the value of K provides a measure of how the reactants and products are distributed relative to each other at a given temperature and can be theoretically and mathematically connected to ΔG°_{rxn}, which is defined as the Gibbs energy change between reactants and products when each species is in its standard state (pure component) at the given temperature. It's this value of ΔG°_{rxn} and thus the equilibrium constant K that can be placed inside Equation 39.2 along with calorimeter data for ΔH_{rxn} to quantify how equilibrium shifts with temperature. Again, the depth of details involved here go beyond the scope of this book, but the origin of these details is entropy. Without it, neither Gibbs nor Helmholtz could have derived such powerful and enabling theories.

* * *

The fact that the use of Gibbs energy equations is based on constant temperature-pressure conditions shifted the conversation away from *the entropy of an isolated system increases to a*

[19] For a good overview of the use of the Gibbs–Helmholtz equation in the chemical industry, see (Mathias, 2016).
[20] (Lewis and Randall, 1923) p. 173.

maximum to *the Gibbs energy of a system decreases to a minimum*. Yes, the movement of G at constant temperature and pressure is caused by the dissipation of the chemical potential gradients and this dissipation would normally increase entropy. But in a reversible process operating at constant temperature, as brought about by using a medium, total entropy doesn't change. The entropy of the body can either increase or decrease as a result of energy removal; the entropy of the constant-temperature medium either decreases or increases, respectively, by an identical amount. So this isn't the issue. Instead the issue is this. Is the change in chemical potential in going from reactants to products such that G decreases? This is what tells you whether a reaction will happen or not. This is what tells you whether positive work can be generated from the reaction, or whether you need to add work or energy to the system to make the reaction happen.

*   *   *

The Gibbs–Helmholtz equation is considered a "cornerstone of chemical thermodynamics."[21] As sometimes happens in history, the order of the names doesn't reflect reality as it was Helmholtz who originally derived the equation. More generally, Helmholtz deserves much credit alongside Gibbs for developing such concepts of "free energy" and their application to electrochemical theory. It was Helmholtz who coined the term "free energy," seeing it as the analog to potential energy in mechanics. The *change* in free energy quantifies the energy available to generate work. Subtracting free from total yields energy not available for work ($T\Delta S$), which he called "bound energy." This choice of term was likely influenced by Helmholtz's reliance on Clausius' mechanical theory of heat in which "bound energy" referred to energy required to do internal work, a quantity Clausius focused on when he created his disgregation. And really, this concept wasn't too far off. In mechanical terms, the natural "fall" in chemical potential from reactants to products releases energy that can be used to do both external work (ΔG_{rxn}) and internal work ($T\Delta S_{rxn}$) at constant T-P. The calorimeter measures the sum of both. This understanding helps explain why the heating/cooling requirements of an electrochemical cell aren't correlated with the ΔH_{rxn} measured by the calorimeter but are instead correlated with ΔS_{rxn}.

It was with the guidance of Equation 39.1 that Nernst would later develop the so-called 3rd Law of Thermodynamics by demonstrating through extensive experimentation that the *change* in entropy of matter goes to zero as temperature goes to (absolute) zero. Others including Planck would later take this one step further by showing that the value of entropy itself goes to zero at these conditions for pure crystalline substances.[22] Interestingly enough, Nernst himself never fully embraced the concept of entropy and instead relied on the changes in values of the Helmholtz energy term (A) with temperature to conduct his analysis, showing the convergence of this term with internal energy (U) as temperature approaches absolute zero. In his theory, for an isothermal process, $dA/dT = \Delta S$. Interpreted in the context of entropy and $\Delta G = \Delta H - T\Delta S$, as temperature goes to absolute zero, ΔG and ΔH converge, meaning that ΔS goes to zero.

Gibbs' and Helmholtz's collective work helped solve the problems inherent to the theory of affinity and thus helped chemists become much more accepting of classical thermodynamics based on energy and entropy. Eventually, free energy replaced chemical affinity in such textbooks as Lewis and Randall's *Thermodynamics*.

[21] (Kragh, 1993) p. 421.
[22] (Nernst, 1969) p. 85.

> Classical Thermodynamics
>
> –
>
> Concluding Thoughts

Entropy would eventually overcome the barriers that rose against it. It offered too much to be ignored. It served as one of the critical building blocks of Gibbs' work on equilibria and also facilitated the quantified analysis of irreversible change in chemistry. As the works of Gibbs and Helmholtz spread, many began to recognize that entropy, while not solely central, was still the final property of matter that enabled the first fundamental equation to be written and from which all chemical thermodynamics would follow. Entropy completed the puzzle. It was a property of matter that was directly linked to issues of mechanical, thermal, and chemical equilibrium, and while it was also a property that would eventually characterize chemical equilibrium using probabilistic arguments, what it meant in the late 1800s was irrelevant. The only thing that was relevant was that it enabled completion of the path, based on primitive properties that chemists could work with, that led from an initial state to a final state.

There was no specific date when entropy was fully accepted by the chemists. It happened over time, in fits and starts, with its inclusion in the textbooks being the best measures of true acceptance. But if one were forced to pick a date, the 1923 publication date of Gilbert Lewis and Merle Randall's *Thermodynamics* would be as valid as any. Yes, this is the same Lewis who had resisted entropy's central role in thermodynamics. But his resistance was not extreme. While he and Randall focused on their own developments involving fugacity and activity, concepts which also took time to be accepted, they additionally spent considerable time and care to explain entropy and all of its implications, such as its role in the calculation of "free energy." Such growing acceptance of entropy simply reflected its fundamental role in nature, not as a rule-of-thumb or an empirical correlation or a mathematical curiosity, but as a fundamental property of matter, just like temperature and pressure but for the unfortunate fact that it didn't lend itself to either direct measurement or ready understanding.

This now brings us to the end of classical thermodynamics but not to the end our journey with entropy. Classical thermodynamics worked perfectly, the equations all locked together in a mathematically precise way, all effectively deduced from the first (differentiated) fundamental equation of nature, Clausius' $dU = TdS - PdV$. One simply cannot help but be awed by how much could be generated from such a singular starting point.

While this new science rapidly grew and enabled great breakthroughs in science and industry, the curious couldn't simply sit by and not ask, *why?* The equations worked, yes, but why did they work? What was happening at the molecular scale that explained why the equations worked and what they meant? While the atomic theory had not yet achieved breakthrough, this didn't stop the curious from thinking, *well, I don't know if atoms exist or not, but let's assume they do exist. Let's create a mathematical model for such a system and see what it predicts. If the predictions agree with nature, well then maybe our starting assumption was indeed correct.* And this is indeed what the curious did. The mathematical density of the resulting work was very high. But the results were amazing and in the end strongly supported the existence of atoms. I share these final concluding steps through the stories of Clausius and his development of the kinetic theory of gases, Maxwell and his inclusion of statistics in physics, and finally

Boltzmann and his inclusion of probability theory in physics. All of these efforts led to the rise of statistical mechanics, statistical thermodynamics, and the fundamental definition of entropy based on probability.

*   *   *

The largest obstacle to entropy's acceptance into the scientific community was undoubtedly the fact that no one knew what it meant when Clausius first introduced it, not even Clausius himself. A few may have perceived what it meant. But no one really knew. As the road to understanding entropy necessarily leads through the world of probability and statistics, we now turn toward the first significant step on this road, the kinetic theory of gases.

40 Clausius: the kinetic theory of gases

The deepest understanding of thermodynamics comes, of course, from understanding the actual machinery underneath

– Richard Feynman[1]

We shall find that we can derive all kinds of things—marvelous things—from the kinetic theory, and it is most interesting that we can apparently get so much from so little

– Richard Feynman[2]

We *sense* the air around us. We feel wind. We feel hot air. We see our sealed empty water bottles collapse in the airplane seat during descent. We can't help but wonder about such phenomena. We can't help but ask *why is this happening?* Our instinct drives us to figure out the answer.

We haven't changed as a species over the past many thousands of years. We're drawn to understand. Darwinian evolution rewarded us for this. So it was regarding air. We were and continue to be fascinated by the "sea of air"[3] we live in. Otto von Guericke's experiment probably captured this fascination the best. To witness the failed attempt of a team of horses to pull apart two hemispheres held together by air pushing against a vacuum likely thrilled many. One can't witness such things without wanting to know *why?*

Daniel Bernoulli was arguably the first to start probing the *why*. He envisioned "minute corpuscles [moving] hither and thither with a very rapid motion." He brought calculations to such movements and so was the first to propose the kinetic theory of gases. Unfortunately his work went unnoticed for many years, thus leaving others to start the process all over again, only later discovering that Bernoulli had already traveled the path they were on. And this brings us back to the second half of the nineteenth century when the kinetic theory of gases was rediscovered. Energy, entropy, and the laws of thermodynamics successfully rose without the need to answer *why*. Indeed, a large part of the all-important community acceptance that led to this success was due to the fact that the main players, especially Clausius and later Gibbs, intentionally stayed away from speculating about the inner workings of nature. The laws of thermodynamics didn't require such speculation. But having said this, the same main players couldn't help but address their own need to understand.

---

[1] (Feynman et al., 1989a) Volume I, p. 39–2.
[2] (Feynman et al., 1989a) Volume I, p. 41–1.
[3] (Brush, 2003a) p. 1.

Block by Block: The Historical and Theoretical Foundations of Thermodynamics,
Robert T. Hanlon, Oxford University Press (2020). © Robert T. Hanlon.
DOI: 10.1093/oso/9780198851547.001.0001

The first step in understanding any natural phenomenon is construction of a physical model. Clausius, for example, strongly embraced such an effort as he believed it helped "facilitate the comprehension by giving something definite to the imagination"[4] It is rather remarkable how far one can go with even a simple model. The physical model affords a foundation of stone on which subsequent high-powered mathematics can reside. Absence of such a model risks building such computational structures on sand.

This step is typically much harder than one would think. Scientists and engineers, students and seasoned professionals, all are confronted with this step. It's rather fascinating how seemingly complex problems can be reduced to a simple physical model; typically, the simpler the better. It is so very difficult to achieve "simpler" but when it is done, many stand back and ask *now why didn't I think of that?!*

We've all been exposed to the billiard-ball model that serves as the basis for the kinetic theory of gases. We look at this model and say, *of course – this makes perfect sense.* But we forget that we've been raised in the world of atoms. The model only makes sense because we *know* that atoms exist as spheres and move "hither and thither," no speculation involved. But such was not the case in the second half of the nineteenth century.

To understand the challenge involved at that time, one arguably needs to go back to Newton. As I wrote before, Newton unintentionally hindered progress in the kinetic theory of gases by exploring an incorrect model. When he addressed air pressure in the *Principia*, he suggested, "Air acts as if it were composed of particles that exert repulsive forces on neighboring particles, the magnitude of the force being inversely proportional to the distance between them."[5] He was applying his laws of motion to explain the "springiness" of air discovered by the work of Boyle who showed that PV is a constant for constant temperature. Thus, the physical model that rose from his work consisted of static particles pushing against each other and so creating pressure. The thing is that while Newton was quite clear that he was not proposing that this simple model reflected reality but instead simply wanted to mathematically show what would happen if one were to *assume* so, this message was lost somewhere along the way. The model survived while the caveat didn't.

Newton's static-atom concept wasn't the only model hindering progress. Others contributed, such as Rankine's vortex model and all the models trying to account for the proposed existence of caloric or ether within or between the particles. Such models were rather complex, contained embedded paradigms of what heat was and ultimately led many down wrong paths.

As Brush well summarized,[6] what was needed was not a connection between heat and molecular motion, because even with a static-atom model one could make that connection, but a connection saying that heat is *nothing but* molecular motion and that regarding heat of a gas, this motion is the free movement of molecules through space. A huge leap was needed.

Bernoulli was the first published scientist to make this leap. He assumed heat to be nothing but the motions of atoms [not his word] moving freely through space until they collided with other atoms with the "before" and "after" of each collision obeying both energy and momentum conservation laws. In a complex world of Newton's repulsive interactions between atoms and of caloric and ether, such a "billiard-ball" model of heat was too simple to be taken seriously. But others, unaware of Bernoulli, would later propose the same model. John Herapath

[4] (Clausius, 2003b) p. 139.
[5] (Brush, 1983) p. 21.
[6] (Brush, 2003a) p. 14.

(1790–1868) did so when he submitted a comprehensive paper on this topic to the Royal Society in 1820,[7] where it was rejected by Humphry Davy as being "too speculative."[8] While his paper did see the light of day the following year in *Annals of Philosophy*, a lesser journal than the Royal Society's *Philosophical Transactions*, the recognition and acknowledgement just wasn't the same. Had it been, he might have been the one most regarded as the founder of the kinetic theory. He almost received a boost of attention when Joule developed his own model based on Herapath's foundation but this too received little to no attention.

John James Waterston (1811–1883) also developed an elemental kinetic theory of gases but his submission to the Royal Society suffered the same fate as Herapath's. It too was rejected, this time based on recommendations from two referees whose reasoning reflected the depth of the paradigm that they lived in. To one, Waterston's theory was "very difficult to admit, and by no means a satisfactory basis for a mathematical theory" and to the other it was "nothing but nonsense, unfit even for reading before the Society."[9] Ultimately, Waterston's paper was read to the Royal Society in 1846 but to little effect. The paper itself remained in the RS's archives until it was later discovered in 1892 by Lord Rayleigh. For both Herapath and Waterston, the Royal Society proved to be an unfortunate experience and "a formidable obstacle to progress."[10]

With his 1856 publication *Elements of a Theory of Gases*, August Karl Krönig (1822–1879) could be credited with reviving the kinetic theory after 1850. The physical models he presented reflected an intelligent vision for what the kinetic theory of gases should be based on. For example, included in his paper is one of the first references to the physical cause for heating when a gas is compressed. "If a gas is compressed... then the gas atoms must bounce back... with greater velocity than they used when they arrived. Therefore, the gas must heat up."[11] But unfortunately such visions were not followed up with sufficient calculations and comparisons with experimental data. Despite such lack of substance, this paper served a higher purpose. It lit a competitive fire within Clausius.

Rudolf Clausius

While Clausius successfully completed his Nine Memoir *Mechanical Theory of Heat* without having to rely on any assumptions regarding the inner mechanical nature of heat, this didn't mean that he wasn't thinking about such assumptions. It's just that his scientific philosophy and rigor kept this as a separate concept so as not to taint this specific work. Fortunately his need to understand *why* led him to simultaneously develop on a separate and parallel path his famed publication, *The Nature of the Motion which we call Heat*, on the kinetic theory of gases. The publication date of 1857 was somewhat misleading as he noted: "Before writing my first memoir on heat, which was published in 1850, and in which heat is assumed to be a motion, I had already formed for myself a distinct conception of the nature of this motion... In my former memoirs I intentionally avoided mentioning this conception." Clausius acknowledged that he

[7] (Herapath, 1836) Herapath furthered this work by using his kinetic theory of gases to determine of the speed of sound.
[8] (Brush, 2003a) p. 14
[9] (Brush, 2003a) p. 17
[10] (Brush, 2003a) p. 18
[11] (Krönig, 1856) p. 322.

wasn't the first to this topic. Others, and specifically Krönig, had already published on the subject and indeed were the reason, or more likely the competitive motivation, for him to publish and earn some priority for his technical advances in the process. Of note, no acknowledgement of Bernoulli was given, likely due to the simple fact that Clausius didn't see his study.

When reading Clausius, one realizes how sound and logical his thinking was. There was a reason why the great mind of Gibbs was drawn to Clausius. When it came to developing a physical model of a system before employing mathematics, Clausius surely ranked as one of the best. In his papers, Clausius very carefully and thoroughly laid out all the relevant assumptions he was making regarding the invisible physical phenomena inside air.

While others before him employed a billiard-ball mechanical model of air, Clausius employed a more generalized and thus a more powerful model, because who really knew the shape of the hypothesized atoms? Instead of assuming elastic spheres, Clausius proposed more complicated molecules that embodied three different types of motion, "hither and thither" translational motion and also rotational and vibrational motions. As will be discussed more shortly, such an approach enabled him to use the ratio of heat capacities C_p/C_v to probe the inner structure of gas molecules. This and other insightful assumptions served to establish Clausius' paper as the first strong step toward the complete development of the kinetic theory of gases and eventually statistical mechanics. Indeed, Daub cited it as "the watershed between the primitive and sophisticated versions of the kinetic theory of gases."[12]

Clausius even broached the topic of intermolecular force laws, writing "I imagine...two forces,"[13] and then went on to describe the shift between them as the distance between passing molecules closes: the effects move from no effect, to a change in direction (attractive force) to a collision (repulsive force) and rebound. After sharing this insight, he simplified the mathematics by saying that by focusing on forces at very small distances, one can simply assume a "sphere of action" and calculate how far on average a molecule can move before entering another's "sphere of action." While he never attempted to incorporate a force law into his model, he was still able to start a powerful conversation going in the world of physics, one involving not only the existence of molecules but the fact that they're constantly colliding, a hypothesis that would have significant implications later on.

The Math behind the Kinetic Theory of Gases

So let's now look at how Clausius started this journey. I'll walk you through the essence of his logic by modifying his text with some of Feynman's own explanations.[14]

Clausius started by assuming the gas to be *ideal*, meaning that only a small space relative to the whole is filled by molecules and that intermolecular forces are insignificant. He also assumed the molecules to all have the same velocity even though he realized that this was not actually true, writing "There is no doubt that actually the greatest possible variety exists amongst the velocities."[15]

[12] (Daub, 1970c) p. 106.
[13] (Clausius, 2003b) p. 138.
[14] (Feynman et al., 1989a) Volume I, Chapter 39, *The Kinetic Theory of Gases.*
[15] (Clausius, 2003a) pp. 126–127.

This assumption served to greatly simplify the mathematics and so enabled Clausius to reach an answer and test his hypothesis. (Note: Velocity and speed are used somewhat interchangeably in the literature. Going forward I will use velocity to mean speed unless otherwise stated.)

Clausius' first goal was to determine the relationship between the translational motion of the molecules and the pressure of the gas. Pressure was a well-understood property to start with, better understood than temperature as it readily lent itself to a mechanical calculation. The force acting against the wall is caused by molecules hitting the wall. The important variables to consider are the mass and velocity of the molecules and then the rate at which they hit the wall.

For a single molecule and a single hit, the force generated against the wall is the change in momentum of the molecule. If the collision is perfectly elastic and the molecule is aimed directly at the wall, the change in momentum of the molecule is

Before: mv_1
After: mv_2

But we know that for elastic collisions v_2 equals $-v_1$ and so the change in momentum of the molecule is simply $2mv_1$. From Newton's 2nd Law of Motion,

Force = rate of change of momentum
　　= d(mv)/dt
　　= (change in momentum of single molecule) ×
　　　(rate of collision of molecules against wall)
　　= 2mv × (rate of collision of molecules against wall)

To calculate the collision rate, we need to know how many molecules simultaneously strike the wall at each moment in time. Assume the molecules travel an infinitesimal distance (dx) for an infinitesimal time (dt). The molecules striking the wall during an interval of dt will be those dx distance away from the wall. In other words, a range of molecules with different velocities will strike at the same moment in time if their relative distance from the wall is proportional to their relative velocity.[16] The total number of molecules thus striking the wall is equal to the density of the gas (N/V) times that volume of the gas that will strike the wall, which is the area of the wall (A) times the "striking" distance, which as discussed above is v_x. Note the "x" subscript on the velocity. We're only interested in the velocity of the molecule in the direction of the wall. Since the molecules move equally in all three x, y, and z directions—the system as a whole is stationary—the squares of the mean velocities in each direction are equal to each other and thus equal to the square of the mean velocity. But we need to calculate the total number of molecules moving in just one of these three directions, specifically the direction aimed directly at the wall. This number is one-third of the total, and actually one-half of this number since the other half is heading directly *away* from the wall. So let's see what we have:

Force = $(2mv_x)(1/2)[(N/V)\times(Av_x)]$
　　= $A(N/V)mv_x^2$
　　= $A(N/V)1/3\,mv^2$

[16] Note that this calculation is based on continuous hitting of the wall by waves of molecules. The calculation is independent of time. Regardless of the time chosen, the calculation tells you how many molecules are striking the walls at any given moment in time. The change in momentum is thus continuously applied.

where "v" now represents the mean velocity. Re-arranging some gives us

$$\text{Pressure} = \text{Force/Area} = (N/V) \times (2/3) \times (1/2\ mv^2)$$
$$PV \qquad = 2/3\ N\ (1/2\ mv^2)$$

for which the last term, $\frac{1}{2}\ mv^2$, is naturally the kinetic energy of the gas molecule moving at the mean velocity.[17] Clausius arrived at this answer from a more circuitous pathway involving angles of collisions.

Once Clausius arrived at this equation, he compared it against the available experimental data in the form of the gas laws of Boyle (PV = k at constant T), Gay-Lussac (P = kT at constant V, V = kT at constant P) and Clapeyron who tied the two together in the ideal gas law,

$$PV = T \times \text{constant}$$

and concluded that the kinetic energy, or *vis viva* as he referred to it, of the molecules is proportional to absolute temperature. He then took this one step further by calculating the densities (V/Nm) at a given pressure for a range of ideal gases, which enabled determination of their mean velocities:

Oxygen =   461 m/s
Nitrogen =   492 m/s
Hydrogen = 1844 m/s

These values represent the translational speed of the molecules. Clausius next sought to quantify the relative proportion of total energy ($C_v T$) caused by translational motion ($1/2\ mv^2$), referencing Rankine's view that C_v represents the "true specific heat." The unstated assumptions behind this logic were that U is a function of T only and that C_v is a constant that quantifies the total energy associated with thermal motion—translational, rotational, vibrational—in a gas. This was one of the early versions of the equipartition theory, which is discussed further in Chapter 41.

$$\text{Translational Energy} = N\ \tfrac{1}{2}\ mv^2 = 3/2\ PV$$

$$\text{Total Energy} = C_v T = C_v\ PV/R$$

$$\text{Translational Energy / Total Energy} = 3/2\ R/C_v = 3/2\ (\gamma - 1)$$

where

$$R = C_p - C_v$$

and

$$\gamma = C_p / C_v$$

for which the quantity γ is a useful number to have in certain calculations involving gases.[18]

---

[17] Note that the force balance involves change in momentum, not kinetic energy. Kinetic energy falls out of the mathematics from the time element: total momentum change equals change per atom × number of atoms striking at one moment in time; the additional velocity component comes in with this last variable.

[18] $\gamma = C_p/C_v$ is a useful ratio because 1) it is independent of units used, 2) it tells you something about the complexity of the molecules, and 3) it appears in calculations involving adiabatic processes, including sound. See (Brush, 2003a) p. 13.

At the end of his paper, seeking to gain insight into the structure of atoms, Clausius inserted some heat capacity data obtained by Regnault for air and arrived at

Translational Energy / Total Energy = 0.63

and thus concluded that translational motion alone does not account for the total energy of motion contained in an ideal gas, in this case air. As he wrote, "We must conclude, therefore, that besides the translatory motion of the molecules as such, the constituents of these molecules perform other motions, whose *vis viva* also forms a part of the contained quantity of heat."[19]

Heat Capacity and the Monatomic Gas

This concluding calculation was enlightening for what Clausius didn't show, specifically that a γ value of 5/3 results in a ratio of exactly 1, meaning that the only energy present is the translational energy of atoms moving across space. Clausius believed in creating the most general model possible; hence, his inclusion of all possible thermal motions: translation, rotation, and vibration. So it's interesting that in this creation he never seemed to consider the possibility that molecules could occur as hard spheres with no rotational energy nor the possibility—in later evaluations by him and others—that diatomic molecules could occur with no axial-spin energy. This challenge reflected the very deep challenge of envisioning the actual physical shape of molecules. The mental block was somewhat understandable. How could Clausius or really anyone have possibly envisioned the existence of electron orbitals that would provide characteristics of a solid sphere of a specific radius but without the concomitant mechanical spinning component? To us today, this model is easy to imagine, as we know that an atom's mass is concentrated in the center with negligible mass—and thus negligible spin energy—at the perimeter. But to the theoretical physicists yesterday? They really didn't have any idea what the physical shape of the atom was. There were only limited data available to guide them and only the beginning of data to suggest that molecules themselves even existed. I wonder if they would have been deeply surprised to learn that atoms indeed behave like billiard balls—and molecules like billiard balls connected by a spring—and can be modeled as such. They had no idea how close they were to reality with their assumptions. This single issue of trying to imagine what couldn't be seen proved very troublesome in the late 1800s as it led to a discrepancy between the atomic theory and experiment, leaving many in doubt and uncertainty about the theory.

Mean Free Path

Clausius' paper in and of itself was a solid piece of work but it was his subsequent theoretical work that multiplied its influence, and it all started as a response to a great question raised by Dutch meteorologist C. H. D. Buys-Ballot[20] (1817–1890) who asked (to paraphrase), *if the speed of the molecules is so high, then why does it take on the order of minutes to smell the release of an odorous gas from across the room?* What a great question to ask. It forced Clausius to take a step back and

[19] (Clausius, 2003a) p. 134.
[20] (Clausius, 2003b) p. 136.

do a reality check. If the speed *is* so fast, why is the diffusion so slow?[21] Clausius rose to the occasion—"I rejoice at the discussion of this point by M. Buijs-Ballot"[22]—and set about to add a layer of complexity to his physical model that incorporated collisions between molecules, proposing this phenomenon as the mechanism impeding diffusion. The question he pondered was, "How far on average can the molecule move, before its center of gravity comes into the sphere of action of another molecule?"[23] for which the "sphere of action" was the radius for which intermolecular forces are significant around a molecule. Again, not having any real evidence for what a molecule looked like and what kind of forces it contained, and clearly there were forces to consider in light of the fact that gases condense (attraction) and that two bodies cannot exist at the same spatial coordinates (repulsion), Clausius assumed each molecule to have a specific radius and all molecules except one to be spread evenly throughout a given volume with the one exception moving at a mean velocity through this structure. Based on his calculated probability of how far a moving sphere would pass through evenly spaced obstacles before collision, Clausius determined that

$$\frac{\text{Mean Length of Path}}{\text{Radius of Spheres}} = \frac{\text{Total Volume of System}}{\text{Volume Occupied by Gas Molecules Spheres}}$$

and based further on his educated guess that the second ratio is about 1000:1—which is consistent with the approximate increase in volume that occurs when a liquid becomes a vapor— he concluded that a molecule is likely to travel about 1000 times its radius before colliding with another molecule. This result has since held up. Looking at the numbers, the radius of either nitrogen or oxygen is around 8×10^{-9} cm and the mean free path is about 1000 times higher at around 8×10^{-6} cm.

Interesting Consequences of the Kinetic Theory of Gases

Clausius' push to quantify the atomic world helped open the door for others to consider the "If they exist, then…" startling implications of the theory. Of this group, I share the work of three here in this short little detour. As we'll shortly see, Maxwell determined a fundamental equation for gas viscosity based on Clausius' mean free path theory. He then used Stokes' viscosity data in this equation to estimate that each gas particle makes about 8 billion collisions per second.[24] As we'll also shortly see, Boltzmann developed theories to quantify the rate at which a given population distribution moves toward the equilibrium Maxwell distribution. He showed that because of such high collision frequencies, a distribution in which all gas molecules start with the same velocity would achieve a Maxwell distribution after only a hundred-millionth of a

---

[21] (Brush, 2003b) pp. 421–450. Brush wrote about the importance of such science "gadflies" as Buys-Ballot to the progress of science. It was this group who asked the difficult penetrating questions that motivated others to excellence while unfortunately not being later acknowledged by historians for their critical role. "Clausius gets the credit for introducing the mean free path, while Buys-Ballot is forgotten." Interesting to read in (Clausius, 2003b) p. 147 how defensive Clausius was at the end of his "mean free path" paper regarding this situation, suggesting that he had clearly stated simplifying assumptions that did not lend themselves to determining a mean free path. He did not regard "motion as it really occurs" and used this as his defense for not considering the problem Buys-Ballot developed.

[22] (Clausius, 2003b) pp. 136–7.

[23] (Clausius, 2003b) p. 139.

[24] (Maxwell, 2003a) p. 166.

second.[25] But perhaps the award for the most striking calculation goes to Josef Loschmidt (1821–1895).[26] In 1865 he started looking at Clausius' work and those of others and noted two relationships. Molecular volume depends on Nd^3 while Clausius' mean free path depends on Nd^2 as it's the density of circumference-based surface area (and not volume) that hinders atomic motion. Thus, he reasoned that if he could quantify both terms, he could solve for the combined variables in each. He estimated Nd^3 by considering the volume change of gas liquefaction: the total volume of molecules in a given volume of gas is equal to the total volume of condensate generated from this mixture. He then estimated Nd^2 by considering the combined gas viscosity studies of Stokes, Maxwell, and O.E. Meyer for which viscosity was mathematically linked to "mean free path" and thus to Nd^2. With two equations and two unknowns, Loschmidt was able to estimate d of about 10×10^{-8} cm. While he didn't further focus on this, one could use the same data to then calculate N and, using his data, this turns out to be 2×10^{18} molecules for 1 cc volume at 0 °C and 1 atm pressure. This number was later revised to 2.687×10^{19} and became known as the Loschmidt number; it also later evolved into the similar Avogadro's number of 6.02×10^{23} molecules per gram-mole based on the standard volume of a perfect gas being 22420.7 cc atm mole^{-1}. Such calculations as these and others, being based on the implications of moving and colliding atoms, were rather fascinating to consider and the fact that they started to make the invisible visible started to enhance the credibility of the atomic theory itself.

* * *

So where does this leave us now? Clausius' two papers arguably launched a revival of the kinetic theory of gases by laying out a strong physical model together with some intelligent simplifying assumptions to make certain calculations that shed light on the invisible "sea of air" around us. He brought new attention to the speed of molecules and their fundamental connection to temperature. He showed how heat capacity data could help us to better understand the shapes and structures of molecules. He developed his "mean free path" theory that opened up a whole a new discussion about molecular interactions. All in all, Clausius' build-up of arguments was quite impressive and certainly laid the groundwork for the explosion of mathematics that was to follow in the hands of Maxwell and Boltzmann. In short, Clausius' content and credentials added *gravitas* to the kinetic theory of gases and solidified its initial standing in the science community.

But Clausius' work fell short in one key way. It failed to link his predictions to some measurable macroscopic property. It failed the ultimate test of the scientific method: does prediction match reality? This was clearly a difficult task. How could one measure molecular speed, radius or "mean free path"? How could one measure the invisible? Clearly a bridge was needed to cross from micro to macro. He himself didn't see this bridge. But Maxwell did.

---

[25] (Boltzmann, 2003c) p. 402.
[26] (Brush, 1986a) pp. 75–76 and (Brush, 2003b) pp. 429–430.

41

Maxwell: the rise of statistical mechanics

The greatest mathematical physicist the world had seen since Newton

– Sir James Jeans (1931)[1]

We now return once again to James Clerk Maxwell (1831–1879). While his earlier presence in this book was needed to bring clarity to the rise of classical thermodynamics, his presence here is more importantly needed to bring clarity to the rise of the kinetic theory of gases and statistical mechanics. It was really Maxwell who, with his advanced mathematical skills, raised Clausius' work to a much higher level of theoretical power, paving the way for the future contributions of Boltzmann and others. This effort along with his other works, especially including his development of electromagnetic theory, his own experiments on gas viscosity, and his opening of the famed Cavendish Laboratory in 1874, established Maxwell as one of the founding fathers of both theoretical and experimental physics.

Early Life and Saturn's Rings

Born in Edinburgh and raised nearby in Glenlair, Maxwell was another of the Scottish scientists/physicists in the mode of Watt, the Thomson brothers, and the adopted Joule: self-confident yet modest, ready to tackle any task.[2] With a childhood curiosity that never left him, Maxwell sought truth with both mind and hands ("*mens et manus*"), pursuing the fundamental questions of nature while employing experimental techniques as his validation. Like Gibbs, he used an organized and thorough thought process to develop and present foundational theories in long, extended papers.

One of Maxwell's first scientific accomplishments was his analysis of the rings of Saturn. In 1855 the organizers of the Adams prize, an award given by Cambridge University to a mathematician for distinguished research, issued a technical challenge that loosely read, *why are the rings of Saturn stable? Why don't they crash down into the planet or fly off into space?* Maxwell spent between 1855 and 1859 answering this question by using Newtonian mechanics to analyze the gravitational forces acting between ring and planet and also within the ring itself. He concluded that the rings could be neither solid nor fluid but must instead be comprised of

---

[1] Sir James Jeans quote cited in (Everitt, 1975) p. 11.
[2] (Mahon, 2003) p. 23.

Block by Block: The Historical and Theoretical Foundations of Thermodynamics,
Robert T. Hanlon, Oxford University Press (2020). © Robert T. Hanlon.
DOI: 10.1093/oso/9780198851547.001.0001

a large number of small bodies. As the only contestant who completed this challenge, Maxwell won the award and at the same time gained a strong interest in the motions of large numbers of colliding bodies.

Maxwell's experience with Saturn primed him to favorably respond to Clausius' work. In April of 1859 he read Clausius' 1858 "mean free path" paper and saw a connection between the moving bodies in Saturn's rings and those in gas-phase systems. While he had earlier considered the technical challenges involved in modeling Saturn's rings as being "hopelessly complicated,"[3] his attitude shifted when confronting Clausius' gas theories, reflecting the strength of Clausius' ideas, his own mathematical fortitude, and a recently read review of statistics, which we'll get to in a moment.

From Micro to Macro

Maxwell was well suited to develop Clausius' work further, more so than Clausius himself. Clausius had derived an expression for "mean free path" but didn't attempt to validate the hypothesis, which isn't meant to pass judgment but simply meant to emphasize the size of the challenge involved. There was no simple way to do this. But Maxwell figured out a way. Being very familiar with the experiments of Thomas Graham (1805–1869) on the diffusion of gases and Sir George Gabriel Stokes (1819–1903) on the viscosity of gases, including Stokes' experimental approach to studying gas viscosity by quantifying the dampening of pendulums, Maxwell realized that if molecules indeed existed as Clausius proposed, then their hindered movement across space caused by collisions should correlate in some way to the macroscopic properties dependent on this movement, such as the aforementioned diffusion and viscosity. Such familiarity combined with his mathematical firepower and experimental experience—again demonstrating the power of working at interfaces—enabled Maxwell to see the bridge from micro to macro and fostered in him the self-confidence he needed to start his own challenging journey across the bridge.

Testing for Absurdity

What drove Maxwell to attack such a complicated problem? The rather interesting answer is found in his letter of May 30, 1859 to Sir George Gabriel Stokes. Wrote Maxwell in regards to Clausius' paper,

> *I thought that it might be worthwhile examining the hypothesis of free particles acting by impact and comparing it with phenomena which seem to depend on this "mean path"...I do not know how far such speculation may be found to agree with facts...I have taken to the subject for mathematical work lately and I am getting fond of it and require to be snubbed a little by experiments...[4]*

---

[3] (Everitt, 1975) p. 131.
[4] Maxwell quote cited in (Brush, 2003a) pp. 26–27.

Maxwell was intrigued by Clausius' hypotheses. But being a firm adherent of the scientific method, he fully understood the need to "follow out the consequences of the hypothesis"[5] and validate, or "snub" as he termed it, the resulting predictions by experiments. As stated before, the problem was that Clausius' work offered no clear means of doing this. But Maxwell had an insight. He realized that, assuming that molecules actually existed and behaved as Clausius hypothesized, then certain measurable macroscopic properties of gas, such as viscosity, thermal conductivity, and diffusion, all of which involved the transport of molecules from one zone to another, should depend on the hypothesized "mean free path" of the molecules. Using his strong mathematical skills together with a healthy skepticism of Clausius' claims—Maxwell had been taught Newton's static theory of gases as a student at Edinburgh, which was the prevailing opinion at that time, and so could be forgiven for questioning the kinetic theory of gases—he determined how to predict these bulk properties based on the "mean free path," believing that this approach would lead to "absurd"[6] predictions and so disprove Clausius. Little did he know that this interesting "exercise in mechanics"[7] would lead to the validation of the kinetic theory of gases and the rise of statistical mechanics. His vigorous and unsuccessful attempt to kill Clausius' hypotheses was what ultimately made them survive. A more powerful example of the power of the scientific method could hardly be found. He followed this philosophy to a "t" and arrived at a prediction that had not yet been observed and that he had not expected.

Maxwell's Two Publications on the Kinetic Theory of Gases

We assert that the gross properties of matter should be explainable in terms of the motion of its parts.

–Richard Feynman[8]

So many of the properties of matter, especially when in the gaseous form, can be deduced from the hypothesis that their minute parts are in rapid motion, the velocity increasing with temperature, that the precise nature of this motion becomes a subject of rational curiosity.

– James Clerk Maxwell (1860)[9]

Maxwell published two major papers on the kinetic theory of gases, his first in 1860 and his second in 1866 to correct technical mistakes found in the first by Clausius. This theoretical effort combined with his own experimental program resulted in both the first integration of statistics into physics and also in a new understanding of the properties of gases based on the atomic theory of nature.

As was the case with Gibbs, Maxwell didn't write short notes. These two papers were long and thick with equations requiring tutorials in both calculus and statistics to get through. I have chosen to focus on three major developments from this effort, namely 1) the rise of

---

[5] (Maxwell, 2003a) p. 150.
[6] (Brush, 2003a) p. 27.
[7] (Brush, 2003a) p. 27.
[8] (Feynman et al., 1989a) Volume I, p. 40-1.
[9] (Maxwell, 2003a) p. 149.

statistics in physics to characterize the velocity distribution of atoms and molecules in gaseous system, 2) the rise of the equipartition theory in physics, and 3) the rise of transport equations for gas properties of viscosity, thermal conductivity, and diffusion. I cover each in the order listed.

> Maxwell's Velocity Distribution

Picture a box containing a large number of very small hard balls in motion continually colliding with each other, each collision being perfectly elastic such that the combined momentum and combined kinetic energy before and after are equal. Further, picture observing the motions of this large number of colliding bodies. What would you see? Well, it would be difficult to really see anything, wouldn't it, other than a mess of motion? But one thing you would notice would be that the density of the balls would seem about the same throughout the box, simply reflecting that the motions of the balls have no preference for direction. None. The x, y, and z directions are all equal to the colliding balls. In fact, if you had enough computing power and could model the motion of each and every ball, assuming simple two-body collisions and the two conservation laws (momentum and energy), your equations would show no bias in any of the directions. This makes sense since, in the absence of an external force such as gravity, nature has no concern for direction inside the system.

What else would you see? This is where things get difficult. You'd see some slow-moving balls, some fast-moving balls and some medium-moving balls. Some moving left, some right, up / down, front / back. If you could follow any one individual ball, which would be easy if it were bright red and all others white, you'd see this ball changing velocity—both speed and direction—constantly and randomly. Given this, how would you ever come to think that a structure could possibly exist inside the chaos? And even if you did, how would you even approach describing such a system of motions mathematically?

Well, first, you could start with a simple assumption. Since the entire system is stationary, then the average of all the velocities (speed and direction) would have to equal zero since there is no net velocity in any given direction, as there would otherwise be in a flowing system. But while the average velocity would be zero, the average speed, being the absolute value of the velocity vector, would not. (As mentioned earlier, the terminology regarding velocity and speed gets somewhat confusing. Again, I'll continue to use velocity to mean speed unless otherwise stated, such as I've just done here.)

Then what? Well, you could do what Clausius did. As a great mathematical simplification, he assumed that all molecules had one and the same mean velocity (speed) even though both he and Maxwell realized that the velocities really spanned a range. But if you did this, you would limit the power of the model. Alternatively, you could start to mentally work your way through this. Say all of the molecules started out moving equally in all directions at the same speed or mean velocity, as Clausius assumed. Could this last? Only if they never collided. In his first paper, before he started considering collisions, this is what Clausius assumed. For his original calculation of pressure against a wall, he assumed that the molecules moved parallel to each other (no collisions) and perfectly perpendicularly to the walls. But this assumption quickly disappeared by Clausius' own hand when he created his "mean free path" theory. After this, both he and the rest of the world realized that the molecules, if they existed, must be constantly colliding. It was this paper that Maxwell first read. He made constant collision of all molecules one of his main starting assumptions.

Colliding Balls – Physical Model

So, going back to the ideal assumption of all spheres starting out with the same mean velocity, what happens when these spheres start colliding with each other? Well, it's conceptually simple but mathematically complex. Each collision between two hard balls—and only binary collisions were assumed by Maxwell as the likelihood of ternary collisions in such a rarified system was considered highly improbable—must conserve both momentum and energy. As discussed earlier in this book, the science of collision theory and laws of impact had been established by the likes of John Wallis, Christopher Wren (1632–1723), and Christiaan Huygens[10] well before Clausius and Maxwell began their work. Each was fully conversant in this subject matter.

So let's walk through this. When all was said and done, for an initial condition in which all balls had the same velocity, after all balls experienced their *first* collision, unless the collision were exactly head-on, then half would have higher velocities and half lower. It would have to be this way since total energy is conserved. For any collision other than exactly head-on, one must gain, the other must lose.

Now what about the second set of collisions? And the third? And the millionth? You can picture a spreading of the velocities from the starting point, some going lower, some higher, most remaining in the middle. But why like this? Why not, for example, an even spreading over the entire range of velocities? Because such distributions would be highly improbable in a world where energy is conserved. Here's why. When two balls hit, the faster one typically slows down and the slower one speeds up. The energies move toward each other, which is the fundamental reason behind why temperatures move toward each other. Can the energies move apart? Yes, if they hit each other at the perfect angle. Possible? Yes.[11] Probable? No. In fact, it's highly improbable, for the same reason it's highly improbable that temperatures move apart, which is more than highly improbable since many of these rare collisions would have to happen at the same exact time to affect a macroscopic property like temperature. Now given this, when a molecule gains velocity, what happens to it? What does it see? As its velocity increases, it sees less and less faster molecules in its neighboring community and more and more slower

---

[10] (Brush, 2003a) p. 3.

[11] Say two atoms (same mass) having the same mean velocity hit each other exactly head-on. They would rebound with the same exact mean velocity. But say that they didn't hit each other exactly head-on, which is much more likely. What then? A whole range of results could possibly happen, each dependent on the angle of collision. In one extreme, say that Ball 1 was moving in the y-direction and Ball 2 in the x-direction. And say that Ball 1 hit Ball 2 exactly perpendicularly at the exact point in time when the center of Ball 2 was at the origin of the x-y axis. Since Ball 2 had zero velocity in the y-direction, then it would be as if Ball 1 (moving in the y-direction) hit a stationary ball. Ball 1 would thus stop dead in its tracks while Ball 2 would gain a velocity component in the y-direction. The velocity of Ball 1 would become zero (motionless) and that of Ball 2 would become the square root of 2 greater. Note in this scenario that even though Ball 2's initial speed could have been higher than Ball 1's, it would have become even higher after collision. Possible? Yes. Probable? No. Most but not all binary collisions cause kinetic energies to move toward the mean.

ones. The probability that its next collision is with a slower molecule increases and so increases the probability that its velocity will decrease. In effect, as a molecule's speed increases or decreases, it's pulled with increasing probability back toward the mean, which is sometimes termed "regression to the mean." If you consider this from a probability standpoint, the probability that a molecule will have a certain velocity decreases the further that velocity deviates from the mean. This fact is what gives the distribution its shape.

But exactly what shape? Is it a straight downward line on either side away from the mean? A curved line? As it turns out, nature's behavior in this context is exactly described by a specific distribution named after Carl Friedrich Gauss (1777–1855). The Gaussian distribution was based on the distribution of errors from a mean value, hence giving it the name "law of errors."[12] The critical assumption in this derivation was that large errors are less likely than small errors. This is simply common sense. If you throw many darts at the dartboard (and are competent in throwing them), many more darts will be near the bulls-eye than far from it. Gauss' error in this situation being the distance between the throw and the bulls-eye means that large errors are less probable. Substitute the difference between velocity and the mean velocity for errors and you arrive at the same mathematics. The probability of a molecule moving at a certain velocity decreases as that velocity is further removed from the mean. Large errors are less likely than small. Because of this, probability says that the distribution of velocities must spread out from a starting point of identical speeds in all direction but that it can't spread out too far. The spread is constrained.

The above step-by-step progression of thought helps us understand why there is a naturally occurring distribution of velocities—the Gaussian distribution[13]—that reflects a balance between the competing tendencies to spread and get reined in. It makes sense to us, each step building on the one before. So it's rather interesting that in his original 1860 paper, Maxwell didn't describe his path to the Gaussian distribution in this manner, thus prompting the question from historians: How did he arrive there? Did he deduce this distribution *a priori* or was there another path?

Maxwell's Path to the Gaussian Distribution

If a great many equal spherical particles were in motion in a perfectly elastic vessel, collisions would take place among the particles, and their velocities would be altered at every collision; so that after a certain time the vis viva will be divided among the particles according to some regular law.

– James Clerk Maxwell[14]

Maxwell deeply believed that while collisions amongst a large number of moving bodies are random, "some regular law" must describe the end result. Just as each coin flip is random, the

---

[12] (Gillispie, 1963)

[13] To be exact, the Gaussian distribution is of the form $\exp(-x^2)$ for $x = -\infty$ to $+\infty$ and applies to the velocity component of a specific direction such as u_x. The related distribution for speed in which $u = (u_x^2 + u_y^2 + u_z^2)^{1/2}$ is $u^2 \exp(-u^2)$. The details of the mathematical derivation are presented in, for example, (McQuarrie and Simon, 1997) pp. 1106–1112.

[14] (Maxwell, 2003a) p. 152.

distribution between heads and tails after a large number of flips isn't. The statistics involved are clear and powerful and show how chaos with small numbers can lead to structure with large numbers. In the world of physics, it was Maxwell who first led us in this direction.

The history surrounding Maxwell's introduction of statistics into physics—"a new epoch in physics"[15]—is very interesting. When reading his 1860 paper, there's a moment where he lays out some rather simple-sounding, "these-make-sense" assumptions which had nothing to do with his preceding discussion on collision theory and the conservation of energy, and then *bam!*, seemingly out of nowhere he states, "Solving this functional equation, we find"[16] which is followed by the Gaussian distribution. When I first read this, eager to see Maxwell's rationale for his distribution, eager to finally learn its origin, my reaction was, "*Say what?*"

While Maxwell wrote as if he were following a true deductive process, historians suggested an alternative and more likely version of events, one in which he saw the answer before he started the process. They suggested that Maxwell knew where he needed to go and just had to figure out how to get there.

In his 1860 paper Maxwell stated, "It appears from this proposition that the velocities are distributed among the particles according to the same law as the errors are distributed among the observations in the theory of the 'method of least squares,'"[17] a method that he first learned about in an 1850 review by John Herschel (1792–1871) of Adolphe Quetelet's (1796–1874) 1846 book, *Letters on the Theory of Probability*. As a student in Edinburgh, Maxwell read Herschel's review the year it was published and was deeply impressed, writing about it to his friend Lewis Campbell, "…the true logic for this world is the Calculus of Probabilities."[18] This was an interesting statement to make during a time when determinism held sway. But, in fact, there was no conflict here. The elastic collision of spheres *is* deterministic. But when multiplied by the reality of millions of spheres, such determinism falls apart, not because it's no longer there but instead because one simply can't keep track of it all. Rather than simply giving up, Maxwell realized that statistics could provide a pathway out based on his belief that nature's seemingly random behavior at the small scale gives way to a structured behavior at the large scale that could be captured by the statistical laws of large numbers.

In considering this history in greater depth, here's what we know. We know that Maxwell not only knew of Herschel's essay but was likely influenced by it in his 1860 derivation,[19] which would explain the lack of physics in the derivation. Turning now to what we don't know, we don't know what Maxwell's exact path of discovery was. He didn't describe it in his 1860 paper. So here's what we believe based on the historical detective work of, among others, Francis Everitt, Stephen Brush, and Michael Strevens.

Maxwell had no precedent to follow in addressing this problem. He was on his own. Statistics had not yet been used in physics. But his intuition into the physical nature of "many bodies in constant collision" told him that his answer existed in statistics. Because he was well read in the theories on collisions and energy, he knew that the velocities of the spheres in his mechanical system must be distributed about a mean. His thought process led him down a path toward a

[15] (Everitt, 1975) p. 134
[16] (Maxwell, 2003a) p. 153.
[17] (Maxwell, 2003a) p. 154.
[18] (Brush, 1983) p. 59.
[19] (Brush, 1986a) p. 185; (Everitt, 1970) p. 219; (Everitt, 1975) pp. 136–7; (Strevens, 2013) pp. 15–17.

shape that had earlier resonated with him, namely the Gaussian distribution as described in Herschel's review. He knew that this was the answer he wanted.[20] He just didn't have the mathematical means in 1860 to arrive at this answer based on physics. And he likely didn't want to wave his hands around and present an answer he believed to be true without a sound deductive process supporting it. So he relied on Herschel's derivation by converting assumptions about the distributions of darts on a dartboard to distributions of colliding spheres in a box.

Read carefully Maxwell's description of his derivation.[21] He envisioned a group of his moving spheres clustered at the x-y-z origin and then at an initial time (t = 0) allowed them to disperse, almost like allowing a small volume of gas atoms to experience free expansion. After a given unit of time, the fastest ones would naturally have traveled further than the slower ones and the resulting population density as a function of distance or radius from the origin would reflect the velocity distribution contained in the original system. The similarity between the distribution of spheres around the origin after a certain time and the distribution of darts around the bulls-eye is telling.

Likely realizing his need to anchor his derivation in sound physics, Maxwell revisited this problem in his 1867 paper, this time using more elaborate mathematics based on collision theory and conservation of energy while also addressing an important problem with one of his assumptions. In his 1860 paper, he assumed that the velocity distribution of his elastic spheres was independent of direction. However, in reality, such is not the case. For example, the inherent resistance to flow in a gas is caused by a velocity gradient. In this non-equilibrium transport system, the perpendicular motion of slower moving molecules into the faster moving stream creates an effective resistance to flow as quantified by viscosity. Direction clearly must be accounted for in this scenario. Maxwell addressed this by modifying this critical assumption to allow the velocity to be variable for each of the x, y, and z directions.

Included in Maxwell's mathematics was use of the calculus term, d/dt. Maxwell began considering how the velocity distribution would evolve over time, which took him in a theoretical research direction that would later inspire Boltzmann. The fact that the distribution changes over time by itself is interesting, but it's the *direction* of change that's more interesting. As will be discussed later, the concept of entropy's increase to a maximum over time in an isolated system is, at the molecular level, the concept of the velocity distribution changing over time toward the one of maximum probability, which is the Gaussian distribution. Velocity distribution isn't the only determinant of entropy in an ideal gas system; location is also important. But the complexities associated with the changing velocity distribution, complexities that Maxwell began to address and that Boltzmann spent his lifetime addressing in his famed H-theorem, were the ultimate key bridge between the entropy of classical thermodynamics and the entropy of statistical mechanics. It all began in a rough form in Maxwell's 1867 paper.[22]

---

[20] In an interesting aside, this scenario played out twice in history. The Gaussian distribution and its variations are ones that occur in nature. Maxwell recognized this when considering the distribution of molecular velocities and Planck recognized this when considering the shape of the blackbody radiation curve as we discussed previously. The shape of this Gaussian-based curve, arising from natural phenomena, resonated with both physicists but perhaps for different reasons. For Maxwell, it made physical sense whereas for Planck, it provided him, at least initially, with a mathematical form that accurately characterized the BBR data.

[21] (Maxwell, 2003a) p. 153.

[22] Of interest is the thought that since Maxwell's distribution is correct then the hypotheses on which it was based must in some sense be justifiable. See (Everitt, 1975) pp. 137–138 and also (Kac, 1939) for a mathematical proof.

Regarding his distribution, Maxwell didn't know if he was correct or not[23] and didn't have a ready experiment in mind to find out. But it made intuitive sense to him, and later to Boltzmann, and the subsequent development of the kinetic theory of gases and transport property predictions based on this theory and the further development of statistical mechanics and its subsequent validation showed no contradictory information. Nothing pointed to failure. But still, without a validating experiment, his distribution would have to remain more hypothesis than fact. This would change.

A Beautiful Confirmation Experiment of the Maxwell Distribution

Some 60 to 70 years after Maxwell first proposed his velocity distribution and after much effort, scientists landed on a beautiful experimental validation.[24] Employing a modified experimental procedure of the one laid out in the early 1920s by Professor Otto Stern (1888–1969) in Germany, scientists first created a beam of gas atoms, not of common gases like air but of metal gases such as mercury and bismuth which could be easily vaporized and deposited upon contact with a suitable surface. This beam was aimed at a spinning hollow cylinder containing a slit. As the slit rotated into the beam, a pulse of atoms entered the spinning cylinder and deposited on the inside. This pulse was a representative sample of the gas and thus contained atoms moving at different velocities according to Maxwell's hypothesized population distribution. (In a way, this experiment followed Maxwell's thought experiment of placing gas atoms at the center of a dart board at t = 0 and then letting them spread.) The moving slit allowed the gas-pulse to enter and hit the inside wall of the cylinder opposite the slit. If the cylinder were moving too slowly, a single blob of deposit would result. But as the cylinder spun faster and faster, the distance separating the fast moving atoms from the slow moving atoms started to increase. By around 1930, scientists had been able to tune the cylinder speed such that a beautiful distribution of deposit density (i.e., population) vs. speed appeared. It was the distribution predicted by Maxwell.

| Equipartition of Energy |

Of all the questions about molecules which Maxwell puzzled over during this period the most urgent concerned their structure. His uneasiness about the discrepancy between the measured and calculated specific heat ratios of gases... increased after 1868 when Boltzmann extended the equipartition theorem to every degree of freedom in a dynamical system composed of material particles; and it turned to alarm with the emergence of a new area of research: spectrum analysis.

– Francis Everitt[25]

[23] (Darrigol, 2018) p. 365. "as late as 1895, Maxell's derivation of his distribution law could still be deeply misunderstood and... the assumptions needed in this derivation were just beginning to be understood."
[24] (Chalmers, 1952) pp. 154–5.
[25] (Everitt, 1970) p. 223.

Heat capacity is such a seemingly nondescript physical property. Add heat, measure temperature change, divide the two, and get heat capacity. What's so special about this, right? But nondescript couldn't be further from the truth. Indeed, heat capacity has played a major role several times in the rise of physics, arguably starting with Mayer's discovery of energy and its conservation and continuing with Clausius' and Maxwell's development of the kinetic theory of gases. Whereas Mayer focused on the difference between C_p and C_v for an ideal gas since this enabled his calculation of the mechanical equivalent of heat, Clausius and Maxwell focused on the ratio of the two since this enabled them to probe the inner mechanical workings of molecules. Recall Clausius' fundamental question: *where does the heat go?* This line of inquiry was what ultimately led to the equipartition of energy theory.

When Clausius summarized, formalized, and built on the work of others, he showed that if gas is indeed comprised of rapidly moving molecules, then the translational kinetic energy associated with this motion can be calculated using the kinetic theory of gases, while the total energy, which additionally accounts for rotational and vibrational energies, can be calculated using heat capacity. Based on this kinetic theory, the translational energy should be less than or equal to the total energy.

Maxwell furthered Clausius' work, not so much by bringing more mathematics to the situation, but by contemplating more deeply what the results meant. Recall that Maxwell's main objective in his 1860 paper was to challenge Clausius' hypothesis. He built a mechanical model that operated per the known laws of motion and applied very advanced mathematics in assessing the consequences. Should the model predict the real world, in this case the ideal gas, then that would strongly support the kinetic theory of gases. But going into this, he believed that he would find "absurd" predictions of his mechanical system relative to ideal gases and so was seeking the scientific proof that would disprove Clausius. To Maxwell, the ratio of specific heats, C_p/C_v, which falls out of the mathematics, became one of the most significant pieces of evidence in this trial. His deep inquiry into the numbers and their significance opened the doors to the equipartition theory of energy and eventually to quantum physics.

The Physical Meaning of γ

Before proceeding, I want to simplify things. Clausius focused on the ratio of translational energy to total energy, Maxwell focused on the ratio of total energy to translational energy, and both focused on the ratio C_p/C_v since this ratio was central to both. This ratio gained prominence in earlier classical thermodynamic development because of its appearance in adiabatic processes such as compression for which PV^γ remains constant (ideal gas only) and in speed of sound calculations (adiabatic wave compressions). From here on out, I'll focus solely on the value of γ as this ratio has endured while the others have not.

As discussed in the previous chapter, according to the kinetic theory of gases, the value of γ for a monatomic ideal gas should exactly equal 5/3, reflecting the fact that all internal energy of the gas is accounted for by the kinetic energy of the rapidly moving spheres and that the atoms as such exhibit no rotational energy, meaning that collisions between monatomic ideal gases are perfectly elastic. Both Clausius and Maxwell had the mathematics in front of them to address the possibility of a γ value of 5/3 but neither did. Again, they likely could not envision how a solid sphere could move across space but not spin while doing so; how was it possible to have translational energy but no spinning energy?[26] This discussion is clearly not meant to criticize either. It's solely meant to provide insight into how challenging the world of physics was in the days before the atomic theory was established. Compounding the challenge was the lack of data. One might have asked, *but surely someone must have observed a γ value of 5/3 for some monatomic gas?* And had this data point been available then, things might have turned out differently. But as it was, the value of 5/3 wouldn't be measured until 1875 by August Kundt (1839–1894) and Emil Warburg (1846–1931) for mercury vapor, and then again in the 1890s once William Ramsay[27] (1852–1916) discovered the inert gases of helium, krypton, and, with Lord Rayleigh (1842–1919), argon.[28] In fact, Ramsay even used the value of 5/3 to support his belief that helium is monatomic and to also support analysis of his future monatomic noble gas discoveries: neon, krypton, xenon. But even with these findings, the electron-shell structure of the atom that accounts for zero rotational energy was not even on the horizon.[29]

Maxwell pushed his advanced mathematics forward to calculate what the value of γ should be given his assumption that molecules possess both translational and rotational energy. He made significant progress leading to the larger theory of the equipartition of energy, a theory that built on the earlier work of Waterston who stated that there is an equal partition of energy between gaseous molecules of different mass, meaning that they all had the same kinetic energy, which was connected with absolute temperature.[30] Maxwell showed this theory to be more than this.[31] He proved that, given a system of atoms continually colliding, the average *vis viva* of each of the components of translational energy and the average *vis viva* of each of the components of rotational energy must all be equal. The underlying theory says that, regardless of whether the involved atoms are monatomic or bound to other atoms, they experience one and the same energy and momentum conservation laws at the moment of collision. The summary conclusion is what Maxwell called his "equipartition theorem."[32] While he assigned a value of 3 for the translational DOF he also (incorrectly) assigned a value of 3 for the rotational DOF and none for vibrational DOF, thus giving him a total DOF of 6, a value of γ of 1.33 (relevant equation coming shortly) and a value of 2.0 for the ratio of total energy to total translational energy. The γ value Maxwell had to work with was 1.408 for air. To him, the difference

---

[26] (Maxwell, 2003a) pp. 166–167. Maxwell wrote, "When two perfectly smooth spheres strike each other, the force which acts between them always passes through their centres of gravity; and therefore their motions of rotation, if they have any, are not affected by the collision, and do not enter into our calculations." In this way, Maxwell did envision the possibility of no rotational energy for atoms, although his reasoning was different than that which was later learned based on the atom's electron-orbital structure.

[27] Sir William Ramsay (1852–1916) received the Nobel Prize in Chemistry in 1904 "in recognition of his services in the discovery of the inert gaseous elements in air." His work in isolating the noble gases argon, helium, neon, krypton, and xenon led to the development of a new section of the periodic table. See (Ramsay, 1904)

[28] (Brush, 2003a) p. 29.

[29] (Brush, 1986b) pp. 353–356. Brush summarizes well the range of theory-versus-data issues involved with specific heats for gases.

[30] (Brush, 2003a) p. 17.

[31] (Maxwell, 2003a) pp. 170–171.

[32] (Maxwell, 2003a) p. 149.

between the two—1.33 theory vs. 1.408 experiment—was significant, very significant, so significant in fact that he concluded his 1860 paper by writing that the consequences of his mechanical system "could not possibly satisfy the known relation between the two specific heats of all gases." He was so certain of his mathematical development that he was willing to jettison the whole kinetic theory of gases based on this single discrepancy.

So what was going on here? Let's look at the known physics.[33] For a monatomic gas, the value of γ should indeed be 5/3. For a diatomic gas, things get complicated. The molecules can rotate and they can vibrate. Maxwell's equipartition theory is indeed correct, *each* degree-of-freedom (DOF) contains the same amount of energy, later determined to be ½ kT. For a given temperature, total internal energy (U) is thus distributed equally to each physical DOF. This falls out of Clausius' mathematics for ideal gases. Since

$$\text{Total Energy} = U = C_v T = C_v PV / R = C_v PV / Nk$$

$$PV = NkT = (\gamma - 1) U \text{ wherein N = number of molecules; k = Boltzmann's constant}$$

$$U = NkT / (\gamma - 1)$$

and separately

$$U = DOF \times \tfrac{1}{2} kT$$

then

$$NkT / (\gamma - 1) = DOF \times \tfrac{1}{2} kT$$

$$\gamma = (DOF + 2) / DOF \qquad\qquad [41.1]$$

For monatomic gases, each molecule can move in three dimensions, thus DOF equals 3 and γ equals 5/3. For diatomic gases, you have three DOF numbers for the translational motion and then another two for rotational motion (which Boltzmann correctly pointed out in correcting Maxwell), thus DOF equals 5 and γ equals 7/5. (Note: no DOF for spinning around axis since again, there's negligible mass at the perimeter.) And if the diatomic molecule can also vibrate, this contributes two more DOF numbers, one for the kinetic energy of vibration and then another for the potential energy in an oscillating molecule. Maxwell's equipartition theorem applies to all forms of energy, including potential energy, and so in an oscillating molecule, the average kinetic energy must equal the average potential energy. Thus for a vibrating diatomic molecule, DOF equals 3 + 2 + 2 = 7 and γ equals 9/7.

So what we have is the following for ideal gases (no intermolecular forces):

Monatomic	$\gamma = 5/3 = 1.67$	no spin energy (all mass in nucleus)
Diatomic/rotation	$\gamma = 7/5 = 1.40$	vibration energy "frozen out"
Diatomic/rotation/vibration	$\gamma = 9/7 = 1.29$	2 DOF for vibration (both K.E. and P.E.)

---

[33] (Feynman et al., 1989a) Volume I, pp. 40-7 to 40-10.

Not to state the obvious, but the central issue in all of this was the lack of knowledge about atomic structure.[34] Yes, prior to the definitive conclusion that atoms are real, chemists and physicists *believed* they were real and even started proposing hypotheses about their structure. By the 1860s they had decided that most gaseous molecules were diatomic,[35] especially since evidence to prove otherwise wouldn't be available until Kundt and Warburg's research with monatomic mercury vapor in 1875, which might explain why Maxwell, for example, focused on physical models involving diatomic structures. But more details beyond diatomic were needed and Maxwell for one continued to challenge the scientific community about this. While Boltzmann resolved the issue of the DOF value for diatomic molecules by pointing out that a rigid connection between two mass-points yields a rotational DOF of two instead of Maxwell's assumed three and thus showed that the predicted value of γ for air of 1.408 was in good agreement with the theoretical value of 1.4, Maxwell challenged his central assumption as summarized by Brush: "How could an atom be elastic (so that kinetic energy and momentum would be conserved in collisions) and also unchangeable in form (so that it can be an ultimate atom in the philosophical sense)?"[36] Confounding this whole situation even more were two issues, the first relating to early experimental results showing some semblance of internal atomic structure based on the existence of spectral lines and yet not showing any correspondence between these lines and the value of DOF, and the second relating to the role of ether. The elimination of ether by Einstein from science wouldn't occur until well after this discussion. The attempts to account for the presence of ether in the motion of gaseous atoms remained in the background. How did the molecules exchange energy with ether and how did this impact the kinetic theory of gases assumptions? Such questions as these and many more—*what exactly do those spectral lines mean?*—made the approach to physical modeling extremely difficult. These issues were deeply significant to Maxwell—and also to Boltzmann who, whenever new specific heat data on gases appeared in the literature, "felt obliged to worry about the distribution of energy among the internal motions of polyatomic molecules."[37] Maxwell's adherence to the scientific method was paramount. To him, the failure of theory to predict nature required the theory to be discarded. As simple as that. In 1877 he wrote regarding the challenge of accurately predicting specific heats that nothing remained but to adopt the attitude of "thoroughly conscious ignorance that is the prelude to every real advance in knowledge."[38]

There was one additional factor at work here in creating confusion in the data. Quantum physics. Molecules need to get to a certain energy level before they vibrate. At low enough temperature, they don't vibrate and so there is no contribution to the DOF value due to vibration. As Jeans presciently observed, at low enough temperature, certain kinds of motions "freeze out."[39] With increasing temperature, vibration becomes possible, DOF increases, heat capacity increases (there's a new place to store energy), and γ decreases. Rotational states are

---

[34] (Brush, 1986b) pp. 353–363; (Feynman et al., 1989a) Volume I, pp. 40-7 to 40-10.

[35] (Brush, 1986a) p. 46. As part of his molecular theories of gases, Amedeo Avogadro (1776–1856) was one of the first to propose that oxygen and hydrogen were diatomic molecules. Stanislao Cannizzaro (1826–1910) revived Avogadro's theories at a conference in Karlsruhe in 1860.

[36] (Brush, 1986b) pp. 354–5.

[37] (Brush, 1970) p. 266.

[38] (Everitt, 1970) p. 223.

[39] (Feynman et al., 1989a) Volume I, p. 40-9.

quantized as well but the energy spacing between the states is too close together to have a significant impact on γ. The important point here is that it was the study of heat capacities that helped inform physicists that there were some unexplained phenomena within the theories of classical physics and the kinetic theory of gases. This was a positive result, not a negative one, because such unexplained findings as these served to motivate physicists to correct the problem and so helped lead to the discovery of quantum physics.

$$\boxed{\text{Transport Equations}}$$

Maxwell's above two developments were significant. His proposed velocity distribution was truly groundbreaking and pulled statistics into physics, where it would remain, while also bringing attention to the fact that a distribution *must* exist and moreover must be the one he hypothesized. While his work to understand heat capacity data via the kinetic theory of gases was not as groundbreaking since it built on Clausius' work, his focus on the discrepancies between theory and experiment and his attempt to resolve them using his newly coined "equipartition theorem" helped bring new insight to the internal mechanics of molecules. But arguably his most significant development was his embrace of the interplay between theory and experiment in the development and validation of the first transport equations based on the atomic theory of matter. It was through this effort that Maxwell sought to disprove Clausius' "mean free path" theorem and ended up doing just the opposite.

Molecules in gas move naturally in three dimensions. In an equilibrated system, no direction is favored and so Maxwell's velocity distribution applies equally to all three directional components of velocity. But in the world of gas-phase transport phenomena, by definition, equilibrium does not exist since a gradient must exist to cause the transport. Concentration gradients cause diffusion, temperature gradients cause heat conduction, and flow gradients cause transfer of momentum between velocity zones as quantified by viscosity. Each of these phenomena occurs in a gas because of molecular motion. While the motion across space is hindered by collisions, net motion does result when gradients exist; odorous gases eventually do make it across the room. Maxwell figured that if he could use his mathematical skills to extend Clausius' theories of motion and collision to such transport phenomena, he could then test his resulting predictions against experimental data.

The details involved here are "fiendishly complicated."[40] Read Maxwell's papers and you'll understand. The mathematics are heavy and dense, especially in two areas, the first involving the fact that velocity is no longer the same in all directions during transport. For example, in the flow of gas, the value of the bulk mass flow-rate must be added to the normal mean molecular velocity component in the direction of flow (but not in the other directions). The second involved the need to account for intermolecular forces. This was extremely difficult since 1) the force laws were not known then, and 2) regardless of its form, a force law would have to be a function of distance, thus necessitating integration of the force law, in whatever form it took, through the entire distance-varying encounter between the two interacting molecules for all possible angles and velocities, and thereby converting motion from straight-lines to complex curves. "Fiendish" indeed.

---

[40] (Mahon, 2003) p. 135.

Gas Viscosity – Theory Followed by Experimental Validation

> *In the whole range of science there is no more beautiful or telling discovery than that gaseous viscosity is the same at all densities.*
>
> – Lord Rayleigh[41]

The single most important result out of this effort was Maxwell's equation for gas-phase viscosity. The phenomenon of viscosity is caused by the natural motion of molecules in a flowing gas from the slow zone—say, for example, next to a stationary pipe wall—into the fast zone—say, for example, in the middle of the pipe—thus necessitating a compensating force to maintain flow, the magnitude of the force being a measure of gas viscosity. Viscosity quantifies the rate of molecular transport, which in turn is governed by both the speed of the molecules and the distance between collisions. Maxwell's mathematics accounted for this "mean free path" flow and arrived at two counter-intuitive and rather stunning predictions, one suggesting that viscosity *increases* with temperature, which didn't make sense since everyone knew that viscous fluids like oil flow faster with increasing temperature, and the other suggesting that viscosity is independent of density, which also didn't make sense since higher density surely led to more friction. Wrote Maxwell regarding this latter result, "Such a consequence of a mathematical theory is very startling."[42]

In fact this consequence was so "startling" that Maxwell turned to Stokes, a fellow physicist of Cambridge and an expert in fluid flow and viscosity, and asked, "Have you the means of refuting this result?"[43] Maxwell believed that his model of a mechanical system of colliding spheres was correct, as were his predictions based on this model. He acknowledged that this system indeed showed that transport of his mechanical spheres from a slow zone into a fast zone was independent of the density of the spheres. The rate stays the same since the presence of more spheres is exactly canceled by the presence of more obstacles (shorter mean free path). This all made sense to Maxwell. But what didn't make sense to him was the comparison of his ideal model predictions to the real world. It simply didn't make sense that gas viscosity should not vary with density and should increase with temperature. Not helping matters was that by 1860 not many experimental studies had been conducted on gas viscosities, thus leaving Maxwell stranded without a good reality check. This was one of those classic and rare moments in science when someone predicted something that hadn't yet been observed. Fortunately, being the consummate scientist that he was, Maxwell couldn't just let his prediction sit there. He had to know. So he took matters into his own hands.

What a truly remarkable moment in the history of the atomic theory. Clausius' updated kinetic theory of gases led Maxwell to predict two "absurd" consequences. To validate the theory (or not), Maxwell rolled up his sleeves and between 1863 and 1865 did the experiments, but not alone. Instead, he had help from an unlikely source, namely his wife Katherine (1824–1886), as Maxwell himself acknowledged in a postcard to Tait: " My better ½...did the real

---

[41] Lord Rayleigh quote cited in (Lewis, 2007) p. 125.
[42] (Maxwell, 2003a) p. 166.
[43] (Brush, 2003a) p. 27.

work of the kinetic theory."[44] The two of them working together oversaw the installation in their attic of a large isolated chamber containing a swinging pendulum, an experimental apparatus and concept which had been in use for many years starting around 1660 with Boyle[45] and Hooke, and also later with others including Stokes. After some technical challenges, they were eventually able to start the pendulum swinging back and forth and then monitor the impact of the presence of air in the chamber on its decay rate. They varied both pressure and temperature and generated "their most useful contribution to experimental physics"[46] by verifying that Maxwell's predictions were indeed correct. Viscosity was independent of pressure[47] and increased with temperature.

The verification was not perfect, however. While Maxwell had predicted a linear dependency of viscosity with $\sqrt{T}$, his experiments showed a linear dependency with T. This was *the key* discrepancy that caused Maxwell to incorporate a force function into his 1867 mathematics.[48] The original prediction made in 1860 was based on the simple collision of spheres of definite radius with no force function. Motivated by this discrepancy and also by Clausius' very constructive feedback on his 1860 paper, Maxwell re-visited his assumptions and added a general force function.[49] The complexity of his generalized equations went up significantly, so much so that neither he nor Boltzmann after him could solve them. It wouldn't be until 1911–1917 that general solutions would be solved independently by S. Chapman (1888–1970) and D. Enskog (1884–1947).[50]

That gas viscosity increases with temperature can be understood from the physical model. Viscosity results from the transport of molecules from slow to fast zones. Higher temperature leads to higher speeds leads to faster transport leads to higher viscosity. Maxwell predicted dependency with absolute temperature only because, through a stroke of good mathematical fortune, he found an exact mathematical solution for his equations when assuming that force varies with distance between centers of gravity to the negative 5th power. (Note that the term for molecular radius was not included in Maxwell's 1867 mathematics; there were no contact collisions in Maxwell's model, only strong short-range repulsion.) Later experiments by others determined the dependency to be somewhat less than first order. But the bottom line for Maxwell was that based on his force function assumptions, he showed theory and experiment to agree for viscosity dependency on temperature. His success in predicting both the pressure and the temperature dependencies of gas viscosity helped cement the kinetic theory of gases in

[44] (Everitt, 1975) p. 185, note 71.
[45] Per Brush in (Boltzmann, 1995) p. 5: Boyle placed pendulum in an evacuated chamber and discovered, to his surprise, that the presence or absence of air makes hardly any difference to the period of the swings or the time needed for the pendulum to come to rest.
[46] (Everitt, 1975) p. 141.
[47] (Everitt, 1970) p. 219. Gas viscosity's independence of pressure only breaks down when density is so high that mean free path becomes comparable with molecular diameter or so low that it becomes comparable with dimensions of apparatus.
[48] Force function also enabled shift from collision by contact to collision by strong repulsion
[49] (Cercignani, 2006) Maxwell eliminated direct use of "mean free path" in his 1867 transport equations; the concept was "inadequate" as a foundation for the kinetic theory. But note that he was able to do this by incorporating the conclusions of his "mean free path" analysis—density independence, dependent on absolute temperature—into the viscosity coefficient.
[50] (Everitt, 1975) p. 151.

science and helped establish the atomic theory of matter, on which the theory was based, as a viable hypothesis.

Summarizing Maxwell

It is thoroughly characteristic of the man that his mind could never bear to pass by any phenomenon without satisfying itself of at least its nature and causes.

– P. G. Tait on Maxwell[51]

Maxwell's 48 years were spent being fascinated with the natural world. Perhaps fascinated is the wrong word. Too passive. A better phrasing would be, Maxwell spent his life attempting to understand how the natural world worked. As a child, Maxwell started asking "what's the go o' that?"[52]—loosely translated as, how does that work?—and never stopped. Throughout his life, Maxwell embraced this curiosity and more importantly acted on it. His contributions to physics were like the blocks of limestone at the base of the pyramids. His powerful influence on, among others, the rise of thermodynamics, the kinetic theory of gases, statistical mechanics, and the electromagnetic theory, provided a firm base on which others could build. His ideas were new, almost too new. When he predicted the existence of radio waves, his contemporaries were "bemused."[53] Yet 25 years later, Heinrich Hertz (1857–1894) produced and detected the waves. Maxwell was an extremely creative genius whose life was cut short by cancer. Had he lived until 90, he would have experienced the discovery of the atom and the rise of quantum physics. He could have interacted with Gibbs; one can excitedly imagine what that conversation among equals would have been like. He could have interacted with so many during the excitely turbulent time of physics in the late 1800s. One can only wonder at the additional contributions he would have made. Such an engaging question: what if he had lived?

So how does one end a section on this man? I am left to share this quote from Oliver Heaviside,

A part of us lives after us, diffused through all humanity—more or less—and through all nature. This is the immortality of the soul. [The soul] of Shakespeare or Newton is stupendously big. Such men live the best part of their lives after they are dead. Maxwell is one of these men. His soul will live and grow for long to come, and hundreds of years hence will shine as one of the bright stars of the past.

– Oliver Heaviside[54]

*   *   *

Does the Meaning of Entropy Lay inside the Mechanical Models?

The kinetic theory of gases provided a means to test the hypotheses that matter consists of atoms, that the atoms are in constant and colliding motion and that macroscopic properties

---

[51] Tait quote cited in (Cercignani, 2006) p. 83.
[52] (Mahon, 2003) p. 3.
[53] (Mahon, 2003) p. 2.
[54] Heaviside quote cited in (Mahon, 2003) p. 185.

such as temperature, pressure, heat capacity, and viscosity are directly related to these phe-nomena. Maxwell's success in predicting and experimentally verifying that gas viscosity is independent of density and proportional to absolute temperature helped bolster the theory and led to its early adoption by some. It would take many more years until the science commu-nity at large followed suit, reflecting the long time required to change the community para-digm created by those who regarded the atomic theory of matter as a "convenient fiction"[55] and not a genuine reality.

Given this, it was only a matter of time before someone started asking, if we can connect *some* bulk properties to molecular motion, why not connect *all* bulk properties to molecular motion and especially why not entropy? Entropy was difficult enough to comprehend to begin with, so why not see if its meaning lay in the mechanical models? The journey toward the mechanical theory of entropy started with Clausius, went through Maxwell, and then ended for all intents and purposes with Boltzmann, who took Maxwell's mathematics, turned them toward the issue of change with time and then ultimately toward the world of probability.

*   *   *

Maxwell created an ideal mechanical world of hard, elastic spheres, set them in motion and then brought high-powered mathematics toward connecting this motion with the macro-scopic properties of gases, asking whether the behavior of the ideal system matched the natural world. If so, then the hypothesis that the natural world is comprised of such moving spheres would be strengthened. Maxwell sought truth and, as a healthy skeptic, continually challenged all theories purporting to have discovered this truth, including and perhaps especially his own. While his mathematical model accurately captured certain key features of this idealized sys-tem, its "faith shaking"[56] failure to accurately predict heat capacities[57] limited his full embrace of the gas theories involved despite his and others' demonstrated successes with these theories elsewhere. He was a scientist's scientist and wasn't given to embracing theories that weren't soundly tested by the scientific method.

Not many knew what he was doing. The mathematics were difficult and the assumptions confronting—air comprised of colliding spheres?? Recall the pushback against Waterston's paper on the kinetic theory of gases. This is what Maxwell and others faced with their own advances of this theory. But while few understood him and fewer still agreed with him, all it took was one. How many single threads of connection have we come across so far in this book? Well, here's one more. There was one who was "entranced"[58] by Maxwell's work, understood what it was all about, and agreed with it deeply and passionately. One who used it as a starting point for even more complicated mathematics and theoretical connections, leading in direc-tions Maxwell may have foreseen but not fully addressed.[59] His name? Ludwig Boltzmann.

---

[55] (Cercignani, 2006) p. vii.
[56] (Mahon, 2003) p. 113.
[57] (Gibbs, 2008) p. x. Even Gibbs held strong to 6 DOF vs. experimental 5.
[58] (Mahon, 2003) p. 135.
[59] (Cercignani, 2006) p. 118.

42

Boltzmann: the probabilistic interpretation of entropy

[The universe] is written in the language of mathematics
– Galileo Galilei[1]

Ludwig Boltzmann (1844–1906) entered the University of Vienna in 1863 to a world abuzz with the new atomic thinking. In 1866 he obtained his PhD degree and launched his career— Assistant Professor at Vienna University in 1867 and then Chair of Mathematical Physics at Graz University in 1869—based on this new thinking by publishing his famous 1872 paper *Further Studies on the Thermal Equilibrium of Gas Molecules* that directly built on Maxwell's work. Who was this wunderkind who could join Maxwell, 13 years his senior, in such a complex wrestling match with nature? While Boltzmann did perform well in school, there aren't any interesting stories to engage you with here—no signs of genius in his youth, no tales of academic glory to recount, no #1 prize ribbons pinned to his shirt. "Boltzmann, it appears, was a man whose individual genius was to awaken later."[2]

A Mathematical Tour de Force

Boltzmann's work on gas theory could be described as a mathematical *tour de force*. He sensed where he wanted to go and then went there, Point A to Point B, regardless of any obstacles in the way, solving each problem as he came upon it, believing he would find a final answer at the end of the journey. "It never occurred to him to doubt whether he would succeed."[3] His strong mathematical skills and sound intuition for physics and the natural world enabled him to take the already complicated starting point of Maxwell and greatly extend it. His *Lectures on Gas Theory* (1896–8) neatly summarized his extensive work on this subject and became an "acknowledged masterpiece of theoretical physics"[4] and "one of the greatest books in the history of exact sciences."[5] As a conscientious and thoughtful teacher who "never exhibited his superiority"[6] and whose lectures, recalled Lise Meitner, "were the most beautiful and most

---

[1] Galileo quote cited in (Popkin, 1966) p. 65.
[2] (Lindley, 2001) p. 34.
[3] (Lindley, 2001) p. 81.
[4] (Boltzmann, 1995) p. 1.
[5] (Boltzmann, 1995) p. 2.
[6] (Cercignani, 2006) p. 37.

Block by Block: The Historical and Theoretical Foundations of Thermodynamics,
Robert T. Hanlon, Oxford University Press (2020). © Robert T. Hanlon.
DOI: 10.1093/oso/9780198851547.001.0001

stimulating that I ever heard,"[7] Boltzmann wrote to provide understanding. Perhaps he wrote more than was needed. But he felt this important, almost as a counter weight to Maxwell, who in Boltzmann's mind at least tended to gloss over the details.[8] Maxwell captured both sides of this relationship in a note to Tait, "By the study of Boltzmann I have been unable to understand him. He could not understand me on account of my shortness, and his length was and is an equal stumbling block to me."[9]

Regarding Boltzmann's work and the high-density mathematics behind it, I don't have the room here to go into his level of detail. As was the case for Gibbs, reading Boltzmann is not for the faint of heart. Isn't it interesting though that it was Maxwell who read and understood both? (Yes, he eventually did understand Boltzmann.) Gibbs worked alone at Yale, and Boltzmann was largely ignored in Germany. Fortunately for the world of science, Maxwell's genius recognized the genius in each, helped raise them from their isolation to positions of influence and so helped accelerate the rise of both classical and statistical thermodynamics. Of specific relevance to this discussion, Maxwell's active interest in Boltzmann's publications encouraged Boltzmann to bring more rigor and clarity to his work. Oh, that Maxwell had lived longer!

While I have chosen not to go into the depths of Boltzmann's mathematics here, I do want to address the critical assumptions behind his physical model as understanding them helps clarify the connection of his work to natural reality and ultimately to the mechanical interpretation of entropy.

Boltzmann's Shift into the World of Probability

If one does not merely wish to guess a few occasional values of the quantities that occur in gas theory, but rather desires to work with an exact theory… [then] one must find the number of molecules out of the total number whose states lie between any given limits.
– Ludwig Boltzmann (1872)[10]

In his 1872 paper, building on Maxwell's work, Boltzmann assumed the existence of an initial distribution $f(x,t)$ of "single point mass"[11] gas molecules, which enabled a simpler mathematical approach than would otherwise have been encountered with polyatomic atoms. In theory, if you took two snapshots of the moving and colliding molecules, separated by an infinitesimal amount of time, and from this calculated the speed and kinetic energy of each, you could graph a population distribution of the total number of molecules $f(x)$ on the y-axis against infinitesimal ranges of kinetic energies x on the x-axis for each dt time increment. (The focus on kinetic energy aided his energy-conservation mathematics.) Such ranges, or buckets as I call them, are

---

[7] (Lindley, 2001) p. 194. Also, per the recollection of Meitner's nephew, Otto Frisch in (Frisch, 1978) p. 426, "And to the end of her life she remembered the lectures of Ludwig Boltzmann, torch bearer for the kinetic theory of gases, who gave her the vision she never lost, of physics as a battle for final truth."

[8] (Brush, 2003b) p. 431.

[9] (Lindley, 2001) p. 83.

[10] (Boltzmann, 2003a) p. 264.

[11] Point mass – mathematical construct of point having inertia but no rotational energy expressed in a quantitative way by the concept of mass.

needed since populations for an exact and single energy can't be defined.[12] Boltzmann assumed that molecules in any given kinetic energy bucket were continually colliding with molecules in each of the many other buckets, the frequency of collision being proportional to the product of the populations in each bucket. He also assumed that both kinetic energy and momentum were conserved for each collision. Next he accounted for the angle at which the collisions occurred. Did the molecules hit head-on or at an angle? Did one hit the other from behind or vice versa? While both total energy and momentum remained constant for each two-body collision (only two-body collisions were considered), the distributions of energies and momenta between the two molecules *after* collision varied because of such issues. Inherent to these phenomena was the additional need to account for intermolecular forces. Boltzmann tied all of these variables together and wrote a population balance for each kinetic energy bucket to account for the accumulation and depletion caused by collisions. A molecule could leave a given bucket by colliding with a molecule from another bucket and a molecule could enter this same given bucket as a result of a collision between two molecules from other buckets. Needless to say, Boltzmann critically relied on his strength in advanced calculus and statistics and on his keen intuition with physics to create simplifying assumptions and so arrive at a reasonably tractable problem.

In the build-up of his assumptions, Boltzmann ventured into the world of probability. As he acknowledged in the beginning of his 1872 paper, the hypothesized molecules of gas are in constant and ever-changing motion and yet the averages of these motions as quantified by such macroscopic properties as temperature and pressure remain constant at our level of perception. He likened such averages to other averages provided by statistics in which each number comprising the average is "determined by a completely unpredictable interaction with many other factors."[13] Consider how unpredictable is the age or height of a random person relative to the average age or height of many persons. The variability in all of the factors combines to make it difficult to predict the height of a specific person in a group of people but easier to predict the average for that group.[14] Boltzmann wrote, "The determination of average values is the task of probability theory. Hence, the problems of the mechanical theory of heat are also problems of probability theory."[15] Here he was simply acknowledging the impossibility of following individual molecules and the need for statistical averaging to attempt any mathematical modeling. In so doing, he was careful to point out that the use of statistics in no way suggested a lack of deterministic behavior in the collisions of molecules. Instead it provided a means to probe nature by making assumptions, to develop mathematical predictions based on the assumptions, and to assess the validity of the assumptions based on the predictions' agreement or lack thereof with experiment. This is what led him (and what earlier led Maxwell) to propose the existence of a distribution $f(x,t)$ that is governed by some "regular law."

While Boltzmann used the word "probability" in 1872, he did not seem to be "fully aware"[16] of the many subtleties involved and where this would then lead him. As Cercignani noted,

---

[12] A probability based on an exact velocity is impossible because you would need an infinite number of significant figures to qualify as "exact." This is why probability based on ranges is used instead.

[13] (Boltzmann, 2003a) p. 264.

[14] (Pierce, 1980)

[15] (Boltzmann, 2003a) p. 264.

[16] (Cercignani, 2006) p. 8.

Boltzmann felt his use of $f(x,t)$ left him within the realm of mechanics, but in actuality the large numbers involved caused his population distribution to morph into a probability distribution since probability is directly tied to frequency of occurrence. Over time, Boltzmann's eventual realization of this fact contributed to his full venture into the world of probability theory based on permutations in 1877, which we'll get to shortly.

Given a population distribution, Boltzmann made a most critical subsequent step by differentiating it with respect to time: $\partial f(x,t)/\partial t$. While Maxwell did not rigorously address the change in distribution with time, Boltzmann made this a center of his study. Specifically, he was interested in how an initial distribution at t = 0 changed after a differential amount of time. He wanted to know, at the first step out of the gate before the collisions themselves changed the distribution, which direction the distribution would move in. This was the first equation ever written to describe the time evolution of a probability and would become particularly useful in statistical mechanics for developing transport equations and describing non-equilibrium systems.[17]

Two significant results fell out of his math. The first was this. When Boltzmann set "$\partial/\partial t$" equal to zero, signifying the point at which the distribution no longer changed and equilibrium was achieved, he confirmed that the solution for the resulting distribution was indeed Maxwell's. With this step, Boltzmann was able to "prove" Maxwell's distribution at a more general level than Maxwell himself was able to do previously. But Boltzmann wanted to know more about this. He wanted to know what would happen according to this equation if he started with a "completely arbitrary"[18] distribution of energy. How would it change over time? This then led to his second result.

The second result addressed the issue of what "direction" even means in the context of a distribution's motion. Through a very (very) complicated mathematical process, perhaps easy for Boltzmann as indicated by his use of such phrases as "one sees easily" or "one can see immediately" but tortuous for the rest of us, Boltzmann was able to show that "whatever may be the initial distribution of kinetic energy, in the course of a very long time it must always necessarily approach the one found by Maxwell."[19] In an amazing feat of sheer intellectual might, he did this by developing $f(x,t)$ into another mathematical function, for which he used the German E but which others[20] mistook for an H, a choice later adopted by Boltzmann,[21] and proved that the time derivative of this H-function must be either negative or—once the distribution equaled Maxwell's distribution—zero. In other words, he found that the molecular collisions that cause $f(x,t)$ to change toward Maxwell also cause H to decrease to a minimum. In essence, by linking the "change toward Maxwell" to a decreasing H, Boltzmann found the means to both identify and quantify the direction of change caused by collisions.

The significance of this second result was this. The mathematical form of the H-function agreed up to a constant factor with Boltzmann's separate 1871 derivation of δQ/T,[22] which he neatly summarized in his *Lectures on Gas Theory*. In this derivation he combined use of the

[17] (Cercignani, 2006) p. 97.
[18] (Boltzmann, 2003a) p. 280.
[19] (Boltzmann, 2003a) p. 291.
[20] (Cercignani, 2006). By others, I mean S. H. Burbury of England, p. 129.
[21] (Brush, 2003a) p. 182.
[22] (Boltzmann, 2003a) p. 291.

kinetic gas theories plus Maxwell's and his statistical averaging methodology, specifically quantifying δQ as an infinitesimal increase in molecular energy and external work. The fact that the H-function resulting from setting *f(x)* to Maxwell's distribution equaled the separately derived δQ/T term (to within a constant) suggested a connection between H and S.[23] Based on this connection, he concluded the 2nd Law to be a probability law and that since H is "closely related to the thermodynamic entropy in the final equilibrium state, our result is equivalent to a proof that the entropy must *always* [my italics] increase or remain constant, and thus provides a macroscopic interpretation of the second law of thermodynamics."[24]

But was this truly a proof? If it were only that simple. Little did Boltzmann realize the battlefront he had opened with this statement and especially his use of the word "always."

A Challenge Presented to Boltzmann: the Reversibility Paradox

Boltzmann showed that for any given distribution of a large number of hard spheres, the initial set of collisions causes the distribution to shift toward Maxwell's distribution, thus causing the value of H to decrease. He also showed a mathematical connection between H and δQ/T that suggested that the decrease in H to a minimum corresponded to the increase in S to a maximum. As the same phenomena of motions and collisions were at the root of each, he felt the connection to be sound and thus led him to conclude the connection to be a mechanical proof of the 2nd Law.

But Boltzmann wasn't the only person engaged in this discussion. Others—few as they were—were seeking a deeper understanding of entropy as well. It was in this spirit that Maxwell, Tait, and Thomson[25] in Britain and later (independently) Josef Loschmidt in Vienna (a university colleague of Boltzmann's) proposed a strong counter argument in the form of a great thought experiment,[26] not as naysayers but more as "gadflies"[27] as defined by Brush: "outspoken critics who challenge the ideas of geniuses, forcing them to revise and improve those ideas, resulting in new knowledge for which the genius gets the credit while the gadfly is forgotten."[28] While I use the word gadflies here for the group, I'm really referring more to Loschmidt, as he was the main critic who motivated Boltzmann and also the one who got little credit for doing so. In this thought experiment, Loschmidt said that after the initial collisions

---

[23] (Boltzmann, 1995) pp. 68–75.

[24] (Boltzmann, 2003a) p. 263.

[25] (Brush, 1986b) p. 602. It was from this counter-argument approach that Maxwell also created (1867) his famed "Demon" as a thought experiment to challenge the thought that the 2nd Law is explained by his and Boltzmann's mechanical system. His "Demon" was a creature that sat at a door between two systems of gas molecules both at the same temperature and selectively allowed only the fast "hot" molecules to move from one system to the other through the door, thus enabling a increase in temperature difference over time and a means to break the 2nd Law.

[26] (Brush, 1986b) pp. 598–615.

[27] (Brush, 1986b) pp. 605–7. Loschmidt was probably the main "gadfly" in this discussion. It was really his critique that stimulated Boltzmann toward his ultimate major advance in providing a statistical interpretation of entropy based on permutation theory. And it was really he who never received the fame for doing so. Boltzmann was largely unaware of Thomson's prior published work on the "reversibility paradox."

[28] (Brush, 2003b) pp. 421–450.

occurred in Boltzmann's scenario, if for each point mass you instantaneously flipped the signs on all velocities to reverse each direction while keeping the same speed, this entire situation would reverse itself and you would have identified an *f(x)* distribution that would move *away* from Maxwell's distribution, resulting in an *increase* in H and a *decrease* in S.

The 2nd Law, as written, was an absolute law based on many years of experience. Heat *always* flows from hot to cold, or as influenced by Clausius, entropy *always* increases to a maximum in an isolated system. But Boltzmann's initial belief that H *always* decreases was not based on experience. Instead it was based on a mathematical model of an idealized, mechanical system that purported gas to be made of moving atoms. The issue that Loschmidt and others had was not with the 2nd Law (it was "law" after all) but with Boltzmann's use of *always* in describing the decrease of H with time. They saw a flaw in his argument—*always* and *never* are two very strong words to use in science; say them and you'll invite challenge. It may have seemed like a minor mathematical issue but, in fact, it was a major philosophical contention since it took Boltzmann's mechanical system from *always* to *almost always*, thus breaking the link between it and the 2nd Law and in so doing challenging the hypothesis at the center of it all: the atomic theory of matter.

This insightful thought experiment, soon to be called the "reversibility paradox," became a gauntlet thrown before Boltzmann. He needed to respond. Little did the gadflies know that it was Boltzmann who was correct and not the 2nd Law as written.

Boltzmann's Response to the Challenge

Some of Boltzmann's best work resulted not from isolation, as was the case for Gibbs, but from intellectual engagement with others. Maxwell's publications initially motivated Boltzmann toward the use of statistical averaging in the kinetic theory of gases. Their subsequent back-and-forth spurred Boltzmann even further, all the way through 1879, the year of Maxwell's death and his last publication on this topic. The "reversibility paradox" and Loschmidt's role in it strongly motivated Boltzmann's deep foray into probability theory and the resulting mechanical proof of the 2nd Law. Others motivated him to write his famed *Lectures on Gas Theory*, but this may have been more in reaction to their attacks[29] than to their healthy intellectual engagement. For the most part, Boltzmann enjoyed such engagements and debates, indeed thrived on them, even if they critically focused on his own work. But after he committed suicide in 1906, many attributed the cause to the emotional distress brought about by these debates. The situation was a little more complicated than this as other possible causes, which we'll get to in short order, were also present.[30] The bottom line is that Boltzmann welcomed active intellectual debate—so long as the debaters were reasonable and rational.

[29] (Cercignani, 2006) p. 25. (Darrigol, 2018) p. 420.
[30] (Cercignani, 2006) pp. 34–37.

Boltzmann's Shift to an Even Deeper World of Probability

Perhaps the greatest triumph of probability theory within the framework of nineteenth-century physics was Boltzmann's interpretation of the irreversibility of thermal processes.

– Ernest Nagel[31]

In reading Clausius we seem to be reading mechanics; in reading Maxwell, and in much of Boltzmann's most valuable work, we seem rather to be reading in the theory of probabilities.

– J. Willard Gibbs[32]

One could even calculate, from the relative numbers of the different state distributions, their probabilities, which might lead to an interesting method for the calculation of thermal equilibrium. In just the same way one can treat the second law.

– Ludwig Boltzmann (1877)[33]

Recall how Boltzmann first approached his theoretical development upon reading Maxwell's papers? He first hypothesized that gas is comprised of moving molecules that collide with each other, thereby accounting for the pressure and temperature of the gas as well as certain macroscopic properties dependent on the collision-hindered diffusion of molecules across space. He then hypothesized that the population of gas molecules distributes itself based on velocity or energy as originally proposed by Maxwell and proved this distribution to be valid based on the fact that it rendered his $\partial f(x,t)/\partial t$ equal to zero. He also showed how both the kinetic gas theories and statistical averaging provided the means via his H-theorem to explain the entropy of the gas from a mechanical standpoint. While together these accomplishments amounted to a very powerful step forward in physics, they weren't sufficient to resolve the "reversibility paradox."

When Loschmidt initially highlighted the paradox to Boltzmann, Boltzmann believed such a reversibility scenario to be "extraordinarily improbable" and "impossible for practical purposes,"[34] suggesting it to have the same improbability as that of air spontaneously un-mixing into pure oxygen and pure nitrogen. He thus felt that the paradox provided insufficient reason for rejection of the mechanical proof of the 2nd Law. But in this intellectual debate, Boltzmann naturally had to admit that "highly improbable" does not mean "absolutely impossible," thus marking the transition from *always* to *almost always* for both him and the world of physics. This was such a significant shift in thinking. As a scientist of high integrity, Boltzmann realized he could no longer hold onto the use of *always*. But as a scientist of high intelligence, Boltzmann also realized that the problem didn't reside in his mathematics but instead resided inside the 2nd Law itself. It required a rewrite. But one can't just simply say that a law is wrong. One must prove it. And this is what Boltzmann set out to do.

What Boltzmann was zeroing in on was obvious only in hindsight. At the time, it was revolutionary. In essence, as he considered the highly improbable reversibility paradox, he started to consider its opposite: the highly probable state of equilibrium. He started to realize the

---

[31] Nagel quote cited in (Brush, 2003d) p. 526.
[32] Gibbs quote cited in (Cercignani, 2006) p. 84.
[33] (Boltzmann, 2003b) p. 366.
[34] (Boltzmann, 2003b) p. 366.

significance of the fact that nature moves toward the most probable state, especially when large numbers are involved. This was the essence of irreversibility, moving from improbable to probable. How could nature move anywhere else? And at the end, once the moving stops, wherever it is that nature ends up must then be, by definition, the most probable state. Thus, one means to validate Maxwell's hypothetical distribution would be to determine if it was indeed the most probable distribution. But how?

How Many Different Ways to Put Balls in Buckets?

How many different ways can you arrange a certain number of objects? The answer is to be found in the mathematics of permutations as first addressed in the world of gambling where determination of effective betting strategies for rolling dice or playing cards was highly valued. If you want to learn why seven is the most probable result of rolling two dice, work out all the different results you could possibly have and you'll find that there are more combinations that lead to seven than to any other number.

As Boltzmann considered his need to quantify the probabilities associated with Maxwell's distribution and the 2nd Law, his mind went toward permutations. Recall that for large numbers, frequency of occurrence equals probability. In 1872 he considered $f(x)$ to represent the population distribution across energy buckets, but in 1877 he revised his thinking to view $f(x)$ as the *probability* distribution across the same energy buckets. Same ball-in-bucket numbers but with a profound shift in paradigm attached.

In his new way of seeing things, Boltzmann envisioned a certain number of objects (atoms, molecules; Einstein later also envisioned photons) and a certain number of buckets to place them in (infinitesimally sized volumes based on some combination of location, velocity, momentum, and energy) and so sat down and worked out the math. It was a fascinating change in direction for him. All of the gas theory assumptions used in his previous work about moving molecules, angled collisions, and changes with time were surprisingly eliminated from the discussion. These variables became like the variables affecting the height of a person. Even though each variable is part of the deterministic process governing the individual outcome, they became lost when considering group statistics. The only things that mattered were the constraints: fixed energy (U), fixed volume (V), fixed mass (N). He asked, given these U-V-N constraints, how many ways could a given number of molecules be distributed into a given number of buckets? He summarized his answer in his famous 1877 paper *On the Relationship between the Second Fundamental Theorem of the Mechanical Theory of Heat and Probability Calculations Regarding the Conditions for Thermal Equilibrium*[35] and so became the first to use permutation theory in a "very unique"[36] and much more complex way than his statistical-averaging approach in 1872 to re-interpret the meaning of entropy.

---

[35] (Sharp and Matschinsky, 2015)
[36] (Sharp and Matschinsky, 2015) p. 1975.

One Approach to Understanding Statistical Mechanics

In approaching the use of probability theory, we must let go of deterministic thinking for there is no rigorous mathematical path that takes one from determinism to probability. In the ideal world, you don't need probability because you're accounting for everything. But in the real world, you can't account for everything. You can't account for the motions of 10^{23} molecules. You simply don't have the capacity. But rather than giving up, you have recourse with statistics. You can make reasonable assumptions—based on your beliefs and intuition—on which to build statistical models; if the models predict what nature does, then your assumptions may be reasonable. If not, then you change your assumptions.

Cutting up a High-Speed Film – Each Frame Equals a Unique "Complexion"

Boltzmann developed a rather ingenious approach to solving the problem in front of him. In essence, using an analogy that we understand today but that wasn't available then, he envisioned taking a high-speed movie of a system of gas molecules, the frames separated from each other by infinitesimal increments of time. From the still frames or snapshots at each moment, the location and velocity (including direction) of each molecule could be measured. From these data he could then prepare population distributions based on the three dimensions of location and the three components of velocity (x,y,z). He could also naturally convert velocity into energy and make another population distribution.

In this way, each frame in the movie could be converted into a population distribution based on any of several parameters: spatial coordinates, velocity, momentum, energy. He called each frame a "complexion," which we refer to as a "microstate" in today's terminology. He considered the population distribution, which itself was comprised of a tremendous amount of data, to be but a single data point for a given set of U-V-N constraints.

Boltzmann then proposed the following. After steady state has been achieved, the movie continues to show motion at the microscopic level even though the macroscopic properties remain largely constant, at least to our level of perception. The motion fluctuates around averages and so doesn't lead anywhere; maximum entropy has been established. To model this steady-state system, one could use Newtonian mechanics to follow all the billions and billions of molecules as they move from one frame to the next and so generate the sequence of resulting population distributions as a function of time. But given the large numbers involved, Boltzmann realized the impossibility of this path and so proposed an alternative. He proposed that one could cut up the entire film into individual frames, put them all into a large bowl and then mix them up. Each individual frame would then become a stand-alone complexion, independent of how it came about, thus eliminating the variable time and all $\partial/\partial t$'s from the discussion. In this bowl would exist a tremendously large but still finite set of unique complexions—no repeats, in theory—each having a population distribution that was put together or arranged uniquely from the available molecules to work with and each fulfilling the U-V-N constraint.[37] In this physical model, the cut-out frames in the bowl, one frame for each possible complexion, would represent *all* possible complexions. Now many arguments challenging all the assumptions I just listed populated the literature for a long time after Boltzmann's proposals, including many from the pure mathematicians, but I don't want to get into these here as they are well discussed by others, such as Cercignani[38] and Brush.[39] The bottom line is that with all these assumptions, the predictions Boltzmann later ended up with agreed with experiment and that's what mattered.

In looking at this collection of complexions, Boltzmann realized he had no reason whatsoever to say that any one was more or less probable than any other. In the spirit of Laplace's statement, "When probability is unknown, all possible values of probability between zero to one should be considered equally likely,"[40] Boltzmann decided to make each resulting distribution just as probable as any other.[41] Strange as it seems, even a given single complexion showing a distribution of all molecules having the same speed (Clausius' original assumption) was just as likely to occur as a given single complexion showing the Gaussian distribution as predicted by Maxwell. Clarity of this strange situation results from the realization that many, many, many more complexions fall into the Maxwell-distribution category, making its occurrence much more frequent and thus, by definition, much more probable. So Boltzmann had reason to propose that each and every complexion had one and the same probability, meaning that for J possible complexions, the probability of occurrence for each would be $1 / J$.

Boltzmann interpreted each frame as the result of a ball-in-bucket process and based his buckets on some combination of location, velocity, and energy. It all depended on the point he was trying to make. The issue of velocity or energy became somewhat complicated as Boltzmann found that velocity "as is" provided the equiprobable mathematics while energy required a correction.[42] He worked with both. He hypothesized that molecules could populate all of the velocity (or, more accurately, momentum) buckets with equal probability to create all possible complexions, again with the U-V-N constraint. He then rolled up his equiprobable ball-in-bucket events into his equiprobable complexion events with each frame of the movie being just as probable as any other frame.[43]

In this large collection of J possible complexions, many showed identical population distributions of the number of balls in buckets. While the distributions were identical, the ways that

---

[37] Each microstate would also have other properties such as temperature and pressure based on considering the various kinetic energies present. These properties would not be constant for constant U-V-N. Important to note is that entropy is NOT dependent on any given microstate; instead, entropy is an inherent property of U-V-N and defined by the number of possible arrangements of N molecules with the U-V constraint. Unlike temperature and pressure, entropy is not a property that depends on in the instantaneous locations or velocities of the particles.

[38] (Cercignani, 2006)

[39] (Brush, 2003e)

[40] (Ben-Naim, 2008) p. 59.

[41] (Lewis, 2007) Boltzmann's assumption of equal probability for all complexions "seemed to imply that any given system would over time pass through all possible configurations or microstates. Such an assumption became known as the 'Ergodic' hypothesis."

[42] (Dugdale, 1966) p. 90.

[43] He mentioned an alternative approach in which he started with all possible frames resulting from no energy constraint (infinite temperature) and then subtracting out all those frames not complying with the energy constraint of U.

they were arranged from the given molecules were unique, ignoring here a lengthier discussion around the issue of indistinguishability.[44] The complexions showing the same distributions could be grouped together; each distribution encompassed a specific group of complexions. Boltzmann labeled the number of complexions P in each distribution i as P_i for i = 1 to the total number of distributions. Thus, $\sum P_i = J$.

Finally, Boltzmann proposed that out of this collection of many distributions, there would be one single distribution that would encompass the maximum number (P^{max}) of complexions. And not only would it be maximum, but because of the "Law of Large Numbers" numbers involved, it would be maximum by an overwhelming margin such that P^{max} effectively equaled J. Technically, the natural world embodies all possible distributions, not solely those aligned with P^{max}. But for large systems, the existence of these other distributions is insignificant and can be ignored.[45]

With this physical model and set of assumptions, Boltzmann was ready to apply his mathematical skills around permutation theory to determine an equation to quantify P and then, by setting the first derivative of this equation equal to zero and solving, to quantify P^{max} and the distribution that encompasses it.

How Many Ways Can You Place 7 Balls in 8 Buckets?

There's a great illustrative example that Boltzmann used in this 1877 paper to explain the mathematics involved in these calculations (Figure 42.1). In this example he showed how 7 balls[46] can be arranged into 8 energy buckets, ranging in equal increments from zero to the final and highest-energy bucket while keeping total energy constant at an amount equal to a single ball being in the final bucket. This naturally becomes one possible distribution of these balls: place one ball in the final bucket, achieve the necessary total energy of the system, and then place all other balls into the zero-energy bucket. By rotating which of the 7 balls goes into the final bucket one finds that there are 7 different ways to create this specific distribution, each being a unique complexion of the same identical distribution.

It turns out that there are 1716 different complexions possible for this example and these can be grouped into any of 15 possible distributions. Permutation theory can be used to write an equation quantifying the number of complexions (P) in each of the 15 distributions

$$P = n! / \left[n_0! \, n_1! \, n_2! \, n_3! \, n_4! \, n_5! \, n_6! \, n_7! \right]$$

[42.1]

where the n_i terms represent the number of molecules in the bucket having energy i and n the total number of balls. Each set of n_i values characterizes a specific distribution that fulfills the fixed energy constraint. So for the example used above in which n = 7, setting $n_7 = 1$ and $n_0 = 6$ results in $P = 7$. In another example, you could put all the balls in bucket n_1. Since n_1 has 1/7th the energy of n_7, then this distribution meets the fixed total energy requirement. Plug these numbers into the above equation and you'll see that there's just one single way to put all balls

[44] (Sharp and Matschinsky, 2015) See Sharp's discussion on indistinguishability.
[45] (Dugdale, 1966) p. 113.
[46] Boltzmann actually used molecules in his example. I chose to use balls.

THE BOLTZMANN DISTRIBUTION

in Boltzmann's 1877 publication that established the probabilistic basis for entropy, he used the following illustrative example *

how many different ways can you assign 7 balls to 8 different energy buckets for a fixed total energy of 7?

energy buckets   #balls = 7

energy → 0 1 2 3 4 5 6 7   fixed energy

$$\text{total energy} = \frac{\text{energy}}{\text{of bucket}} \times \frac{\text{\#balls}}{\text{in bucket}} \qquad \frac{\text{\#balls}}{\text{in bucket}} = 7$$

by doing this exercise in probability, a total of 1716 arrangements can be categorized into 15 distributions:

distribution number	0	1	2	3	4	5	6	7	# of ways to arrange 7 balls
1	6							1	7
2	5	1					1		42
3	5		1			1			42
4	5			1	1				42
5	4	2				1			105
6	4	1	1		1				210
7	4	1		2					105
8	4		2	1					105
9	3	3			1				140
10	3	2	1	1					420
11	3	1	3						140
12	2	4		1					105
13	2	3	2						210
14	1	5	1						42
15		7							1

(bucket energy →)

the most probable distribution (→ 420)

* SHARP 2015 + CERCIGNANI 2010 p.122

---

the most probable distribution looks like this:

#balls in bucket vs bucket energy (points at 0 1 2 3 4 5 6 7)

advanced mathematics → Large #balls Large #buckets

#balls   exponential decrease

$$\text{\#balls in bucket} \propto e^{-E_{bucket}}$$

bucket energy

from this work came the **BOLTZMANN DISTRIBUTION**

$$\frac{N_i}{N} \propto e^{-E_i/kT}$$

T = temperature
N_i = the number of particles having energy E_i

The **BOLTZMANN DISTRIBUTION** says that the maximum number of ways to arrange particles occurs when lower energy states are more populated than higher energy states.

This distribution can be applied to a gas of non-interacting atoms to yield Maxwell's distribution of velocity for a single coordinate

0   v_x v_y or v_z

$$\text{\#atoms with } v_x \propto e^{-v_x^2/RT}$$

Maxwell arrived at this equation by assuming a Gaussian distribution

---

The **BOLTZMANN DISTRIBUTION** applies to each degree of freedom (DOF) for dispersing energy

For molecular speed $v = (v_x^2 + v_y^2 + v_z^2)^{1/2}$

advanced mathematics leads to the **MAXWELL-BOLTZMANN DISTRIBUTION**

#atoms   increasing T →

v

$$\text{\#atoms with } v \propto v^2 e^{-v^2/RT}$$

It was the shape of this mathematical relationship that later inspired Planck in his theoretical research on black body radiation

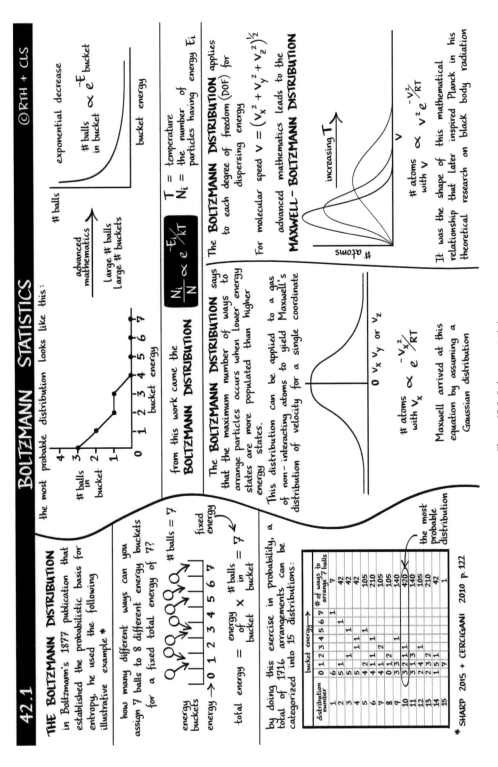

Figure 42.1 Boltzmann statistics

into one bucket. Remember, it's only whether or not a given ball goes into a given bucket that matters; the order in which they are placed there doesn't.

Continuing with Equation 42.1, since n is fixed at 7, the goal of finding the distribution with the most number of complexions is to minimize the denominator. In this example, n is rather small which makes the mathematics somewhat complex, but for the tremendously large values of n for a system of molecules in a given system, the Law of Large Numbers comes into play and makes the mathematics somewhat easier since the limit of x! is x^x for large values of x, thus leaving the optimization process to become the *minimization* of the product of the $n_i^{n_i}$ terms or, in the more popular logarithmic version, the sum of $n_i \ln (n_i)$ terms.[47] The caveat of fixed total energy adds complexity to the optimization process, but Boltzmann shows the reader how to work through the mathematics.[48] Again, this is a very nice mathematics tutorial for those who are interested. As Boltzmann noted, when the goal is the minimization of these terms without constraint, the solution is an equiprobable distribution of n_i terms. This occurs when looking at how molecules spread through space in the absence of any external force. The solution simply says that molecular density is spread equally throughout the constrained volume V. With the energy constraint though, the n_i terms occur as equiprobable only for infinite energy; when energy is constrained, the Gaussian shape results.

After many pages of extensive mathematics, Boltzmann arrived at the solution. The distribution that encompasses the maximum number of complexions is none other than Maxwell's distribution. Starting from an entirely new direction in physics, that of pure probability and having nothing to do with collisions or time, Boltzmann demonstrated that Maxwell's is indeed the most frequently occurring and thus the most probable distribution.

Of course Boltzmann did not stop there. He went on to demonstrate how the mathematics could be performed for both discrete and continuous energy distributions and then additionally went on to demonstrate how the mathematics could be performed for the distribution involving *both* location (three coordinates) and momentum (three velocity components), resulting in location–momentum buckets of six dimensions. This opened the door to understanding the impact of such events as fluid flow and external forces such as gravity on the spatial and momentum coordinates of a collection of molecules, including dissimilar ones. The inclusion of location also opened the door to incorporating the variable of volume into his mathematics, which enabled the dependency of entropy on *both* energy and volume to be captured, leading to state function relationships: S (U,V).[49]

The Arrival of S = k log W

After a discussion on how his mathematics applied to polyatomic molecules (very complicated) and some other subjects, Boltzmann concluded this paper by establishing for the first

---

[47] When converted into the world of logarithms, the sequence of $n_i^{n_i}$ terms transforms into the familiar form that some have used $\sum n_i \ln (n_i)$ and the minimization process thus focuses on this term. Since for large numbers, frequency of occurrence can be transformed to probability to yield another familiar form of this term, $\sum p_i \ln (p_i)$, which will re-appear in the discussion around Claude Shannon and Information Theory.

[48] (Dill and Bromberg, 2011) Dill demonstrates the need of Lagrange multipliers to solve the mathematics involved. Although Boltzmann did not specifically reference as such, this is the approach he used.

[49] (Sharp and Matschinsky, 2015) See commentary regarding the importance of including volume together with energy—S(U,V)—in Boltzmann's statistical mechanics.

time a solid link between entropy and probability. During his H-theorem work of 1872 Boltzmann developed mathematics around the population distribution that resulted in his use of the term $\int f(x) \ln f(x) \, dx$. In his work of 1877, starting with permutation mathematics, he developed new mathematics that resulted in his use of the term $\sum n_i \ln (n_i)$. As the size of the buckets assumed in 1877 became smaller this summation could be written in terms of integrals, thus bringing him to the same $\int f(x) \ln f(x) \, dx$ term he used in 1872, thereby uniting the two publications and enabling him to employ the mathematics from 1872 in 1877 to quantify the natural log of the total number of complexions—$\ln(P)$—contained in a given distribution. He called this number the "permutability measure" and denoted it Ω. Assuming a monatomic, ideal gas (no interactions) with no external forces, he derived an equation for Ω and compared this with a separately derived equation for $\int dQ/T$ that he had previously developed based on classical thermodynamics and the ideal gas law. The resulting equation (to within an additive constant)

$$\int dQ / T = 2/3 \ln\left(P^{max}\right) = 2/3 \, \Omega$$

demonstrated a direct link between absolute entropy (S) and Boltzmann's permutability measure. With this result, Boltzmann quantified the entropy of an ideal, monatomic gas as the logarithm of the number of different ways (P^{max}) a system could be arranged for the distribution that yields this maximum value. In sum, starting with probability based on permutations, Boltzmann found a path to his H-theorem and then connected this result with the classically defined entropy based on his separate derivation $\int dQ/T$, thus leading to his simple but powerful statement, "[in this way] the calculus of probabilities is introduced."[50]

As an interesting historical fact regarding the confusing use of symbols, the equation that Boltzmann is best known for, the equation written on his tombstone, is not this one. The fact that $\int dQ/T$ can be written as S is clear (assuming S_o equal to zero[51]). But a careful reading of Boltzmann's paper shows that Ω itself is the natural log of the number of permutations—$\ln(P^{max})$—in the optimal distribution. In his paper of 1900, Planck applied Boltzmann's discrete permutation mathematics to his study of blackbody radiation and in so doing derived an equation for entropy using the term W for the number of permutations as opposed to Boltzmann's P. Boltzmann had used W to denote the number of complexions P for a given distribution i divided by the total number of complexions for all distributions—$W = P_i / J$. Thus Boltzmann's W was a probability between zero and 1 while his P (and thus Ω) was an absolute number that captured the extensive nature of entropy. So when Planck wrote out what would become the equation carved into Boltzmann's tombstone in 1933, the following equation (using log as opposed to natural log),

$$S = k \log W$$

---

[50] (Boltzmann, 1995) p. 58.
[51] (Dugdale, 1966) p. 158. "Per the 3rd Law, the entropy of all systems and of all states of a system is zero at absolute zero. In this state the important thing is not that the value of the entropy at 0 K is *zero* but that it is the *same* for all systems or states of a system. You could give this entropy any value you pleased but since it would have to be the same for all systems and since its value would be entirely arbitrary, it is obviously simplest to put it equal to zero."

one can see where the confusion comes from. Planck's use of W stuck. Perhaps Boltzmann would have argued for P^{max} instead.

*   *   *

Probability is not explicitly featured in the above fundamental equation linking entropy to the number of microstates; it's implicit in the factorial-based mathematics that led to the answer. Even though it's difficult to grasp, this number, which today we define as W, is an inherent property of the system. Once U-V-N conditions are fixed, S and thus W are also fixed. As U-V-N conditions vary, S and W also vary, exactly as spelled out by Gibbs. Statistical mechanics quantifies how W varies and thus provides a means to link changes in molecular distributions to corresponding changes in macroscopic properties.

The Meaning of Entropy from Different Perspectives

Let's take a step back at this point and consider where we are regarding the story of entropy and our understanding of this "ghostly quantity." Clausius' analysis of the Carnot cycle showed that $\oint \delta Q/T$ equals zero for the reversible closed-cycle operation and thus led him to conclude that $\delta Q/T$ equals an infinitesimal change in a newly discovered state property he named entropy: $dS = \delta Q/T$. Boltzmann then provided the mechanical definition of entropy by showing its proportionality to the logarithm of the number of possible microstates comprising the system: $S = k_B \ln(W)$. Together, the two definitions enabled the completion of classical thermodynamics and the rise of statistical thermodynamics. While neither field requires a deeper physical understanding of entropy than the equations shown, the primary motive for this book was to provide this understanding. To this end, let's explore the meaning of entropy from some different perspectives.

Constraints decrease entropy

Given Boltzmann's clarification about what entropy is, one can understand why the maximum entropy for a given number of atoms occurs when the atoms experience no other constraints than fixed energy and volume. For fixed energy, the atoms naturally can't populate all energy buckets equally. The number of accessible buckets is limited by the constraint and the atoms spend most of their time around the population distribution defined by Maxwell and Boltzmann. For fixed volume, the atoms spend equal time in all spatial locations within the dimensions allotted by the constraint. The entropy of this system described here is the highest possible for the given the fixed U-V-N conditions.

Intermolecular interactions decrease entropy

Now what would have to happen to decrease entropy of the above system? Add additional constraints. For example, in the above discussion, the implied assumption is that the atoms are non-interacting. This is the only way for each to spend equal amounts of time in all spatial locations available. But say you started adding interactions and even chemical bonding. The atoms would start spending more time around other atoms rather than in the full volume provided, thus shifting their spatial movements away from an equiprobable distribution, thus decreasing entropy. When one atom is bound to another, it certainly can't freely access the entire volume available; it's tied down.

Energy gradients decrease entropy

Beyond the total energy constraint, what additional energy-related constraints could be added to decrease entropy? Energy gradients themselves. Should any energy gradients related to pressure, temperature, or chemical potential exist inside the system, the number of possible arrangements decreases. For example, some have more fast-moving atoms on one side than the other, reflecting a temperature gradient. Each such distribution is just as probable as any other. It's just that there're many less of them relative to the number comprising the most probable distribution of uniform location and Gaussian velocity. So if you were to create a system with such a gradient, you would do so by effectively removing all the many, many microstates corresponding to the most probable distribution, thus making the energy-gradient distribution become more probable. The act of removing microstates to achieve this, by definition, decreases entropy.

Thoughts on the mechanical meaning of entropy

Let's now shift the discussion to another perspective. The usefulness of entropy to thermodynamics manifests itself in three characteristics of the property, the first being that, building on the above discussion, the entropy of an isolated system increases to a maximum. It seems like a simple concept, the movement of nature toward her most probable state. Common sense, right? But it's remarkable that such a property as entropy exists to quantify this movement and even more remarkable that Clausius caught the initial glimpse of this inside a steam engine. It was this concept that proved critical to Gibbs' development of chemical equilibria, especially as regards his creation of the concept of chemical potential. He showed that it is the dissipation of all energy gradients—temperature, pressure, and chemical potential—that lies at the heart of equilibria.

I now come to the other two important characteristics of entropy. Carnot built his cycle based on isothermal heating and adiabatic volume change. For the former, entropy increases by exactly dQ/T, while for the latter entropy doesn't change at all. The presence of these two characteristics inside Carnot's heat engine enabled both the means to create the first fundamental equation, $dU = TdS - PdV$, and also the ability to use this equation to solve real problems. Knowing how entropy changes due to reversible heat and work enables calculation of property change between states.

It remains a mystery to me why the above is so. Why is dS *exactly* equal to dQ/T and why does entropy not change during reversible adiabatic volume change? To attempt to answer this, let's talk about $\delta Q/T$ first and let's start with Clausius' concept of disgregation that rose from his famed question, where does the heat go? As thermal energy (δQ) enters a system, the moving parts move faster. Sometimes this ends up as faster motion: faster translation, faster vibration, faster rotation. And sometimes this ends up as the movement of atoms and molecules away from each other as the conducted thermal energy is transformed into higher potential energy: attracting atoms stop clinging to each other, bound atoms break away from each other, phases change from solid to liquid to gas, reactions happen, and energy equipartition governs. The sum of all of these responses is a net increase in total energy, both kinetic and potential. This is where the heat goes. And from this, one can see why entropy increases when heat enters the system. There's more kinetic motion and this provides the energy for atoms and molecules to more freely access the full volume of the system and thus make more configurations possible.

But from a quantitative standpoint, we're still left with the challenge of trying to understand why entropy increases *exactly* as $\delta Q/T$ and why entropy doesn't change at all during adiabatic volume change. Adding to this challenge is the fact that both phenomena are absolutely independent of the contents of the system. Realized Carnot, "the working substance doesn't matter." It amazes me that such an exact relationship between entropy, thermal energy, and absolute temperature exists and that the relationship has zero dependence on what's inside. The solution to this challenge resides with Boltzmann.

Per Boltzmann, entropy quantifies the total number of ways a certain number of particles can be placed into a certain number of location–energy buckets[52] while maintaining fixed U-V-N. And in this situation, it's the system's temperature that determines which energy buckets are accessible. There's a property called the partition function, which we don't need to get into the details of here, that quantifies the number of energy buckets in play along with the extent of their involvement; this function is a singular function of temperature. Fix the temperature, fix the number of buckets. This makes sense since at the molecular level, it's the particles' temperature-related kinetic energies that are involved in the collisions that cause the spread in the population distribution as regards both kinetic and potential (which is captured within the location variable) forms. It's temperature that quantifies how many buckets are in play.

So what happens when you add an infinitesimal amount of thermal energy δQ to the system? That incremental energy increases the number of buckets in play by an incremental amount as can be quantified using Boltzmann's mathematics by imposing a new energy constraint, $U + \delta Q$, which we won't delve into here but which Boltzmann did in seeking the mechanical explanation of entropy.

In the end, when you divide δQ by T, you're dividing the *increase* in the number of accessible buckets by the number of accessible buckets already there. You're calculating the relative increase in the number of buckets that the particles can access. Whether it's increasing the number from 10 to 11 or from 100 to 110, it's the same relative increase as quantified by the same increase in entropy. This is why higher values of δQ are needed at higher temperatures to increase entropy by the same amount.

It is fascinating though that in the end the relationships involved here—along with the dS = 0 for adiabatic volume change—are indeed exact. But perhaps the relationships are exact only because we defined entropy based on this exactness.

---

[52] The energy bucket comprises all forms of energy (including potential, which is embedded in the location variable) manifested in a particle, not just translational kinetic energy.

Sackur–Tetrode Validation

I reiterate here that all of the mathematics involved in Boltzmann's work and really in the greater work of all the gas theories were based on an idealized mechanical system. The connection between this system and nature was a hypothesis that needed to be tested. Maxwell, Boltzmann, and later Gibbs were adamant about maintaining this distinction. In the introduction to his *Elementary Principles in Statistical Mechanics* (1901) which arguably established the basic structure of statistical mechanics by re-casting Boltzmann's work in a slightly different form, Gibbs wrote, "Certainly, one is building on an insecure foundation, who rests his work on hypotheses concerning the constitution of matter."[53] But given this constraint, the fact is that these mechanical models matched up well with reality, thus providing validation to the starting assumption that matter is comprised of atoms.

The pinnacle validation of this model arrived with the separate work of two individuals around 1912. One argument against Boltzmann's approach concerned his assumption that the number of complexions was finite. Some at that time argued that nature is a continuum with an infinite number of locations and velocities available for population by the molecules. Indeed, in the classical world, this argument had a point. But in an interesting twist, arrival of the quantum world resolved the dispute by placing Heisenberg limits on location–velocity relationships. Recall that as the precision in measurement of one increases (indicative of a continuum), precision in the other blows up. This is a fact of quantum mechanics. Because of this, nature inherently limits the size and number of available buckets, thus validating Boltzmann's "finite" assumption. Hugo Martin Tetrode (1895–1931) and Otto Sackur (1880–1914) independently used this new understanding to incorporate quantum mechanics into Boltzmann's mathematics and successfully demonstrated the path from statistical mechanics to the ideal gas law for single atoms.[54] The agreement between theory and experiment—the ideal gas law represented many years of experimental work—was one of the great validations in science history and further served to validate the existence of atoms moving "hither and thither."

Boltzmann – Standing Alone on the Battlefield

> *In Germany, there is almost no one who properly understands these matters*
> – Ludwig Boltzmann describing his work to Hendrik Antoon Lorentz (1886)[55]

Returning to Boltzmann's 1877 publication, why did it meet the resistance it did? Boltzmann showed that entropy is linked to probability. Others believed the 2nd Law to be absolute and not linked to probability and so challenged this finding. While the battle appeared to be about absolute vs. probable, it was actually about something larger: the atomic theory of matter.

---

[53] (Gibbs, 2008) p. x.
[54] (Grimus, 2011)
[55] Boltzmann quote cited in (Darrigol, 2018) p. 323.

One of those leading the resistance was Ernst Mach (1838–1916), whom we met previously. Mach was "violently hostile to an atomic picture of nature"[56] and also unfortunately very influential. He refused to accept the existence of the invisible atoms out of an irrational principle—*scientists should not go beyond the observable facts*—and so refused to accept any theories based on such existence. He and others played the role of the naysaying Old Guard, out to protect their own territories of theories and philosophies, seemingly unwilling to consider new ideas such as Boltzmann's. This was clearly frustrating to Boltzmann as it would naturally be to any scientist in a similar situation.

Granted, in the late 1800s, prior to the arrival of the atomic theory, confusion about atoms reigned. The world of atoms remained the great unknown. No one at that time, including Boltzmann, could really conceive what atoms looked like. As discussed earlier, the interpretation of heat capacities remained a popular means for indirectly determining such structure, but the data weren't making conclusive sense, leaving this situation as "one of the unsolved problems in classical mechanics."[57] On top of this, the science of spectroscopy started revealing all sorts of interesting patterns inside matter, suggesting a more complex internal structure than suggested by the hard sphere model. And on top of this, recall that at that time the ether still existed in peoples' minds, although many didn't know what to do with it and, in fact, some were even asked to conduct their work with the goal of "not invoking the role of ether."[58] So, given this, one may be forgiven for truly wondering, *how is it possible that gas is a collection of moving and colliding hard spheres*? But to not be open to Boltzmann's work, to simply shut it off because of such a strict anti-atom campaign, that was not good science and certainly did not provide Boltzmann the kind of engaged, intellectual debate that he enjoyed.

The challenges Boltzmann faced when confronting such naysayers certainly took a toll on him. But it seems that life itself also took a toll on him. Boltzmann enjoyed a rather "idyllic"[59] beginning of his career. He spent a total of 18 largely happy years (1869–1873; 1876–1890) as a Professor in Graz, married, raised a family, sank deep into theoretical physics, enjoyed the earned honor and respect from the academic community, welcomed visitors, had funding and students, took pleasure in long peaceful walks. Really, all in all, Graz was a great experience. But then a restless desire to move and change came over him, perhaps precipitated by the painful losses of his mother in 1885—he published no scientific papers in 1885—his first son in 1889, and his sister in 1890 as well as by the apparent onset of a manic-depressive syndrome in 1889. Boltzmann left Graz for Munich in 1890, Vienna in 1894, Leipzig in 1900, and finally back to Vienna in 1902. Some of the years, especially those in Munich, treated him well. But some didn't. Mach's 1895 arrival in Vienna didn't help matters. Boltzmann always enjoyed scientific discussions and debates but became a strong adversary of the anti-atom group of which Mach played a strong role. He was a sensitive man, one prone to psychological weakness, so to have a famous colleague openly and strongly disagree with a theory on which he devoted his entire life was very hard to take. Needless to say, his return to Vienna in 1894 was

---

[56] (Cercignani, 2006) p. 26.
[57] (Cercignani, 2006) p. 153.
[58] (Cercignani, 2006) p. 129.
[59] (Cercignani, 2006) p. 17.

not a completely happy one. Not helping matters were the deaths of his best scientific friends, Josef Stefan (1835–1893) and Josef Loschmidt (1821–1895). He left Vienna for Leipzig in 1900 mainly because of his dislike of working with Mach, but in another stroke of bad fortune found himself working next to Wilhelm Ostwald, a personal friend but also a leading figure of the energetics movement which also espoused an anti-atom philosophy. Boltzmann couldn't escape. The resulting scientific debates between them became so bitter than it was Mach of all people who had to interject himself into the situation to calm things down. Unfortunately, Boltzmann was alone on the battlefield, no troops supporting him. Those who could have provided support were sitting in Britain.[60] Those attacking him were sitting next door. While his ideas were actively discussed in Britain, they received little attention or positive engagement on the continent, leaving Boltzmann stranded. While the young students largely loved his lectures and embraced his theories and supported him in his public scientific debates with the energeticists—"We young mathematicians were all on Boltzmann's side"[61]—this wasn't enough. Boltzmann's mind and body began to break down. He experienced an emotional high early in life, achieved renown, but then began a slow descent in an Austrian society where suicide was fairly common. "Beneath the glittering surface [of Austria] was a society whose members were incapable of opening themselves to others."[62] He attempted suicide twice, the first time in Leipzig, after which he spent time recovering in a psychiatric hospital. The second time he tried, he unfortunately succeeded. In 1906, while on summer vacation in Italy, the thought of starting a new school year back in Vienna[63] combined with his own personal unfortunate history and manic-depressive anxieties to send him into a deep depression. While his family was at the beach, he hanged himself, one day prior to his planned return to Vienna.

The timing of Boltzmann's suicide was very unfortunate because acceptance of the atomic theory of matter was right around the corner, with Albert Einstein publishing his theoretical explanation of Brownian motion in terms of atoms in 1905 and then Jean Perrin (1870–1942) experimentally confirming this work in 1908. The rise of such clear evidence naturally turned a spotlight on the reactions of the two main anti-atom protagonists. Ostwald became a convert, writing in 1909, "I am now convinced … of the grainy nature of matter,"[64] while Mach did not, reiterating his disbelief in atoms in 1913 and sharing his view with Einstein around that time that one could use the atomic theory as a good assumption without necessarily accepting the "real existence" of atoms. In his review of this history, Brush aptly described Mach as "the unrepentant sinner."[65]

---

[60] (Darrigol, 2018) p. 568. "British physicists were favorable to molecular theories; they were developing dynamical theories in every domain of physics; and Boltzmann's works were largely a reaction to Maxwell's kinetic theory of gases."

[61] (Cercignani, 2006) p. 27.

[62] (Cercignani, 2006) p. 46.

[63] Mach retired from the University of Vienna in 1901 due to poor health. This improved the atmosphere there for Boltzmann and so helped lead to his return in 1902. But Boltzmann's suicide in 1906 suggests that even in Mach's absence, Vienna was no longer an inspiration for Boltzmann.

[64] (Cercignani, 2006) p. 209.

[65] (Brush, 1986a) p. 295.

Boltzmann's Influence on Those who Followed

I am conscious of being only an individual struggling weakly against the stream of time. But it still remains in my power to contribute in such a way that, when the theory of gases is again revived, not too much will have to be rediscovered.

– Ludwig Boltzmann (1898)[66]

Boltzmann delivered a powerful gift to those who would launch their own careers based on his work, providing them as he did a means to bridge from the classical to the quantum world. Two such individuals stand out: Max Planck (1858–1945) and Albert Einstein (1879–1955). Planck's own story of conversion is worthy of attention. He started as a strong believer in Mach but later became one of his most severe critics. The conversion was that complete. It happened in his own eureka moment discussed earlier when—"after a few weeks, which were certainly occupied by the most dogged work of my life, a flash of lightning lit up the dark in which I was struggling"[67]—he saw Boltzmann's statistical mechanical theories as the path toward solving the blackbody radiation conundrum. Not only did the theories provide the mathematical solution to the radiation curve but they also provided the means to build a bridge to the quantum world. The discrete-energy solution Boltzmann presented in his 1877 paper provided Planck the foundation on which to base his theory of discrete resonators being at cause in the matter of blackbody radiation and, even more importantly, subsequently provided Einstein the foundation on which to base his theory of light quanta. Planck's first-hand experience in applying Boltzmann's work "in such a novel way"[68] to solving real problems plus his eventual endorsement of the statistical view of the second law in 1914[69] contributed to his eventual turning on both Ostwald and Mach. He even went so far as to call Mach a "false prophet."[70] His comment in 1933 on this situation is apropos:[71]

In the eighties and nineties of the last century, personal experience taught me how much it cost a researcher who had had an idea on which he had reflected at length to try to propagate it. He had to realize how little weight the best arguments he exhibited to that end carried, since his voice had not sufficient authority to impose it on the world of science. In those days it was a vain enterprise to try to oppose such men as Wilhelm Ostwald [and] Ernst Mach.

– Max Planck (1933)[72]

---

[66] (Boltzmann, 1995) p. 216.
[67] (Cercignani, 2006) p. 219.
[68] (Pais, 2005) p. 372.
[69] (Darrigol, 2018) p. 410.
[70] (Cercignani, 2006) p. 210.
[71] So as to better balance this analysis, I include here a much appreciated comment received in an 2017 email from Professor Helge Kragh: "The Mach–Ostwald–Helm position was [not] generally accepted in the 1890s. The majority of physicists (and even more so the chemists) did not follow Mach or the energetics school, not even in Germany. There is an excellent discussion in C. Jungnickel & R. McCormmach, *Intellectual Mastery of Nature* (1986), vol. 2, pp. 213–227 (a recommendable work). There is also a recommendable essay on "Atomism versus Thermodynamics" in C. Howson, ed., *Method and Appraisal in the Physical Sciences* (1976), quite detailed and technical, you may appreciate it despite its philosophical framework."
[72] Planck quote cited in (Cercignani, 2006) p. 210.

Turning now to the second beneficiary, having developed his own ideas about distribution theories and then having read Boltzmann's and also Gibbs' works, Einstein became an expert in statistical mechanics and wielded it with some of his own adaptations[73] to resolve a range of very different problems: Brownian motion and the subsequent launch of the atomic theory, radiation, light quanta and the photoelectric effect, specific heats based on discrete quantum states including the "unfreezing" of quantum states. It is rather amazing that Boltzmann had created a statistical mechanics for one situation only to have it become such a powerful source of breakthrough discovery by Einstein in other situations, almost as if someone had designed this whole story as a tragedy followed by a rousing "You were right!" post-mortem epilogue.

Boltzmann's influence went beyond Plank and Einstein. Just look at his students, a who's-who list of famous physicists: Sommerfeld, Nernst, Svante Arrhenius (1859–1927), Meitner, among others. Considering a person's legatees is a good criterion to use when assessing that person's contributions to science. Also look at the other fields of science where his work found a home: the behavior of gas at high altitude where the continuum assumption breaks down, the motion of neutrons in nuclear fission, the motion of charged particles in nuclear fusion, radiation in combustion chambers including his own theoretical deduction of Stefan's T^4 radiation law. All of these fields required Boltzmann's theories to deal with the "grainy nature of matter."

Last but not least, we return to statistical mechanics. Boltzmann's statistical approach to thermodynamics provided a mechanical definition of entropy and so provided a different pathway to calculating entropy than that provided by classical thermodynamics. This opened the door to the whole new world in which the term $k_B \ln(W)$ could be substituted for S in all of Gibbs' thermodynamic equations, which Gibbs himself eventually did.

Gibbs and the Completion of Statistical Mechanics

Boltzmann combined his theories into his seminal 1884 paper on statistical mechanics. Gibbs took this work and expanded it into his own 1902 treatise on the subject, acknowledging the priority of both Maxwell and Boltzmann in so doing. In a note to Lord Rayleigh, Gibbs wrote, "I do not know that I shall have anything particularly new in substance, but shall be contented if I can so choose my standpoint (as seems to me possible) as to get a simpler view of the subject."[74] In the end, Gibbs' version of statistical mechanics won the day.[75] He was the one who, in the same letter to Lord Rayleigh, named this field *Statistical Mechanics*. It was his *ensemble* theory in which each ensemble is comprised of all possible microstates that became famous even though Boltzmann and Maxwell also contributed to the theory. Yes, it is largely Gibbs' version we study today but Boltzmann and Maxwell were there first.

---

[73] (Klein, 1967)
[74] (Cercignani, 2006) p. 143.
[75] (Cercignani, 2006) Chapter 7, pp. 134–152.

Before Leaving Boltzmann – a Final Comment Regarding His Famed Distribution

Before going further, a distinction is needed here as regards distributions. When considering population distributions in statistical mechanics, one can focus either on the energy associated with a specific quantum state or the total energy. For example, the translational motions of non-interacting hard spheres are composed of three components representing the x, y, and z coordinates; each component represents a different quantum state. Thus, total energy is comprised of the contributions of three quantum states. The distribution of energy for a given quantum state follows the Boltzmann Distribution, which is an exponential decay $[\exp(-x^2)]$ of the number of spheres in each energy bucket plotted against the energy associated with that bucket. The distribution of energy for total energy follows the Maxwell–Boltzmann Distribution, which shows the number of particles increasing to a maximum and then decreasing to zero $[x^2\exp(-x^2)]$ as a function of the energy associated with that bucket. The shift from the former to the latter is attributable to the fact that 1) total energy is determined by the sum of the component energies, and 2) the number of ways to achieve a given total energy increases with the energies of the components; hence the addition of the x^2 term to the function. This gets into an area termed *degeneracy*, which effectively adds additional layers of probability calculations to Boltzmann's balls-in-buckets exercise.

With this (very brief) background, we now take a deeper dive into Boltzmann's most important contribution to the new field of statistical mechanics: the Boltzmann Distribution. In his 1877 publication involving permutation theory, Boltzmann derived an equation showing how particles distribute themselves based on energy. Recall his exercise involving 7 balls in 8 energy buckets?[76] He generalized this example to show that the most probable population distribution of particles based on energy levels (E_i) follows an exponential decay curve per the now well-known term $e^{-E_i/kT}$. While Maxwell derived this equation for the distribution of energy in a single velocity coordinate (single quantum state), Boltzmann proved the validity of the equation at a higher theoretical level based on probabilities. Many more particles populate the low energy levels than the high, not because there's any preference of single particles for low-energy states, because there isn't, but instead because many more arrangements are possible for such a distribution, thus making it the most probable distribution. Of all the possible complexions that can exist for a given system, the overwhelming majority aligns with the Boltzmann distribution. Recall earlier that in the extreme, placing one particle at the highest energy level forces all other particles into the lowest energy level and significantly reduces the number of possible arrangements. The occurrence of particles arranging themselves in this way is highly improbable. Far more arrangements exist when most particles populate the low energy levels.[77] You can (roughly) see this by doing the 7 ball / 8 energy bucket example for yourself in which the placing of a ball in a bucket represents a one degree-of-freedom decision. The distribution comprising the most number of different arrangements is 3 (E_0), 2 (E_1), 1 (E_2), and 1 (E_3), and

---

[76] Note: In his exercise, total energy was comprised of a single component, in effect representing a single quantum state and thus a single degree of freedom (DOF).

[77] (Dill and Bromberg, 2011) Chapter 10, p. 173.

no balls in the E_4 through E_7 buckets (see Figure 42.1). This example demonstrates the exponential decay in population relative to energy.

With the Boltzmann distribution as the core building block, statistical mechanics rose to become a powerful means to quantify probability distributions and thus concentrations of particles, atoms, and molecules along with their associated energies. Such information at the microscopic level enables predictions at the macroscopic level in the form of, for example, bulk properties and chemical reaction equilibria, the macroscopic world simply being the average of the microstates present.

Statistical Mechanics Provides the "Why"

For all its immense power, thermodynamics is a science that fails to reward man's quest for understanding. Yielding impressively accurate predictions of <u>what</u> can happen, thermodynamics affords us little or no insight into the <u>why</u> of those happenings.

– Leonard K. Nash[78]

Maxwell's intuition and strong mathematical skills provided the first glimpse of such distributions for simple systems. Boltzmann enhanced, extended, and generalized Maxwell's work, showing the Boltzmann distribution to be at the heart of the Maxwell–Boltzmann Distribution as indicated in the equations above. His generalized approach opened the door to applications in more complicated systems. As the phase changes from gases to liquids to solids, and as the molecules change from mono-atoms to poly-atoms that can rotate and vibrate, and as the energies involved in each system of phases and molecules are discretized due to quantum mechanics, statistical mechanics becomes very complicated. Complicating matters even further are systems in which increasing energy levels result in certain new accessible arrangements such as occurs with the unfreezing of vibration states or the unfolding of proteins. For all systems, complicated or not, the microstates associated with each energy level are accounted for, with each microstate being equiprobable with all others *at that specific energy level* and with the total energy at each level being equal to the population of microstates at that level times the energy of the level. The population of all energy states, being comprised of their respective microstates, is finally distributed per the Boltzmann Distribution with adjustment to account for the change in number of microstates. Higher energy states become more populated as the number of accessible microstates increases. Yes, it gets complicated. For the ideal monatomic gas, this whole process becomes tremendously simplified since each energy level is equivalent to just the translational kinetic energy of the atom. This simplification leads to the Maxwell–Boltzmann Distribution based on velocities, which is a special case of the more general Boltzmann Distribution. A final complication in this whole discussion, as if there weren't enough, concerns the macroscopic properties constraining the system. Recall that statistical mechanics involves the substitution of the mechanical definition of entropy, $S = k_B \ln W$, into Gibbs' classically derived thermodynamic equations, which were developed based on assumed constraints. The collections of microstates corresponding to each constrained system are called

---

[78] (Nash, 2006) pp. 1–2.

ensembles per Gibbs. The U-V-N ensemble is called microcanonical and the T-V-N ensemble canonical. The mathematical construct used to model the differently constrained ensembles takes into account the fact that certain properties vary or fluctuate when other properties are constrained. For example, when sitting in a constant temperature bath, a body's total energy can fluctuate.

Given such complexity, the end result is very powerful. Indeed, the rise of statistical mechanics was a tremendous accomplishment and its historical and theoretical foundations shared here serve as a sound conclusion to this section on entropy and the laws of thermodynamics. Most undergraduate students aren't exposed to this subject as many other courses, such as classical thermodynamics, take priority. But the learning of this subject opens doors to a better understanding of nature and thus provides a powerful tool for probing nature, for predicting what she will do and ultimately for controlling her to accomplish productive tasks. Reflecting on Leonard Nash's above quote, statistical mechanics provides an explanation of not just *what* happens in thermodynamics but also *why* it happens. Even though it's a mathematical construct based on an idealized mechanical system of particles and interactions and stands on its own as an exact science, it still provides a path toward testing assumptions about the microscopic behavior of assemblies of atoms and molecules since these assumptions can be incorporated into the mathematics. If the resulting predictions match reality, then perhaps the assumptions are valid. And this is the heart of science. Many validations of this field of science have populated history's timeline, starting with Maxwell's work on gas viscosity. To me, the first and most visually powerful validation remains Otto Stern & Co.'s Maxwell–Boltzmann spray pattern of atoms on a rotating cylinder. They made the invisible visible.

Gibbs – a Great Capstone to His Life

> *All this Maxwell and Boltzmann have explained; but the one who has seen it most clearly, in a book too little read because it is a little difficult to read, is Gibbs, in his Elementary Principles of Statistical Mechanics*
>
> – Poincaré[79]

To close this section, I now return to Gibbs to say a little bit more about his completion of Boltzmann's work. One cannot fault Gibbs for the fact that his treatise obscured the contribution of Maxwell and Boltzmann to statistical mechanics. One can only praise Gibbs for what he did. It is admittedly very difficult to wrap one's head around the machine that Gibbs created. But just because Gibbs was a genius does not mean that the task he challenged himself with was easy on him. In fact, it was just the opposite. This effort, which summed up much of his life's work, ran from the fall of 1900 to the summer of 1901. During this time, Gibbs compiled results 14 years in the making into a book that can best be characterized as "prolific in the smallest possible space."[80] Try reading the book and you'll understand. How he did this, no one knows. No notes with indications of his thought process or his false starts were left behind. It's

---

[79] Poincaré quote cited in (Rukeyser, 1988) p. 336.
[80] (Rukeyser, 1988) p. 343.

as if this magnificent book just appeared. During this time of writing, he carried on a full load of teaching at Yale while working on the book through the afternoons and into the evenings. The machine he created provided immense usefulness to those who followed despite—or because of—his intentional exclusion of the atomic theory. The machine was pure and absent contamination by hypotheses about nature. It is rather remarkable that his statistical approach required almost no modifications upon the arrival of quantum theory. But all this work, all of the immense physical and intellectual effort that went into this work, took a toll on Gibbs. As one of his closest friends commented, after such a sustained effort to complete the book Gibbs was "a worn-out man and never fully came back."[81] He passed away in 1903 at the age of 64.

---

[81] (Rukeyser, 1988) p. 343.

43 Shannon: entropy and information theory

"My greatest concern was what to call it. I thought of calling it 'information,' but the word was overly used, so I decided to call it 'uncertainty.' When I discussed it with John von Neumann, he had a better idea. Von Neumann told me, 'You should call it entropy, for two reasons. In the first place your uncertainty function has been used in statistical mechanics under the name, so it already has a name. In the second place, and more important, no one knows what entropy really is, so in a debate you will always have the advantage.'"

– Claude Shannon as captured by Myron Tribus[1]

This conclusion to Part IV is about communication. The reason will take some explaining.

What is communication? This sounds like an easy concept but is in fact rather complicated. An idea or thought appears in our mind, we use our vocal chords to convert the message we want to convey into small pressure waves of sound that move through air across space where they are detected by the ears of another person and converted back into the original message (Figure 43.1). This sounds simple. And yet how often do the sent and received messages match? For simple messages, frequently, but for complex ones, not so frequently. Misunderstanding is common. We lack a language in which an exact correspondence exists between words and their meanings. Thus, the receiver, is frequently uncertain and sometimes left wondering, *now what did he mean by that?* So the sender uses more descriptive words, adds in some facial expressions, body language, and volume and pitch variation, and the match between sender and receiver improves. But still, for many reasons, it's very difficult to achieve perfect communication.

Adding to the challenge of perfect communication is distance. How does one communicate with another who isn't standing within shouting distance? Well, communication evolved in different ways over time to meet this challenge. For example, a message was told to a runner or a rider, they memorized it, traveled to the receiver and then repeated it. One of the best examples of this occurred when Pheidippides ran the original marathon in 490 BC from Marathon to Athens with the news of the great victory by his people over the Persians at Marathon. This message took about 3 hours to cover 26 miles, which is quite fast and may have been why he immediately died after delivering the message. But in urgent times, such as war, even faster communication was needed as it could be used to great advantage. Sound generated by drums and horns were great for this. But light was even faster. Flags and bonfires provided near-

---

[1] (Tribus and McIrvine, 1971)

Block by Block: The Historical and Theoretical Foundations of Thermodynamics,
Robert T. Hanlon, Oxford University Press (2020). © Robert T. Hanlon.
DOI: 10.1093/oso/9780198851547.001.0001

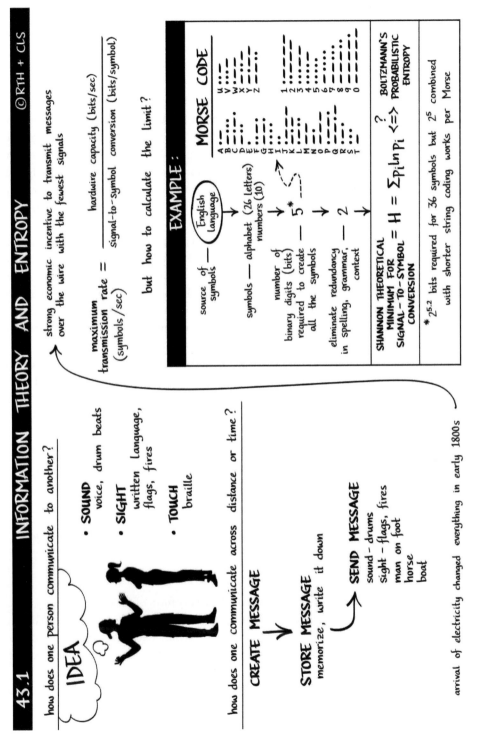

43.1 INFORMATION THEORY AND ENTROPY © RTH + CLS

how does one person communicate to another?

IDEA

- **SOUND**
  voice, drum beats
- **SIGHT**
  written language,
  flags, fires
- **TOUCH**
  braille

how does one communicate across distance or time?

CREATE MESSAGE

STORE MESSAGE
memorize, write it down

SEND MESSAGE
sound - drums
sight - flags, fires
man on foot
horse
boat

arrival of electricity changed everything in early 1800s

strong economic incentive to transmit messages
over the wire with the fewest signals

maximum
transmission rate = hardware capacity (bits/sec) / signal-to-symbol conversion (bits/symbol)
(symbols/sec)

but how to calculate the limit?

EXAMPLE:

source of ── English language
symbols

symbols — alphabet (26 letters)
numbers (10)

number of
binary digits (bits)   5*
required to create
all the symbols

eliminate redundancy   2
in spelling, grammar,
context

SHANNON THEORETICAL
MINIMUM FOR
SIGNAL-TO-SYMBOL = H = $\sum p_i \ln p_i$? **BOLTZMANN'S**
CONVERSION $\Longleftrightarrow$ **PROBABILISTIC**
ENTROPY

* $2^{5.2}$ bits required for 36 symbols but 2^5 combined
with shorter string coding works per Morse

MORSE CODE

A ·—	U ··—
B —···	V ···—
C —·—·	W ·——
D —··	X —··—
E ·	Y —·——
F ··—·	Z ——··
G ——·	
H ····	
I ··	
J ·———	1 ·————
K —·—	2 ··———
L ·—··	3 ···——
M ——	4 ····—
N —·	5 ·····
O ———	6 —····
P ·——·	7 ——···
Q ——·—	8 ———··
R ·—·	9 ————·
S ···	0 —————
T —	

Figure 43.1 Information theory and entropy

instant messages beyond sound's reach. In the twelfth century BC, the Greeks used fire beacons that could be seen 20 miles distant to announce the fall of Troy to Clytemnestra four hundred miles away in Mycenae. And who can forget the great scene in *The Lord of the Rings* when warning beacons were lit on mountaintops to tell the Rohirrim that Minas Tirith was under siege. Naturally, in these long-distance circumstances, the correlation between message sent and message received naturally had to be defined in advance. Seeing a fire from a distance is only valuable if you know in advance that the fire means HELP!

What about for more complicated messages, and not just those crossing distance but also those crossing time, such as the histories passed down from generation to generation. For these, a strong memory was needed, aided by the strategic use of rhythm and rhyme such as was done in ancient Greece to preserve history in poems. This legendary skill lasted right up until the advent of writing. The transcription of Homer arguably represented the something-lost/something-gained transition from an oral-based life to one of fixed symbols on a papyrus sheet.

About 30 000 years ago *Homo sapiens* began to scratch and paint images of hunters and prey on stone walls, and over time this use of images for art developed into their use for communication with different peoples in different regions coming up with different approaches. At one extreme, the Chinese created many symbols, one for each word. At the opposite extreme, the Phoenicians along the eastern Mediterranean developed the alphabet, with each minimal sound component receiving one written letter. The development of written numbers and mathematics soon followed.

Writing enabled communication of increased size and complexity. But the speed of sending such communications remained limited by the speed of the foot or hoof or the wind in the ship's sail. Written messages did not lend themselves to transmission via sound or light, although one could argue that the evolution of drum-language in Africa or semaphore flag signaling did achieve high levels of complexity.

Then came electricity and everything changed.

Samuel F. B. Morse

> It would not be difficult to construct a system of signs by which intelligence could be instantaneously transmitted.
>
> – Samuel F. B. Morse [2]

In 1800 Italian physicist Alessandro Volta invented an early form of the electric battery, called a voltaic pile, and thus provided civilization with a source of a steady electric current. The later invention of the electromagnet by William Sturgeon in 1825 demonstrated the concept of using electricity to move objects from a distance. It didn't take long before people realized the possibilities.

While returning by ship from Europe in 1832, Samuel F. B. Morse observed a demonstration of the electromagnet in action, set aside his then vocation of painting, and launched himself

[2] (Gleick, 2011) p. 142.

in an entirely new direction. He was familiar with much that had been discovered in the world of electricity; the possibility for a "eureka" moment was set. When he witnessed the electromagnet in action, he received "the inspiration which was to come to him like a flash of the subtle fluid which afterwards became his servant."[3] Duly impressed, Morse soon developed the keypad system that closed circuits in one location and caused motion of the electromagnet in another location, resulting in an audible "click." His development of the hardware, though, was surpassed in complexity by the development of the software. What message could be sent by clicking? Morse initially thought he could send numbers via clicks, the number of clicks naturally corresponding to the intended digit with the space between separating the digits. He proposed assigning each number to a word. But this approach proved to be a dead-end as the receiver had to spend inordinate amounts of time looking up the number-word pairs. He then realized a more eloquent, less brute-force approach, one made available to him with his hardware. A short click could be termed a "dot" and a long click—hold the key down 3× longer than for a dot—could be termed a "dash." He also had both short and long spaces during which the circuit was open. So all told, he had four different signals to work with. In the end, he set to work using dot-dash combinations to transmit letters and spaces to separate letters and words.

Morse's dot-dash system could be viewed as a binary system based on zeros and ones. He determined that combinations of five dot-dash symbols enabled 32 different combinations ($2^5 = 32$). Including some shorter-string combinations enabled him to cover all 26 letters plus 10 numbers. In this way he developed a new language. Experienced keypad operators could "talk" with each other through the clicking sounds that the inexperienced couldn't even begin to understand.

Market forces soon drove the design of Morse's code toward improved efficiency. The less time it took to send a message, the lower the cost, the happier the customer. Morse realized that the best code would thus assign faster coding to the more frequently used letters. How did he determine letter usage frequency? He went to a printer's shop and looked at the type cases. (Obvious perhaps only in hindsight.) The letter E contained the most types and so was assigned to a single dot. The letter T also contained many types and so got a single dash. The less populated letter Z was assigned two dashes and two dots. In this way the statistics of English were embedded into the final structure of Morse's code (Figure 43.1). Intelligent customers employed additional efficiency-improving statistics as they learned to shorten their messages and so lower their costs by using abbreviations, such as "ymir" – your message is received.[4]

The concept of "now" became very real upon arrival of the telegraph. Information that previously had taken days to travel now took seconds. "Now" in one location became the same "now" in another location. Instant knowledge across distance became possible, all the more so when large landmasses were connected by the laying of cable underwater, starting with the English Channel (1851) and later across the Atlantic Ocean (1858), a feat that in 1852 had been authoritatively declared to be "utterly impracticable and absurd."[5]

---

[3] (Morse, 1914) p. 3.
[4] (Gleick, 2011) p. 154.
[5] (Gleick, 2011) p. 146. Note that William Thomson's valuable contributions to the successful installation of the transatlantic cable led to his becoming Lord Kelvin in 1892 by Royal decree.

Disruptive innovation occurred again in 1875 with the arrival of Alexander Graham Bell's (1847–1922) telephone. By the turn of the century, the telephone surpassed the telegraph. The reason was not difficult to understand. Everyone could use the telephone. Soon new hardware sprung up all over to support the desire of everyone to use the phone. By 1948 more than 125 million conversations passed daily through AT&T's 138 million miles of cable and 31 million telephone sets.[6] As the complexity grew—how does one efficiently connect caller to receiver?—AT&T's research division, Bell Laboratories, started hiring mathematicians to design the logic required to meet the growing challenges efficiently and effectively. As a result, Bell Labs' culture rose into a powerful center of practical mathematics, second to none, and one characterized by a creative tension between the ideal mathematician and the practical engineer.

As Bell Labs' mathematicians waded into this new field, breakthroughs started to happen. Of note in the context of this book's conclusion were the contributions of two. The first, Harry Nyquist (1889–1976), showed how continuous waves could be converted into discrete or "digital" data by strategic sampling of the waves at intervals. This advance enabled the digitization of sound, thus complementing the original digitization of sight in the form of such visuals as writing and images. The second, Ralph Hartley (1888–1970), a colleague of Nyquist's, brought new theories and definitions to address capacity issues resulting from the rapid growth. In so doing he started using the word *information* to encompass the use of symbols and signals in transmitting messages and also suggested a new approach to quantifying the size of messages by taking the $\log_2$ of the number of symbols available to use (more on these mathematics later). With many such new ideas circulating through the hallways of Bell Labs, the stage was set for one to absorb them all, to take a step back from the specifics, to seek a more generalized place to stand and from that place to see the essence at its theoretical core. John Pierce wrote, "we can often see men of an earlier day stumbling along the edge of discovery but unable to take the final step."[7] Many came close. But in the end, it was Claude Shannon who took the final step.

Claude Shannon

The fundamental problem of communication is that of reproducing at one point either exactly or approximately a message selected at another point.

– Claude Shannon[8]

The story of Claude Shannon's (1916–2001) journey toward creating *information theory* strikes similar chords with the stories of others' shared in this book. As was the case for the others, an innate intelligence inhabited Shannon. He achieved dual B.S. degrees (Electrical Engineering, Mathematics) at the University of Michigan and then both Master's (1937) and Doctorate (1940; Mathematics) Degrees at MIT, where he worked on Vannevar Bush's differential analyzer, an early version of the analog computer. While at MIT he spent two summers at Bell Labs in New York City and then in 1940 spent a post-doctoral year as a National Research Fellow at the Institute for Advanced Study in Princeton, where he encountered such influential mathematicians as Hermann Weyl (1885–1955) and John von Neumann (1903–1957)

---

[6] (Gleick, 2011) p. 5.
[7] (Pierce, 1980) p. 20.
[8] (Shannon and Weaver, 1998) p. 31.

while pursuing his interest in the transmission of intelligence. When WWII broke out he returned to Bell Labs and worked on improved weapons firing and cryptography. While there he read the work of Nyquist and Hartley in addition to the works of others outside Bell Labs such as Norbert Wiener (1894–1964) at MIT. In sum, Shannon absorbed a broad and expansive experience involving both hardware and software, was exposed to the transmission problems that needed solving especially regarding capacity and also noise, filled his mind with a range of ideas, like Galileo saw opportunities at the interface, and then, like Newton, Carnot, and Gibbs among others, isolated himself within Bell Labs and thought. His managers left him alone, not being certain what it was he was working on, but nonetheless willing to support him, reflecting Bell Labs laissez-faire approach to managing their highly talented research group back then.

The focus of Shannon's thinking was communication. The end result was "The Mathematical Theory of Communication," Shannon's famed 1948 paper, made very readable by a "light touch"[9] that delivered communication theory to the science and mathematics community fully formed. Similar to the approach employed by Maxwell, Boltzmann, and Gibbs, Shannon's paper outlined his creation of an exact mathematical model that stood on its own with no assumptions about the real world and only then turned toward the real world for comparison and validation. The paper wove the ideas of others such as Nyquist, Hartley, Wiener, and Andrey Kolmogorov (1903–1987) into his own to reveal the larger picture of what they all touched but only Shannon saw. The impact was significant. As suggested at that time by Warren Weaver (1894–1978), the Director of Natural Sciences for the Rockefeller Foundation, what Shannon had done for communication theory "was what Gibbs did for physical chemistry."[10] While he never became famous to the public, he became "iconic"[11] to those in the field he formed.

So what did Shannon see? He saw the possibility of a new paradigm, a unifying theory, about how communication works. As Carnot did with the steam engine, Shannon did with communication. He stripped away all except what mattered, separated the relevant from the irrelevant and sought a generalized approach to account for any and all forms of communication. He saw the process of communication as originating from a source of symbols, such as the written letters or words of any language, the sounds of a voice or a symphony, and the images in a photograph or a motion picture. Each source provided a different means by which minds can affect other minds. A sender creates a message by selecting a sequence of symbols from the source and these are then converted or *encoded* in some way into signals, such as dots-and-dashes or zeros-and-ones, suitable for transmission. Shannon had a preference for using zeros and ones and labeled them *binary data*, which he then shortened to the famed word *bits*, each bit being either a zero or a one. In this way, all symbols in the source can be converted into a string of bits. The receiver on the other end takes the string of bits and *decodes* them back into the original symbols and so re-creates the message.

In addition to unifying the field of communication under a single theory, Shannon also wanted to address two significant problems: capacity and noise. Regarding noise, during

---

[9] (Shannon and Weaver, 1998)

[10] (Gleick, 2011) p. 221. Warren Weaver, the director of natural sciences for the Rockefeller Foundation, wrote this to Shannon in 1949. Weaver would later write an essay about Shannon's theory and the two papers would be combined into a book, *The Mathematical Theory of Communication*, in 1949.

[11] (Gleick, 2011) p. 262.

transmission of signals corresponding to a specific message, noise enters the channel from the external environment via natural and unwanted processes and overlays randomness on top of structure. When he considered this problem, Shannon reflected back on his work on cryptography. "From the point of view of the cryptanalyst, a secrecy system is almost identical with a noisy communication system."[12] Both coding and noise place a veil over pattern, one intentionally, the other not. Shannon's increased understanding of the statistical structure of language enabled him to understand how this works and thus how to account for the impact of noise on the capacity of a communication system.

Turning now to capacity, Shannon recognized that the capacity issue was not with the message itself but with the message's encoded signals, because it was these signals that were transmitted. Building on Hartley's original inquiry, Shannon reasoned that a commercially viable system must be designed to handle all possible messages created from the source and that messages would emanate from the source at a certain rate, say number of symbols per second. The symbol flow would then be converted into a signal flow using the ratio of signals per symbol, thereby resulting in a transmission flow of signals per second. In the case of zeros-and-ones, the corresponding units of transmission rate are bits per second. The maximum message transmission rate can then be calculated.

$$\text{Maximum Transmission Rate} \left(\text{symbols/sec}\right) = \frac{\text{Hardware capacity} \left(\text{bits/second}\right)}{\text{Signal-to-Symbol Conversion} \left(\text{bits/symbol}\right)}$$

[43.1]

Hardware limitations set the maximum transmission capacity of bits per second and thus, working backwards, the maximum capacity for sending messages. The critical parameter in these calculations then is the conversion of symbols to bits. The conversion factor can depend on a range of issues, especially the skill level of the person doing the coding. But Shannon took this to a higher level, seeking the limit by asking, "is there a measure which will tell us the *minimum* [my italics] average number of binary digits per symbol?"[13] His goal was to shorten the number of signals required to transmit the messages and thereby maximize the message transmission rate for the given hardware limitations. By providing the means to quantify the minimum for a given information source, encoders would have a goal to shoot for.

The critical point was that, following Hartley's lead, Shannon wasn't concerned with any given message, per se, but instead was concerned with *all* possible messages that could originate from the source. Furthermore, he wasn't concerned with the *meaning* of the messages—meaning was irrelevant to his approach—but instead was concerned with defining a readily quantifiable number based on mathematics and not human judgment. This became a stumbling block for some. *How can information ignore meaning?*

Shannon started his journey using the word "communication" but eventually adopted Hartley's "information" as this was the word more in use at the time. To him, information was that which connected symbols to signals. It focused on the encoding process. The sequences of

---

[12] (Gleick, 2011) p. 216.
[13] (Pierce, 1980) p. 79. It's interesting that Carnot sought to identify maximum work, Shannon sought to identify minimum signal-to-symbol and both *what's the limit?* inquiries ended up at entropy.

bits in the signal were not binary numbers but instead identification tags matching symbol to signal.

So the question before Shannon was, how to calculate the minimum value of bits per symbol? Morse arguably started addressing this question in the way that he developed his famed code. His information source consisted of letters and numbers. He encoded most but not all of these 36 symbols into five-bit strings of dots and dashes, reserving the right to use shorter strings for more frequently used letters. Customers later reserved the right to save money by using intelligent abbreviations to achieve even more brevity. In this way, scientists and engineers realized that the transmission capacity issue had to focus not on the symbols but the resulting signals, which was approximately $\log_2$ of the number of symbols in the source.[14] Shannon started gravitating toward this function on account of his previous experience with analyzing the electronic circuitry. Because circuit switches could be in either of two different positions, he developed his abstract circuitry mathematics for the sequencing of such switches around base-two. When he approached information theory, he realized he could use the same approach.

In his search to quantify the minimum value of bits-per-symbol, Shannon recognized that the sequence of symbols in the message could not be totally random since such a message would be no message at all. Indeed, messages must contain structure. Q is most always followed by u. The sequence "*men is*" is not allowed. We recognize "*cn u rd ths?*" as "*can you read this?*" We guess that the word at the end of "*it ain't over 'til the fat lady...*" is *sings*.[15] We know this and much more because we know English. The math had to account for all of the spelling, grammar, and contextual relationships between the symbols, relationships that represented a certain amount of *redundancy* in the source, which refers to the amount of text in a given language that can be removed without losing any information. Shannon then demonstrated that such redundancy is tied to probability. For example, it is highly probable that u follows Q and highly improbable that "is" follows "men".[16] This embrace of probability gave Shannon a path to his mathematics.

How can one measure the probabilities inherent to a system of symbols? Morse did this in a crude but highly effective fashion by visiting the print shop. Shannon took this concept to a higher level. He took a book off the shelf and started counting the occurrence of many different sequences of letters and words, realizing that as the numbers become large, frequency turns into probability. From his "experiments", he determined, for example, that the redundancy of ordinary English is about 50%, meaning that about 50% of the letters and words can be removed without loss of meaning. Whereas Morse showed that strings of 5 binary digits for each symbol sufficed for transmitting English, Shannon showed that by including probability into the calculations and shifting the source of information from letters to words, which is facilitated by the fact that English words have much redundancy, the number of binary digits

---

[14] $2^{5.2} = 36$. Morse achieved use of five-bit coding by assigning some symbols to shorter strings

[15] The statistical structure of language is embedded in the search functions you use in Google or Siri when each completes words and sentences for you.

[16] The many languages around the world naturally evolved to have some redundancy as such structure helps to provide context and ensure accuracy of both message transmission and also message understanding. In fact, one could argue as (Gleick, 2011) did (p. 353), that most communication in life lies in the middle between predictability and randomness where the two "interlace." What fun is a predictable movie or a randomly constructed song?

per letter could be reduced to about 2.[17] The reduction from Morse's 5 to Shannon's 2 for English demonstrates the compression available when using probabilities to account for all types of redundancy, e.g., spelling, grammar, and context. Again, such calculations show only what is possible but not how to get there. Good coders are still needed.

It is now time to share why the above discussion is relevant to this book. The most prominent equation Shannon derived in his paper was [43.2] below. It quantifies the signal-to-symbol ratio (H) in equation [43.1] in units of average number of bits per symbol as follows:

$$H = -\sum p_i \log p_i \qquad\qquad\qquad [43.2]$$

for i = 1 to n, n being the number of symbols in the source and p_i being the probability of symbol i being chosen.

As you can see, this equation is rather familiar. Boltzmann arrived at a similar equation.

Shannon Embraces Statistical Mechanics Terminology to Describe Information Theory

> *Several physicists and mathematicians have been anxious to show that communication theory and its entropy are extremely important in connection with statistical mechanics. This is still a confused and confusing matter… [and] efforts to marry communication theory and physics have been more interesting than fruitful.*
>
> – John R. Pierce[18]

> *It is probably dangerous to use this theory of information in fields for which it was not designed, but I think the danger will not keep people from using it.*
>
> – J. C. R. Licklider (1950)[19]

Maxwell's demon is a good place to start this final discussion. If you'll recall, around 1870, in his continual pursuit of ways to challenge the colliding-atoms definition of entropy, Maxwell devised the now classic thought experiment involving the presence of a "demon" sitting at a gate in a wall separating two volumes of gas, selectively allowing fast atoms to go through but not slow ones, thus leading to diverging temperatures between the two volumes. This proved to be a very powerful challenge to the mechanical definition of entropy since it led to the decrease in entropy of an isolated system. It took until the 1920s for a Hungarian physicist, Leó Szilárd (1898–1964), to solve this puzzle by pointing out that the information required by the demon to make the *fast-or-slow?* decision does not come without a cost. When the demon and its decision-making actions are included as part of the system, the decrease in entropy caused by the diverging temperatures is offset by the increase in entropy of the demon as a result of the actions of the information system, thus demonstrating that the 2nd Law is not violated in this thought experiment. More arguments around this line of thinking would continue and it's not

---

[17] 9.14 binary digits per word or about 2 bits per letter (this assumes an average of 4.5 letters per word – per (Pierce, 1980) p. 75 and 88.
[18] (Pierce, 1980) p. 24
[19] Licklider quote cited in (Gleick, 2011) p. 233.

my intent to go further. But the point of my including this history is the following link Szilard made between thermodynamics and information theory.

Shannon did not read German and so was not exposed to the work of Boltzmann. But Szilárd did and was and shared ideas with von Neumann, who in turned shared ideas with Shannon. So it was that Shannon learned that in preparing a mathematical approach to information theory he had re-invented Boltzmann's mathematical approach to statistical mechanics. Once he realized this, he pulled terminology from Boltzmann into his work and so caused some confusion. Compounding the confusion was his use of terminology from probability mathematics to describe events in everyday life. Language can serve to unify and clarify or sow seeds of confusion. In Shannon's case, his language arguably tipped the scale more toward the side of confusion.

I intentionally left out some of the terminology that Shannon used so that you could experience for yourself the confusion. I'll now share some examples of them. Instead of creating a message, the sender *chooses* the message from a list of available symbols, each having a *discrete probability*. Information is a measure of one's *freedom of choice* in this *statistical process* and a message with *varying probability* the outcome. The receiver is *uncertain* of what the message says until the message is received and decoded. The information source is an *ergodic* system— a term that Boltzmann created—that continually transmits all possible messages. It's no coincidence that Shannon labeled the size of information with the letter H in light of Boltzmann's H-theorem. He called H the *size of information* which itself is confusing as H is really a ratio of two numbers. And as a final example, in seeking a tighter one-word term for H, he employed the word *entropy*. H is the entropy of the information source. While the end result was powerful and foundational, reading the language and terminology, such as the above-italicized words, is rather confusing, especially the use of the word *entropy*.

Boltzmann and Shannon started from two very different points and ended up at the same equation, but this doesn't necessarily mean that the two starting points are related. Boltzmann's entropy quantifies the number of different ways that discrete entities can be arranged based on location and momentum. Shannon's entropy quantifies the minimum number of binary digits necessary to encode the symbols used to transmit messages. While I leave it to others[20] to argue about the similarity (or not) between these two entropies, my intent in presenting it here is simply to point out that, for many, Shannon started a paradigm shift from thinking about energy to thinking about information, even going so far as to suggest that information is more fundamental than energy.

> To do anything requires energy. To specify what is done requires information.
> – Seth Lloyd (2006)[21]

---

[20] Arieh Ben-Naim has probably done as much as anyone to outline statistical thermodynamics based on information theory, being the first to reformulate Shannon's work into physical terms. To learn more, I highly recommend his book, *Farewell to Entropy*. (Ben-Naim, 2008)

[21] (Gleick, 2011) p. 355.

Afterword

Happy the man, who, studying nature's laws,
Thro' known effects can trace the secret cause.

Virgil

My journey over the past many years in writing this book has naturally left me with a greater understanding of how and why thermodynamics came to be what it now is. I very much enjoyed this journey, and especially the more challenging parts of it, as the struggle to understand the depths of both the science and history eventually gave way to the rewarding moment of understanding *cause*. Now that my journey is complete, I take a step back to share a few other insights and observations I unexpectedly gathered along the way, starting with my reflections on the science itself.

The Science

The disconnect between the physical world and classical thermodynamics was by design. Classical thermodynamics' founders, unsure of what the atomic world looked like, adhered to sound scientific principles and kept the two separate. The result was an exact science acclaimed by Einstein as "the only physical theory of universal content which I am convinced will never be overthrown." The downside is that it is just this disconnect that makes thermodynamics a challenge to learn. To many, thermodynamics occurs as an indecipherable black box. The opportunity in front of us is to open the box to see what's going on inside, and the first step in this process is to connect the physical world with the foundations of thermodynamics, namely the 1st and 2nd Laws. A quick review of the material in Parts III and IV will make this clear.

In making this connection, consideration is given to the fact that classical thermodynamics comprises both science and language. As the famed physicist John Wheeler (1911–2008) once said, "We humans are so easily trapped in our own words."[1] So the opportunity is broader than just the science. It includes the opportunity to "untrap" us from the words by learning their historical origins. When we bring both a physical and historical understanding to the foundations of thermodynamics, we gain a deeper understanding of the subject and so strengthen our ability to creatively use it to solve real problems.

---

[1] (Wheeler, 2008)

Block by Block: The Historical and Theoretical Foundations of Thermodynamics,
Robert T. Hanlon, Oxford University Press (2020). © Robert T. Hanlon.
DOI: 10.1093/oso/9780198851547.001.0001

Revisiting the 1st Law of Thermodynamics

When Rudolf Clausius considered Sadi Carnot's ideal heat engine in light of James Joule's discovery of work–heat equivalence, he concluded that

$$dU = \delta Q - \delta W \tag{5.12}$$

This equation, which arguably represents the first quantified energy balance around a system and which later became known as the 1st Law of Thermodynamics, says that the internal energy (U) of a system only changes due to either heat or work. Energy became the umbrella concept under which heat and work operate. If no heat and no work, then internal energy doesn't change.

We trace the concept of work back to the operation of the lever and its measure of performance: weight (mg) multiplied by the change in vertical height (Δh). While this concept is relatively easy to grasp—*How many rocks can we lift? How much water can we pump?*—the concept of heat isn't. Heat is one of the terms we are "trapped" by, since this term originated from a time when scientists felt that heat was a substance. As we've since learned, heat is not a *thing*. It simply quantifies a *change* in thermal energy. Physical height is a property. The difference in heights between two people is not. It's a difference.

The 1st Law of Thermodynamics as embodied by Equation 5.12 represents an energy balance around a body for which the following equation is valid: accumulation = in − out. If no "in" and no "out," then no "accumulation," and energy remains constant. From a historical perspective, Rudolf Clausius based this equation on Sadi Carnot's piston-in-cylinder assembly where the only means by which the energy in this closed system could change were by heat and work processes, which are both based on the same atom–atom collision phenomena (Figure 13.1). The equation was subsequently generalized to account for other means by which internal energy could change, such as by opening the system to allow for the exchange of mass; Gibbs showed how to mathematically handle this scenario through use of his newly created property of "chemical potential" (μ). Additional terms were later added to Equation 5.12 to account for other energy-changing processes, such as those involving surface tension, electrical charge, and magnetic systems.

Revisiting the 2nd Law of Thermodynamics

While the 1st Law of Thermodynamics is necessary to set limits on what is and is not possible, it is not sufficient for this task, for it allows such phenomena as diverging temperatures between two bodies upon contact with each other. This is why we need the 2nd Law, to prevent such things from happening. Entropy tells you where nature stops when energy says you could keep going.

One reason why the 2nd Law is difficult to learn is that, unlike the 1st, it comes in many forms, which reflects the fact that its discovery occurred deep inside the consequences of the law (Figure 32.3). At the various stages along this journey different versions of the 2nd Law appeared, each moving further up the tree of consequences toward the top where the probabilistic form of entropy sits. The lower branches reflecting the time period around Sadi Carnot included such statements as "perpetual motion is impossible" and "heat cannot flow from cold to hot," then higher up were William Thomson's statements on "energy dissipation", while higher still were statements by Rudolf Clausius and J. Willard Gibbs such as "entropy increases to a maximum in an isolated system," which served as the basis for Gibbs' work on chemical equilibrium. Higher still, on top of the tree, arrived Boltzmann's statements involving probability, such as "nature moves to the most probable state." Each of these statements in the branches is true, and each is also a consequence of the fact that nature moves to the most probable state, regardless of starting point, as defined by the number of ways that a fixed number of entities—atoms, molecules, electrons, nuclei, and even photons as Einstein demonstrated—can be arranged based on momentum and location in a system constrained by total energy and volume: $S = S(U,V)$.

The many different versions of the 2nd Law, with only the later ones including the concept of entropy itself, remain in our lexicon and operate as Wheeler "traps." For example, consider such statements as

> *The 2nd Law marks a distinction between heat and work. Heat from a single source whose temperature is uniform cannot be completely converted into work in any cyclic process, whereas work from a single source can always be completely converted into heat.*[2]

and

> *The second law of thermodynamics, unlike the first, recognizes that work readily dissipates into heat, but heat does not readily turn back into work.*[3]

Such statements can be confusing. They have their origins in Carnot's ideal heat engine, which was designed to continuously and reversibly extract thermal energy from a source and convert it into work for which the following statement is true: $Q_{in} = Q_{out} + W$. But a single Carnot heat engine is but one part of a larger energy conversion process. For example, starting with a continuous energy source at a given temperature and pressure, such as a carbon-based fuel and oxygen at ambient conditions, one can generate a flowing hot gas stream and then reversibly generate work via shaft-work machines, e.g., turbines, and an infinite number of Carnot heat engines, while bringing the temperature and pressure of the gas stream back down to the starting temperature and pressure (Figure A.1). While it appears that calculating the maximum work, being the sum from all machines and engines, would be quite difficult, Gibbs showed that it's actually quite easy. The maximum work equals the change in Gibbs energy between the reactants and products. This is the fact. Trying to generalize this fact by creating heat-to-work and work-to-heat conclusions from a single Carnot heat engine can be misleading and confusing, especially to students attempting to learn the subject.

Generally speaking, trying to learn about entropy and the 2nd Law from Carnot's heat engine is challenging. Clausius' analysis of the heat engine revealed entropy as a new property of state. This was an amazing find buried inside this engine. Clausius had the persistence, focus, and stamina—fifteen years' worth—to find it there. It was buried deep, so deep in fact that teaching students today about entropy based on the heat engine is arguably not a best

[2] (Keenan, 1970) p. 289.
[3] (Munowitz, 2005) p. 261.

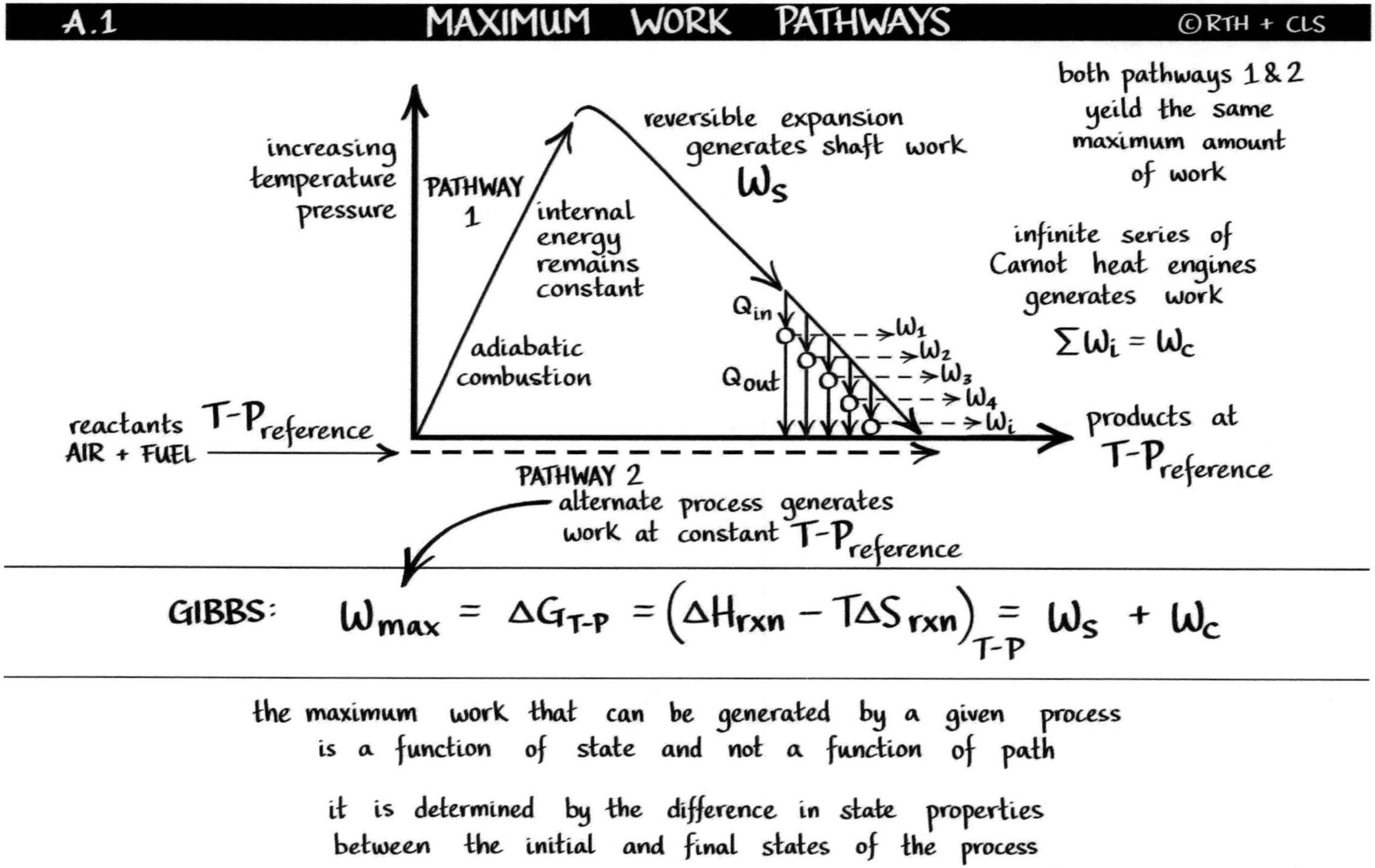
A.1
MAXIMUM WORK PATHWAYS
©RTH + CLS
both pathways 1 & 2 yeild the same maximum amount of work
increasing temperature pressure
PATHWAY 1
internal energy remains constant
adiabatic combustion
reversible expansion generates shaft work
Ws
Qin
Qout
W1
W2
W3
W4
Wi
infinite series of Carnot heat engines generates work
ΣWi = Wc
reactants T-P reference
AIR + FUEL
products at T-P reference
PATHWAY 2
alternate process generates work at constant T-P reference
GIBBS: Wmax = ΔG T-P = (ΔHrxn − TΔS rxn) T-P = Ws + Wc
the maximum work that can be generated by a given process is a function of state and not a function of path
it is determined by the difference in state properties between the initial and final states of the process

Figure A.1 Maximum work pathways

practice, especially since the engine is complicated enough as it is. Carnot's history is important to understand, but perhaps not as the first step in introducing entropy.

*   *   *

The first solid bridge built between the atomic world and classical thermodynamics occurred with the development of the kinetic theory of gases by Clausius and Maxwell and then with the further development of statistical mechanics by Maxwell, Boltzmann, and Gibbs. These events, based on colliding hard spheres, brought physical meaning to temperature, energy, and most importantly the obscure entropy. While at the level of classical thermodynamics based on the macroscopic properties of matter such meaning isn't required, its absence leaves students saying, *I never understood entropy*. The material provided in this book, in which I intentionally intertwine history and science as the two are inseparable for thermodynamics, is intended to support the curious student's quest to understand the foundations of thermodynamics, including the probabilistic entropy. Understanding Boltzmann's work and really all of the foundational works presented in this book and involving both history and science provides access to the deep power offered by thermodynamics.

The History

Good stories engage, educate, and inspire. The history of thermodynamics has many good stories. I chose to tell those that best explain the roots of the physical world on which thermodynamics is based, selecting four themes to do so. Each theme has its own historical timeline, some short, some long, but all interesting. When we look at these timelines collectively and explore what they have in common, several observations stand out.

The Nature of Scientific Discovery – Thrills, Hard Work, Emotional Hardship

The histories that led to the advances in each of the four themes illuminate the nature of scientific discovery. The stories are just as rich as those in other fields, such as exploration, politics, war, entertainment, sports, and so on. Told in the right way, they serve to both excite and inspire. We can't help but feel that some would make for good educational movies. I can easily imagine a movie of Galileo rolling balls down the inclined plane, hearing the tick as they hit each fret in constant time increments and at increasing distance increments according to the Law of Fall, and a movie of Newton discovering that the falls of the Moon and of the apple toward the Earth are governed by the same law. Would Joule's successful efforts leading up to the falling weight and the spinning paddle be interesting? I don't know. But it's worth asking the question and using his story to teach students (in the classroom or on the job) how to conduct good science.

Some stories reveal the thrill of discovery. Galileo and Newton experienced such moments, as did many others. Who wouldn't have wanted to have been Rudolf Clausius when he calculated that both ideal gas and liquid/vapor water perform identically as working substances in Carnot's engine, or Vesto Slipher seeing that most all the stars are moving away from us, or Edwin Hubble seeing that the speeds of the stars' outward motions increase with distance?

Or how about Ernest Rutherford at the moment the alpha particle bounced back, or George Gamow when he calculated the temperature at the beginning of time?

Other stories, while less thrilling, are no less educational, especially those that highlight the reality that good science can be hard and without reward. Look at Sadi Carnot's story. At the end of his life he had hit a brick wall and was left hopeless and adrift. Look at Ludwig Boltzmann's story. At the end of his life he was suffering, at least in part, from an unwelcoming and resistant world. And yet the perseverance of each in the face of such challenges established for us the foundations of thermodynamics and statistical mechanics, accomplishments that neither lived to joyfully experience.

The emotional pain Boltzmann encountered was not uncommon. Consider the stories of Newton versus Hooke, Mayer versus the establishment, Tyndall versus Thomson and Tait, Galileo versus the Roman Catholic Church, or really of anyone standing alone in their fight against consensus. The pain involved with such stories is of the type that keeps us up at night, stomach clenched, blood pressure high. We are human beings, after all. We are driven to be heard, to be first, and to be acknowledged; failure to reach these goals can be crushing. Consider the gadflies whom Stephen Brush wrote about. They asked great questions and received little recognition. Their challenges reside deep in obscure books, while the discoveries they motivated still flap on top of the scientific flagpoles. Yes, the world of science can exact emotional hardship.

A different kind of emotional hardship is also evident in the stories of those involved with deep inquiries into theory. J. Willard Gibbs aged during the creation of his treatise on statistical mechanics, as did Einstein during his development of his General Theory of Relativity. Living day in, day out, deep in sustained thought, creating new theories where no one had yet travelled, can be exciting but also exhausting, at times irreversibly so.

And then there's the physical hardship involved in obtaining data. But such hardship can be compensated by joy, especially if success is achieved. Joule seemed to enjoy the many hours spent obtaining the sixth significant figure in his data. Maxwell and his wife also likely enjoyed the team-based construction and operation of their equipment to measure gas viscosity. And somehow Madame Curie enjoyed the hours spent stirring the pitchblende, perhaps because she too was part of a team effort with her husband in the pursuit of science. The life of an astronomer is also physically difficult. Whether sitting behind a telescope on top of a mountain on a cold dark night like Slipher and Hubble, or working in an office trying to make sense of library plates, like Henrietta Leavitt, physical hardship was present. But again, the success each achieved likely provided joy. Naturally it's helpful to remember that we aren't reading the stories of those who experienced similar hardships without achieving success. I'm guessing that there wasn't the same amount of joy for them.

But I emphasize that without this group of individuals who were willing to roll up their sleeves and obtain data, we wouldn't have gotten to where we are today. The power of a single data point cannot be overemphasized.[4] So thank you, Galileo, Tycho Brahe, Joseph Black,

---

[4] A former colleague once told me a story about a technical problem in one of our facilities. A certain process hadn't been behaving as expected. Many office discussions involving many hypotheses couldn't resolve the situation. So my colleague finally decided to do what hadn't been done. Get a data point. He went out into the facility where the process was situated, shoved a thermocouple through the thick insulation and found the reactor wall temperature to be far

Henri Regnault, James Joule, and all the others, successful or not, for rising from comfort, enduring hardship, conducting the experiments, and obtaining data. Thermodynamics couldn't have progressed without you.

Creative Thinking – Solitude Versus Group

There doesn't seem to be one successful approach to thinking. On the one hand you had J. Willard Gibbs thinking in what appeared to be complete solitude, while on the other hand you had William Thomson thinking within the Scottish community of scientists. All others seemed to fall between these two extremes. My own belief is that the achievement of true "eureka!" insight requires the sustained thought offered by solitude in order to follow the connective threads of logic. But I have also been in situations where back-and-forth banter can generate insights. Look at the interactions between Richard Feynman and Hans Bethe, among others, in Paul Halpern's *The Quantum Labyrinth*, with Feynman often shouting out "No, no, you're crazy!"[5] to Bethe's novel ideas. Unorthodox? Yes. Effective? Also, yes.

Creative Thinking – Power of Working at an Interface

As manifested in the below table, many of those responsible for contributing to the rise of thermodynamics achieved their respective successes by working at the interface between at least two different fields of study. The success of such a strategy here and elsewhere has led to the encouragement of many scientists (and others) to explore the interface between different "silos" of science, business, art, and so on. It's at the interface where creative opportunity exists.

Scientist	Interface	Scientist	Interface
Galileo	Academia – Craftsmanship	Sadi Carnot	Academia – Engineering
Newton	Terrestrial – Celestial	William Thomson	Physics (Theory) – Practical Application
Bernoulli	Medicine – Physics	Gibbs	Mathematics – Thermodynamic Theory
Euler	Calculus – Mechanics	Maxwell	Mathematics – Thermodynamic Theory – Experimentation
Lazare Carnot	Theory (Mathematics) – Practical Engineering	Claude Shannon	Hardware – Software
Mayer	Physiology – Physics		

removed from where it should have been. That single data point disappeared many hypotheses in an instant and served to support my former boss's belief that "one data point is worth one thousand opinions."

[5] (Halpern, 2017) p. 97.

The Impact of Paradigms

The concept of paradigm played a big role in many of these stories. Galileo lived in Aristotle's world, where the speed of a falling body increases with weight. John Dalton lived in Newton's world, where gaseous atoms vibrate but are otherwise static in space. Sadi Carnot and William Thomson lived in Lavoisier's world, where the phenomenon of heat is due to the flow of an element called caloric that could not be decomposed. I pay tribute here to those who started out deeply entrenched in such false paradigms and successfully climbed their way out. Galileo and Thomson did; Dalton and Carnot didn't. I don't pay tribute lightly, for such paradigms swayed entire communities and maintained strangleholds on progress. It's hard to even conceive that we today may be living deep down in paradigms we're not even aware of. But remember that it wasn't all that long ago when many, including Einstein himself, believed the evening stars to be stagnant. Who would have thought that they are all racing away from us?

It was rather interesting to learn that some of the major theories in science occurred from within incorrect paradigms, large or small. So when you have a difficult time understanding the theories, just remember that the contemporaries of those who came up with them were sometimes just as confused. Consider that much of our language around heat formed at a time when many thought heat to be a physical, weightless substance that flowed from one object to another. In reality, we now know that nothing actually does flow during heat transfer, and yet we continue to ask such questions as, "how much heat is flowing out of that vessel?"

As the paradigm manifests itself in one's assumptions, the lesson learned here is the importance of continuously challenging those assumptions. The scientific method is an excellent vehicle for doing this, which brings me to my next observation.

The Scientific Method

Thermodynamics came into being because the scientists involved adhered to sound science as embodied in the scientific method. There are great historical examples demonstrating the effectiveness of this method. Perhaps the epitome was the work of James Clerk Maxwell as regards the kinetic theory of gases. He hypothesized colliding hard spheres, calculated consequences, built equipment to test the predictive power of the consequences, and then experimentally confirmed the predictions: ideal gas viscosity is independent of density! And this validation of a deduced consequence lent credibility to the induced hypothesis it came from: gas is comprised of atoms that behave as colliding hard spheres. What a great story. What a great teaching vehicle.

The interesting thing about the scientific method is that its power is only available to those willing to engage with it. Science doesn't occur in nature. It occurs in the thoughts and actions of human beings. It's an intentional and well-proven process. When considering the stories involved with the founding of thermodynamics, it's enlightening to hold them up against this yardstick of what good science looks like. Some compare well; others do not. The stories thus serve to teach students about both the science itself and also the process of conducting the science.

The Human Drama of Science

Human drama can't be separated from the science. In the end, this is what makes the stories so engaging to us. We can picture ourselves in the situations described in this book and we can wonder, what would I have done in that situation? And from such inquiries we can become inspired to work hard, to strive for excellence, to continue our journey after getting knocked down or criticized or mocked by others, following our own instinct in the pursuit of understanding nature, the scientific method as our judge.

Final Thoughts

Thermodynamics does not live solely in the textbook. It lives in the worlds of heat and work. It lives in the worlds of machines, chemistry, biology, power generation, radiation, and more.

In the end, thermodynamics provides the means by which we can answer a critical question. *Is what I'm doing worth the effort?* While sources of energy are all around us—sunshine, waterfall, buried coal—money is required to corral these sources for productive use. You have to build the solar cells to generate electricity from the Sun and the turbines to generate electricity from the falling water. The high cost of mining coal was one of Carnot's key motivators as he contemplated how to get more energy from a steam engine for a given bushel of coal. The concepts of maximum work, minimum work, and energy efficiency and their respective impacts on capital costs and operating expenses lie at the heart of many economic evaluations involving return on investment. As the grandfather of thermodynamics, Carnot was the first to start bringing forth the theory needed to address such issues. Gibbs completed the task with his use of state properties. *Here's the starting point, there's the end point, what will it take to get from here to there? Is it worth it? Is it even possible?*

*   *   *

In approaching the use of thermodynamics to guide progress, the more we understand the physics behind the foundations, the better able we'll be able to creatively and powerfully approach and solve the problems encountered. While today's computers can handle much of the mathematics involved, identifying the physical essence of the problem will still depend on the person. To bring this book to a fitting end, a quote I introduced in Chapter 3 bears repeating here:

> *All things are made of atoms – little particles that move around in perpetual motion, attracting each other when they are a little distance apart, but repelling upon being squeezed into one another. In that one sentence, you will see, there is an enormous amount of information about the world, if just a little imagination and thinking are applied.*
>
> – Richard Feynman[6]

---

[6] (Feynman et al., 1989a) Volume I, p. 1-2.

Bibliography

Allen, Peter. 1976. "Francis Arthus Freeth. 2 January 1884 – 15 July 1970." *Biographical Memoirs of Fellows of the Royal Society* 22 (November): 104–18.

Ben-Naim, Arieh. 2008. *A Farewell to Entropy: Statistical Thermodynamics Based on Information: S = logW*. Hackensack, NJ: World Scientific.

Berry, Charles William. 1913. *The Temperature-Entropy Diagram*. 3rd Edition. New York: John Wiley & Sons.

Bevilacqua, Fabio. 1993. "Helmholtz's Ueber Die Erhaltung Der Kraft: The Emergence of a Theoretical Physicist." In *Hermann von Helmholtz and the Foundations of Nineteenth-Century Science*, edited by David Cahan. California Studies in the History of Science 12. Berkeley: University of California Press.

Boltzmann, Ludwig . 1995. *Lectures on Gas Theory*. Translated by Stephen G. Brush. Dover ed. New York: Dover Publications.

———. 2003a. "Further Studies on the Thermal Equilibrium of Gas Molecules." In *The Kinetic Theory of Gases: An Anthology of Classic Papers with Historical Commentary*, edited by Stephen G. Brush and Nancy S. Hall, 262–349. History of Modern Physical Sciences 1. London; River Edge, NJ: Imperial College Press; Distributed by World Scientific Pub.

———. 2003b. "On the Relation of a General Mechanical Theorem to the Second Law of Thermodynamics." In *The Kinetic Theory of Gases: An Anthology of Classic Papers with Historical Commentary*, edited by Stephen G. Brush and Nancy S. Hall, 362–67. History of Modern Physical Sciences 1. London; River Edge, NJ: Imperial College Press; Distributed by World Scientific Pub.

———. 2003c. "Reply to Zermelo's Remarks on the Theory of Heat." In *The Kinetic Theory of Gases: An Anthology of Classic Papers with Historical Commentary*, edited by Stephen G. Brush and Nancy S. Hall, 392–402. History of Modern Physical Sciences 1. London; River Edge, NJ: Imperial College Press; Distributed by World Scientific Pub.

Bondi, Hermann. 1988. "Newton and the Twentieth Century—A Personal View." In *Let Newton Be! A New Perspective on His Life and Works*, edited by R. Flood, J. Fauvel, M. Shortland, R. Wilson. New York: Oxford University Press.

Boyer, Carl B. 1959. "Commentary on the Papers of Thomas S. Kuhn and I. Bernard Cohen." In *Critical Problems in the History of Science*, edited by Marshall Clagett, 384–90. Madison, WI: The University of Wisconsin Press.

Brooks, Michael. 2008. *13 Things That Don't Make Sense: The Most Baffling Scientific Mysteries of Our Time*. New York: Vintage Books.

Brush, Stephen G. 1970. "Boltzmann, Ludwig." In *Dictionary of Scientific Biography (1970–1990)*, 260–68. New York: Charles Scribner's Sons.

———. 1983. *Statistical Physics and the Atomic Theory of Matter from Boyle and Newton to Landau and Onsager*. Princeton, NJ: Princeton University Press.

———. 1986a. *The Kind of Motion We Call Heat: A History of the Kinetic Theory of Gases in the 19th Century. Book 1: Physics and the Atomists*. 3rd print. North-Holland Personal Library.

———. 1986b. *The Kind of Motion We Call Heat: A History of the Kinetic Theory of Gases in the 19th Century. Book 2: Statistical Physics and Irreversible Processes*. 2nd print. North-Holland Personal Library.

———. 2003a. "Introduction." In *The Kinetic Theory of Gases: An Anthology of Classic Papers with Historical Commentary*, edited by Stephen G. Brush and Nancy S. Hall, 1–42; 179–96. History of Modern Physical Sciences 1. London; River Edge, NJ: Imperial College Press; Distributed by World Scientific Pub.

———. 2003b. "Gadflies and Geniuses in the History of Gas Theory." In *The Kinetic Theory of Gases: An Anthology of Classic Papers with Historical Commentary*, edited by Stephen G. Brush and Nancy S. Hall, 421–50. History of Modern Physical Sciences 1. London; River Edge, NJ: Imperial College Press; Distributed by World Scientific Pub.

———. 2003c. "Interatomic Forces and Gas Theory from Newton to Lennard-Jones." In *The Kinetic Theory of Gases: An Anthology of Classic Papers with Historical Commentary*, edited by Stephen G. Brush and Nancy S. Hall, 451–79. History of Modern Physical Sciences 1. London; River Edge, NJ: Imperial College Press; Distributed by World Scientific Pub.

———. 2003d. "Statistical Mechanics and the Philosophy of Science: Some Historical Notes." In *The Kinetic Theory of Gases: An Anthology of Classic Papers with Historical Commentary*, edited by Stephen G. Brush and Nancy S. Hall, 524–54. History of Modern Physical Sciences 1. London; River Edge, NJ: Imperial College Press; Distributed by World Scientific Pub.

———. 2003e. "Part III: Historical Discussions." In *The Kinetic Theory of Gases: An Anthology of Classic Papers with Historical Commentary.*, edited by Stephen G. Brush and Nancy S. Hall, 421–553. History of Modern Physical Sciences 1. London; River Edge, NJ: Imperial College Press; Distributed by World Scientific Pub.

Brush, Stephen G., and Ariel Segal. 2015. *Making 20th Century Science: How Theories Became Knowledge*. New York: Oxford University Press.

Burbidge, E. Margaret, G. R. Burbidge, William A. Fowler, and F. Hoyle. 1957. "Synthesis of the Elements in Stars." *Reviews of Modern Physics* 29 (4): 547–650.

Burchfield, Joe D. 1974. "Darwin and the Dilemma of Geological Time." *Isis (The History of Science Society)* 65 (3) (September): 300–321.

Burton, A. 2002. *Richard Trevithick: Giant of Steam*. London: Aurum.

Callen, Herbert B. 1960. *Thermodynamics: An Introduction to the Physical Theories of Equilibrium Thermostatics and Irreversible Thermodynamics*. New York: Wiley.

———. 1985. *Thermodynamics and an Introduction to Thermostatistics*. 2nd edition. New York: Wiley.

Caneva, Kenneth L. 1993. *Robert Mayer and the Conservation of Energy*. Princeton, NJ: Princeton University Press.

Cardwell, D. S. L. 1971. *From Watt to Clausius; the Rise of Thermodynamics in the Early Industrial Age*. Ithaca, NY: Cornell University Press.

———. 1989. *James Joule: A Biography*. Manchester: Manchester University Press; Distributed exclusively in the USA and Canada by St. Martin's Press, New York.

Carnot, Sadi. 1986. *Reflexions on the Motive Power of Fire*. Edited and translated by Robert Fox. Manchester: Manchester University Press.

Carnot, Sadi, E. Clapeyron, and R. Clausius. 1988. *Reflections on the Motive Power of Fire by Sadi Carnot and Other Papers on the Second Law of Thermodynamics by E. Clapeyron and R. Clausius*. Edited with an Introduction by E. Mendoza. Mineola, NY: Dover.

Carr, Matthew A. 2013. "Thermodynamic Analysis of a Newcomen Steam Engine." *The International Journal for the History of Engineering & Technology* 83:2 (July): 187–208.

Carslaw, H. S., and J. C. Jaeger. 1986. *Conduction of Heat in Solids*. 2nd edition. New York: Oxford University Press.

Cercignani, Carlo. 2006. *Ludwig Boltzmann: The Man Who Trusted Atoms*. Oxford: Oxford University Press.

Chalmers, Thomas Wightman. 1952. *Historic Researches. Chapters in the History of Physical and Chemical Discovery*. New York: Charles Scribner's Sons.

Chandrasekhar, Subrahmanyan. 1998. *Truth and Beauty: Aesthetics and Motivations in Science*. Repr. Chicago, IL: University of Chicago Press.

Chang, Hasok. 2007. *Inventing Temperature: Measurement and Scientific Progress*. Oxford Studies in Philosophy of Science. Oxford: Oxford University Press.

Clagett, Marshall, ed. 1969. *Critical Problems in the History of Science*. Madison, WI: University of Wisconsin Press.

Clausius, R. 1867. *The Mechanical Theory of Heat: With Its Applications to the Steam-Engine and to the Physical Properties of Bodies*. Edited by T. Archer Hirst. London: John Van Voorst.

———. 1879. *The Mechanical Theory of Heat (1879)*. Translated by Walter R. Browne. London: Macmillan.

———. 2003a. "The Nature of the Motion Which We Call Heat." In *The Kinetic Theory of Gases: An Anthology of Classic Papers with Historical Commentary*, edited by Stephen G. Brush and Nancy S. Hall, 111–34. History of Modern Physical Sciences 1. London; River Edge, NJ: Imperial College Press; Distributed by World Scientific Pub.

———. 2003b. "On the Mean Lengths of the Paths Described by the Separate Molecules of Gaseous Bodies." In *The Kinetic Theory of Gases: An Anthology of Classic Papers with Historical Commentary*, edited by Stephen G. Brush and Nancy S. Hall, 135–47. History of Modern Physical Sciences 1. London; River Edge, NJ: Imperial College Press; Distributed by World Scientific Pub.

———. 2003c. "On a Mechanical Theorem Applicable to Heat." In *The Kinetic Theory of Gases: An Anthology of Classic Papers with Historical Commentary*, edited by Stephen G. Brush and Nancy S. Hall, 172–78. History of Modern Physical Sciences 1. London; River Edge, NJ: Imperial College Press; Distributed by World Scientific Pub.

Clibbens, Douglas A. 1920. *The Principles of the Phase Theory*. London: MacMillan and Co. (reprinted from the University of Michigan Libraries collection).

Cohen, I. Bernard, and Richard S. Westfall, eds. 1995. *Newton: Texts, Backgrounds, Commentaries*. 1st edition. A Norton Critical Edition. New York: W.W. Norton.

Cook, G., and R. H. Dickerson. 1995. "Understanding the Chemical Potential." *Am. J. Phys.* 63 (8) (August): 737–42.

Coopersmith, Jennifer. 2010. *Energy, the Subtle Concept: The Discovery of Feynman's Blocks from Leibniz to Einstein*. New York: Oxford University Press.

Copernicus, Nicolaus. 1995. *On the Revolutions of Heavenly Spheres*. Buffalo, NY: Prometheus Books.

Crombie, A. C. 1969. "The Significance of Medieval Discussions on Scientific Method for the Scientific Revolution." In *Critical Problems in the History of Science*, edited by Marshall Clagett, 79–101. Madison, WI: University of Wisconsin Press.

Cropper, William H. 1986. "Rudolf Clausius and the Road to Entropy." *Am. J. Phys.* 54 (12) (December).

———. 2001. *Great Physicists: The Life and Times of Leading Physicists from Galileo to Hawking*. New York: Oxford University Press.

Crosland, M. P. 2008a. "Gay-Lussac, Joseph Louis." In *Complete Dictionary of Scientific Biography*, 5:317–27. New York: Charles Scribner's Sons.

———. 2008b. "Avogadro, Amedeo." In *Complete Dictionary of Scientific Biography*, 1:343–50. New York: Charles Scribner's Sons.

Crowther, J. G. 1937. *Famous American Men of Science*. New York: W.W. Norton & Company, inc.

Darrigol, Olivier. 2018. *Atoms, Mechanics, and Probability: Ludwig Boltzmann's Statistico-Mechanical Writings—an Exegesis*. First edition. Oxford: Oxford University Press.

Daub, E. E. 1970a. "Clausius, Rudolf." In *Dictionary of Scientific Biography (1970–1990)*, 303–11. New York: Charles Scribner's Sons.

———. 1970b. "Entropy and Dissipation." *Historical Studies in the Physical Sciences (University of California Press)* 2: 321–54.

———. 1970c. "Waterston, Rankine, and Clausius on the Kinetic Theory of Gases." *Isis (The History of Science Society)* 61 (1): 105–6.

———. 1975. "The Hidden Origins of the Tait–Tyndall Controversy." *Proc. XIV Int. Congr. Hist. Sci.* 2: 241–44.

———. 1976. "Gibbs Phase Rule: A Centenary Retrospect." *J. Chem. Educ.* 53 (12) (December): 747–51.

———. 1978. "Sources for Clausius' Entropy Concept: Reech and Rankine." *Human Implications of Scientific Advance: Proceedings of the 15th International Congress of the History of Science, Edinburgh, 10–15 August 1977*, 342–58.

Deltete, Robert J. 2011. "Josiah Willard Gibbs (1839–1903)." In *Philosophy of Chemistry, Volume 6 of Handbook of the Philosophy of Science*, 89–100. Amsterdam: Elsevier.

Deprit, Andre. 1984. "Monsignor Georges Lemaître." In *The Big Bang and Georges Lemaître*, edited by A. Berger, 363–92. Dordrecht: Springer.

Dickinson, Henry Winram, and Rhys Jenkins. 1927. *James Watt and the Steam Engine: The Memorial Volume Prepared for the Committee of the Watt Centenary Commemoration at Birmingham 1919*. Oxford: Clarendon Press.

Dijksterhuis, E. J. 1969. "The Origins of Classical Mechanics from Aristotle to Newton." In *Critical Problems in the History of Science*, edited by Marshall Clagett, 163–84. Madison, WI: University of Wisconsin Press.

Dill, Ken A., and Sarina Bromberg. 2011. *Molecular Driving Forces: Statistical Thermodynamics in Biology, Chemistry, Physics, and Nanoscience*. 2nd edition. New York: Garland Science.

Dirac, P. A. M. 1963. "The Evolution of the Physicist's Picture of Nature." *Scientific American* 208 (5) (May): 45–53.

Donnan, F. G. 1925. "The Influence of J. Willard Gibbs on the Science of Physical Chemistry." *Journal of the Franklin Institute* 199 (4): 457–83.

Drake, Stillman. 1973. "Galileo's Discovery of the Law of Free Fall." *Scientific American* 228 (5): 84–92.

———. 1975. "The Role of Music in Galileo's Experiments." *Scientific American* 232 (6): 98–104.

Dugas, René. 2012. *A History of Mechanics*. New York: Dover.

Dugdale, John Sydney. 1966. *Entropy and Low Temperature Physics*. London: Hutchinson.

Dunham, William. 1999. *Euler: The Master of Us All*. Washington, DC: Mathematical Association of America.

Durant, Will, and John R. Little. 2002. *The Greatest Minds and Ideas of All Time*. New York: Simon & Schuster.

Emsley, John. 1990. *The Elements*. Reprinted (with corrections). Oxford: Clarendon Press.

England, Philip. 2007. "John Perry's Neglected Critique of Kelvin's Age for the Earth: A Missed Opportunity in Geodynamics." *GSA Today* 17 (1) (January): 4–9.

'Espinasse, Margaret. 1956. *Robert Hooke*. Berkeley, CA: University of California Press.

Everitt, C. W. F. 1970. "Maxwell, James Clerk." In *Dictionary of Scientific Biography (1970–1990)*, 198–230. New York: Charles Scribner's Sons.

———. 1975. *James Clerk Maxwell: Physicist and Natural Philosopher*. DSB Editions. New York: Scribner.

———. 1984. "Maxwell's Scientific Creativity." In *Springs of Scientific Creativity. Essays on Founders of Modern Science*, edited by Rutherford Aris, H. Ted David and Roger H. Stuewer, 71–141. Minneapolis: University of Minnesota Press.

Feynman, Richard P., Michael A. Gottlieb, Ralph Leighton, and Matthew L. Sands. 2013.

Feynman's Tips on Physics: Reflections, Advice, Insights, Practice: A Problem-Solving Supplement to the Feynman Lectures on Physics. New York: Basic Books.

Feynman, Richard Phillips, Robert B. Leighton, and Matthew L. Sands. 1989a. *The Feynman Lectures on Physics. Volume I. Mainly Mechanics, Radiation, and Heat*. Vol. 1. The Feynman Lectures on Physics 1. Redwood City, CA: Addison-Wesley.

———. 1989b. *The Feynman Lectures on Physics. Volume II. Mainly Electromagnetism and Matter*. Vol. 2. The Feynman Lectures on Physics 2. Redwood City, CA: Addison-Wesley.

———. 1989c. *The Feynman Lectures on Physics. Volume III. Quantum Mechanics*. Vol. 3. The Feynman Lectures on Physics 3. Redwood City, CA: Addison-Wesley.

Fourier, Joseph. 1952. "Theory of Heat." In *Great Books of the Western World. 45. Lavoisier, Fourier, Faraday*, 163–251. William Benton, Encyclopedia Britannica, Inc.

Fox, Robert. 1971. *The Caloric Theory of Gases: From Lavoisier to Regnault*. Oxford: Clarendon Press.

Freese, Barbara. 2003. *Coal: A Human History*. Cambridge, MA: Perseus Pub.

Frisch, Otto Robert. 1973. *The Nature of Matter*. New York: Dutton.

———. 1978. "Lise Meitner, Nuclear Pioneer." *New Scientist*, November, 426–28.

Fromkin, David. 2000. *The Way of the World: From the Dawn of Civilizations to the Eve of the Twenty-First Century*. New York: First Vintage Books.

Galilei, Galileo. 1974. *Two New Sciences, Including Centers of Gravity & Force of Percussion*. Translated by Stillman Drake. Madison, WI: University of Wisconsin Press.

Galloway, Robert Lindsay. 1881. *The Steam Engine and Its Inventors: A Historical Sketch*. London: Macmillan.

Gibbs, J. Willard. 1889. "Rudolf Julius Emanuel Clausius." *Proceedings of the American Academy of Arts and Sciences* 24: 458–65.

———. 1902. *Elementary Principles In Statistical Mechanics: Developed With Especial Reference To The Rational Foundation of Thermodynamics (1902)*. New York: Charles Scribner's Sons.

———. 1993. *The Scientific Papers of J. Willard Gibbs. Volume One Thermodynamics*. Woodbridge, CT: Ox Bow Press.

———. 2008. *Elementary Principles In Statistical Mechanics: Developed With Especial Reference To The Rational Foundation Of Thermodynamics (1902)*. Whitefish, MT: Kessinger Publishing, LLC.

Gillispie, Charles Coulston. 1963. "Intellectual Factors in the Background of Analysis by Probabilities." In *Scientific Change*, edited by A. C. Crombie, 431–53. London: Heinemann.

———. 1990. *The Edge of Objectivity: An Essay in the History of Scientific Ideas*. 10. Paperback printing and first printing with the new preface. Princeton, NJ: Princeton University Press.

Gillmor, C. Stewart. 1982. "Review of: L'Accueil Des Idées de Sadi Carnot: Et La Technologie Française de 1820 à 1860-de La Légende à l'histoire by Pietro Redondi (Paris: Librairie Philosophique, 1980)." *Technology and Culture* 23 (No. 1) (January): 102–4.

Gingerich, Owen. 2005. *The Book Nobody Read: Chasing the Revolutions of Nicolaus Copernicus*. London: Penguin Books.

Gleick, James. 2003. *Isaac Newton*. 1st edition. New York: Pantheon Books.

———. 2011. *The Information: A History, a Theory, a Flood*. 1st edition. New York: Pantheon Books.

Grant, Andrew. 2013. "Atom & Cosmos: Proton's Radius Revised Downward: Surprise Measurement Could Lead to New Physics," *Science News*, 23 February, 183(4): 8.

Griffiths, Robert B., and John C. Wheeler. 1970. "Critical Points in Multicomponent Systems." *Physical Review A* 2 (3) (September): 1047–64.

Grimus, W. 2011. "On the 100th Anniversary of the Sackur–Tetrode Equation." *ArXiv:1112.3748*

[Cond-Mat, Physics:Physics, Physics:Quant-Ph], December.

Guerlac, Henry. 2008. "Lavoisier, Antoine-Laurent." In *Complete Dictionary of Scientific Biography*, 66–91. New York: Charles Scribner's Sons.

Guillen, Michael. 1996. *Five Equations That Changed the World: The Power and Poetry of Mathematics.* New York: Hyperion.

Guth, Alan H. 1997. *The Inflationary Universe: The Quest for a New Theory of Cosmic Origins.* Reading, MA: Perseus Books.

Hall, Marie Boas. 2008. "Boyle, Robert." In *Complete Dictionary of Scientific Biography*, 2:377–82. New York: Charles Scribner's Sons.

Halpern, Paul. 2017. *The Quantum Labyrinth: How Richard Feynman and John Wheeler Revolutionized Time and Reality.* 1st edition. New York: Basic Books.

Harper, Douglas. 2007. "Energy." In *Online Etymology Dictionary.* Archived from the Original on October 11, 2007. Retrieved May 1, 2007.

Harriman, David. 2000. "Galileo: Inaugurating the Age of Reason." *The Intellectual Activist* 14 (nos. 3-5) (May).

———. 2008. "Isaac Newton: Discoverer of Universal Laws." *The Objective Standard*, 53–82.

———. 2010. *The Logical Leap: Induction in Physics.* New York: New American Library.

Harris, T. R. 1966. *Arthur Woolf. The Cornish Engineer. 1766-1837.* Truro, Cornwall: D. Bradford Barton Ltd.

Hawking, Stephen W. 1988. *A Brief History of Time: From the Big Bang to Black Holes.* Toronto: Bantam Books.

Hawley, John Frederick, and Katherine A. Holcomb. 2005. *Foundations of Modern Cosmology.* 2nd edition. New York: Oxford University Press.

Heath, Chip, and Dan Heath. 2008. *Made to Stick: Why Some Ideas Survive and Others Die.* Hardcover edition. New York: Random House.

Heisenberg, Werner. 2007. *Physics & Philosophy: The Revolution in Modern Science.* 1st Harper Perennial Modern Classics edition. New York: HarperPerennial.

Helmholtz, Dr. H. 1853. "Art. IV. - On the Conservation of Force." In *Scientific Memoirs, Selected from The Transactions of Foreign Academies of Science and from Foreign Journals. Natural Philosophy.*, edited by John Tyndall and William Francis, 114–62. London: Taylor and Francis.

Helprin, Mark. 2012. *In Sunlight and in Shadow.* Boston: Houghton Mifflin Harcourt.

Herapath, J. 1836. "Exact Calculation of the Velocity of Sound." *The Railway Magazine* 1: 22–28.

Heywood, John B. 2011. *Internal Combustion Engine Fundamentals.* New Delhi: McGraw Hill Education (India) Private Limited.

Hiebert, Erwin N. 1981. *Historical Roots of the Principle of Conservation of Energy.* The Development of Science. New York: Arno Press.

———. 1982. "Developments in Physical Chemistry at the Turn of the Century." Edited by Bernhard, Crawford, and Sorbom. *Science, Technology and Society in the Time of Alfred Nobel (Oxford, 1982)*, 97–115.

Holbrow, C. H., and J. C. Amato. 2011. "What Gay-Lussac Didn't Tell Us." *American Journal of Physics* 79 (1): 17–24.

Hosken, Philip M. 2011. *Oblivion of Richard Trevithick.* Cornwall, England: Trevithick Society.

———. 2013. *Genius, Richard Trevithick's Steam Engines.* Launceston, Cornwall: Footsteps Press.

Hutchison, Keith. 1981. "W. J. M. Rankine and the Rise of Thermodynamics." *The British Journal for the History of Science* 14 (1): 1–26.

Isaacson, Walter. 2007. *Einstein: His Life and Universe.* New York: Simon & Schuster.

Jaffe, Bernard. 1934. *Crucibles: The Lives and Achievements of the Great Chemists.* London: Jarrolds.

James, Henry. 1901. *The Sacred Fount*. London: Methuen & Co.

Jensen, William B. 2010. "Why Are q and Q Used To Symbolize Heat?" *Journal of Chemical Education* 87 (no. 11) (November): 1142.

Johnson, George. 2009. *The Ten Most Beautiful Experiments*. New York: Vintage Books.

Joule, James Prescott. 1845. "On the Changes of Temperature Produced by the Rarefaction and Condensation of Air." *Philosophical Magazine* 26 (174): 369–83.

Kac, M. 1939. "On a Characterization of the Normal Distribution." *American Journal of Mathematics* 61 (3) (July): 726–28.

Kant, Immanuel. 1896. *Immanuel Kant's Critique of Pure Reason: In Commemoration of the Centenary of Its First Publication*. London: Macmillan.

Kaznessis, Yiannis Nikolaos. 2012. *Statistical Thermodynamics and Stochastic Kinetics: An Introduction for Engineers*. Cambridge: Cambridge University Press.

Keenan, Joseph H. 1970. *Thermodynamics*. Cambridge, MA: MIT Press.

Kerr, Philip. 2015. *Prayer: A Novel*. New York: G.P. Putnam's Sons.

Klein, Martin J. 1967. "Thermodynamics in Einstein's Thought." *Science* 157 (August): 509–16.

———. 1969. "Gibbs on Clausius." *Historical Studies in the Physical Sciences (University of California Press)* 1: 127–49.

———. 1983. "The Scientific Style of Josiah Willard Gibbs." In *Springs of Scientific Creativity: Essays on Founders of Modern Science*, edited by Rutherford Aris, H. Ted Davis, and Roger H. Stuewer, 142–62. Minneapolis: University of Minnesota Press.

Kragh, Helge. 1984. "Julius Thomsen and Classical Thermochemistry." *The British Journal for the History of Science* 17 (No. 3) (November): 255–72.

———. 1993. "Between Physics and Chemistry: Helmholtz's Route to a Theory of Chemical Thermodynamics." In *Hermann von Helmholtz and the Foundations of Nineteenth-Century Science*, edited by David Cahan, 403–31. California Studies in the History of Science 12. Berkeley, CA: University of California Press.

———. 1996. *Cosmology and Controversy: The Historical Development of Two Theories of the Universe*. Princeton, NJ: Princeton University Press.

———. 2000. "Max Planck: The Reluctant Revolutionary." *Physics World*, December, 31–35.

———. 2001. "Van't Hoff and the Transition from Thermochemistry to Chemical Thermodynamics." In *Van't Hoff and the Emergence of Chemical Thermodynamics: Centennial of the First Nobel Prize for Chemistry, 1901–2001: An Annotated Translation with Comments of L'équilibre Chimique Dans Les Systèmes Gazeux Ou Dissous a l'état Dilué*, edited by J. H. van 't Hoff, Willem J. Hornix, S. H. W. M. Mannaerts, and Koninklijke Nederlandse Chemische Vereniging, 191–211. Delft, the Netherlands: DUP Science.

———. 2016. "The Source of Solar Energy, Cz. 1840–1910: From Meteoric Hypothesis to Radioactive Speculations." *The European Physical Journal H* 41: 365–94.

Kragh, Helge, and Stephen J. Weininger. 1996. "Sooner Silence than Confusion: The Tortuous Entry of Entropy into Chemistry." *Historical Studies in the Physical and Biological Sciences* 27 (1): 91–130.

Krauss, Lawrence Maxwell, and Cormac McCarthy. 2012. *Quantum Man: Richard Feynman's Life in Science*. New York: W.W. Norton.

Krönig, August. 1856. "Grundzüge Einer Theorie Der Gase." *Annalen Der Physik* 175 (10): 315–22.

Kuhn, Thomas S. 1958. "The Caloric Theory of Adiabatic Compression." *Isis (The History of Science Society)* 49 (2) (June): 132–140.

———. 1959. "Energy Conservation as an Example of Simultaneous Discovery." In *Critical

Problems in the History of Science, edited by Marshall Clagett, 321–56. Madison, WI: University of Wisconsin Press.

———. 1987. *Black-Body Theory and the Quantum Discontinuity, 1894–1912*. Chicago, IL: University of Chicago Press.

———. 1996. *The Structure of Scientific Revolutions*. 3rd edition. Chicago, IL: University of Chicago Press.

Lewis, Christopher J. T. 2007. *Heat and Thermodynamics: A Historical Perspective*. Greenwood Guides to Great Ideas in Science. Westport, CT: Greenwood Press.

Lewis, Gilbert Newton. 1901. "The Law of Physico-Chemical Change." *Proceedings of the American Academy of Arts and Sciences* 37 (3) (June): 49–69.

Lewis, Gilbert Newton, and Merle Randall. 1923. *Thermodynamics and the Free Energy of Chemical Species*. New York: McGraw-Hill Book Company, Inc.

Lightman, Alan P. 2005. *The Discoveries*. 1st edition. New York: Pantheon Books.

Lindauer, Maurice W. 1962. "The Evolution of the Concept of Chemical Equilibrium from 1775 to 1923." *Journal of Chemical Education* 39 (8) (August): 384–90.

Lindley, David. 2001. *Boltzmann's Atom: The Great Debate That Launched a Revolution in Physics*. New York: Free Press.

———. 2004. *Degrees Kelvin: A Tale of Genius, Invention, and Tragedy*. Washington, D.C: Joseph Henry Press.

———. 2007. *Uncertainty: Einstein, Heisenberg, Bohr, and the Struggle for the Soul of Science*. 1st edition. New York: Doubleday.

Livio, Mario. 2013. *Brilliant Blunders: From Darwin to Einstein – Colossal Mistakes by Great Scientists That Changed Our Understanding of Life and the Universe*. First Simon & Schuster hardcover edition. New York: Simon & Schuster.

Mach, Ernst. 1911. *History and Root of the Principle of the Conservation of Energy (1910)*. Translated from the German and Annotated by Philip E. B. Jourdain, M.A. (Cantab.). Chicago, IL: The Open Court Publishing Co.

Mackenzie, R. C. 1989. "George Martine, M.D., F.R.S. (1700–1741): An Early Thermal Analyst?" *Journal of Thermal Analysis* 35: 1823–36.

Mahon, Basil. 2003. *The Man Who Changed Everything: The Life of James Clerk Maxwell*. Chichester, West Sussex, England: Wiley.

Mahoney, Michael S. 2008. "Mariotte, Edme." In *Complete Dictionary of Scientific Biography*, 9:114–22. New York: Charles Scribner's Sons.

Mathias, Paul M. 2016. "The Gibbs–Helmholtz Equation in Chemical Process Technology." *Industrial & Engineering Chemistry Research* 55: 1076–87.

Maxwell, James Clerk. 2003a. "Illustrations of the Dynamical Theory of Gases." In *The Kinetic Theory of Gases: An Anthology of Classic Papers with Historical Commentary*, edited by Stephen G. Brush and Nancy S. Hall, 148–71. History of Modern Physical Sciences 1. London; River Edge, NJ: Imperial College Press; Distributed by World Scientific Pub.

———. 2003b. "On the Dynamical Theory of Gases." In *The Kinetic Theory of Gases: An Anthology of Classic Papers with Historical Commentary*, edited by Stephen G. Brush and Nancy S. Hall, 197–261. History of Modern Physical Sciences 1. London; River Edge, NJ: Imperial College Press; Distributed by World Scientific Pub.

Maxwell, James Clerk, and Peter Pesic. 2001. *Theory of Heat*. Mineola, NY: Dover Publications.

McQuarrie, Donald A., and John D. Simon. 1997. *Physical Chemistry: A Molecular Approach*. Sausalito, CA: University Science Books.

Michelson, Albert. 1896. "Annual Register…with Announcements For…," 159.

Morris, Robert J. 1972. "Lavoisier and the Caloric Theory." *The British Journal for the History of Science* 6 (21): 1–38.

Morse, Samuel Finley Breese. 1914. *Samuel F. B. Morse: His Letters and Journals*.

Edited by Edward Line Morse. New York: Houghton-Mifflin.

Munowitz, M. 2005. *Knowing: The Nature of Physical Law*. New York: Oxford University Press.

Nagel, Ernest. 1969. "Commentary of the Papers of A. C. Crombie and Joseph T. Clark." In *Critical Problems in the History of Science*, edited by Marshall Clagett, 153–61. Madison, WI: University of Wisconsin Press.

Nash, Leonard Kollender. 2006. *Elements of Statistical Thermodynamics*. 2nd edition. Mineola, NY: Dover Publications.

Nernst, Walther. 1969. *The New Heat Theorem: Its Foundations in Theory and Experiment*. New York: Dover Publications.

Newton, Isaac. 2006. *The Mathematical Principles of Natural Philosophy: Volume 1*. Translated by Andrew Motte. London: H. D. Symonds, 1803.

———. 2017. *The Mathematical Principles of Natural Philosophy: Volume 2*. Translated by Andrew Motte. London: H. D. Symonds, 1803.

Noakes, G. R. 1957. *A Text-Book of Heat*. London: Macmillan.

Ostwald, Wilhelm. 2016. *Wilhelm Ostwald: The Autobiography*. Edited by Robert Smail Jack and Fritz Scholz. Cham, Switzerland: Springer International.

Pais, Abraham. 1988. *Inward Bound: Of Matter and Forces in the Physical World*. Oxford: Clarendon Press.

———. 2005. *"Subtle Is the Lord": The Science and the Life of Albert Einstein*. New York: Oxford University Press.

Pauli, Wolfgang. 2013. *Writings on Physics and Philosophy*. Edited by Charles P. Enz and Karl von Meyenn. Berlin/Heidelberg: Springer Science & Business Media.

Peebles, P. J. E. 1993. *Principles of Physical Cosmology*. Princeton Series in Physics. Princeton, NJ: Princeton University Press.

Perry, John. 1895. "On the Age of the Earth." *Nature* 51: 224–27, 341–42, 582–85.

Pierce, John R. 1980. *An Introduction to Information Theory: Symbols, Signals & Noise*. 2nd edition. New York: Dover Publications.

Pimentel, George C, and Richard D Spratley. 1969. *Chemical Bonding Clarified through Quantum Mechanics*. San Francisco: Holden-Day.

Planck, Max, and Alexander Ogg. 1990. *Treatise on Thermodynamics*. 3rd edition. Transl. from the 7th German edition. New York: Dover.

Platt, John R. 1964. "Strong Inference." *Science* 146 (3642): 347–53.

Popkin, Richard Henry, ed. 1966. *The Philosophy of the Sixteenth and Seventeenth Centuries*. New York: The Free Press.

Pupin, Michael Idvorsky. 1898. *Thermodynamics of Reversible Cycles in Gases and Saturated Vapors*. New York: J. Wiley & Sons.

———. 1927. *Josiah Willard Gibbs: Exercises in Celebration of the Fiftieth Anniversary of Publication of His Work on Heterogeneous Substance : With Addresses at the Graduates Club in New Haven Connecticut in June, MCMXXVII*. New Haven, CT: The Graduates Club.

Ramsay, William. 1904. "Sir William Ramsay – Nobel Lecture." December 12, 1904.

Rankine, William John Macquorn, William J. Millar, and Peter Guthrie Tait. 1881. *Miscellaneous Scientific Papers*. London: Charles Griffin and Company.

Reilly, Conor. 1969. *Francis Line, S. J.: An Exiled English Scientist, 1595–1675*. Rome: Institutum Historicum.

Renn, Jürgen, Peter Damerow, Simone Rieger, and Domenico Giulini. 2000. "Hunting the White Elephant: When and How Did Galileo Discover the Law of Fall?" *Science in Context* 13 (3–4): 299–419.

Reynolds, Terry S. 1983. *Stronger than a Hundred Men: A History of the Vertical Water Wheel*. Baltimore: Johns Hopkins University Press.

Roe, J. W. 1916. *English and American Tool Builders*. New Haven, CT: Yale University Press.

Rolt, L. T. C., and John Scott Allen. 1977. *The Steam Engine of Thomas Newcomen.* Nottingham, England: Moorland Pub. Co.

Romer, Alfred. 1997. "Proton or Prouton?: Rutherford and the Depths of the Atom." *American Journal of Physics* 65 (8): 707–16.

Rowlinson, J. S. 2010. "James Joule, William Thomson and the Concept of a Perfect Gas." *Notes and Records. The Royal Society Journal of the History of Science* 64 (1): 43–57.

Rukeyser, Muriel. 1988. *Willard Gibbs.* Woodbridge, CT: Ox Bow Press.

Sagan, Carl. 1980. *Cosmos.* New York: Random House.

Schopenhauer, Arthur. 2009. *Collected Essays of Arthur Schopenhauer.* Radford, VA: Wilder Publications.

Seiler, Drederick. 2012. "The Role of Religion in the Scientific Revolution." *The Objective Standard*, Fall: 43–55.

Settle, Thomas B. 1983. "Galileo and Early Experimentation." In *Springs of Scientific Creativity*, edited by Rotherford Aris, H. Ted Davis, and Roger H. Stuewer, 3–20. Minneapolis, MN: University of Minnesota Press.

Shachtman, Tom. 1999. *Absolute Zero and the Conquest of Cold.* Boston, MA: Houghton Mifflin Co.

Shannon, Claude E., and Warren Weaver. 1998. *The Mathematical Theory of Communication.* Champaign-Urbana: University of Illinois Press.

Shapin, Steven, and Simon Schaffer. 2011. *Leviathan and the Air-Pump: Hobbes, Boyle, and the Experimental Life.* Princeton, NJ: Princeton University Press.

Sharp, Kim, and Franz Matschinsky. 2015. "Translation of Ludwig Boltzmann's Paper 'On the Relationship between the Second Fundamental Theorem of the Mechanical Theory of Heat and Probability Calculations Regarding the Conditions for Thermal Equilibrium' Sitzungberichte Der Kaiserlichen Akademie Der Wissenschaften. Mathematisch-Naturwissen

Classe. Abt. II, LXXVI 1877, Pp 373–435 (Wien. Ber. 1877, 76:373-435). Reprinted in Wiss. Abhandlungen, Vol. II, Reprint 42, p. 164–223, Barth, Leipzig, 1909." *Entropy* 17 (4): 1971–2009.

Sibum, Heinz Otto. 1995. "Reworking the Mechanical Value of Heat: Instruments of Precision and Gestures of Accuracy in Early Victorian England." *Studies in History and Philosophy of Science Part A* 26 (1): 73–106.

Simhony, M. 1994. *Invitation to the Natural Physics of Matter, Space, and Radiation.* Singapore: World Scientific.

Singh, Simon. 2004. *Big Bang: The Origin of the Universe.* 1st US edition. New York: Fourth Estate.

Smith, Crosbie. 1998. *The Science of Energy: A Cultural History of Energy Physics in Victorian Britain.* Chicago, IL: University of Chicago Press.

Smith, Crosbie W., and Matthew Norton Wise. 2009. *Energy and Empire: A Biographical Study of Lord Kelvin.* Cambridge: Cambridge University Press.

Smith, J. M., and H. C. Van Ness. 1975. *Introduction to Chemical Engineering Thermodynamics.* 3rd edition. McGraw-Hill Chemical Engineering Series. New York: McGraw-Hill.

Sobel, Dava. 2000. *Galileo's Daughter: A Historical Memoir of Science, Faith, and Love.* New York: Penguin Books.

Stachel, John. 2009. "Bohr and the Photon." In *Quantum Reality, Relativistic Causality, and Closing the Epistemic Circle*, edited by W. C. Myrvold and J. Christian. The Western Ontario Series in Philosophy of Science 73. Berlin/Heidelberg: Springer Science & Business Media.

Strevens, Michael. 2013. *Tychomancy: Inferring Probability from Causal Structure.* Cambridge, MA: Harvard University Press.

Stukeley, William. 1936. *Memoirs of Sir Isaac Newton's Life (1752).* London: Taylor and Francis.

Suisky, Dieter. 2009. *Euler as Physicist*. Berlin: Springer.

Taleb, Nassim Nicholas. 2005. *Fooled by Randomness: The Hidden Role of Chance in Life and in the Markets*. 2nd edition, updated. New York: Random House.

——. 2010. *The Black Swan: The Impact of the Highly Improbable*. Revised edition. London: Penguin Books.

Tester, Jefferson W., and Michael Modell. 1997. *Thermodynamics and Its Applications*. 3rd edition. Prentice-Hall International Series in the Physical and Chemical Engineering Sciences. Upper Saddle River, NJ: Prentice Hall PTR.

Thackray, Arnold. 2008. "Dalton, John." In *Complete Dictionary of Scientific Biography*, 3:537–47. Detroit: Charles Scribner's Sons.

Theobald, D. W. 1966. *The Concept of Energy*. London: E. & F. N. Spon Ltd.

Thompson, Silvanus Phillips. 1910. *The Life of William Thomson, Baron Kelvin of Largs*. London: Macmillan.

Thomson, James. 1849. "XXXVII. Theoretical Considerations on the Effect of Pressure in Lowering the Freezing Point of Water." *Transactions of the Royal Society of Edinburgh* 16(5): 575–80.

Thomson, William. 1848. "On an Absolute Thermometric Scale Founded on Carnot's Theory of the Motive Power of Heat and Calculated from Regnault's Observations." *Philosophical Magazine [from Sir William Thomson, Mathematical and Physical Papers, Vol. 1 (Cambridge University Press, 1882), Pp. 100-106.]*, October.

——. 1849. "An Account of Carnot's Theory of the Motive Power of Heat." *Transactions of the Edinburgh Royal Society* XVI: 541.

——. 1850. "The Effect of Pressure in Lowering the Freezing Point of Water Experimentally Demonstrated." *Proceedings of the Royal Society of Edinburgh*, January.

——. 1851. "On the Dynamical Theory of Heat." *Transactions of the Royal Society of Edinburgh*, March, 1851, and Philosophical Magazine IV. 1852. Compiled in Transactions of the Royal Society of Edinburgh, Vol. 21, Dickson, 1857.

——. 1852. "XLVII. On a Universal Tendency in Nature to the Dissipation of Mechanical Energy." *Philosophical Magazine* 4 (25): 304–6.

——. 1856. "On the Discovery of the True Form of Carnot's Function." *Philosophical Magazine* 11 (Series 4) (74): 447–48.

——. 1864. "XXIII. On the Secular Cooling of the Earth." *Transactions of the Royal Society of Edinburgh* XXIII (April): 167–69.

——. 1874. "Kinetic Theory of the Dissipation of Energy." *Nature*, April, 441–44.

Todd, A. C. 1967. *Beyond the Blaze. A Biography of Davies Gilbert*. Truro, Cornwall: D. Bradford Barton Ltd.

Tolstoy, Leo. 1894. *The Kingdom of God Is Within You*.

Trevithick, F. 1872. *Life of Richard Trevithick: With an Account of His Inventions*. E. & F.N. Spon.

Tribus, Myron, and Edward C. McIrvine. 1971. "Energy and Information." *Scientific American*, 179–88.

Tunkelang, Daniel. 2009. *Faceted Search*. San Rafael, CA: Morgan & Claypool Publishers.

Vonnegut, Kurt. 2009. *Player Piano*. New York: Random House Publishing Group.

Waller, John. 2002. *Einstein's Luck: The Truth behind Some of the Greatest Scientific Discoveries*. Oxford/New York: Oxford University Press.

Ward, John. 1843. *The Borough of Stoke-upon-Trent*. S.R. Publishers.

Weinberg, Steven. 1993. *The First Three Minutes: A Modern View of the Origin of the Universe*. Updated edition. New York: Basic Books.

——. 2003. *The Discovery of Subatomic Particles*. Revised edition. Cambridge, UK: Cambridge University Press.

Westfall, Richard S. 1971. *Force in Newton's Physics: The Science of Dynamics in the Seventeenth Century*. New York: Elsevier.

———. 1996. *Never at Rest: A Biography of Isaac Newton*. Cambridge/New York: Cambridge University Press.

Wheeler, John. 2008. "John Wheeler (1911–2008) Quantum Theory Poses Reality's Deepest Mystery." *Science News*, May, 32.

Wheeler, Lynde Phelps. 1998. *Josiah Willard Gibbs: The History of a Great Mind*. Woodbridge, CT: Ox Bow Press.

Wilczek, Frank. 2004a. "Whence the Force of F = Ma? I: Culture Shock." *Physics Today* 57 (10): 11–12.

———. 2004b. "Whence the Force of F = Ma? II: Rationalizations." *Physics Today* 57 (12): 10–11.

———. 2005. "Whence the Force of F = Ma? III: Cultural Diversity." *Physics Today* 58 (7): 10–11.

Wohlwill, Emil. 2000. "The Discovery of the Parabolic Shape of the Projectile Trajectory." *Science in Context* 13 (3–4): 645–80.

Wordsworth, William. 1850. *The Prelude. Book Third. Residence at Cambridge.*

Young, Thomas, M.D. 1807. *A Course of Lectures on Natural Philosophy and the Mechanical Arts*. London: Joseph Johnson.

Index

Note: Figures are indicated by an italic *f* following the page number.